Planning and Reporting in BI-supported Controlling

Dietmar Schön

Planning and Reporting in BI-supported Controlling

Fundamentals, Business Intelligence, Mobile BI, Big Data Analytics and AI

Springer

Dietmar Schön
Fachhochschule Dortmund
University of Applied Sciences and Arts
Dortmund, Germany

ISBN 978-3-658-41043-8 ISBN 978-3-658-41044-5 (eBook)
https://doi.org/10.1007/978-3-658-41044-5

This book is a translation of the original German „Planung und Reporting im BI-gestützten Controlling“ by Schön, Dietmar, published by Springer Fachmedien Wiesbaden GmbH in 2022. The translation was done with the help of an artificial intelligence machine translation tool. A subsequent human revision was done primarily in terms of content, so that the book will read stylistically differently from a conventional translation. Springer Nature works continuously to further the development of tools for the production of books and on the related technologies to support the authors.

Planning/Editorial: Vivien Bender
This Springer imprint is published by the registered company Springer Fachmedien Wiesbaden GmbH, part of Springer Nature.
The registered company address is: Abraham-Lincoln-Str. 46, 65189 Wiesbaden, Germany

Preface to the 4th edition

Digitization is advancing. Artificial intelligence is supposed to help companies to align themselves even better with the needs of their customers and to significantly improve their corporate management. With the help of ever larger amounts of available data, it should be possible to identify opportunities and risks earlier and to significantly improve the target orientation and the forecasts. The advertising for artificial intelligence (AI) suggests that companies could soon be controlled autonomously. But we are still far from that, because in corporate practice there are still many construction sites in data processing and information use. They range from significant data quality deficiencies to heterogeneous software landscapes to confusing control systems with insufficiently transparent analysis options. Therefore, it is important not to be seduced by the technical highlights, but to evaluate the differently mature BI technologies for corporate management from the respective status quo of the company and to select and use the appropriate technologies. The principle still applies: "The operational requirements determine the technical implementation". And this is exactly where companies need support. Especially with young companies in new markets, but also with long-established companies with historically grown structures, we very often find inadequate planning and reporting solutions. *"Report after report is developed anew, without really replacing old ones. Contents overlap. Departments generate a shadow reporting. The planning takes too long and is quickly outdated. New technologies are only used selectively."* Many more reasons could be listed. The question arises: What do companies lack to build a good planning and reporting system for management and controlling? Often there is a lack of a good planning and reporting concept and the knowledge to implement it technically and organizationally. The extensions in the 4th edition consistently address this. To get ideas for a concept, you need suitable implementation examples. These can be found in literature, on the internet, at consulting and IT companies and in other sources in partial sections. But usually only individual report examples or suggestions for specific areas of application are shown. Therefore, in the new edition, the comprehensive controlling cockpit example is extended by the analysis areas "strategic management" and "production". Thus, the operational and strategic management for the executive board has been sensibly supplemented. Many analysis and planning examples for the most important functional areas "sales, purchasing, production" as well as the "project management" are available for the reader.

The controlling cockpit contains over 85 report suggestions with about 30 key performance indicators and well over 100 control variables. The strategic planning is particularly noteworthy here. While in many companies the strategic planning is carried out without technical support, the cockpit example shows a proposal of BI-supported implementation. The companies can use the strategic planning areas of the environmental, market, corporate, SWOT and portfolio analysis to guide their strategic orientation within the framework of the Balanced Chance and Risk Card and consistently pursue their strategic projects through the roadmap control.

The new production analysis area closes the gap of the previously covered functional areas of sales and procurement. The production control ranges from the production program planning to the production cost analysis to detailed analyses such as the inventory analyses and the productivity analyses for employees and plants.

The cockpit is characterized by economically sound controlling content and standardized design suggestions for tables and graphs that are used in modern BI-supported controlling. Suitable visualization types, e.g. sparklines, colour-coding or bullet charts, help executives to quickly recognize trends, outliers or deviations from targets. In addition, the controlling cockpit offers comfortable analysis, planning and recording possibilities.

The proven structure of the first two editions of the book remains unchanged by integrating professional, organizational, process-related and IT-related aspects of planning and reporting in BI-supported controlling. Innovations of the last years have been integrated into all chapters. This enables the reader to approach the topic from the basics to traditional, proven standards to the new approaches in BI-supported controlling.

The innovations of the 4th edition cover the most important new technological trends for the further development of BI-supported controlling systems.

First, the importance of artificial intelligence is scientifically classified, and important basics are explained. Machine learning, deep learning and neural networks are information technology methods that provide improved analyses and planning for specific delimited tasks by evaluating collected data sets and using statistical methods. They offer interesting development perspectives for business management control tasks, especially for identifying relationships and dependencies in the analysis and for improving the forecast quality, aspects which are mainly used in the context of data mining and predictive analytics. With the use of data discovery, modern frontend solutions help to link third-party data ad hoc with data from the data warehouse and to evaluate those with the help of statistical algorithms and forecasting models. With the support of data visualization, new forms of visualization are used that help with the analysis of the data. For this purpose, modern graphic types, such as sunburst, treemap and bullet charts can be used.

Another important technological advance for controlling is evident in the development of chatbots. Chatbots are computer-based dialogue systems that communicate with humans in a chat, which can also be based on artificial intelligence. Via voice assistants or text recognition, individual questions of the employees and the management can be answered based on the available data from the data warehouse and beyond. This results in new possibilities of a faster, more targeted and resource-efficient information supply.

With RPA (Robotic Process Automation), there is another IT approach that can be used as an intermediate technology in controlling. It makes sense where, despite the use of software (such as ERP, CRM and SCM systems), many complex manual activities are still necessary to operate the systems because these software systems are maintenance-intensive and not optimally coordinated. RPA accesses the surface of these software products as software and can automatically perform well-structured processes and thus replace complex manual maintenance work.

These new technical possibilities should not be understood as a panacea for replacing established tools. Rather, it is shown for which controlling tasks humans or proven as well as new technologies are best suited. For companies that want to deal with the new technological possibilities of AI, data mining, predictive analytics, chat bots or RPA, the necessary basics and prerequisites are explained and their sensible use in operational application is presented.

As an alternative to conventional menu functions and navigation controls, "app" technology (application technology) is increasingly used for interface design. Apps are accessed via tiles. Important tile types, with which the software operation can be designed in the tile look, are shown.

In the section on cloud computing, alternative combinations of cloud variants are highlighted in the context of planning and reporting. These are IaaS, PaaS or SaaS in connection with a public, a private or a hybrid cloud. In addition, this section focuses especially on IT security.

The section on mobile computing concludes the book. The use of mobile devices for planning and reporting continues to grow. Executives today expect that reporting and planning can be supported regardless of location and device. Mobile devices such as smartphones and tablets complement the classic use of PC and laptop. The development and use of further "wearables" and "gadgets" (glasses, watches, etc.) remains exciting. Special requirements for mobile BI-supported reporting and their data storage are shown. Other important topics in this section are mobile device management and mobile security.

The topic of data quality management has also been updated and expanded. Due to the intensive development of data warehouse and BI-supported controlling systems in recent years, data has been growing more and more, but the data quality in the upstream systems and in the data warehouse lags behind. Therefore, in the 4th edition, the legal foundations (e.g. compliance with the German or European GDPR) as well as data quality enhancing measures, such as data profiling, data governance, BI monitoring and especially authorization systems, were shown.

The 4th expanded edition continues to serve as a valuable reference work for practitioners and researchers who deal with the topic of planning and reporting in

BI-supported controlling. They receive many suggestions and extensive application examples that provide valuable assistance in the development and also further development of existing systems. New aspects and innovations on the important development trends of recent years have been completed in the 4th edition. These include especially the topics of AI, big data, data mining, predictive analytics, self-service BI, data discovery, data visualization, app technology, cloud computing and mobile computing.

Finally, I would like to express my sincere thanks especially to Kirsten Wächter and Guido Schön for their professional support and suggestions in preparing the English translation of the 4th edition of the book. In addition, thanks go to my dear family, who made it possible for me to have the necessary support for this project.

Dortmund, Germany, February 2022 Prof. Dr. Dietmar Schön

Trademarks

Almost all hardware and software names and other product brands that are mentioned in this book are also registered trademarks or should be considered as such, e.g.:

- Balanced Chance and Risk Card®is a registered trademark of Prof. Dr. Thomas Reichmann, CIC GmbH.
- Board is a software of Board International SA.
- Cognos PowerPlay and Cognos Visualizer are trademarks of Cognos. Inc.
- Corporate Planning®is a registered trademark of CP Corporate Planning AG.
- Cubeware Solutions Platform C8, C8 Cockpit are trademarks of Cubeware GmbH.
- Diamant/4 and Diamant®/3 IQ are trademarks of Diamant Software GmbH & Co. KG.
- EVA®is a registered trademark of Stern Stewart & Co.
- Evidanza is a software of evidanza AG.
- HTML and XML are trademarks or registered trademarks of the W3C®, World Wide Web Consortium, Massachusetts Institute of Technology.
- Infor®is a registered trademark of the company Infor.
- Intel®is a registered trademark of Intel Corporation.
- InSight & dynaSight (now arcplan Enterprise) are registered trademarks of the company arcplan.
- IBM®, DB2®, Informix®and PASW (formerly SPSS a statistics software of SPSS Inc.) are registered trademarks of IBM Corporation in the USA and other countries.
- Jaspersoft®is a registered trademark of the company TIBCO Software Inc.
- JAVA®is a registered trademark of Sun Microsystems, Inc., JAVASCRIPT®is a registered trademark of Sun Microsystems, Inc., used under the license of the technology developed and implemented by Netscape.
- Jedox®is a registered trademark of Jedox AG.
- Lotus®is a registered trademark of Lotus Development Corporation, USA.
- LucaNet is a software of LucaNet AG.

- MICROSOFT®; WINDOWS®; MS PowerPoint®and MS SQL Server®, MS SharePoint, MS Analysis Services®, MS Office®, MS Excel®, MS Word®, MS Access®, MS Query®, Microsoft Reporting Services® and Microsoft Power BI® are registered trademarks of Microsoft Corporation, USA.
- MicroStrategy®is a registered trademark of MicroStrategy Deutschland GmbH.
- ORACLE®, Oracle Hyperion Financial, Oracle Hyperion Planning, Oracle Hyperion Essbase are a registered trademark of ORACLE Corporation.
- Qlik View®and Qlik Sense®are registered trademarks of QlikTech International AB.
- SAP®, R/3®, SAP ECC®(ERP Central Component), ABAP/4®SAP BW®, SAP BO®, SAP SEM®, SAP Hana®, SAP S4/Hana®, SAP Analytic Cloud®, SAP Lumira, Management Cockpit®, SAP NetWeaver and other SAP products and services mentioned in the text as well as the corresponding logos are trademarks or registered brands of SAP AG in Germany and other countries.
- The trademarks, illustrations and symbols of the iPhone and iPad are the exclusive property and trademarks of Apple Inc.
- SAS-Software®is a registered trademark of the company SAS Institute.
- Tableau®is a registered trademark of the company Tableau Software.
- Targit®is a registered trademark of the company Morton Systems.
- Unit4 Prevero® is a registered trademark of the company Unit4 (formerly Prevero).
- The trademarks, illustrations and symbols of the RIM- and Blackberry family are the exclusive property and trademarks of Research in Motion Limited.

All other names of products and services are trademarks of their respective companies.

The information in this work is published without regard to any possible patent protection. Product names are used without guarantee of free usability. Publisher and author cannot assume any legal responsibility or liability for incorrect information and its consequences. Publisher and author are grateful for suggestions for improvement, hints and erroneous information on factual and product information.

Contents

Abbreviations

AC	Actual
ACID	Atomicity, Consistency, Isolation, Durability
ACT	Actual
AI	Artificial Intelligence
ANN	Artificial Neural Networks
App	Application
APS	Advanced Planning and Scheduling
BARC	Business Application Research Center
BASE	Basically, Available, Soft state, Eventually consistent
BBRT	Beyond Budgeting Round Table
BC	Best Case
BCR-Card	Balanced Chance and Risk Card
BDSG	Federal Data Protection Act (deutsches Bundesdatenschutzgesetz)
BEA	Business Economic Analysis
BEP	Break-Even-Point
BHI	Boots Healthcare International
BI	Business Intelligence
BICC	BI Competence Center
BPM	Business-Performance-Management
BSC	Balanced Scorecard
BU	Business Unit
BYOD	Bring our Own Device
B2B	Business to Business
B2C	Business to Consumers
B2G	Business to Government
CAP	Consistency, Availability, Partition tolerance
CEP	Complex Event Processing
CFROI	Cash Flow Return On Investment
CIS	Chief Information System

CM	Contribution Margin
CMS	Content Management Systems
CO_2	Carbon dioxide
CPI	Cost Performance Index
CPU	Central Processing Unit
CRM	Customer Relationship Management
CSR	Corporate Social Responsibility
CSV	Comma-Separated Values
CYOD	Choose Your Own Device
DB	Database
DCF	Discounted Cash Flow
DDE	Dynamic Data Exchange
DDL	Data Definition Language
Dim.	Dimension
DIN	German Industry Standard (Deutsche Industrie Norm)
DMC	Direct manufacturing costs
DMI	Deloitte Institute for Medium-sized Enterprises
DML	Data Manipulation Language
DMS	Document Management Systems
DRS	German Accounting Standards (Deutsche Rechnungslegungsstandards)
DSS	Decision Support-System
DWH	Data Warehouse
EBIT	Earning Before Interest and Taxes
EBITDA	Earning Before Interest, Taxes, Depriciation and Amortization
EIS	Executive Information System
EKAM	European Competence Center for Applied SME Research
ELT process	Extraction, Load, Transformation Process
ERP	Enterprise Resource Planning
ETL process	Extraction, Transformation and Load Process
EU	European Union
EUS	Decision Support Systems
EVA	Economic Value Added
EXP	Expert Systems
FASMI	Fast Analysis of Shared Multidimensional Information
FC	Forecast
GAN	Generative Adversarial Networks
GDPR	General Data Protection Regulation
GPRS	General Packet Radio Service
GPS	Global Positioning System
GQ	Good Quantity
GSM	Global System for Mobile communications
GSP	Gross sales price

GUI	Graphical User Interface
HDFS	Hadoop Distributed File System
HOLAP	Hybrid OLAP
HR	Human Ressources
hrs.	hours
HSPA	High Speed Packet Access
HTML	Hypertext Markup Language
IaaS	Infrastructure as a Service
IC	Intercompany or Individual Company
ID	Identification number
IDC	International Data Corporation
IEC	International Electrotechnical Commission
IIRC	Integrated International Reporting Council
IfM	Institute for SME research (German Institut für Mittelstandsforschung)
IFRS	International Financial Reporting Standards
ISO	International Organization for Standardization
IT	Information Technology
IoT	Internet of Things
KDD	Knowledge Discovery in Databases
KPI	Key Performance Indicators
KVD	Key Value Drivers
LDM	Local Device Managements
LME	London Metal Exchange
LTE	Long Term Evolution
MDM	Mobile Device Management
MDX	Multidimensional Expressions
MIS	Management Information Systems
ML	Machine Learning
MOLAP	Multidimensional OLAP
mths.	monthly
MQE	Managed Query Environments
MRP	Material Resource Planning
MSS	Management Support Systems
MVCC	Multiversion Concurrency Control
NFC	Near Field Communication
NN	No Name or Net-Net
NOPAT	Net Operating Profit After Taxes
NoSQL	Not only Structured Query Language
ODBC	Open Database Connection
ODS	Operational Data Stores
OEE	Overall equipment effectiveness
OEM	Original Equipment Manufacturer

OLAP	Online Analytic Processing
OLE	Object Linking and Embedding
OLTP	On-Line Transaction Processing
PaaS	Platform as a Service
PASW	Predictive Analysis Software
PC	Personal Computer
PD	Product Development
PDA	Personal Digital Assistant
PDF	Portable Document Format
PHP	Hypertext Preprocessor, originally Personal Home Page Tools
PL	Plan
P&L	Profit and Loss statement
PY	Previous year
QR-Code	Quick-Response Code
RAM	Random-Access Memory
R&D	Research and Development
RDA	Robotic Desktop Automation
RDBMS	Relational Database Management System
RFID	Radio-Frequency Identification
ROCE	Return on Capital Employed
ROLAP	Relational OLAP
ROI	Return On Investment
RPA	Robotic Process Automation
SaaS	Software as a Service
SBU	Strategic Business Unit
SCM	Supply Chain Management
SG	Subgroup
SIGDSS	Special Interest Group on Decision Support, Knowledge and Data Management Systems
SLA	Service Level Agreement
SME	Small and medium-sized enterprises
SPoT	Single Point of Truth
SPI	Schedule Performance Index
SPSS	Statistical Package for the Social Science, later also: Superior Performing Software System
SSBI	Self-Service BI
SQL	Structured Query Language
SWOT	Strengths, Weaknesses, Opportunities and Threats
TA	Target achievement
TCP/IP	Transmission Control Protocol/Internet Protocol
TO	Total output
TOM	Technical and Organizational Measures (in the context of the GDPR)

UMTS	Universal Mobile Telecommunications System
URL	Uniform Resource Locator (uniform identifier for Internet sources)
Var.	variable
VBA	Visual Basic for Applications
VPN	Virtual Private Network
WACC	Weighted Average Cost of Capital
WC	Worst Case
WLAN	Wireless Local Area Networks (also known as Wi-Fi)
WPA2	Wi-Fi Protected Access 2
WWW	World Wide Web
XBRL	eXtensible Business Reporting Language
XML	Extensible Markup Language
XPS	Expert Systems
YARN	Yet Another Resource Negotiator
YTD	Year to date
YTE	Year to end

List of Figures

List of Tables

Introduction

1

Contents

1.1 Problem Statement

In many medium-sized enterprises, the existing planning and reporting solutions in controlling are mapped in simple spreadsheet programs and report generators within the existing application programs. This is very time-consuming and costly, the information is not integrated and meaningful information from other areas and systems is not or only insufficiently included.[1] What applies to medium-sized enterprises is also partly true for large enterprises. Although they have the advantage of being able to draw on larger IT and personnel resources, there is also a proliferation of heterogeneous software landscapes here which are unsuitable for today's planning and reporting tasks in controlling.

For management support, powerful planning and reporting systems based on data warehouse technology and modern cockpit solutions are available; these are discussed in research under the umbrella term business intelligence (short BI). In recent years, a lot of progress has been made in the field of BI-supported controlling.[2] New topics, including AI-supported analysis and planning features in the context of big data and predictive

[1] Cf. Schön (2011, p. 1–47).

[2] Cf. e.g. the BARC surveys (2020) and the Lünedonk market sample (2016).

D. Schön, *Planning and Reporting in BI-supported Controlling*,
https://doi.org/10.1007/978-3-658-41044-5_1

analytics, robotic process automation (RPA), chatbots, data discovery, data visualization, expand the forecasting and analysis capabilities of enterprises and need to be sensibly integrated into existing BI systems or complement them. In addition, the existing BI systems are growing enormously, so that data supply, data quality management, and authorization systems still need to be given high attention. The design of planning and reporting systems in its basic requirement profiles (content, organization, processes and IT support) is not comprehensively and coherently considered in the literature and on the Internet, despite numerous new technical trends. Also, the scientific research still often treats the topic of planning and reporting separately, and there is only little integrative relation between planning and reporting.[3]

According to the joint study "The State of Business Intelligence in Academia 2010" by the Teradata University Network and the Special Interest Group on Decision Support, Knowledge and Data Management Systems (SIGDSS), universities (worldwide) are not able to produce graduates with the appropriate qualifications, although the companies need employees who have both knowledge of business intelligence and business processes.[4] This is where the book comes in. For business practice and science, existing basics as well as current trends for planning and reporting in BI-supported controlling are presented.

1.2 Objective

The aim of the book is to reveal current deficits in planning and reporting and to develop a requirement profile for an integrated concept of planning and reporting in BI-supported controlling, a concept which fundamentally shows content-related, organizational, process-related and IT-specific design options. This scientific concept is enriched with numerous practical examples, so that a benefit for teaching and research as well as for practice is created.

The core questions are:

- What characterizes good planning and good reporting?
- How can reporting and planning be meaningfully integrated?
- How should planning and reporting be effectively designed and efficiently used?
- How are the new technologies in BI-supported controlling best used for the purposes of planning and reporting?

There can be no generally valid answers to these questions, as companies have individual and industry-specific characteristics and therefore may have different requirements and

---

[3] Cf. e.g. Horváth (2008), Küpper (1995), Reichmann (2011), Weber et al. (2005, 2008, 2009), Weber and Linder (2008), Klein and Gräf (2014) and Gladen (2014).

[4] Cf. Terradata (2011).

priorities for planning and reporting in certain areas. However, basic requirement profiles and design recommendations for planning and reporting and their integration can be developed. In addition, the IT infrastructure in the companies is very differently set up and requires individual development recommendations regarding the new BI technologies and their extensions, e.g. in the topic areas of AI- and BIG-Data-supported analytics.

1.3 Approach

For the integration of planning and reporting in BI-supported controlling, it makes sense to consider the **four perspectives "content, organization, processes and their IT support"** in the overall context. For this reason, there is a step-by-step structure in this book. It treats the four perspectives systematically in individual chapters. Here, planning and reporting are not analyzed in isolation, but in integration.

Chapter 2 defines the basic terms "planning and reporting" and provides an integrated analysis profile for them.

First, the special importance of the integration of planning and reporting in the context of corporate management and controlling is worked out. The interlocking and important linking function of planning and reporting in the management control cycle is shown, supporting the following thesis:

▶ An isolated view of planning and reporting leads to inconsistencies in the management process and can thus cause mismanagement in the company.

For example, if responsibilities and objects of planning and reporting are not aligned, overlaps and gaps can arise which lead to certain areas being not or only inadequately controlled. Differently designed contents, e.g. key figures, lead to misinterpretations and controversial opinions, something which should be avoided.

In corporate practice, however, it can be observed that planning and reporting projects were rarely carried out together in the past, but separately due to the complexity and time commitment of the resources. This has the disadvantage that the connection between planning and reporting in controlling is given too little attention. Therefore, the following project recommendation applies:

▶ Ideally, a planning project should also always be a reporting project or vice versa. If a connection between planning and reporting project cannot be realized due to the scarce resources, at least the connection to planning or reporting should be intensively worked on in the isolated project.

After the basic definitions for planning and reporting, the perspectives listed above are systematized in an investigation framework. This results in an integrated analysis profile.

The **four perspectives** (**technical content, organization, processes and IT support**) to be examined represent the main part of the book and are divided into the following chapters:

- Chap. 3: Technical and content design
- Chap. 4: Organization and processes
- Chap. 5: IT support

In Chap. 3 the special relation of the technical and content design to the **corporate strategy** and to the **value-creating factors of the business model** of a company is highlighted at the beginning. These are the most important drivers for good reporting and planning.

Subsequently, important hints and practical design suggestions are given for the structuring, navigation and analysis paths in planning and reporting. An overview of the reporting and planning objects (dimensions, hierarchies, attributes and values) as well as types of reports serves as an introduction for the presentation of the most important basic forms of reporting such as the target-actual comparison and the ABC analysis. In the context of report design, besides general recommendations, especially the filter and selection functions, the layout and the main components of reports (tables, diagrams etc.) are discussed. In addition to well-known graphic types such as pie, line and bar charts, new visualization possibilities, e.g. with sunburst, treemap and bullet charts, are introduced. Here, the readers can get many suggestions and tips for their individual report design. Moreover, the peculiarities of planning forms and the alignment of planning and reporting contents are addressed. Based on exemplary practical examples, many ideas for the design of special reporting and planning areas are shown. Here, a comprehensive example of a BI-supported controlling cockpit for corporate management is presented. It covers the control areas "corporate management (including operational and strategic controlling), sales, production, purchasing and project management" and provides valuable suggestions for the design of a BI-supported controlling cockpit in corporate practice. Companies that have questions and need assistance in this area may contact the author.

Chap. 4 deals with the organizational integration and the design of the processes in connection with planning and reporting. It examines the influence of corporate connections, the organizational structure and the leadership style on planning and reporting. Here, among other things, the peculiarities of medium-sized and large companies are worked out. In addition, the tasks and relationships of the involved roles are distinguished. These are, on the one hand, the recipients and, on the other hand, the creators or senders and coordinators. For planning and reporting, the following processes were differentiated: The introduction process, the cyclical process (continuous execution), and the quality assurance process (e.g. to ensure the information or data quality). By systematically dividing the process steps, the companies are given many impulses that are important for the practical implementation of planning and reporting solutions.

In Chap. 5 the history and the basic hardware components of planning and reporting solutions are briefly discussed. Concerningt the software solutions for the planning and reporting tasks, we distinguish between special ERP systems, spreadsheet programs, special software programs (based on relational database technology), and data warehouse or business intelligence supported systems and highlight their advantages and disadvantages. Numerous examples from practice provide suggestions for the practical design of planning and reporting solutions. Due to the rapid development and the increasing importance of information technology, the basics and definitions of data warehouse, business intelligence and BI-supported controlling are worked out in Sect. 5.5 and 5.6. Here, among other things, OLAP data modeling, OLAP storage concepts, the ETL processes, the different analysis tools such as cockpit and dashboard solutions, as well as portals and BI monitoring are in focus. In addition, planning tools and other possible uses for management tasks such as the Balanced Scorecard and risk management are shown. New technological trends in the form of artificial intelligence (AI), robotic process automation (RPA) and chatbots (conversational UX) are comprehensively treated in Sect. 5.7. Based on the basics of AI (among others, machine learning, deep learning and neural networks), meaningful applications and challenges in the field of data mining, predictive analytics, data discovery and data visualization are shown. These new technologies are not considered as a replacement, but as a potential complement for existing BI-supported planning and reporting systems. For their specific task assignment in planning and reporting, criteria are shown that have to be weighed sensibly in terms of benefits and profitability. Sect. 5.8 deals with the basics of big data. It is worked out for which applications in planning and reporting big data analytics are suitable and to what extent they can usefully complement classical BI systems. Big data is characterized by a high processing speed and the processing of large heterogeneous data sets. Here, important technical basics such as in-memory computing, NoSQL databases and Hadoop are explained. Furthermore, the terms business intelligence and business analytics are delimited and classified within the framework of traditional and exploratory BI. In Sect. 5.9 different models of cloud computing are presented and discussed with regard to their use for planning and reporting. Data warehouse, business intelligence, big data and cloud computing also form the basis for the final topic area "Mobile Reporting and Mobile Planning", which is addressed in Sect. 5.10, Mobile Computing and Mobile BI. Mobile devices will be increasingly used for planning and reporting tasks in the future. Phrases such as "pop up on the screen" become mobile and virtual reality through the use of smartphones, smartpads or tablet PCs. Therefore, content requirements, technical possibilities and the problems and potentials of mobile applications are presented. Modern application technologies with different app or tile types are addressed as well as mobile device management.

At the end, the book summarizes the most important findings and future trends regarding planning and reporting in BI-supported controlling in the four perspectives of the presented analysis profile.

References

BARC. 2020. BARC Guide Data, BI & Analytics 2020. BARC-Surveys. 2016. https://bi-survey.com/. Accessed: 10 June 2017.

Gladen, W. 2014. *Performance Measurement. Controlling mit Kennzahlen*, 6th edn. Wiesbaden: Springer Gabler.

Horváth, P. 2008. Grundlagen des Management-Reportings. In *Management-Reporting – Grundlagen, Praxis und Perspektiven*. Editors P. Horváht, R. Gleich and U. Michel. Munich: Haufe.

Klein, A., and J. Gräf. 2014. Reporting und Business Intelligence – update. In *Der Controlling-Berater*. Editors R. Gleich and A. Klein, Vol. 32. Freiburg: Haufe.

Küpper, H. U. 1995. *Controlling, Konzeption, Aufgaben und Instrumente*. Stuttgart: Schäffer-Poeschel.

Lünedonk-Marktstichprobe. 2016. Der Markt für Business Intelligence und Business Analytics in Deutschland. http://luenendonk-shop.de/out/pictures/0/lue_bi-studie_f201016_fl.pdf. Accessed: 10 June 2017.

Reichmann, T. 2011. *Controlling mit Kennzahlen, Die systemgestützte Controlling-Konzeption mit Analyse- und Reportinginstrumenten*, 8th edn. Munich: Vahlen.

Schön, D. 2011. Ergebnisse zur empirischen Untersuchung: Business Intelligence für Reporting und Planung im Mittelstand – April 2011. The complete results of the study are available for download via the following link. http://www.fhdortmund.de/de/studi/fb/9/personen/lehr/schdie/103020100000206873.php. Accessed: 1 June 2011.

Terradata. 2011. Hochschulen haben Business Intelligence zu wenig im Visier. http://www.beyenetwork.de/view/15350. Accessed: 14 July 2011.

Weber, J., and S. Linder. 2008. Neugestaltung der Budgetierung mit Better und Beyond Budgeting? In *Series of publications Advanced Controlling*, Editor J. Weber, Vol. 64. Weinheim: Wiley.

Weber, J., S. Schaier and O. Strangfeld. 2005. Berichte für das Top-Management. In *Series of publications Advanced Controlling*, Editor J. Weber, Vol. 43. Weinheim: Wiley.

Weber, J., R. Malz, and T. Lührmann. 2008. Excellence im Management-Reporting. In *Series of publications Advanced Controlling*, Editor J. Weber, Vol. 62. Weinheim: Wiley.

Weber, J., P. Nevries, D. Breiter, et al. 2009. Operative Planung. In *Series of publications Advanced Controlling*, Editor J. Weber, Vol. 71. Weinheim: Wiley.

Fundamentals

2

Contents

2.1 Planning and Reporting in Relation to Corporate Management and Controlling

Companies today face an extreme field of tension: On the one hand, they have to successfully sell existing and new products and services in the market; on the other hand, they have to do everything to keep costs and performance under control if they want to survive in the market. In doing so, they are constantly challenged by the competition to build up new potentials and to use them effectively and efficiently. Planning and reporting are essential tools for management and controlling to recognize the dynamic changes in the economic environment and in the company and to successfully align the course of the company. Controlling supports management in its leadership tasks by helping to create transparency about the company and its environment. Management needs this for the formation of corporate goals, the creation of framework conditions, the making of

D. Schön, *Planning and Reporting in BI-supported Controlling*,
https://doi.org/10.1007/978-3-658-41044-5_2

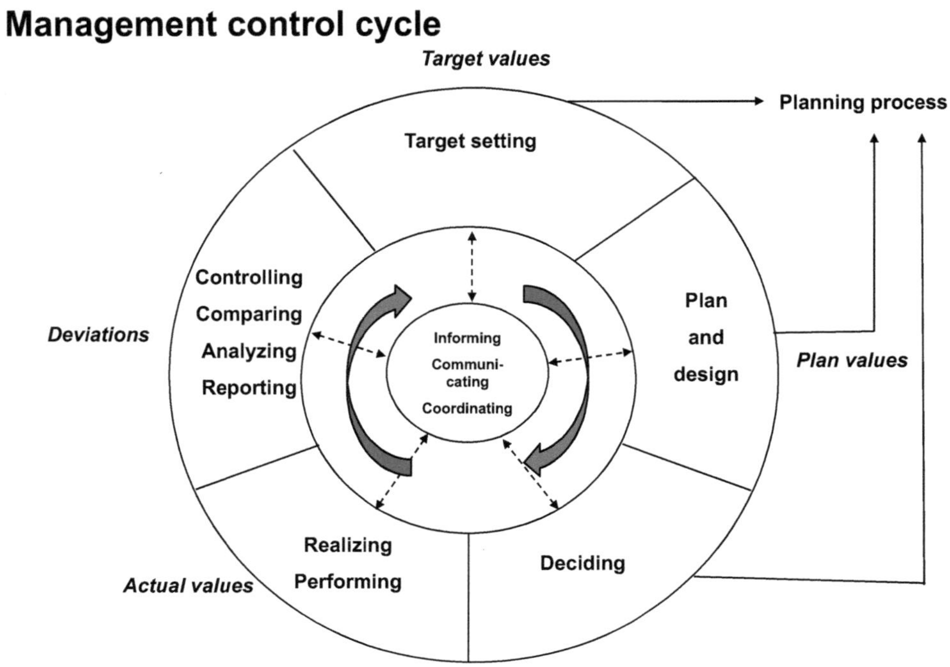

Fig. 2.1 Planning and reporting in the management control cycle (Own representation. Cf. similar representations of the management control cycle used in cybernetics and in management literature, e.g. Wild 1981, p. 40. As synonyms of the term management control cycle, the terms cybernetic control cycle or control cycle are also used.)

decisions and the initiation of their implementation. Important subtasks of controlling are planning and reporting for performance control.[1]

Planning helps to align with previously set goals and standards, which are developed in the context of a planning process. It shows in advance the scope of action of the company and helps later with a change of course which has become necessary due to occurring deviations.

Reporting[2] should transparently reflect the company's activities and show the degree of goal achievement and their deviations. It is therefore indispensable for the control of the company.

Planning and reporting functions are to be considered in the context of the management control cycle, which represents the functions of leadership that are supported by controlling (see Fig. 2.1).

[1] Cf. Schön and Irmer (2010, p. 49–56) and Reichmann et al. (2017, p. 4–19).

[2] The term reporting is used synonymously with the term reporting system.

Based on the visions and targets of the company, the planning of the entrepreneurial action takes place. Decisions for shaping the future of the business are made and implemented by measures. With the help of reporting, states and developments can be monitored and analyzed so that it is possible for the decision-makers to initiate feedback processes for the correction of target specifications, plans, decisions and measures. The basis of these management control cycle functions is the exchange of information as well as the communication and coordination with each other, all supported by controlling in the company.

How complex this simply presented management control loop is in reality can be quickly seen from the temporal dimension of corporate management (see Fig. 2.2).

Strategic, tactical and operationaltasks have to be coordinated, and this often across the group and worldwide. Dynamic market changes, corporate restructuring and shorter product cycles require constantly new, decision-relevant information. Nevertheless, the transparency and accuracy of the information should not suffer and it should be provided in a user-friendly way. Looking at the software market, one can quickly get the impression that there are a variety of powerful planning and reporting tools that make analysis a walk in the park with a mouse click. Especially with business intelligence and big data analytics, the IT world suggests that management tasks could be solved in the future almost by click and algorithm (see also chapter 5). But this is not the case. If you ask the management itself, many feel over-supplied with information without being properly informed. Planning and information deficits are still present. The problems for this are manifold, but they can be mastered with the help of BI-supported controlling systems.

Fig. 2.2 Time dimensions of corporate management

2.2 Basics of Planning and Reporting

Based on the target setting, planning and reporting have always been the central coordination and communication instrument for management and controlling. Surprisingly, only in recent years there have been increased contributions and books that deal more intensively either with planning or reporting. In addition, there are publications of implemented practical solutions that point out especially the peculiarities, problems and implementation proposals of the respective reporting and planning systems.[3] For this reason, the study will systematically examine the integration aspect of both sides, starting from the definition, goals and tasks of planning and reporting. Subsequently, a common analysis profile with the four main perspectives "content, organization, processes and IT support" is developed, in which planning and reporting are examined together and not in isolation. In many cases, the statements apply to both planning and reporting. In cases where planning and reporting differ, this is particularly highlighted.

2.2.1 Planning Definition

To evaluate the performance, productivity and profitability, orientation and comparison standards are required in all areas of the company.

A future-oriented corporate management needs decision-relevant information that has a prognostic character. This applies to all subareas of corporate planning. In accordance with Wild's definition, corporate planning is defined as follows:

> Corporate planning comprises the systematic, future-oriented thinking and defining of corporate targets as well as measures, means and ways to achieve them.[4]

Corporate planning can be differentiated according to the criteria of planning object, planning content and planning horizon.[5]

Objects are the considered subjects and areas of responsibility that underlie the planning, such as companies, cost centers, cost units, customers, suppliers, employees, facilities etc. The content comprises the quantity and value structures as well as other properties that are forecasted and budgeted with regard to the planning objects.

---

[3] Cf. Horváth (2008, p. 17).

[4] Cf. Wild (1981, p. 13).

[5] Cf. Homburg (1991, p. 18).

In view of the planning horizon, we distinguish the general corporate target planning, the strategic and the operational corporate planning. The planning horizon indicates the period (short-, medium-, long-term) for which planning should be done.

In Fig. 2.3 an exemplary planning model for the entire company is sketched, in which the essential sub-plans of corporate planning and their relationship to each other are listed. To better distinguish the plan data level, monetary or value-based plans are shaded in grey.[6]

In the general target planning, the corporate principles (guidelines) and fundamental material, social and value targets are formulated. In the strategic planning, the success potentials of all important business areas of the company as well as of the entire company are planned. This includes, besides strategic targets of the business areas, also general business-shaping aspects, such as the organizational structure, the choice of location and legal form. The operational planning is characterized by a multitude of planning areas starting from the sales and revenue planning to various functional and resource plans up to the value-based result plans.

The planning of costs and services allows a preview of how future corporate activities will affect the operating result; thus, it provides useful decision bases for corporate management.[7]

Many information requirements for management decision problems, e.g. make-or-buy decisions, process selection decisions, production program and production quantity decisions, are future-oriented, i.e. they require success-oriented plan or forecast data, mainly in the form of monetary or quantitative data.[8] Due to increasing qualitative information requirements of management, associated with new management concepts (among others, Balanced Scorecard, benchmarking, customer relationship management), non-monetary information such as quality characteristics, customer assessment and employee satisfaction are increasingly supplemented.

The existing interdependencies between the upstream quantity-oriented sub-plans and the downstream value-based success and financial plans require, from a theoretical point of view, a simultaneous overall planning.[9] However, since a simultaneous overall business planning, which takes into account as many of the complex business relationships as possible, is considered not feasible, a coherent and consistent corporate planning can only be based on a successive planning of the business sub-plans.[10] The process is usually oriented to the business process chain and its bottleneck area, starting from the sales planning and inventory planning, through the further process and potential plans, to the

[6] Cf. Schön (1999, p. 23).

[7] Cf. Krause and Fröhling (1991, p. 275).

[8] Cf. Hummel (1992, p. 76 ff.).

[9] Cf. Horváth (2009, p. 163 ff.).

[10] Cf. Scherrer (1991, p. 143).

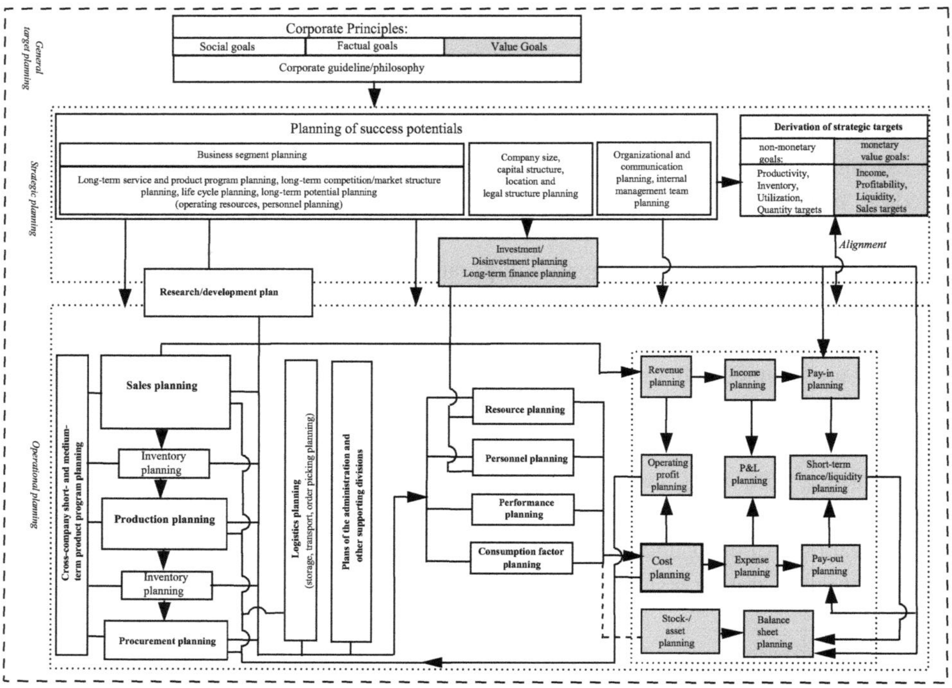

Fig. 2.3 Interdependencies of the operational sub-plans. (Taken from Schön 1999, p. 23)

valued monetary plans (financial, success and balance sheet planning) including the condensed expressions in key figure systems, management and group reports.

The process of the planning calculation is carried out in individual steps, which have to be repeated several times according to the system of the management control cycle (see Sect. 2.2). Theoretically, the partial plans of the company are adjusted again and again until the results of the operational planning are compatible with the targets demanded by the overall company. In this process, the strategic targets derived from the management (e.g. profit and productivity targets) can be adapted step by step with the operational planning figures of the individual business areas in the sense of the counter-current method. In business practice, such planning cycles are hardly feasible, as they would take too much time and resources, so that one limits oneself to a few planning rounds and rough plan corrections. In business practice, one also finds one-sided planning processes that are either only top-down or only bottom-up oriented (see Sect. 4.2.2.1). Besides the annual operational budgeting, intra-year forecasts (forecasts) in the cyclical planning process (see Sect. 4.2.2) play a greater role in corporate management today. Here, updated assessments of the management responsible for planning changes are needed. In addition, cost- and value-driver based planning with BI technology can be set up in which planning and forecasting can be calculated with the help of defined calculation algorithms (value driver models) by changing a few success and cost drivers (see Sect. 5.4.5). For the improvement of the BI-supported forecasting models, forecasts can be supplemented with the help of AI-based models, e.g. in the context of predictive analytics, for specific tasks (see Sect. 5.7). With the in-memory technology, extensive planning processes with complex calculation algorithms can provide the planning and forecasting calculations much faster (see Sect. 5.8). This technology allows large amounts of data to be processed in a timely and fast manner, so that computationally intensive iterative planning processes can benefit from it. Coordination processes remain unaffected. The big data technology, on the other hand, can be used for processing large amounts of data, among other things, from external data sources, such as social networks or online market platforms (see Sect. 5.8). In the run-up to the planning, this method can partially identify, for example, changed customer needs that are taken into account in the planning. However, there will be no full automation in planning. The planning managers retain control over the planning process. The technology helps where it makes sense.

The planning process cannot be seen in isolation from decision, implementation and control, since planning without control is meaningless, and control without planning does not work. For this reason, the authority must be defined for different system applications of planning and reporting tools in which application the analysis of deviations takes place. If different systems are used for planning and reporting, import and export interfaces must also be taken into account. The phases of the planning and control process and thus also the connection between planning and reporting are indicated in Fig. 2.4. The reporting takes over the task of analysis and control. In addition, the reporting is also a supplier of information in the target formation and planning process and thus supports the entire corporate management.

2.2.1.1 Excursion on Budgeting, Better and Beyond Budgeting

At this point, a small excursion on the discussion regarding (traditional) budgeting as well as better and beyond budgeting will be made.

Often, the term budgeting is equated with the term cost planning. However, unlike this imprecise conceptual distinction, budgeting is understood as the concrete numerical determination of the planned value and possibly quantity sizes for a partial area of the company within the framework of operational cost and performance planning.[11] The budget is the monetary target agreement between two parties (e.g. the general management and the division management) and forms the final link in the chain of planned value generation. The budget specifications delegate the decision-making and action freedom to decentralized areas, something which leads to the relief of the executives. Budgeting is criticized, among other things, because of the bureaucratic effort associated with it and

[11] Cf. Wild (1981, p. 40).

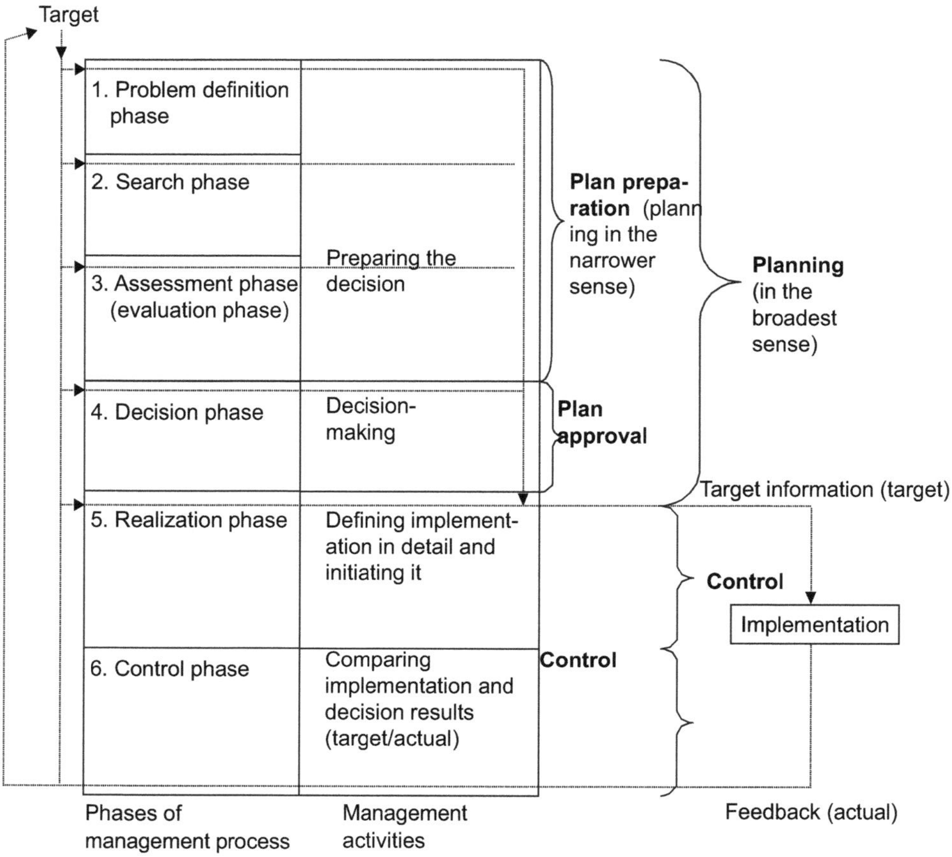

Fig. 2.4 Phases of the planning and control process. (Based on: Hahn 1994, p. 42)

the possibly dysfunctional behaviors that arise. For example, unused budgets are simply cut in the following year (budget wasting) or buffers are built in to make the targets easier to achieve (budgetary slacks).[12]

Newer concepts of "better budgeting" are intended to eliminate bureaucratic disadvantages of budgeting by systematically accelerating, simplifying and making the traditional budgeting more flexible. This instrumental change is essentially achieved by improved reporting and planning tools. This book also follows this approach. In addition, however, the improvement is also considered in relation to the technical content, the organizational integration and the process support.

"Beyond budgeting", on the other hand, wants to dispense with the traditional budgeting and replace it with a new control instrument. This is a fundamental change in

[12] Cf. Wala and Haslehner (2009, p. 308).

the management approach and dispenses with operational planning and budgets. Protagonists of beyond budgeting are Jeremy Hope, Robin Fraser and Peter Bunce, who propagate a comprehensive management approach based onprime examples such as Svenska Handelsbanken or Boots Healthcare International (BHI); instead of budgets, this approach steers with target-oriented key value drivers (20% of the information (KVD) = 80% of the success), internal and external benchmarks and flexible strategic initiatives.[13] Here, this concept shows great similarities with the Balanced Scorecard approach. A management forum is set up as a so-called BBRT (Beyond Budgeting Round Table). The new control instrument does not use any operational, plan-based accounting systems with budgets, but steers with forward-looking forecasting values. The section of pure operational planning and budgeting is thus not needed for "beyond budgeting"; instead, strategic tools and forecasting move into the foreground.

However, operational planning and budgeting cannot be dispensed with. The "beyond budgeting" approach is **not** tenable for the following reasons:

- Forecast is a variant of planning, in which the year-to-date actual is supplemented by an operational residual planning (possibly rolling). Planning is therefore still needed.
- A sound planning of a new business field cannot be done without new planning.
- For the control of the companies, calculations, price determinations and tariff determinations are necessary, which are generated on the basis of an operational planning and budgeting. These would not be possible without an operational planning and budgeting.

2.2.2 Targets and Tasks of Planning

The central task of corporate planning is to quantify future developments that arise in the course of operational business processes, usually under uncertainty (forecasting function) and to establish a benchmark for the analysis of the actual development (information and documentation function). The task of corporate planning is to determine the desired performance and the resources to be used for this purpose, as well as the resulting profitability of future periods (design function). In doing so, the basis for decisions and dispositional purposes for controlling the company should be established by providing information. The design function also subsumes the flexibility, security and optimization function. Possible changes and risks should be detected early, and options for change and countermeasures should be shown. In decisions such as the selection of different alternative possibilities, the decision quality should be increased so that the targets of the company are optimally achieved. Deviation, comparison and simulation analyses help in planning and controlling the effectiveness and efficiency, as well as in creat-

[13] Hope and Fraser (1999).

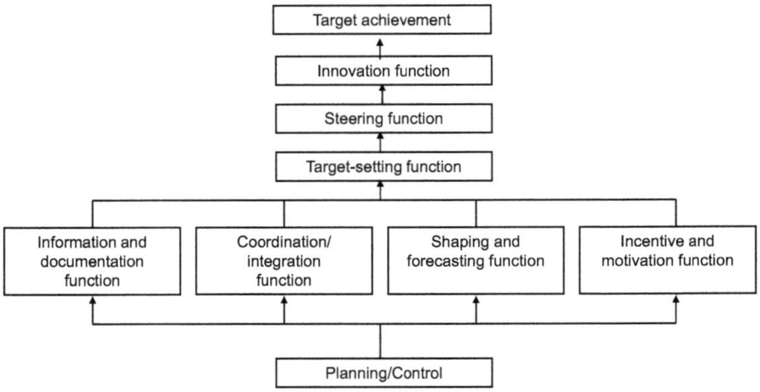

Fig. 2.5 Targets and tasks of planning. (Based on Töpfer 1976, p. 97)

ing entrepreneurial scope for action and exploiting alternative design possibilities (see Fig. 2.5).[14]

Further essential characteristics of corporate planning are:

- It takes place in a multi-level, decision-oriented process.
- It is a systematic, rational and partly subjective information processing process (coordination and integration function), which is determined by legitimate planning agents.
- It is carried out with the help of information technologies due to its complexity (see Chap. 5).
- It is target-oriented, future-oriented and requires the formation of will of the participants (target-setting function).
- It fulfills specification and orientation functions by trying to create an image of the future desired reality (target-setting, control, target-achievement function).
- It should motivate people and provide incentives for improving the company situation and profitability. Economization possibilities should be uncovered (innovation function, incentive and motivation function).

2.2.3 Reporting Definition

Of operational[15] reporting (reporting system) everyone quickly has an idea. One imagines various paper- or IT-based management reports and analyses in which different kinds of information are presented with tables, diagrams and comments. If we take a closer look at the definitions, we can find different definitions of reporting.

[14] Cf. Gabele and Fischer (1992, p. 126).

[15] The statements on operational reporting and planning can be transferred to public and non-profit organizations as well as to private-sector organizations. See Pook and Tebbe (2002, p. 159).

Weber, for example, understands the term reporting as the totality of reports addressed to internal addressees of a company.[16] Other authors like Taschner distinguish internal and external addressees. Among the internal addressees, managers and employees without management tasks are further differentiated.[17]

In the *Gabler-Wirtschaftslexikon* (Gabler Economic Dictionary), reporting is understood as all systematically prepared information relevant to, decisions and managementin written or electronic form for internal and external addressees.[18]

Küpper conceives reporting more broadly. It encompasses the entire operational information system, but also the associated data processing.[19]

A broad definition is provided by Blom, who includes in the operational reporting system "all facilities, means and measures of a company (...) for the preparation, forwarding, processing and storage of information about the operation and its environment".[20] Like Blom, Reichmann also includes the information supply process as an important component of reporting.[21]

Horváth, on the other hand, defines management reporting much more concisely. As part of the operational reporting system, management reporting according to Horváth covers the phases of information provision and transmission as well as information use, but not the phases of information need determination, procurement and generation.[22]

The dimensions of the reporting definition can therefore be delimited according to the recipient group of the information, the content scope of the information and the process of information supply (see Fig. 2.6).

In contrast to Horváth, I believe that the early information process is inseparable from the reporting system, as the determination of information needs and the procurement of source data are important for reporting and thus difficult to be separated from it.[23] In addition, the preparation and management of the source data is a special task for reporting. With the establishment of a data warehouse in BI-supported controlling, an information level is created that contributes to the improvement of the management's decision quality in the sense of a single point of truth (see Sect. 5.5, 5.6 and 5.7).

In summary, the reporting definition used here is broader:

[16] See Weber et al. (2005, p. 13).

[17] See Taschner (2013, p. 2 f.).

[18] See Gablerlexikon (2004, p. 364).

[19] See Küpper (1995, p. 148 ff.).

[20] Cf. Blom (1975, p. 1924–1930).

[21] Cf. Reichmann (2011, p. 12).

[22] Cf. Horváth (2008, p. 18–20).

[23] Cf. Horváth (2008, p. 18).

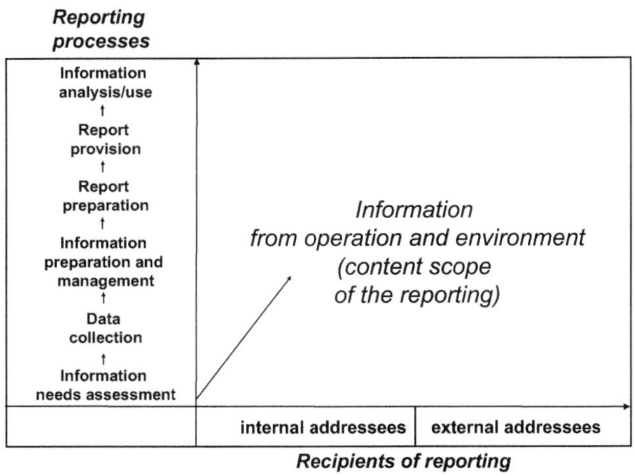

Fig. 2.6 Dimensions of the reporting definition

Operational reporting in the broadest sense includes the determination, procurement, preparation, provision, use and analysis of all information relevant for control and decision making of the operation and its environment for external and internal addressees of the company in the form of reports, whereby these ideally are prepared in a reporting system tailored to the addressees.

Since planning and reporting are closely related to the management control cycle, the general planning definition can be usefully extended.

Corporate planning comprises the systematic, future-oriented thinking through and setting of corporate targets as well as measures, means and ways to achieve them.[24] This requires a planning-relevant information need determination, procurement, processing, provision and use of all control- and decision-relevant information of the company and its environment for the planning and control managers in the company. Planning tools, such as planning forms and functions, are ideally provided in a planning system according to the needs of the addressees.

Planning and reporting should be integrated and coordinated with each other.

2.2.4 Targets and Tasks of Reporting

The aim of reporting is to provide the respective report recipients with current, correct and relevant information (information function) as quickly as possible for the pending

[24] Cf. Wild (1981, p. 13).

decisions and control issues (control function). In doing so, reporting supports the management at every level of leadership (leadership function) of the company and helps the responsible employees to better assess the current situation and the achievement of targets (control and analysis function) and to anticipate future situations (forecast preparation function). In the interaction of decision-makers from different areas of the company, reporting helps with communicating and coordinating their tasks (communication and coordination function). It provides suggestions for the design of the company and motivates economic actions (innovation and motivation function) possibly supported by incentive systems (incentive function). Reporting should answer the following basic questions as best as possible:[25]

- Why should reporting be done? (Report purpose/benefit)
- What should be reported? (Content, level of detail)
- How should reporting be done? (Design, medium)
- Who should report to whom? (Creator and recipient)
- When and where should reporting be done? (Cycle of reports and location of presentation)

The general reporting purpose was already addressed in the context of presenting the management cycle (see Sect. 2.1). Reports serve as a basis of information for the control of the company, i.e. the influence of the employees, the processes and organizational framework conditions of the company by the management and leadership staff which are directly or indirectly initiated in the process of target formation, design, planning, realization and control of the management cycle by the management and the executive staff.

The most important **tasks of reporting** can be summarized in bullet points as follows:

- Documentation and information of operationally relevant facts, achieving
- improved communication
- improved coordination
- improved analysis and control
- improved basis for decision-making
- improved management and leadership support
- improved basis for target-setting, design, forecasting and planning

In summary, reporting creates the necessary transparency for corporate management. In order to have a transparent information base, the following basic requirements of information must be met:

- Extensive completeness of information
- Fast availability of information

[25] Cf. Antony et al. (1972).

- Complexity reduction[26] or detail level of information for the respective analysis and decision making
- Accuracy of information regarding data consistency, comprehensibility, relevance and consistency.

Especially the completeness and the complexity reduction entail a trade-off, which can be resolved by designing the reporting in stages. Depending on the recipient, for example, an overview of the available reports and information should be provided first, before the report recipient can analyze step by step from condensed to detailed information.

The trade-off between the fast availability of information and the accuracy and completeness of information can be improved mainly by modern IT (see Chap. 5) and organizational measures (see Chap. 4).

Ideally, corporate reporting should be able to cope with these requirements. However, an increasing complexity of the company in a constantly more dynamic economic environment leads to an information dilemma regarding the divergence of information generation, processing and demand.[27] On the one hand, the decision-makers are overwhelmed by a flood of mostly outdated and backward-looking data from various sources and systems, while on the other hand they need fast, current, relevant, reliable and mostly forward-looking information for their decisions. The growth of the additionally relevant structured data, such as success and financial data, as well as unstructured data, e.g. in the form of sensor data from machines and parts or in the form of videos and blog entries on the internet, is also increasing considerably.

2.2.5 General Influencing Factors of Planning and Reporting

The **scope and degree of reporting** and planning is shaped by many influencing factors. General influencing factors are:

- The **dynamics of the environmental and contextual conditions** of a company require constant realignment and adaptation to the changing circumstances.
- The **strategic target orientation of a company** is to be steered by means of planning and reporting (see Sect. 3.1).
- The **value-creating factors of the business** (see Sect. 3.2) are central to be reflected in planning and reporting.
- If there are **corporate connections**, planning and reporting requirements arise for the superior corporate unit, e.g. the holding company requires planning and reporting from its subsidiary. In addition, there are consolidation requirements and the associated consideration of intercompany relationships (see Sect. 4.1.2).

[26] Cf. Mehrmann (2004, p. 55 f.).

[27] Cf. Nölken (2002, p. 25).

- The **organizational structure** and the responsibility and decision areas (see Sect. 4.1.3) as well as **the management style** of the general management (see Sect. 4.1.4) influence the content and addressees of planning and reporting.
- The **company size:** the larger the company, the more extensive usuallythe planning and reporting solutions (see Sect. 4.1.1).
- An **internationally oriented company** with several foreign locations and representations may need multilingual system solutions beyond the corporate language. In addition, cultural and legal regulations of the respective country have to be taken into account.
- **Industry-specific characteristics** have to be considered in planning and reporting. For example, in hospital management, mandatory statistics (E1–E3) have to be created on specific medical services. Other industries such as insurance have other peculiarities with risk reporting.

Therefore, no recommendation can be given here on the number of online reports and the page number of print reports. In corporate practice, it is not uncommon to find companies that produce more than 30 reports per month with a total number of up to 50–100 pages. However, when designing the reporting, it is important to achieve a balanced scope of reporting and to ensure that the information density meets the requirements of the decision-makers and report recipients as optimally as possible.

When designing the planning, it is important, as in reporting, to create a balanced planning. On the one hand, it should not be too detailed, as this would be very costly for those involved in the planning and, in addition to being uneconomical, would also cause frustration. On the other hand, it should not be too sketchy, as otherwise the analysis and control functions of the planning would not work.

In addition, the form (e.g. online, mobile, print) of the report output or the planning input should be aligned with the report recipient. Here, simplicity and user-friendliness are expected.

2.3 Integrated Analysis Profile for Planning and Reporting

In most sources on planning and reporting, the analysis is completely detached, focusing only on planning or only on reporting. This is understandable in principle due to the complexity of these analysis areas, but certainly more advantageous isthe connection of both topics. Also in business practice, planning and reporting projects are often initiated independently of each other although a planning project should always also consider the reporting, and a reporting project should also always consider the planning. Otherwise, coordination and alignment problems such as overlaps and gaps or misinterpretations of objects and contents can arise, leading to faulty management. For example, planned values for an important deviation analysis are missing or the granularity of the planning and reporting objects is not consistent.

The importance of the connection between planning and reporting in the management control cycle was already explained in Sect. 2.1. The interlocking was also shown in the development of the definition of planning and reporting in Sect. 2.2. An analysis of the set targets and target manifestations without planning specifications does not work, and planning without control would be a waste of time. Planning and reporting should therefore be considered more closely in relation to each other in research and practice than has been done so far. For this reason, reporting and planning topics are considered integrally in the following chapters. Similarities can be shown in direct relation. Where there are differences, such as in the planning process, these are pointed out and integration aspects are worked out.

As a common analysis profile, criteria are sought that allow planning and reporting to be examined together. As a framework, four fields offer themselves: the technical content, the organization, the processes and the IT support (see Fig. 2.7).

Since the analysis of the four fields does not have a mandatory order, but rather is to be understood in the overall context, cross-references to connections within the four fields are made in the following chapters.

Management-specific content: • Relation to corporate strategy • Value drivers of the business model • Planning/report content (structure and navigation, report and planning objects) • Planning form/report design (report types, basic forms, filter/selection options, layout, special features of planning forms)	**IT support:** • Hardware • Software • ERP-supported systems • Spreadsheet programs • Relational database systems • Data warehouse and BI-supported systems (requirements/criteria: data connectivity, data modeling, harmonization and quality, analysis and planning functionality, flexibility and design options, speed, etc.). • Mobile BI • AI-powered and Big Data analytics • RPA and chatbots
Organization: • Company connections • Organizational structure • Management style • Company size • Addressees/recipients of planning/reports • Senders/creators/coordinators of planning/reports (central and decentralized responsibilities)	**Processes:** • Implementation processes (framework conditions, information needs and actual state analysis, best practice comparison, blueprint, target concept, IT selection, IT concept & implementation, coaching/training) • Cyclical implementation processes • Cyclical planning process (preparation (including data preparation), implementation, coordination and approval) • Cyclical reporting process (information procurement, preparation, report preparation, analysis preparation, report provision, information analysis and control) • Quality assurance processes (including support)

Fig. 2.7 Four-field framework for reporting and planning

References

Antony, R. N., J. Dearden and R. F. Vancil. 1972. *Management control system – Text, cases and readings*. London: R.D. Irwin.

Blom, H. 1975. Informationswesen, Organisation. In *Handwörterbuch der Betriebswirtschaft*, Editors E. Grochla and W. Wittmann, Vol. 2, 4th edn. 1924–1930. Stuttgart: Poeschel.

Gabele, E. and P. Fischer. 1992. *Kosten- und Erlösrechnung*. Munich: Vahlen.

Gablerlexikon. 2004. *Stichwort Berichtswesen*, 16th edn. Wiesbaden: Gabler.

Hahn, D. 1994. *Planungs- und Kontrollrechnung, – PuK Controllingkonzepte*, 4th edn. Wiesbaden: Springer Gabler.

Homburg, C. 1991. *Modellgestützte Unternehmensplanung*. Wiesbaden: Gabler.

Hope, J. and R. Fraser. 1999. *Beyond budgeting round table*. Poole: Dorset.

Horváth, P. 2008. Grundlagen des Management-Reportings. In *Management-Reporting – Grundlagen, Praxis und Perspektiven*, Editors P. Horváht, R. Gleich and U. Michel, 15–42. Munich: Haufe.

Horváth, P. 2009. *Controlling*, 11. Aufl. Munich: Vahlen.

Hummel, S. 1992. Die Forderung nach entscheidungsrelevanten Kostenrechnungsinformationen. In *Handbuch Kostenrechnung*, Editor W. Männel, 76–83. Wiesbaden: Gabler.

Krause, H. and O. Fröhling. 1991. PC-gestützte Kostenplanung. Intelligenter Baustein eines computergestützten Unternehmens-Controlling. In *conference volume 6. Deutscher Controlling Congress*, Editor T. Reichmann, 273–310. Dortmund: Gesellschaft für Controlling e. V.

Küpper, H. U. 1995. *Controlling, Konzeption, Aufgaben und Instrumente*. Stuttgart: Schäffer-Poeschel.

Mehrmann, E. 2004. *Controlling in der Praxis*. Wiesbaden: Springer Gabler.

Nölken, D. 2002. *Controlling mit Intranet- und Business Intelligence Lösungen*. Frankfurt a. M.: Peter Lang.

Pook, M. and M. Tebbe. 2002. Berichtswesen und Controlling. In *Die neue Kommunalverwaltung*, Editors H. Bals, H. Hack and C. Reichard, Vol. 6. Munich: Jehle.

Reichmann, T. 2011. *Controlling mit Kennzahlen, Die systemgestützte Controlling-Konzeption mit Analyse- und Reportinginstrumenten*, 8th edn. Munich: Vahlen.

Reichmann, T., M. Kißler and U. Baumöl. 2017. *Controlling mit Kennzahlen – Die systemgestützte Controlling-Konzeption*, 9th edn. Munich: Vahlen.

Scherrer, G. 1991. *Kostenrechnung*, 2nd edn. Stuttgart: Lucius & Lucius.

Schön, D. 1999. *Neue Entwicklungen in der DV-gestützten Kosten- und Leistungsplanung*. Frankfurt a. M.: Lang.

Schön, D. and K. H. Irmer. 2010. Effiziente Steuerung mit Forecasting und Integrierter Unternehmensplanung bei der GRAMMER Gruppe. *Controlling* 22(1): 49–56.

Taschner, A. 2013. *Management Reporting. Erfolgsfaktor internes Berichtswesen*. Wiesbaden: Springer Gabler.

Töpfer, A. 1976. *Planungs- und Kontrollsysteme industrieller Unternehmungen*. Berlin: Duncker und Humblot.

Wala, T. and F. Haslehner. 2009. *Kostenrechnung, Budgetierung und Kostenmanagement*. Vienna: Linde.

Weber, J., S. Schaier and O. Strangfeld. 2005. *Berichte für das Top-Management*. Weinheim: Wiley.

Wild, J. 1981. *Grundlagen der Unternehmensplanung*, 3rd edn. Opladen: Westdeutscher Verlag.

Professional Content Design

3

Contents

D. Schön, *Planning and Reporting in BI-supported Controlling*,
https://doi.org/10.1007/978-3-658-41044-5_3

The technical and content design of the planning and reporting are largely determined by the information needs and actual analysis (see Sect. 4.2.1.2) and the target or functional concept for planning or reporting (see Sect. 4.2.1.3).

In this chapter, the focus is on the technical and content design of the reports and planning forms. The following questions are answered:

- What influence do the corporate strategy and the business model have on the planning and reporting?
- How should planning and reporting be structured?
- What analysis and navigation paths should be designed?
- What types of reports and planning forms can generally be used?
- What basic forms of reports can be used?
- What factors influence the scope, and what areas are to be distinguished in planning and reporting?
- What design options can be used for the planning forms and reports?

At the beginning of this chapter, the content relation of planning and reporting to the corporate strategy and the value-creating factors of the business model of a company is established in advance. The focus here is on the integration of the entire corporate planning and control.

3.1 Relation to Corporate Strategy

The starting point for the conception, implementation, and further development of planning and reporting is always the connection to higher-level goals and strategies of the company. All strategy-relevant factors are to be taken into account in the planning and reporting process. To support strategic planning, reporting should be able to provide supportive information for the situation analysis, the strategic goal-setting, and the action design.

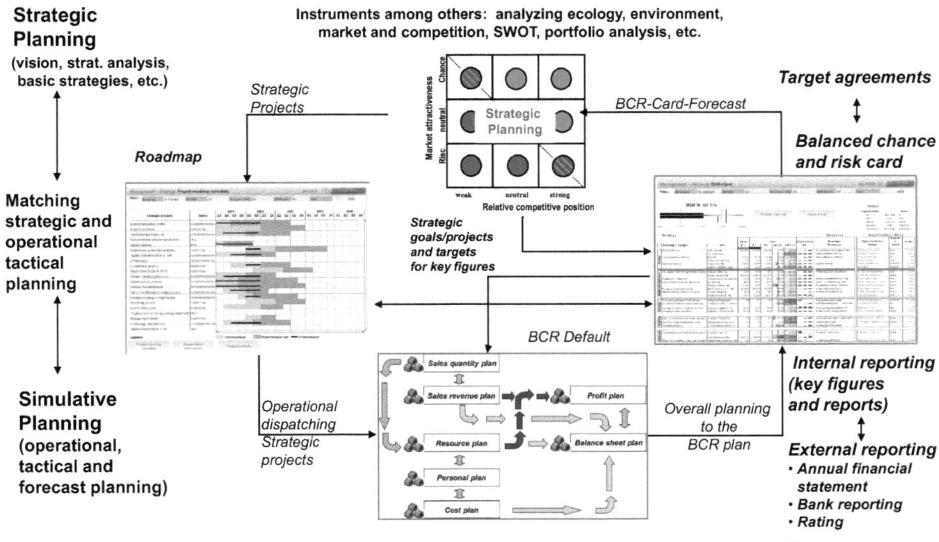

Fig. 3.1 Overall model overview for integrated corporate planning and control

3.1.1 Integration of Strategic, Tactical and Operational Control

In most practical applications, there is no consistent link between planning and control instruments from the strategic to the operational level or the reporting. The concept of integrated corporate planning and control pursues an integrative controlling approach and integrates planning, control and monitoring in the sense of a holistic management control loop in the temporal horizon (see Sect. 2.2). Strategic, tactical and operational control components are thus interlocked.[1] Important elements of the integrated corporate planning and control are shown in Fig. 3.1.

The formulation of overarching goals and strategies as well as the application of the already established strategic instruments, such as market, competitive, environmental, context, SWOT analyses and portfolio techniques, are integrated into the concept to find a strategic orientation. The detailed description of the instruments for strategic planning in a cockpit is given in Sect. 3.9.2. Further explanations on strategic planning are provided by the following literature sources.[2] The result of the strategic planning is the definition of the strategy, the strategic goals, and the necessary projects and metrics to

[1] See Schön and Irmer (2007, p. 245–255).

[2] See, for example, Reichmann et al. (2017, p. 580–601), Bauer (1995, col. 1653–1668, esp. col. 1660) or Köhler (1993, p. 21 f.).

achieve them. For the strategic goal-setting, planning, control and steering, the Balanced Scorecard[3] or the extended "Balanced Chance and Risk Card®" are suitable.[4]

Balanced Scorecards can be set up for the strategic business units, functional areas, divisions and/or regional areas, all of which are consolidated in the overall company, up to the corporate group. For the integration of the strategic and operational/tactical planning, trends and rough targets as orientation values are set up for selected top indicators (also called KPI, Key Performance Indicators) of the Balanced Scorecard (e.g. net sales, ROCE etc.).

Furthermore, for the integration of the strategic planning with the operational-tactical level, planning premises and top-down targets are given which are discussed in more detail in Sect. 3.1.2.

The Balanced Scorecard is a management concept that links the rather long-term oriented corporate strategy with the short-term control of the operational business and, in addition to the financial perspective, also includes the customer, market, product perspective, the learning and development perspective, and the internal process perspective.[5]

The Balanced Scorecard can be interpreted as a balanced goal system that takes all perspectives into account in a balanced way.

The starting point of Balanced Scorecard management is the development and definition of strategic goals as well as the derivation of suitable indicators, targets, actual data and strategic projects for each strategic goal.[6]

The BCR-Card by Reichmann extends the classic Balanced sSorecard by integrating risk management.[7] The BCR-Card forms the central control instrument in the overall concept; this instrument signals the achievement/failure of the strategic goals based on the indicator measurement and controls risk-prone fields. An example of a BCR-Card is shown in Fig. 3.2. The BCR-Card shows in the upper area the top indicator and a button with a jump functionality to the strategic planning (see Sect. 3.9.2); there, the management exclusively defines the BCR fields (perspectives), the strategic goals and projects. Via another button, the cause and effect relationships of projects on indicators of the strategic goals can be forecasted or simulated with value driver tree-like models.

In the main table below, the BCR fields with their strategic goals, indicators and projects as well as their chances and risks are to be analyzed. The example of the BCR-Card extends the four basic perspectives of the Balanced Scorecard (finance, market,

[3] See Horváth (2008, p. 229 ff.).

[4] The Balanced Chance and Risk Card® is a registered trademark of Prof. Dr. Thomas Reichmann, CIC GmbH & Co. KG (see Reichmann 2011, p. 590 ff.).

[5] Cf. Reichmann (2011, p. 550 ff.) and Horváth (2008, p. 229–232).

[6] Cf. Blumenschien and Dick (2004, p. 659–678).

[7] Cf. Reichmann (2011, p. 550 ff.).

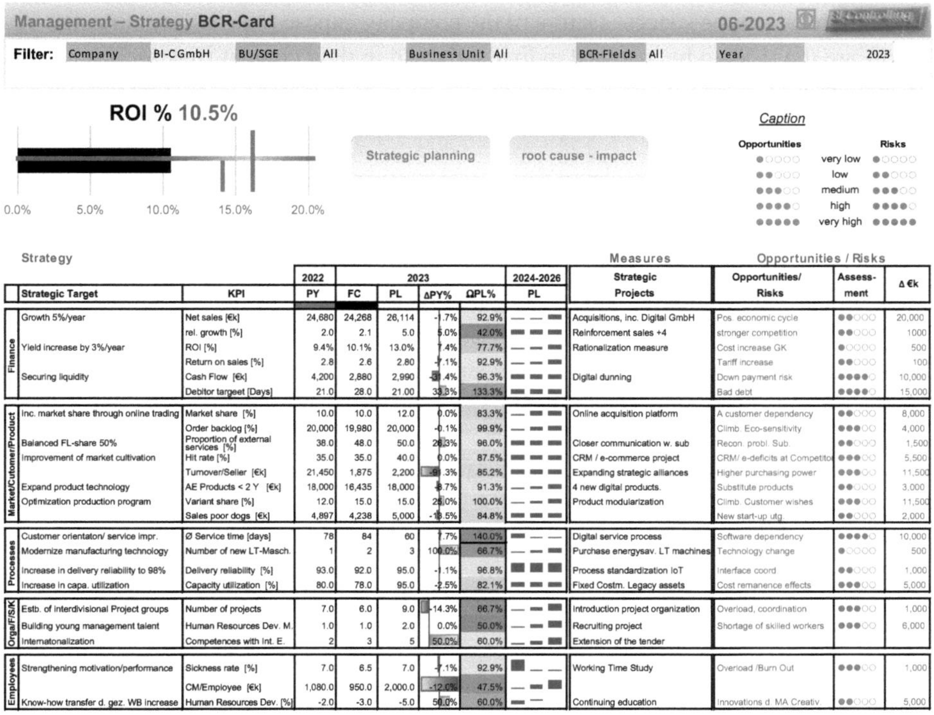

	Strategic Target	KPI	2022 PY	2023 FC	2023 PL	2023 ΔPY%	2023 ΩPL%	2024-2026 PL	Strategic Projects	Opportunities/ Risks	Assess-ment	Δ €k
Finance	Growth 5%/year	Net sales [€k]	24,680	24,268	26,114	-1.7%	92.9%		Acquisitions, inc. Digital GmbH	Pos. economic cycle	●●○○○	20,000
		rel. growth [%]	2.0	2.1	5.0	5.0%	42.0%		Reinforcement sales +4	stronger competition	●●○○○	1000
	Yield increase by 3%/year	ROI [%]	9.4%	10.1%	13.0%	7.4%	77.7%		Rationalization measure	Cost increase GK	●○○○○	500
		Return on sales [%]	2.8	2.6	2.80	-7.1%	92.9%			Tariff increase	●●○○○	100
	Securing liquidity	Cash Flow [€k]	4,200	2,880	2,990	-31.4%	96.3%		Digital dunning	Down payment risk	●●●●○	10,000
		Debitor targeet [Days]	21.0	28.0	21.00	33.3%	133.3%			Bad debt	●●●●○	15,000
Market/Cutomer/Product	Inc. market share through online trading	Market share [%]	10.0	10.0	12.0	0.0%	83.3%		Online acquisition platform	A customer dependency	●●○○○	8,000
		Order backlog [%]	20,000	19,980	20,000	-0.1%	99.9%			Climb. Eco-sensitivity	●●●○○	4,000
	Balanced FL-share 50%	Proportion of external services [%]	38.0	48.0	50.0	26.3%	96.0%		Closer communication w. sub	Recon. probl. Sub.	●●○○○	1,500
	Improvement of market cultivation	Hit rate [%]	35.0	35.0	40.0	0.0%	87.5%		CRM / e-commerce project	CRM/ e-deficits at Competitor	●●●○○	5,500
		Turnover/Seller [€k]	21,450	1,875	2,200	-91.3%	85.2%		Expanding strategic alliances	Higher purchasing power	●●●○○	11,500
	Expand product technology	AE Products < 2 Y [€k]	18,000	16,435	18,000	-8.7%	91.3%		4 new digital products.	Substitute products	●●○○○	3,000
	Optimization production program	Variant share [%]	12.0	15.0	15.0	25.0%	100.0%		Product modularization	Climb. Customer wishes	●●●○○	11,500
		Sales poor dogs [€k]	4,897	4,238	5,000	-13.5%	84.8%			New start-up utg.	●●○○○	2,000
Processes	Customer orientaton/ service impr.	Ø Service time [days]	78	84	60	7.7%	140.0%		Digital service process	Software dependency	●●●●○	10,000
	Modernize manufacturing technology	Number of new LT-Masch.	1	2	3	100.0%	66.7%		Purchase energysav. LT machines	Technology change	●○○○○	500
	Increase in delivery reliability to 98%	Delivery reliability [%]	93.0	92.0	95.0	-1.1%	96.8%		Process standardization IoT	Interface coord.	●●○○○	1,000
	Increase in capa. utilization	Capacity utilization [%]	80.0	78.0	95.0	-2.5%	82.1%		Fixed Costm. Legacy assets	Cost remanence effects	●●●○○	5,000
Orga/F/S/K	Estb. of interdivisional Project groups	Number of projects	7.0	6.0	9.0	-14.3%	66.7%		Introduction project organization	Overload, coordination	●●●○○	1,000
	Building young management talent	Human Resources Dev. M.	1.0	1.0	2.0	0.0%	50.0%		Recruiting project	Shortage of skilled workers	●●●○○	6,000
	Internatonalization	Competences with Int. E.	2	3	5	50.0%	60.0%		Extension of the tender			
Employees	Strengthening motivation/performance	Sickness rate [%]	7.0	6.5	7.0	-7.1%	92.9%		Working Time Study	Overload /Burn Out	●●●○○	1,000
		CM/Employee [€k]	1,080.0	950.0	2,000.0	-12.0%	47.5%					
	Know-how transfer d. gez. WB increase	Human Resources Dev. [%]	-2.0	-3.0	-5.0	50.0%	60.0%		Continuing education	Innovations d. MA Creativ.	●●○○○	5,000

Fig. 3.2 Exemplary Balanced Chance and Risk Card. (Slightly modified from Schön and Irmer 2007, p. 251. The Balanced Chance and Risk Card® is a protected trademark of Prof. Dr. Thomas Reichmann, CIC GmbH & Co. KG (see Reichmann 2011, p. 590 ff.))

processes and knowledge) into the five BCR fields finance, market/customer/products, processes, organization/leadership/culture and employees.

The integration of risk management is achieved by assigning possible hazards and opportunities to the individual strategic goals. Only in this way can a risk-adequate success control of the strategic projects be achieved. The BCR-Card can be cascaded according to the corporate organizational structure, e.g. for business units, divisions, functional areas or regions.

Risks are defined as possible positive and negative deviations from a target value determined under uncertainty. Already during the strategic planning, risks are identified that could endanger the set goals. Here, various questions have to be answered (see Gleißner 2000, p. 1625):[8]

[8] See Gleißner (2000, p. 1625), Gleißner and Romeike (2005, p. 260 f.) and Denk et al. (2008, p. 124).

- What threats are the success factors of the company exposed to (risk identification)?
- How are the risks to be assessed individually and in their context (risk assessment and aggregation)?
- Does the existing equity capital provide sufficient risk coverage potential (risk coverage)?
- Which "core risks" must the company necessarily bear itself, and which risks can be avoided, reduced or transferred (risk management and transfer of risks)?
- Which key figures take risks into account when determining the company's success (risk monitoring and development of value-oriented control systems)?

The definition of the strategic targets and subtargets must always be linked to concrete strategic projects that contribute to the direct or indirect achievement of the target. Here, a suitable connecting instrument often lacks in practice. The roadmap planning and control in which strategic projects are concretely planned and thus operationalized is to be favored here (see Figs. 3.3, 3.4 and 3.5).

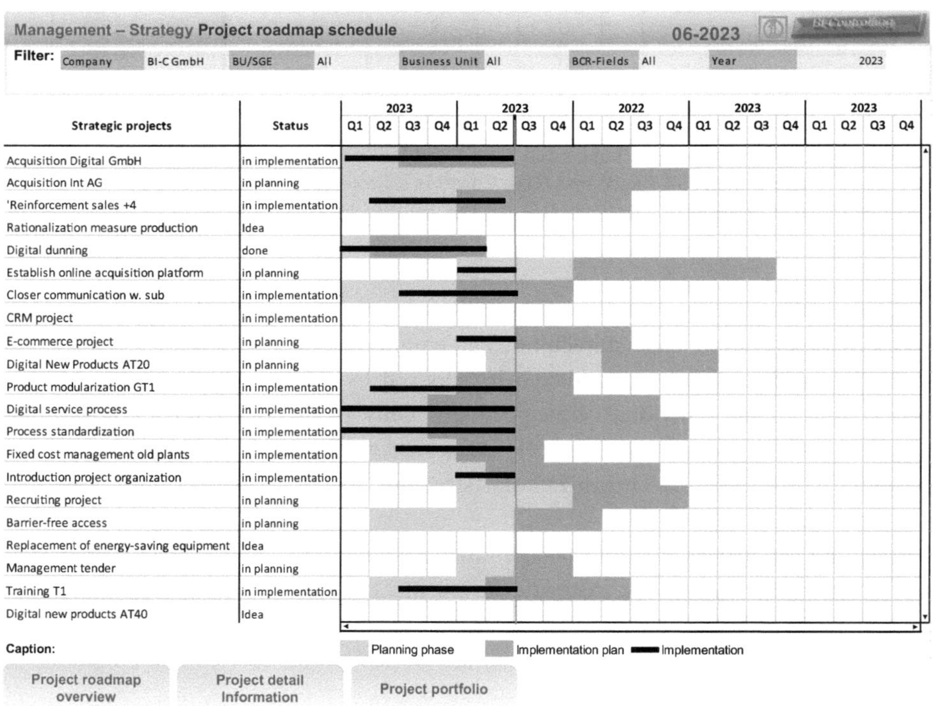

Fig. 3.3 Exemplary roadmap (schematic representation). (Based on the source: Schön and Irmer 2007, p. 248)

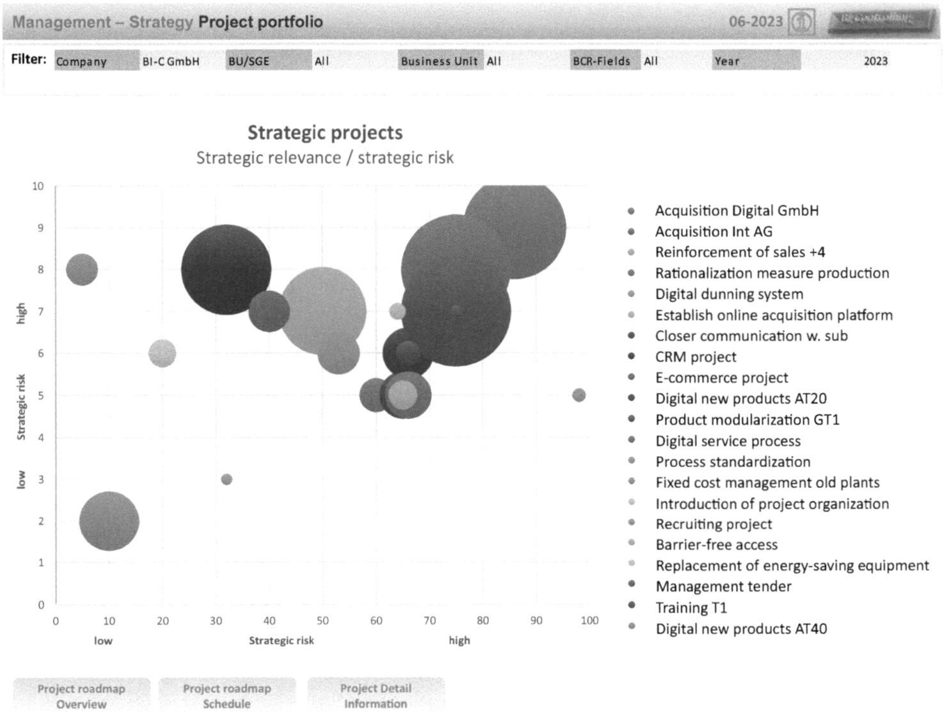

Fig. 3.4 Strategic project portfolio

The **Roadmap Management** aims at bundling strategic projects in a project portfolio from various areas of activity of the enterprise, e.g.:

- Strategic projects of the corporate management
- Long-term projects of the functional areas
  - e.g. measures in the area of supply chain management
  - e.g. compensation of capacity bottlenecks or vacancies in production
- Research and development projects
- Investment projects
- Projects to avoid hazards from risk management
- …

The roadmap is the result of a process that collects, evaluates, selects and prioritizes project ideas. As a complement to the project roadmap, it is advisable to systematize the projects in the company according to their strategic relevance and their strategic risk

Management – Strategy Project roadmap overview 06-2023

Filter: Company BI-C GmbH | BU/SGE All | Business Unit All | BCR-Fields All | Year 2023

Strategic risk	Status	Responsibility	Plan costs	Cost variance	Net present value	Utility (1-10)	Completion degree	Deadline-delay	Risk index
Acquisition Digital GmbH	in implementation	Mr Müller	4,720,000	472000	2.36	8.00	28.6%	-28	86
Acquisition Int AG	in planning	Mrs Maier	5,340,000	267000	5.34	9.00	0.0%	-10	75
Reinforcement sales +4	in implementation	Mr Schulz	3,430,000	171500	3.43	8.00	82.3%	-6	50
Rationalization measure production	Idea	Mrs Dehner		0	2.30	7.00	0.0%	0	
Digital dunning	done	Mr Sieger	56,300	-5630	ohne	5.00	100.0%	-15	32
Establish online acquisition platform	in planning	Mrs Tama	1,030,000	103000	1.03	8.00	0.0%	-61	80
Closer communication w. sub	in realization	Mr Sodlest	960,000	96000	0.96	4.00	76.4%	-14	65
CRM project	in realization	Mrs Konrad	3,600,000	360000	3.60	2.00	53.6%	2	32
E-commerce project	in planning	Mr Schwer	750,000	75000	0.75	3.00	0.0%	-30	40
Digital New Products AT20	in planning	Mrs Leicht	1,150,000	115000	1.15	5.00	30.0%	0	66
Product modularization GT1	in realization	Mr Tal	5,400,000	270000	5.40	2.00	64.2%	7	75
Digital service process	in realization	Mrs Schwarz	960,000	96000	0.96	5.00	75.3%	-6	66
Process standardization	in realization	Mr Schön	800,000	400000	0.80	4.00	42.3%	-5	53
Fixed cost management old plants	in realization	Mrs Zepper	1,600,000	160000	1.60	8	91.2%	-35	10
Introduction project organization	in realization	Mr Schröder	120,000	12000	ohne	4	65.0%	14	64
Recruiting project	in planning	Mrs Koslowski	80,000	8000	ohne	7	0.0%	-12	98
Barrier-free access	in planning	Mr Schubilski	360,000	-72000	ohne	2	0.0%	-26	65
Replacement of energy-saving equipment	Idea	Mrs Lotweg		0	1.06	3.00	0.0%	0	
Management tender	in planning	Mr Garten	55,000	5500	ohne	4.00	0.0%	0	75
Training T1	in realization	Mrs Morgen	263,000	26300	ohne	8.00	15.3%	-4	66
Digital new products AT40	Idea	Mr West		0	2.30	3.00	0.0%	0	

Project roadmap schedule | Project detail information | Project portfolio | Investment plan

Fig. 3.5 Project roadmap overview

(e.g. concretized by the complexity or the amount of resource commitment and project costs) in a project portfolio, so that it becomes more transparent which projects have a special strategic importance, are particularly urgent or have certain dependencies on each other. Projects with high strategic importance and low risk are called "quick wins" and are to be implemented with the highest priority (see Fig. 3.4). An important criterion for project evaluation is also the net present value or value contribution of the considered project.

When designing the roadmap management, the consistent control of the strategic projects through an appropriate project management and controlling is important (see Sect. 3.9.6). The following project plans and reporting information are to be used for the control:

- Compact project structure plans and their changes
- Schedule and process plans
- Resource and cost plans

- Success and financing plans
- Risk plans
- Project reports including degree of goal achievement and impact of project results (measured by the indicators in the Balanced Scorecard)

The instruments are to be designed appropriately in view of the economic importance of the strategic projects.[9] In the sense of the multi-project controlling and the "simultaneous planning", the management bodies of the company take into account all important information of the strategic projects at an early stage with regard to the partial and overall objectives of the company. The advantage is that difficulties and risks can be detected and controlled sooner. By specifying the resource use, costs, results and finances for a project, the possibility arises to take them into account directly in the operational and tactical planning. Ideally, the most important strategic projects are linked to the Balanced Scorecard, in order to be analyzed and monitored in the management cycle. Exemplary reports for an adequate project management are shown in Sect. 3.9.6. A slightly adapted overview report for the strategic planning can be found in Fig. 3.5. In a project roadmap overview, the most important strategic projects are listed with their key indicators (e.g. planned costs, cost variance, net present value, utility value, completion rate, schedule delay and risk index). The project roadmap schedule, the project roadmap portfolio, and detailed information on the individual projects and the investment plan can be accessed via the navigation buttons.

Both the roadmap planning and the Balanced Scorecard requirements result in an integration of the strategic planning with the operational and tactical planning. For the simulative, operational and tactical planning and reporting, a new system level is created that is detached from the previous systems but integrated. Business intelligence applications in the form of data warehouse-based planning and reporting systems are suitable as planning applications.[10] These enable, in interaction with the ERP systems or the merchandise management systems and the commercial systems as well as further upstream systems, to build a powerful and integrated planning and control environment for all areas of the company,[11] an environment which enables the necessary data aggregation and integration for planning and control.[12] Various planning cycles can be realized, e.g.:

---

[9] See Schön (2003).

[10] Cf. e.g. for SAP-BI and other IT solutions Egger et al. (2005) and Schön (2004a, p. 287–337).

[11] Cf. e.g. for SAP-BI solutions Heuser et al. (2003).

[12] Cf. Fischer (2003).

- The operational budget planning is carried out for 12 months.
- The tactical planning (multi-year planning) complements the operational planning up to e.g. the 3rd or 5th subsequent year.
- The forecast planning is usually carried out until the end of the year or on a rolling basis for 12 or more months.

The core task of operational planning is the preparation and coordination of the operational sub-plans, starting from the sales planning to the profit, finance and balance sheet planning for the entire company or the consolidation result for the group of companies. In contrast to a separate (tactical) medium-term planning, the supplementary medium-term planning is recommended after the operational budget planning. It is faster and more economical, if it can be carried out in an integrated planning system. The first year of the (tactical) medium-term planning can be automatically taken over from the budget planning.

In contrast to highly aggregated planning models, the integrated planning model is very strongly connected with the structures and values of the operational control systems of the ERP systems used as well as further upstream systems. This avoids the mistake of performing planning simulations in condensed planning models that are not feasible in reality in terms of resources and performance. Later operational corrections lead to multiple costs and sources of loss that are avoidable.

For the system of integrated corporate planning, the most significant business value creation processes of the companies have to be modeled and then implemented with IT support. In Fig. 3.6 a shortened model is sketched in which the essential sub-plans of corporate planning and their relationship to each other are listed.

The critical success factors of the companies and the sub-plans have to be integrated from the perspective of corporate planning in a comprehensive model. The granularity of the information has to decrease and give way to condensed information without significantly reducing the transparency and relevance for control.

The internal management control loop closes with the control and analysis of deviations, trends and other information of the reporting, which can be used for the internal steering. The balanced scorecard stands here as a representative for the entire reporting. The content and IT-supported implementation as well as the process and organizational anchoring will be comprehensively addressed in the following chapters.

If one complements the internal view of management control with the external reporting, the reporting should ideally also cover the external reporting requirements at the same time. This includes, among other things, the requirements of an integrated accounting system, in which external and internal accounting are linked, mainly affecting the income statement and the short-term income statement. In external reporting, the German Accounting Standards (DRS) and the International Financial Reporting Standards

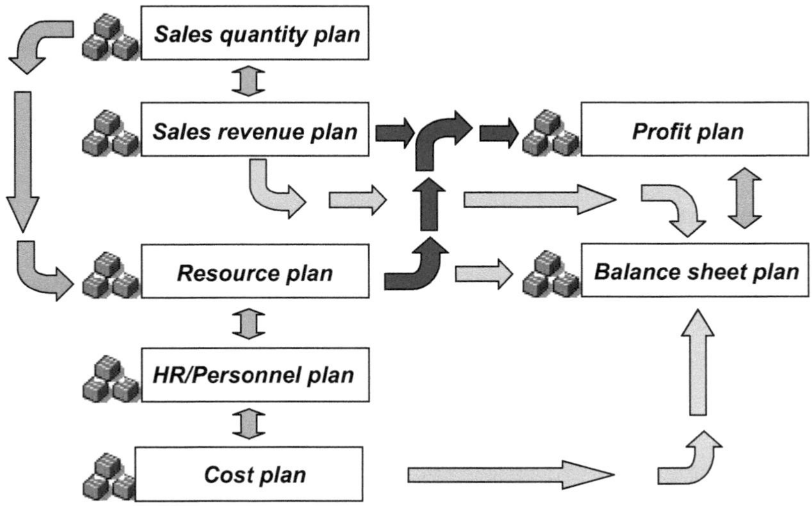

Fig. 3.6 Interdependencies of the operational sub-plans. (Slightly modified from Schön and Irmer 2007, p. 250)

(IFRS) as well as the reporting requirements of the Integrated International Reporting Council (IIRC) must also be taken into account.[13] Stakeholders such as customers, suppliers, banks, investors, employees, business partners, associations, and the state or society in general have an interest in being informed about the condition and development of the enterprise, and this interest must be adequately met. According to the framework concept of the IIRC, in addition to the establishment of guidelines for report preparation and presentation, the following components, which are mainly oriented towards the stakeholders, should exist:[14]

- Overview of the environment and the organization,
- Governance,
- Business model,
- Opportunities and risks,
- Strategy and resource utilization,
- Performance (output/success),
- Outlook.

[13] Cf. IIRC (2013, p. 4) and Weißenberger (2014, p. 441).

[14] Cf. IIRC (2013, p. 24).

From the perspective of the integrated model for strategic, tactical and operational control, these external reporting requirements should be met without creating unnecessary duplication of work. It is thus a complementary set of requirements that is simultaneously fulfilled in integrated reporting.

3.1.2 Planning Premises and Top-Down Specifications

Linking strategic planning with the tactical-operational level is done on the one hand by planning strategic projects in the roadmap as described in the previous chapter, and on the other hand by the specifications and premises that are defined in the context of planning.

The strategic framework conditions are given as premises in a central planning area in the planning process. Examples of such premises are typically

- Price increases of important raw, auxiliary and operating materials such as metals,
- Pay changes,
- Market assessments,
- Exchange rates,
- Operating and plant calendars

and many other planning premises.

An example of a premise planning is shown in Fig. 3.7, in which central key values for inflation rates, economic growth etc. can be set for a standard variant and possibly for further planning variants. In addition to the general planning premises, there are planning area-dependent variables, e.g. for personnel planning such as special shift models, pay scales and premiums.

The planning speed, especially the avoidance of many planning loops and coordination processes, can be avoided if key performance indicators (KPI) and key values of the result plan are given top-down and broken down and reconciled to the areas and individual result objects. This can already be done in the context of the BSC planning within the framework of the target values but also independently of this with the help of key performance indicators. Clear target specifications replace time-consuming budget discussions in this case.

An example of the key performance indicator sales is shown in Fig. 3.8. The total sales are broken down here to the result objects of the subsidiaries and the business areas and major customers. The counter-current planning shows whether the target value has been reached or exceeded or fallen short of.

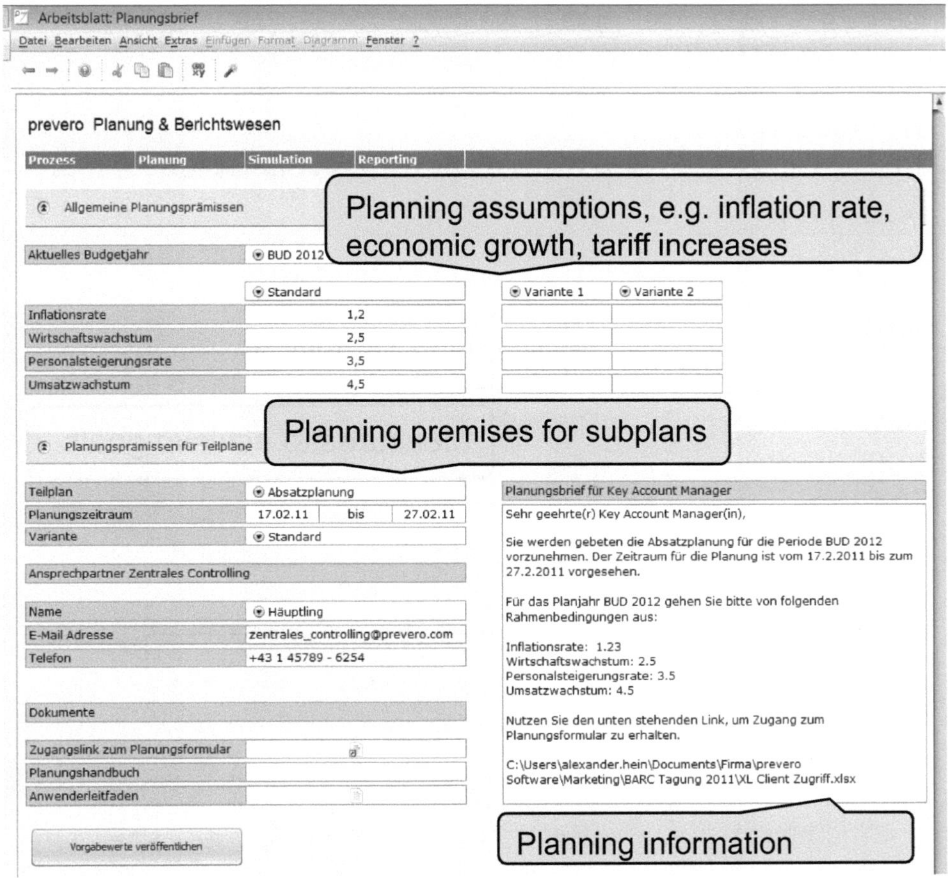

Fig. 3.7 Exemplary premise planning. (Provided from the lecture notes of Hein 2011, p. 39)

3.2 Value-Creating Factors of the Business Model

The identification of the control-relevant and value-creating factors of the business model of a company forms, in addition to the strategic orientation, another important component for the design of the planning and reporting. Here, the operative control understanding in connection with the business model has to be worked out with the management.

For example, if the operative business consists of the marketing and execution of project services on the one hand, and the sale and production of different components on the other hand, these two business directions have different requirements for the planning

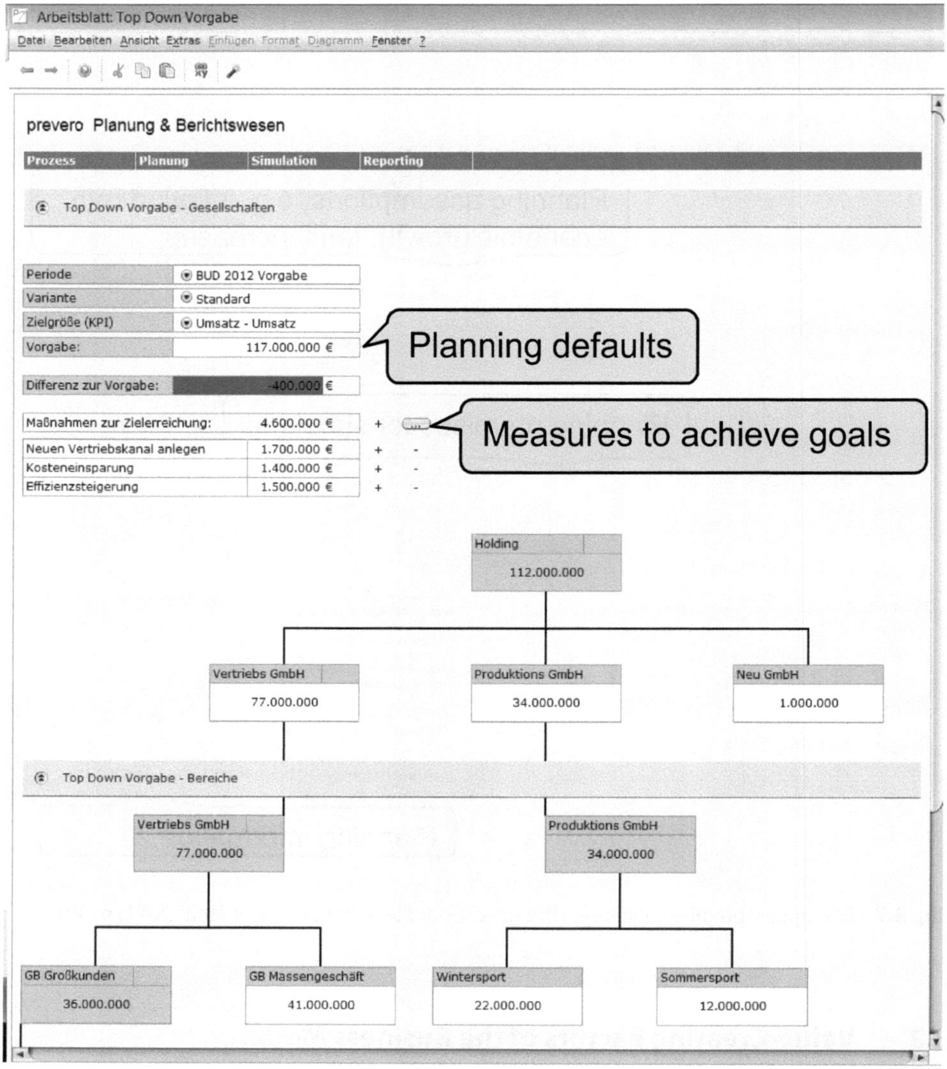

Fig. 3.8 Exemplary top-down specifications and reconciliations. (Provided from the lecture notes of Hein 2011, p. 41)

and reporting. In business practice, this often leads to the dominant part of the business model asserting its requirements in the planning and reporting and the systems being mostly aligned to it. However, a reasonable approach would be the adequate representation of all business parts. This means, for example, that the result analysis should be set up once from the project perspective and once from the product perspective.

Examples of central value-creating factors in the company are listed below:

- Sales volumes, prices and conditions
- Purchase volumes, prices and conditions
- Personnel hours performed and rates
- Times for productive and indirect activities
- Temporary worker ratio
- External service ratio
- Other cost variables
- Payment behavior
- ...

The variables have to be examined according to their influenceability. Non-influenceable variables have to be accepted as framework conditions and should be analyzed with regard to their deviations. Influenceable variables are the leverage variables that have to be positively changed in favour of the company goals by targeted project and measure derivation. If the company considers the value-creating factors in the planning, this is also called a driver- or value-driver-based planning.

3.3 Structure and Navigation in Reporting

The **structure in reporting**should follow the basic principle that a more in-depth analysis must be possible from the general company analysis in the top management over the next management levels of the company to the decentralized units. When looking at a corporate group, for example, the structure of the group would be considered starting from the overall group, over the subgroups, to the individual companies.

Figure 3.9 shows, for example, the alignment of the reporting from the central corporate analysis of a single company to the three management levels business units, locations and functional areas.

Depending on the organizational structure (see Sect. 4.1.3), the specifics of the divisional, regional, and functional organization or the matrix organization have to be taken into account. Each level receives a central top-level indicator report from which one can branch into the detailed reports. Because of the integration with the strategic orientation, the Balanced Scorecard is suitable as a top-level indicator report (see Sect. 3.1.1) which has to be cascaded to the respective level. In medium-sized companies, where no Balanced Scorecard (BSC) has been established yet, alternative indicator reports are available which provide the most important control variables of the respective areas for analysis.

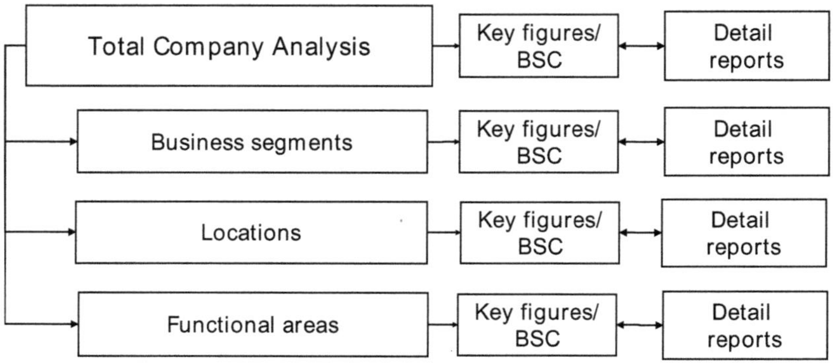

Fig. 3.9 Example of structure in reporting

An example taken from corporate practice shows how this basic principle can be mapped in the form of analysis paths in reporting (see Fig. 3.10).

The start cockpit displays the top-level indicators of the entire company. Here, the most important indicators for controlling the individual companies of the company are highlighted as target-actual deviations.

For the **navigation**, a flexible model is implemented. From the start cockpit, one can either switch to a start navigation, from which the detailed analyses can be accessed, or directly jump to linked detailed reports via linking of the top-level indicators.

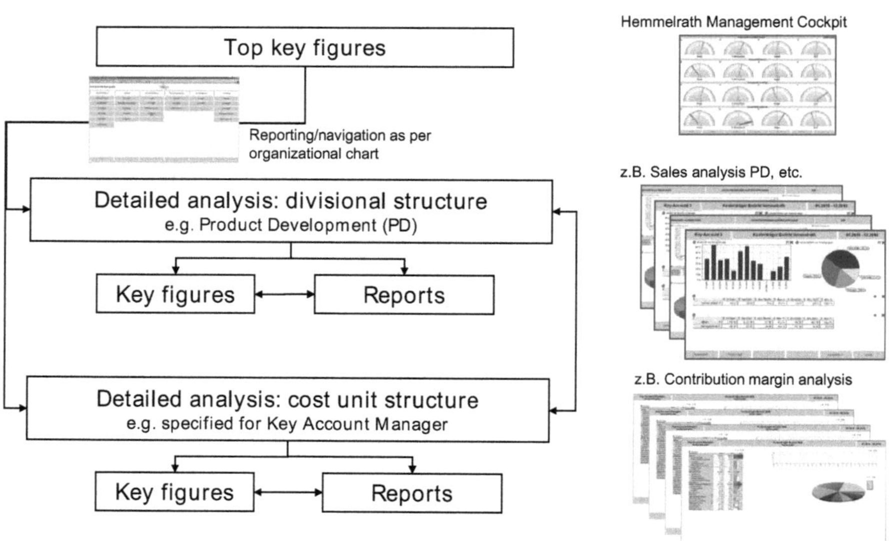

Fig. 3.10 Analysis paths in reporting. (Taken from Schön et al. 2011, p. 232)

The start navigation allows an entry into the entire reporting from the perspective of the organizational areas (see Fig. 3.11) or the cost center areas (without illustration). In the area navigation (e.g. for production and sales), the corresponding top-level key figure reports and further individual reports can be accessed. Top-level key figure reports are to be designed as entry-level reports. At first glance, the most important control-relevant facts of the management area are to be analyzed here. Due to the IT support in today's time, these entry-level reports are available in a timely manner, at least daily and independent of location (mobile). In the cost center navigation, the most important key figures and reports for the customer, plant and item view can also be analyzed.

The structuring possibilities of the reports depend, as stated above, very much on the organizational structure of the company. Since divisions, product groups, and regional structures are very company-specific, the following will list structuring possibilities for the top management level and the functional areas. Here, reports that are used in several areas are deliberately assigned multiple times:

- Corporate reporting/Top management
  - Top-level key figures/Balanced Scorecard
  - Sales/revenue reporting
  - Order intake/order backlog reporting
  - Income statement
  - Contribution margin-oriented project/product income statement

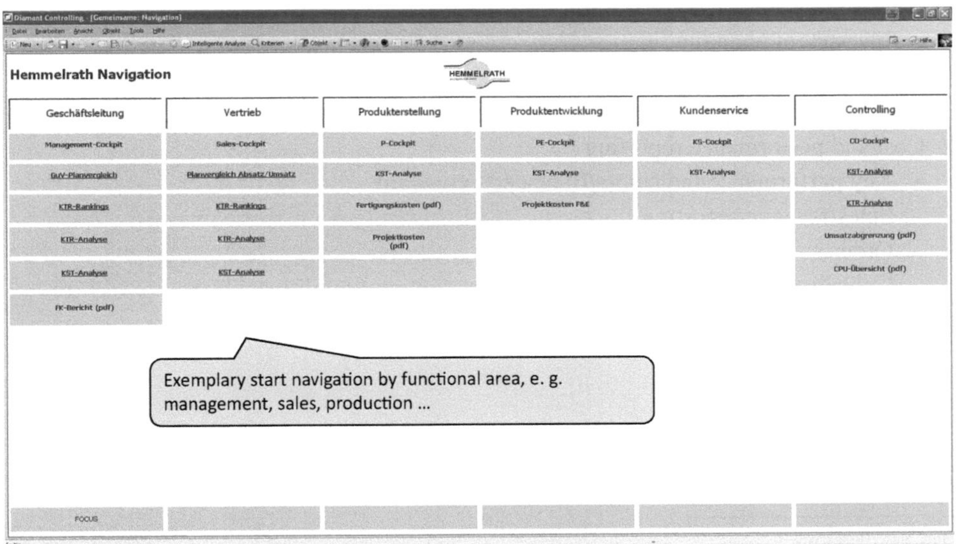

Fig. 3.11 Example of start navigation. (Taken from Schön et al. 2011, p. 233)

  - Profit center reporting
  - Cost center reporting
  - Balance sheet
  - Income statement/Profit & loss statement/business economic analysis
  - Asset schedule
  - Equity development
  - Management report
  - Development reports
  - Strategic reporting (environmental/competitive/SWOT analysis etc.)
  - Competitor and benchmark analyses
  - Opportunity and risk reports
  - Corporate governance reports
  - Overview of company and environment
  - Overview of organizational structures
  - Business model and business plan
  - Sustainability reports[15]
  - Supervisory board reporting
  - Reporting for the advisory board and external recipients
- Financial and Investment reporting
  - Key figures/Balanced Scorecard
  - Open items/payment statistics
  - Liquidity development
  - Financial planning and capital requirement determination
  - Cashflow statement
  - Investment planning
  - Bank reporting
  - Rating reporting
- Cost and performance reporting
  - Key performance indicators/Balanced Scorecard
  - Sales/revenue reporting
  - Order intake/order backlog reporting
  - Contribution margin-oriented project/product profitability analysis
  - Income statement
  - Profit center reporting
  - Cost center reporting

[15] In sustainability reporting, reports on the ecological, economic and social sustainability of the company are required for the stakeholders of the company. Sustainability management is also dealt with under the term Corporate Social Responsibility (CSR). See regulations of the Commission of the European Communities (2001, p. 7).

- Sales reporting
  - Top indicators/Balanced Scorecard
  - Sales/revenue reporting
  - Order intake/order backlog reporting
  - Contribution margin-oriented product profitability analysis
  - Contribution margin-oriented customer profitability analysis
  - Cost center reporting
  - Profit center reporting
  - Product analyses
  - Customer analyses (incl. key account analyses)
  - Offer and sales opportunity statistics
  - Seller analyses
  - Competitor and benchmark analyses
  - Price and condition analyses
  - Warehouse and shipping analyses
  - Marketing analyses
  - Contract analyses
  - Service analyses
  - Employee analyses
  - Complaint analyses
- Production reporting
  - Key performance indicators/BSC
  - Cost center reporting
  - Capacity analyses
  - Hourly analyses
  - Downtime analyses
  - Maintenance analyses
  - Error analyses
  - Quality reports
  - Employee analyses
  - Accident statistics
- Procurement reporting
  - Key performance indicators/BSC
  - Supplier analyses
  - Purchasing analyses
  - Order and delivery analyses
  - Price and condition analyses
  - Employee analyses
  - Supplier income statement
  - Buyer analyses

- Delivery and storage analyses
- Cost center reporting
- Research and development
  - Top indicators/BSC
  - Innovation reporting
  - Cost center reporting
  - Employee analyses
  - R&D project reports
- Project reporting
  - Top indicators/BSC
  - Project income statement
  - Project schedule and progress overview
  - Project payment overview
  - Earned-value analysis
  - Employee analyses
  - Project risk reports
- Human resources reporting
  - Top indicators/BSC
  - Employee analyses
  - Hour analyses
  - Qualification reports
  - Personnel cost analyses
  - Cost center reporting
  - Collaboration reports with tasks, notes and messages
- Logistics reporting
  - Delivery analyses
  - Transport analyses
  - Warehouse analyses
  - Cost center reporting
- IT-Reporting
  - CPU statistics
  - Cost center reporting
  - Employee analyses
  - Service analyses
  - IT cost analysis
  - Downtimes
- ...

Not all areas and individual reports have to be relevant for the company. For smaller companies, the number of reports will be significantly lower.

The structuring can be done in parallel or individually or in combination. Here, the "right classification system" has to be designed individually for each company. In principle, the classifications should be **non-overlapping** as well as **complete** and contain **similar elements** on the same structuring levels. Non-overlapping does not mean that reports from different perspectives cannot also occur twice, but that one should avoid using different report names for the same or similar contents. Once the structure is fixed, it should be communicated to the decision makers and report recipients in the company, so that the access can be done quickly. The structures are then created in parallel in the case of different IT systems (portal, BI frontend, ERP system etc.) or media (print reports). It is advisable to use the structure as uniformly as possible and to fix it for a period of time.

For the order and access of the reports, there are various data-processing possibilities, such as single-level or multi-level lists, folder structures, tabs, report folders (briefing books), storyboards or navigation reports (see Sect. 5.5.5.7). Through these, the individual reports can be accessed directly. In a **storyboard**, a complete report is composed of individual reports, additional information, comments and hints that are in a thematic context. In this case, one also speaks of a **storytelling**.

In the individual reports, further individual reports can be accessed or branched back to the central navigation report via menu functions, context menu functions (right mouse button), links or again via graphical buttons (e.g. similar to a status bar at the bottom of the report layout), so that the user can quickly move through the entire available reporting system.

A limitation is made here by the authorization system and the user profile stored here. If the report structure is built from the outset according to the user role, the navigation can also be a kind of pre-filtering of the possible reports to be accessed. Reports and data for which no permissions exist are then not listed or displayed at all.

3.4 Structure and Navigation in Planning

The structure and navigation of the planning is primarily process-oriented. The planning process can also be very well integrated with the reporting phases. An example of a process-oriented overview of the reporting and planning steps using a calendar is shown in Fig. 3.12.

Here, the monthly, quarterly and annual financial statements are displayed in the planning calendar next to the strategic planning, the medium-term planning and the budget planning with all their planning areas.

A shortened process model of a planning in which the connections between individual planning areas can be seen is shown in Fig. 3.13. It shows, besides the strategic planning, important planning areas that are typical of an industrial manufacturer.

The planning is based on the budgeting according to the sub-planning areas. Starting from the sales plan, it follows primarily the value creation process of the company up to the income, balance sheet and financial plans.

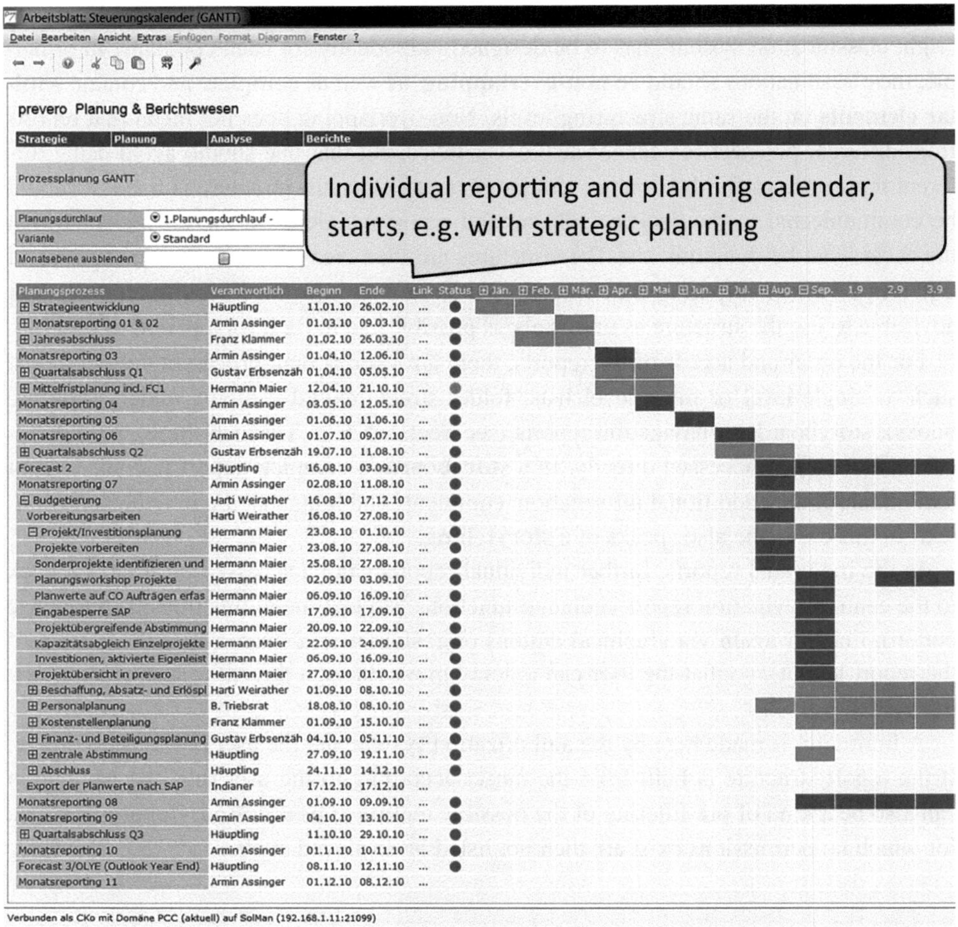

Fig. 3.12 Reporting and planning calendar with prevero. (Provided from the presentation materials of Hein 2011, p. 38)

Further criteria for dividing the planning into areas, which depend strongly on the company structure and organizational structure as in the reporting, are discussed in Sect. 4.1.3. Criteria for dividing the planning can be, e. g., the group structure with its companies, the business fields, but also the branches, locations and functional areas.

An example of a possible division of the planning according to the planning process and the sub-planning areas shows the following structure:

- Strategic planning
  - Vision
  - Strategic goals

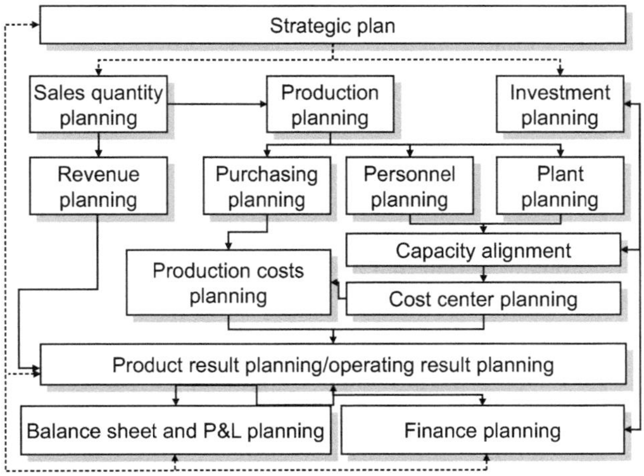

Fig. 3.13 Model of the sub-planning areas. (Slightly adapted from: Schön and Irmer 2010, p. 51)

  - Strategic project planning/roadmap/project portfolio planning
  - Balanced Scorecard/top indicators planning
  - Risk planning
  - Top-down specifications
- Medium-term planning (Tactical Planning)
  - Sales/revenue planning
  - Profit planning
  - Financial planning
  - Investment planning
  - Income statement and balance sheet planning
  - BSC/key performance indicator planning
- Operational Planning (Budgeting/Annual planning)
  - Premise planning
  - Sales/Revenue planning
  - Production planning
  - Material/purchasing planning
  - Resource planning
    Direct Personnel planning
    Plant planning
  - Indirect personnel planning
  - Product success planning/contribution margin planning
  - Project planning
    Customer order-related projects
    R&D projects
    Internal projects

  Other projects
  Multi-project planning
  - Capital budgeting
  - Income or Profit planning
  - Cost center planning
    Central primary cost type planning
    Remaining primary cost and performance planning
    Secondary cost planning (internal cost allocation)
    Rate determination
  - Profit center planning
  - Other profit planning (taxes, financial result, non-periodic and extraordinary result)
  - P&L planning
  - Balance sheet planning
  - Financial/cash flow planning
  - Key figure planning
  - Operative risk planning
- Forecast (extrapolations/predictions)
  - Based on the end of the year
  - Rolling for a period of e.g. 12 months in advance.

Individual planning areas can be further subdivided, such as the project planning with regard to:

- Project schedule planning
- Project resource planning
- Project outcome planning
- Project payment planning
- Project risk planning

When planning, planning loops, coordination and approval rounds, e.g. between top-down specification and bottom-up planning, as well as simulations and scenario planning in different planning versions have to be taken into account, whereby a final version has to be released as planning version (see also Sect. 4.2.2).

3.5 Reporting and Planning Objects (Dimensions, Hierarchies and Values)

The most important reporting and planning objects of the company can be differentiated into dimensions, hierarchies, attributes, and values, so-called key figures (measures).

Values (key figures, measures) are the parameters to be planned or reported, such as quantities (e.g. piece numbers and working hours), amounts in monetary units (e.g.

revenue and cost values, contribution margins) but also stock values (fixed asset values) and percentage values (e.g. profitability ratios).

The values can be evaluated for dimensions of the company. Dimensions (e.g. cost centers, products and customers) can occur as hierarchies (e.g. the cost center hierarchy) but also as attributes with only one or a few characteristics (e.g. the gender of the employees).

The objects can be analyzed or planned from the overall, group or individual object perspectives or vice versa via the hierarchies. This is why the object hierarchies are so interesting for analysis and planning.

Important hierarchies in the company can be found in the following list:

- Corporate hierarchy
- Customer hierarchy
- Product hierarchy
- Profit center hierarchy
- Cost center hierarchy (functional areas, departments, branches)
- Account or revenue and cost type groups
- Cost object hierarchy
- Item/material hierarchy
- Project hierarchy
- Process hierarchy
- Supplier hierarchy
- Resource component hierarchy
- Hierarchies for cooperations and networks
- Time hierarchy
- Quantity hierarchy (e.g. for units of measure)
- Value hierarchy

Through the hierarchy or by further derivation rules, additional attributes for planning and reporting can be identified, among others:

- Industries (e.g. through the attributes of customer or supplier)
- Regions (e.g. through the attributes of customer or supplier)
- Product groups (e.g. through the item or material hierarchy or the attributes of sellable products or resource components)
- Segments (e.g. through the profit center hierarchy or cost object hierarchy)
- Branches (e.g. through the profit center or cost center hierarchy)

If the account model and not the key figure model is used for data modeling (see Sect. 5.5.4.1), value and quantity hierarchies can be derived from the value dimensions of the account model. In particular, the value hierarchies are very useful in business reporting as many reports have standardized line schemes, such as the income statement, the liquidity analysis, the cost center report or the balance sheet (see Sects. 3.9.1.1, 3.9.1.2, 3.9.1.3 and 3.9.1.4). If the schemas are arranged as value hierarchies in one dimension, individual result levels can be dynamically analyzed in the analysis by folding and unfolding the result levels. It is quasi a dynamic drill-down or bottom-up analysis along the value hierarchy.

Some of the hierarchies are already predefined in many programs as they are the same for all companies. These include especially the hierarchies for time and quantity units.

Hierarchies and attributes for planning and reporting should meet the following criteria:

- Completeness
- Uniqueness/consistency/non-overlapping
- Uniformity

The basic problem with the hierarchies and evaluation attributes is the determination of the assignment at the data source for the movement data. If information or derivation possibilities (via the master data) are missing or ambiguous already at the booking documents and the movement types, the planning and reporting cannot be properly supplied. In addition, the master data problem and the associated data quality is a basic problem in many companies (see Sect. 4.2.4.4). The correct evaluation and forecast of data starts already with the correct classification. Furthermore, new requirements of the master data assignment are systematically ensured by a change management process.

3.6 Types of Reports

For the information processing, the following types of reports are distinguished for operational use:[16]

- Standard reporting
- Exception reporting (exception reports/triggered deviation reports)
- Analysis reporting *and*
- Ad-hoc reporting (individually requested reports)

Standard reporting takes place at fixed dates and cycles to exactly specified recipients. The content and form of the standard reports are fixed and cannot be changed spontaneously by the recipient. Therefore, the contents and design of these reports must also

[16] Extended in comparison to: Küpper (2004, p. 171 ff.) and Göpfert (2006, p. 695).

be defined as far as possible in the context of the information needs analysis, so that as few as possible or no information gaps can arise for the recipient. As practice shows, however, the standard reports are also subject to a certain shelf life and are eventually changed at intervals for various reasons (e.g. restructuring, new influences).

Exception reports or **triggered reports** are generated when predefined tolerance sizes, deviation or threshold values are reached. They are automatically presented to the recipients when predefined limits are exceeded for information analysis and help to draw attention to particularly relevant issues for control. Threshold values for early warning indicators (so-called sentinels = Sentinels) can be determined manually or in connection with statistical functions such as setting a confidence level. In the form of Desktop Notifications, threshold exceedances can also be displayed directly on the desktop without the application having to be open.

Analysis reporting summarizes the structured research and search for new insights based on the existing data basis of the information systems by the report recipient. In contrast to standard reporting and triggered reports, the entry point is partly predefined, but the further analysis of the information is not predetermined; instead, it is created interactively between the analyst and the information system by querying a provided data set. The advantage is that the analyst is not limited by predefined report structures, but can extend and change their information needs individually as far as the provided data set allows.

Individually requested demand reports or **ad-hoc reporting** are created based on a professional authority and their specific information needs. They often result from the occurrence of events (e.g. significant deviations or negative results) and the goal of creating transparency about them, as well as from new information needs that cannot be derived from standard reporting. Ad-hoc reports can be generated quickly with the analysis reporting as far as the data set allows. In some cases, the existing data set has to be enriched with the data that is not yet available, and the design of the data query has to be specially adapted.

Demand reports are explicitly requested by the recipient to cover the additional need for individual information in the short term. Consequently, the demand report should only contain the information requested by the recipient. With the development of information and communication technology, this form of reports is gaining in importance. They increasingly complement the standard reports. The recipient assumes, by means of direct access to central databases, the active role of information producer (see Sects. 4.1.5.4 and 5.7.3, keyword Self-Service BI).

While a standard reporting system is now technically implemented at the level of the so-called accounting systems (e.g. cost accounting), the information technology implementation of problem-oriented special reports usually requires the use of special analysis and reporting systems (report generators) that are based on different database technologies (e.g. relational databases, OLAP databases or NoSQL databases) and storage techniques (hard disk versus in-memory). The IT support is discussed extensively in Chap. 5. The extension and classification of the analysis and forecasting spectrum in BI-supported controlling is shown in Sect. 5.7.3.

In addition to this controlling and management-related reporting system, further reporting requirements are also necessary for the operational execution of the business; these requirements can partly be created with tools of office communication or with the same tools of report generators. This includes, among other things, the form system, e.g. for the creation of delivery notes and invoices. These are not the subject of further investigation.

3.7 Basic Forms of Reporting

The number of basic forms of reporting for comparisons and analyses is very diverse and specific in business practice. Nevertheless, important basic patterns for reporting can be worked out. For this reason, important basic forms for reports as well as their advantages and disadvantages are elaborated in the following sections. The different basic forms of reporting can be applied both alone and in combination.

3.7.1 Actual-Actual Comparisons

Actual-actual comparisons consider a result object with respect to a simple time comparison. Here, identical periods such as years, quarters or months are compared (see Fig. 3.14).

The example report is structured in **T-shape format**. In the middle, the report lines are listed. In the left part of the table, the comparison of the accumulated actual values up to the current month in comparison to the previous year is shown. In the right part of the table, the monthly comparison is shown. Absolute deviation values are often

ACT YTD	PY YTD	Δ PY	Δ PY %	Report line	ACT mths.	PY mths.	Δ PY	Δ PY %
1,000	900	100	11.1%	Sales revenues	133	110	23	21.2%
200	250	-50	-20.0%	Other revenues	17	19	-2	-11.5%
400	400	0	0.0%	Changes in inventories	37	33	4	12.0%
1,600	**1,550**	**50**	**3.2%**	**Total output**	**187**	**162**	**25**	**15.5%**
500	600	-100	-16.7%	Cost of materials	42	54	-12	-22.8%
100	200	-100	-50.0%	External services	8	17	-8	-50.0%
600	500	100	20.0%	Personnel costs	50	42	8	20.0%
100	50	50	100.0%	Depreciation	9	4	5	124.0%
100	150	-50	-33.3%	Other costs	8	11	-2	-20.6%
1,400	**1,500**	-100	-6.7%	**Total costs**	**118**	**127**	-9	-7.3%
200	**50**	**150**	**300.0%**	**Profit (before taxes)**	**70**	**35**	**35**	**98.1%**

Fig. 3.14 Actual-actual comparisons

supplemented by percentage deviation values. Comments on the causes and consequences for the future development should be added to significant deviations.

Advantages of actual-actual comparisons are:

- Changes compared to the reference period are quickly recognizable.
- The data basis of the comparison periods can usually be determined well.
- Experience values and orientation values can be derived.

Disadvantages of actual-actual comparisons are:

- The comparison periods have different company states, so that the values are difficult to compare without adjustment.
- In extreme cases, there are no historical comparison data, e.g. for a new business field.
- Inefficiencies in the past are not recognizable.
- The analysis of the target achievement of plan values is not possible due to the lack of plan values.
- Forecasts about the future development are not taken into account.
- Trends in terms of a detailed temporal development of the individual period values are not recognizable in pure period comparisons but can be evaluated in combination with time series comparisons.

3.7.2 Target-Actual or Planned-Actual Comparisons

Target-actual comparisons or Plan-actual comparisons compare the actual values achieved by a result object with the set target values, e.g. the target or plan value for a considered period. The deviations show whether the set target values were met, exceeded or missed (Fig. 3.15).

ACT YTD	PL YTD	Δ PL	Δ PL %	Report line	ACT mths.	PL mths.	Δ PL	Δ PL %
1,000	1,000	0	0.0%	Sales revenues	133	118	15	12.7%
200	250	-50	-20.0%	Other revenues	17	19	-2	-11.5%
400	400	0	0.0%	Changes in inventories	37	33	4	12.0%
1,600	**1,650**	**-50**	**-3.0%**	**Total output**	**187**	**171**	**17**	**9.9%**
500	500	0	0.0%	Cost of materials	42	46	-4	-8.8%
100	150	-50	-33.3%	External services	8	13	-4	-33.3%
600	450	150	33.3%	Personnel costs	50	38	13	33.3%
100	100	0	0.0%	Depreciation	9	8	1	12.0%
100	100	0	0.0%	Other costs	8	6	2	31.6%
1,400	**1,300**	**100**	**7.7%**	**Total costs**	**118**	**110**	**7**	**6.6%**
200	**350**	**-150**	**-42.9%**	**Profit (before taxes)**	**70**	**60**	**9**	**15.8%**

Fig. 3.15 Target-actual or plan-actual comparisons

Plan values are the original planned values, which were e.g. fixed as budget values in the operational planning. Target values are plan values adjusted to the actual activity. In the cost center accounting, this is done e.g. by separating the fixed and variable plan costs, whereby the variable plan costs are adjusted to the actual activity and the fixed costs, as the name implies, are retained, as they do not react to changes in activity.

The example report is again in T-shape format. In the middle, the report lines are listed. In the left part of the table, the comparison of the accumulated actual values up to the current month with the plan is shown. In the right part of the table, the monthly comparison with the plan is shown. Absolute deviation values are supplemented by percentage deviation values.

Advantages of target-actual or plan-actual comparisons are:

- Inefficiencies compared to the set plan are quickly recognizable.
- Control needs to achieve the set target objectives are faster to identify.

Disadvantages of target-actual or plan-actual comparisons are:

- A pronounced planning is necessary.
- The planning can already be outdated by current events.
- Trends in terms of a detailed temporal development of the individual period values are not recognizable in pure period comparisons but can be evaluated in combination with time series comparisons.
- Forecasts about the future development remain unconsidered.

To better visualize deviations, table presentations can be coupled with graphical elements. For example, important negative outliers are highlighted with a red traffic light, and deviations are displayed in the form of a bar chart. In addition, actual, plan and previous year values are abbreviated and graphically highlighted in a uniform notation, so that they can be distinguished immediately in any report (see Fig. 3.16).

3.7.3 Planned-Forecast-Comparisons (Feedforward/Extrapolations/Predictions)

Since planned-actual comparisons and target-actual comparisons do not show an updated forecast of the future development, plan-will comparisons are often prepared as a complement. "expected actual values" stand for forecasts regarding the development of the future actual values considered for a target period, usually the end of the fiscal year or, in the case of a project report, the planned project end. These "expected actual values" are

German states		PY	PL	AC	ΔPY	ΔPL	ΔPL%	ΔPL
NRW		21.2	18.2	24.8	+3.6	+6.6	+36%	
Hessen		15.0	13.4	15.1	+0.1	+1.7	+13%	
Bayern		5.6	5.9	5.4	-0.2	-0.5	-8%	
Baden-Württemberg		4.6	4.8	5.3	+0.7	+0.5	+10%	
Niedersachsen	•	84.0	91.7	81.5	-2.5	-10.2	-11%	
Berlin	•	18.4	18.4	13.5	-4.9	-4.9	-27%	
Hamburg		8.4	7.7	9.9	+1.5	+2.2	+29%	
Bremen		25.1	25.4	25.4	+0.3	+0.0	+0%	
Schleswig-Holstein		5.8	2.8	3.2	-2.6	+0.4	+14%	
Rest		2.6	2.4	2.8	+0.2	+0.4	+17%	
Deutschland		190.7	190.7	186.9	-3.8	-3.8	-2%	

Fig. 3.16 Graphically supported table presentation

also referred to as forecast (preview) or extrapolations or projections. They are composed of the accrued actual values and the future updated remaining plan values until the considered target date. For the generation of the remaining plan values, different methods can be applied:

- Adoption of the old periodic plan values
- Manual (analytical) replanning of the periodic plan values
- Trend calculations using statistical and mathematical methods (often based on the past actual values). If statistical methods are used in connection with data mining or predictive analytics (see Sect. 5.7.4) for the forecast calculation, this is also called a "digital or predictive forecast". In this case, in addition to the data from, for example, accounting and the ERP system, structured and unstructured data from external sources (websites, IoT, etc.) can also be used to create forecast models (e.g. value driver models),[17] compare also the possibilities of using AI (artificial intelligence) for forecast models (see Sect. 5.7)

While pure extrapolations are based on the trend calculation of historical data, forecast values or projections can also include updated analytical plan values regarding the future.

The example report in Fig. 3.17 complements a plan-actual comparison in the left part of the table with the plan-will comparison in the right part of the table. The comparison value is presented here as a forecast.

[17] See Mehanna (2016, p. 22–25) and Möller et al. (2017, p. 509–518).

Report line	ACT YTD	PL YTD	Δ PL	Δ PL %	Forecast	PL total	Δ PL	Δ PL %
Sales revenues	1,000	900	100	11.1%	1,300	1,200	100	8.3%
Other revenues	200	250	-50	-20.0%	350	400	-50	-12.5%
Changes in inventories	400	400	0	0.0%	700	600	100	16.7%
Total output	**1,600**	**1,550**	**50**	**3.2%**	**2,350**	**2,200**	**150**	**6.8%**
Cost of materials	500	500	0	0.0%	800	750	50	6.7%
External services	100	150	-50	-33.3%	250	300	-50	-16.7%
Personnel costs	600	450	150	33.3%	700	600	100	16.7%
Depreciation	100	100	0	0.0%	200	150	50	33.3%
Other costs	100	100	0	0.0%	150	120	30	25.0%
Total costs	**1,400**	**1,300**	**100**	**7.7%**	**2,100**	**1,920**	**180**	**9.4%**
Profit (before taxes)	**200**	**250**	-50	-20.0%	**250**	**280**	-30	-10.7%

Fig. 3.17 Plan-will comparison (forecast)

Advantages of planned-forecast comparisons are:

- Inefficiencies compared to the set plan are quickly recognizable.
- Control needs to achieve the set target objectives are faster to identify.
- Forecasts about the future development are taken into account.

Disadvantages of planned-forecast comparisons are:

- A pronounced updating of the planning is necessary.
- Trends in terms of a detailed temporal development of the individual period values are not recognizable in pure period comparisons but can be evaluated in combination with time series comparisons.

3.7.4 Target Achievement Reports

Target achievement reports compare the current and actual values achieved by a result object with the set target values for the entire period under consideration. This is intended to show, for example, how many sales still need to be generated after the third quarter to achieve the planned annual sales. The deviation thus shows the open values that need to be achieved or exceeded in the remaining time.

The sample report in Fig. 3.18 shows for various business indicators on the left how high the target achievement is compared to the annual plan (per annum) and in the rear part compared to the previous year's value. The target achievement rate is shown as a ratio and supplemented by the open absolute values in the following column.

Advantages of target achievement reports are:

- Control needed to achieve the set target values are quickly recognizable by the target achievement rate.

Business figures (in €m)	PL p.a.	ACT YTD	PL in %	to go	PY p.a.	PY in %	to go
Operating income	8,980	8,709	97.0%	-271	8,688	100.2%	21
Operating expenses	8,168	7,988	97.8%	-180	7,953	100.4%	35
Profit	812	721	88.8%	-91	735	98.1%	-14
Return on sales	9.0%	8.3%	91.6%	-0.8%	8.5%	97.9%	-0.2%
Capital expenditure	516	431	83.5%	-85	415	103.9%	16
Free cash flow	6,840	5,950	87.0%	-890	5,760	103.3%	190
Balance sheet total	71,603	84,676	118.3%	13,073	82,659	102.4%	2,017
Equity	2,857	3,534	123.7%	677	2,745	128.7%	789
Number of employees (FTE)							
Group (excluding temporary staff)	44,178	44,803	101.4%	625	44,698	100.2%	105
Headquarters (excluding temporary staff)	32,919	30,863	93.8%	-2,056	30,699	100.5%	164
Temporary staff	1,571	1,690	107.6%	119	1,602	105.5%	88

Fig. 3.18 Target achievement report

- The open values that still need to be achieved in the remaining time for the entire period considered are quickly recognizable.

Disadvantages of target achievement reports are:

- A pronounced planning is necessary.
- The planning can already be overtaken by current events.
- Trends in terms of a detailed temporal development of the individual period values are not recognizable in pure period comparisons, but can be evaluated in combination with time series comparisons.

3.7.5 Time Series Analysis

Comparisons between individual values for individual periods do not reveal trends. Outliers below or above in individual periods are also hardly identifiable. Therefore, time series analysis is often used as a complement.

The sample report in Fig. 3.19 shows a graphical and a tabular time series for sales figures of certain products in the upper and lower part, respectively. Graphical time series analysis are more intuitive for the viewer than table representations. Important cyclical fluctuations and outliers should be explained with comments.

Advantages of time series analysis are:

- Trends in terms of a detailed temporal development of the individual period values are very well recognizable.
- Outliers in the course of time above and below are very well recognizable.

Disadvantages of time series analysis are:

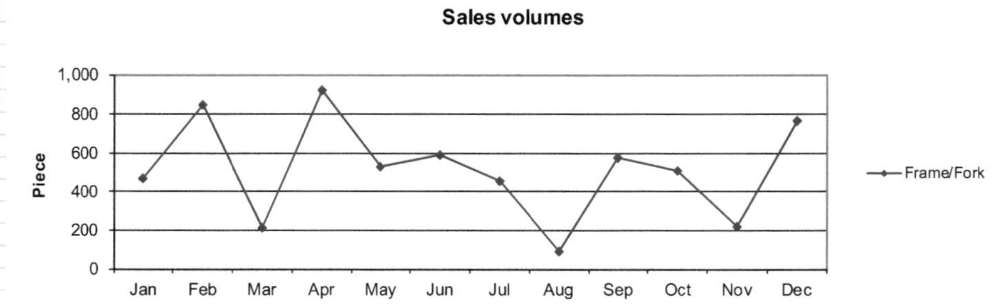

Products	Jan	Feb	Mar	Apr	May	Jun	Jul	Aug	Sep	Oct	Nov	Dec	Total
Frame/Fork	471	847	213	920	529	590	457	91	577	508	218	770	**6,189**
Wheels/h.	93	199	15	80	250	343	336	435	29	703	532	98	**3,113**
Wheels/v.	416	454	989	376	286	405	247	964	522	148	822	921	**6,550**
Bottom bracket	769	525	987	895	71	791	412	697	839	923	4	339	**7,253**
Handlebar	187	471	961	4	683	461	292	495	795	305	231	722	**5,605**
Total	**1,936**	**2,497**	**3,164**	**2,274**	**1,819**	**2,589**	**1,744**	**2,682**	**2,761**	**2,587**	**1,807**	**2,850**	**28,710**

Fig. 3.19 Time series analysis

- Due to the abundance of data of individual period values, possibly confusing columns of numbers are created, especially in the table representation. Important cyclical fluctuations and outliers should be explained with comments.
- If only one or a few periods are considered, the analytical capability of the time series analysis is limited.

3.7.6 ABC, Flop/Top and Class Analysis

If you want to know which result object contributed the most or the least to the success, it is advisable to sort the result objects considered for selected key figures for a period of time. This way, you get, e.g. by an ascending or descending sorting, the best or the worst result objects. If you limit the view to a certain number of result objects, you get a list of the top or flop result objects.

The sample report in Fig. 3.20 shows the top ten customers in table and graphical form for the sales values from the current and the previous year.

Another form of sorting is the ABC analysis which can be used to rank the considered result objects according to their importance and economic significance. This allows to focus especially on the result objects that have an above-average share of the value of a key figure. The classification shown in the result indicates which result objects need to be particularly examined. Often, the so-called *Pareto* curve is confirmed, whereby with only about 20% of the result objects more than 80% of the total volume of the selected

	Top Ten Customers	Revenues ACT €k	Revenues PY €k
1	Allianz	164,674	162,357
2	R+V	95,212	99,516
3	Generali	85,304	83,232
4	Zurich	73,767	74,290
5	Debeka	61,359	74,810
6	Würtembergische	55,128	55,637
7	Bayern Versicherung	55,075	58,943
8	Nürnberger	48,854	48,233
9	Hamburg-Mannheimer	48,302	52,128
10	AXA	46,650	34,402

German insurance providers

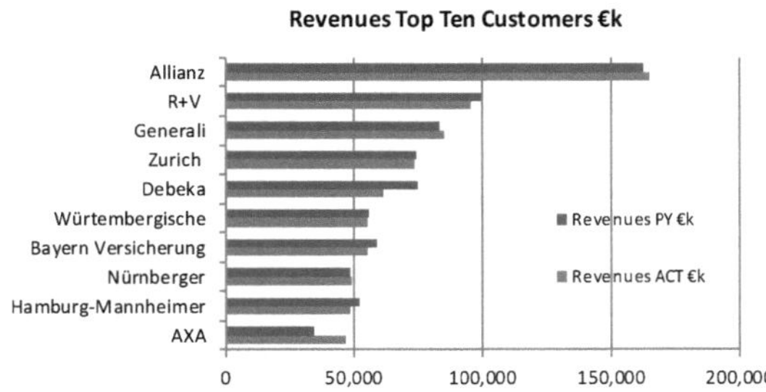

Fig. 3.20 Top ten analysis (example MS Excel)

key figure value is achieved (A-class). With about 50% of the result objects, more than 90% of the total volume of the selected key figure value is achieved. The result objects added here are assigned to the B-category. The remaining 50% of the result objects often contribute only with less than 10% to the total volume of the selected key figure value. They are assigned to the C-category. Other classifications, as the following examples show, are also possible.

To display the ABC analysis, all result objects are sorted in descending order of value and shown in a table. In addition, the key figure values are accumulated continuously. The class division is done by setting threshold values, e.g. percentage or absolute value limits. If percentage threshold values are used, the A, B and C result objects can be formed based on the percentage of accumulated shares, starting from the total value of 100%.

For the formed classes, standard strategies are often derived in the respective functional areas. In sales, for example, marketing measures or care intensities are developed and defined for the customers in their respective class.

The sample report in Fig. 3.21 shows an ABC analysis from the sales area for the key figure sales. The threshold values are set for the A-class between 0–70%, for the B-class between 70–90%, and for the C-class between 90–100% of the accumulated sales share. The *Pareto* curve is shown in the graph to the right of the table.

Another example of an ABC analysis is shown in Fig. 3.22, in which a different class division was chosen: 0–60% (A), 60–80% (B) and 80–100% (C).

The ABC analysis divides the analysis objects into three classes. In addition, the objects can be divided into any number of classes. From a business perspective, the result objects of the companies are represented in hierarchies, as already mentioned, e.g. product or customer hierarchies. Such hierarchies and group divisions can be interesting for the analysis when it comes to distinguishing significant from insignificant classes. For the visualization of such classes, besides traditional pie, column and bar charts, also modern chart types such as treemap, sunburst or bubble charts are suitable (see Sect. 3.8.4).

Designation	Revenues	Revenues %	Cumulative revenues	Cumulative revenues %	Class
Allianz	1,850,000	19.4%	1,850,000	19.4%	A
R+V	1,510,000	15.8%	3,360,000	35.3%	A
Generali	1,265,000	13.3%	4,625,000	48.5%	A
Zurich	1,115,000	11.7%	5,740,000	60.2%	A
Debeka	965,000	10.1%	6,705,000	70.4%	A
Würtembergische	660,000	6.9%	7,365,000	77.3%	B
Bayern Versicherung	550,000	5.8%	7,915,000	83.1%	B
Nürnberger	490,000	5.1%	8,405,000	88.2%	B
Hamburg-Mannheimer	200,000	2.1%	8,605,000	90.3%	B
AXA	190,000	2.0%	8,795,000	92.3%	C
HDI-Gerling	170,000	1.8%	8,965,000	94.1%	C
Provinzial	150,000	1.6%	9,115,000	95.7%	C
Swiss	105,000	1.1%	9,220,000	96.8%	C
Victoria	68,000	0.7%	9,288,000	97.5%	C
Iduna	60,000	0.6%	9,348,000	98.1%	C
Alte Leipziger	48,400	0.5%	9,396,400	98.6%	C
Cosmos	36,000	0.4%	9,432,400	99.0%	C
DBV	35,000	0.4%	9,467,400	99.4%	C
Gothaer	30,000	0.3%	9,497,400	99.7%	C
Volkswohl Bund	20,400	0.2%	9,517,800	99.9%	C
neue Leben	10,500	0.1%	9,528,300	100.0%	C
Total	9,528,300	100.0%			

German insurance providers

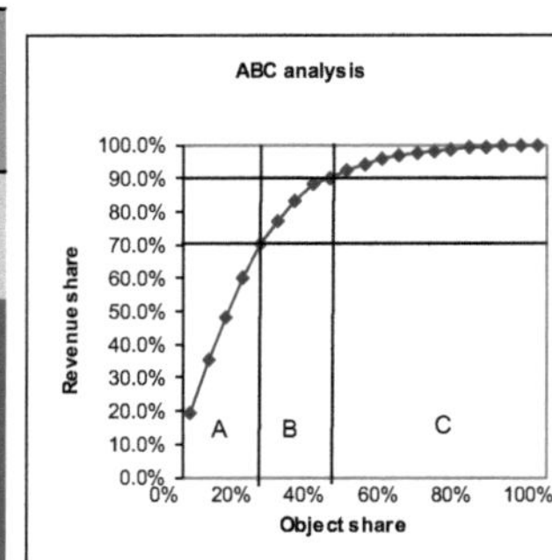

Fig. 3.21 ABC analysis (example MS Excel)

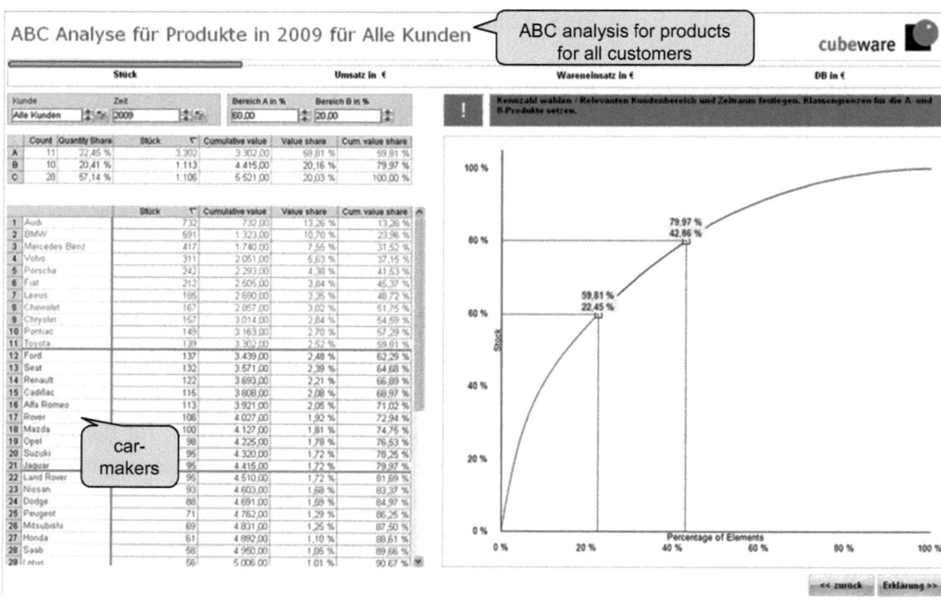

Fig. 3.22 ABC analysis (example Cubeware). *Source* Cubeware image gallery (2011)

Advantages of the ABC, top/flop and class analyses are:

- The result objects are quickly identifiable with respect to their economic significance.
- Outliers to the top (tops) and bottom (flops) are very easy to recognize.
- Standard strategies can be applied to the formed classes.

Disadvantages of the ABC, top/flop and class analyses are:

- The determination of the threshold values for the ABC and class analyses is often difficult at the boundaries. Possibly, result objects at the boundaries fall only slightly out of the formed class.
- With many result objects, the lists can become very long.
- Result objects that today still belong to the B or C class can belong to the A class in the future (and vice versa).

3.7.7 Portfolio Analysis

The portfolio analysis is characterized by its classification and analysis capability of result objects in a matrix formed by criteria. It is used in many economic areas, e.g. in financial economics for planning a balanced investment portfolio or in strategic corporate planning for optimizing business areas, product areas or projects. The most well-known portfolio analyses for strategic corporate analysis and planning are the four-field matrix of the Bosten Consulting Group and the 9-field matrix of the consulting firm McKinsey.

The portfolio analysis also finds increased application in other functional areas of the company, such as assessing suppliers in a supplier portfolio or a customer portfolio for analyzing customer groups.

Figure 3.23 shows the four-field matrix of the Bosten Consulting Group. Here, the most important report characteristics can be illustrated. To display the portfolio, three data categories must be determined for the relevant result objects (here the strategic business areas) with which the portfolio is to be formed. In the four-field matrix, these are the key figures market growth, relative market share and sales share of the considered strategic business areas. For the formed fields, classifications and standard strategies can be defined which are used for control. In the nine-field matrix of the consulting firm McKinsey, in contrast to the single criteria of the four-field matrix, many criteria are condensed to the axis values market attractiveness and competitive strength using a scoring model (see Sect. 3.7.10).

Advantages of portfolio analyses are:

- By forming the comparative value levels (relative, nominal, ordinal, cardinal), different result objects can be measured with the same standards.
- Standard strategies can be applied to the formed portfolio fields.
- The graphical representation provides a high communication and analysis value.

Business Unit	Own market share	Market share strongest competitor	Relative market share	Market growth	Sales in € million	Sales share
A	3.20%	7.60%	0.4	4.2%	250	23.15%
B	7.40%	2.90%	2.6	4.9%	120	11.11%
C	9.40%	5.30%	1.8	2.4%	310	28.70%
D	4.30%	3.10%	1.4	4.2%	280	25.93%
E	1.40%	3.30%	0.4	1.5%	120	11.11%

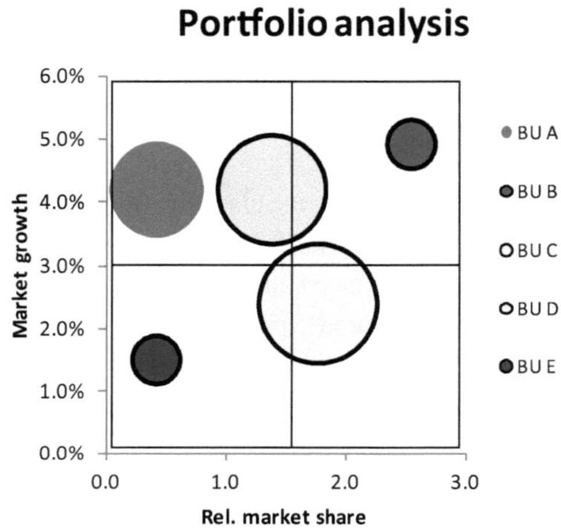

Fig. 3.23 Portfolio analysis

Disadvantages of portfolio analyses are:

- In the reduction of complexity to three value levels (X-axis, Y-axis and size representation of the result objects), important factors can be lost.
- Dependencies and synergy effects may not be recognized.
- General standard strategies are not applicable to specific problems.

3.7.8 Object and Benchmark Comparisons

In object comparisons two or more objects are compared with each other within a considered period of time. The objects can be both internal and external objects. For the external objects, the objects can be from the same industry (industry comparisons) or even across industries. The aim is to better assess the performance of the objects to be compared and to derive control measures from this.

Figure 3.24 shows a simple internal object comparison of two locations which are compared based on different indicators. Absolute and percentage deviation values help to capture the differences faster.

Business figures (in € million)	Location Hamburg	Location Munich	Δ	Δ %
Operating income	10,980	10,709	271	2.5%
Operating expenses	8,168	7,988	180	2.3%
Profit	2,812	2,721	91	3.3%
Return on sales	25.6%	25.4%	0.2%	0.8%
Capital expenditure	516	431	85	19.7%
Free cash flow	6,840	5,950	890	15.0%

Fig. 3.24 Object comparison

Similar to object comparisons are benchmark comparisons. Here, selected indicators are compared based on the best comparison objects or based on the best achieved values. For orientation, alternatively or additionally, comparisons can also be made to the average of the comparison objects or to the average achieved values. The comparison objects can again be chosen internally, across companies or even across industries.

Figure 3.25 shows a benchmark comparison of one location to the best location (in the left part of the table) and to the average of all locations (in the right part of the table).

Advantages of object and benchmark comparisons are:

- Better performance assessment of the objects to be compared.
- Orientation towards the "best" helps in the systematic search for improvement possibilities in the company.
- Orientation towards the average value helps in assessing one's own position in comparison to the average.

Disadvantages of object and benchmark comparisons are:

- It may not be possible to find comparison objects.
- Determining the comparison and benchmark objects and generation of the benchmark values (best values, average values) is often complex.

Business figures (in €m)	Location Hamburg	best Location	Δ +L	Δ +L %	Locations on average	Δ ØL	Δ ØL %
Operating income	10,980	11,709	-729	-6.2%	9,540	1,440	15.1%
Operating expenses	8,168	7,988	180	2.3%	7,950	218	2.7%
Profit	2,812	3,721	-909	-24.4%	1,590	1,222	76.9%
Return on sales	25.6%	31.8%	-6.2%	-19.4%	16.7%	8.9%	53.7%
Capital expenditure	516	531	-15	-2.8%	451	65	14.4%
Free cash flow	6,840	6,950	-110	-1.6%	4,950	1,890	38.2%

Fig. 3.25 Benchmark comparison

- Comparisons are sometimes difficult to interpret as the comparison objects may have different and specific conditions in the comparison period which are hard to recognize or eliminate.

3.7.9 Break-Even Analysis

In the case of declining employment, the company management asks itself whether the profit target aimed at for the coming year can be achieved at all and when or from which sales quantity or achieved turnover this can be the case. Here it is helpful to analyse the Break-Even-Point.

The instrument of break-even analysis determines the sales quantity or the corresponding sales revenues at which the total fixed costs and the sales quantity-dependent (variable) costs are fully covered. This point of coverage is called break-even point. It defines the revenue-quantity combination from which the company makes profits. Formally, one determines this critical value by equating the incurred accumulated total costs (K_g) of the turnover with the achieved accumulated revenues (U).

The determination of the break-even sales quantity (BEP_M) is obtained by dividing the fixed costs (K_f) of the total period by the unit contribution margin, which results from the difference between the unit revenues (p) and the proportional unit costs (k_v). If the break-even sales quantity is then multiplied by the unit revenue, the break-even sales are obtained.

$$\mathrm{BEP_M} = \frac{K_\mathrm{f}}{p - k_\mathrm{v}}$$

Figure 3.26 shows a break-even analysis for the case of a single-product consideration. The premise structure of the break-even analysis is very restrictive:[18]

- one product is produced
- costs, prices, capacities are fixed and known
- prices are independent of quantity
- no discounts on purchases
- no interval-fixed costs
- no parameter change during the observation period
- no cost remanence (costs of this period are independent of costs of the previous period)
- produced quantity = sold quantity → no inventory

[18] Cf. Heigl (1989, p. 128 f.) and Poensgen (1981, col. 308).

BEP (initial data)	Jan	Feb	Mar	Apr	May	Jun	Jul	Aug	Sep	Oct	Nov	Dec
Quantity mths.	1,000	1,000	1,000	1,000	1,000	1,000	1,000	1,000	1,000	1,000	1,000	1,000
Quantity YTD	1,000	2,000	3,000	4,000	5,000	6,000	7,000	8,000	9,000	10,000	11,000	12,000
Variable costs € mths.	250	250	250	250	250	250	250	250	250	250	250	250
Variable costs € YTD	250	500	750	1,000	1,250	1,500	1,750	2,000	2,250	2,500	2,750	3,000
Fixed costs € Full Year	1,150	1,150	1,150	1,150	1,150	1,150	1,150	1,150	1,150	1,150	1,150	1,150
Revenues € mths.	375	375	375	375	375	375	375	375	375	375	375	375
Revenues € YTD	375	750	1,125	1,500	1,875	2,250	2,625	3,000	3,375	3,750	4,125	4,500
Total costs € YTD	1,400	1,650	1,900	2,150	2,400	2,650	2,900	3,150	3,400	3,650	3,900	4,150

Break-Even-Point-Calculation	
BEP sales volume	9,200
BEP sales revenues €	3,450

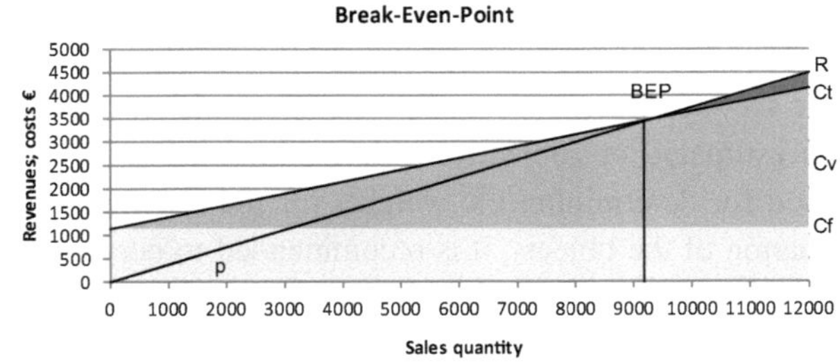

Fig. 3.26 Break-even analysis

The very restrictive premises of the simple model of the break-even analysis can be largely eliminated by various extension possibilities. For a multi-product company with different product contribution margins, it makes sense to perform the break-even analysis first at the product or, if possible, at the product group level with the associated product or product group fixed costs, and then to determine the profit threshold for the entire sales mix.[19]

Advantages of the break-even analysis are:

- The BEP is a good reference value for the past and future performance of the company.
- If the input data is available in the BEP model, it helps to quickly estimate the break-even point.
- Differentiated BEP analyses per product and product group provide transparency regarding the specific break-even points for each product or product group.

Disadvantages of the break-even point analysis are:

- Restrictive premise structure of the model assumption.
- Stepwise reducible fixed costs (cost remanence effects) are not considered in the basic model.
- Preparation of the input data for multi-product companies with different revenue and contribution margin structures (regular case) requires complex planning.

[19] Cf. Kilger (1988, p. 802 f.).

3.7.10 Scoring or Utility Analyses

To complement the analysis of quantitative values, scoring or utility analyses are often used to assess the qualitative values. The terms scoring and utility analysis are used synonymously here. For the scoring analysis, in the first step, evaluation criteria of quantitative and qualitative nature are developed. They should be non-overlapping and complete. It is also advisable to form groups of criteria. The individual criteria are finally weighted in the second step according to their different importance. In practice, these weights are often estimated or subjectively determined by the creators or experts. A more precise method for determining the weights for the criteria is the pairwise comparison. For the evaluation of the objects, it is recommended to normalize the total score to 100 or 1000. The advantage is that the evaluation object can be classified more quickly by the analyst using a percent or per mille scale. In the third step, the result objects to be compared are evaluated. Ideally, cardinal evaluations with precise delimitation criteria can be derived, in which, for example, a criterion such as the price (in a predefined price range) receives a number of points. But also nominal evaluations (e.g. good/bad) or ordinal evaluations (rank classes without distance definition) are used.

Figure 3.27 shows a simple scoring analysis of two comparison products with a weighted point rating.

Advantages of the scoring or utility analysis are:

- Simple evaluation method in which individual user criteria can be compiled and weights can be assigned for the evaluation of the criteria.
- Takes into account qualitative evaluation criteria in addition to other economic considerations and thus creates good transparency about them.
- The scoring model can be easily adapted individually to different company requirements.

		Own product		Competitor product	
Comparison criteria	Weighting	Points (0-10)	weighted points	Points (0-10)	weighted points
Price	30	8	240	9	270
Performance	20	7	140	9	180
Innovation	20	6	120	8	160
Quality	20	9	180	7	140
Service	10	5	50	6	60
Total	**100**	**35**	**730**	**39**	**810**

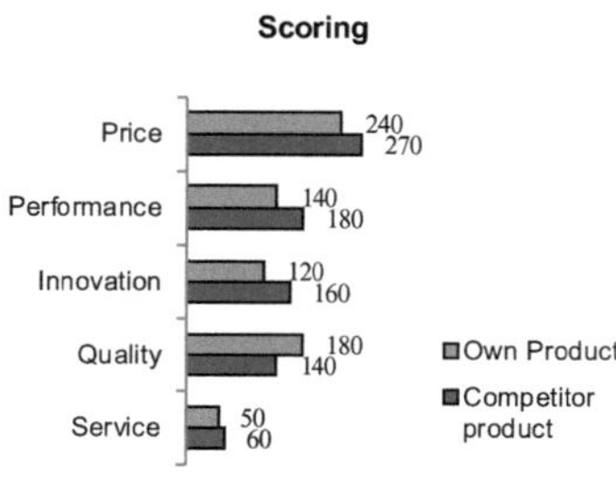

Fig. 3.27 Scoring analysis

Disadvantages of the scoring or utility analysis are:

- Subjective determination of the evaluation criteria and weights.
- The measurement of the evaluation criteria is not always possible.
- The evaluation of the objects is often subjective and can be easily manipulated.

3.8 Report Design

In the context of designing individual reports, the following topic areas are of particular importance: the filter or selection criteria, the layout, and the evaluation area, especially with the tables, comments, diagrams and other visualization aids used here. In addition, basic hints and recommendations for the design of reports are given.

3.8.1 Filter or Selection Criteria

When selecting or filtering report information, one must generally distinguish between a pre-selection for the report and a detailed selection in the report of key figures and dimensions.

The pre-selection allows the analyst to call up the report already with selected data material, in which e.g. only certain dimension values such as a company, a region, a cost center etc. are addressed. If a pre-selection is made, only the pre-selected data material can be used for analysis or planning when calling up the report. Further detailed analyses are then only possible in the pre-selected data set, e.g. the research of revenues and costs for a pre-selected region.

To expand the data view of a pre-selected report online, there is usually no other way than to exit the report and extend the pre-selection accordingly. The benefit of the pre-selection is the limitation of the data material to be analyzed. A pre-selection and narrowing of the data material is also done by the restricted view of the available data set in the user's authorization profile. For example, a cost center manager only sees data from his cost center.

The detailed information in the individual report can still be selected by various tools, whereby the following are the most common tools:

- Filter and menu functions in which e.g. different dimensions can be selected.
- Context menu functions with which individual functions can be performed using the right mouse button. This selection function can be accessed here in separate filter and menu bars but also directly in objects (e.g. in graphics and tables) by marking graphic elements (e.g. column) or table values.
- Drop-down list, e.g. for existing master data such as e.g. personnel groups.

- Hierarchy selection of all available master data according to maintained hierarchy (e.g. cost center hierarchy).
- Radio button or boxes that can be checked for selectable dimensions or master data (e.g. the regions North, West, East, South).
- Selection via Boolean operators such as "and" (conjunction), "or" (disjunction), "not" (negation) and "XOR" (exclusive or).
- Set bookmarks for already analyzed reports or contents.

A further form of data preparation is the **sorting** in which the key figure values in the report are displayed in ascending or descending order. This can direct the analysis to the most important (e.g. largest and smallest) values.

3.8.2 Layout

3.8.2.1 General Design Guidelines

The following design guidelines and recommendations are to be observed in principle for the content and layout of the management reporting.[20]

Criterion	Design guidelines/recommendations
Expressiveness	Key statements, important messages, relationships, dependencies, classifications, correlations and backgrounds of statements are to be highlighted as best as possible by appropriate combination of tables, texts and graphics and their design (e.g. by class formation, trend presentation or multidimensional evaluations).
Balance	Information contents should not be overloaded but also not too content-poor. An appropriate redundancy-free information density should be chosen. Distracting information should be avoided.
Quality/Correctness	The contents of a report should be checked and qualitatively correct.
Structure	The reports should be structured in such a way that a quick overview of the most important information is possible. If reports consist of several report pages, a logical structure at the beginning of the report is useful. The weighting of the chapters should be balanced in relation to the importance of the chapter.

[20] Cf. for design tips among others Frontline-Consulting (2011), Hichert+Partner AG (2008, p. 1–17), Bissantz et al. (2013, p. 439–462), Drews and Schilling (2014, p. 125–144), Klein (2014, p. 100 ff.), Schöneberg (2014, p. 330), Müller and Lenz (2013, p. 245), Kohlhammer et al. (2013, p. 41–80) and Gerths and Hichert (2011, p. 15 ff.).

Criterion	Design guidelines/recommendations
Uniformity (in general)	The reports should be designed uniformly, so that a quick capture of the most important information is possible. A similar report structure helps to quickly find one's way around the contents. A constant change confuses the viewer.
Completeness	The contents of a report should be complete, e.g. all costs of a cost center are listed.
Font size	Use max. 3 font sizes per report.
Font type	Use a uniform font type. It should be considered that rare, uncommon fonts are not available on every computer and in this case the font type is replaced by a similar one. Therefore, standard fonts are advantageous here.
Line widths	Use max. 3 line widths per report.
Colors/Patterns/Hat chings	Colors, patterns and hatchings should be used sparingly (e.g. 3–5 different ones), otherwise the desired highlighting will no longer be perceived. Use the same highlighting for similar factual contexts, e.g. actual values or target values should always be the same color. The colors, patterns and hatchings used should be well differentiated from each other, so that a distinction is easily visible.
White spaces	Sufficient white spaces and distances should be left between the layout elements such as tables, graphics etc. The available space should be used sensibly. Unfilled white spaces are useful to direct the attention to the essential.
Avoidance of unnecessary elements	The focus should be on the content of the reports and not on certain special effects, such as 3D or Flash effects. Core information should be highlighted visually. Unnecessary elements and information should be avoided in the report.
Headings	Appropriate report titles should reflect the main message of the report. Parts of the report (tables, diagrams etc.) should also receive appropriate headings.
Emphasis	Important things should be emphasized (e.g. by frames, bold fonts, arrows etc.). They should, like the colors and patterns, be used sparingly, so that they are perceived. Underlines should not be used as emphasis because of their use for links (e.g. web link). Marginalia on the page margin help to highlight important information or assign comments in bullet points.
Text additions/bullet points	Additional important text information (e.g. comments) should be summarized with a few keywords in bullet points. Only in exceptional cases should complete texts be added.
Visual elements	Using or supplementing with visual elements such as diagrams helps to better illustrate core contents.
Division	The report should be divided into meaningful report sections which ideally have a similar structure. This way, the viewer can quickly find and absorb the information, e.g. the right part of the report always shows the comments.

Criterion	Design guidelines/recommendations
Uniform notation and legends	Uniform notation and legends regarding used symbols and periods, e.g. the abbreviation "h" for hour (Latin hora) or the use of uniform currency symbols, such as €k for thousand euros. Important dimensions such as product groups, data types (such as plan, forecast, actual data) should also get a uniform notation.
Uniform value and quantity representation	Value and quantity representations should be uniform e.g. with regard to the decimal representation.
Uniform elements	Used elements such as icons, symbols, abbreviations, or words should be used in a similar way.

3.8.2.2 Basic Grid

The basic grid of a report has different elements and areas, which are illustrated by Fig. 3.28.

Navigation Menu

The navigation menu helps the user to navigate in the reporting system (see also Sect. 3.3), to access individual reports or detailed information faster. Alternatively to the navigation menu, objects from different reports are cross-referenced via so-called **links**. If such a "linked" object is clicked, the viewer is taken to the linked report.

General Report Information

The general report information includes the following elements of the report:

- the title (report headline or name)
- the author
- the recipient
- the creation time (day, time, possibly loading time/status of the data)
- the reporting period (e.g. quarter 1)
- the central report object(s) (e.g. cost center(s))

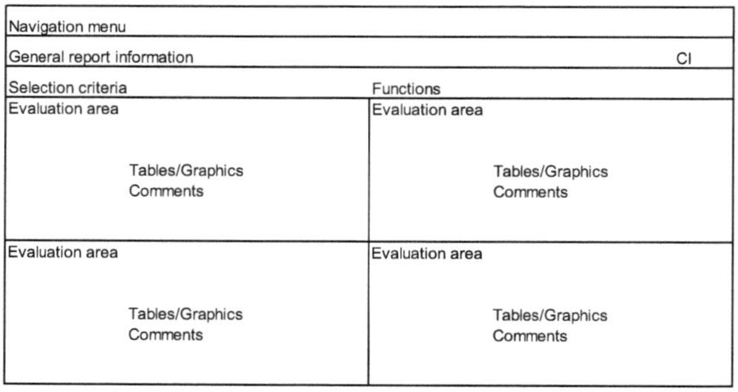

Fig. 3.28 Exemplary basic grid of a report

The following report components can also be part of the general information:

- Report cover sheet (useful for central printers to find your report)
- Table of contents/structure of the report for extensive multi-page report compilations (often in print reporting)
- Report header/footer information: page number, date, author, title/heading, time, company logo etc.
- Report margin information: marginalia, short comments etc.

Corporate Identity
With the corporate identity, a company strives for a distinctive identity and high recognition. In reporting, these are style elements, such as the use of a **Logo** or a **company slogan** on the report pages. This also includes the use of a uniform report design, i.e. the complete design, starting from the font, to the use of colors, to the full design of the page layout.

Selection Criteria
Often the reports are based on more extensive data material, so that the report recipient has the possibility to limit the data to certain data contents. A distinction must be made between a pre-selection before the report construction and a selection in the report itself (see Sect. 3.8.1).

Functions
In the report, functions can be built in that perform an action. This is common in planning forms, in which, for example, plan values are distributed, extrapolated or saved. For reporting, however, actions such as printing and sending or entering comments and attachments are also possible.

Evaluation Area
The evaluation area represents the largest part of the report and contains the core information. Important elements of the evaluation area are tables, diagrams and comments, which include, for example, status information and planned measures. Due to the better visual presentation of information, table, text and graphic information are also displayed in combination; therefore, it is advisable to divide the evaluation area into meaningful sections. Due to the landscape format of the screen resolution, different grids are possible, such as the division into four quadrants (Fig. 3.29). Here it is assumed that the 1st quadrant receives the most (approx. 35%) and the 4th quadrant the least (approx. 15%) attention of the analyst. The 2nd and 3rd quadrant are balanced in between at approx. 25% of the attention.[21]

[21] Cf. Schmid-Gundram (2014, p. 46).

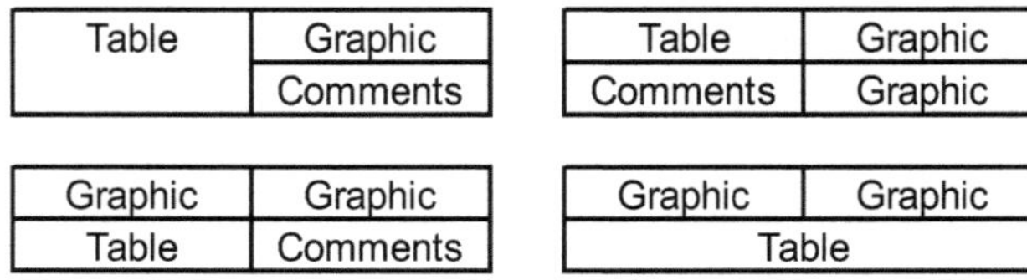

Fig. 3.29 Exemplary division possibilities of the evaluation area

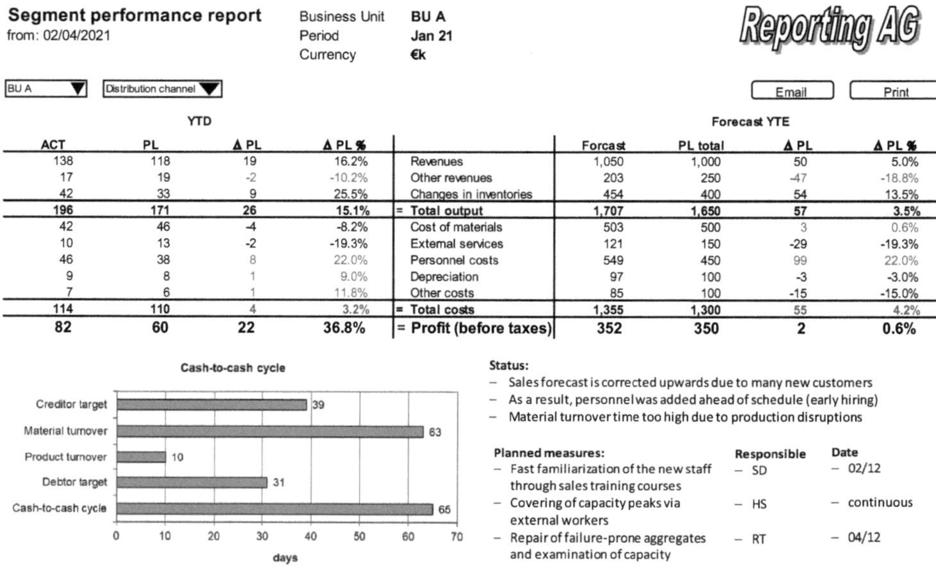

Segment performance report
from: 02/04/2021

Business Unit BU A
Period Jan 21
Currency €k

Reporting AG

BU A | Distribution channel | Email | Print

YTD ACT	YTD PL	YTD Δ PL	YTD Δ PL %		Forecast YTE Forcast	PL total	Δ PL	Δ PL %
138	118	19	16.2%	Revenues	1,050	1,000	50	5.0%
17	19	-2	-10.2%	Other revenues	203	250	-47	-18.8%
42	33	9	25.5%	Changes in inventories	454	400	54	13.5%
196	171	26	15.1%	= Total output	1,707	1,650	57	3.5%
42	46	-4	-8.2%	Cost of materials	503	500	3	0.6%
10	13	-2	-19.3%	External services	121	150	-29	-19.3%
46	38	8	22.0%	Personnel costs	549	450	99	22.0%
9	8	1	9.0%	Depreciation	97	100	-3	-3.0%
7	6	1	11.8%	Other costs	85	100	-15	-15.0%
114	110	4	3.2%	= Total costs	1,355	1,300	55	4.2%
82	60	22	36.8%	= Profit (before taxes)	352	350	2	0.6%

Status:
- Sales forecast is corrected upwards due to many new customers
- As a result, personnel was added ahead of schedule (early hiring)
- Material turnover time too high due to production disruptions

Planned measures:	Responsible	Date
– Fast familiarization of the new staff through sales training courses	– SD	– 02/12
– Covering of capacity peaks via external workers	– HS	– continuous
– Repair of failure-prone aggregates and examination of capacity	– RT	– 04/12

Fig. 3.30 Exemplary report based on the basic grid

Various combinations are conceivable here.[22] An example of a clear report with navigation menu, general report information, selection criteria, functions, evaluation areas with table, graphic and comments is shown in Fig. 3.30.

Since tables often have many entries in rows or columns downwards or sideways, a connection of two quadrants makes sense, whereby for larger tables a **scroll bar** (**scroll bar**) can also be used. For longer tables that can only be viewed completely using the scroll function, it is advisable to list not only the attribute headings, but also the sum formation at the beginning of the table (i.e. at the top or left). Figure 3.31 shows, for example, in the right part a table in which the utilization and turnover of many areas and blocks of a stadium are shown. The sums are aggregated upwards. While costs, quantities, revenues and other values can be easily accumulated, other key figures such as prices, utilizations, etc. may have to be determined on average for the considered consolidation area. In the left part, the evaluation is supported by the geo-reporting in which

[22] Cf. e.g. the controller's target-actual form with expectation calculation—form "4 windows" by Deyle (2001, p. 31).

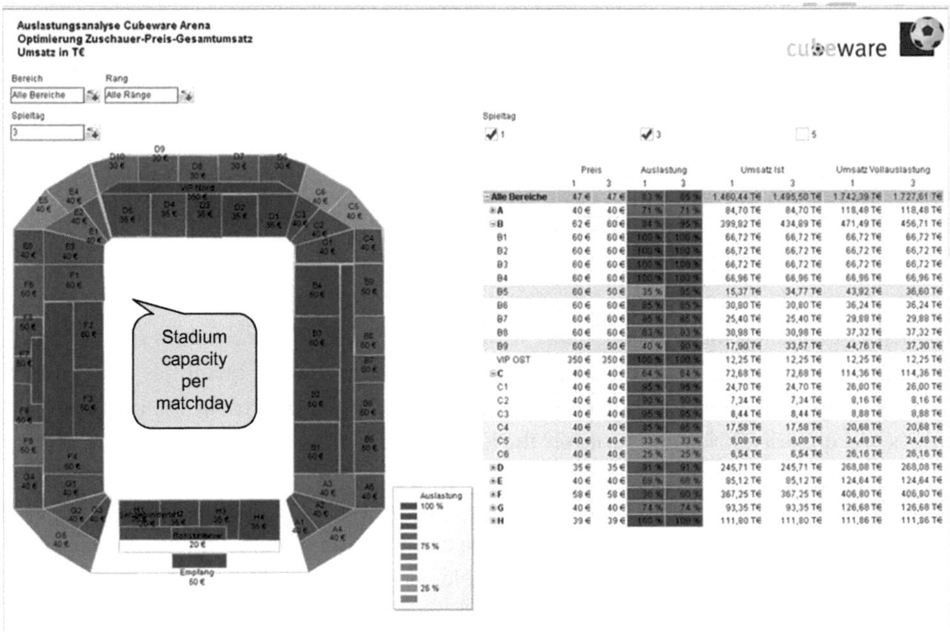

Fig. 3.31 Table with a sum shown at the top. *Source* Cubeware Picture Gallery (2011)

the utilization of the individual stadium areas and blocks is differentiated in a graphic from green to red according to the level of utilization (see also the geo-visualization in Sect. 3.8.4).

Depending on the report content, one should choose a suitable report structure. Similar reports should be structured in the same way, so that a quick recognition and analysis is possible.

3.8.3 Tables

As already shown in Sect. 3.7, Basic Report Forms, tables are the most frequently used form for presenting management information. Here, flat tables, cross tables and pivot tables are distinguished.

Flat Tables

The flat table is the simplest form of a table (see Table 3.1). In terms of relational databases, it consists of the relation of attributes (columns) and tuples (rows). Attributes stand for the properties or names of fields and the tuples for a record with the data contents for the fields. Of course, the table can also be displayed tilted by 90 degrees; in this case the attributes are in the rows and the tuples in the columns. For the fields, one dis-

Table 3.1 Flat table

Order number	Article number	Quantity	Price	Revenue
10001	20101	2	10	20
10001	20201	1	20	20
10002	20101	3	10	30
10002	20201	4	20	80
10002	20203	1	15	15

tinguishes characteristics (master data) with the respective attribute values (e.g. number, description and further assigned properties) and key figures (e.g. quantities and values). In the example of Table 3.1, order number and item number represent the characteristic values and quantity, price and turnover the key figures.

Cross Tables

The cross tables represent tables that contain a combination of data contents from at least two or more feature values. A simple cross table, as shown in Table 3.2, shows the values for the key figure "sales quantity" for the feature value "customer" and "product".

One or more feature dimensions or additional key figures can be achieved by further subdivisions in the columns or rows, as Table 3.3 shows using the product group as an example. If additional feature dimensions are inserted, the limits of the cross table become easily recognizable as it becomes increasingly opaque due to further nesting.

Pivot Tables

The pivot table gives the user the possibility to flexibly evaluate feature dimensions by optionally adding or removing feature dimensions and key figures for the analysis without changing the source data of any source. The pivot table represents a consolidation of the data fields of the source data, based on the selected feature dimensions and key figures. The advantage is that larger data sets can be reduced to manageable feature dimensions in evaluations.

In the exemplary pivot table, the source data are condensed to the sales of different material groups and sales organizations (see Fig. 3.32). Further dimensions such as the distribution channel, the time dimension, and the materials, and also further key figures, such as the net sales or discounts, can be flexibly evaluated.

Due to the evaluation of numerous feature dimensions, the pivot table is also called multidimensional analysis. This means that the evaluation possibilities are similar to those that are presented for multidimensional data cubes (cubes), which are based on the technology of OLAP (On-Line Analytical Processing) (see Sect. 5.5.4). In contrast to the pivot table, the OLAP cubes are a multidimensional data storage. Pivot tables are about the multidimensional analysis function, which often only accesses one or more flat tables that contain multidimensional data.

Table 3.2 Simple cross table

Paragraph	Product 1	Product 2	Product 3	Product 4	Total
Customer 1	200	100	100	0	**400**
Customer 2	100	50	200	100	**450**
Customer 3	50	100	50	0	**200**
Customer 4	100	300	100	50	**550**
Sum	**450**	**550**	**450**	**150**	**1600**

Table 3.3 Extended cross table

Paragraph	Product group A		Product group B		
	Product 1	Product 2	Product 3	Product 4	Total
Customer 1	200	100	100	0	**400**
Customer 2	100	50	200	100	**450**
Customer 3	50	100	50	0	**200**
Customer 4	100	300	100	50	**550**
Int.-Sum	**450**	**550**	**450**	**150**	**1600**
Sum	**1000**		**600**		**1600**

The most important basic operations (rotation or pivoting, slice and dice, drill-down and roll-up, drill-through and drill-across, as well as split and merge) of a multidimensional pivot table will be presented below using a simple data cube (see Fig. 3.33) with three dimensions (product, time, region) and key figures (such as sales, quantity, contribution margin).[23]

Rotation or Pivoting

Due to the visual limitation of a paper or screen report, usually only a two-dimensional section of a multidimensional data set is analyzed. Therefore, the analyst's view is directed to the front view of the data cube, for example, the analysis of product sales by time (see Fig. 3.34 bottom left).

Rotation or pivoting means to turn the cube around its own axis in order to make other combinations of dimensions visible, e.g. the product sales in the regions (bottom right) or the annual sales of the regions (see Fig. 3.34 top).[24]

[23] See also Voß and Gutenschwager (2001, p. 269 ff.).

[24] See Schmidt-Volkmar (2008, p. 23 f.) and Kemper et al. (2010, p. 96).

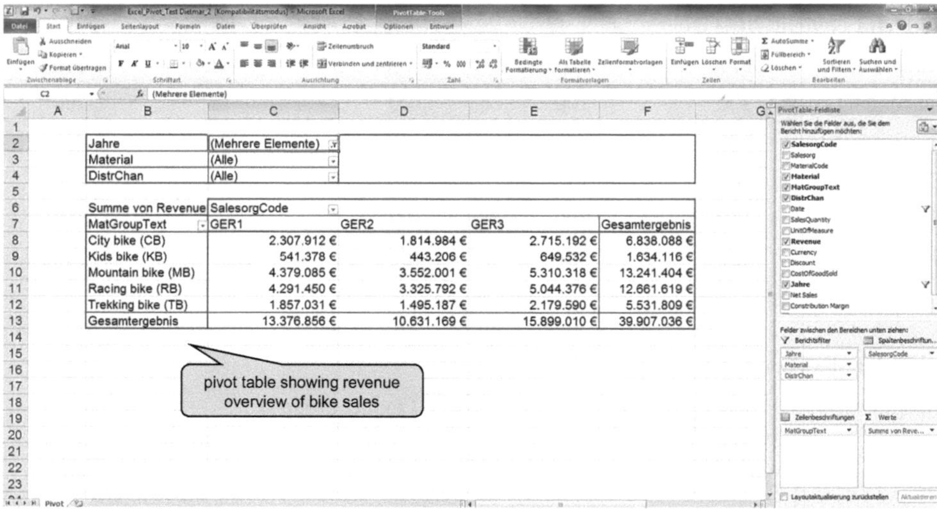

Fig. 3.32 Pivot table (MS Excel)

Fig. 3.33 Simple data cube

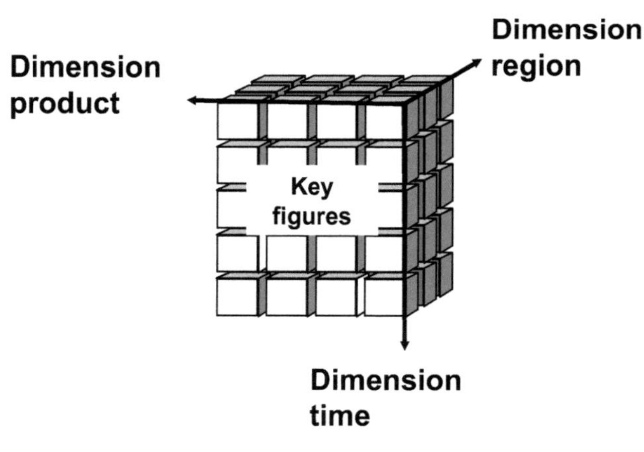

Slice and Dice

To filter data according to the needs, there are two operations: slice and dice.[25] With slice, only a data slice is viewed in the multidimensional cube. Practically, this is achieved by restricting a dimension to a value. For example, the regional manager has the possibility to view all product sales only of his region (see Fig. 3.35).

[25] Cf. Oehler (2006, p. 27) or Linden (2016, p. 150).

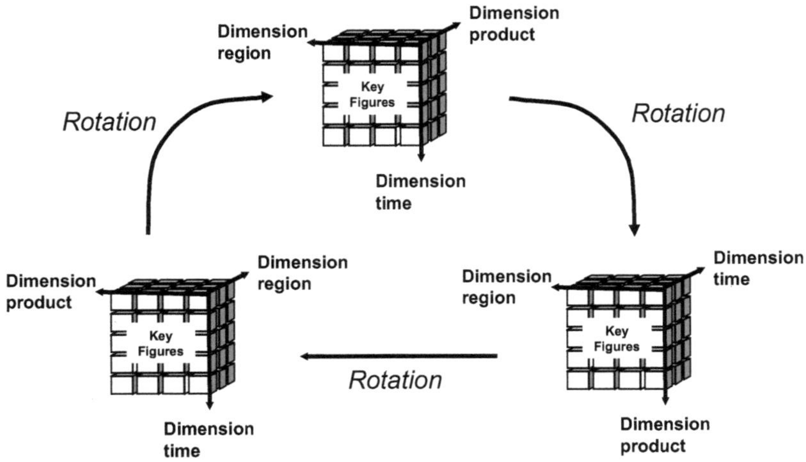

Fig. 3.34 Rotation or pivoting

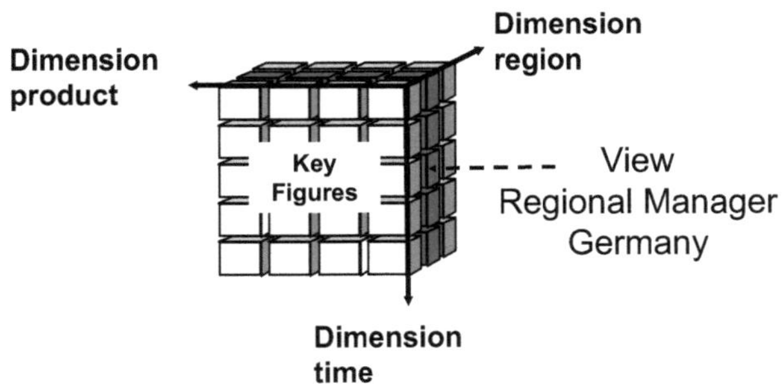

Fig. 3.35 Slice

With dice, a subcube is cut out of the cube (see Fig. 3.36). In this case, several dimensions are each restricted by a dimension element or a set of dimension elements. This results in a new (smaller) cube that is limited to a specific data area.

Roll-Up and Drill-Down

With the help of the operations roll-up and drill-down, one can navigate within the existing dimension hierarchies.[26] By roll-up, key figure values are further aggregated or condensed, but this operation reduces the level of detail along the dimension hierarchy. In contrast, the operation drill-down provides a more detailed view of the key figure values. Figure 3.37 illustrates this process visually using the time (here months to quarter 1) and product hierarchy (sum and individual products).

[26] Cf. Kemper et al. (2010, p. 96 f.).

Fig. 3.36 Dice

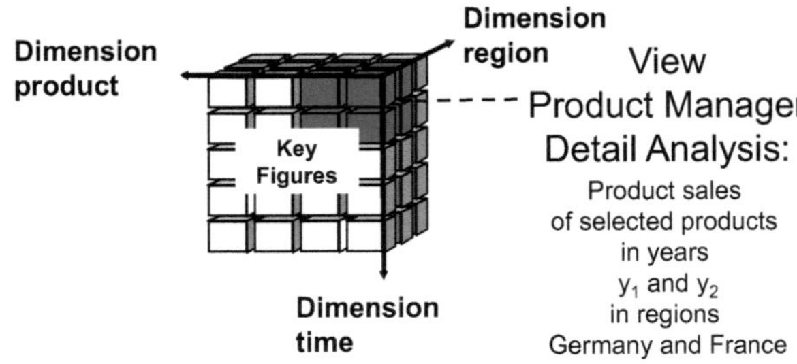

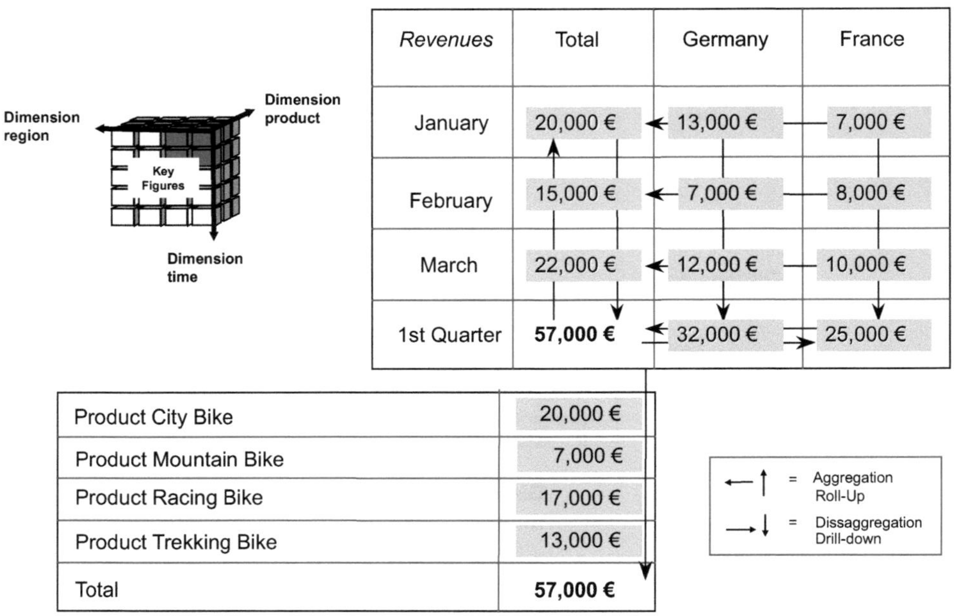

Fig. 3.37 Drill-Down and Roll-Up

Drill-Through and Drill-Across

With the operations drill-through and drill-across, navigation beyond the original cube is possible.[27] When one reaches the highest level of detail with a drill-down operation, no further refinement is actually possible. However, a drill-through allows to switch to another data source, usually the original physical data source, and thus gets access to further, more detailed data. The switch takes place in the background and does not affect the user interface. In other words, the drill-through enables an extended vertical search. In the example of Fig. 3.38, the booking receipts from financial accounting for the revenues from the product Mountain Bike are read.

[27] Cf. Kemper et al. (2010, p. 97).

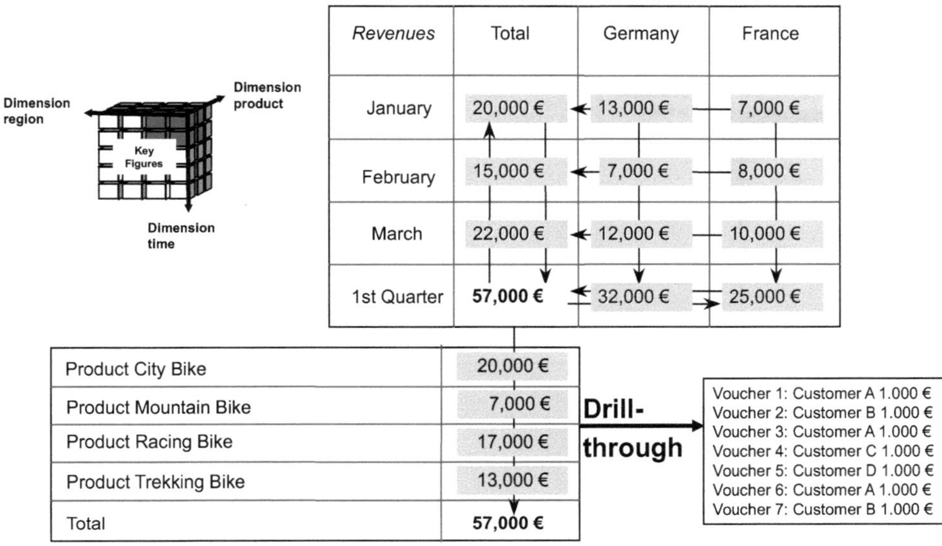

Fig. 3.38 Drill-Through

In contrast, the drill-across operation extends the horizontal search possibilities by enabling the switch between multidimensional cubes (see Fig. 3.39). For example, in a company, two data marts (smaller data cubes) could exist for the areas of product management and sales, both of which use the product dimension. By the operation

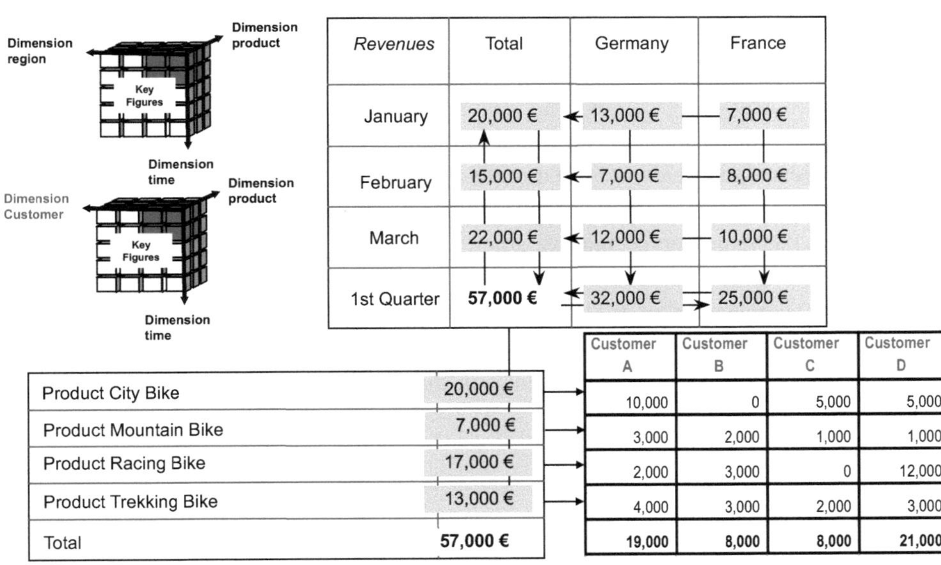

Fig. 3.39 Drill-Across

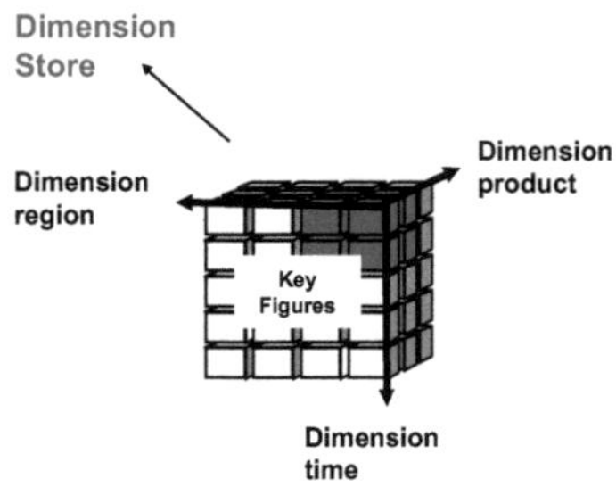

Revenues	Total	Germany	France
January	20,000 €	13,000 €	7,000 €
February	15,000 €	7,000 €	8,000 €
March	22,000 €	12,000 €	10,000 €
1st Quarter	**57,000 €**	32,000 €	25,000 €

Fig. 3.40 Split and Merge

drill-across, the product management could access the sales data, for example, to analyze the customers in detail.

Split and Merge
With the help of the split operation, a value can be broken down by elements of another dimension and thus, a further detail of a value is possible. In contrast, the merge operation can remove the additional dimension, thereby reducing the level of detail.[28] For example, the sales of a branch for a certain amount of products can be displayed or hidden. Unlike drill-across, the dimension is already contained in the multidimensional data cube (see Fig. 3.40).

3.8.4 Charts and Other Visualization Aids

The visualization of business management information with the help of charts and other graphical representation aids are often more illustrative than value tables and data lists, and they provide a faster overview of the significant information. Basically, the following rules apply to the visual representation, among others:[29]

1. Consistent explanation of the contents of diagrams with headings, axis labels and legends. However, the explanations should not overload the diagram. Repetitions, such as the value type € in the heading, in the axis label or in the legend, should be avoided.

[28] Cf. Oehler (2006, p. 27).
[29] Cf. Hichert+Partner AG (2011, pp. 1–29, 2008, pp. 1–17) and Klein (2014, pp. 126 ff.).

2. Do not omit or distort anything: extreme values should also be shown to support the credibility of the presentation. Distortions and other distortions should not call into question the value expressions of the presentation.
3. Comparisons help to classify the facts and create transparency. If many objects can be compared with each other, they should not be shown on different pages, but rather on one page, even with a smaller resolution, to use the direct visual comparison.
4. Sortings help to analyze the ranking of the objects faster.
5. Diagrams should bring the viewer closer to the causes, relationships and explanations for facts.
6. Comments and Messages should address explanations, backgrounds, recommendations for action and measures directly in connection with the facts in the graphic.

Figure 3.41 shows avoidable visualization errors (top) that can be easily corrected (bottom).

In the following part, the most important types of diagrams and other visualization aids that are used in management reporting will be presented. Here, all considered objects can be analyzed according to different dimensions, attributes and values.

Pie Charts, Donut, Sunburst and Treemap Charts

Pieorcharts show subsets of a total amount in the form of circular sectors (see Fig. 3.42). The larger the subset of a considered object is, the larger is the share of the circular sector in the total amount (100% correspond to 360° of the full circle). Legends and comments on the circular sectors can be added to improve the illustration of the object names and percentage values. The use of colors, patterns or hatching should also be appropriate (see Sect. 3.8.2.1).

Pie charts are only suitable for management reporting if the number of objects considered is not too large. If the number of objects exceeds 10, the number of circular sectors becomes unclear. The legends then often become too small in font size, so that the transparency is lost. In this case, objects with a smaller share should be grouped together if this does not affect the analysis. Otherwise, another chart form, e.g. the column or bar charts, should be used.

A variation of the pie charts are semi-circle charts. They segment the 100% distribution on half (usually the upper) area of the circle.

Pie charts are generally not suitable for comparisons, as the changes and deviations are only poorly visible. The following errors should also be avoided when using pie charts:

- The segments of the pie charts are not sorted by their importance or size.
- The totality of the segments does not correspond to the considered population, but only to a selected section.
- Only the absolute or the relative values are displayed.

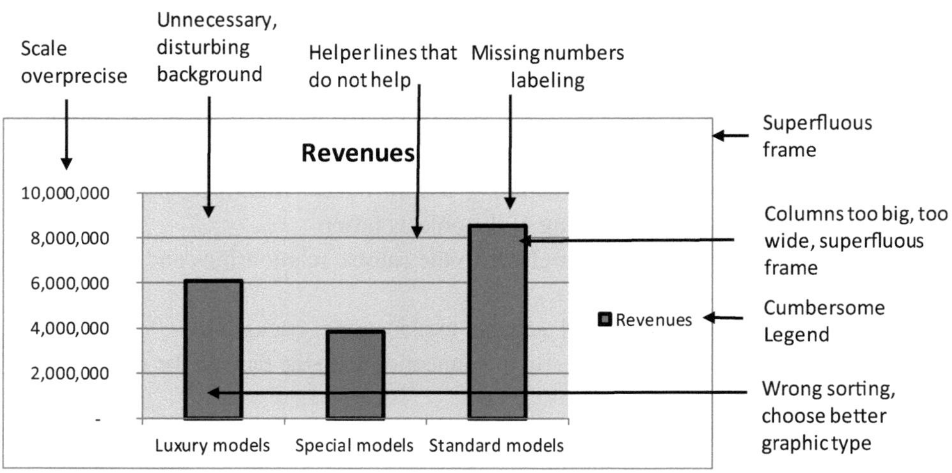

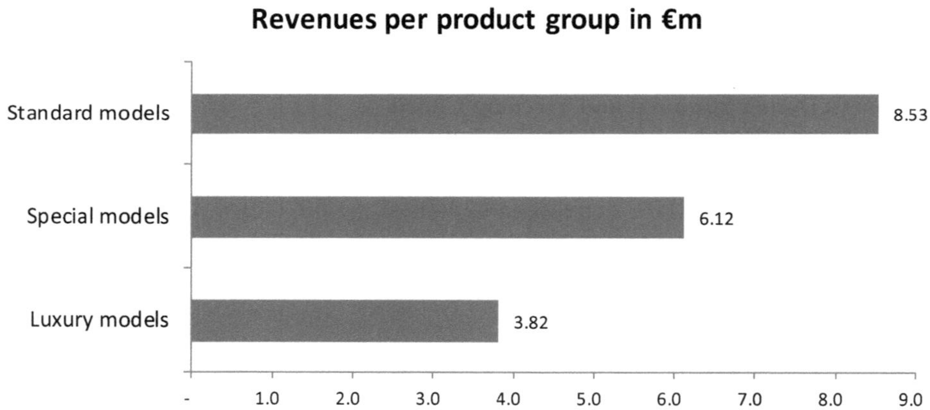

Fig. 3.41 Avoiding visualization errors

- Small color differences lead to a longer search in the legend.
- The pie chart contains too many small segments that are no longer readable.

If several pie charts are shown in parallel, e.g. in a coordinate system or a portfolio (see Sect. 3.7.7), the area size can also be used to display the volume of a selected value. This chart type is also called a bubble chart and is discussed in the following topic section together with portfolios. Here, for example, the sales of a country by several SBU can be displayed and compared with other countries.

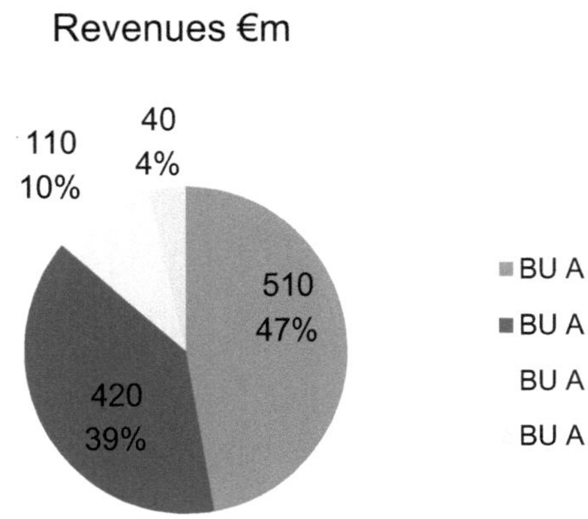

Fig. 3.42 Pie charts

As an alternative to full circle diagrams, the representation of donut charts can be used, in which the segments are displayed in the form of a ring. The informative value of the donut chart is identical to that of a circle diagram. The empty area in the ring circle can be used to display the total volume (see Fig. 3.43).

For the visual representation of different size classes within circle diagrams, the sunburst chart is used as an extension of the circle diagram. Here, the values can be grouped by different classes and highlighted according to their weights. An example of a sunburst chart can be seen in Fig. 3.44. Except for the grouping, the main advantages and disadvantages of a circle diagram remain, especially if there are too many groups and subgroups.

Circle diagrams and their extended forms have the disadvantage that value and text entries in circles or segments are often not well placed. As an alternative for the visualization of size classes, treemap charts can be used (see Fig. 3.45). They weight the size

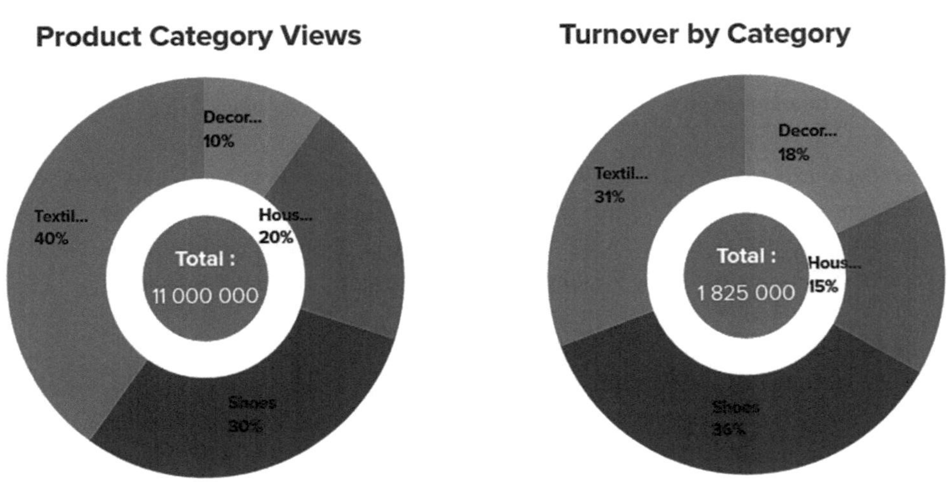

Fig. 3.43 Donut chart. *Source* bime Showcase (2017)

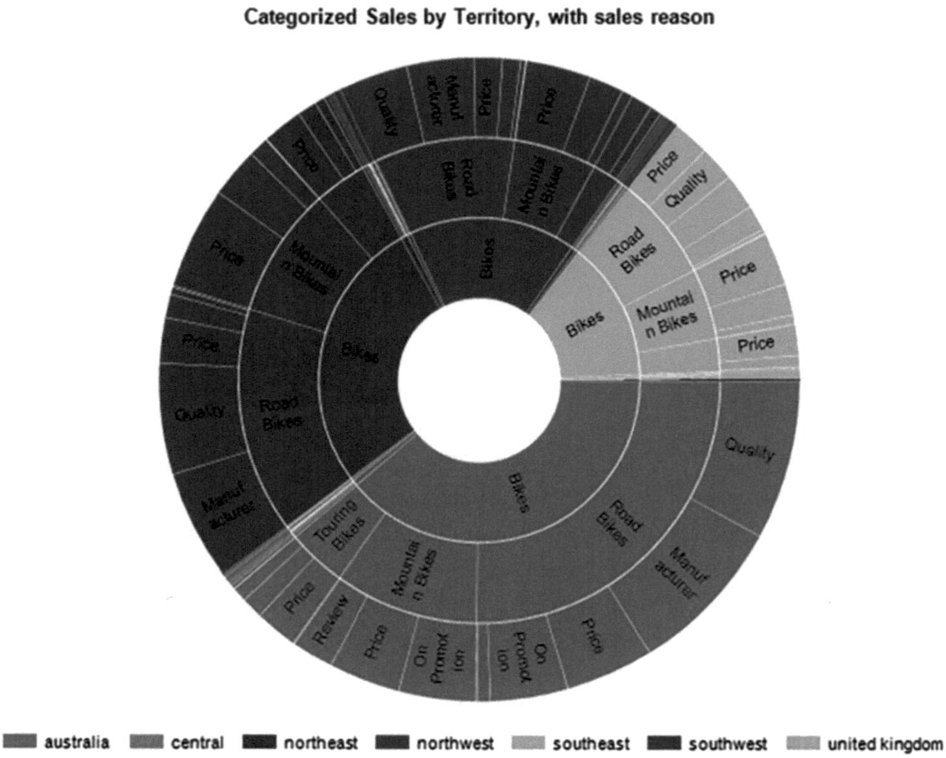

Fig. 3.44 Sunburst chart. *Source* Microsoft (2017a)

classes in the form of rectangles or tiles, something which has advantages for the labeling. The most important (e.g. volume-strongest) analysis objects are displayed in the top left, the least important in the bottom right. Groups can be formed by the color scheme, which can in turn be subdivided into parts. A too fine segmentation also leads to confusion in the treemap charts. However, more segments can be displayed than in the circle diagrams.

Sunburst and treemap charts, but also other charts such as bubble charts and word clouds, are popular in the context of BI systems in the topic areas of data mining and predictive analytics (see Sect. 5.5.5.3) as well as data discovery and data visualization (see Sect. 5.7.3), because they visualize the weightings of different classes well.

Portfolio/bubble Charts

Portfolio or.Bubble charts position selected objects, such as products, projects, strategic business units, funds or securities according to certain criteria in the portfolio. These criteria are defined in a coordinate system on the X and Y axis, a system which is usually represented as a matrix diagram with quadrants (e.g. the four-field matrix or the nine-

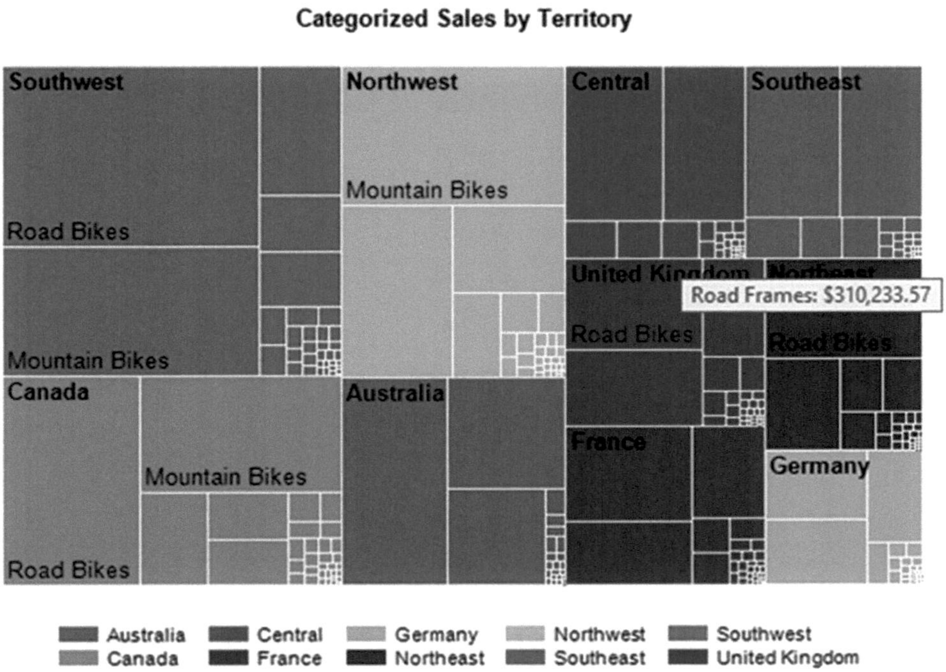

Fig. 3.45 Treemap chart. *Source* Microsoft (2017b)

field matrix) (see Fig. 3.46). For the fields formed, classifications and standard strategies can be defined, which can be used, for example, to develop the product program or the portfolio. The values of the objects to be analyzed are differentiated by the size of the bubbles, e.g. the sales volume or the securities volume. Further explanations and examples of portfolio analyses have already been given in Sect. 3.7.7.

Size class ratios can also be displayed in bubble charts with or without a coordinate system. Here, the positioning on the two criteria of the coordinate system and the importance of the objects are indicated by the size of the circle area. Figure 3.47 shows on the left a bubble chart, in which, for example, the risks and opportunities of selected business categories are positioned in a four-field matrix. The right diagram of the figure shows a bubble chart without a coordinate system. Important trends and highlights can be shown by this. If the bubble charts are viewed over time, changes in the positioning and importance of the analysis objects can also be analyzed.

Column/bar Charts and Bullet Charts

Column charts represent the objects under consideration using vertically aligned lines, rectangular areas or other animated area types (e.g. 3D columns) that are placed more or less wide on a horizontal axis (see Fig. 3.48).

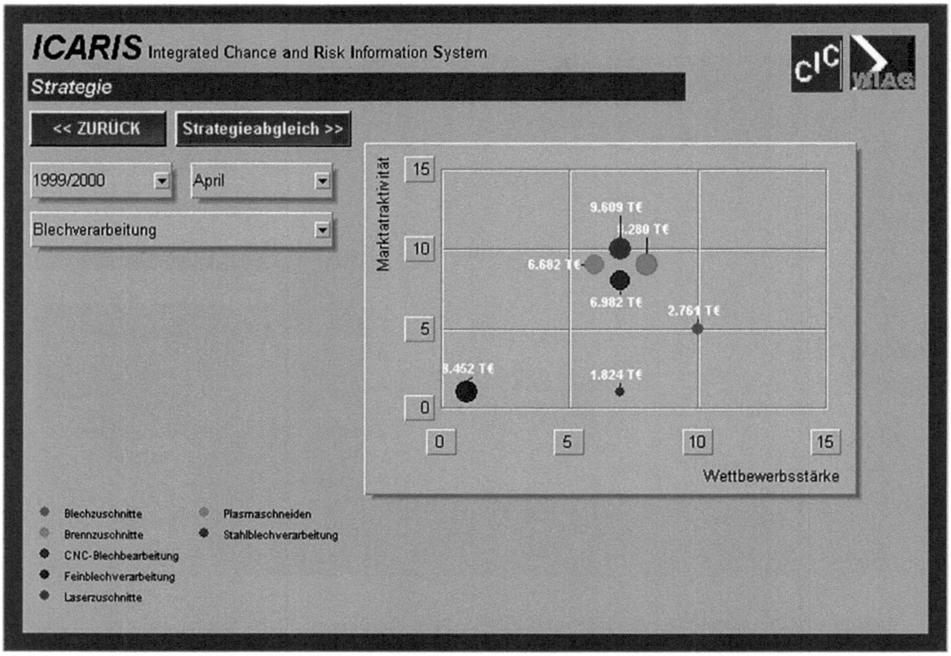

Fig. 3.46 Example of a portfolio/bubble chart. (Image excerpt taken from Schön et al. 2001, p. 386)

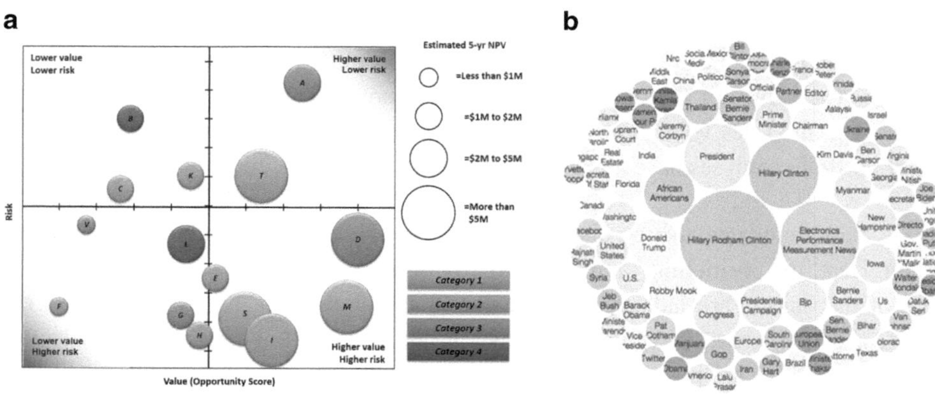

Fig. 3.47 Bubble chart. *Source* **a**: PPM Execution (2017), *Source* **b**: react (2017)

Bar charts are column charts rotated by 90°, where the lines, rectangular areas or columns lie on the vertical axis (see Fig. 3.49).

The width of the areas for a measure or value type for the same dimension expression (e.g. month) must be identical in order not to create any deception when viewing as otherwise the areas and not the height dominate the perception.

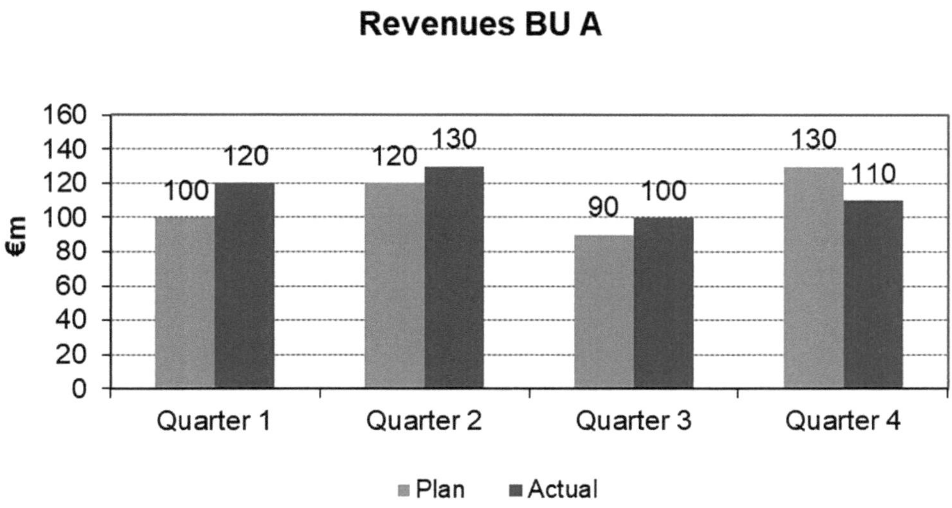

Fig. 3.48 Column chart

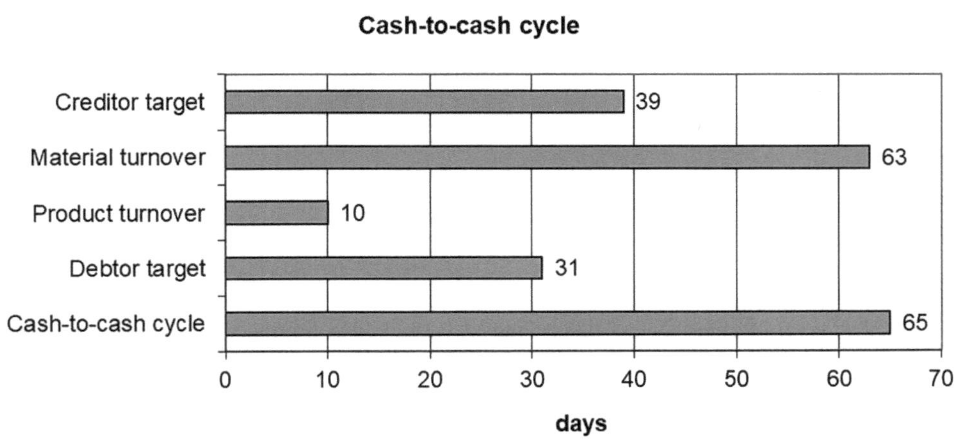

Fig. 3.49 Bar chart

Bar and column charts can also be displayed in the negative number range, like line and point charts. This is particularly suitable for displaying positive and negative deviations as well as results, see Fig. 3.50.

Comparisons can be well represented with column and bar charts as the difference of the compared objects can be well estimated by the different column or bar height or represented by an additional column or bar.

With the help of column charts, trends, development tendencies and changes over time can be well recognized, as with line and curve charts. The value expressions can be well displayed above the columns or next to the bars.

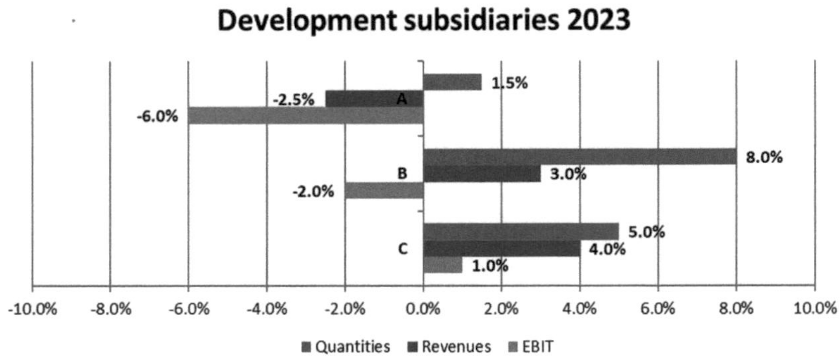

Fig. 3.50 Bar chart with negative and positive values

Fig. 3.51 Normal stacked columns

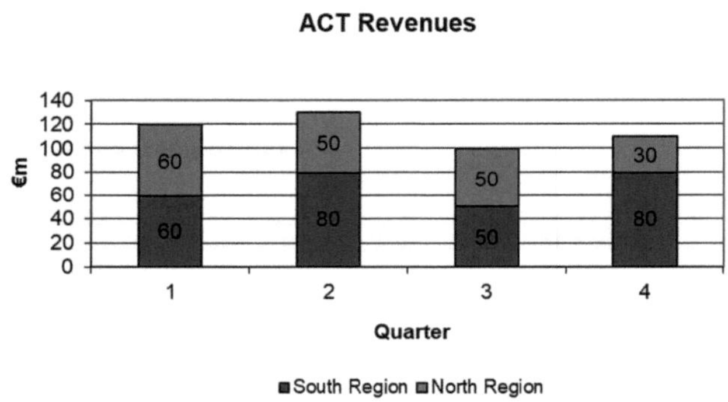

Another level of analysis can be obtained for stacked columns or bars by dividing the rectangular total area into subareas (see Fig. 3.51). However, if the number of subareas is too large, the diagram becomes confusing again.

An alternative to the normal stacked column is the percentage-weighted stacked column (see Fig. 3.52). Here, the subareas are normalized to 100% in the sense of a breakdown number, which allows a better comparison of the object value ratios that are represented in the subareas. The disadvantage is the loss of transparency of the absolute total value of the objects under consideration (here, for example, the temporal development of sales).

Furthermore, additional information can also be included in the columns or bars, such as values, attributes or deviations. This inclusion increases the information content, but may also contribute to greater confusion.

For column or bar charts, as with other types of charts, it should be avoided that the chosen legend (section and scale intervals) distorts or falsifies the statement, as the following example shows (see Fig. 3.53). Due to the cut-off legend and the larger legend

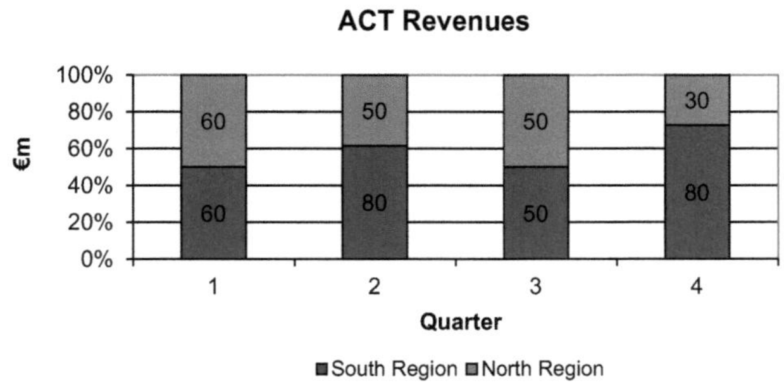

Fig. 3.52 Percentage-weighted stacked columns

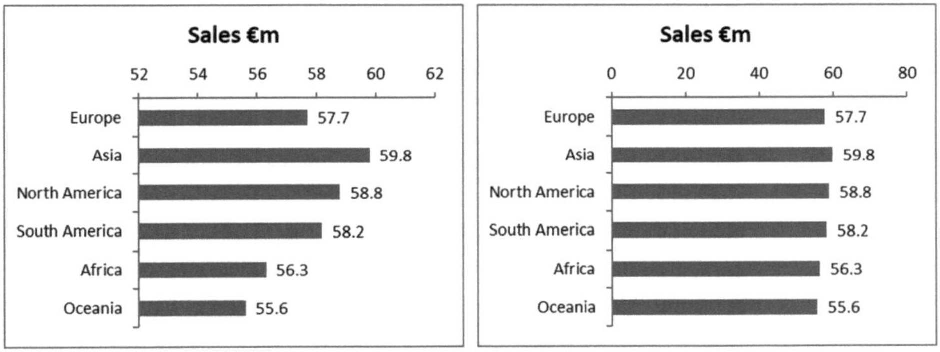

Fig. 3.53 Avoidance of distortions

scale (left), marginal sales differences of the regions are shown which actually have no great significance (right).[30]

An advanced visualization technique of column or bar charts are so-called bullet charts, sometimes also called target graphs[31] (see Fig. 3.54). With this chart, value expressions of a key figure can be compared to reference values based on a scaling division. Reference values are e.g. target or plan values which are compared with the current actual values. For this purpose, e.g. a line is used as a orientation boundary. Previous year values can be visualized e.g. with the help of a point or a half-stroke (see the bullet chart representation of the key figures in the example of the BI-supported controlling cockpit in Sect. 3.9). As a scale division, threshold values are suitable to which categories are assigned. These categories can then be visually highlighted e.g. with fixed colors (top) or with soft color transitions (bottom). In particular, traffic light colors are suitable for this. The bullet chart is very similar to the speedometer (see further below). But unlike the speedometer, it uses a flat bar instead of a semicircle.

[30] Further examples can be found in Kohlhammer et al. (2013, p. 72–77).

[31] Cf. Kohlhammer et al. (2013, p. 133).

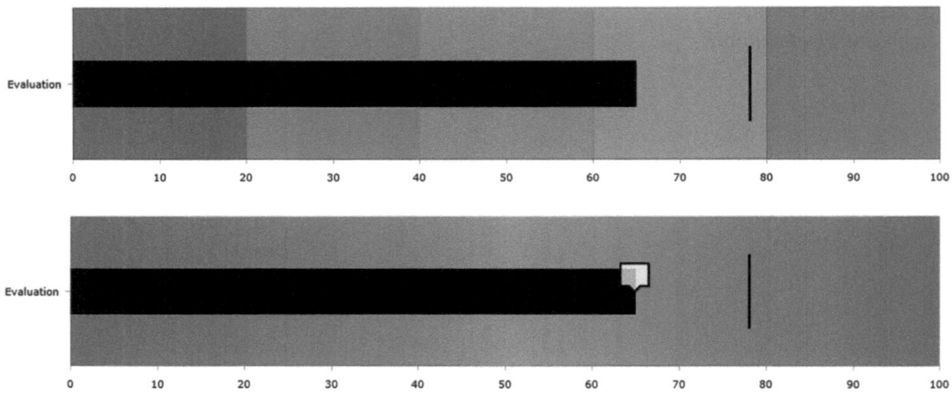

Fig. 3.54 Bullet chart. *Source* Online Chart Makers—amCharts (2017b)

Waterfall Charts, Histograms, Box Plots and Funnel Charts

Waterfall charts are actually extended column charts, which show in particular the positive and negative changes of key figures over time (see Fig. 3.55). The color support helps to analyze e.g. the change of direction, and the line connection of the columns helps to analyze, the initial level of the considered period. Waterfall charts are especially suitable for the analyzing key figures in their temporal development with regard to their change of direction and intensity.

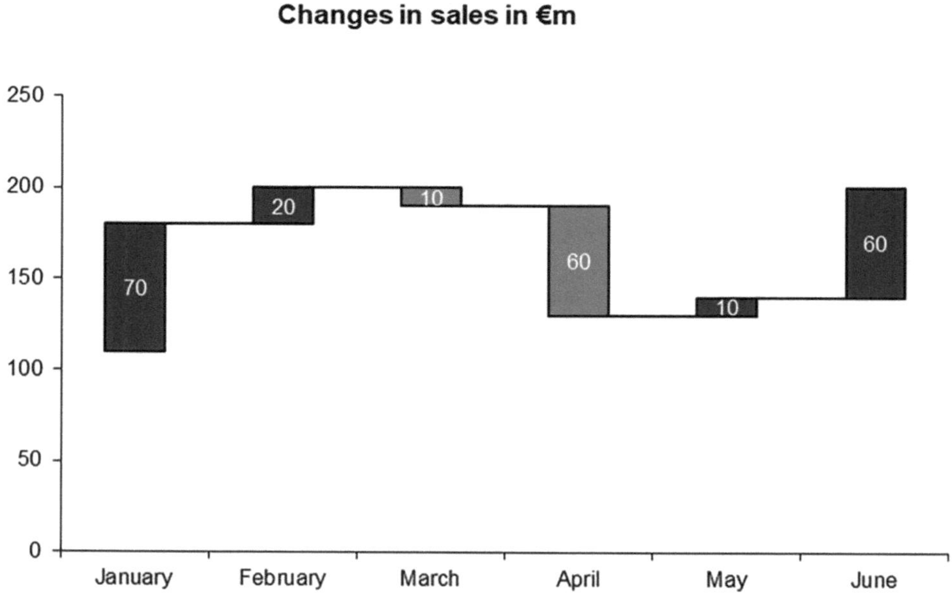

Fig. 3.55 Waterfall chart with MS Excel

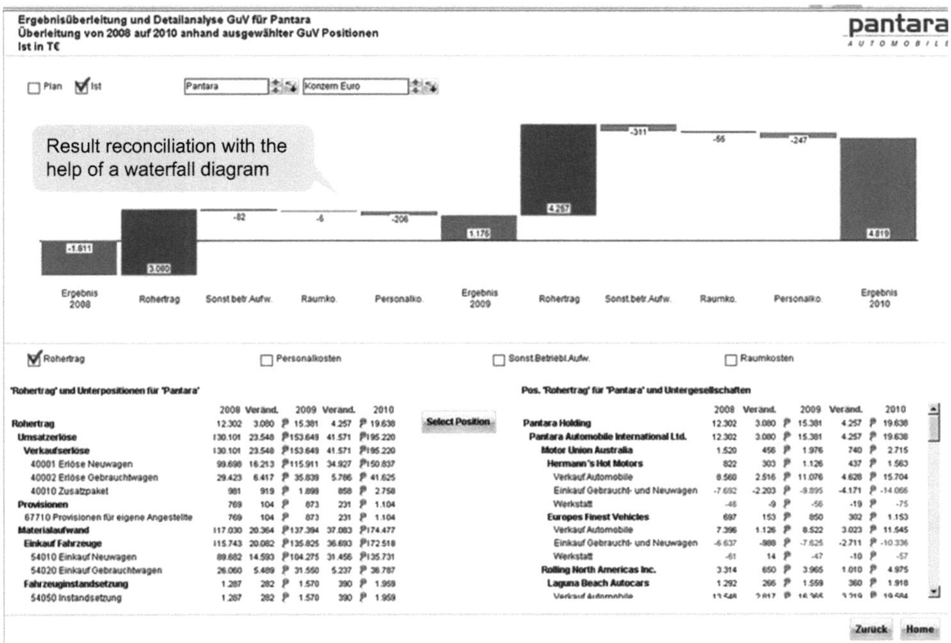

Fig. 3.56 Waterfall diagram with Cubeware. *Source* Cubeware image gallery (2011)

Furthermore, waterfall diagrams can be used to show changes of values in important stages, as shown by example (based on gross profit and various cost items up to the result) in Fig. 3.56.

Column and bar charts are also very suitable for mathematical and statistical evaluations in combination with line charts. Popular chart types are the histogram and the box plot (see Fig. 3.57). A histogram shows the frequency distribution of cardinal scaled features, whereby a class division of the data has to be made. The class division can be done with fixed or variable width. The box plot or the box-whisker plot shows the following statistical values for value expressions of an analysis object in a period of time:

- Lower and upper outlier (smallest and largest data value)
- Median (50% of the values are smaller or larger than this value)
- Lower and upper quartile (25%/75% of the values are smaller or larger than this value)
- Range (total value range of the data)
- Interquartile range (value range in which the middle 50% of the data are between the quartiles)

The term box-whisker refers to the appearance where the quartile values indicate the box and the hair-like antenna the outlier values (minimum and maximum). The median is

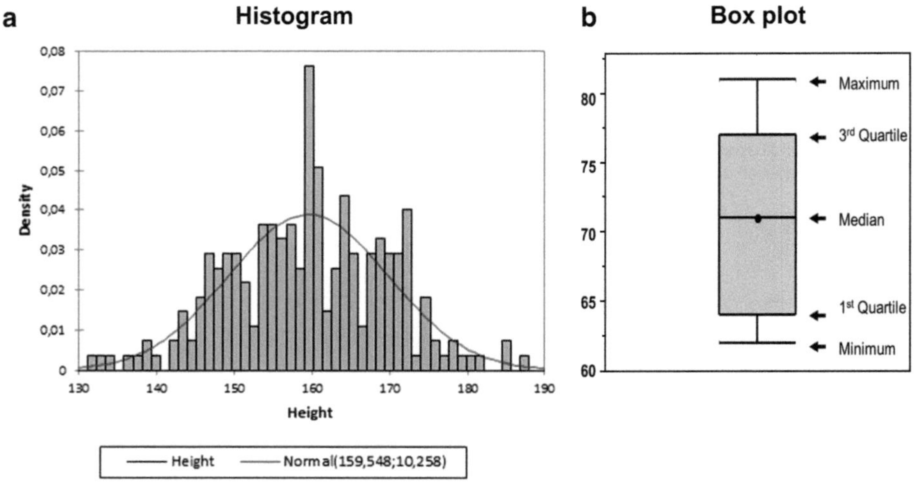

Fig. 3.57 Histogram and box plot. *Source* **a**: XLSTAT (2017), *Source* **b**: Sullivan and LaMorte (2017)

visualized as a line in the box. If the median is more up or down in the box, the distribution is skewed as opposed to a normal distribution. The quartiles are replaced for other purposes (e.g.: risk considerations) by other sizes, e.g. 2.5%-quantile, where especially the outliers are considered. The box plot can be used sensibly, for example, for price developments or risk analyses in companies.

If one centers the bar or column deviations, a funnel chart can be displayed. Here, the value expressions of the analysis objects usually become smaller from level to level. They are popular for presenting the acquisition analysis in sales when inquiries and prospects become qualified leads, offers and customer orders. Figure 3.58 shows an example of the acquisition success of a website from the view to the order. The size of a surface section in the funnel is determined by the value as a percentage of the sum of all values and decreases from top to bottom like the passage of a funnel.

Curve, Line, Point and Area Charts

For the representation of temporal developments and seasonal effects, the curve or line chart is best suited (see Fig. 3.59). Outliers, stable trends or dynamic changes of courses are quickly recognizable. For a better overview, point values can be displayed in the coordinate system. A normalization in relation to index numbers is also possible. For example, the target value is normalized to 100%, and one can see how far the target achievement degree of the actual value is compared to the target value.

Comparisons of value series can be well represented in the same coordinate system in order to determine special differences and similarities or parallel or opposite developments.

If the number of lines in a chart is very high (e.g. more than 6 lines) and the lines overlap several times, the curve charts become more and more unclear. In order not to

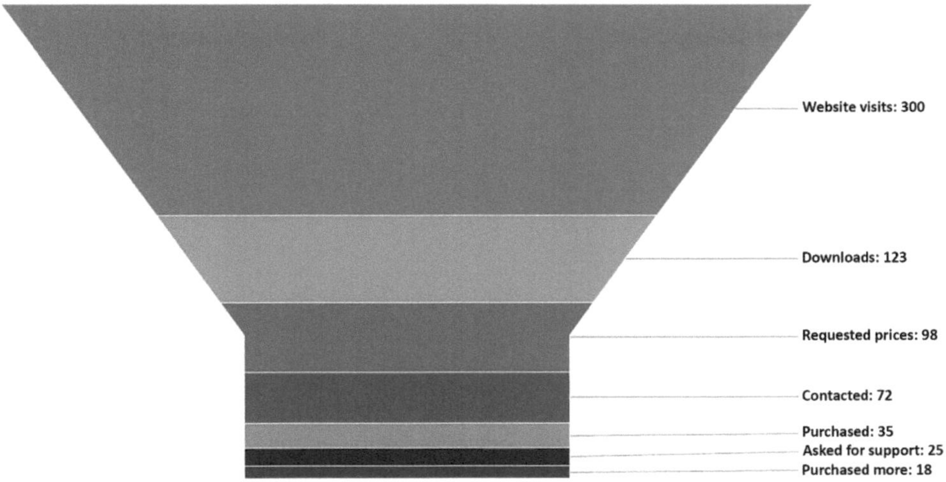

Fig. 3.58 Funnel chart. *Source* Online Chart Makers—amCharts (2017a)

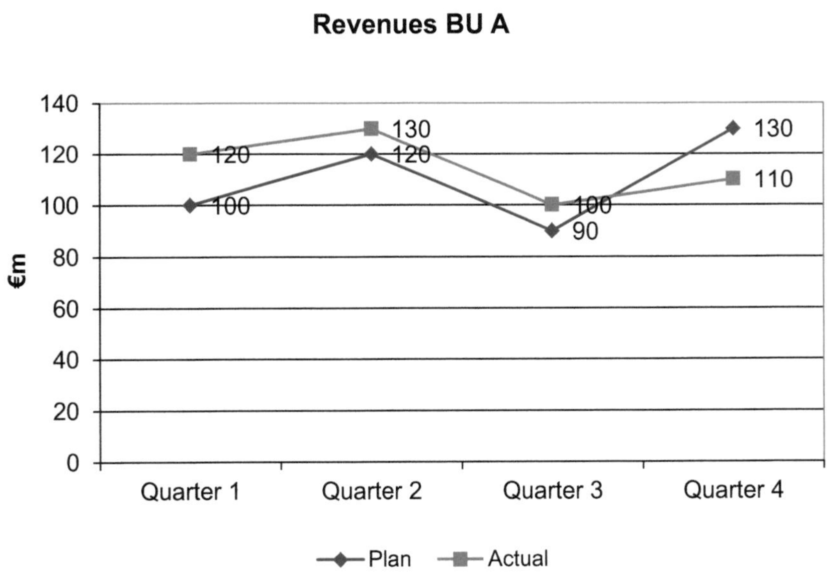

Fig. 3.59 Normal curve or line chart

deceive the visual perception, the distances of the observed periods are to be represented evenly. Changes in the scaling to e.g. see overlaps better lead to erroneous and distorted perception and should not be done.[32]

[32] Kohlhammer et al. (2013, p. 57).

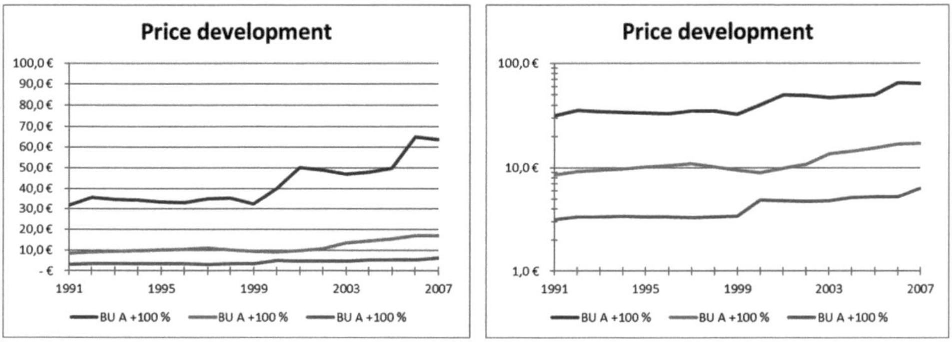

Fig. 3.60 Distorted comparisons of price developments

If a series of values is fixed as a basis, a normalized curve or line diagram is to be displayed; this diagram shows the changes of other series of values relative to this basis series. The advantage here is the easier comparison to the basis series. The disadvantage is the poorer transparency of the absolute value development.

For line diagrams, as with other types of diagrams, it should be avoided that chosen legends (section and scale intervals) distort or falsify the statement (here price developments) as the following example shows (see Fig. 3.60). In a price comparison of different product segments with different prices, the price elasticity of the expensive price segment A is presented as if it were a price explosion compared to the other two segments (left). If the comparison of the price development is to be in the focus of the analysis, a logarithmic representation should be chosen, as here the angles of the price increases starting from the respective base value are more transparent (right).

Another alternative to the line diagram is the dot diagram (see Fig. 3.61). Unlike the line diagram, it has the advantage that the values are only displayed as dots and thus the connecting line does not suggest a linear value progression. However, the disadvantage is the poorer transparency of the development of the value; here, the intersecting lines in particular would act as optical aids to quickly recognize changes.

If two key figures, which have a uniform notation (e.g. percentage representation), e.g. growth rate and profit increase, are compared in a line or dot diagram, it makes sense to use a rectangular coordinate system as a representation. The coordinate axes divide the potential diagram area into four quadrants. Here, the positive upper right quadrant is used most frequently. If negative values are also shown, further quadrants must also be displayed (see Fig. 3.62).

For the representation of the utilization of resources and capacities and similar problems, area diagrams are suitable, which, in contrast to line diagrams, color the underlying areas (cf. Fig. 3.63).

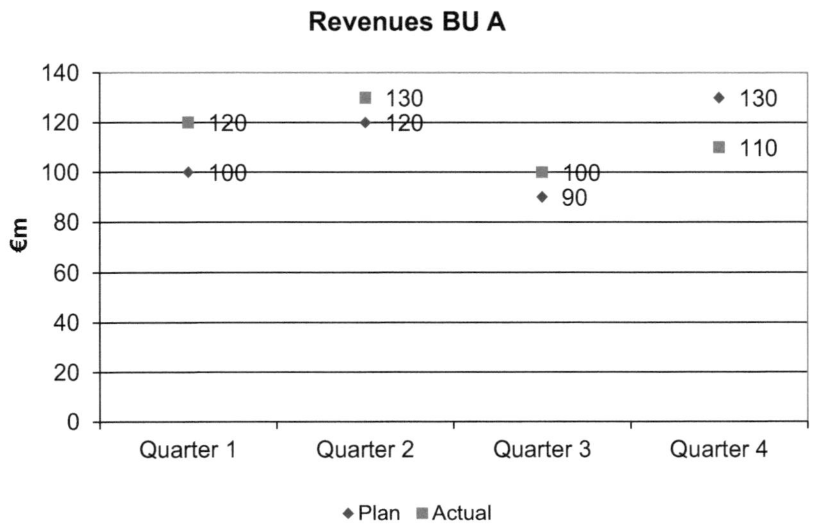

Fig. 3.61 Dot diagram

Traffic Signals and Heatmap

Traffic signals or Traffic lights are intended to show the viewer three categories, using a familiar warning symbol from road traffic; the categories represent whether good (green), bad (red) or only first warnings (yellow) have been achieved, symbolized by the corresponding color.

The classic three categories can be further differentiated by subdividing the color scheme into light to dark colors (e.g. light green to dark red).
The traffic signals can be used in very different ways in diagrams (cf. Fig. 3.64). For example, only the signal colors are shown, without the viewer knowing the corresponding detail values at first glance, or they can be used as field-specific area or font color to support the value display. A simple traffic light function is familiar to many users of spreadsheet programs by the red representation of negative values. Often, the planned values are also used as threshold values, whereby a predetermined percentage deviation then determines the color change. For the Geo-visualization, the "Color-Coding" of different areas is suitable, too.

A prerequisite for the definition of **signal colors** is the previous determination of tolerance or threshold values. This gives the information a predefined control function, which the creator and/or the recipient of the report have previously determined. Alternatively, the threshold values, so-called **Sentinels**, can also be determined by the programs using statistical models and calculations (see Sect. 5.5.5.3 Data Mining and Predictive Analytics). Then the threshold values are automatically determined based on the evaluated data range. The user can choose which sentinels he wants to use. To facilitate the

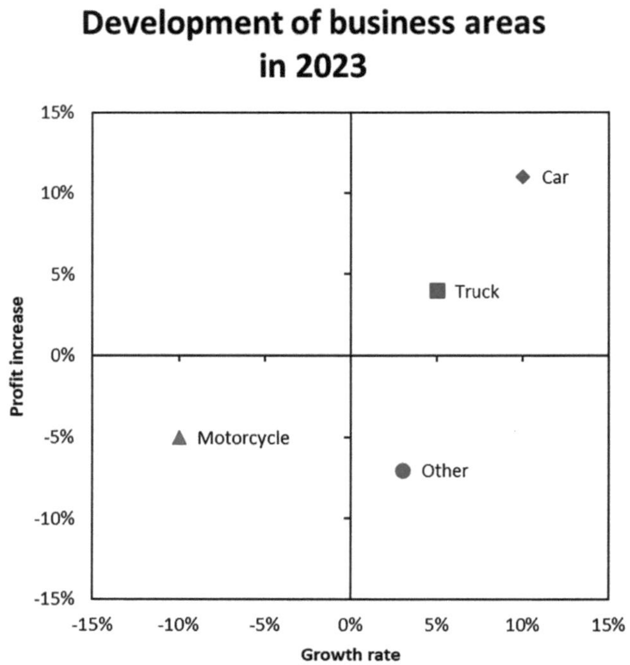

Fig. 3.62 Coordinate system with four quadrants

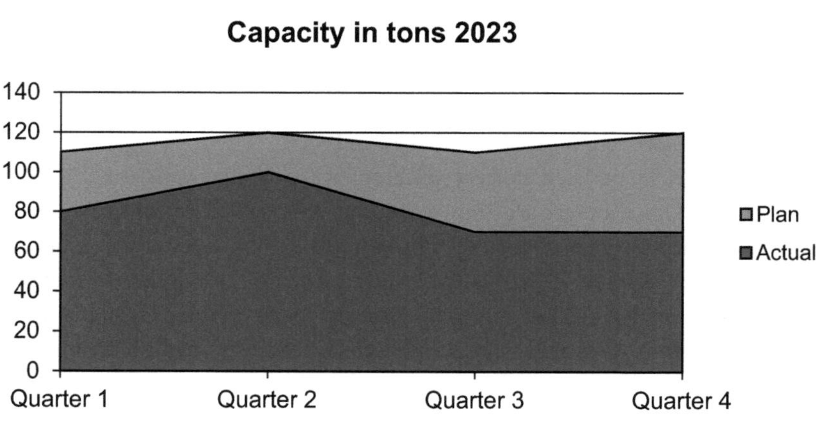

Fig. 3.63 Area diagram

selection, e.g. the confidence level is displayed. The higher this level, the more reliable the rule.

The advantage of the traffic light signals lies in the faster visual perception of starting or already existing positive or critical states. Moreover, the person who sets the threshold values has to think in advance about the control intentions of the signal colors, some-

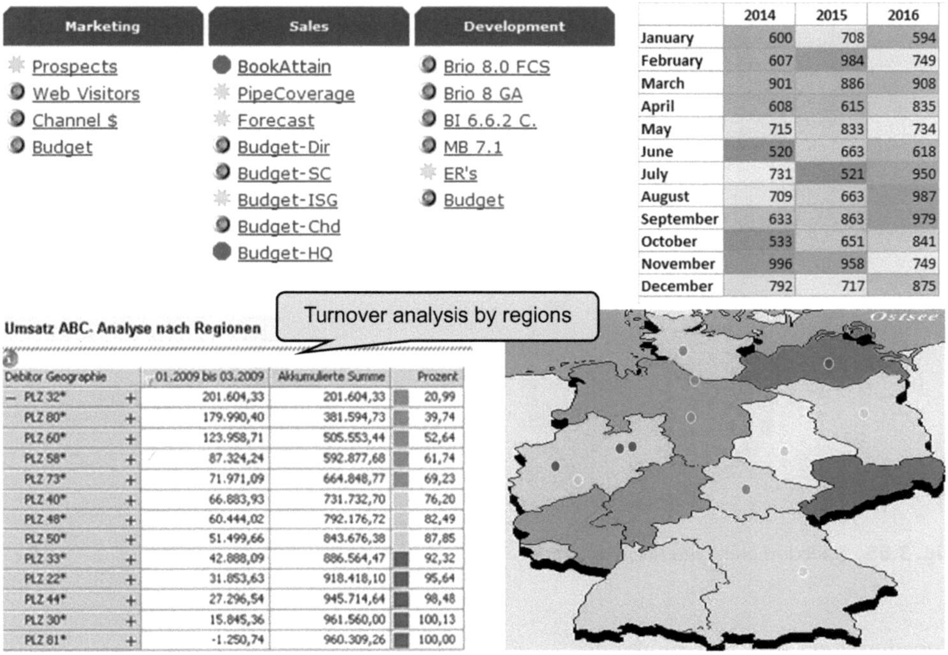

	2014	2015	2016
January	600	708	594
February	607	984	749
March	901	886	908
April	608	615	835
May	715	833	734
June	520	663	618
July	731	521	950
August	709	663	987
September	633	863	979
October	533	651	841
November	996	958	749
December	792	717	875

Debitor Geographie		01.2009 bis 03.2009	Akkumulierte Summe	Prozent
– PLZ 32*	+	201.604,33	201.604,33	20,99
PLZ 80*	+	179.990,40	381.594,73	39,74
PLZ 60*	+	123.958,71	505.553,44	52,64
PLZ 58*	+	87.324,24	592.877,68	61,74
PLZ 73*	+	71.971,09	664.848,77	69,23
PLZ 40*	+	66.883,93	731.732,70	76,20
PLZ 48*	+	60.444,02	792.176,72	82,49
PLZ 50*	+	51.499,66	843.676,38	87,85
PLZ 33*	+	42.888,09	886.564,47	92,32
PLZ 22*	+	31.853,63	918.418,10	95,64
PLZ 44*	+	27.296,54	945.714,64	98,48
PLZ 30*	+	15.845,36	961.560,00	100,13
PLZ 81*	+	-1.250,74	960.309,26	100,00

Fig. 3.64 Traffic signals/Heatmap. (Report elements among others taken from Schön and Müller 2010, p. 123–165 and Schön 2004b, p. 328)

thing which in turn contributes to the control and decision quality in the company. The disadvantage is the higher effort that is tied to the fact that the threshold values have to be checked and maintained over time. A minor disadvantage may consist in the lack of transparency of the "not" reported detail information. This information is usually only possible by calling up another report, something which corresponds to the drill-down principle.

In summary, the traffic light signals are very positive to evaluate as the analyst is quickly directed to those areas where there is a need for action and where he can now look for further detail information.

In Fig. 3.64, the example at the top left shows a list of management-relevant subareas in marketing, sales and product development. The traffic light signals indicate in which subareas which signal states are currently present. Thus, the analyst can relatively quickly see in which areas there is a need for control. At the top right is a table as a heatmap. Heatmap charts range from high-resolution thermal imaging analysis to simple representations, e.g. with the help of conditional formatting in MS Excel. The table at the bottom left shows a an ABC analysis with sales by debtor classes, whereby the classes for A, B and C customers (green, yellow and red) are color-coded (see also Sect. 3.7.6). The figure at the bottom right shows regional sales values displayed on a map where the colors draw attention to problem regions.

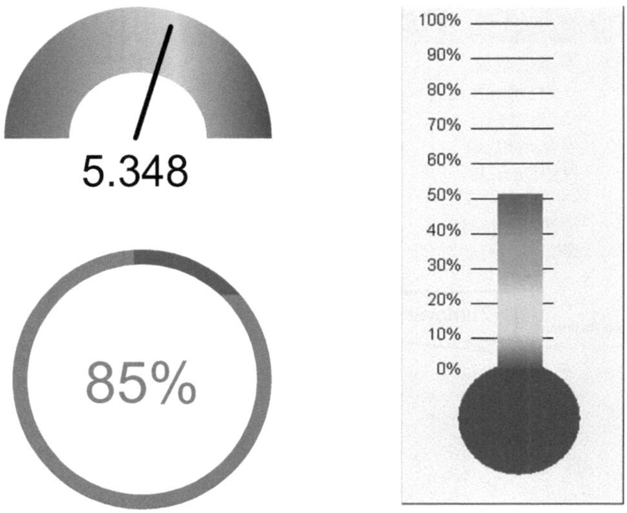

Fig. 3.65 Speedometers and thermometers

Speedometers and Thermometers

Another possibility to quickly inform the information recipient about a state and, at the same time, to point out possible disturbance signals is given by the use of scale-related measuring instruments such as speedometers or thermometers (see Fig. 3.65, which have great similarities with bullet charts (see above).). As intended for their actual use, e.g. the speed or temperature display, these chart types can be used to visualize the state of individual key figures. They have the advantage, similar to the traffic light signals, of using display areas for this purpose and highlighting, e.g. by color or by a scale, where the key figure reaches good, medium or bad values. For the division of the scale, as with the traffic light signals, threshold values have to be defined in advance. In contrast to the traffic light, the viewer can even see whether the state of the achieved key figure is in the upper or lower range between the defined threshold limits. The display scale usually has equally large cardinal definable measurement distances, in which the signal areas can be placed arbitrarily.

The disadvantage, as with the traffic light signals, is that often the underlying detailed information, such as the absolute values and the temporal development, are not recognizable. However, the signal function gives the viewer the incentive to look at the detailed analysis for very good or very bad values. Thus, he can direct his analysis time sensibly to the areas where there is a need for action.

In the first speedometer example at the top left, the key figure is given in terms of value under the speedometer and displayed on the speedometer scale at the top. The color scale can be oriented very well, for example, to the target or planned value. At an achievement level of 100%, for example, the pointer is exactly vertical. For all states that are better, the pointer then tilts to the right, and for states that are worse, it tilts to the left.

In the thermometer example, a deviation from the target value is shown. The greater the deviation, the more the value in the thermometer rises.

Both with the speedometer and with the thermometer, visual irritations can arise if the key figures in the scale values do not improve from right to left or from top to bottom in analogy to the speed or temperature. Examples of this are the cost level, which is usually assessed better when it is low. Therefore, it makes sense to mark the evaluation areas with the traffic light signal color in the presentations. In the speedometer at the bottom left, the scale is divided into three threshold areas. In the speedometer at the bottom right, the target achievement level in percent is in the foreground of the analysis.

Geo-visualization

From the point of view of management reporting, geo-visualization or geo-reporting means the adapted visual presentation of management information (such as sales and profit statistics) that are embedded in flat maps (see Fig. 3.66) or other spatial representations (e.g. layout of a stadium occupancy, see Fig. 3.36). They serve the user to quickly analyze the selected regional data and to derive control conclusions and recommendations for action from them.

They are especially helpful for the regional result control of plants, districts, business units and strategic business units, especially when the regional result view is of greater importance for the company. Due to the international orientation of the economy in all sectors, the regional view for the global corporate control is indispensable.

In the example of Fig. 3.66, several pieces of information have been linked and displayed. The areas of the individual American states are color-coded here for the different market potentials. The columns show the sales change of the last quarters per region. With the help of the traffic light signals, good, average and bad changes can be seen here. In addition, the points indicate the respective locations of the sales units.

The potential and the operational evaluation possibilities of geo-data have increased rapidly in the last years, and it can be assumed that this trend will continue. Examples from the economy are diverse, e.g. the regional shopping behavior of customers or the maintenance-needy networks of an energy supplier.

Scoring Mapping

With the help of the scoring mapping, the evaluations of criteria to be analyzed can be displayed in a scoring profile (see Sect. 3.7.10). Here, both quantitative and qualitative criteria can be analyzed. Graphics for a scoring or utility analysis show the criteria evaluations of selected objects in a grid. Alternative representations are possible, e.g.:

- Comparison with one or more object values
- Comparison with an average value
- Comparison with a best or worst value
- Comparison with a minimum or maximum value.

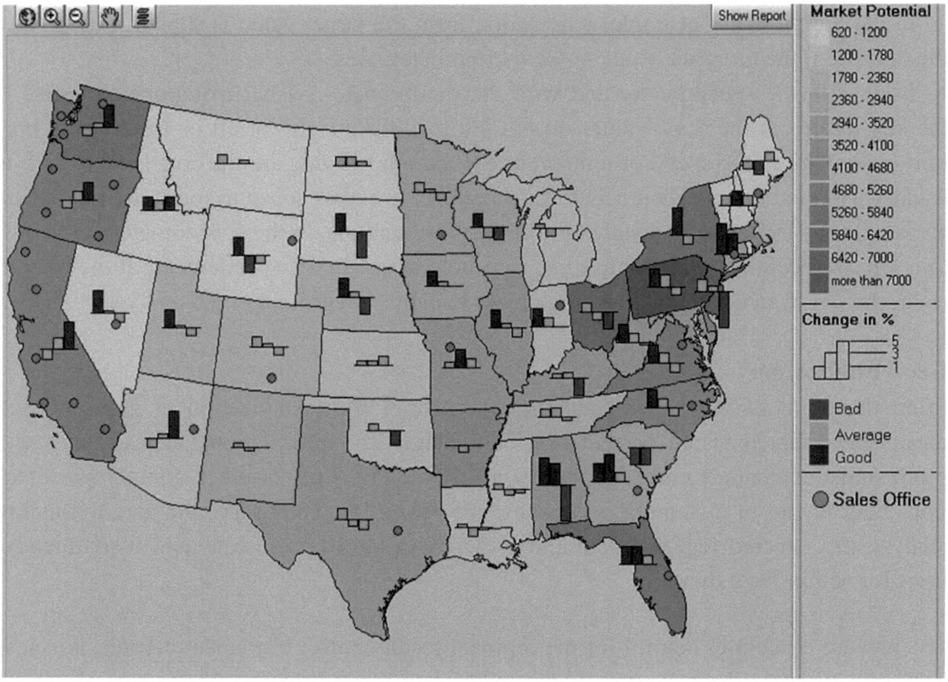

Fig. 3.66 Geo-visualization. *Source* SAP AG (2003)

Scoring mappings are also based on a rectangular coordinate system. The evaluation of the criteria is displayed on the main coordinate (usually horizontal). If the criteria are evaluated with points, for example, the coordinate distances correspond to the point value ranges (e.g. 1–10 or 1–100). Alternatively to point values, individual characteristics of the criteria can also be displayed. However, this is only clear if the evaluations are done with uniform units, e.g. currencies (for sales). If different criteria with different evaluation scales and units are compared, a point value assessment with separate indication of the units of the individual criterion is recommended. A scoring example for a supplier comparison with threshold value is shown in Fig. 3.67.

Spider Web or Network Diagrams

Spider web or network diagrams help to carry out comparisons for a selected number of evaluation criteria, whereby the scale for evaluation should be as normalized as possible (e.g. by uniform grades, point values or percentage values). The values are plotted in the network diagram with data points on the axes, starting from the center of the network. The farther a value is from the axis, the better (e.g. goal achievement) or worse (e.g.

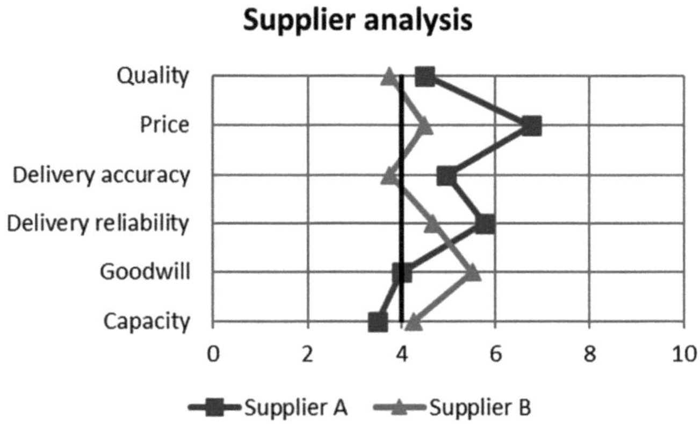

Fig. 3.67 Supplier comparison with threshold value

risks) it is evaluated. It is important that the evaluation criteria swing in the same direction and are as normalized as possible in the swing, as e.g. with school grades.

Often, the data points are connected with lines from axis to axis. This creates an area in the spider diagram. With normalized value scales, a larger area thus means a better or worse overall assessment across all criteria, whereby the individual criteria must then be equally weighted.

However, if the number of criteria is too large, the spider diagram becomes easily confusing. Also, plotting several key figures (e.g. time-dependent) leads rather to a visual overload, as Fig. 3.68 shows for a radar.

The base year in the example is the planning year 2009, which specifies the target value 100%. Based on this, the years 2010 and 2009 in actual and the planning years 2010 and 2011 in comparison for all key figures of the balanced scorecard are displayed. However, the normalization to percentage values results in the loss of absolute values. Also, a weighting of the individual values, which distinguishes the importance of individual criteria, is no longer recognizable here.

Therefore, the spider diagram representation is only recommended if few evaluation criteria, which are normalized and equally weighted, can be displayed.

Sparklines or Microcharts

By sparklines or microcharts one understands miniature diagrams that complement data and number series in tables and text information without having to insert larger diagrams in parallel. The sparklines are only as large as the cells of the tables or the text line, so that they can be integrated directly. The original concept of sparklines was developed by

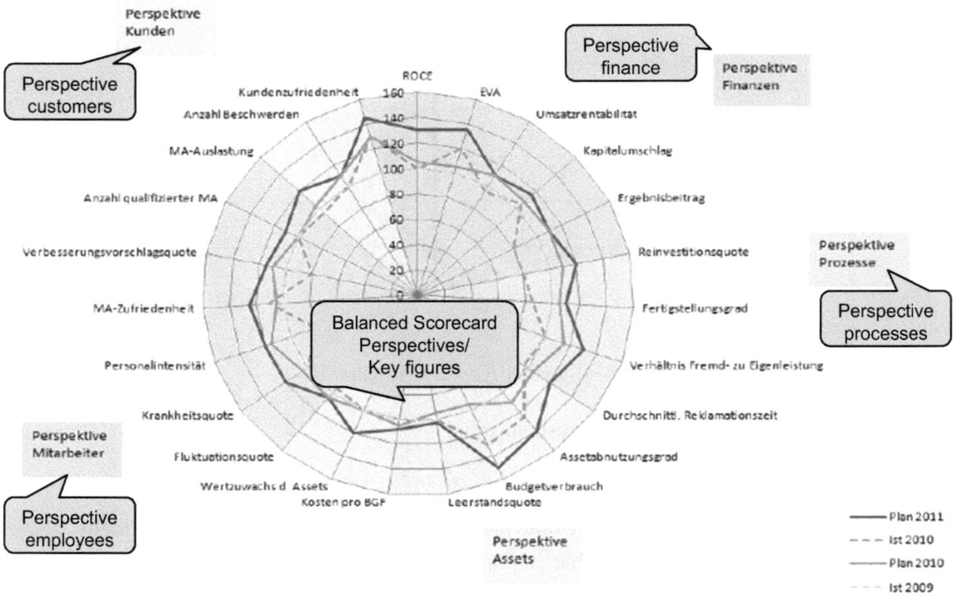

Fig. 3.68 (Spider) spider diagram. (Own representation based on Gehringer and Michel 2000, p. 28)

Edward Tufte.[33] The advantage of sparklines is that they visually illustrate trends and effects of more extensive data by small miniature diagrams line by line; this illustration, among other things, also makes comparisons between data rows easier. In addition, the use of sparklines can visually supplement data with their periodic view. For example, the distribution of the period values in the miniature diagram can be displayed in addition to the absolute annual value (value representation).

In principle, many chart types that have already been mentioned above are suitable as diagram types for the sparklines. However, the most suitable use for the sparklines are usually the charts that are best suited to represent trends over time, such as line and column charts (see Fig. 3.69). Due to the sharper contours, column charts are to be preferred here.[34]

Course Charts/Chart Analysis Charts

Course charts or chart analysis charts are used to display stock prices or similar prices for securities, commodities, etc. (see Fig. 3.70). They can be used for any time periods

---

[33] Cf. the information from Tufte (2011).

[34] See Bissantz et al. (2013, p. 451 ff.).

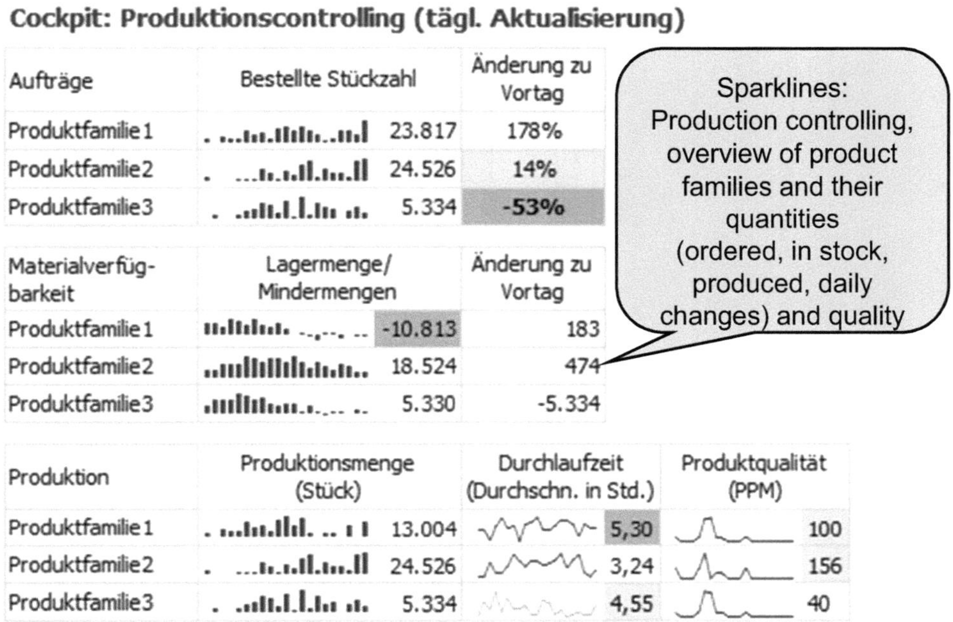

Fig. 3.69 Sparkline example. (The example was taken from the web presentation pages of Bissantz & Company GmbH 2011)

(days, weeks, months, years, etc.) and visualize the fluctuations and developments of the prices. Typically, closing prices, and high and low points of the price development are shown. In addition, the traded volume can be depicted.

Alternatively, price movements can also be represented in the form of line charts in which trend lines, scenarios or other chart analysis aids can also be incorporated. The banks, stock exchanges and other institutions offer numerous toolsets for chart analysis (see Fig. 3.71).

Further Visualization Aids and Animations
For most of the shown diagram types, there is the possibility to use them also in combination with other diagram types. For example, the pie chart can be supplemented by a bar chart for a selected sector area (e.g. the largest partial section) containing the most important shares of the largest group values (see Fig. 3.72).

On the other hand, the use of **3D effects** in graphics may be appealing to the eye, but the 3D representation is rather hindering for the data analysis, as value ratios are distorted depending on the view of the 3D object (see Fig. 3.73). It is even possible that data is hidden by other 3D objects in the foreground and thus not recognized at all. Therefore,

Fig. 3.70 Example course chart

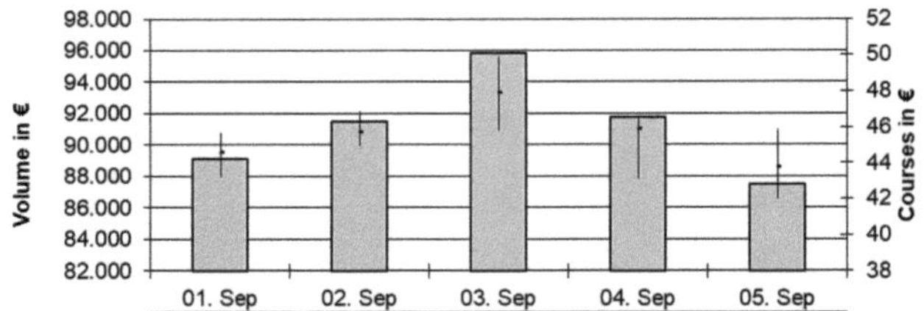

	01. Sep	02. Sep	03. Sep	04. Sep	05. Sep
◻Volume	89.120	91.460	95.820	91.780	87.520
High	45,67	46,89	49,88	46,47	45,86
Low	43,22	44,98	45,77	43,11	42,01
· Close	44,56	45,73	47,91	45,89	43,76

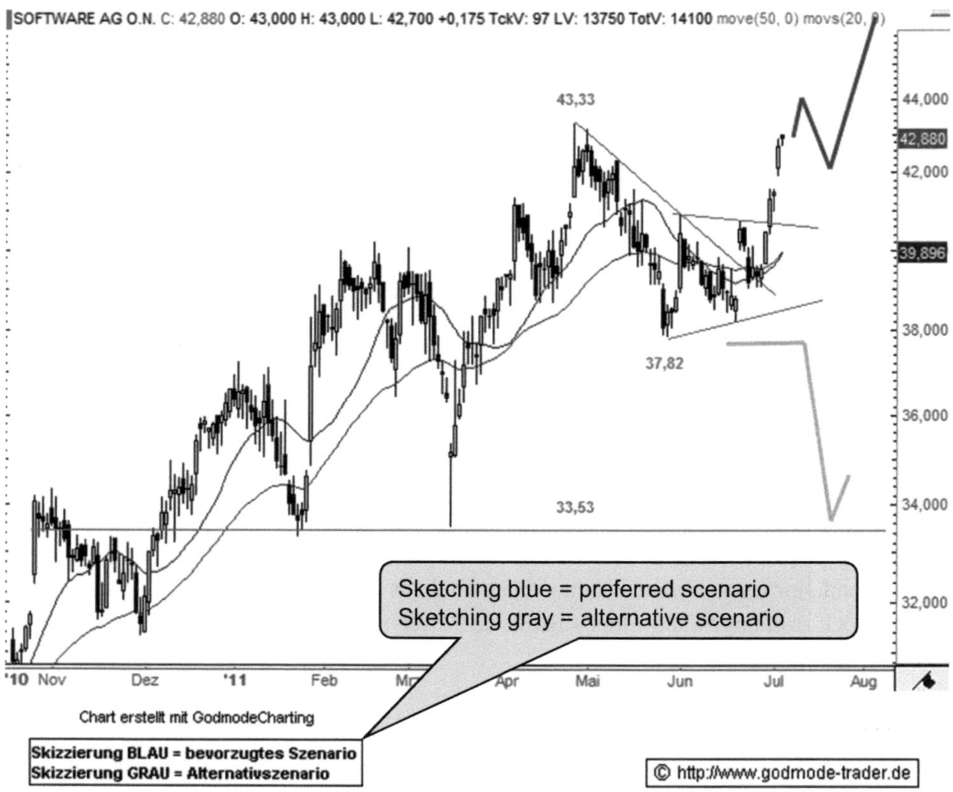

Fig. 3.71 Example chart analysis diagram. *Source* godmode-trader (2011)

one should avoid 3D effects, unless the number of indicators to be displayed is very small and the distortion of the view does not play a big role for the analysis.

Negative aspects in this example are also the repetitions in the time dimension, in the legend, and in a category axis, as well as the vertical object labels in the category axis for

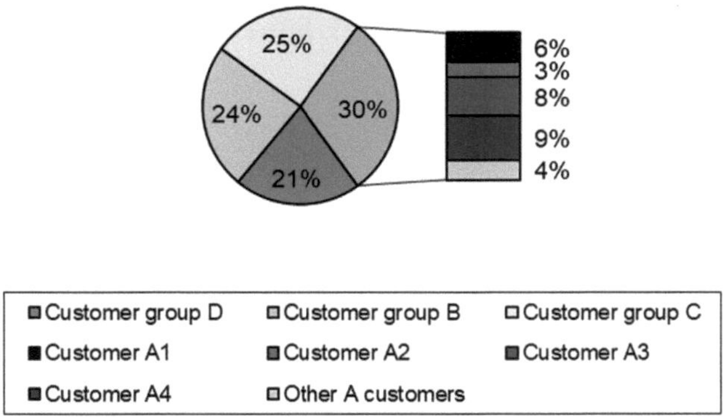

Fig. 3.72 Combination pie-bar chart

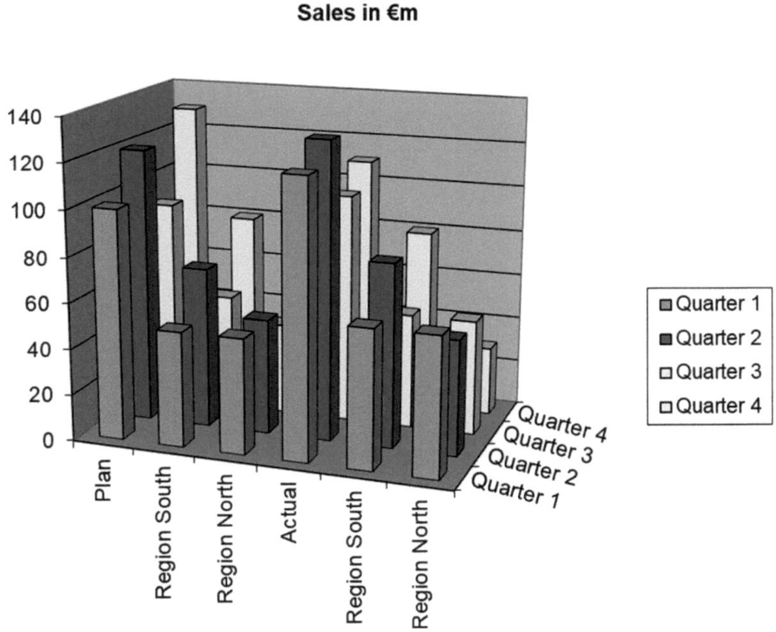

Fig. 3.73 Distortions by 3D effects

the regional view. In addition, plan and actual values are not clearly separated from each other.

Also further animation effects, such as **flash effects**, usually do not serve the analysis capability, but rather the show of the presentation. With flash, parts or even whole diagram contents are only created by calling the report or by selecting certain functions. Here, flowing fade-ins of different kinds can be set, e.g. from above, from below, from right or left and random fade-ins. They should therefore only be used, if a particularly important piece of information is to be highlighted in a presentation.

Fig. 3.74 Dynamic diagrams that can be changed by selection—before (QlikView). *Source* QlikView-Demo Market Share (2015)

Better than 3D and flash effects is the use of changeable **dynamic diagrams**, which are controlled by selection. For example, by selecting a row in a table, the diagram can change its presentation and show only the selected objects, e.g. a customer group or an individual customer. The diagram then shows, for example, the sales, contribution margin and cost development only for the selected objects. This is advantageous when the diagrams would otherwise become too confusing due to many objects.

In Fig. 3.74 (before), several segments and categories have been selected. By selecting the category Smart Phones and segment Consumer, the values in the tables and graphs above adjust (Fig. 3.75 after).

The selection can be done not only by choosing the selection filters, but also directly in the graphic, as Fig. 3.76 shows for a specially selected period of time. In the graphic, for example (top right), the selection for reprocessing the report elements and key figures is triggered by marking the time period. The result is shown in Fig. 3.77below. The example shows full-time staff statistics for a healthcare facility, which was built with the software HEARTBEAT based on the technology of Qlik-Sense. For the data basis of

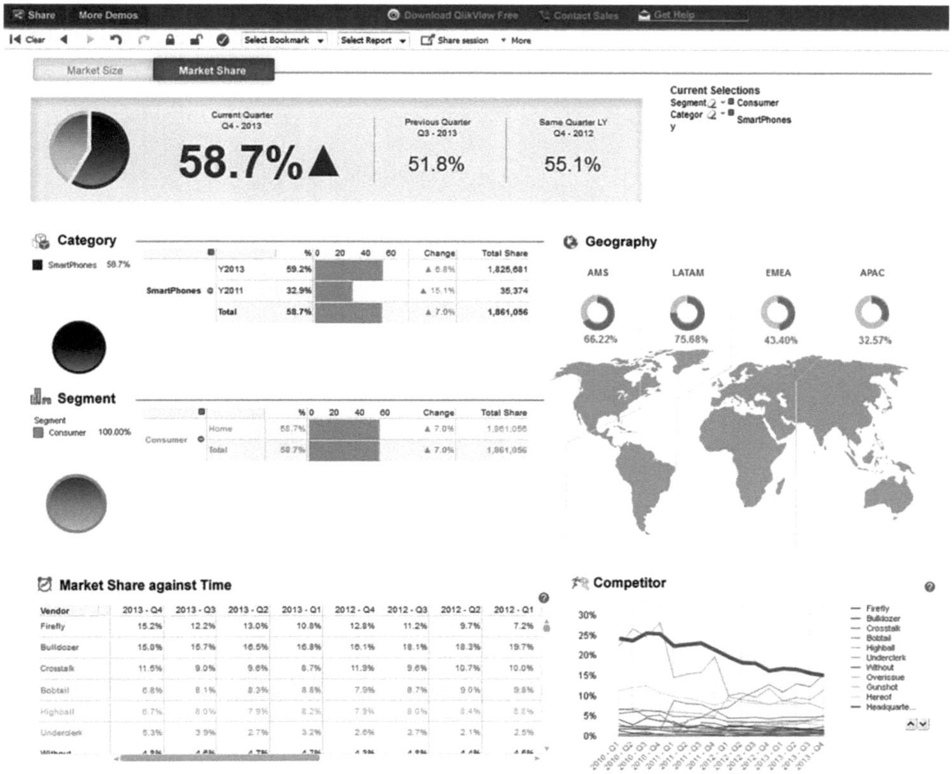

Fig. 3.75 Dynamic diagrams that can be changed by selection—after (QlikView). *Source* QlikView-Demo Market Share (2015)

HEARTBEAT, mainly pre-systems from the personnel area are connected. The full-time staff statistics show the converted total number of available full-time workforces, a number which can be differentiated and analyzed according to the part-time rate by service types, organizational units, etc. HEARTBEAT is a special management and personnel management system of Pradtke GmbH in Bochum, a provider of personnel management software with a focus on healthcare.

A popular function of the software applications is also the **drag & drop functionality**, in which the selected objects are marked, held with the mouse button and moved into another object. This changes, for example, a graphic or table and adapts it to the selected objects.

Special types of diagrams, which are especially used for the visualization of data pattern recognition in the context of data mining and predictive analytics, are exemplarily listed in Sect. 5.5.5.3 in Fig. 5.37. They are also often used in BI systems that are particularly strong in the area of data discovery and data visualization in the context of exploratory BI (see Sect. 5.7.3).

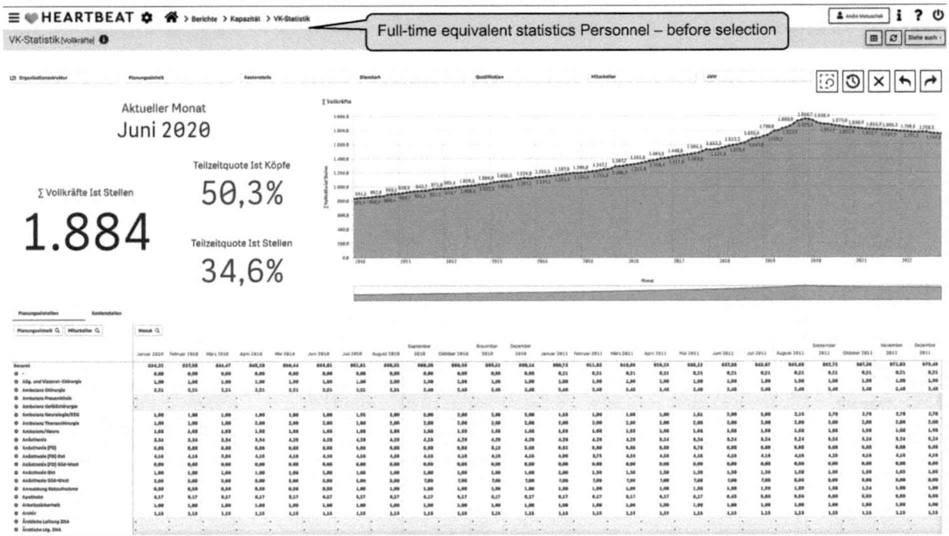

Fig. 3.76 Dynamic full-time staff statistics with HEARTBEAT—before. *Source* Pradtke GmbH (2021)

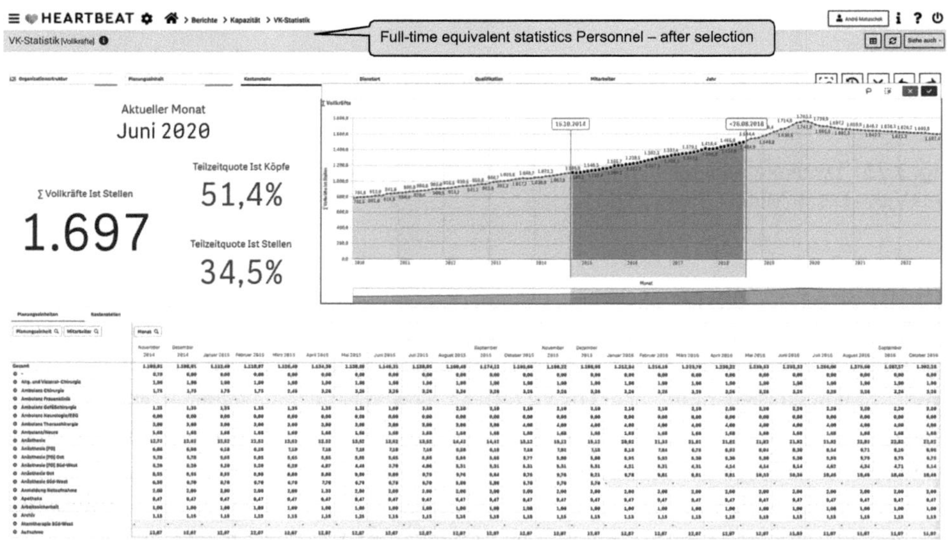

Fig. 3.77 Dynamic full-time staff statistics with HEARTBEAT—after. *Source* Pradtke GmbH (2021)

As further information aids for the report, the following objects can be distinguished in particular:

- photos
- logos
- symbols
- audio files
- movies/videos
- text files
- links to websites or similar

For the enhancement of reports, it is advisable to supplement important information with **photos** (see Fig. 3.78). For example, photos of the individual products or the faces of the salespeople can be shown next to the product results or to the sales statistics. When using photos and foreign images, the copyright and personality rights must be respected.

Logos serve mainly to support the corporate identity of the company and are often a general part of a report page (often as header or footer information). In addition, logos can also be used sensibly to distinguish between different companies in the participation reporting in order to better separate them from each other.

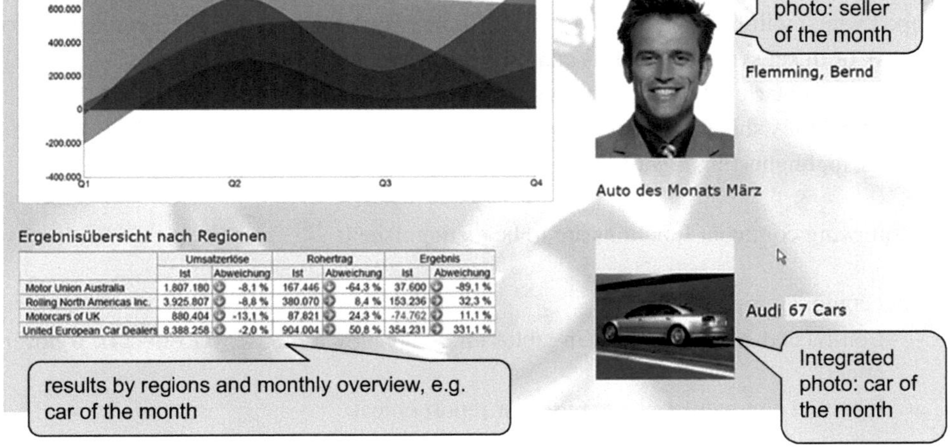

Fig. 3.78 Integrated photos and logos (example Cubeware). *Source* Cubeware image gallery (2011)

Symbols such as trend arrows or smileys can be used sensibly to define tendencies or evaluations in predefined classes.[35] Similar to sparklines, they have the advantage that they can be integrated directly into the table or text lines of the report.

For digital reporting, the addition of other media forms is also possible. Here, one should think especially of **Audiofiles** and **Movies** or **Videosequences**. They can be attached as supplementary information to the reports or partial information, for example, short comments on particularly important contents of the reports or general company information for shareholders and potential investors in the context of external reporting or in sales at customer presentations.

3.8.5 Comments, Messages and Management Summary

Management reports often include a lot of detailed information, so that important contents may be overlooked. Here, commentaries and short texts can help to capture the most important facts and findings of the report in a condensed way.

Good commentaries can be recognized by the following profile:

- They are written in bullet points and short.
- Important statements and information (also in the tables and graphs) are highlighted.
- They do not describe the contents already shown, but emphasize analytical results. (Not: The sales dropped in the second quarter, but e.g. the insolvency of the main customer XY led to the strong sales decline in the second quarter of …) They show exceptional developments and important special effects as well as economically relevant impacts.
- Ideally, they already integrate recommendations for action and suggestions for measures that can be followed up (e.g.: "The sales declines are to be compensated by increased new customer sales. For this purpose, a marketing offensive has been developed which will be implemented in the following quarter."). If there is a connection to a task or ticket system, the tasks can be created and taken over directly.
- The commentaries should be structured and sorted by importance.
- Commentaries should be picked up by the recipient at first glance. In tables and diagrams, highlights (highlighting) and comment hints (e.g. numbers) can be placed.

The following comment functions are to be distinguished:

- Text commentaries
  - Cell-related comments (e.g. in tables and graphs by a mouseover effect or rollover effect)
  - Free text commentaries per report or report object

[35] See Unrein (2016, p. 217 f.).

   - Summarized comment reports
   - Jump by link to a comment document
- Audio commentaries
- Video commentaries

It would be desirable to aggregate commentaries by levels. Here, it is defined whether a detailed commentary is also displayed on higher condensed levels. Such consolidation functions for commentaries are, however, rarely found.

Before the overall report or at the end, it is advisable to prepare summaries of the most important contents and analysis results. These summaries are also called "executive summary" or "management summary".

Important contents and messages should be presented in short sentences. They are the focus of the report and show explanations, recommendations for action and measures that result from the planning and analysis results.[36]

> **Important**
> So do not write:
> The material costs increase by 3% (**pure statement**).
> But rather:
> Due to the rising energy costs, the material costs increase by 3% (**explanation**).
> By substituting energy-intensive input materials with less energy-intensive materials in product development, a cost improvement is to be achieved (**recommendation for action**).
> In addition, alternative cheaper sources of supply are already being sought in the project "New Suppliers in the Next Decade" by the purchasing department without significant quality losses (**measure**).

At the end of an overall report, in addition to the summary and a conclusion, an outlook should also be described. Hints on the next steps or the further procedure round off the report. This underlines the objective and sustainability of the reporting.

3.9 Control in the BI-Supported Controlling Cockpit

The design of individual reports and planning forms must be accompanied by the design of a comprehensive controlling system. With the data warehouse and frontend technology in the context of business intelligence, the controlling department has the possibility to design "BI-supportedcontrolling cockpits".[37]

[36] Cf. Hichert+Partner AG (2011, p. 1–29, 2008, p. 1–17).

[37] Compare the technical possibilities and the terms data warehouse, business intelligence and BI-supported controlling in Chap. 5.

In aviation, the term cockpit (dashboard)[38] refers to the pilot's cabin, which is used to control the aircraft. The display instruments in the aircraft cockpit as well as in other vehicles help the pilot or vehicle operator to keep an eye on the effects of his or her control. This idea was first transferred to the management area by Patrick M. Georges and his team through the concept of the management cockpit.[39] It was a specially equipped room with numerous screens on the walls, which were supposed to provide the decision-makers with differentiated information.[40] The resulting management cockpit® is a trademark of SAP AG and was installed in Swiss banks at the end of the 1980s.[41] Over the next decades, the hardware and software technology as well as the business background evolved continuously. Alternative names to the cockpit such as Dashboard are used synonymously.[42] Following Reichmann, "all control-relevant internal and external information should be made available to the decision-maker in a clear, concise and interdependent manner by means of a controlling cockpit."[43]

In this book, the main focus will be on the visual and economically sensible design of a controlling cockpit. Using a fictitious model company, it will be shown how a BI-supported controlling cockpit can be built in the sense of navigation or analysis paths. Analysis paths are to be developed "top-down", while the operational data supply is done "bottom-up", starting from undistorted base data up to information of the highest level of aggregation. The focus of the control support of the example below is on the operational and strategic controlling for the following areas:

- Management
  - Operational controlling
  - Strategic controlling
- Sales
- Production
- Purchasing
- Projects

The example is structured in such a way that further functional areas such as logistics, human resources, service, IT etc. can be easily added.[44] The example shows important key figures and reports for each of these areas, which are intended to provide guidance and inspiration for the implementation and design. It becomes clear that a meaningful

[38] Alternatively, the term dashboard (overview page) is also used.

[39] Cf. Schutt (2014, p. 25), Georges (2000, p. 131) and Meier et al. (2002, p. 134).

[40] Cf. Reiterer et al. (2000, p. 10).

[41] Cf. Georges (2000, p. 131).

[42] Lux (2006, p. 29).

[43] Cf. Reichmann et al. (2017, p. 31).

[44] An example of a cockpit for the IT area is shown by Growe et al. (2020, p. 80–85).

business-oriented structuring and design significantly increases the usability of a cockpit. The controlling department is in charge as the coordinating institution for the development. The implementation error of letting BI systems grow report by report leads to confusing report collections and thus to unused data piles.

After successful login to the system, e.g. via the one-time identification "Single Sign On", the user enters the start menu of the cockpit. The users start their analyses here. Via the navigation menu bar or tiles, they select their analysis entry point. The authorization settings only make those menu items visible to the user for which he has access rights.

For a relatively large number of analysis reports, their structure is of high importance for navigation.[45] This structure should be as simple and intuitive to use as possible to avoid unnecessary clicks and searching for reports. The steering committees of the company, such as e.g. management, division and department heads, often shape the structure of the navigation of a BI-supported controlling cockpit leading to a stakeholder-oriented division of the cockpit. As an alternative structuring option, functional divisions according to tasks or topics are possible. For example, the controlling modules could include, among others, "sales controlling, production controlling, financial controlling, etc." as a division. Also, mixtures of stakeholder and functional divisions can be useful, if e.g. special analyses are to be provided only to certain persons or groups and other topics are to be made available in general (see Fig. 3.79). In order not to take up too much space for the report presentation on the screen surfaces, it should be possible to hide or display this bar, e.g. via an icon (e.g. arrows or triangles). The arrow behind the areas symbolizes here that further areas can be reached via a sub-control. With the help of a 'home' button, jumps to the start cockpit can be set. Further functions of the navigation are, among others:

- Drill-pad function: Navigate forward or backward in the last navigation steps.
- Undo function: Reset the last input or filtering and restore the previous view.
- Back-to-start function: Reset a report or a planning form to the initial situation.

The position of the current navigation should be visible throughout the controlling cockpit. In the example, the header shows the area and the report or the planning form. Furthermore, the analysis date is visible and can be changed here, if necessary, by the analyst. Further supplementary but not central information on the reports and planning forms can be accessed via an information function, in which e.g. the data source, the last update of the data provision, the report responsible or the last planner can be queried.

The start cockpit provides a similar entry area for all users of the cockpit as a portal entry and should therefore be tailored to the entire user group. The start cockpit (as well as later intermediate dashboards) is also referred to as a landing page to which the analyst can always jump back and continue the analysis during the research. For motivation and

[45] Cf. Schön et al. (2013, p. 254 ff.).

Fig. 3.79 Start menu of the BI-supported controlling cockpit

orientation for the company, general information is suitable for the central start cockpit. In the example, bullet points of the mission statement[46] of the company should help to continuously remind the employees and managers of it.[47]

The following questions should be answered by the start cockpit:

- What does our company stand for?
- What are the values of our company?
- What analysis paths can be used?
- What important key figures has our company achieved?
- How do I motivate my managers and employees in planning and analysis?

Key data on the company, such as the development of total performance or top indicators such as productivity[48] or accident-free days, can inspire and generate attention. Sensitive top indicators, e.g. on the result, must not be displayed here for reasons of data access

[46] On the term mission statement see Al-Laham et al. (2017, p. 199 ff.).

[47] See Mentzel (2015, p. 13 ff.).

[48] On the indicator productivity see Krause and Arora (2008, p. 1–5).

rights as authorization systems usually work on a report basis. Further important top indicators therefore belong in the start cockpits of the respective control areas.

3.9.1 Cockpit Executive Board: Operational Controlling

The executive board and the controlling department need important information on the profitability, success and financial situation of the company for steering. In addition, they have the access rights for all other functional areas and can analyze steering-relevant information from other areas such as sales, production and personnel in the entire cockpit. It is therefore a privileged protected analysis area.

From a steering perspective, it makes sense to divide the cockpit into the following segments (e.g. via a menu control bar):

- Operational controlling
- Strategic controlling (only for executives with strategic relevance)
- Planning (budget/forecast/medium-term planning)

The entry area is operational controlling. Here, the most important current information on the company's situation can be analyzed with key figures and management reports. In addition to the main areas of success and finance, the functional areas of sales, production and purchasing as well as cross-sectional areas such as controlling for human resources, logistics and quality are among the analysis areas in the cockpit.

Strategic controlling deals with the strategic status analysis as well as the strategic goal setting and long-term planning. The strategic status analysis includes tools such as the environmental and context analysis, the SWOT analysis and the portfolio analysis. From the status analysis, strategic plans can be derived which result, for example, in roadmap planning and investment planning. As a link to the operational management, the use of a Balanced Scorecard or the Balanced Chance and Risk Card (BCR Card) is recommended.[49] Based on the strategic goals of the perspectives of corporate management used, key figures are collected, based on which the effect of strategic projects can be measured. Furthermore, risk management and reports on sustainability can be placed in this strategic sector.

In the area of planning, the entry is made into those parts of the short- and medium-term planning that are relevant for the user group. Here, the planning areas and forms of the budget planning, the forecast and the medium-term planning can be accessed. A status and tracking report shows the current processing status of the individual planning areas and approval rounds as well as their deadlines. The management deals here, for

[49] For the term balanced scorecard, compare Sects. 3.1.1, 3.12.1 and, among others, Al-Laham et al. (2017, p. 843 ff.).

example, with the specifications and premises as well as the coordination and approval processes of the operational, tactical and strategic planning. In addition, alternative plans such as the scenario planning (worst/best/middle case) and the forecast are part of their tasks.

In the entry image of the operational controlling, the management or the controlling should get a quick overview of the current situation of the company based on top key figures and graphics. This image and all other intermediate start cockpits of the individual analysis areas are also landing pages into the overall navigation. They include the most important key performance indicators (KPIs) on a highly aggregated level of the respective analysis area. In the start cockpit of the management, central questions and important fields of action should be covered, for example:

- Where does the company stand at the moment?
- Will the company achieve the set values of the planning, e.g. regarding sales, profit and return?
- When will the company reach the break-even point?
- How are key control indicators, such as the return on investment, the contribution margin, the productivity and the utilization, developing?

The view on the individual indicators should always be the same or similar. For this purpose, speedometers, tiles, and column or bar charts are suitable for visualization as they depict the state of the current situation (actual values). In the present example, actual values are uniformly represented by a black bar (so-called bullet charts or bullet graphs). The target or plan value is indicated by the "blue" line and the previous year value by the grey bars. The threshold ranges can be quickly analyzed by the traffic light color of the orientation line. The direction of the positive and negative threshold values depends on the respective indicator. For the sales indicator, higher values are better (green), while e.g. for the complaint rate, higher values are worse (red). The peak indicator in the start cockpit example is the **return on investment (ROI)**.[50] Ideally, the **operating (business) ROI** should be displayed here adjusted for extraordinary and non-operating results. Business practice has shown that too many indicators are confusing. Nevertheless, a large number of detailed indicators are needed for control. Therefore, it makes sense to develop a graded indicator system for the individual business control. In addition to the content assignment of the individual indicator, the visual location of the indicator is important in the BI-supported controlling cockpit. The model company has decided in the start cockpit (see Fig. 3.80) in favor of success and return indicators but against a liquidity indicator, as this plays a rather subordinate role for the company; the financing and solvency are covered as important secondary objectives by the financial and cash

[50] For the ROI indicator, see Reichmann et al. (2017, p. 83).

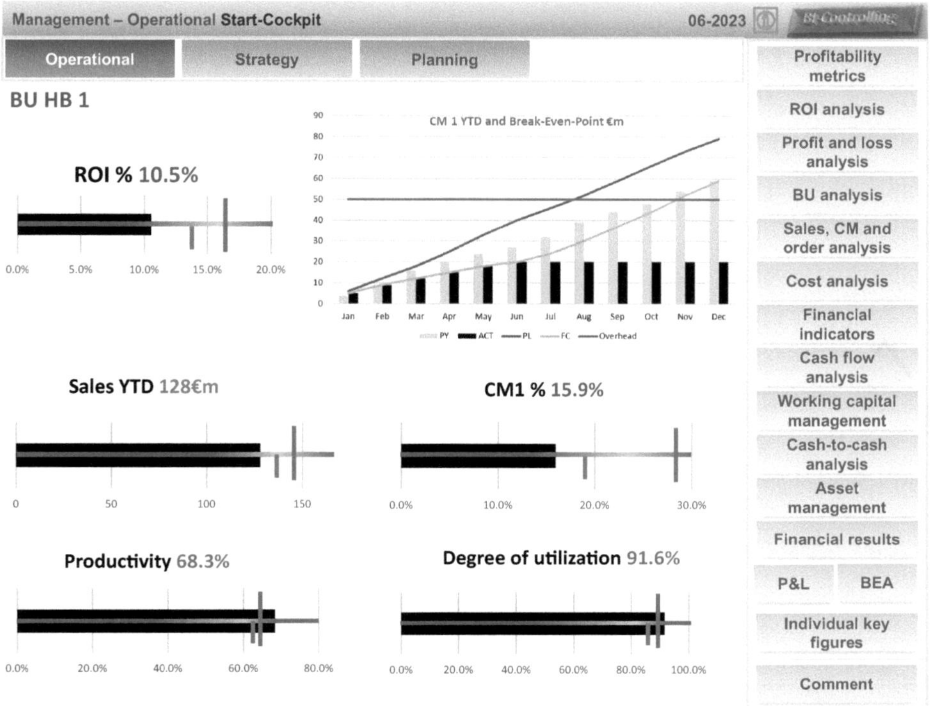

Fig. 3.80 Start cockpit for the management (operational controlling)

flow analysis[51] in the finance area. But there are other companies, e.g. from the area of "private equity", which rather see cash flow indicators at the first place.

As an alternative to the shown start cockpit view, modern systems use the tile look which has been inspired by the pioneers Apple and Microsoft and which can be used primarily to launch an application (app) but also to select a report (see Sect. 5.10.3; Fig. 3.81). The menu and header lines are comparable except for the "look". For example, the currently selected menu item is underlined in bold and white. The tiles are directly linked to the individual reports or another intermediate dashboard is reached. In the intermediate dashboard, for example, further tiles can be triggered to allow access to deeper report hierarchies. The tiles can generally have different types:[52]

- Standard tile with pure text information and icons
- Dynamic tiles with simple key figure values

[51] See Krause and Arora (2008, pp. 70–91).

[52] See also Sect. 5.10.3 as well as Krüger (2015, p. 204 ff.) and Keist et al. (2016, p. 109–113).

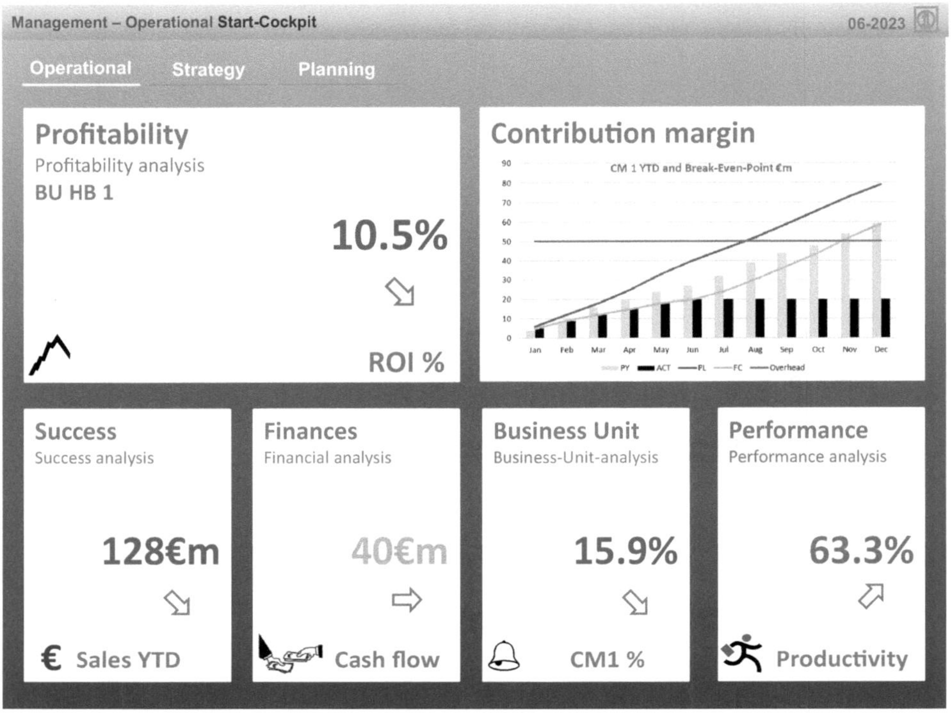

Fig. 3.81 Tile look for the start cockpit

- Dynamic tiles with key figure values, warning symbols and colors based on set tolerance values
- Dynamic tiles with charts

The advantages of the tile look are its appealing look and overview. The symbols and longer text blocks help to find analysis areas. The display of the individual reports in another menu bar (e.g. line) was omitted here. The tiles are to be built in accordance with a report hierarchy so that the individual reports, programs, web pages etc. can be reached via intermediate tile pages. The more extensive the report hierarchy is, the more confusing the intermediate pages with the tile look can become. In contrast, the reports could be selected directly from the start cockpit via the "classic" report menu bar. This structure saves a few clicks.

For the conceptual development of a performance measurement system, the RL-performance measurement system is particularly suitable [53] as it is modular and can

[53] Cf. Reichmann et al. (2017, p. 86 ff.).

be individually adapted and expanded for the company's purposes. The **Break-Even-Point**[54] is visualized with the help of **Contribution Margin 1**[55] using variable production costs in connection with planned fixed costs. The accumulated sales and the relative contribution margin 1 in percent of the total output are further important performance indicators of the company. In the model case, the planned and previous year values are not reached. The **Productivity** and the **Capacity Utilization** are positive and show better values than planned.

Positive and negative states or developments of the key figures have to be analyzed in detail with regard to their opportunities and risks. This can be done, for example, by means of trigger functions that branch into the detailed analysis when activated (e.g. by selecting the key figure or a button). These trigger functions have to be defined and documented when designing the reports.

A basic selection of detailed analysis reports for the management and the controlling department can be accessed via a menu control bar (in the example on the right of Fig. 3.80). There are the following reporting areas within the scope of the company management, of which selected areas are described below:

- Profitability analysis
  - Profitability ratios
  - ROI analysis
- Success analysis
- Cost analysis
- Financial and liquidity analysis
- Working capital management
- Annual financial statement analysis
- Analysis of individual key figures

Alternatively to the menu control, connected analyses can be accessed by clicking on the top key figures of the respective start cockpit (landing pages) via trigger functions,[56] for example, the top key figure ROI leads to the further profitability analysis.

3.9.1.1 Profitability Analysis

Profitability and profit are the central target variables for the business in many companies. The profitability and success analyses[57] answer important control questions in the company:

[54] For the key figure Break-Even-Point cf. Reichmann et al. (2017, p. 199–203).

[55] For the key figure Contribution Margin cf. Reichmann et al. (2017, p. 469–470).

[56] Cf. Gleich and Klein (2016, p. 188 f.).

[57] On the term success analysis cf. Lutz et al. (2011, p. 69).

- Will the company achieve the desired profit and return on the capital employed? How does this return look compared to other investment alternatives and other companies?
- How high is the share of the return and success solely contributed by the operational activity or the core business? Were there extraordinary effects? How much are the earnings burdened by debt? Can the positive or negative leverage effect be used by capital restructuring?
- Where is there a need for action to improve the profitability and success situation of the company?
- How high is the capital commitment? Do we use the capital provided efficiently? Where are the best levers for improving the result and the capital utilization?

For deeper detailed analyses, the user branches into further analysis areas, such as success, finance, sales, etc. and researches on further questions:

- What are the sources of profit and loss of the company? Which result objects (customers, products, regions …) are particularly successful and which are not?
- Which revenue and cost components are mainly responsible for the result?
- What is the liquidity situation? Are we solvent in the short term or do we have to close financing gaps?

Various reports are offered. The report profitability key figures (see Fig. 3.82) shows key ratios and success ratios in their periodic development up to the current (yellow) forecast compared to the plan.

The first column shows the most important return ratios:[58]

- The operating **Return on Investment** shows the operating result generated in relation to the assets employed for operational purposes. Thus, the ROI can be used to analyze the purely operational earning power of the capital employed.
- The **Return on Equity** shows the shareholders how successfully their capital was used.[59] For investors, it serves as a benchmark for the attractiveness of investing in the company compared to alternative investments.
- The **Return on (Total) Capital** shows all stakeholders the total success of the company for the entire invested equity and debt capital.[60] It is thus a benchmark for the performance of other companies in terms of time or industry comparison. If the return on total capital is greater than the interest burden from the debt capital, the leverage effect can be exploited.

[58] For an overview of profitability ratios cf. Reichmann et al. (2017, p. 132).

[59] For the key figure return on equity see, among others, Laier (2011, p. 54) and Reichmann et al. (2017, p. 112).

[60] For the key figure return on total capital see Reichmann et al. (2017, p. 111–112).

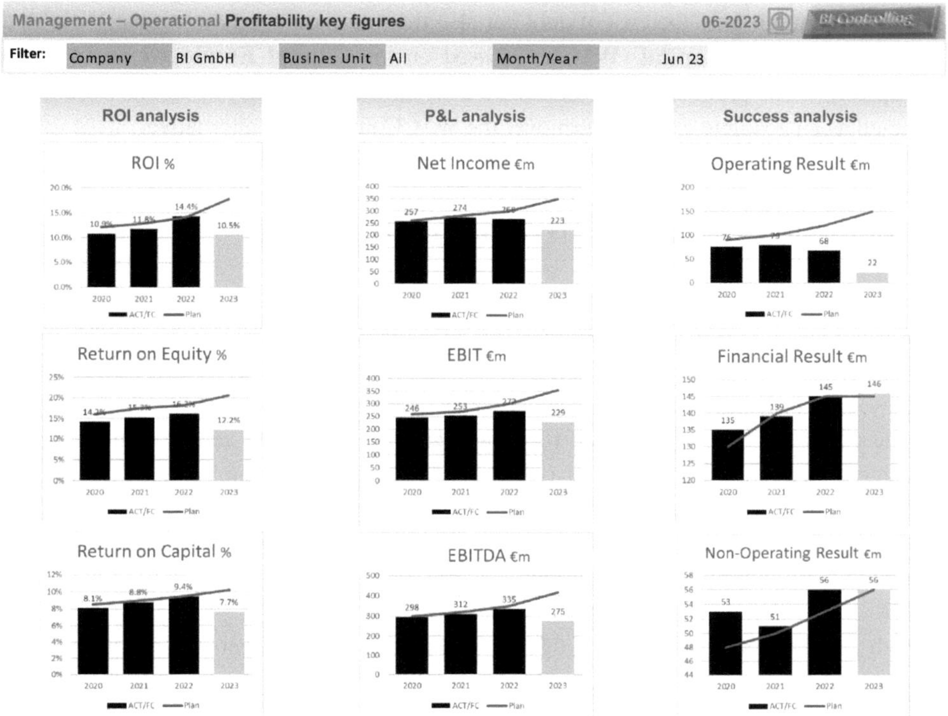

Fig. 3.82 Profitability key figures

For smaller medium-sized businesses, these return figures are the right entry-level figures. For larger companies, corporations and corporate holdings, these can be supplemented and expanded with the key figures of value-oriented corporate management (shareholder value approach). The central figure here is the Economic Value Added,[61] which can be determined by two different variants:

- Variant 1: The **Economic Value Added (EVA)** represents a book value-oriented and single-period measure of success over the total operating and capital costs (residual income).[62] Depending on whether a pre- or post-tax view is used as the basis for calculation, either the **Net Operating Profit After Taxes (NOPAT)**[63] or the **Earnings Before Interest and Taxes (EBIT)**[64] is used as the measure of success. This earning power of the company is contrasted with a risk-oriented benchmark return, which takes

[61] See Weber et al. (2017, p. 42 ff.). EVA® is a registered trademark of Stern Stewart & Co.

[62] Cf. Steger (2014, p. 159).

[63] For the NOPAD ratio, see Hostettler (2000, p. 150 ff.).

[64] For the EBIT ratio, see Krause and Arora (2008, p. 16–18).

into account not only the cost of debt but also the return on equity using the **Weighted Average Cost of Capital (WACC)**[65] based on the **Capital Employed or Invested Capital)**.[66] The product of this difference with the capital employed is thus to be interpreted as an absolute excess return over the WACC investment alternative.
- Variant 2: The EVA is determined using the return on invested capital with the ratio **Return On Capital Employed (ROCE)**. This return is a relative measure and results from dividing the profit (e. g. EBIT) by the capital employed.[67]

The NOPAT corresponds to the accounting operating profit before deduction of capital costs (interest) and after deduction of a tax adjustment. Furthermore, so-called conversions (adjustments) are to be considered when calculating the NOPAT, EBIT and Capital Employed in order to transform balance sheet items into economically relevant items, e.g. the adjustment of depreciation for non-operating assets. The calculation steps and adjustments are often complex and rarely justify the relevance and control quality for smaller medium-sized businesses.[68]

The EVA can be used for value driver management similar to the ROI. However, the calculations are more complex, and therefore this model gives priority to the ROI analysis.[69]

With regard to success, the key figures of the income statement and the operational success analysis are distinguished.

- The **net income** shows the absolute profit contribution of the company from performance and financial activities as well as extraordinary effects; based on these one can derive the increase in equity of the company.
- The key figure **EBIT** (Earnings Before Interest and Taxes) is a measure of success before capital costs (interest) and taxes. It is therefore often used for cross-sectional, sometimes international, comparisons of companies as this key figure is independent of the capital costs of the underlying capital structure and the tax conditions.
- The key figure **EBITDA** (Earnings Before Interest, Taxes, Depriciation and Amortization)[70] represents the result of the business activity before interest, taxes and depreciation are taken into account. This figure is intended to enable the cross-sectional comparability of the earning power for companies independent of their capital costs (capital structure) as well as the different tax conditions and the different investment

[65] For the WACC, see Reichmann et al. (2017, p. 298 f.) and Krause and Arora (2008, p. 115–117).

[66] Cf. Copeland et al. (1998, p. 261), Seppelfricke (2005, p. 24), Knorren (1998, p. 43), Black et al. (1998, p. 61).

[67] Cf. Reichmann et al. (2017, p. 145–147) and Hostettler (2000, p. 19).

[68] Cf. Hostettler (2000, p. 19 and p. 150 ff.).

[69] Cf. Weissman et al. (2014, p. 55 ff.).

[70] For the key figure EBITDA see Krause and Arora (2008, p. 19–21).

behavior of the companies. For example, in private equity companies, the EBITDA is also used as an approximation for the cash flow.

Unlike the income statement-oriented key figures, the key figures of the profitability and liquidity system for success analysis show which sector is responsible for the annual surplus:

- The **operating result** shows the operationally relevant part of the periodic result. Here, unlike according to the German Commercial Code, only items that are really operationally relevant are included in the result. These include, among others, according to the profitability and liquidity key figure system, the cost of debt (interest), without which a business existence of the company would not be possible, and other ordinary operating income and expenses.[71]
- The **non-operating result** contains, in addition to extraordinary income and expenses, also period-unrelated income and expenses, which are hidden in the other expense and income items. For example, period-unrelated sales of assets above the book value increase the non-operating result.
- The **financial result** indicates, adjusted for the cost of debt, which earning power is attributable solely to the operational financial activities. These are, for example, income and expenses from investments as well as interest income for the excess financial assets.

As navigation alternatives, the individual key figure analysis (see Sect. 3.9.1.7) or the navigation buttons for jumping to detail reports are available for all key figures.

The return analysis can be intensified by using the ROI analysis button, which allows the user to break down the value drivers of the return step by step.

The **ROI analysis** (see Fig. 3.83) represents a mathematically derivable development concept for return and performance analysis, something which can be determined from the annual financial statements. It is based on the ROI indicator system (Du-Pont indicator system), which was developed as early as 1919 by the American chemical company Du Pont de Nemours and Co.[72] Meanwhile, there are variants of the original scheme which have been adapted to different legal norms or corporate purposes. The ROI analysis splits the top indicator ROI mathematically into its components. The ROI results from the multiplication of the return on sales and the capital turnover frequency:[73]

---

[71] See Reichmann et al. (2017, p. 109 ff.) Here, instead of operating result, the term "ordinary operating result" is used.

[72] See Coenenberg (2009, p. 1155), Reichmann et al. (2017, p. 82–83).

[73] See e.g. Lachnit and Müller (2012, p. 292–294).

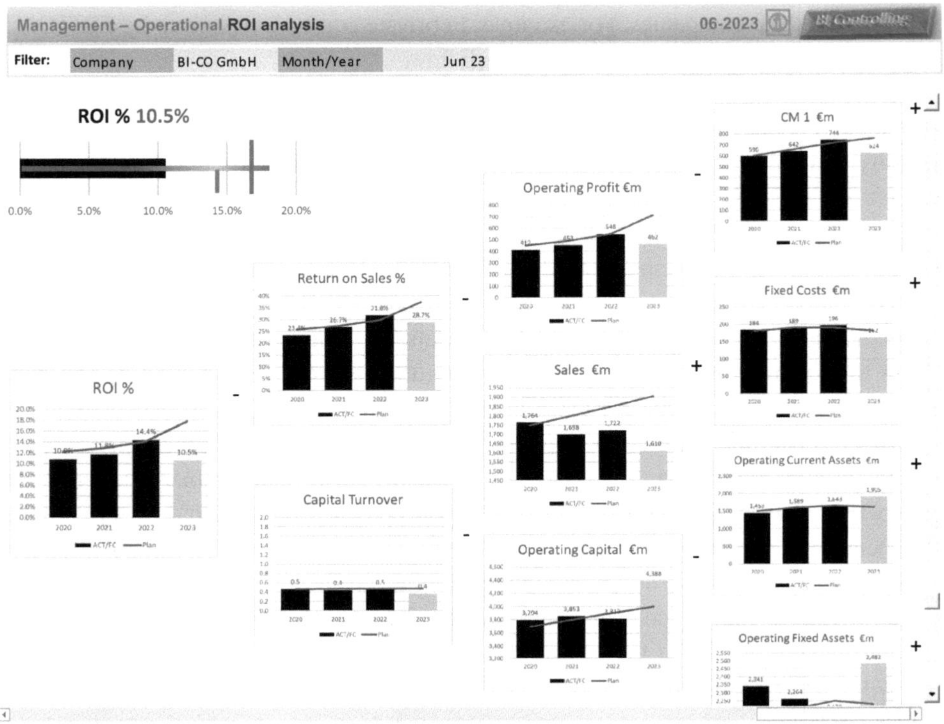

Fig. 3.83 ROI analysis

- The **return on sales** (or also Return on Net Sales)[74] illustrates, on the one hand, how successfully the company can enforce its prices for its products and services in the market, and, on the other hand, how well it succeeds in producing and selling these products and services profitably. It shows the profit margin for every euro of sales or net sales without taxes. The larger the profit margin, the more leeway there is to absorb price declines and cost increases. If the cash flow is used instead of the result in relation to sales, the so-called "cashflow margin" is obtained. This indicator shows the internal financing power that is generated with every euro earned.[75]
- The **capital turnover** indicates to the company how often the employed operating assets or capital have been "turned over" (used) by the sales.[76] This shows the utilization intensity of the employed assets, which can vary greatly depending on the industry and business type. While e.g. in the trade sector very high capital turnover

[74] For the return on sales indicator see Krause and Arora (2008, p. 37–39).

[75] See Weissmann et al. (2014, p. 143).

[76] See Steger (2014, p. 36).

frequencies of 20 and more can be achieved, they are usually rather low for project-related manufacturers and can e.g. be around the factor 1. These two types of companies behave inversely concerning the return on sales.

The individual levels of key figures of the ROI analysis can be expanded and collapsed by the key figure hierarchy. This function and the scroll function keep the analyst's overview in the entire ROI tree scheme. For each management, there is the possibility to model the planned changes in the business or to analyze the achieved business figures. The origin areas for profit and loss can be segmented or aggregated. Starting from the return on sales, the following sections will examine the profitability and performance indicators of a company. The operating result and the turnover are each broken down into their detailed components.

The operating result can be analyzed in detail using the profitability analysis (see Sect. 3.9.1.2) starting from the turnover via the total output, the gross profit to the contribution margins and the result lines. In the context of the cost analysis (see Sect. 3.9.1.3), cost blocks can be broken down into individual or overhead costs or variable and fixed cost components. The entrepreneur recognizes, for example, the need for cost reduction in the variable sales or production costs or the fixed-cost-intensive overhead areas.

The turnover can be analyzed from the gross to the net turnover to the total output. Here, sales tax, pricing and revenue reductions in the form of trade discounts, cash discounts and rebates as well as the analysis of inventory changes play an important role.

The ROI and the capital turnover can be increased by the more efficient use of the capital employed in addition to an increase in turnover. To do this, one navigates from the operating fixed assets to the asset management analysis (without example) or from the operating current assets to the working capital management analysis (see Sect. 3.9.1.5).

3.9.1.2 Profitability Analysis

In the context of **profitability analysis**, the focus is on the income statement of the company in relation to its result objects. Besides products, any result objects such as customers and customer groups, divisions, profit centers,[77] branches, sales areas or sales channels can be defined as profitability objects (see, for example, Fig. 3.84 with a result scheme from the trade sector). Due to the numerous and interrelated result objects, the possible result object combinations are multiplied. This means that when implementing the income statement for all conceivable combinations, the corresponding revenue and cost information must also be planned and recorded multidimensionally in the cost accounting system.

In business practice, such a complex income statement with many dimensions is often costly and time-consuming without OLAP technology. In order to be able to evaluate the income statements multidimensionally efficiently, the following requirements, among others, must be observed:

---

[77] For the term profit center, see Lutz et al. (2011, p. 191).

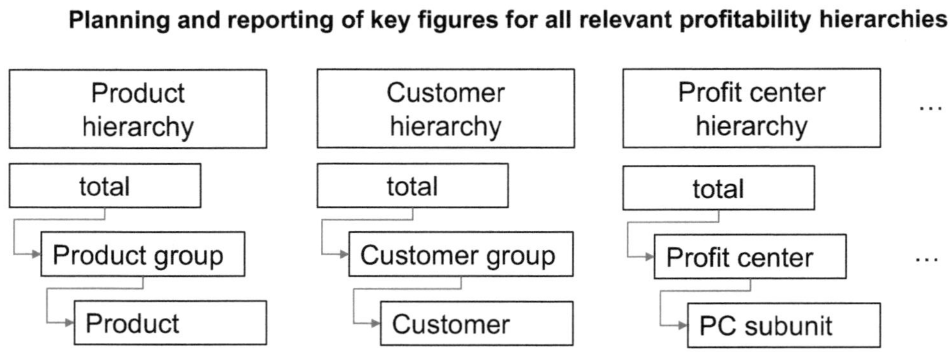

	€m	PY	PLAN	ACT	ΔPY	ΔPL	ΔPY%	ΩPL%
Total Output	+	**136**	**142**	**132**	**-4**	**-10**	**-2.9%**	**93.0%**
Discounts	+	7	4	6	-1	2	-14.3%	150.0%
Bonuses	+	13	14	12	-1	-2	-7.7%	85.7%
Outgoing freight	+	5	5	6	1	1	20.0%	120.0%
Net Sales	-	**111**	**119**	**108**	**-3**	**-11**	**-2.7%**	**90.8%**
Material costs	+	38	35	39	1	4	2.6%	111.4%
External services	+	25	22	28	3	6	12.0%	127.3%
Gross Profit	+	**48**	**62**	**41**	**-7**	**-21**	**-14.6%**	**66.1%**
Var. production costs	+	21	22	20	-1	-2	-4.8%	90.9%
CM 1	-	**27**	**40**	**21**	**-6**	**-19**	**-22.2%**	**52.5%**

Fig. 3.84 Result object hierarchies. (Based on Schön 1999, p. 63)

- The result object hierarchies are to be defined and maintained in the upstream systems in a structured, complete and non-overlapping way (e.g. customer hierarchy, product hierarchy, regional hierarchies, business unit hierarchy, etc.). Problems arise mainly from the continuous master data quality in the upstream systems (inventory management, accounting, CRM, PPS, ERP systems, etc.).
- The scheme of the profitability or contribution margin accounting is optimally designed for the analysis and decision requirements of the business model. The success scheme is best implemented as a value hierarchy in the data model, so that the individual result levels (via the plus and minus signs) can be expanded and collapsed (see Fig. 3.84). Important variables of the income statement are:
  - gross sales (before taxes and before revenue reductions)
  - net sales after taxes and after direct revenue reductions
  - total output (including changes in inventory)
  - net-net sales after further revenue reductions or conditions
  - gross profit after material costs and external services
  - contribution margins after variable sales and production costs
  - further contribution margins after overhead or fixed cost levels
  - total result

- Performance indicators complete the profitability analysis, such as:
  - ontribution margin relative to the total output or the sales
  - growth indicators to previous periods e.g. for sales and contribution margin
  - target achievement indicators to planning e.g. for sales and contribution margin
  - sales or contribution margin for used resources, e.g. for employees and plants[78]
- The cost accounting must be consistent in the value flow. This means that the assignment of value types (accounts/cost types) in the income statement schema is clear, and the postings (movements) to the result objects can be directly allocated.
- Direct movement rates (e.g. revenues and costs), imputed cost allocations and variance re-allocations from the cost center to the cost unit accounting must be integrated into the value flow of the cost accounting and can be assigned to the result object hierarchies. The core dimensions/hierarchies represent those of the product and the customer. Ideally, further dimensions can be derived from the core dimensions by means of derivation rules. For example, the industry and regional assignment results from the unique assignment of the customer in the customer master.
- Amounts that cannot be assigned to final objects of the result hierarchy pose a special challenge. Certain performance and cost items, such as rebates and advertising costs,[79] may not be allocatable to products, but only to customers. However, the multidimensionality of the data storage requires the aggregation of the individual values and demands an allocation. One solution is to distribute the node amounts to the end objects of a dimension (e.g. by means of a splash function, which corresponds to a keying that, for example, the partial cost accountants want to avoid, while the full cost accountants accept). Alternatively, these amounts can be assigned to residual items (e.g. to dummies "without allocation"), so that, for example, a stepped fixed cost coverage calculation for customer groups can be depicted even if costs cannot be clearly allocated to them.

The income statement shows the comparison with the previous year and the planned-actual comparison for the underlying contribution margin scheme (see Fig. 3.85). The top performance indicator is the performance measure "contribution margin 1 (after variable costs) in relation to total output" visualized. Ideally, the row and column structures of the income statement are uniformly structured. In the example, the contribution margin scheme of the income statement reflects the row scheme. It can ideally be further unfolded or aggregated based on the value hierarchy of the income statement up to the account allocation. For this purpose, the frontend solutions often use the plus or minus sign (+ or −). Higher-level hierarchy nodes should be highlighted and, for example, marked in bold. If entire hierarchy nodes or the entire hierarchy are unfolded, the income

[78] See Krause and Arora (2008, p. 188–190).

[79] Cf. Lutz et al. (2011, p. 256).

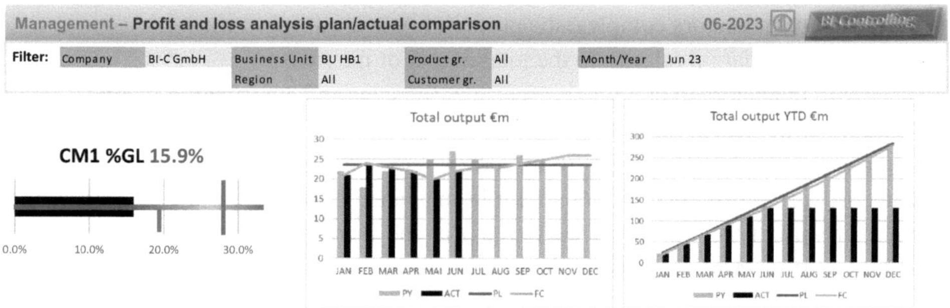

€m	PY	PL	ACT	ΔPY	ΔPL	ΔPY%	ΩPL%	% Total Output
Total output	**136**	**142**	**132**	**-4**	**-10**	**-2.9%**	**93.0%**	**100.0%**
Discounts	7	4	6	-1	2	14.3%	150.0%	**4.5%**
Bonuses	13	14	12	-1	-2	-7.7%	85.7%	**9.1%**
Outgoing freight	5	5	6	1	1	20.0%	120.0%	**4.5%**
Net sales	**111**	**119**	**108**	**-3**	**-11**	**-2.7%**	**90.8%**	**81.8%**
Material costs	38	35	39	1	4	2.6%	111.4%	**29.5%**
External services	25	22	28	3	6	12.0%	127.3%	**21.2%**
Gross profit	**48**	**62**	**41**	**-7**	**-21**	**14.6%**	**66.1%**	**31.1%**
Var. production costs	21	22	20	**-1**	**-2**	-4.8%	90.9%	**15.2%**
CM 1	27	40	21	**-6**	**-19**	**22.2%**	**52.5%**	**15.9%**
Material overheads	3	3	2	-1	-1	33.3%	66.7%	**1.5%**
Production overheads	9	10	8	-1	-2	11.1%	80.0%	**6.1%**
CM 2 after production cost	**15**	**27**	**11**	**-4**	**-16**	**26.7%**	**40.7%**	**8.3%**
Administrative overheads	3	2	2	-1	0	33.3%	100.0%	**1.5%**
Sales overheads	4	5	4	0	-1	0.0%	80.0%	**3.0%**
Calc. result	**8**	**20**	**5**	**-3**	**-15**	**37.5%**	**25.0%**	**3.8%**
Deviation BU overheads	-10	0	-12	-2	-12	20.0%	0.0%	**-9.1%**
CM Business Unit	18	20	17	-1	-3	-5.6%	85.0%	**12.9%**
Deviation other overheads								**0.0%**
Operating result	18	20	17	-1	-3	-5.6%	0.85	**12.9%**
CM1 % Output	**19.9%**	**28.2%**	**15.9%**	**-3.9%**	**-12.3%**	19.9%	56.5%	
CM BU % Output	**13.2%**	**14.1%**	**12.9%**	**-0.4%**	**-1.2%**	-2.7%	91.4%	

Fig. 3.85 Profitability analysis

statement can only be completely analyzed using the scroll function. The start hierarchy should therefore be optimized and adapted to the screen size.

The row scheme of the income statement or alternative income statements of the company is to be set up for the entire company as a value hierarchy in which all relevant revenue and cost types are clearly assigned. These row schemes are to be established as part of the cost accounting concept in terms of an optimal value flow presentation and control.

The column scheme always shows, in the same order, the previous year values, the planned values, the actual values, the deviations from the previous year and planned values in absolute terms. The table column of the actual values can be recognized by the black bar. The previous year values are slightly grayed out. Visually highlighted are the last three columns of the table. First, the growth from the previous year is shown positively or negatively using the sparkline column. The degree of target achievement is signaled by means of color-coding in traffic light colors. Degrees of target achievement above 100% are marked green, for example, below yellow and significantly below red. The percentage share of the row values measured against the base size (here total output) shows the economic importance of the row value. In the example above, the material cost share is almost 30%.

The success analysis is supported by dynamic charts. Depending on the row selection, the selected key figure is shown in its temporal and accumulated development in bar charts. Thus the analyst has all the possibilities to analyze the success and its cost contribution in detail.

The numerous filtering options in the filter bar allow the view of the data to be sensibly restricted and expanded, here, for example, by business unit and by product or customer groups or regional aspects. The time selection allows the periods under consideration to be limited or extended. For forecast periods, the forecast values are used instead of the actual values.

Since the success analysis requires different views, it makes sense to provide alternative reports. In addition to the classic planned-actual comparison, an object comparison could be useful, for example.

In the business unit analysis, selected business units are compared side by side in the structure of the contribution margin accounting. The abbreviations HB, TB and SB in Fig. 3.86 stand for abbreviated BU names. This allows to identify successful and less successful business units more quickly. Cost allocations that cannot be assigned to a business unit or not at all are shown separately (without BU).

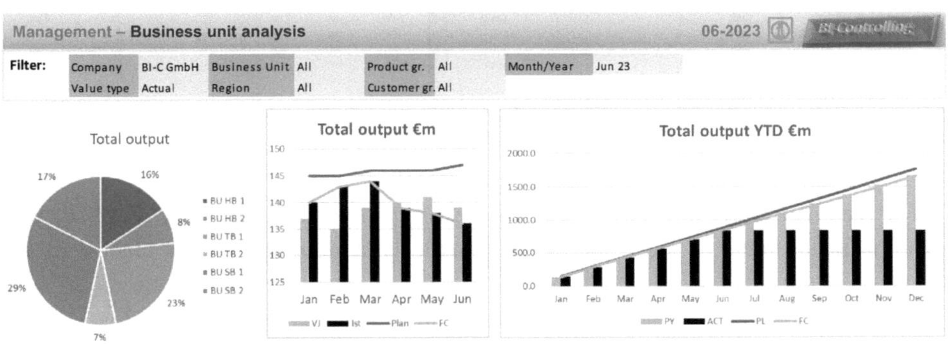

€m		BU HB 1	BU HB 2	BU TB 1	BU TB 2	BU SB 1	BU SB 2	Without BU	Total	% TO
Total output	+	**132**	**65**	**196**	**56**	**245**	**146**		**840**	100%
Discounts	+	6	3	9	5	2	1		26	3%
Bonuses	+	12	4	3	6	6	1		32	4%
Outgoing freight	+	6	6	8	7	4	3		34	4%
Net sales	-	**108**	**52**	**176**	**38**	**233**	**141**		**748**	89%
Material costs	+	39	28	35	25	20	18		165	20%
External services	+	28	12	26	9	12	8		95	11%
Gross profit	+	**41**	**12**	**115**	**4**	**201**	**115**		**488**	58%
Var. production costs	+	20	6	32	2	53	63		176	21%
CM 1	-	**21**	**6**	**83**	**2**	**148**	**52**		**312**	37%
Material overheads	+	2	1	3	1	5	4		16	2%
Production overheads	+	8	4	10	2	21	25		70	8%
CM 2 after production cost	-	**11**	**1**	**70**	**-1**	**122**	**23**		226	27%
Administrative overheads	+	2	1	4	1	12	9		29	3%
Sales overheads	+	4	2	8	2	20	16		52	6%
Calc. result	-	**5**	**-2**	**58**	**-4**	**90**	**-2**		**145**	17%
Deviation BU overheads	+	-12	-8	10	-10	-2	-3		-25	-3%
CM Business Unit	-	**17**	**6**	**48**	**6**	**92**	**1**		**170**	20%
Deviation other overheads	+							-61	-61	-7%
Operating result	-	**17**	**6**	**48**	**6**	**92**	**1**	**61**	**231**	28%
CM1 % Output		**15.9%**	**9.2%**	**42.3%**	**3.6%**	**60.4%**	**35.6%**		**37.1%**	
CM BU % Output		**12.9%**	**9.2%**	**24.5%**	**10.7%**	**37.6%**	**0.7%**		**20.2%**	

Fig. 3.86 Business unit analysis

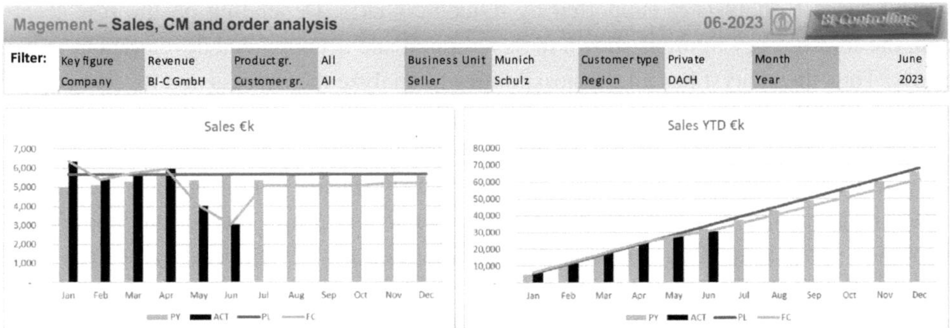

Selection hierarchy	June €k							YTD €k						
Product Hierachy	PY	PLAN	ACT	ΔPY	ΔPL	ΔPY%	ΩPL%	PY	PLAN	ACT	ΔPY	ΔPL	ΔPY%	ΩPL%
+Total	**956**	**1,046**	**1,093**	**137**	**47**	**14.4%**	**104.5%**	**4,832**	**5,044**	**5,131**	**298**	**87**	**6.2%**	**101.7%**
+Product group A	**441**	**476**	**503**	**62**	**27**	**14.2%**	**105.8%**	**2,582**	**2,594**	**2,741**	**158**	**147**	**6.1%**	**105.6%**
+Article group A1	**250**	**290**	**330**	**80**	**40**	**32.0%**	**113.8%**	**1,500**	**1,550**	**1,580**	**80**	**30**	**5.3%**	**101.9%**
Item A1.1	100	110	90	-10	-20	-10.0%	81.8%	600	650	720	120	70	20.0%	110.8%
Item A1.2	90	100	150	60	50	66.7%	150.0%	540	500	560	20	60	3.7%	112.0%
Item A1.3	60	80	90	30	10	50.0%	112.5%	360	400	300	-60	-100	-16.7%	75.0%
+Article group A2	**191**	**186**	**173**	-18	-13	**-9.3%**	**93.2%**	**1,082**	**1,044**	**1,161**	**78**	**117**	**7.2%**	**111.2%**
Item A2.1	6	6	3	-3	-3	-46.6%	53.5%	32	34	31	-2	-3	-4.9%	89.8%
Item A2.2	100	120	100	0	-20	0.0%	83.3%	600	610	620	20	10	3.3%	101.6%
Item A2.3	85	60	70	-15	10	-17.6%	116.7%	450	400	510	60	110	13.3%	127.5%
+Product groupe B	**515**	**570**	**590**	**75**	**20**	**14.6%**	**103.5%**	**2,250**	**2,450**	**2,390**	**140**	-60	**6.2%**	**97.6%**
+Article group B1	**250**	**290**	**330**	**80**	**40**	**32.0%**	**113.8%**	**1,110**	**1,200**	**1,180**	**70**	-20	**6.3%**	**98.3%**
Item B1.1	100	110	90	-10	-20	-10.0%	81.8%	610	600	650	40	50	6.6%	108.3%
Item B1.2	90	100	150	60	50	66.7%	150.0%	250	300	210	-40	-90	-16.0%	70.0%
Item B1.3	60	80	90	30	10	50.0%	112.5%	250	300	320	70	20	28.0%	106.7%
+Article group B2	**265**	**280**	**260**	-5	-20	**-1.9%**	**92.9%**	**1,140**	**1,250**	**1,210**	**70**	-40	**6.1%**	**96.8%**
Item B2.1	80	100	90	10	-10	12.5%	90.0%	350	400	390	40	-10	11.4%	97.5%
Item B2.2	100	120	100	0	-20	0.0%	83.3%	590	600	610	20	10	3.4%	101.7%
Item B2.3	85	60	70	-15	10	-17.6%	116.7%	200	250	210	10	-40	5.0%	84.0%

Fig. 3.87 Sales, contribution margin and order analysis

The pie chart allows the sizes of the business units to be assessed very well. The row selection in the table of the result scheme has a dynamic effect on the change of the charts.

Further alternative column and row schemes are potentially conceivable. However, the controlling concept should set standards here and not allow any proliferation. Another example of the profitability analysis according to the product hierarchy is presented in the sales area (see Sect. 3.9.3). Similar to the following sales analysis, the analysis of the success components can be done according to the selected hierarchy.

With the help of the sales analysis (see Fig. 3.87), those product areas and individual products can be identified based on the product hierarchy in which sales growth compared to the previous year was stronger or weaker for the considered month or accumulated. In addition, the degree of target achievement from the planned value can be analyzed again. The available hierarchies, e.g. the customer hierarchy, can be selected by changing the hierarchy selection.

The report is formulated as a sales analysis at the entry by the default setting of the key figure "sales" in the filter area. However, this analysis can be changed to other key

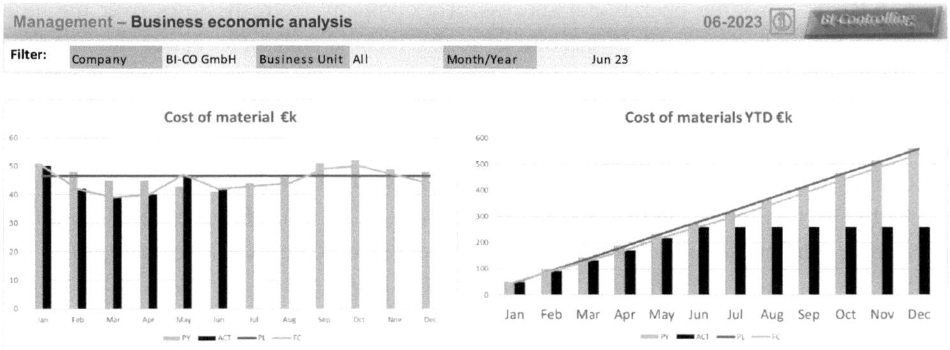

		PY	PL	ACT	ΔPY	ΔPL	ΔPY%	ΩPL%	% Total Output
Total output	-	**883**	**973**	**825**	**-58**	**-148**	**-6.6%**	**84.8%**	**100.0%**
Other operating income	+	1	1	1	0	0	0.0%	100.0%	0.1%
Cost of materials	-	-273	-280	-260	13	20	-4.8%	92.9%	-31.5%
Gross profit	-	**611**	**694**	**566**	**-45**	**-128**	**-7.4%**	**81.6%**	**68.6%**
Personnel expenses	+	-173	-173	-178	-5	-5	2.9%	102.9%	-21.6%
Depreciation and amortization	+	-63	-64	-46	17	18	-27.0%	71.9%	-5.6%
Rental and lease expenses	+	-18	-18	-19	-1	-1	5.6%	105.6%	-2.3%
External repairs and maintenance	+	-18	-15	-22	-4	-7	22.2%	146.7%	-2.7%
Energy	+	-14	-13	-15	-1	-2	7.1%	115.4%	-1.8%
Leasing	+	-5	-5	-6	-1	-1	20.0%	120.0%	-0.7%
Insurance	+	-3	-3	-4	-1	-1	33.3%	133.3%	-0.5%
Vehicle expenses	+	-15	-13	-9	6	4	-40.0%	69.2%	-1.1%
Advertising costs	+	-5	-5	-4	1	1	-20.0%	80.0%	-0.5%
Freight/Packaging	+	-6	-5	-7	-1	-2	16.7%	140.0%	-0.8%
Commissions	+	-3	-2	-3	0	-1	0.0%	150.0%	-0.4%
IT and telecommunications	+	-10	-18	-15	-5	3	50.0%	83.3%	-1.8%
Legal and consulting costs	+	-3	-3	-5	-2	-2	66.7%	166.7%	-0.6%
Training costs	+	-1	-1	-1	0	0	0.0%	100.0%	-0.1%
Other operating expenses	+	-1	-1	-2	-1	-1	100.0%	200.0%	-0.2%
Operating result	-	**273**	**355**	**230**	**-43**	**-125**	**-15.8%**	**64.8%**	**27.9%**

Fig. 3.88 Business Economic Analysis

figures, such as e.g. **order backlog** and **contribution margins,** by using the key figure selection.

Furthermore, the filter options can be used for detailed views, e.g. by organization, customer type or salesperson.

Again, the charts behave dynamically to the selected rows. Thus, individual selected items or item groups can be analyzed in detail with regard to their monthly development and in their accumulated view.

The profitability and return analyses contain a powerful scope of analysis with the selected reports and functions to be able to answer the questions posed by the management and controlling in detail.

A simple presentation of the profitability analysis without illustration and use of a cost accounting system is done with financial accounting means via the **Business Economic Analysis**, short “BEA” (see Fig. 3.88). It is especially popular among smaller medium-sized companies and is used by many tax accountants that support the monthly reporting for their corporate clients on behalf of them.

The BEA shows the operational success of a company from the sales over the total output to the operating result.[80] In contrast to the income statement, fewer intermediate results are shown up to the gross profit. Contribution margins are missing in the BEA. The presentation is often done without cost accounting adjustments and imputed items based purely on account information from the financial accounting. Sometimes, financial accounting adjustments are used to distribute expense items periodically, which are only booked at certain times, such as insurance premiums, Christmas bonuses, etc. For a first overview of the profitability, individual expense and income items can be analyzed.

The best-practice techniques of presentation are again:

- Hierarchy analysis of the expense and income items down to the account level using the plus and minus signs of the value hierarchy (here BEA hierarchy). There are no legal requirements regarding the structure here.
- Uniform column structure with visualization aids for growth, target achievement and size classification in relation to the base size "total output" or "sales".
- Dynamic charts showing the temporal development of selected rows for the months of the current year and their accumulated development. This makes it easy to see the target achievement compared to the plan and the forecast of the target achievement (forecast).

3.9.1.3 Cost Analysis

The cost analysis for the selected business areas complete the profitability analysis. Here, either special cost blocks (e.g. operating costs such as maintenance costs, energy costs, etc.) or cost centers as the location or cause of the cost occurrence are analyzed. Important questions around the cost analyses are:

- What does the cost structure of the company look like?
- How have the costs developed?
- Could the cost targets be met?
- Which cost deviations are particularly significant?
- Are the costs appropriate to the services provided?

This is shown here for the sales area as an example for all other business areas.

The cost center report as a partial section of the operating statement serves as the main source of analysis for the overhead cost analysis (see Fig. 3.89). The row structure of the cost center report is based on certain classification rules as shown in the example:

- Primary costs
- Secondary costs

[80] Cf. Varnholt and Dagit (2009, p. 191 ff.).

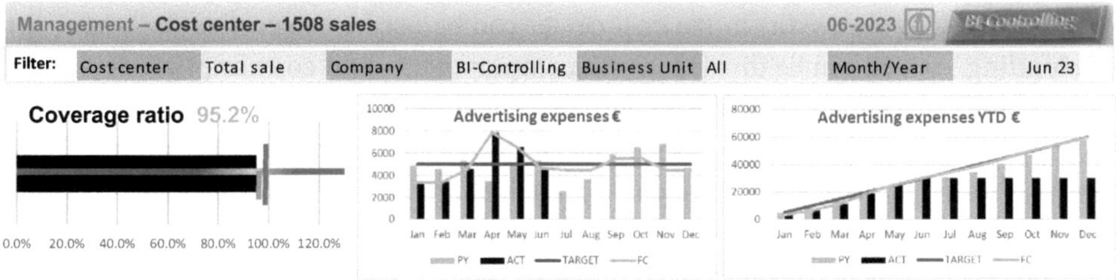

		June					YTD				
		PY	TA	ACT	ΔPY%	ΩTA%	PY	TA	ACT	ΔPY%	ΩTA%
Supplementary wages	+	1,527	1,500	1,638	7.3%	109.2%	9,467	9,000	9,992	5.5%	111.0%
Salaries	+	28,480	30,000	14,560	-48.9%	48.5%	176,576	180,000	88,816	-49.7%	49.3%
Non-wage personnel costs	+	2,848	2,600	2,567	-9.9%	98.7%	17,658	15,600	15,659	-11.3%	100.4%
Personnel costs	-	***32,855***	***34,100***	***18,765***	***-42.9%***	**55.0%**	***203,701***	***204,600***	***114,467***	***-43.8%***	***55.9%***
Advertising expenses	+	5,631	5,000	4,678	-16.9%	93.6%	33,786	30,000	30,407	-10.0%	101.4%
Office expenses	+	5,364	5,100	4,956	-7.6%	97.2%	32,184	30,600	28,745	-10.7%	93.9%
Calc. depreciation	+	3,265	3,000	3,187	-2.4%	106.2%	19,590	18,000	18,166	-7.3%	100.9%
IT costs	+	7,341	7,000	6,898	-6.0%	98.5%	44,046	42,000	41,388	-6.0%	98.5%
Primary costs total	-	**54,456**	**54,200**	**38,484**	**-29.3%**	**71.0%**	**333,307**	**325,200**	**233,172**	**-30.0%**	**71.7%**
Energy costs		1,170	1,200	1,123	-4.0%	93.6%	7,020	7,200	6,738	-4.0%	93.6%
Maintenance		917	800	723	-21.2%	90.4%	5,502	4,800	4,338	-21.2%	90.4%
Vehicle fleet		725	900	762	5.1%	84.7%	4,350	5,400	4,572	5.1%	84.7%
Secondary costs total	-	**2,812**	**2,900**	**2,608**	**-7.3%**	**89.9%**	**16,872**	**17,400**	**15,648**	**-7.3%**	**89.9%**
Total costs after internal activity allocation	-	**57,268**	**57,100**	**41,092**	**-28.2%**	**72.0%**	**350,179**	**342,600**	**248,820**	**-28.9%**	**72.6%**
Relieving overhead costs	+	54,700	57,100	39,100	-28.5%	100.0%	333,670	342,600	242,420	-27.3%	70.8%
Shortfall / Surplus	-	**2,568**	**0**	**1,992**			**16,509**	**0**	**6,400**		
Measures	-										
Number of employees		9	10	9	0.0%	94.7%	9	10	9	0.0%	94.7%
Number of orders		4,320	4,500	3,505	-18.9%	77.9%	26,352	27,000	20,680	-21.5%	76.6%
Cost of goods sold		1,145,360	1,268,889	747,127	-34.8%	58.9%	6,643,088	7,613,333	4,408,051	-33.6%	57.9%
Key figure	-							-			
Total costs / Cost of goods sold		5.0%	4.5%	5.5%	10.0%	122.2%	5.3%	4.5%	5.6%	7.1%	125.4%
Total costs / Employee		127,262	133,567	83,014	-34.8%	62.2%	123,020	133,567	81,631	-33.6%	61.1%
Total cost / Order		13.26	12.69	11.72	-11.6%	92.4%	13.29	12.69	12.03	-9.5%	94.8%

Fig. 3.89 Cost center analysis

- Determination of under- and overfunding
- Measures and indicators

In the area of **primary costs**, the most important cost type groups or cost types of the cost center are listed.[81] These are mainly personnel costs and other costs that are borne by the company from outside, for example in the sales cost center these are costs for advertising, office supplies, imputed depreciation and IT costs. For other cost centers, e.g. from production, other cost types are relevant, such as energy costs, maintenance costs, auxiliary and operating materials.

In the area of **secondary costs**, the internal services are presented using the internal service allocation. These are services of the preliminary or auxiliary cost centers that are allocated to the final or main cost centers. In addition, allocated services for cost objects, the so-called "allocated costs", are shown in the area of relief for the final or main cost centers.

[81] For the terms primary and secondary costs, see Müller (1995, p. 35 ff.).

The debits and the credits in the cost center can be used to determine the **under- or overfunding** , which can be divided into the **employment- and consumption deviation**. For this, the determination of the target cost function is required, which presupposes a cost splitting into fixed and variable standard costs. It shows to what extent the cost center is utilized by its service recipients. For a final production cost center, the under- or overfunding corresponds to the total deviation from employment and consumption deviation, something which is explained in detail in the production controlling cockpit (see Sect. 3.9.4.2). If the allocated standard costs are related to the actual costs, the **coverage ratio** is obtained; this is shown as a peak indicator. It shows how well the costs are covered by the service creation and allocation of the cost center. If it is greater than 100%, more fixed costs were allocated in the actual than planned. If it is less than 100%, not all costs could be passed on. In the case of the sales cost center, it is assumed that the overhead costs have mainly fixed cost character. The standard costs correspond in this case to the target costs.

In the area of **measures** and **key figures**, the services or reference variables can be used for the cost planning and allocation or for an additional key figure formation. Reference variables are mainly used for the calculation as well as the planning and the cost control of the cost center performance. Typical examples of final production cost centers are production, machine and setup times, based on which the allocation rates (e.g. hourly rates) are determined for the valuation of the services (e.g. times) which have arisen or were planned for the product calculation and order valuation. The calculatory valuation with allocation rates is carried out per order for the affected operation per assigned workplace for the respective cost center.

For indirect overhead areas such as sales or purchasing, only allocation variables for the overhead surcharge calculation are often used in cost accounting. Alternatively, process services, e.g. the number of processed orders for the process cost rate determination for the overhead allocation, could be formed. Furthermore, measures for the internal service allocation (e.g. for allocations) can be depicted. Alternatively, measures for the determination of transfer prices or for statistical purposes are used. Examples of measures are the number of employees (heads, full-time equivalents), the square meters, the number of PCs etc., which are used among other things in the context of cost accounting for simple cost allocations.

For the entire line scheme of the operating cost sheet or the cost center reports, it is recommended to build a value hierarchy for the entire company in which all relevant revenues, cost types, measures and key figures are clearly assigned and the corresponding calculation lines are defined. This line scheme is to be set up in the context of the cost accounting concept in the sense of an optimal value flow presentation and control.

The analysis of the costs, services and key figures can be carried out in the well-known column scheme. Monthly or accumulated values can be analyzed in terms of growth to the previous period or as target achievement to the planning. This is visually supported by the miniature bars and the traffic light colors.

For the selected line of the operating cost sheet, the periodic and accumulated development can be displayed. Thus, cost peaks and peculiarities can be quickly detected, here for the advertising costs.

Cost analysis can be used in all areas of the company, so that a repeated explanation in the following sections is omitted except for production.

3.9.1.4 Financial and Liquidity Analysis

Securing the ability to pay and avoiding overindebtedness are important secondary conditions for maintaining the business, in contrast to the main goal of profitability and profit increase. The financial and liquidity analysis answers important control questions in the company:

- How high are the cash balances and payment reserves for the upcoming short- and medium-term payments?
- What is the share of cash flow from operating activities, from investments or divestments, from financing activities and from non-operating and extraordinary effects?
- How long does the internal financing power last to cover upcoming business and investments?
- Is the financing structure balanced in terms of long-term investments and short-term needs from the ongoing business?
- Does the company have sufficient security through equity, reserves and hidden reserves?
- …

The financial ratio analysis provides an entry point into the dynamic and structural financial analysis (see Fig. 3.90).

The most important ratios for the dynamic financial analysis can be tracked in their temporal development and target achievement:

- The **Cash Flow** represents the cash surplus of a period; this is generated from the own strength (internal financing power) of the business operation without being dependent on third parties.[82] It can be determined retrogradely from the annual surplus minus important non-cash-relevant expense and income items or directly from all cash-relevant expense and income items. In the frequently used retrograde determination, the cash flow is determined in the simple form by adding the depreciation and changes in provisions to the annual surplus. Furthermore, changes in working capital can be taken into account. The cash flow can be used for any investments, current business expenses, debt repayments, dividend payments and increases in liquidity reserves in

[82] Cf. Reichmann et al. (2017, p. 117 f.).

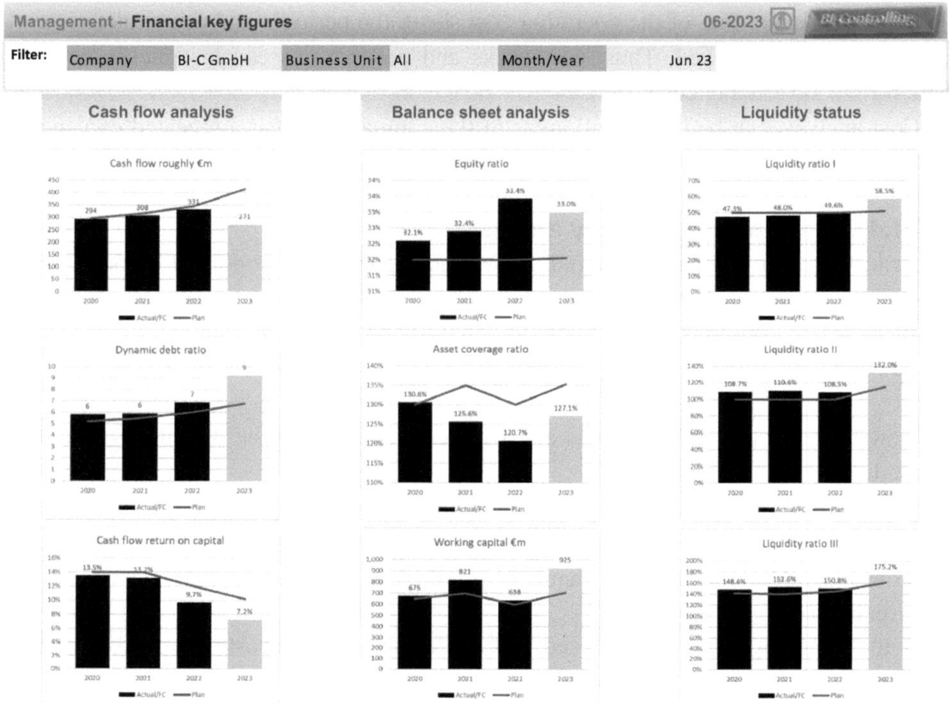

Fig. 3.90 Financial ratio analysis

the company, just like financial resources from third parties. In connection with the repayment ability in the form of repayments and interest on borrowed capital, it is a key figure that serves to determine the debt repayment capacity of the company. In connection with securing the ongoing solvency, it is a central variable for liquidity planning. For the short-term, partly weekly to monthly, planned/actual comparison of the key figure "cash flow", the preparation of a sectoral cash flow statement, which can be accessed via the detailed cash flow analysis, is recommended.

- The **Dynamic Debt Ratio**[83] indicates as a financial risk ratio how often the cash flow of a period must be earned in order to theoretically repay the entire debt. The larger the ratio, the higher the financial risk of overindebtedness and insolvency. Loans must be extended, expanded or replaced. Rating agencies and banks consider a dynamic debt ratio greater than 6 years to be very critical and and a high risk for insolvency. Alternative calculations consider only the net debt, which deducts the liquid funds from the debt.

[83] Cf. Reichmann et al. (2017, p. 118 f.).

- The **Cash Flow Return on Capital** indicates the percentage ratio between the cash flow and the balance sheet total. This shows how large the share of capital return through the internal financing power of the cash flow is. The capital return ratio should be clearly positive in order not to deter investors and lenders.[84]

In the context of the classical balance sheet analysis, the most important key figures for the structural financial analysis can be determined:

- The **equity ratio**[85] shows the proportion of equity in relation to total capital. It is one of the most significant financial risk indicators and included in every rating. The corresponding key figure to this is the **debt ratio** (debt ratio = 1 − equity ratio). With regard to the debt risk, the equity ratio is divided into sections, e.g. less than 10% as very critical, less than 20% as critical, greater than 20% as good and greater than 30% as very good.
- The **asset coverage**[86] shows the ratio between the long-term capital, consisting of long-term debt and equity, and the long-term fixed assets. It is thus a key figure for analyzing the balance of the long-term financing of the long-term assets (maturity-congruent asset financing). It should be clearly above 100%, as otherwise extensions and substitutions for loans become necessary and difficult. If the key figure falls below 100%, parts of the fixed assets have to be financed short-term. If this is not possible, insolvency is often imminent unless other sources of financing are available.
- The **working capital**[87] gives the difference between current assets and short-term liabilities. Due to the potential danger that creditors demand the repayment of their liabilities in the short term, the working capital represents a security measure that shows how large the net current assets are to cover further liabilities. The key figure should be positive, so as not to have to cover the short-term liabilities over longer-term external financing. The relations of the liquidity of the individual components of the current assets to cover the short-term liabilities show the liquidity ratios of the golden balance sheet rule:
  - liquidity ratio I only relates the liquid funds (bank balances, checks and securities) to the short-term liabilities
  - liquidity ratio II relates the liquid funds and the receivables to the short-term liabilities in addition to the liquid funds (here, a percentage value greater than 100% should already be achieved by good companies)

[84] Cf. Gleißner (2011, p. 328).

[85] For the key figure equity/debt ratio see Krause and Arora (2008, pp. 96–98).

[86] See Reichmann et al. (2017, p. 120).

[87] See Reichmann et al. (2017, p. 119).

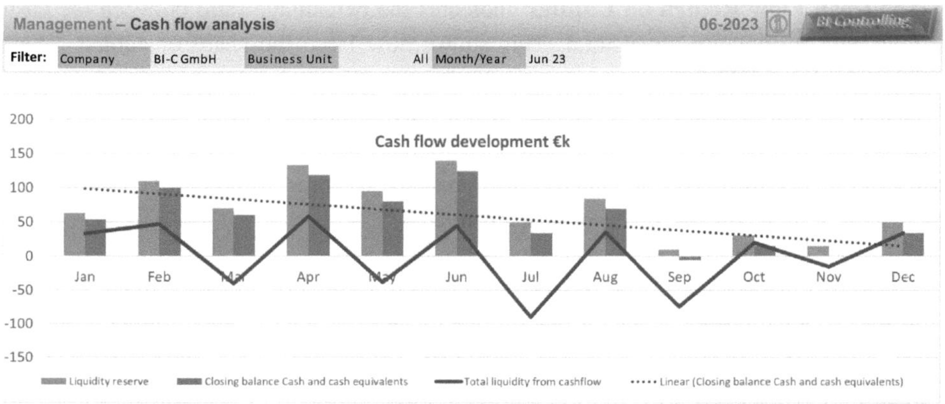

		Jan	Feb	Mar	Apr	May	Jun	Jul	Aug	Sep	Oct	Nov	Dec
Liquidity reserve	-	63	110	70	133	95	139	49	84	10	30	15	49
Credit lines	+	10	10	10	15	15	15	15	15	15	15	15	15
Closing balance Cash and cash equivalents	+	**53**	**100**	**60**	**118**	**80**	**124**	**34**	**69**	**-5**	**15**	**0**	**34**
Opening balance Cash and cash equivalents	+	20	53	100	60	118	80	124	34	69	-5	15	0
Total liquidity from cashflow	-	**33**	**47**	**-40**	**58**	**-38**	**44**	**-90**	**35**	**-74**	**20**	**-15**	**34**
Cash fow from operating activities	-	**32**	**32**	**45**	**33**	**36**	**27**	**24**	**30**	**25**	**19**	**33**	**32**
Cash flow from operating activities before Interest and Taxes	-	**43**	**43**	**54**	**49**	**45**	**39**	**34**	**40**	**39**	**41**	**43**	**45**
Operating cash inflows	+	***139***	***139***	***149***	***148***	***147***	***135***	***140***	***143***	***142***	***141***	***141***	***139***
Operating cash outflows	+	***96***	***96***	***95***	***99***	***102***	***96***	***106***	***103***	***103***	***100***	***98***	***94***
Cash flow from debt payments and taxes	+	-11	-11	-9	-16	-9	-12	-10	-10	-14	-22	-10	-13
Cash flow from non-pperating activities	+	**8**	**8**	**7**	**9**	**10**	**10**	**10**	**10**	**6**	**6**	**7**	**6**
Cash flow from investing activities	+	**-4**	**0**	**-98**	**0**	**-80**	**1**	**-120**	**0**	**-100**	**0**	**-50**	**1**
Cash flow from financing activities	+	**-3**	**7**	**6**	**16**	**-4**	**6**	**-4**	**-5**	**-5**	**-5**	**-5**	**-5**

Fig. 3.91 Cash flow analysis

- liquidity ratio III relates the entire current assets (including inventories) to the short-term liabilities (here, a percentage value significantly greater than 100% should be achieved by good companies)

Especially the sectoral Cash flow analysis shows the company continuously how the cash flow changes, which needs are planned and where possible liquidity bottlenecks arise and additional financing needs exist. In contrast to the cash flow analysis, the pure **liquidity status** focuses on the final balances of the individual accounts. In addition, potential safety stocks and reserves, e.g. overdraft facilities, are shown.

In the cash flow analysis (see Fig. 3.91) the liquidity reserves and balances form the starting point of the financial flow statement. As already mentioned concerning the income statement, the detail lines of the cash flow analysis can be expanded or aggregated using the plus or minus sign. The main part of the cash flow analysis divides the cash flow development into four sectors:

- **Sector I:** Cash flow from operating activities
  - Cash flow from operating activities before interest and taxes

    - Operating inflows
        - Net sales (inflows)
        - Advance payments
        - Other operating inflows
    - Operating outflows
        - Material purchase
        - External services
        - Personnel outflows
        - Pension outflows
        - Other operating outflows
  - Cash flow from debt service and taxes
    - Payments for interest on debt
    - Dividends
    - Tax liability from value-added/tax
    - Payment for income tax
    - Subsidies
- **Sector II:** Cash flow from non-operating activities
  - Non-operating inflows
  - Non-operating outflows
- **Sector III:** Cash flow from investing activities
  - Inflows from divestments
  - Outflows for investments
  - Outflows for fixed assets and intangible assets
  - Outflows for financial assets
- **Sector IV:** Cash flow from financing activities
  - Inflows from borrowing
  - Outflows from loan repayment
  - Inflows from capital increase

With the help of liquidity planning, forecasts can be planned for the change in the cash flow ratio in each sector. If liquidity bottlenecks arise, for example, planned investments in the forecast (yellow period) are to be postponed or new financing arrangements are to be set up. In contrast to some sectors, the operating cash flow can also be determined for result objects such as business units. Generally, financing for most companies is a cash pool view for the entire company.

3.9.1.5 Working Capital Management

The change in working capital is part of the working capital management and should be broken down into the analysis of inventories, receivables and cash (see Fig. 3.92).

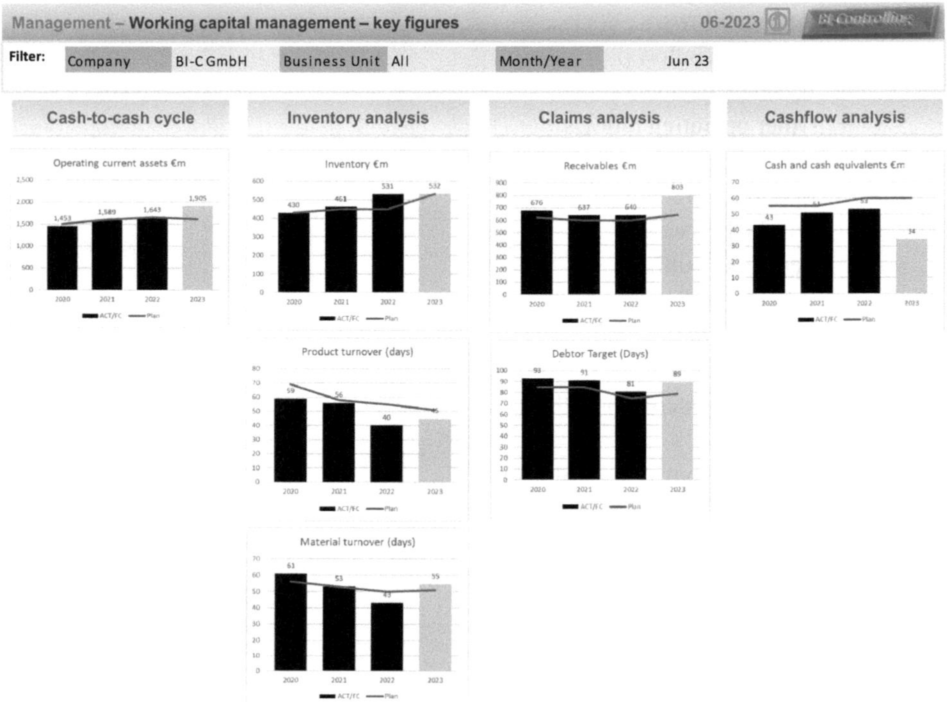

Fig. 3.92 Working capital management analysis

In the context of the inventory analysis, the following key figures should be reduced tendentially due to the associated capital commitment costs (storage and interest):

- The **inventory development of stocks** shows the capital commitment[88] of materials, semi-finished and finished products. Detailed analyses are to be carried out as part of the inventory analysis (see Sect. 3.9.5.5).
- The **product turnover time**[89] shows the period in days that is required to turn over the product inventory (product stock) with the help of sales (revenue).
- The **material turnover time**[90] shows the period in days that is required to turn over the material inventory (input goods stock) by means of material consumption in the production process.

[88] Cf. Lutz et al. (2011, p. 127).

[89] Cf. Reichmann et al. (2017, p. 114).

[90] For the key figure material turnover time, see Krause and Arora (2008, 243–246).

The recommendation to reduce the inventories and thus the capital commitment also applies to the key figures in the receivables analysis:

- The **receivables balance**[91] shows the capital commitment of outstanding receivables. It raises awareness, among other things, for the risks of bad debts and provides indications for the payment behavior of customers. Detailed analyses are to be carried out as part of the receivables analysis (see Sect. 3.9.3.6).
- The **debtor target**[92] shows the period in days that the customers (debtors) need on average[93] to pay the outstanding receivables in the considered period. With the help of debtor management, e.g. by granting selective discounts and consistently following up payments (reminders, dunning, collection options, etc.), the debtor target is to be reduced.

The liquid funds show the bank and cash balances that have to be monitored weekly as part of a continuous liquidity status. Payment bottlenecks and needs are to be identified and covered by financing measures (e.g. in the context of external financing, the extension of expiring loans or the borrowing of loans) with the support of the ongoing liquidity planning (see Sect. 3.9.1.4).

As a complement to working capital management, the analysis of the entire cash-to-cash cycle including the creditor target is recommended (see Fig. 3.93); this target can be examined in detail through the analysis of the liabilities and the payment behavior (without illustration).

This cycle results from the addition of the material and product turnover time plus the debtor target minus the creditor target. Based on the cash-to-cash cycle analysis, improvements over the previous year and deviations from the planned value can be examined. Detailed analyses of the inventory, the own payment behavior and the receivables analysis can be accessed via detailed reports.

3.9.1.6 Annual Financial Statement Analysis

Every business analysis includes the preparation and examination of the business reports and annual financial statement information in the form of balance sheet, income statement and notes including management report. Depending on the size and group perspective of the company, consolidated financial statement information, quarterly financial statements, the cash flow statements, the segment reporting, the risk reporting as well as information on compliance and corporate governance are also added. The preparation of this information is the core task of accounting and controlling. For this purpose, various tools and IT solutions are used, in which this information can be determined and

---

[91] For the term receivables, see Lutz et al. (2011, p. 84).

[92] The debtor target is also called receivables turnover time, see Reichmann et al. (2017, p. 115).

[93] Cf. Lutz et al. (2011, p. 183).

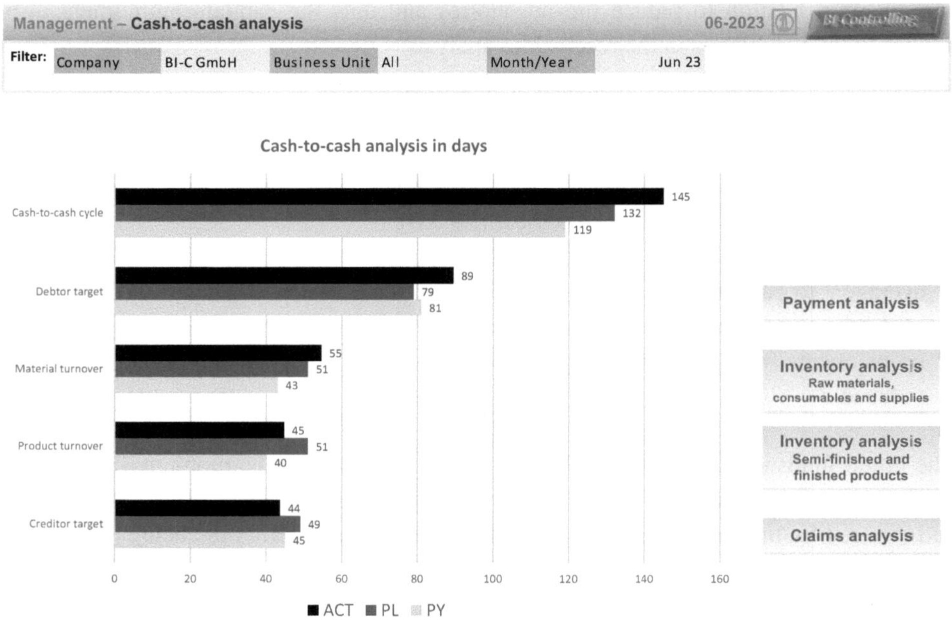

Fig. 3.93 Cash-to-cash cycle analysis

reported. Nevertheless, this information is also interesting for the BI-supported controlling cockpit, but does not have to be stored redundantly in a database. Glossy documents such as the annual report including annual financial statement information can be stored, for example, via DMS systems, which the cockpit accesses.

Important questions around the financial statements are:

- What is the asset, income and debt situation of the company or the group of companies?
- How is the company funded and secured?
- What developments, opportunities and risks are to be expected for the company?
- What commercial result was achieved?

The central reporting elements of the financial statements are balance sheet (see Fig. 3.94) and income statement (see Fig. 3.95).

By presenting the balance sheet in the staggered form, a view of the current situation and development of the balance sheet is possible, unlike the rigid financial statements in the annual report.[94] Compared to the profitability analysis, this analysis is not done con-

[94] See Lutz et al. (2011, p. 337).

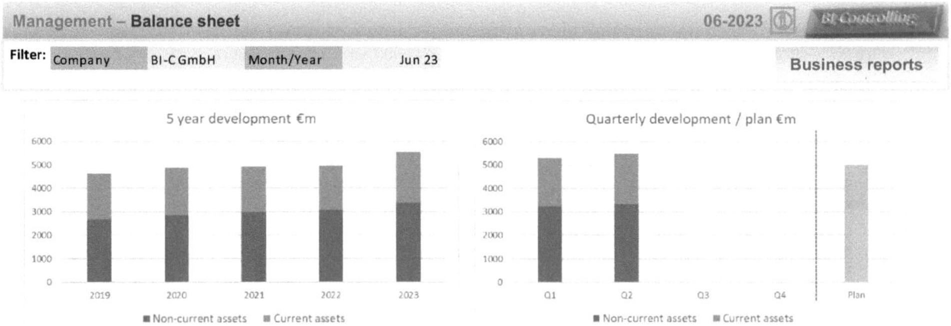

	PY	PL	ACT	ΔPY	ΔPL	ΔPY%	ΩPL%	% Balance Sheet Total
+Assets	4,963	5,389	5,483	520	94	10.5%	101.7%	
+Non-current assets	3,070	3,130	3,342	272	212	8.9%	106.8%	61.0%
+Intangible assets	5	5	4	-1	-1	-20.0%	80.0%	0.1%
+Property, plant and equipment	2,165	2,225	2,438	273	213	12.6%	109.6%	44.5%
+Financial assets	900	900	900	0	0	0.0%	100.0%	16.4%
+Current Assets	1,893	1,860	2,140	247	280	13.0%	115.1%	39.0%
+Inventories	531	531	532	1	1	0.2%	100.2%	9.7%
+Receivables and other assets	740	740	903	163	163	22.0%	122.0%	16.5%
+Securities	250	250	235	-15	-15	-6.0%	94.0%	4.3%
+Cash and cash equivalents	372	339	470	98	131	26.3%	138.6%	8.6%
Prepaid expenses and deferred charges	1	1	1	0	0	0.0%	100.0%	0.0%
+Liabilities	4,963	5,389	5,483	520	94	10.5%	101.7%	
+Equity	1,659	1,701	1,820	161	119	9.7%	107.0%	33.2%
Share capital	394	394	505	111	111	28.2%	128.2%	9.2%
+ Revenue reserves	772	772	872	100	100	13.0%	113.0%	15.9%
Profit / Loss carried forward	225	185	220	-5	35	-2.2%	118.9%	4.0%
Net income / Loss for the period	268	350	223	-45	-127	-16.8%	63.7%	4.1%
'+Provisions	2,827	3,205	2,958	131	-247	4.6%	92.3%	53.9%
+Liabilities	476	482	704	228	222	47.9%	146.1%	12.8%
Deferred income and accrued expenses	1	1	1	0	0	0.0%	100.0%	0.0%

Fig. 3.94 Balance sheet

tinuously, but at larger intervals. Due to the basic accounting principle of double-entry bookkeeping "debit to credit", the assets and liabilities of the balance sheet are always balanced.

Best-practice techniques of presentation are again:

- Hierarchy analysis of the asset and liability items down to the account level by means of the value hierarchy (plus and minus signs). This allows the legally prescribed reporting structure to be slightly adapted with regard to presentation freedoms and options.
- Uniform column structure with visualization aids for growth, target achievement and size classification in relation to the base size "balance sheet total".
- Dynamic charts show the temporal development (5-year and quarterly development) of selected rows. The default setting shows, for example, the ratio of fixed assets to current assets.

Further analyses of the balance sheet are possible, for example, via the key figure analyses already presented in the areas of profitability/success and finance/liquidity. They form the basis for the internal and external company rating.

€m		PY	PL	ACT	ΔPY	ΔPL	ΔPY%	ΩPL%	% Total Output
Sales	+	861	952	805	-56	-147	-6.5%	84.6%	97.6%
Inventory change	+	21	20	19	-2	-1	-9.5%	95.0%	2.3%
Own work capitalised	+	1	1	1	0	0	0.0%	100.0%	0.1%
Total output	-	**883**	**973**	**825**	**-58**	**-148**	**-6.6%**	**84.8%**	**100.0%**
Other operating income	+	1	1	1	0	0	0.0%	100.0%	0.1%
Material expenses	-	-273	-280	-260	13	20	-4.8%	92.9%	-31.5%
Exp. for raw materials, supplies and purchased goods	+	-176	-186	-165	11	21	-6.3%	88.7%	-20.0%
Exp. for purchases services	+	-97	-94	-95	2	-1	-2.1%	101.1%	-11.5%
Gross profit	-	**611**	**694**	**566**	**-45**	**-128**	**-7.4%**	**81.6%**	**68.6%**
Personnel expenses	+	-183	-173	-178	5	-5	-2.7%	102.9%	-21.6%
Depreciation	+	-63	-64	-46	17	18	-27.0%	71.9%	-5.6%
Other operating income	+	-91	-101	-111	-20	-10	22.0%	109.9%	-13.5%
Operating result	-	**274**	**356**	**231**	**-43**	**-125**	**-15.7%**	**64.9%**	**28.0%**
Financial and investment result	+	**-3**	**-3**	**-5**	**-2**	**-2**	**66.7%**	**166.7%**	**-0.6%**
Result from ordinary activities	-	**271**	**353**	**226**	**-45**	**-127**	**-16.6%**	**64.0%**	**27.4%**
Income taxes	+	-2	-2	-2	0	0	0.0%	100.0%	-0.2%
Profit after income taxes	-	**269**	**351**	**224**	**-45**	**-127**	**-16.7%**	**63.8%**	**27.2%**
Other taxes	+	-1	-1	-1	0	0	0.0%	100.0%	-0.1%
Net income / Loss for the period	-	**268**	**350**	**223**	**-45**	**-127**	**-16.8%**	**63.7%**	**27.0%**

Fig. 3.95 Profi and loss statement

The examination of the profit situation from the perspective of the legal annual financial statement requirements is carried out via the income statement.

The income statement is presented either in the form of the total cost method or the cost of sales method, depending on the choice of the company.[95]

In contrast to the income statement, the structure of the profit determination is bound to legal requirements and not to control-relevant issues. Nevertheless, a monthly look at the current situation and development of the income statement is useful.

The best-practice techniques of presentation are again:

- Hierarchy analysis of the expense and revenue items down to the account level via the plus and minus signs of the value hierarchy. This allows the legally prescribed reporting structure to be slightly adapted with regard to presentation freedoms and options.
- Uniform column structure with visualization aids for growth, target achievement and size classification in relation to the base size "total output" in the total cost method or "sales" in the cost of sales method.

[95] Cf. Lutz et al. (2011, p. 96).

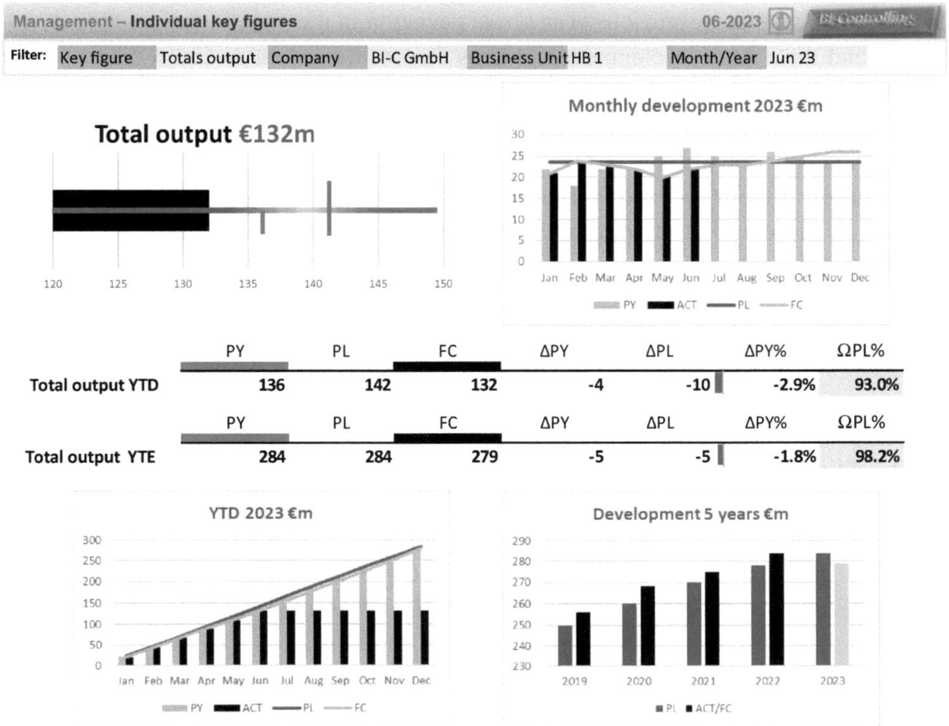

Fig. 3.96 Single key figures analysis

- Dynamic charts show the temporal development of selected rows for the months of the current year and their accumulated development. This allows the target achievement to the plan and the forecast of the target achievement to be quickly visible.

3.9.1.7 Single-Indicator Analysis

The report Single-indicator analysis (see Fig. 3.96) allows for a comprehensive detailed analysis of each defined and available indicator. The following questions are generally to be answered:

- Have the target values been achieved, failed or exceeded?
- How has the indicator developed compared to the previous period?
- How does an indicator develop over different observation periods?

The single-indicator analysis can be accessed via trigger functions on report objects from upstream reports. Alternatively, the single indicator can be selected directly in the filter area. Via a drill pad or return function, you can jump back to the last viewed report.

Unlike the previously presented indicator reports, the single-indicator report allows different views on the indicator selected in the filter area. Starting from the already known graphical plan or previous year-actual deviation in the top left, the monthly development and the 5-year development including forecast (e.g. year-to-end = YTE) can be viewed. For both the monthly and the accumulated view (year-to-date = YTD), absolute deviations as well as the growth compared to the previous year and the target achievement compared to the planned value are displayed in the value table.

Compared to the other reports, the single-indicator analysis eliminates the disadvantage that in those an indicator can usually only be viewed in one form of presentation with limited visibility due to space constraints.

Just like with the cost analyses, indicator analyses can be used in all business areas, so that a repetitive explanation in the following sections (e.g. for sales, purchasing, etc.) is omitted.

3.9.2 Executive Board Cockpit: Strategic Controlling

There are many good reasons why strategic controlling is indispensable. The long-term survival of a company depends on building and defending strategic success potentials. Companies are subject to constant change, which they have to face, such as:[96]

- The advancing globalization and the change of international trade relations with international trade conflicts and tariffs, but also mergers of new markets. The international orientation is not limited to the expansion of the sales markets, but also extends to the provision of capital as well as the procurement of raw materials and personnel. Economic, ethical and legal risks of the respective countries have to be taken into account.
- The technological progress in various directions, e.g. in the form of production, information, communication and transport technologies.
- The growth and change of the world population with constant changes in the needs of the market participants on the one hand and resource scarcity on the other hand.
- The change of political, social and ecological conditions, such as political change, social justice and environmental protection.
- The changes caused by a political crisis or other uncertainties, such as the oil crisis, the financial crisis, the real estate bubble, corporate insolvencies, Brexit, the refugee flows, the Covid-19 crisis, the climate change etc.

To this end, it is important to know one's strengths and weaknesses as well as to predict one's opportunities and risks, to formulate one's strategic goals and to develop a suitable action plan with which these goals are to be achieved.

[96] Cf. Bea and Haas (2013, p. 10) and Reichmann et al. (2017, p. 580 ff.).

Central questions of strategic controlling are therefore:

- What does the company stand for in general?
- Where is the company today?
- What are the strengths and weaknesses of the company and the competition?
- What opportunities and risks does the company have?
- What changes will have an impact on the company? Where can constancy be expected?
- What are the strategic goals of the company?
- With what measures does the company want to achieve the strategic goals?

To better answer these questions, a lot of information is necessary that is not directly available internally, but has to be estimated, forecasted and collected. Here it is important to identify and combine the right information from the multitude of data sources and to draw the right strategic conclusions from them. In contrast to the individual use of strategic instruments, the cockpit has the advantage of carrying out the strategic control in a more structured and systematic way on an integrated data base. Advantages of the digital visualization and analysis possibilities are especially to recognize new opportunities and dangers sooner and to get various suggestions for good strategic project ideas. The following existing deficiencies of the strategic control should be avoided:

- The strategic planning is partly or completely neglected, one focuses on the operating business processing.
- The strategic planning takes place only sporadically, e.g. in the form of retreats.
- For the strategic planning, only selected information sources and known instruments are used.
- Strategic decisions are made by gut feeling.
- Strategic plans disappear in the drawers of the executives.
- There are hardly any IT-supported tools for a holistic strategic controlling.

Due to the sustainable economic importance and sensitivity of the strategic data, the strategic cockpit must be particularly protected from misuse and competitive espionage. Compare Sect. 5.5.3 on authorization system and access control.

While cockpit examples of operational controlling tasks are more frequently found in literature and practice, research for strategic controlling cockpit solutions is still at its beginning.[97] The following example shows how many of the known strategic instruments can be integrated and how also non-monetary data can be included.

[97] Cf. Panke (2015, p. 11).

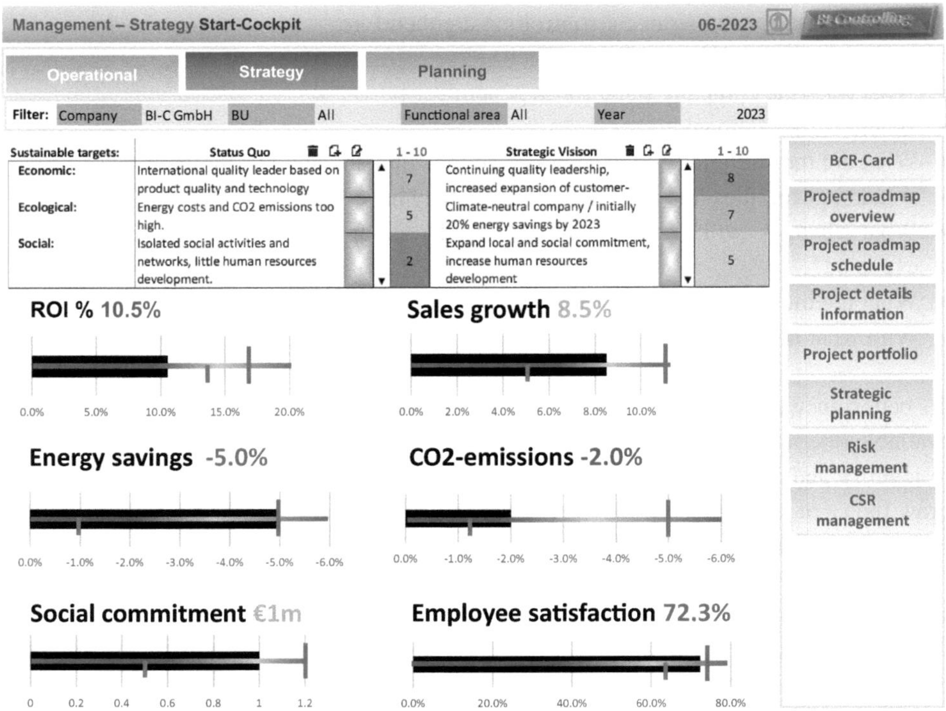

Fig. 3.97 Strategic planning - start cockpit

3.9.2.1 Strategic Start Cockpit/Sustainability

The strategic start cockpit serves as a point of orientation for the central goals of the enterprise and their achievement. While in the past the purely economic goals regarding profitability and liquidity were in the foreground, today three pillars of goal achievement are targeted under the keyword "sustainability":[98]

- Economic goals
- Ecological goals
- Social/Societal goals

These should be planned, prioritized and controlled in accordance with the mission statement of the enterprise. For this purpose, the start cockpit offers an entry point into the individual planning, in which the status quo and the strategic vision can be assessed or monitored (see Fig. 3.97). For the entry, simple evaluations, e.g. from 1–10 or percentage estimates, are sufficient.

[98] Cf. Tremmel (2003, p. 56).

Since the first pillar of economics dominates in most companies, Corporate Social Responsibility (CSR) should serve as a basis to voluntarily integrate social issues and environmental concerns into their business activities and into their interactions with the stakeholders of the company.[99] The requirements in CSR are limited to two of the three sustainability dimensions, i.e. the ecological and the social concerns of the company. Social responsibility in the company mainly concerns the employees on issues such as occupational health and safety or investments in human capital. Environmentally conscious action mainly takes place in production and in the resources used for it. The main dimensions of CSR are: human resource management, occupational health and safety, adaptation to change, environmental impacts and management of natural resources, local communities, business partners, suppliers and consumers, human rights and global environmental protection.[100]

In the start cockpit, the most important key figures can be analysed in accordance with the three pillars of sustainability management (see Fig. 3.97). For the area of economic objectives, the ROI and the sales growth were selected. For the ecological objectives, the key figures for energy saving and the change rate of CO_2 emissions were shown. The social commitment and the employee satisfaction are shown in the social objectives.

The individual planning and analysis fields of strategic controlling can be selected via the menu at the edge.

The BCR-Card as an extension of the Balanced Scorecard and the roadmap planning serves as a link between the strategic, tactical and operational control and was already explained in detail in Sect. 3.1.1. The strategic planning with its various instruments will be described in the following section. In addition, the risk management with its control process consisting of risk identification, risk analysis and assessment, risk control and monitoring can be called up for the general risk management control.[101] The CRS management for the detailed control of the ecological and social objectives can also be selected here.

3.9.2.2 Strategic Planning

An important historical event that impacted strategic planning is certainly the oil crisis of 1973, associated with the question of whether such special effects and the associated trend breaks are predictable.[102] Furthermore, slight changes in the market, e.g. due to new technologies, globalization, changes in buying behavior etc. are reasons for a continuous strategic adjustment. Instead of transferring the past into the future, the concept

[99] Cf. Green Paper of the Commission of the European Communities (2001, p. 7).

[100] Cf. Green Paper of the Commission of the European Communities (2001, p. 2–7).

[101] Cf. Hornung et al. (1999, p. 318 ff.).

[102] Cf. Bea and Haas (2013, p. 12–15).

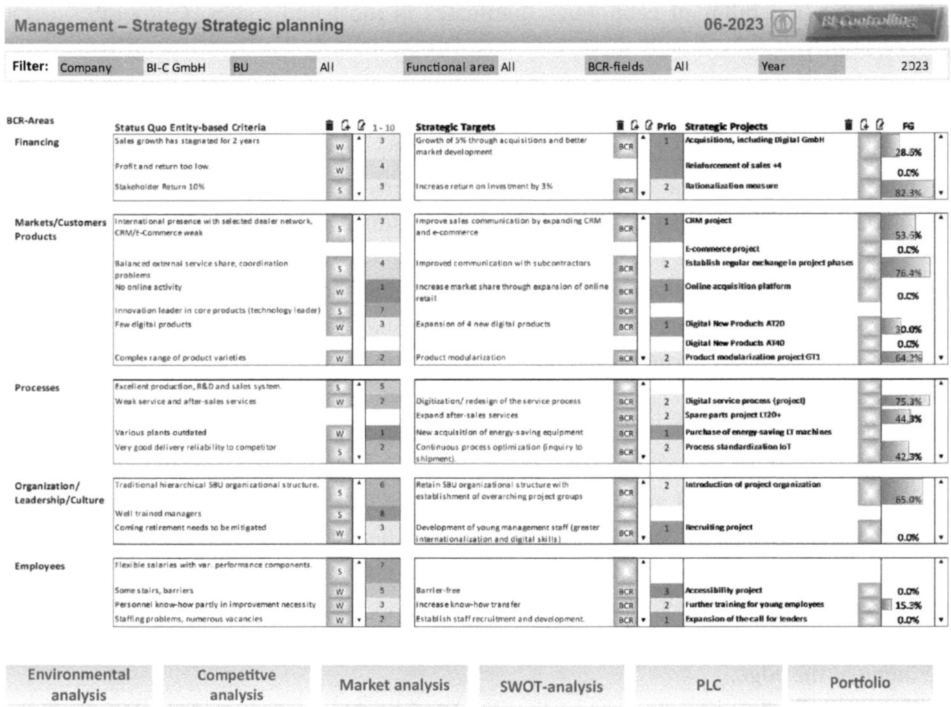

Fig. 3.98 Strategic planning

of **environmental analysis** emerged to systematically analyse future opportunities and risks in the company's environment.[103]

In addition to a general environmental analysis, specific **market analyses** and **company analyses** as well as **various portfolio analyses** and the **experience curve concept** have been established to develop a status quo positioning and an adaptation strategy. This involves both an external and internal orientation of the company, which are often summarized in the instrument of the **SWOT analysis** in a strengths-weaknesses and opportunities-threats profile.

In the cockpit overview of strategic planning, the final strategic status quo positioning and the final prioritization of the strategic objectives and projects, which are assigned to the respective BCR fields (perspectives) of the BCR -Card, are carried out (see Fig. 3.98). The status quo positioning can be assigned from the further strategic analyses. Strengths and weaknesses are displayed in categories. In further menus of the cockpit, key figure assignments and risk assignments can be added. By using the system-

[103] Cf. Ansoff and McDonnell (1990, p. 3 ff.), Bea and Haas (2013, p. 12–15) and Reichmann et al. (2017, p. 584).

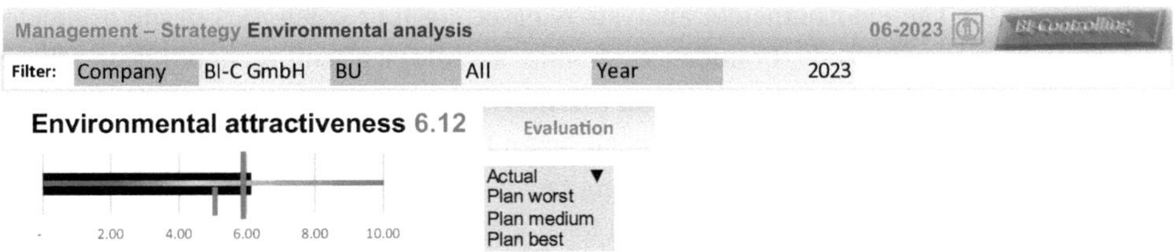

Criteria environmental analysis	Status quo		critical 1	2	3	4	5 (Evaluation)	6	7	8	9	favorable 10	Weithting	weighted scenario 2023	Plan scenario Ø Scenario	Δ Ø Scenar.	worst	middle	best
+Political													0.28	1.72	1.69	0.03	1.47	1.64	1.96
Political stability	high	O									•		0.07	0.63	0.60	0.03	0.50	0.60	0.69
Legislation trade	no changes							•					0.03	0.18	0.20	-0.02	0.16	0.17	0.27
Tax legislation	Tax privileges	T				•							0.07	0.28	0.28	0.00	0.25	0.27	0.31
Environmental legislation	higher limits	T			•								0.03	0.09	0.08	0.01	0.06	0.09	0.09
Capital regulation	low	O			•								0.03	0.09	0.09	0.00	0.08	0.09	0.10
Labor market	very good	O									•		0.05	0.45	0.44	0.01	0.41	0.43	0.50
+Economic													0.23	1.17	1.18	-0.01	1.06	1.11	1.38
GDP trend	slightly increasing	O					•						0.03	0.15	0.15	0.00	0.14	0.14	0.17
Unemployment	very low	O									•		0.04	0.36	0.35	0.01	0.32	0.34	0.40
Interest rate development	very low	O		•									0.04	0.08	0.09	-0.01	0.08	0.08	0.11
Monetary policy	little influence						•						0.03	0.15	0.17	-0.02	0.14	0.14	0.23
Exchange rates	to $ rather increasing	T			•								0.05	0.15	0.15	0.00	0.14	0.14	0.17
Available purchasing power	high	O								•			0.04	0.28	0.28	0.00	0.25	0.27	0.31
+Socio-cultural													0.17	1.17	1.15	0.02	1.05	1.11	1.29
Demographic development	stable							•					0.04	0.24	0.24	0.00	0.22	0.23	0.26
Population growth	low	T				•							0.03	0.12	0.12	0.00	0.11	0.11	0.13
Education level	very high	O									•		0.04	0.36	0.35	0.01	0.32	0.34	0.40
Culture/Lifestyle	west									•			0.03	0.24	0.24	0.00	0.22	0.23	0.26
Career expectations	high	O							•				0.03	0.21	0.21	0.00	0.19	0.20	0.23
+Technological													0.32	2.06	2.05	0.01	1.76	1.96	2.43
Innovation potential	very high	O									•		0.06	0.54	0.53	0.01	0.49	0.51	0.59
Degree of automatization	very high	O									•		0.03	0.27	0.23	0.04	0.14	0.26	0.30
Degree of digitalization	moderate						•						0.06	0.30	0.35	-0.05	0.27	0.29	0.48
New developments/patents	low	T			•								0.04	0.12	0.12	0.00	0.11	0.11	0.13
Patent protection	high	O							•				0.03	0.18	0.18	0.00	0.17	0.17	0.21
Government spending on research	moderate						•						0.05	0.25	0.25	0.00	0.23	0.24	0.28
Private sector R&D spending	high	O								•			0.05	0.40	0.39	0.01	0.36	0.38	0.44
Environmental attractiveness													1.00	6.12	6.07	0.05	5.51	5.81	6.73

Fig. 3.99 Environmental analysis in the form of the PEST analysis

supported planning of the strategic objectives and projects, the management gets a better overview and can act according to priority. The link to the BCR-Card improves the strategy implementation and operational execution, as the control by the executives on different organizational levels is bundled by the BCR-Card. In the sub-menu to the BCR-Card, cause and effect relationships of projects on key figures of the strategic objectives can be forecasted or simulated with value driver tree-like models.

In the data model for strategic planning, it is important to model BCR fields, strategic objectives, key figures and projects independently, so that not a rigid, but a flexible strategic planning model is set up, in which new objectives etc. can be created and links between objects can be made. In this way, the strategic planning in connection with the BCR-Card can be designed flexibly and lived dynamically, unlike in separate Excel solutions. The subordinate strategic analyses and planning areas are briefly described below:

- **Environmental analysis**

The starting point for the general environmental analysis is the PEST analysis (also STEP analysis). It was developed by Fahey and Narayanan in 1986 in *Macroenviromental Analysis for Strategic Environment* and is still a popular method for examining and presenting the external business environment (macro-level) from the perspective of the company.[104] The environmental analysis is based on four categories of environmental factors (see Fig. 3.99):[105] **P**olitical, **e**conomic, **s**ociological and **t**echnological ; these flow into the top indicator **environmental attractiveness** in the cockpit report with their weighted sub-criteria which can vary depending on the region, industry and company.

Changes and extensions of the influencing factors such as e.g. in the PESTEL analysis including the categories **e**nvironmental and **l**egal were not made in the example, since the detailed criteria are already included in the first four categories. Company-specific adaptations are possible. **To increase the clarity in the report, the criteria can be shown or hidden using the plus and minus symbols.** The criteria are to be maintained individually and marked as an opportunity or threat for the takeover in the SWOT analysis (see below). The influencing factors of the company/PEST analysis can usually not be influenced directly by the company. In connection with the **scenario analysis** critical deviations from the status quo can be determined and thus opportunities and threats for the company can be identified. The evaluation is carried out for the current status and for selected planning scenarios using the point scale 1–10. For orientation, an average scenario is formed, which serves as a benchmark. This way, important influencing criteria can be distinguished faster from others. When evaluating the detailed criteria, it should be noted that, depending on the opportunity and threat, the evaluation is critically or favorably exactly reversed. The influencing factors are best sorted and weighted according to their significance.[106] In the author's opinion, the PEST analysis is particularly suitable for market positioning and the development of market strategies such as market entry or exit decisions for companies, business units and regions.[107]

- **Market analysis**

The market analysis is based on the industry structure analysis developed by M. E. Porter in the field of competitive strategy. It focuses on the company's market and thus also on the competitors. The intensity of competition within an industry is influenced by its structure and is a decisive factor for the profitability of an industry and thus also of the companies within the industry.[108]

To analyze the industry structure, Porter distinguishes between five competitive forces: competitors, new entrants, suppliers, buyers and substitute products.[109] This **Five Forces Model** developed by Porter in the 1980s is still one of the most important analysis models of strategic management.[110] The central five forces of competition and the influencing factors acting on them are presented below:

---

[104] Cf. Steuernagel (2017, p. 61) and Paxmann and Fuchs (2010, p. 82).

[105] Cf. Steuernagel (2017, p. 61) and Paxmann and Fuchs (2010, p. 82).

[106] Cf. Paxmann and Fuchs (2010, p. 83 f.) and Eisl et al. (2015, p. 67 f.).

[107] Cf. Steuernagel (2017, p. 68) and Paxmann and Fuchs (2010, p. 82).

[108] Cf. Bea and Haas (2013, p. 109) and Steuernagel (2017, p. 47).

[109] Cf. Bea and Haas (2013, p. 109 ff.) and cf. Steuernagel (2017, p. 47 ff.).

[110] Cf. Bea and Haas (2013, p. 113).

1) **Competitors**
   The rivalry exerted by the already established competitors in an industry arises from the limitedness of the market that the competitors are fighting for. The competitors try to improve their market position by taking measures that other competitors respond to with countermeasures.[111] Typical measures of rivalry are e.g., price discounts, advertising battles or the introduction of new products.[112] The strength of the rivalry depends on a multitude of structural factors that are listed below in a nutshell:[113]
   - Number of competitors: With a large number of competitors, competition tends to be high.[114]
   - Size and position of competitors: Different sizes of companies with different market positions lead to different power relations in the market. The rivalry tends to increase with similar size ratios and similar market positions.[115]
   - Degree of product differentiation and switching costs: The greater the homogeneity of the products offered and the lower the switching costs for the buyers (supplier change), the greater the competitive struggle.[116]
   - Exit barriers: High exit barriers force companies to remain in the market and thus increase the rivalry of an industry.[117]
   - Degree of capacity utilization: The lower the capacity utilization, the higher the fixed cost pressure, something which induces companies to take measures to increase the capacity utilization.[118]
2) **New entrants**
   The lower the entry barriers of a market are, the more new entrants intensify the competition.[119] Important causes for entry barriers are, among others: product differentiation, capital requirements, switching costs, access to distribution channels, government policy, economies of scale and size-independent cost disadvantages such as location costs or learning curve effects but also many other factors such as technological progress, or logistical and digital networking.[120]
3) **Suppliers**
   The stronger the bargaining position of the suppliers, the lower the profit margin for the purchasing companies. Possible reasons for a strong supplier power are, for exam-

---

[111] Cf. Porter (2013, p. 53) and Steuernagel (2017, p. 48).

[112] Cf. Porter (2013, p. 53) and Steuernagel (2017, p. 48).

[113] Cf. Porter (2013, p. 54 ff.).

[114] Cf. Steuernagel (2017, p. 47 f.) and Porter (2013, p. 53 f.).

[115] Cf. Porter (2013, p. 54).

[116] Cf. Porter (2013, p. 55) and Bea and Haas (2013, p. 112).

[117] Cf. Porter (2013, p. 59 f.) and Bea and Haas (2013, p. 112 f.).

[118] Cf. Porter (2013, p. 55 f.) and Bea and Haas (2013, p. 112).

[119] Cf. Steuernagel (2017, p. 52).

[120] Cf. Porter (2013, p. 41–53) and Steuernagel (2017, p. 52 f.).

ple, high switching costs when changing suppliers or a lack of substitute goods for the supplied input goods.[121]

4) **Buyers**
A strong bargaining power of the customers reduces the prices and thus the profitability of a selling company.[122] This is to be expected, among other things, when the concentration of buyers in the market is high, the prices in the market are very transparent and the price elasticity is very high or when the buyers can threaten to take over the upstream value-added stages by backward integration.[123]

5) **Substitute goods**
If the products in the market are interchangeable by substitute goods, the competitive pressure increases. This is especially true when the price/performance ratio of the substitute goods improves in relation to the original products of the market or when technological progress makes older products obsolete.[124]

These five influencing factors are tentatively classified in the market analysis in the cockpit in the categories market structure and product potential and restructured and supplemented with their subcriteria (see Fig. 3.100). For example, the entry barriers for new providers fall into the category competition, and the product potentials are supplemented by the development capability of the product because the greater the development capability of a product, the greater the need for change and thus the competition. All categories and influencing criteria can be adapted and supplemented to the individual needs of the market analysis of the company.

Furthermore, in the market analysis, the criteria market size and market growth are placed in the category market potentials as they form the important starting point for assessing the volume and change of the overall market. These two categories are also used for the market share-market growth portfolio of the Boston Consulting Group (see further below in this section).

The weighted subcriteria go into the evaluation of the top indicator **market attractiveness**, which is the basis for the McKinsey portfolio in comparison to the competitive strength (see further below in this section). **To increase the clarity, the criteria can be expanded or collapsed using the plus and minus symbols.** The criteria of the market analysis are to be maintained individually and marked as an opportunity or risk for the transfer to the SWOT analysis (see below). The criteria are to be prioritized and weighted in advance. The evaluation is done for the current status and for selected planning scenarios using the point scale 1–10 (critical–favorable). A comment on the status quo explains the assessment. For orientation, an average scenario is formed which serves

[121] Cf. Bea and Haas (2013, p. 110) and Lippold (2017, p. 62).

[122] Cf. Porter (2013, p. 164 ff.) and Bea and Haas (2013, p. 111 f.) and Lippold (2017, p. 62).

[123] Cf. Porter (2013, p. 165) and Bea and Haas (2013, p. 111 f.) and Lippold (2017, p. 62).

[124] Cf. Porter (2013, p. 60 f.) and Bea and Haas (2013, p. 112) and Lippold (2017, p. 62).

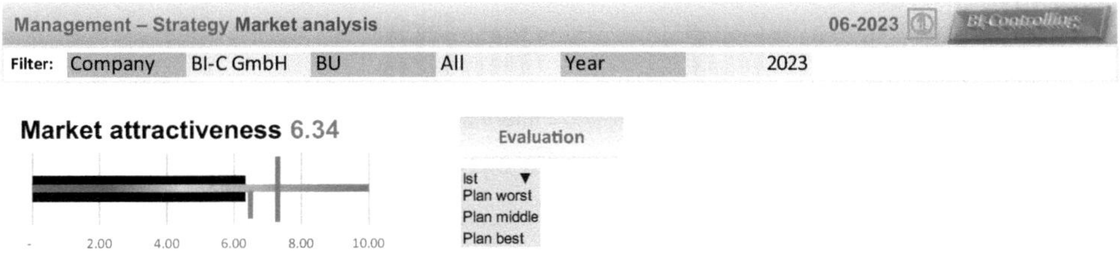

Criteria of the market analysis	Status quo	1 (critical)	2	3	4	5	6	7	8	9	10 (favorable)	Weighting	weighted scenario	Plan scenario: Ø Scenario	Δ Ø Scenar.	worst	middle	best
1. Market potential								•				0.35	2.30	2.90	0.60	2.80	2.95	2.95
+ market size								•				0.20	1.40	1.60	0.20	1.60	1.60	1.60
+ market growth							•					0.15	0.90	1.30	0.40	1.20	1.35	1.35
2. Market structure												0.55	3.34	3.09	-0.25	3.09	3.09	3.09
- Competitors							•					0.25	1.45	1.22	-0.23	1.25	1.25	1.15
Number of competitors	3 big und 10 small T					•						0.10	0.50	0.37	-0.13	0.40	0.40	0.30
Size of competitors	from 100 to 10000 employees T						•					0.10	0.60	0.50	-0.10	0.50	0.50	0.50
Entry barriers	high, own product techn. advanced O							•				0.05	0.35	0.35	0.00	0.35	0.35	0.35
- Suppliers						•						0.12	0.65	0.65	0.00	0.65	0.65	0.65
Negotiating power	weak O						•					0.05	0.30	0.30	0.00	0.30	0.30	0.30
Susceptibility to disruptions	lack of quality increasing T					•						0.05	0.25	0.25	0.00	0.25	0.25	0.25
Factor price development	rising factor prices T					•						0.02	0.10	0.10	0.00	0.10	0.10	0.10
- Buyers								•				0.18	1.24	1.22	-0.02	1.19	1.19	1.29
Negotiation strength	increasing T						•					0.05	0.30	0.25	-0.05	0.25	0.25	0.25
Product commitment	stable customer relationshop O							•				0.10	0.70	0.73	0.03	0.70	0.70	0.80
Price sensitivity	not advanced O								•			0.03	0.24	0.24	0.00	0.24	0.24	0.24
3. Product potential					•							0.10	0.70	0.60	-0.10	0.60	0.60	0.60
+ Substitution potential						•						0.05	0.35	0.30	-0.05	0.30	0.30	0.30
+ Development capability				•								0.05	0.35	0.30	-0.05	0.30	0.30	0.30
Market attractiveness												1.00	6.34	6.59	0.25	6.49	6.64	6.64

Fig. 3.100 Market analysis

as a benchmark. This allows to distinguish important influencing criteria in the strategic market analysis more quickly from others.

- **Business analysis**

The business analysis takes over the internal orientation within the framework of strategic planning. Here, the most important influencing criteria and thus the strengths and weaknesses of the company in its competitive position are analyzed. The criteria can be derived, for example, from the management concept of business value chains developed by M. E. Porter. The business value chain influences the profit margin, return and thus the total value of the company. Here, primary activities for value creation (inbound and outbound logistics, production, sales, customer service) and supporting activities (corporate infrastructure, human resources, technology development, procurement) are distinguished.[125] In the concept approach of "Strategic Management", the basic idea of value chains was refined. Here, the strategic success potentials of the company are divided into performance and leadership potentials (Fig. 3.101).[126]

---

[125] Cf. Bea and Haas (2013, p. 127 f.) and Porter (2014, p. 66 f.).

[126] Cf. Bea and Haas (2013, p. 129 ff.).

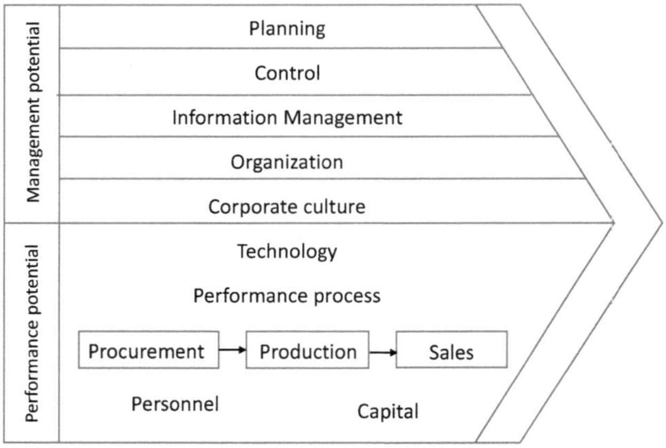

Fig. 3.101 Strategic Management approach according to value chains. *Source* slightly modified from Bea and Haas (2013, p. 129)

The individual criteria can be adopted as a basis for the company analysis and adapted to the specific needs of the company.[127] Thus, in the business analysis presented (see Fig. 3.102), company-specific subcriteria were introduced for the main criteria to allow for a more precise strategic classification and planning; e.g. for the sales area the following subcriteria were developed: product quality, brand name/protective rights, sales organization, product age, distribution channels/online, after-sales service, price/performance, delivery readiness, complaint management, market share, customer loyalty). These individual criteria are again to be prioritized and weighted. **To increase the clarity, the criteria can be hidden or displayed using the plus and minus symbols.** Some criteria (such as product age and price/performance) allow jumps (so-called triggers) to further analyses such as the product life cycle analysis (see later in this section) or the price analyses (see Sect. 3.9.3.3). They can e.g. be displayed with an underlined link like a URL address. This way, the manager has the possibility to verify his or her assessment by deeper analyses of the available cockpit data. A short comment is added for the status quo assessment. The evaluation is based on a direct competitive comparison using the point scale 1–10 (where 1 stands for very poor and 10 for very good). For the criterion technology, e.g. the performance potential can be examined for a strength or weakness in the cost or performance structure (productivity comparison). The results flow into the top indicator **competitive strength**, which, as already mentioned, also forms the basis for the MCKinsey portfolio in comparison to the market attractiveness.

The procurement of the information required for the competitor analysis poses a certain problem. Especially the Internet has an increasing importance for obtaining such information. Other sources of information for the analysis of competitors include, among

[127] Alternative strategic models are e.g. the 7-S model of the consulting firm McKinsey, which distinguishes between hard (strategy, structure, systems) and soft (skills, subordinate goals, staff, style) factors. Cf. Steuernagel (2017, p. 70 f.).

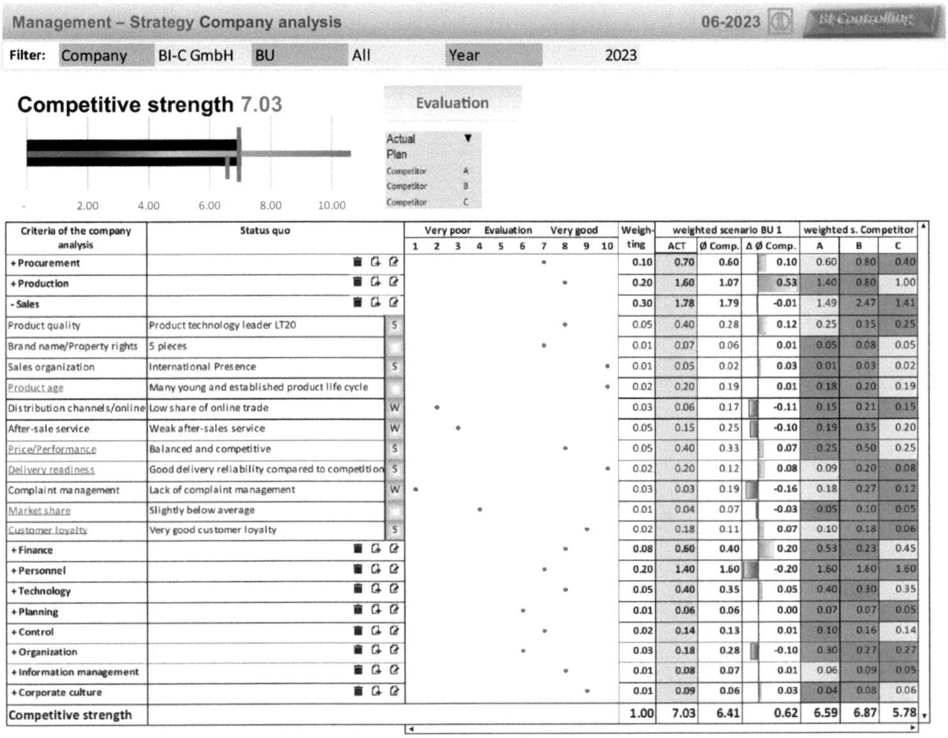

Criteria of the company analysis	Status quo		Very poor 1	2	3	4	5	Evaluation 6	7	8	9	Very good 10	Weighting	weighted scenario BU 1 ACT	Ø Comp.	Δ Ø Comp.	weighted s. Competitor A	B	C
+ Procurement									•				0.10	0.70	0.60	0.10	0.60	0.80	0.40
+ Production										•			0.20	1.60	1.07	0.53	1.40	0.80	1.00
- Sales													0.30	1.78	1.79	-0.01	1.49	2.47	1.41
Product quality	Product technology leader LT20	S								•			0.05	0.40	0.28	0.12	0.25	0.35	0.25
Brand name/Property rights	5 pieces								•				0.01	0.07	0.06	0.01	0.05	0.08	0.05
Sales organization	International Presence	S										•	0.01	0.05	0.02	0.03	0.01	0.03	0.02
Product age	Many young and established product life cycle											•	0.02	0.20	0.19	0.01	0.18	0.20	0.19
Distribution channels/online	Low share of online trade	W		•									0.03	0.06	0.17	-0.11	0.15	0.21	0.15
After-sale service	Weak after-sales service	W			•								0.05	0.15	0.25	-0.10	0.19	0.35	0.20
Price/Performance	Balanced and competitive	S								•			0.05	0.40	0.33	0.07	0.25	0.50	0.25
Delivery readiness	Good delivery reliability compared to competition	S										•	0.02	0.20	0.12	0.08	0.09	0.20	0.08
Complaint management	Lack of complaint management	W	•										0.03	0.03	0.19	-0.16	0.18	0.27	0.12
Market share	Slightly below average					•							0.01	0.04	0.07	-0.03	0.05	0.10	0.05
Customer loyalty	Very good customer loyalty	S									•		0.02	0.18	0.11	0.07	0.10	0.18	0.06
+ Finance										•			0.08	0.60	0.40	0.20	0.53	0.23	0.45
+ Personnel									•				0.20	1.40	1.60	-0.20	1.60	1.60	1.60
+ Technology										•			0.05	0.40	0.35	0.05	0.40	0.30	0.35
+ Planning								•					0.01	0.06	0.06	0.00	0.07	0.07	0.05
+ Control									•				0.02	0.14	0.13	0.01	0.10	0.16	0.14
+ Organization								•					0.03	0.18	0.28	-0.10	0.30	0.27	0.27
+ Information management										•			0.01	0.08	0.07	0.01	0.06	0.09	0.05
+ Corporate culture											•		0.01	0.09	0.06	0.03	0.04	0.08	0.06
Competitive strength													1.00	7.03	6.41	0.62	6.59	6.87	5.78

Fig. 3.102 Company analysis

others, trade fair catalogues, product descriptions, professional publications, annual reports, factory magazines, job advertisements, brochures and advertising material.[128] In the context of big data technology (see Sect. 5.8), the use of an exploratory business intelligence approach (see Sect. 5.8.3) is one option.

The criteria can be compared to the strongest competitor of each individual criterion or to the average. The sum of the weighted scores of all success potentials or success criteria results in the competitive strength. If it is related to the best competitor, the relative competitive advantage is obtained. Alternatively, comparisons can also be made to the planned value or to the average competition.

- **SWOT analysis**

The SWOT analysis developed by the Harvard Business School integrates the previously described analyses (environmental, market and company analysis) into a matrix for deriving strategies for the company and its strategic business units (SBU) or business

[128] Cf. Bea and Haas ((2013, p. 133).

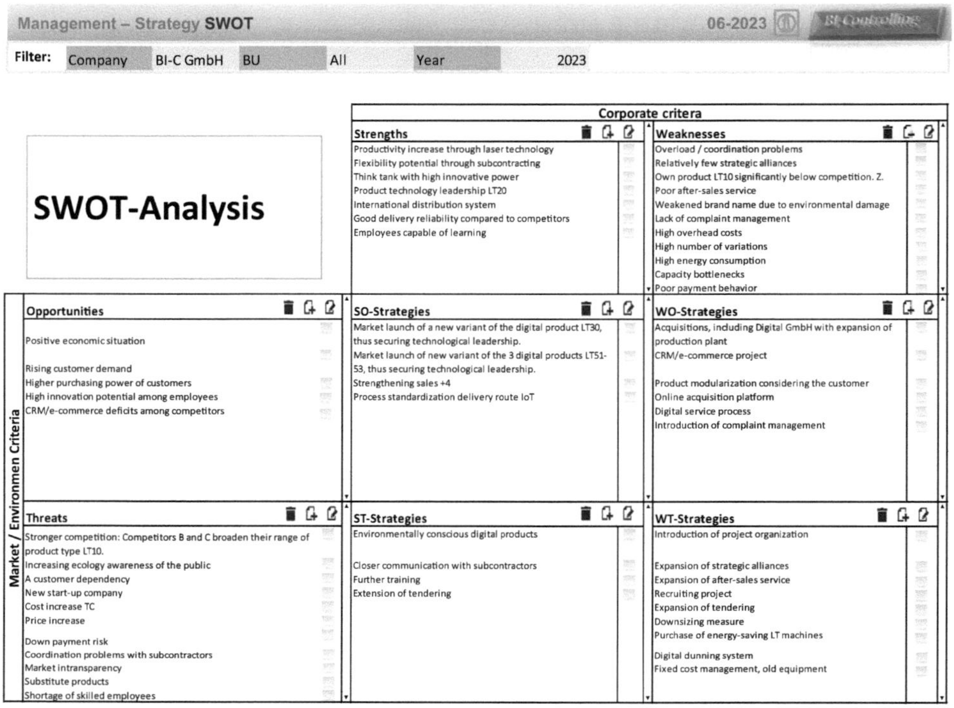

Fig. 3.103 SWOT analysis

areas (BA).[129] The analysis is called SWOT due to the initial letters of the underlying terms Strenghts, Weaknesses, Opportunities and Threats.[130] The juxtaposition of internal strengths and weaknesses and external opportunities and threats within a matrix does not only serve for clarity, but also for deriving standard strategies for individual field combinations of the matrix.[131]

To create the matrix, the opportunities and threats identified in the environmental and market analysis as well as the strengths and weaknesses identified in the company analysis are summarized and sorted and compared according to their importance for the company.[132] In the cockpit, the criteria for the SWOT analysis can be adopted and prioritized (see Fig. 3.103). The recommended standard strategies for the four resulting quadrants of the matrix are strategic directions that are intended to lead to competitive

[129] Cf. Bea and Haas (2013, p. 135) and Steuernagel (2017, p. 77).

[130] Cf. Steuernagel (2017, p. 77) and Krupp (2014, p. 101).

[131] Cf. Steuernagel (2017, p. 77) and Krupp (2014, p. 103).

[132] Cf. Krupp (2014, p. 101 f.).

advantages.[133] They are therefore a guide for strategy formulation and should be adapted to the respective company.[134] The strategic targets individually planned by the company can be adopted in the strategic planning in the cockpit (see above) and used there for the BCR-Card.

With the use of **SO strategies**, opportunities from the environment are to be exploited with the help of one's own strengths.[135] A typical example of this are expansion strategies, in which a core competence of the company, e.g. a strength in technology, is combined with the growth of a foreign market.[136]

With **ST strategies**, risks from the environment are to be fended off by using one's own strengths.[137] An example is the creation of entry barriers, e.g. by occupying distribution channels or registering patents.[138]

WO strategies aim to reduce or transform weaknesses of the company into strengths in order to perceive opportunities.[139] A typical action of this strategic field is the merging of companies through mergers, acquisitions or joint ventures, e.g. to compensate for a low presence on growth markets.[140]

WT strategies are recommendations to reduce weaknesses in order to avoid impending risks from the environment.[141] These include divestment strategies, e.g. the withdrawal from a market with low attractiveness.

- **Portfolio analysis for strategic planning**

In the context of strategic planning, portfolio analyses can be used to position and align individual and condensed influence criteria of the upstream environmental, market and business analyses for the status quo and strategic objectives.[142] The graphical representation of the portfolios belongs to the bubble diagrams which have already been explained in Sect. 3.7.7.

While the SWOT analysis can be used to analyze not only the overall company but also individual strategic business areas or units, the portfolio analysis serves the purpose of strategy formulation. It is often carried out at the corporate level with regard to the strategic business units, but it can also be applied at the business area level, for example,

---

[133] Cf. Krupp (2014, p. 101 f.) and Dillerup and Stoi (2006, p. 234 ff.).

[134] Cf. Steuernagel (2017, p. 77).

[135] Cf. Steuernagel (2017, p. 78) and Krupp (2014, p. 102).

[136] Cf. Steuernagel (2017, p. 78) and Krupp (2014, p. 102).

[137] Cf. Steuernagel (2017, p. 78) and Krupp (2014, p. 102).

[138] Cf. Krupp (2014, p. 102).

[139] Cf. Steuernagel (2017, p. 79) and Krupp (2014, p. 102 f.).

[140] Cf. Steuernagel (2017, p. 79).

[141] Cf. Krupp (2014, p. 103).

[142] Cf. Meier et al. (2002, p. 39).

for products of the business area.[143] Similar to the financial world with investment forms, one strives for an optimal mix of product-market combinations in the form of strategic business units. Individual business units may appear promising when viewed in isolation, but in the portfolio with other business units, the more or less successful positioning in the overall context becomes more visible.[144] In analogy to the financial investment portfolio, for example, the risk diversification should be preferred over the risk concentration by focusing on one industry.[145]

In the portfolio analysis, selected individual or condensed criteria of the environmental or market analysis are plotted on the ordinate, and selected individual or condensed criteria of the company analysis are plotted on the abscissa.[146] This results in four, nine or more quadrants, in which the SBU or products are positioned accordingly.[147] From the resulting portfolio, standard strategies can be derived in turn.[148] By comparing an actual and a target portfolio, a strategic need for action can be identified.[149]

One of the most well-known portfolios used in the cockpit is the market growth-market share matrix (BCG matrix) developed by the Boston Consulting Group (see Fig. 3.104).[150] The ordinate is represented by the criterion of market growth from the environmental or market analysis and the **abscissa by the relative market share from the dimension of the company analysis.**[151] The relative own market share is calculated from the ratio of the own absolute market share to the total market.[152] This allows to infer to what extent the market potential has been exploited compared to the competition.[153] The relative market share is obtained as the ratio of the own relative market share to the market share of the strongest competitor.

The cash flow is the desired target variable. Thus, products that tie up cash flow through investment activities (star and question-mark products) should be supported by products that generate cash flow (cash products).[154] In the portfolio itself, instead of the cash flow, the sales volume is often represented as an upstream indicator with the size of

143 Cf. Steuernagel (2017, p. 85 f.).

144 Cf. Bea and Haas (2013, p. 150).

145 Cf. Bea and Haas (2013, p. 150).

146 Cf. Meier et al. (2002, p. 40).

147 Cf. Meier et al. (2002, p. 40) and Bea and Haas (2013, p. 159 f.).

148 Cf. Meier et al. (2002, p. 40).

149 Cf. Meier et al. (2002, p. 40).

150 See Meier et al. (2002, p. 40) and Bea and Haas (2013, p. 161 f.) and Steuernagel (2017, p. 85 f.).

151 See Meier et al. (2002, p. 40) and Bea and Haas (2013, p. 161 f.).

152 See Eichenberg et al. (2017, p. 153 ff.).

153 See Fischer et al. (2005, p. 104 f.).

154 See Meier et al. (2002, p. 40) and Bea and Haas (2013, p. 161 f.).

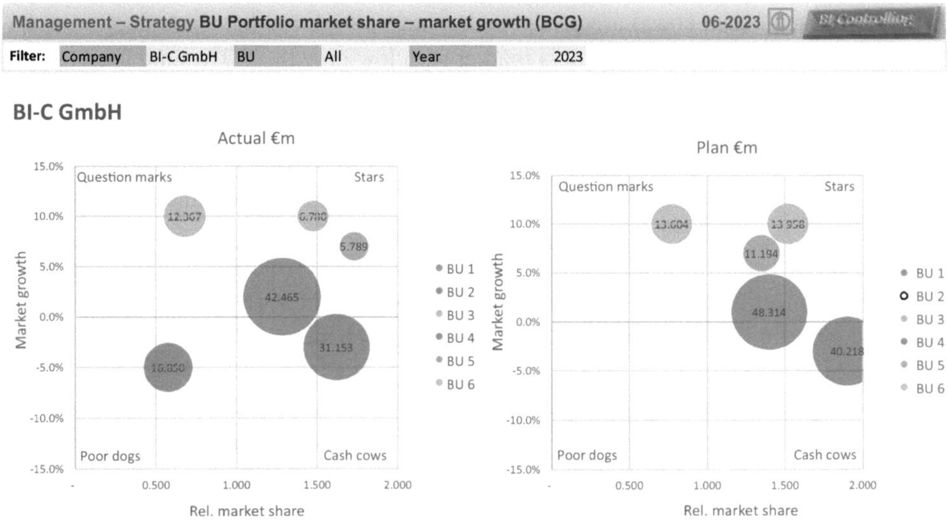

ACT	Sales million	Market growth %	Rel. market share	Rel. own market share	Rel. market share strongest competitor
BU 1	42.465	2.0%	1.284	26.7%	20.8%
BU 2	16.850	-5.0%	0.576	20.5%	35.6%
BU 3	6.780	10.0%	1.476	30.4%	20.6%
BU 4	31.153	-3.0%	1.620	35.0%	21.6%
BU 5	5.789	7.0%	1.730	35.3%	20.4%
BU 6	12.367	10.0%	0.679	15.9%	23.4%

PL	Sales million	Market growth %	Rel. market share	Rel. own market share	Rel. market share strongest competitor
BU 1	48.314	1.0%	1.400	35.0%	25.0%
BU 2	- 0.000	-5.0%	0.333	10.0%	30.0%
BU 3	13.958	10.0%	1.520	30.4%	20.0%
BU 4	40.218	-3.0%	1.900	38.0%	20.0%
BU 5	11.194	7.0%	1.350	27.0%	20.0%
BU 6	13.604	10.0%	0.773	17.0%	22.0%

Product life cycle / learning curve

Fig. 3.104 Market growth-market share matrix (BCG matrix)

the circle diameter. Four product-market combinations are distinguished, with a standard strategy assigned to each of them.[155]

Question marks are new or innovative products with low market share.[156] They are placed on growing markets and require investments to gain market share.[157] Typical standard strategies are offensive strategies or also divestment strategies in view of an impending flop.[158]

Star products (stars) are products that have a high market share on a growing market, but still tie up cash flow.[159] The recommended strategic direction is an investment strategy.[160]

155 See Bea and Haas (2013, p. 161 f.).

156 See Bea and Haas (2013, p. 161 f.).

157 See Meier et al. (2002, p. 40) and Bea and Haas (2013, p. 161 f.) and Steuernagel (2017, p. 85 f.).

158 See Bea and Haas (2013, p. 161 f.).

159 Cf. Bea and Haas (2013, p. 161 f.).

160 Cf. Bea and Haas (2013, p. 161 f.).

Cash products (cash cows) are products that have a high market share due to experience curve effects, but are located on a market with low growth.[161] They are used to generate cash flow for star and emerging products. The strategy recommended here is to use a skimming strategy.[162]

Declining products (poor dogs) are products with low market share, positioned on markets with low growth.[163] A divestment strategy, as in SGE 2, is often the best measure here.[164]

In the market growth-market share matrix, the ordinate (market growth) indirectly contains a connection to the product life cycle analysis and the abscissa (relative market share) a connection to the experience curve effect, which are available as detailed analyses in the cockpit below.

An advanced variant of the BCG matrix is the market attractiveness-competitive strength matrix (McKinsey-**Matrix).**[165] The portfolio offers the advantage, compared to the market growth-market share matrix, of taking into account different and also qualitative environmental and success factors of the enterprise in the analysis.[166] As a disadvantage, the greater data collection and planning effort has to be mentioned. The portfolio consists of nine strategic fields and takes into account the condensed key figures of market attractiveness and competitive strength, figures which result from the weighted evaluation of subcriteria, from the market and company analysis. The diameter of the circle area is backed up with a performance indicator, e.g. the sales (see Fig. 3.105) or the return on investment (ROI) as a target variable for the SBU or products.[167] The planned and actual positioning serves again to derive suitable strategic measures for achieving the objectives. Other well-known portfolio analyses are: competitive position-market life cycle matrix (A.D Little),[168] business field-resources matrix (Albach),[169] and the technology matrix (Pfeiffer et al.).[170]

- **Product life cycle and learning curve**

The product life cycle analysis can be used well for detailed analysis in the cockpit based on the multi-year data stock for the articles of the respective strategic business fields (see

[161] Cf. Bea and Haas (2013, p. 161 f.).

[162] Cf. Bea and Haas (2013, p. 161 f.).

[163] Cf. Bea and Haas (2013, p. 161 f.).

[164] Cf. Bea and Haas (2013, p. 161 f.).

[165] Cf. Bea and Haas (2013, p. 162 ff.) and Steuernagel (2017, p. 85 ff.).

[166] Cf. Bea and Haas (2013, p. 162 ff.).

[167] Cf. Bea and Haas (2013, p. 162 ff.).

[168] For the term competitive position-market life cycle matrix, see e.g. Bea and Haas (2013, p. 166).

[169] For the term business field-resources matrix, see e.g. Bea and Haas (2013, p. 167 ff.).

[170] For the term technology matrix, see e.g. Bea and Haas (2013, p. 169 ff.).

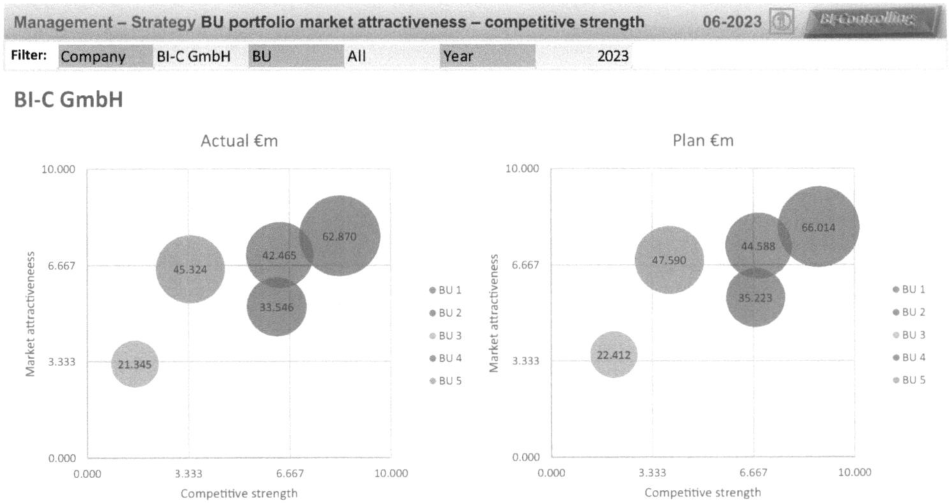

BU	ROI			Market attractivity			Strength of competition			Sales €m		
	PL	ACT	ΔPL	PL	ACT	ΔPL	PL	ACT	ΔPL	PL	ACT	ΔPL
BU 1	13.00%	10.10%	2.90%	6.340	6.710	0.370	7.525	8.123	0.598	42.465	60.342	18
BU 2	9.30%	8.60%	0.70%	5.470	5.530	0.060	4.856	4.980	0.124	33.546	42.343	9
BU 3	10.30%	12.50%	2.20%	2.965	1.200	1.765	2.456	0.000	456	21.345	0.000	-21
BU 4	7.01%	8.90%	1.89%	7.867	7.421	0.446	8.360	8.934	0.574	62.870	60.435	-2
BU 5	5.30%	8.60%	3.30%	8.467	9.487	1.020	5.342	6.929	1.587	45.324	94.210	49
Total										205.550	257.330	

Fig. 3.105 Market attractiveness-competitive strength matrix (McKinsey-Matrix)

Fig. 3.106). For this, sales, contribution margin after operating costs and development costs must be compiled. While sales and contribution margins are easy to determine, the allocation of development costs is more difficult. For this, product-related development costs would have to be separately posted, for example, to internal development projects. Based on the curve profiles, it can be seen well whether the planned profit skimming or the desired "first and follower" strategy was achieved at product launch. The learning curve indicates whether the targeted 20–30% unit cost reduction was achieved when doubling the accumulated output quantity. If the curve is too flat, this is an indication, for example, that too little effort was made in the past to achieve cost savings through economization, and improvements of handling and processes.

3.9.2.3 Comments and Actions

Comments and actions in reporting are often recorded with classic office documents for the monthly, quarterly and other reporting in addition to the individual reports. Here, the overall report is put into a content context that tells, for example, a success story. This approach is often called **storytelling**, which is summarized in a **storyboard** (overall presentation) (see Sects. 3.3 and 5.5.5.8). The software companies in the BI environment of the frontend solutions offer corresponding functions for this. Within the framework of

Management – Strategy **Product life cycle and learning curve** 06-2023

Filter: Company BI-C GmbH BU All Product gr. LT Product LT20

Articel	Turnover	Turnover %	Trend last months	Age	CM	Develop-ment cost	Δ BEP
Total	**9,769,300**	100.0%			**906,440**	**726,010**	**180,430**
LT10	2,150,000	22.0%		7.0	215000	172000	43000
LT20	1,550,000	15.9%		3.5	232500	162750	69750
LT21	1,365,000	14.0%		2.0	68250	88725	-20475
LT22	1,128,000	11.5%		1.5	33840	16920	16920
LT23	936,000	9.6%		5.5	46800	23400	23400
LT24	760,000	7.8%		6.0	76000	38000	38000
LT25	540,000	5.5%		8.0	64800	77760	-12960
LT26	390,000	4.0%		2.6	70200	35100	35100
LT27	198,000	2.0%		1.6	19800	9900	9900
LT28	182,000	1.9%		3.7	18200	36400	-18200
LT29	171,000	1.8%		3.4	17100	8550	8550
LT30	80,000	0.8%		2.4	8000	4000	4000
LT31	51,000	0.5%		1.6	5100	15300	-10200
LT32	28,000	0.3%		0.2	2800	1400	1400
...	...	...	...	...	...	...	...
VT01	60,000	0.6%		10.0	9000	4500	4500
VT02	48,400	0.5%		5.0	4840	24200	-19360
VT03	36,000	0.4%		6.5	3600	1800	1800
VT04	35,000	0.4%		4.0	3500	1750	1750
VT05	30,000	0.3%		3.5	3000	1500	1500
VT06	20,400	0.2%		0.5	3060	1530	1530
VT07	10,500	0.1%		0.1	1050	525	525

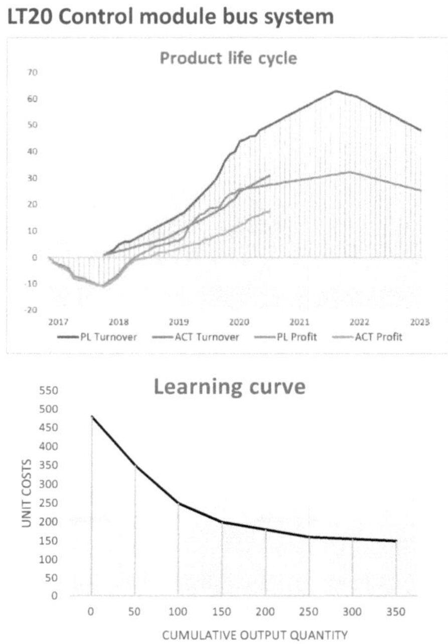

Fig. 3.106 Product life cycle and learning curve

a storyboard, report folders for recipient groups can be compiled here, for example, and supplemented with comments including action recommendations. If the software solutions do not have their own generator for the storyboard, many providers use the interfaces to the well-known Microsoft Office products (MS Excel, MS PowerPoint and MS Word) or create simple print files in the Postscript file format.

Comments and measures can be used in all areas of controlling. In the area of operational controlling, for example, the most important factors for the current and future short-term success are explained, while in strategic controlling the assessment of the status quo and the measures to achieve the strategic targets are described in the BCR-Card (see Fig. 3.107).

If the company wants to place comments and measures in the online cockpit, there are different options.

- Comments and measures can be written in tables and graphs in corresponding editable fields (see, for example, the status quo description in strategic planning in Fig. 3.98) or in separate editable notes (comparable to, for example, MS Excel). In theory, comments can be assigned to any information. A general writing and

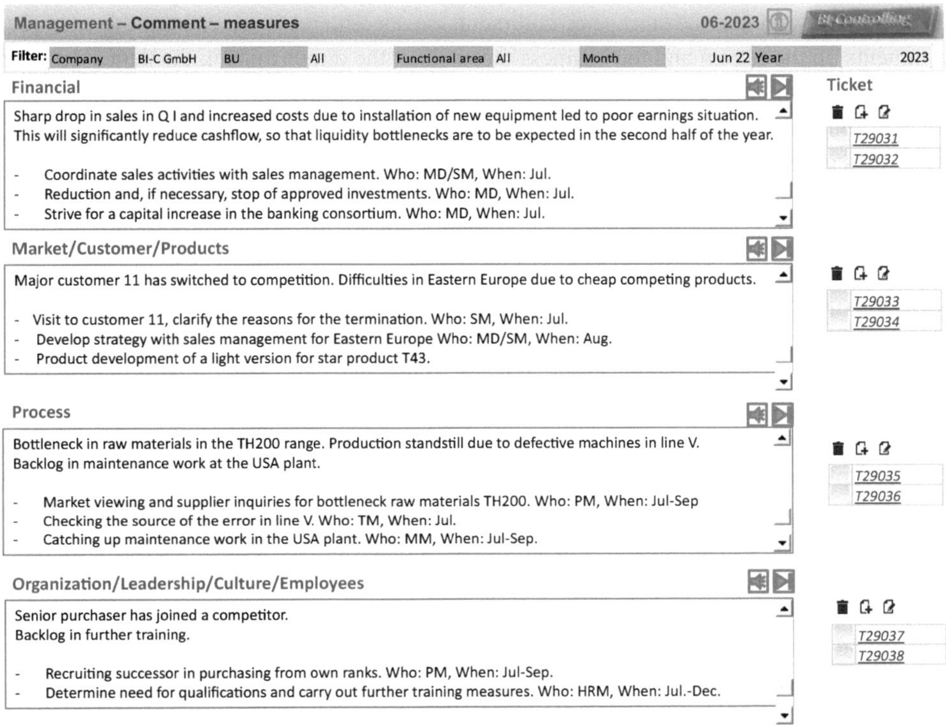

Fig. 3.107 Comments and measures

development right for the pure analysts is not desired for license cost and development reasons. Therefore, editable comments have to be created in the reports. The disadvantage here is the specification and, possibly, limitation of the fields in the data model. In practice, however, only occasional comment placements are desired, and these can change from period to period.

- Separate reports or report areas for comments are developed, which are tailored to the control in the company (see Fig. 3.107). For example, each control body, e.g. executive board or sales management circle, receives structured comment reports for its body, in which comments and measures can be recorded or read for the most important control areas.

Important questions that should be answered with comments and measures are:

- How does the responsible management assess the current situation?
- What are the causes of and reasons for the status quo?
- What developments (opportunities/risks) are expected?
- How does one assess the strengths and weaknesses of the area under consideration?
- What measures and projects are planned to achieve the desired or updated target?

Filter functions enable the storage of comments on specific result objects (in the example, comments on the region and the business unit). Here, the dimensions in the data model for the comments should be kept as low as possible in order to avoid a user overload in maintenance and search of the comments. If too many dimensions are selected for comments, either data garbage is created or many comment fields remain empty during maintenance.

It is also advisable to limit the commenting areas. This gives the user some guidance in his or her inputs, and the reader can search specifically for his or her relevant areas. For the management, perspectives based on the Balanced Scorecard, e.g., finances, customers, processes and employees, are suitable. The scroll function allows the input of longer texts. A possible disadvantage compared to office documents is the limited editing functionality in frontend applications, e.g. for fonts and tabs. Simple editing functions such as dash or bullet lists, underlines and highlights should be possible, however.

Audio and video comments can be placed via additional functions. This makes it even easier to place comments. Measures, on the other hand, should rather be recorded in writing. Contents of the task, addressee or responsible person, as well as deadlines and periods are to be documented. Ideally, a **ticket system** or **ticket management system** in which tasks can be planned and tracked is integrated at such points. The ticket system ensures the correct categorization, assignment and forwarding of tasks in a company. E-mail and chat functions are integrated. The tasks recorded in the ticket can be of any kind and concern, for example, fault reports, hazards, activities and initiatives. A tracking of the ticket task is possible via a ticket history. Affected parties are usually informed in time via e-mails etc. Status quo and changes to the ticket are transparent.

Just as with the cost analyses and individual key figure analyses, comments and measures can be used in all business areas, so that a repetitive explanation in the following sections is omitted. The categories for the comments are to be adapted to the requirements of the functional area, e.g. for

- sales: products, customers, competitors, sellers and sales organization
- purchasing: parts, suppliers, warehouse, buyers and purchasing organization
- production: plant, lines, products, units and employees
- project: projects, customers, competitors, resources and finances

3.9.3 Cockpit Reports for Sales

The sales management and sales staff responsible for controlling need, in their area for sales controlling, mainly information on successful market development. Customer orientation in sales as in marketing is in the foreground of the control in order to achieve the sales market-oriented goals as best as possible.[171] From a control perspective, it makes sense to divide the cockpit via the menu bar into the following areas:

[171] Cf. Bruhn (2010, p. 14).

- Operational controlling
- Strategic controlling
- Planning (budget/forecast/medium-term planning)

The entry area is formed by the **operational sales controlling** (see Fig. 3.108). Here, the most important current information on market development can be analyzed with key figures and management reports.

The **strategic sales controlling** comprises the strategic market analysis, the formation of strategic sales targets as well as the strategic sales planning and control. The strategic market analysis includes the areas of strategic product, customer, regional and competitive analysis as well as portfolio analysis.[172] From the status analysis, strategic targetss and plans can be derived, which, for example, result in a roadmap planning for sales. Strategic sales projects are initiated and controlled. Important key figures in strategic marketing management include, among others:

- The **market share** is a measure of competitiveness.[173]
- The **relative market share** is the market share that the company achieves in comparison to its largest competitor.
- The **brand development index** (BDI) indicates how a specific brand, in a specific customer group, compares with the average of all consumers.
- The **market penetration** is a measure of the popularity of a brand or product line and indicates how much percent of the population has bought a brand or a product of a line at least once in the observation period.
- The **customer satisfaction** is usually determined in surveys in terms of a rating.[174]
- The **market exploitation degree** illustrates to what extent the company has exploited the possible market potential.[175]
- The **market saturation degree** (also maturity) determines the ratio of market volume and market potential, i.e. the market exploitation degree of all companies in an industry. The indicator can be determined by using industry magazines, trade fairs, statistics, expert analyses, etc.[176]
- The value of a customer relationship is measured by the indicator **Customer Lifetime Value** (CLV). In connection with the CLV, the **Prospect Lifetime Value** (PLV), which shows the expected value of a prospect, is also used. The PLV gives the expected value of the customer relationship minus the costs that are necessary to acquire the

172 Cf. Ehrmann (2016, p. 168 ff.).

173 Cf. Lutz et al. (2011, p. 162).

174 Cf. Farris et al. (2007, pp. 36–63, 69–73) and Friedemann and Neumann (2007, p. 142 ff.).

175 Cf. Schneider and Henning (2008, p 233).

176 Cf. Schneider and Henning (2008, pp. 237 f.).

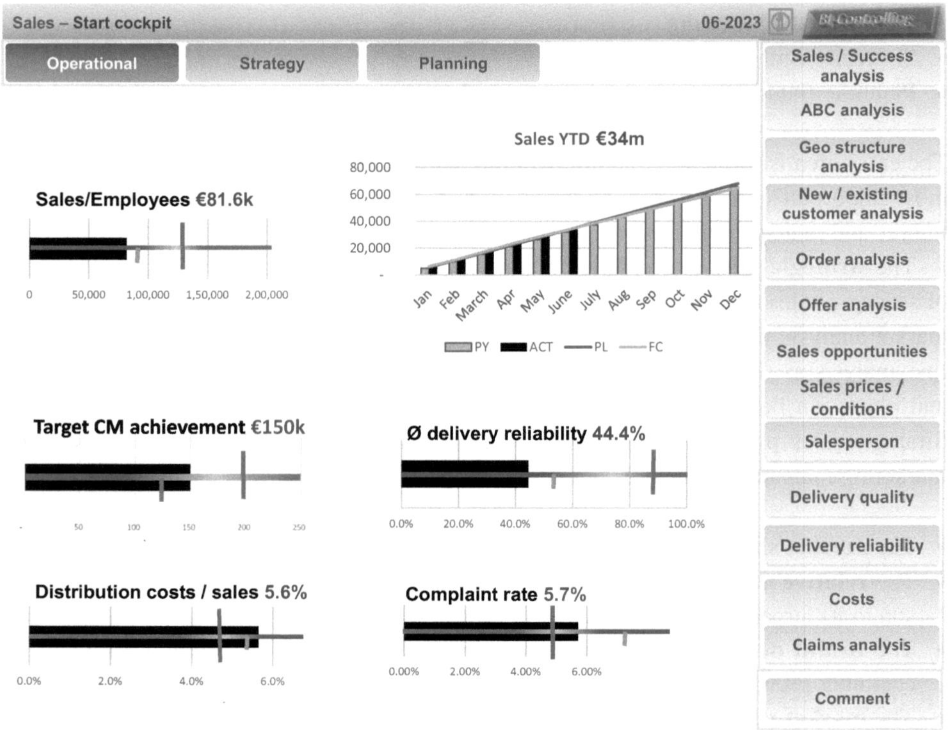

Fig. 3.108 Sales: Start cockpit

prospect as a customer. Only if the indicator is positive, the company should continue its acquisition efforts.[177]

Sales planning is the entry point into the sales-relevant parts of the budget planning, the forecast or the medium-term planning. A status and tracking report shows the current processing status of the individual planning areas and approval rounds as well as their deadlines. For the sales area, the following planning areas are particularly important:

- Sales and revenue planning (incl. conditions) (see Sect. 3.12.2)
- Planning of the special individual costs of sales (especially packaging, freight, commissions and licenses)
- Sales cost planning
- If applicable, upstream planning of order backlogs, offers and sales opportunities

[177] Cf. Farris et al. (2007, pp. 181–204).

The sales and revenue planning takes a high priority for the downstream planning areas as the starting and often as the bottleneck area of the planning. A reasonably derived resource and cost planning is only possible on the basis of a good sales planning.

The following example focuses on the operational sales controlling.

In the introductory image, the sales analyst should get a quick insight into the current situation of the market development of the company based on key figures and charts. It should cover central questions and important areas of action, such as:

- Does the company achieve the set values of the sales planning in terms of revenue, margin, contribution margin and profit?
- How are revenue, margin, contribution margins and profit developing? What is the status quo?
- With which result objects (products, customers, segments, regions, sales organization, industry etc.) do I achieve good or poor values for the sales key figures (sales volume, revenue, contribution margin etc.)?
- How high is the sales cost share?
- How punctually do we deliver to our customers?
- Are our customers satisfied with our services?

The sales cockpit also supports parts of the **marketing mix** with its four dimensions:[178]

- Product policy
- Price policy
- Communication policy
- Sales policy including distribution policy

For control purposes in marketing, however, the cockpit shown must be extended, especially by analyses in the area of communication and distribution policy. In the area of communication policy, for example, the distribution of the budget to the different advertising media or carriers and the success of different communication instruments such as media advertising or direct marketing are to be examined. In the distribution policy, for example, the analysis of the different distribution channels (internet, wholesale etc.) is in the foreground.

The key figures in the sales area are uniformly structured in accordance with the overall system. For this purpose, speedometers, tiles, column or bar charts are suitable for visualization to depict the status of the current situation. In the present example, actual values are always displayed with a black bar. Target values are shown with the help of a "blue" marker line and the previous year's value with a "gray" bar. The threshold ranges can be quickly analyzed using the traffic light color of the orientation line.

---

[178] Cf. Homburg (2017, p. 6–13) and Reinecke and Janz (2007, p. 56).

The top performance indicator in the operational sales cockpit is the performance ratio **revenue/employee**.[179] Based on the number of employees, it serves as a benchmark between companies or parts of companies. A similar ratio would be the **revenue/salesperson**. It focuses the performance more specifically on the number of salespeople in the sales area. The development of the top performance indicator **revenue** is shown graphically in an accumulated way, so that the target achievement and the forecast can be quickly grasped. Alternatively, the ratios **daily, weekly or monthly revenue** could be used, depending on the temporal control intensity of the business of the company.

The sales success can be measured by the ratio **target CM achievement**. If the realized contribution margin is significantly below the planned value, this can be threatening for the company's result. The economic use and the appropriate deployment of sales resources for the overall business can be shown by the ratio **sales costs/revenue**. The **delivery reliability** indicates in percentage how punctually we have delivered to our customers. Alternatively or additionally, the **incorrect delivery rate**[180] can signal wrongly delivered goods. The **complaint rate** expresses the (dis)satisfaction of our customers with our services.

A basic selection of detailed analysis reports for the sales area can be accessed via a menu control bar (in the example on the right). The following areas are distinguished:

- Revenue and profit analyses
- Sales opportunity, offer and order analyses
- Price/condition analyses
- Salesperson analyses
- Shipping date and quality analyses
- Receivables analyses
- Cost analyses
- Comments and actions

The areas cost analysis and comments and actions have already been dealt with in the area of the management board and are not redundantly listed below.

3.9.3.1 Revenue and Profit Analyses

Using the sales analysis (see Fig. 3.109), the result object areas and individual objects can be analyzed based on the selected hierarchy. Important hierarchies are especially the product hierarchy, the customer hierarchy, regional hierarchies and profit center hierarchies (business units). Depending on the business model, these can also be sales channels, order size classes, industry affiliation or other sales-relevant result objects such as salespeople or sales organizations, all of which can be mapped either as a hierarchy or as

[179] Cf. Gladen (2014, p. 304).

[180] For the ratio incorrect delivery rate, see Krause and Arora (2008, p. 221–222).

Sales – Sales analysis 06-2023

Filter:											
	Key figure 1	Sales	Product gr.	All	Branch	Munich	Customer type	Private	Month		June
	Company	BI-C GmbH	Customer gr.	All	Seller	Schulz	Region	BAY	Year		2023

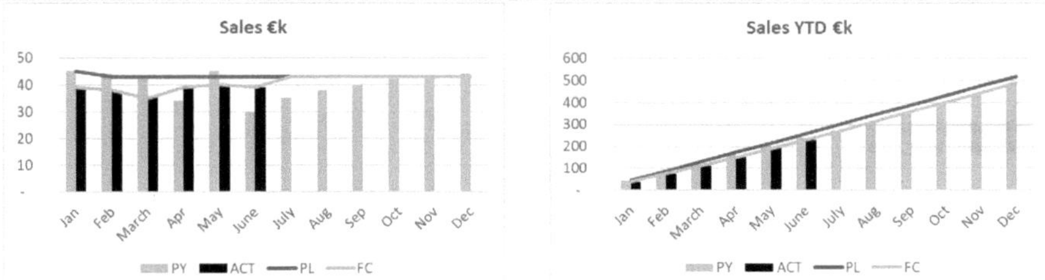

Selection hierarchy	June							YTD						
Producthierachy	PY	PL	ACT	ΔPY	ΔPL	ΔPY%	ΩPL%	PY	PL	ACT	ΔPY	ΔPL	ΔPY%	ΩPL%
+Total	**980**	**1,083**	**1,129**	**149**	**46**	**15.2%**	**104.2%**	**5,040**	**5,270**	**5,330**	**290**	**60**	**5.8%**	**101.1%**
+Productgroup A	**465**	**513**	**539**	**74**	**26**	**15.9%**	**105.1%**	**2,790**	**2,820**	**2,940**	**150**	**120**	**5.4%**	**104.3%**
+Article Group A1	**250**	**290**	**330**	**80**	**40**	**32.0%**	**113.8%**	**1,500**	**1,550**	**1,580**	**80**	**30**	**5.3%**	**101.9%**
Item A1.1	100	110	90	-10	-20	-10.0%	81.8%	600	650	720	120	70	20.0%	110.8%
Item A1.2	90	100	150	60	50	66.7%	150.0%	540	500	560	20	60	3.7%	112.0%
Item A1.3	60	80	90	30	10	50.0%	112.5%	360	400	300	-60	-100	-16.7%	75.0%
+Article Group A2	**215**	**223**	**209**	-6	-14	**-2.8%**	**93.7%**	**1,290**	**1,270**	**1,360**	**70**	**90**	**5.4%**	**107.1%**
Item A2.1	30	43	39	9	-4	30.0%	90.7%	240	260	230	-10	-30	-4.2%	88.5%
Item A2.2	100	120	100	0	-20	0.0%	83.3%	600	610	620	20	10	3.3%	101.6%
Item A2.3	85	60	70	-15	10	-17.6%	116.7%	450	400	510	60	110	13.3%	127.5%
+Product Group B	**515**	**570**	**590**	**75**	**20**	**14.6%**	**103.5%**	**2,250**	**2,450**	**2,390**	**140**	-60	**6.2%**	**97.6%**
+Article Group B1	**250**	**290**	**330**	**80**	**40**	**32.0%**	**113.8%**	**1,110**	**1,200**	**1,180**	**70**	-20	**6.3%**	**98.3%**
Item B1.1	100	110	90	-10	-20	-10.0%	81.8%	610	600	650	40	50	6.6%	108.3%
Item B1.2	90	100	150	60	50	66.7%	150.0%	250	300	210	-40	-90	-16.0%	70.0%
Item B1.3	60	80	90	30	10	50.0%	112.5%	250	300	320	70	20	28.0%	106.7%
+Article Group B2	**265**	**280**	**260**	-5	-20	**-1.9%**	**92.9%**	**1,140**	**1,250**	**1,210**	**70**	-40	**6.1%**	**96.8%**
Item B2.1	80	100	90	10	-10	12.5%	90.0%	350	400	390	40	-10	11.4%	97.5%
Item B2.2	100	120	100	0	-20	0.0%	83.3%	590	600	610	20	10	3.4%	101.7%
Item B2.3	85	60	70	-15	10	-17.6%	116.7%	200	250	210	10	-40	5.0%	84.0%

Fig. 3.109 Sales: Sales analysis

a flat list. Direct sales channels include, for example, field and inside sales, key account management and customer service.[181]

The selected key figure (here sales) can be evaluated monthly and accumulated in the uniform column scheme. Alternative key figures can be changed via the filter. The key figure sales growth compared to the previous year and the target achievement compared to the plan value are highlighted visually.

Numerous filter options allow for more detailed analyses, e.g. by organization, customer type or salespeople. If forecast periods are selected in the filter, then forecast values are analyzed.

The charts behave dynamically to the selected rows. Thus, individual items or selected item groups can be analyzed with regard to their monthly development and in their accumulated view for all value types (actual, plan and forecast).

By changing the key figure in the filter area, the sales analysis can be expanded to a profitability analysis, in which the most important profitability indicators (especially

[181] See Hofbauer and Hellwig (2016, p. 103).

Product Hierachy	Gross sales	Discounts	Discount %	Net sales	Material expences	Third-pary services	Gross profit	Gross profit %	Production costs	CM	CM %
+Total	**7,800**	**720**	**9.2%**	**7,080**	**4,130**	**875**	**2,075**	**29.3%**	**1,400**	**435**	**6.1%**
+Productgroup A	**4,000**	**360**	**9.0%**	**3,640**	**2,200**	**360**	**920**	**25.3%**	**720**	**200**	**5.5%**
+Article Group A1	**2,000**	**180**	**9.0%**	**1,820**	**1,100**	**180**	**510**	**28.0%**	**360**	**150**	**8.2%**
Item A1.1	1,000	100	10.0%	900	500	50	350	38.9%	200	150	16.7%
Item A1.2	200	20	10.0%	180	100	30	50	27.8%	40	10	5.6%
Item A1.3	800	60	7.5%	740	500	100	110	14.9%	120	- 10	-1.4%
+Article Group A2	**2,000**	**180**	**9.0%**	**1,820**	**1,100**	**180**	**410**	**22.5%**	**360**	**50**	**2.7%**
Item A2.1	900	100	11.1%	800	500	50	250	31.3%	200	50	6.3%
Item A2.2	200	20	10.0%	180	100	30	50	27.8%	40	10	5.6%
Item A2.3	900	60	6.7%	840	500	100	110	13.1%	120	- 10	-1.2%
+Product Group B	**3,800**	**360**	**9.5%**	**3,440**	**1,930**	**515**	**915**	**26.6%**	**680**	**235**	**6.8%**
+Article Group B1	**1,600**	**180**	**11.3%**	**1,420**	**800**	**130**	**410**	**28.9%**	**320**	**90**	**6.3%**
Item B1.1	800	100	12.5%	700	400	50	250	35.7%	200	50	7.1%
Item B1.2	200	20	10.0%	180	100	30	50	27.8%	40	10	5.6%
Item B1.3	600	60	10.0%	540	300	50	110	20.4%	80	30	5.6%
+Article Group B2	**2,200**	**180**	**8.2%**	**2,020**	**1,130**	**385**	**505**	**25.0%**	**360**	**145**	**7.2%**
Item B2.1	1,200	100	8.3%	1,100	600	150	350	31.8%	200	150	13.6%
Item B2.2	300	20	6.7%	280	100	135	45	16.1%	40	5	1.8%
Item B2.3	700	60	8.6%	640	430	100	110	17.2%	120	- 10	-1.6%

Fig. 3.110 Sales: Success analysis

gross margin, contribution margins, results) of the income statement can be analyzed. If you want to analyze the result contributions along the income statement in one report at a glance, the following presentation of the profitability analysis is suitable.

In the success analysis, the most important sales-relevant key performance indicators of the income statement are highlighted (see Fig. 3.110). The sales department often focuses on important conditions, net sales, gross profit and contribution margins after variable costs, all of which can be used to measure the sales success for the company. If the key performance indicators can also be reported for other hierarchies, a hierarchy selection is possible, e.g. a switch to the customer hierarchy. Important key performance indicators (here e.g. the contribution margin in percent of sales) are visually highlighted. The traffic light color uniformly indicates the target achievement in a tolerance range. The bar deflections illustrate the percentage coverage. If the hierarchy is changed to a flat object (e.g. individual products), the top or flop products can be sorted into a ranking by using a sorting function.

The top key figure shows the **target CM achievement** for the selected object. The dynamic waterfall chart shows the key figures of the value-added stages of the income statement for the selected object.

A top report for the result analysis is the dynamic ABC analysis (see Fig. 3.111).

The **dynamic ABC analysis** filters the selected key performance indicator (e.g. sales) on several selected result objects, unlike the simple ABC analysis[182]. Unlike in the filter area, the importance (ABC classification) of the objects can be used directly visually for the analysis. Due to the limitation of the screen area, four ABC analysis tables are usually offered. The most important result objects in the example are products and customers (top). They are followed by regions and salespeople (bottom). The result objects can be exchanged if necessary, e.g. for industries, sales organizations and sales channels. An alternative would be the placement of corresponding filters in the menu bar. The order of the objects is usually displayed in the ABC logic from high to low. A change is possible via the radio button from fast seller to slow seller. In addition to the color coding based on set limits for the ABC categories, spark lines help to assess the short-term trend of sales.

The division of the classes is to be done individually by the company. Usually, the following 3 classes (A, B and C) are formed:

- A-items: just 20–30% of all items account for 70–80% of total sales.
- A- and B-items: approx. 50% of all items already account for up to 90% of total sales accumulated.
- C-items: the remaining 50% of all items account for only 10% of total sales.

The dynamic ABC analysis is so well suited for analysis purposes because the selection of selected rows in any ABC analysis table adjusts the other values accordingly. For example, if you select a product in the product ABC table, all values of the other ABC tables are narrowed down to it. If you further limit the view in the salesperson ABC table to a specific salesperson, you can analyze his or her individual product sales to the selected product by customer and sales regions in the other two ABC tables. The four combinations in the analysis provide many views for market analysis. Important sales-related questions can now be easily analyzed:[183]

- In which region is a product very well accepted? In which not?
- What sales do we generate in which regions, with which products and which salespeople?
- Which customers or products are strong or weak in sales?
- Which salesperson has the most success with certain products?

[182] See for the simple ABC analysis e.g. Mehlan (2007, p. 96 ff.).

[183] Cf. Pufahl (2015, p. 208–209).

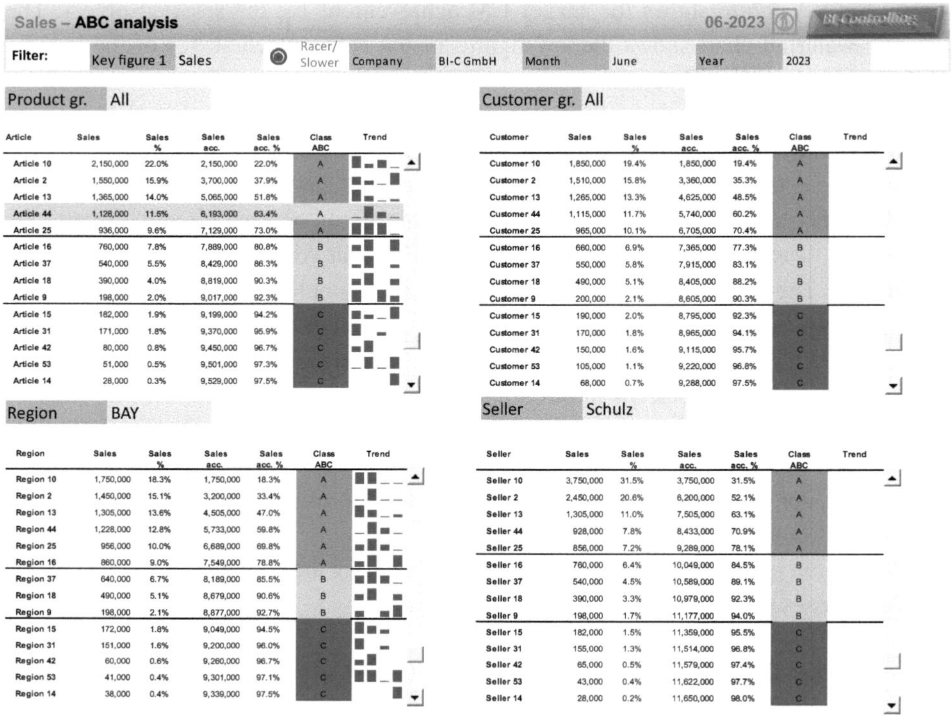

Fig. 3.111 Sales: Dynamic ABC analysis

- Which customers buy certain products most frequently in a certain region?
- In which region is a salesperson most successful with which customers?
- ….

For structural and regional considerations in sales, the geographical and structural analysis is suitable (see Fig. 3.112). Here, the key figures to be analyzed (e.g. sales and contribution margin) are displayed graphically using a map or an object plan (e.g. stadium, cinema hall). Regional differences can be quickly perceived through this. For the structural analysis, pie charts, donut, sunburst and treemap charts as well as column or bar charts are suitable (see Sect. 3.8.4). This allows large and small result object areas such as business units, order size classes, product areas, customer types, etc. to be quickly visually identified. Pie charts, donut, sunburst and treemap charts are suitable for rather few segments. As the number of objects considered increases, column or bar charts are better suitable. The color classifications of the objects should ideally be consistent across the reports. The same colors should not be used for different objects. With many result objects and classification needs, however, the color assignment across all reports is difficult to maintain.

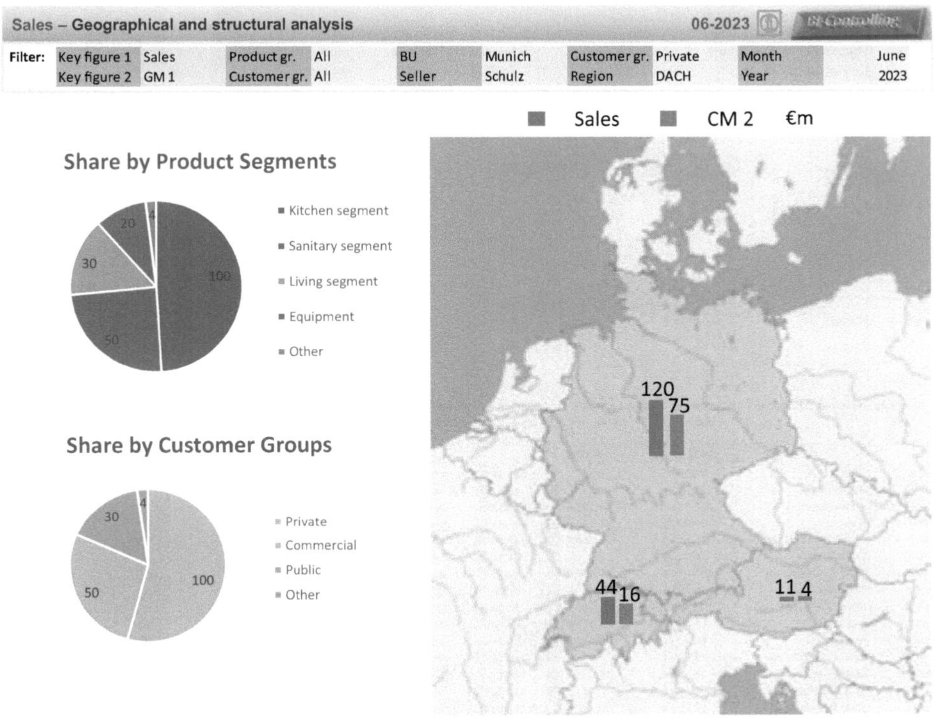

Fig. 3.112 Sales: Geo- and structural analysis

The diagrams react dynamically to the selected filters in the menu bar and to the selected result objects in the diagrams below (e.g. by clicking on a segment in the pie chart or an area/column on the map).

Another very useful analysis area for successful market development is the new and existing customer analysis (see Fig. 3.113). It answers key questions such as:

- How many and which new customers did we acquire?
- How high is the repeat purchase rate among existing customers?
- How many and which customers did we lose?
- Which customers are currently active?

The color of the bars or columns of the top diagrams shows the number of new customers (dark green) and active (light green) and passive (yellow) customers as well as lost (red) customers in the traffic light sequence. The changes in the listed customer categories can be analyzed over time in the last year, so that effects can be recognized and reasons for the loss or gain or activation of customers can be investigated. The key indicators are the **new customer rate**, the **existing customer rate** active and passive and the

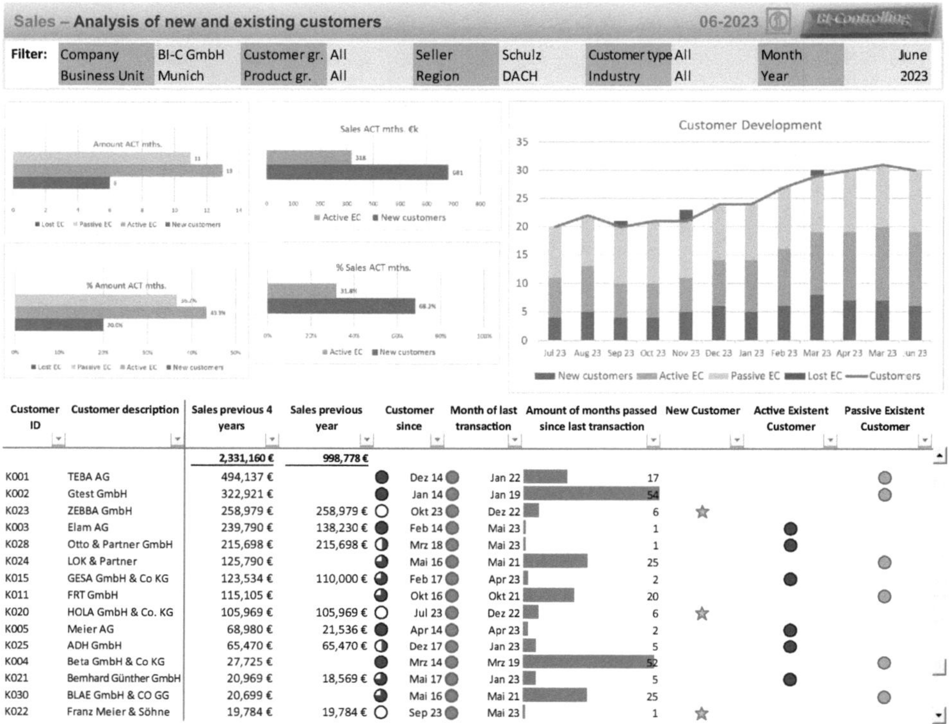

Customer ID	Customer description	Sales previous 4 years	Sales previous year	Customer since	Month of last transaction	Amount of months passed since last transaction	New Customer	Active Existent Customer	Passive Existent Customer
		2,331,160 €	998,778 €						
K001	TEBA AG	494,137 €		Dez 14	Jan 22	17			
K002	Gtest GmbH	322,921 €		Jan 14	Jan 19	54			
K023	ZEBBA GmbH	258,979 €	258,979 €	Okt 23	Dez 22	6			
K003	Elam AG	239,790 €	138,230 €	Feb 14	Mai 23	1			
K028	Otto & Partner GmbH	215,698 €	215,698 €	Mrz 18	Mai 23	1			
K024	LOK & Partner	125,790 €		Mai 16	Mai 21	25			
K015	GESA GmbH & Co KG	123,534 €	110,000 €	Feb 17	Apr 23	2			
K011	FRT GmbH	115,105 €		Okt 16	Okt 21	20			
K020	HOLA GmbH & Co. KG	105,969 €	105,969 €	Jul 23	Dez 22	6			
K005	Meier AG	68,980 €	21,536 €	Apr 14	Apr 23	2			
K025	ADH GmbH	65,470 €	65,470 €	Dez 17	Jan 23	5			
K004	Beta GmbH & Co KG	27,725 €		Mrz 14	Mrz 19	52			
K021	Bemhard Günther GmbH	20,969 €	18,569 €	Mai 17	Jan 23	5			
K030	BLAE GmbH & CO GG	20,699 €		Mai 16	Mai 21	25			
K022	Franz Meier & Söhne	19,784 €	19,784 €	Sep 23	Mai 23	1			

Fig. 3.113 Sales: New- and Existing customer analysis

fluctuation rate (lost existing customers). The table serves for the detailed analysis as customers can be sorted and analyzed according to different criteria. The sales of the last four years, the last year and the month of the last sales as well as the time of customer acquisition and the period in months since the last sales illustrate the activity or loyalty of the customer. Miniature columns, signal symbols and miniature bars help to identify the outliers immediately. In addition, new customers, active and passive existing customers are visually highlighted.

Using the filter options, the market analysis can be meaningfully narrowed down for the research purposes, for example, certain product groups, customer types, sales organizations, industries, etc. can be examined in detail.

The new and existing customer analysis requires, from the perspective of the BI-supported cockpit solution, above all the identification and classification of the new and existing customers in the source data of the CRM system. Rules must be implemented to determine when a customer is a new or existing customer (e.g. sales in the period of the last 18 months, after that existing) or when a customer is active (e.g. sales in the last year).

3.9.3.2 Sales Opportunities, Offer and Order Analyses

The ongoing sales and profit analysis must be supplemented in sales by a forward-looking analysis of the market development. Here, the key figures **order backlog**, **order intake** and **sales opportunity** are in the foreground of the analysis. Central questions are:

- How well is the employment secured in the future?
- How well is the market development by the sales and how does it evolve?
- How high is the success rate, to win prospects and offers?
- How long does the acquisition process take? How successful are the individual phases of the process?
- Why does the company lose prospects and offers?

The order analysis (see Fig. 3.114) focuses on the order end stock (**order backlog** at the end of the period), the **order intakes** and the range of the order backlog (**order range**). A calculated order backlog allows clear statements regarding the future company result. Like any stock size, the order end stock is determined by the initial stock, the inflows

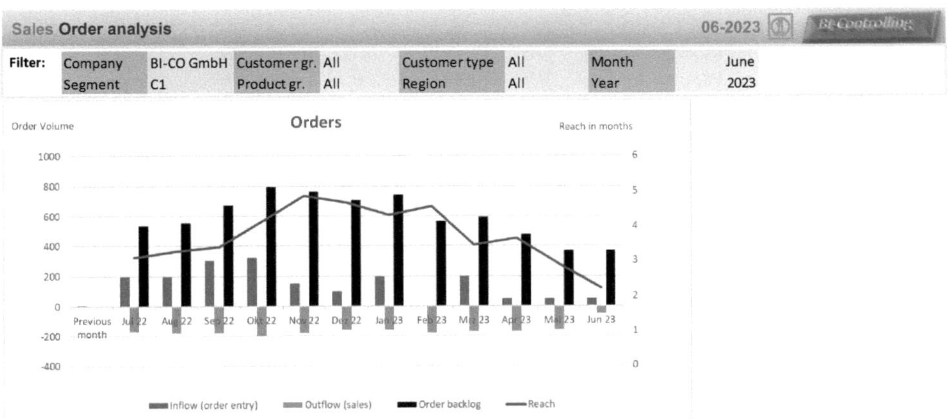

	Sales	Order entry	Order backlog	Ø Sales / Month	Reach in Months
+Total	**2,061**	**2,061**	**25,714**	**7,026**	**3.7**
+Product group A	**779**	**779**	**11,795**	**3,417**	**3.5**
+Article group A1	**373**	**373**	**6,039**	**1,631**	**3.7**
Item A1.1	200	200	3,540	811	4.4
Item A1.2	50	50	365	163	2.2
Item A1.3	123	123	2,134	657	3.2
+Article group A2	**406**	**406**	**5,756**	**1,786**	**3.2**
Item A2.1	136	136	3,100	921	3.4
Item A2.2	34	34	200	100	2.0
Item A2.3	236	236	2,456	765	3.2

Fig. 3.114 Sales: Order analysis

(order intakes) and the outflows (sales) of the considered period. The order analysis can be analyzed in the table for the selected hierarchy by products. Using the average sales per period, the range of the order backlog can be determined and displayed. The order range illustrates how well the company was able to build up business to be able to produce and sell sufficiently in the future. It is an early indicator for the secured utilization of the company.

The table analysis is supported by a dynamic diagram that visualizes the key figures around the order backlog for the selected product line. The order backlog at the end of each period is highlighted in black. The order range is shown as a line with a separate value axis. If order entries are not planned, a rolling view of a past period, e.g. a fiscal year, is useful to identify typical seasonal effects. If order entries and backlogs are planned, the forecast periods (e.g. YTE [Year to End] or the next 3–6 months) should be displayed.

If one goes back one step in the temporal analysis of market development, one arrives at the analysis of the offers (cf. Fig. 3.115).

The **offer backlog** at the end of the period is again determined as a stock indicator from the initial stock of the period, the inflows (**offer entry** = new offers) and the

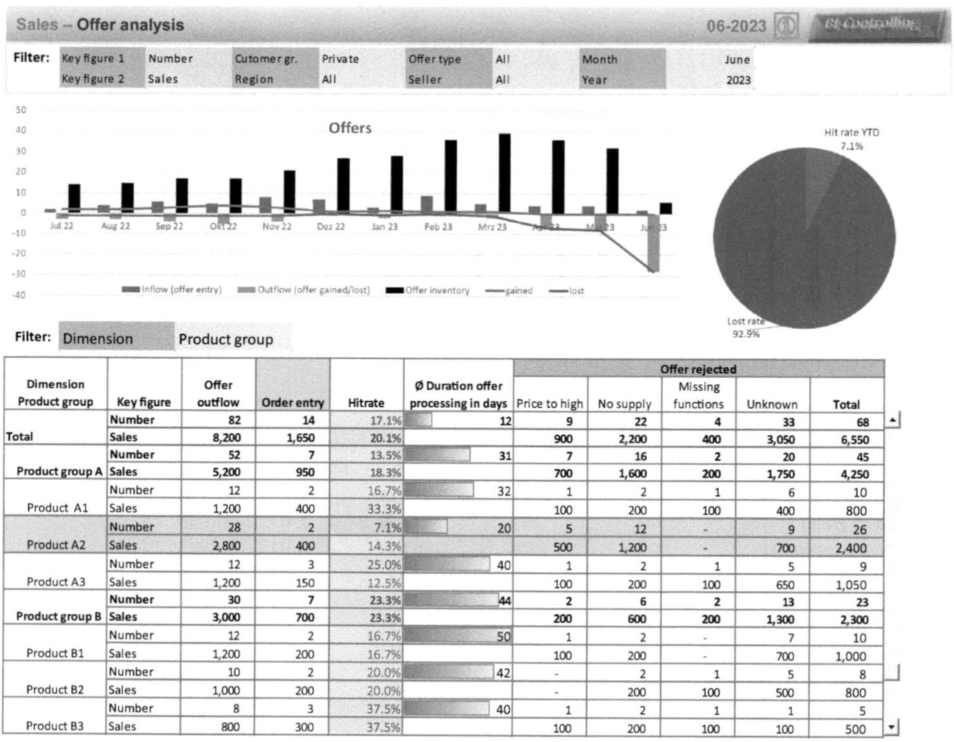

Filter: Dimension Product group

Dimension Product group	Key figure	Offer outflow	Order entry	Hitrate	Ø Duration offer processing in days	Offer rejected: Price to high	No supply	Missing functions	Unknown	Total
Total	**Number**	**82**	**14**	17.1%	12	**9**	**22**	**4**	**33**	**68**
	Sales	**8,200**	**1,650**	20.1%		**900**	**2,200**	**400**	**3,050**	**6,550**
Product group A	**Number**	**52**	**7**	13.5%	31	**7**	**16**	**2**	**20**	**45**
	Sales	**5,200**	**950**	18.3%		**700**	**1,600**	**200**	**1,750**	**4,250**
Product A1	Number	12	2	16.7%	32	1	2	1	6	10
	Sales	1,200	400	33.3%		100	200	100	400	800
Product A2	Number	28	2	7.1%	20	5	12	-	9	26
	Sales	2,800	400	14.3%		500	1,200	-	700	2,400
Product A3	Number	12	3	25.0%	40	1	2	1	5	9
	Sales	1,200	150	12.5%		100	200	100	650	1,050
Product group B	**Number**	**30**	**7**	23.3%	44	**2**	**6**	**2**	**13**	**23**
	Sales	**3,000**	**700**	23.3%		**200**	**600**	**200**	**1,300**	**2,300**
Product B1	Number	12	2	16.7%	50	1	2	-	7	10
	Sales	1,200	200	16.7%		100	200	-	700	1,000
Product B2	Number	10	2	20.0%	42	-	2	1	5	8
	Sales	1,000	200	20.0%		-	200	100	500	800
Product B3	Number	8	3	37.5%	40	1	2	1	1	5
	Sales	800	300	37.5%		100	200	100	100	500

Fig. 3.115 Sales: Offer analysis

outflows (**lost offers +order entries**). The offer analysis is performed for the selected result objects along their hierarchy (e.g. the product hierarchy). In addition to the value size sales, the number of offers is also interesting for the analysis. The success is shown by the won offers, which become order entries (marked in green). The **hit rate** (also called **conversion rate**) shows the success rate of the won offers in relation to the offers made, which is highlighted by the traffic light. The hit rate is additionally graphically highlighted by a pie chart. In the table below, well-running offer placements (e.g. for products or customers) can be identified. The average dwell time in days shows the duration of the offer processing. It is useful for the time analysis of the offer processing. For this, offer entry and exit must be recorded, which can be delayed for a long time for lost offers.

The rejection of the offers should be analyzed according to categorized reasons, which also have to be recorded. If certain reasons dominate, they have to be taken into account in the market development. For example, if the availability of the goods is poor, the inventory management has to be analyzed. For example, if the prices are too high, a price analysis is useful, based on which the pricing policy is adjusted.

The table analysis is again supported by a dynamic diagram that visualizes the most important key figures of the offer analysis for the selected line. The offer backlog at the end of each period is highlighted in black. A dilemma of the offer analysis is the data maintenance, especially for offer exits. If these are not recorded regularly, an unrealistic offer backlog builds up and the dwell times increase. If offer entries are not planned, a rolling view of a past period, e.g. a fiscal year, is useful to identify typical seasonal effects. If offer entries and backlogs are planned, something which is less common, forecast periods should be displayed.

If one goes back one more step in the temporal view of the supply analysis in the market development, one arrives at the analysis of the sales opportunities (see Fig. 3.116).

In the so-called lead management, potential customers (prospects) are identified and optimally analyzed, developed (qualified) and checked for customer acquisition in phases.[184]

The lead management process divides customer acquisition into phases which can vary depending on the business and acquisition process. In the shown example, the phases are recording, selection, planning, negotiation, offer, and order. In each phase, sales opportunities are processed. As an alternative to the complete representation of the initial stocks, inflows, outflows and final stocks, only the won or lost sales opportunities of the phase are visualized here. A "funnel"-like representation is created, as typically a reduction of the potential prospects occurs in each phase. The goal of sales is to increase the customer gains in the overall process and to do this in the shortest possible time span. Non-promising prospects should therefore be sorted out early. If the data of the inflows and outflows can be recorded in the CRM system in a timely manner, specific conclusions can be drawn and measures initiated by analyzing the won and lost sales opportunities as well

[184] Cf. BITCOM (2005, p. 1–36).

Fig. 3.116 Sales: Sales opportunities

as the average length of stay in each individual phase. The hit rate (conversion rate) of the respective phase supports the analysis of the success in the respective phase and shows the percentage success of the won sales opportunities in relation to the total number of processed sales opportunities of the period.

By changing the key figure number to sales, the analysis can be usefully varied. The sales show the value-based customer success compared to the number. What matters are the right customers who should be won. Detailed analyses of market development can be done by restricting the filter options, for example by differentiating regions, salespeople, sales organizations, customer groups and product groups.

3.9.3.3 Price and Condition Analyses

An important task in market development is the pricing and conditions policy. Important questions here are, for example:

- What are the prices at the moment?
- How will the prices develop?
- How large are the deviations from achieved prices to list prices?
- How do the prices change depending on reference prices?

- What conditions can we grant and still be successful?
- Which customers get the most revenue reductions? With which customers do we achieve the best business after deducting the revenue reductions?

To this end, it makes sense to integrate price and conditions analysis in the cockpit (see Fig. 3.117).

The sales prices report shows the price development of selected product-market combinations, in which the articles can be restricted by customer, region, etc. using the filter area. For the sales article, the revenue, the average net sales price and the quantity are displayed. Deviations from the previous year can be analyzed in terms of price, quantity and also total revenue deviations. Visually, the bar lengths of the respective deviation type help here.

The price analysis is again supported by dynamic charts in which the monthly development of gross and net prices to the market average and list price are displayed. Reference prices of important basic materials are compared with the price development. This allows reference price-related changes to be better anticipated and forecasted.

The contribution to success of a company is also influenced by the often neglected conditions policy. Granted discounts, cash discounts and rebates directly reduce the profit, so that here the lever for influencing the result, similar to purchasing materials, is of special importance for the company.[185]

The conditions analysis shows, in its simplest stage, the revenue reductions granted by the customer (see Fig. 3.118). Starting from revenues based on the gross sales price (gross price), the net revenues are shown by deducting the discounts and the net-net revenues by deducting the cash discounts and rebates. If the analysis is extended by special individual costs of distribution, conditions such as customer freight, packaging and commission can be analyzed, for example, with regard to a contribution margin after direct distribution costs. Which customers or customer groups get the largest reduction can be read directly from the deviation of the net-net revenue from the gross revenue.

The dynamic charts allow the analysis of the discount, cash discount and credit rating as well as the absolute change of the conditions in the waterfall chart for each selected result object, e.g. a customer by line selection. Complemented by the filter function, targeted analyses of the result objects in distribution (e.g. salespeople, sales organizations, customer types and regions) can be performed.

3.9.3.4 Seller Analyses

The performance of the sales department to the customer depends on the field service and its salespeople. Due to the direct impact on the employment situation of the company, the sales department must be supported in a targeted way. The seller analysis

---

[185] See image rights: [Copyright by the author]
Hermann (2015, p. 34 ff.). Reproduction rights: Not necessary.

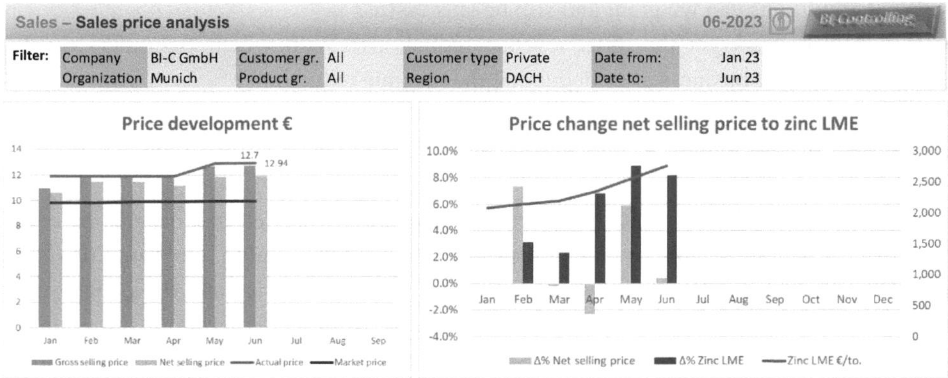

		Jan - Jun					
		PY	ACT	ΔPY	ΔPY% Price	ΔVJ% Quantity	ΔVJ% Sales
+Total	**Sales**	**7,759.68**	**8,262.50**	**502.83**			**6.5%**
+Productgr. A	**Sales**	**3,103.87**	**3,305.00**	**201.13**			**6.5%**
Article 1	Ønet price	10.89	9.48	-1.41	-12.9%		
	Quantity	10	15	5		50.0%	
	Sales	109	142	33			30.6%
Article 2	Ønet price	11.93	12.49	0.56	4.7%		
	Quantity	115	140	25		21.7%	
	Sales	1,372	1,749	377			27.5%
Article 3	Ønet price	115.93	117.85	1.92	1.7%		
	Quantity	14	12	-2		-14.3%	
	Sales	1,623	1,414	-209			-12.9%

Fig. 3.117 Sales: Sales price analysis

offers, especially in combination with the reports of the success analysis (especially the dynamic ABC analysis and the geographical and structural analysis), starting points for an improved market development.

The seller analysis shows the monthly and accumulated sales performance of the salespeople and their development in the current year (see Fig. 3.119). For the performance measurement, the user selects a key figure, e.g. the turnover or the contribution margin. Comparisons with the previous year and the trends of the last months are displayed with miniature bar or column charts. The dynamic charts change when the salesperson selection is changed per row. If target values are recorded for the salespeople, the report can be extended by the target achievement. Often, it is not a monthly planning, but a quota specification which is shown in the charts above as a reference line for the target achievement. The grouping of the salespeople in a hierarchy allows a graded analysis, in which, for example, sales groups, teams or sales organizations can be examined.

The sales management has here a tool at its disposal which should be used very sensitively. Managers receive very individual information here which can be used wrongly. For example, if the salespeople are constantly put under pressure to achieve predefined goals, they may fear not being able to cope with the pressure. Resignation, illness or staff

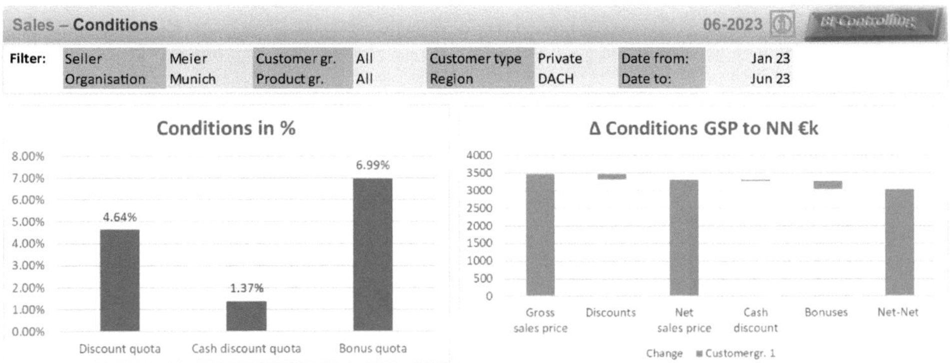

€k	Gross sales price	Discounts	Net sales price	Cash discount	Bonuses	Net-Net	ΔNN-GSP%
+Total	**8629**	**370**	**8259**	**122**	**462**	**7675**	**-11%**
+Customer group. I	**3465**	**161**	**3304**	**45**	**231**	**3028**	**-13%**
Customer 1	727	66	661	13	33	614	-15%
Customer 2	1701	50	1652	17	165	1470	-14%
Customer 3	335	25	310	6	16	289	-14%
Customer 4	160	5	155	3	8	144	-10%
Customer 5	216	6	209	4	6	199	-8%
Customer 6	108	3	105	2	3	100	-8%
Customer 7	89	3	86	0	0	86	-3%
Customer 8	52	2	50	0	0	50	-3%
Customer 9	78	2	76	0	0	76	-3%

Fig. 3.118 Distribution: Conditions

turnover can be a consequence. This is not in the interest of either the employee or the company. This applies of course to all areas of the company. Instead, the tool should be used individually in a motivating and supportive way. For example, the following question patterns could be helpful:

- Which salespeople are currently very successful? By asking questions, sensitive market information and market development techniques can be identified and multiplied.
- Which products are doing very well with the top salespeople? This way, the salespeople can be directed to sensitive customer needs.
- Which customers are doing very well with the top salespeople? This way, the salespeople can be directed to sensitive industries and customer types.
- Which regions are doing very well? This way, the resources can be better positioned for regional reasons.
- How large is the overall target deviation in the sales organization? In the team, one can motivate and encourage better to achieve or even exceed goals.

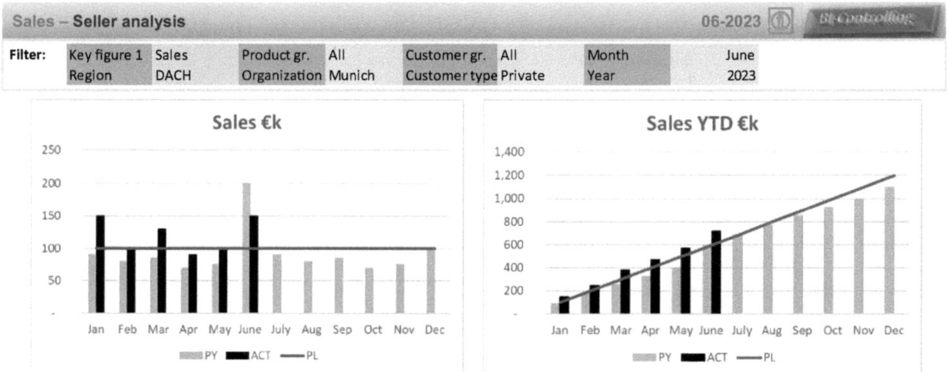

	June				YTD				
Seller groups	PY	ACT	ΔPY	ΔPY%	PY	ACT	ΔPY	ΔPY%	Trend
+Total	**870**	**1,000**	**130**	**14.9%**	**3,000**	**3,160**	**160**	**5.3%**	
+Seller group A	**420**	**430**	**10**	**2.4%**	**2,640**	**2,860**	**220**	**8.3%**	
Seller A1	100	80	-20	-20.0%	1,500	1,580	80	5.3%	
Seller A2	200	150	-50	-25.0%	600	720	120	20.0%	
Seller A3	120	200	80	66.7%	540	560	20	3.7%	
+Seller group B	**450**	**570**	**120**	**26.7%**	**360**	**300**	-60	**-16.7%**	
Seller B1	100	160	60	60.0%	840	850	10	1.2%	
Seller B2	150	180	30	20.0%	240	230	-10	-4.2%	
Seller B3	200	230	30	15.0%	600	620	20	3.3%	

Fig. 3.119 Sales: Seller analysis

3.9.3.5 Shipping and Delivery Quality Analyses

The satisfaction of customer needs is only achieved when the customer receives his goods on time in his expected quality and he is satisfied with the product or service in the long term. The reports delivery reliability in shipping (see Fig. 3.120) and delivery quality (see Fig. 3.121) can be used to check these requirements. Central questions here are:

- Could the delivery time to the customer be met?
- How high was the proportion of incorrect deliveries?
- How long do our deliveries take?
- How satisfied or dissatisfied are our customers? How high are the returns, complaints and credits?
- For what reasons are customers dissatisfied?

The delivery reliability as a key indicator shows how many orders were delivered on time and met the requirements of the customer order. Here, the four “Rs” (right product, right time, right quantity, right place) and possibly also the right quality - as the fifth of the 7

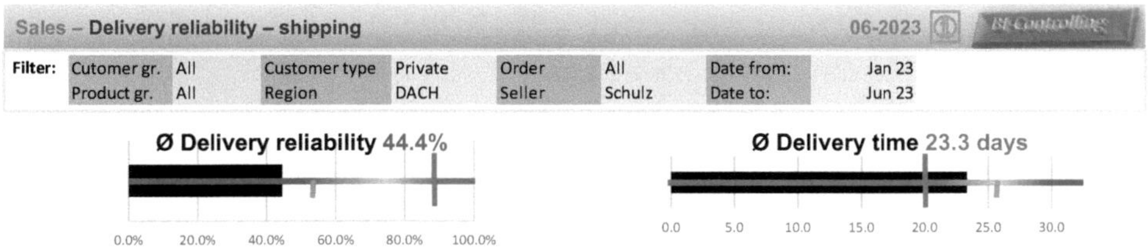

	Order receipt	Desired date	Confirmed date	Delivery date	Delivery in days	Delivery delay in days
Average					**23.3**	**-2.3**
Order 1	03.01.2023	10.01.2023	31.01.2023	06.02.2023	34	-6
Order 2	04.01.2023	20.01.2023	25.01.2023	24.01.2023	20	1
Order 3	05.02.2023	20.02.2023	25.02.2023	28.02.2023	23	-3
Order 4	06.03.2023	25.03.2023	28.03.2023	24.04.2023	49	-27
Order 5	04.03.2023	23.03.2023	28.03.2023	24.03.2023	20	4
Order 6	10.03.2023	20.03.2023	25.03.2023	28.03.2023	18	-3
Order 7	12.03.2023	14.03.2023	01.04.2023	04.04.2023	23	-3
Order 8	04.04.2023	17.04.2023	28.04.2023	24.04.2023	20	4
Order 9	06.04.2023	20.04.2023	25.04.2023	30.04.2023	24	-5
Order 10	14.04.2023	25.04.2023	20.05.2023	15.05.2023	31	5
Order 11	16.04.2023	28.04.2023	30.04.2023	05.05.2023	19	-5
Order 12	04.05.2023	20.05.2023	25.05.2023	25.05.2023	21	0
Order 13	08.05.2023	20.05.2023	28.05.2023	24.05.2023	16	4
Order 14	14.05.2023	26.05.2023	26.05.2023	04.06.2023	21	-9
Order 15	25.05.2023	10.06.2023	15.06.2023	14.06.2023	20	1
Order 16	10.06.2023	20.06.2023	25.06.2023	24.06.2023	14	1
Order 17	10.06.2023	20.06.2023	25.06.2023	open		-5
Order 18	10.06.2023	20.06.2023	25.06.2023	open		-5

Fig. 3.120 Sales: Delivery reliability - Shipping

"Rs" of logistics - are checked.[186] The delivery reliability can also be analyzed in more depth by identifying, for example, only those orders that cannot be delivered on time due to insufficient inventory levels. The **delivery time** and the **delivery delay** of individual orders can be examined in detail according to different result objects. The visual highlighting of the miniature bar charts makes it easy to identify those orders that were delivered with a high delivery delay or those that are still open. With the help of the sorting functions, delivery delay and delivery time can be displayed in a ranking order, so that problem cases become directly visible. In the sum line above, the average delivery times and the average deadline delay are displayed. By limiting the result objects (e.g. customers, products, regions, order types, salespeople) in the filter menu, the sales department gets a powerful analysis tool related to the shipping performance. Critical open orders can be identified and requested immediately.

[186] The sixth "R" (right cost) is checked by analyzing the logistics cost centers or the direct freight, handling and packaging costs of the order. The seventh "R", the right supplementary information to the product, can also be integrated. Synonymously, the indicator delivery reliability in its broadest form is also defined as delivery readiness. See Reichmann et al. (2017, p. 424).

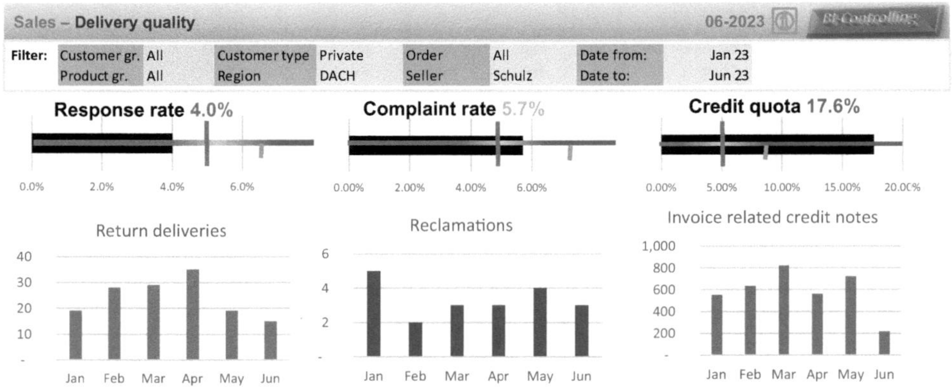

	Quantity			Amount		Sales			Amount
	Delivery	Withdrawal	Return rate	Cancellation delivery note	Complaint	Invoice	Credit note	Credit note rate	Cancellation invoice
Total	**3,610**	**145**	**4.0%**	**16**	**20**	**19,900**	**3,500**	**17.6%**	**17**
Customer 1	100	10	10.0%	5	10	2,000	200	10.0%	5
Customer 2	200	-	0.0%	-	-	4,000	-	0.0%	-
Customer 3	400	10	2.5%	1	-	5,600	200	3.6%	1
Customer 4	200	20	10.0%	1	1	3,500	200	5.7%	1
Customer 5	250	30	12.0%	2	3	2,300	1,250	54.3%	2
Customer 6	150	-	0.0%	-	-	2,500	-	0.0%	-
Customer 7	350	10	2.9%	1	-	5,600	200	3.6%	1
Customer 8	360	20	5.6%	1	1	3,500	200	5.7%	1
Customer 9	230	30	13.0%	2	3	3,500	1,250	35.7%	2
Customer 10	150	-	0.0%	-	-	2,500	-	0.0%	4
Customer 11	150	-	0.0%	-	-	2,500	-	0.0%	-
Customer 12	350	-	0.0%	1	-	5,600	-	0.0%	-
Customer 13	360	5	1.4%	1	1	3,500	-	0.0%	-
Customer 14	360	10	3%	1	1	3,500	-	0.0%	-

Fig. 3.121 Sales: Delivery quality

Compliance with delivery quality is another important criterion for successful market penetration for companies (see Fig. 3.121). On the one hand, a high quality is a guarantee for new customers and the consolidation of existing customers. On the other hand, unnecessary costs of return and subsequent delivery, rework, scrapping, re-production etc. are avoided. Ideally, the company strives for a zero-defect tolerance.

Whether the customer is actually satisfied with the quality of the product can only be determined by direct questioning or usage verification, which often lies beyond the reach of the company's information. Therefore, indirect indicators are useful for quality testing to show how dissatisfied the customers are. At first glance, the peak indicators **return rate**, **complaint rate** and **credit note rate** can be analyzed in a planned-actual-previous year comparison as well as for the last months in their development.

The return rate measures the returns to the number of deliveries. Returns can be motivated in different ways, so it makes sense to analyze the reasons for them: are they quality-related or error-related customer complaints, cancellations of delivery notes by the customer (wrong deliveries, exchange etc.) or self-initiated returns of the company (recall actions, wrong labeling etc.). The value of the returns and the amount of the credit notes indicate the economic importance of the delivery quality. Visual miniature bar charts make it easy to identify those customers where the delivery quality could not be met. Here, for example, quality needs of sensitive customers can be researched, so that the reasons for the quality deficiency can be countered accordingly.

3.9.3.6 Receivables Analysis

The market processing is only successfully completed when the customer pays for the product or the service received. This closes the 360-degree view of the customer. The receivables analysis shows the payment behavior of the customers, the due date of receivables, and the reminder activities of the already due items (see Fig. 3.122). Important questions of the receivables analysis are:

- When do our customers pay?
- Which customers pay on time, which do not?
- Which receivables are due for how long?
- Which open receivables are critical?
- Which measures have already been initiated to collect the receivables?

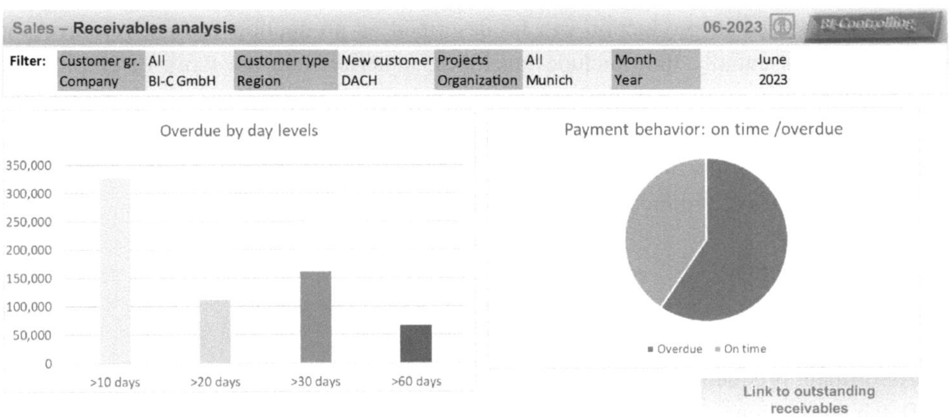

Receivable					Payment		Overdue					
R-Type	Invoice ID	Date	Due Date	Value	Payment	On time	Open	thereof: >10 days	>20 days	>30 days	>60 days	Reminder level
+Total				1,303,247	594,124	469,560	688,537	328,934	112,464	160,621	66,616	
+Customer 1				42,150	5,000	5,000	7,000	15,752	0	14,398	7,000	
Down payment	1001-2345	01.04.2023	15.04.2023	12,000	5,000	5000	7,000				7,000	3
Partial final invoice	1001-2346	01.05.2023	15.05.2023	14,398						14,398		1
Final invoice	1001-2347	01.06.2023	15.06.2023	15,752				15,752				
+Customer 2				73,894	49,564	30,000	33,894	0	0	24,330	9,564	
Partial final invoice	1001-3457	14.04.2023	28.04.2023	9,564	9,564		9,564				9,564	3
Final invoice	1001-3458	05.05.2023	19.05.2023	64,330	40,000	30,000	24,330			24,330		
Customer 3				118,323	39,560	34,560	78,763	56,974	0	21,789	0	
Down payment	1001-4581	17.03.2023	16.04.2023	24,560	24,560	24,560						
Partial final invoice	1001-4582	26.04.2023	26.05.2023	36,789	15,000	10,000	21,789			21,789		2
Final invoice	1001-4583	25.05.2023	24.06.2023	56,974			56,974	56,974				
Customer 4				68,360	0	0	68,360	56,000	12,360	0	0	
Down payment	1001-2415	25.05.2023	08.06.2023	12,360			12,360		12,360			
Final invoice	1001-2416	30.05.2023	13.06.2023	56,000			56,000	56,000				

Fig. 3.122 Receivables analysis

In the overview, the user receives the sum of the due receivables by day classes and proportionally the ratio of the due and timely payments for the filtered result objects.

In the table, the customers can be analyzed in detail along the customer hierarchy with regard to their payment behavior. Issued receivables are displayed by their type (advance payment, partial final invoice, final invoice), their invoice date, their due date and their value. Incoming payment statuses can be analyzed by their timely receipt and due receivables by predefined day classes, e.g. longer than 10, 20, 30 or 60 days. For due receivables, the activities of the reminder system (e.g. payment reminders, reminders by stages) are displayed.

For a direct processing of a customer invoice, a link to the open item accounting in the financial accounting software can be activated. The default settings of the filtering and customer selection are passed on, as far as the pre-system allows.

3.9.4 Cockpit Reports for Production

Production is the focus of value creation in a manufacturing company and thus a decisive success factor.[187] The term "production" refers to the fabrication, manufacture and generation of products within a company by using goods and services.[188] Goods and services that are used in production can be described as production factors, resources or input goods; essentially they include machines, materials and the work of employees. The products that result from the transformation process in production are, among others, finished goods and output goods and can be accompanied by downstream services such as servicing (inspection, maintenance and repair) and spare parts business.[189]

Important types of production in industrial manufacturing are:[190]

- Single production
- Customer order production
- Mass production
- Assortment production
- Batch production
- Variant production
- Lot production

While the single and, partly, the customer order production such as shipbuilding, large-scale plant construction, civil engineering, underground construction etc., are partly

[187] Cf. Gottmann (2016, p. 1–18).

[188] Cf. Bloech et al. (2014, p. 3).

[189] Cf. among others Fandel (2007, p. 1).

[190] Cf. on the types of production among others Weber et al. (2014, p. 177).

controlled in a project-like manner, the other types of production are characterized by a factory control with established production processes and a larger number of similar end products; these will be the main focus here. For project control, compare the Sect. 3.9.6 on the project cockpit.

In a production plant, there can be different types of organization in the production processes:[191]

- Point production
- Job shop production
- Flow production
- Group/cell production

As in other areas within a company, multiple targets are also pursued in production. The targets of production management are also derived from the corporate targets. The main targets of production include, in particular, the cost targets, time targets and result targets, as well as their sub-targets such as the utilization for the costs and the productivity and quality for the result.[192] Furthermore, environmental protection and occupational health and safety targets are to be pursued or complied with.

Within the scope of the **cost targets**, the main target is to keep the production costs as low as possible.[193] The economic principle is applied, which states: With the given production factor (input), the highest possible output should be achieved—maximum principle—or the given output should be produced with the lowest possible input—minimum principle.[194] High cost degression effects can be achieved for capacities if, for example, large quantities can be produced without having to constantly perform setup processes in which no output can be achieved.[195] A high **utilization** of the capacities leads to a better use of the invested costs. Idle costs should be avoided. For this reason, the plant and resource utilization and their availability are of particular importance for meeting the cost target.

With **time targets** in production, the aim is to produce the planned production quantity on time in order to deliver to the customer on schedule. In order to meet deadline targets and avoid shortage costs as much as possible, low lead times of the materials and components in the respective production processes as well as low waiting and lying times between the production processes are necessary. With the help of large capacities, larger quantities can be processed in larger production lots. However, due to the strong

---

[191] Cf. on the types of organization in production, among others, Opresnik and Rennhak (2012, p. 420) and Weber et al. (2014, p. 179).

[192] Cf. Steven (2016, p. 7).

[193] Cf. Herrmann (2010, p. 83 f.).

[194] Cf. Bloech et al. (2014, p. 10).

[195] Cf. Herrmann (2010, p. 84).

product differentiation, the orders of different products compete for the scarce capacities, and setup processes are necessary between the orders. Lead times and capacity utilization are thus partially competing targets. In the sequence planning of the orders in the PPS or ERP systems of the companies, an optimum for all targets of production has to be worked out.

The **objectives** in production lie in the **productivity** and **quality** of the produced goods. The productivity of a company can be measured by the ratio of the produced quantities to the input production factor.[196] To remain competitive in the market, a company must also convince customers with the quality of the available products.[197]

At the forefront of production controlling. and thus also of the production cockpit, there is the comprehensive information supply of the production management for decision making, whose executives are presented below.[198]

In production, there are different executives who take on different roles and responsibilities in production management and are potential users of the production cockpit. These include the plant management, the factory management, the production management, the work preparation, the production foremen, and the supervisors.[199]

The plant manager is the highest leader in production within a company. The tasks of the plant managers vary depending on the company size. Within smaller companies, plant managers are responsible for the management and development of the entire production of a company. In larger operations, there may be separate factory managers for different plants, for example.

The production manager and executives of the work preparation are responsible in production for the economic and technical planning, execution, monitoring, and optimization of the production processes with regard to quality, quantity, costs, and deadlines.

The task of the production manager is the planning of the raw, auxiliary and operating materials that are necessary for production. Here, not only the quantities or the capacity of the production factors are determined, but also the order of the use of the production factors. The production manager monitors the production based on the production data. This ensures quality and productivity, and corrective measures can be initiated, if necessary to continuously improve quality and productivity or avoid waste. In addition, the production manager is responsible for monitoring the environmental protection regulations and ensuring occupational safety.

Foremen or production foremen are technical specialists within an area in production.[200] They are subordinate to the production management and responsible for their own department in production. The main tasks of the production foremen lie above all in

---

[196] Cf. Weber et al. (2014, p. 174).

[197] Cf. Herrmann (2010, p. 84).

[198] Cf. Sejdic (2019, p. 58).

[199] Cf. Ibarra (2005, p. 18) and Haman (2013, p. 19).

[200] Cf. Bühner (2005, p. 216).

the execution, control and monitoring of the production processes.[201] This includes the planning of the material and personnel deployment.

The supervisor is on the lowest level of management in production. As a leader in production, the supervisor is the liaison person between the production foreman and the employees of a group. Depending on the management level, only certain reports and data manifestations in the production cockpit are relevant, and there is a corresponding data access in connection with the access rights. The plant managers usually have complete access to the production data.

From a control perspective, it makes sense again to divide the cockpit into the following areas via the menu bar or navigation:

- Operational controlling
- Strategic controlling
- Planning (budget/forecast/medium-term planning)

The entry area is the operational production controlling. Here, the most important current information on production can be analyzed with key figures and management reports.

The **operational production controlling** is short-term oriented and has the task of ensuring the profitability of the production processes.[202] With the help of various instruments, it aims to implement the strategic production management as best as possible. In contrast to strategic production controlling, operational production controlling is primarily not about effectiveness, but rather about efficiency. Therefore the following aspects are in the foreground: information on cost behavior, adherence to deadlines, productivity, quality assurance, capacity utilization, compliance with environmental protection regulations and occupational health and safety, as well as in connection with the production program on achieved contribution margins. The main tasks of operational production controlling include, among others:[203]

- analysis and control of production performance (quantitative and qualitative)
- analysis and control of production costs
- analysis and control of production orders with regard to process selection, cycle time, lot size and deadline compliance
- analysis and control of capacities (production facilities/machines), as well as their maintenance measures

[201] Cf. Haman (2013, p. 83).
[202] Cf. Steven (2016, p. 9).
[203] Cf. Schnell (2018, p. 15).

- analysis and control of performance especially of the times of the production employees
- analysis and control of inventories for production

Important questions in operational production controlling are:

- How high is the productivity in the entire production and in its subareas? How high are contribution margins and costs? How high are resource use (input) to output quantities (output)?[204]
- How are the plants and capacities utilized?
- Can production be done on schedule? How can production and logistics processes be optimally designed and monitored based on delivery dates, production times (manufacturing and setup times), storage and waiting times, order progress, process alternatives, etc.?
- How effectively are the employees in production deployed?
- How high are the production costs, and where are potential savings that can be achieved, for example, by rationalization, optimal order sequence planning and flexible process use?
- How can inefficiencies such as scrap, 2nd quality, poor quality, repairs, accidents, high setup parts, etc. be avoided?
- How can outage risks and other production risks, such as missing raw materials and purchased parts, be reduced?
- How can the production demand be well planned and controlled? How well can customer needs, storage and logistics and production capacities be coordinated?
- How can internal or externally prescribed quality standards be met in the execution of production and logistics processes?
- How can environmental protection goals and occupational health and safety be achieved or complied with?

From a strategic perspective, it is a management system for the production area that supports decisions, for example, regarding the choice of location, the investments to be made in plants and machines, the choice of technology, the factory layout, the long-term production program planning and the design of the production and logistics processes.[205] The **strategic production controlling cockpit** includes parts of the strategic controlling that are specifically tailored to the production (see Sect. 3.9.2), such as the process perspective in the BCR-Card. Furthermore, it has a strong connection to the investment planning.

[204] Cf. Gottmann (2019, p. 21).

[205] Cf. Steven (2016, p. 1) and Bauer (2017, p. 13).

The main focus of the strategic production controlling are the following questions:[206]

- Where is the optimal production location?
- Which manufacturing concept is optimal for the company? For example, is the implementation of customer demand oriented flow production according to the principles of lean production feasible depending on the order structure?
- What degree of vertical integration do we aimed for? What should be manufactured in-house and what should be outsourced?
- Which technologies should be used in production?
- Which investments in production facilities are to be made in the long term?
- How should the production program be adapted in the long term?
- By which long-term strategies can production failure risks and other production risks be minimized?
- By which measures can long-term environmental protection goals and permanent occupational safety be achieved?
- ...

In the area of **production planning** the entry to the corresponding planning areas of budget planning, forecasting or medium-term planning takes place. A status and tracking report shows the current processing status of the individual planning areas and approval rounds as well as their deadlines. For the production area, the following planning areas are particularly important:

- reconciliation of sales, inventory and production quantities
- new product and variant planning
- quantitative, temporal and process-specific planning of the products to be manufactured using the necessary resources (materials, equipment and staff)
- planning of the special direct costs of production (especially special tools, licenses etc.)
- production cost planning of the production cost centers including transfer prices and surcharges for the calculation of the overhead costs
- maintenance planning
- plant and investment planning

The production planning comprises the quantitative, temporal and process-specific planning of the manufacturing using the necessary repetitive factors of production (raw materials, purchased parts, components etc.) as well as the equipment and staff. Due to the versatility and complexity of these tasks, production planning and control is carried out in many companies with the help of production planning and control systems (PPS sys-

[206] Cf. Schnell (2018, p. 25).

tems). The spectrum of computer-aided production planning and control ranges from medium-term to short-term production planning.[207]

The BI-supported cockpit can support the PPS systems by supporting sales volume, inventory, production, purchasing, personnel and plant planning (cf. Sects. 3.12.2, 3.12.3, 3.12.4 and 3.12.5). While the PPS system is mainly used for daily, weekly and other short-term planning (order planning and scheduling) on an operational level, the BI-supported production planning cockpit is suitable for simulative planning for longer periods (e.g. 3-month forecast, annual planning and multi-year planning). Here, for example, changes in demand behavior in sales planning can be calculated by breaking down the structure data of production (bills of material and work plans) or the calculation structures. Here, new products, new plants, processes and operating models (e.g. shift models), for which no detailed structure data are available yet, can be more easily integrated into the planning model. For new articles, for example, simplified or similar structure data are used.

The following example focuses on the operational production controlling. The entry cockpit (landing page) of the operational production controlling contains important key figures and an overview of the most important analyses (cf. Fig. 3.123).

In the example, balanced key figures from the operational analysis areas were used which cover both cost, result and time targets. The costs are monitored by the plant utilization and the productivity ratio of the production staff. The result is tracked by both the production error rate (quality) and the production result (e.g. contribution margin 1 after variable material and production costs). The key figure deliveries on schedule shows the compliance with the general time target. The overall equipment effectiveness also combines several targets by integrating the machinery availability, the performance rate and the quality rate (cf. Sect. 3.9.4.4).

The following example focuses on selected analysis cockpits in the operational production controlling. A small example of a forecast planning is presented in the context of the production program analysis and planning in the following Sect. 3.9.4.1.

3.9.4.1 Production Program Analysis and Planning

The analysis and planning of the production program is carried out on several levels: the strategic, tactical and operational level of the company.[208] The strategic production program serves the long-term orientation of the product fields; new promising product technologies are targeted, and exit strategies are prepared for critical product fields. The strategic product planning uses the strategic controlling cockpit presented in Sect. 3.9.2.1. Here, for example, the strategic instruments of the product life cycle curve, portfolio management and roadmap management are applied in which research and development projects are planned. The strategic program planning serves to build up future potential for success and to ensure long-term existence. Within the scope of tactical production pro-

[207] Cf. Kistner and Steven (2001, p. 254).

[208] Cf. Bloech et al. (2014, p. 106 f.).

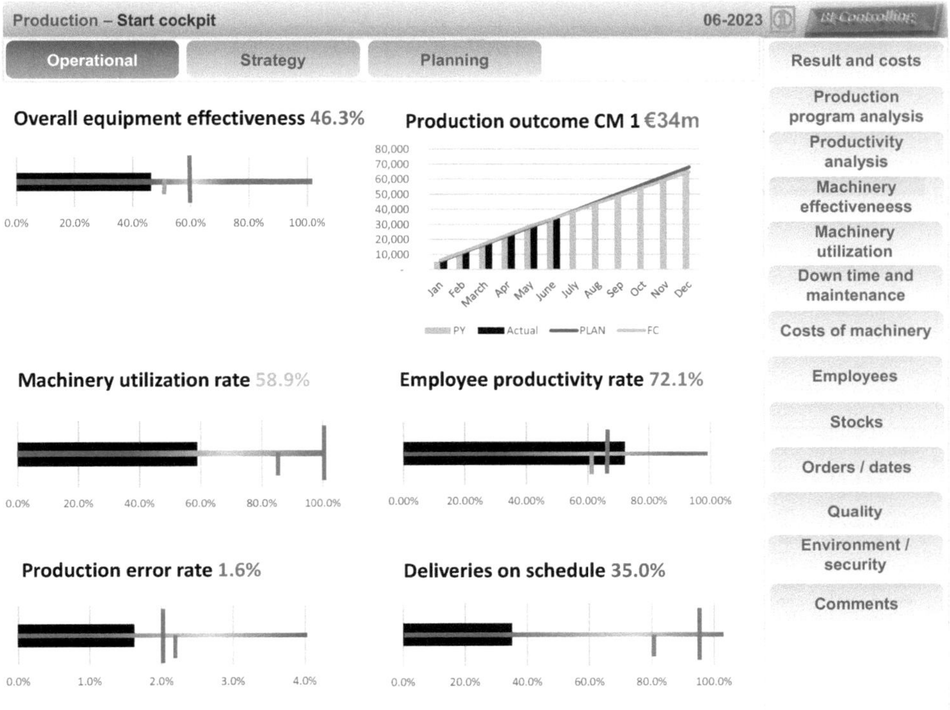

Fig. 3.123 Start cockpit of the operational production controlling

gram planning, the introduction of new products and the variation of existing products are planned. In addition, discontinued products and slow-moving items are eliminated from the product program. In operational production program planning, short-term goals such as profit or profitability maximization are pursued. For all three planning levels, the analysis of the contribution margins of the individual products and the utilization of the required capacities is needed. The cockpit could provide transparent assistance here.

In the production program analysis, the contribution margins and production cost shares with their value-added shares are differentiated for the production quantity, possibly extended by the inventory (final) stocks of the selected period (see Fig. 3.124). The achievement of the target contribution margin indicates as a peak indicator whether the selected product or product area meets the target specification. In the detailed table, the individual products and product areas can be systematically analyzed. The relevant contribution margin I after variable costs can be examined both on the basis of its calculated costs and on the basis of the actual costs (plus deviations) for the most important cost blocks "material, external and production services". The starting point are calculated revenues, so that the sales result remains unaffected. If the deviations become too large, this is a clear indication that the basic quantity and price frameworks for the calculation need

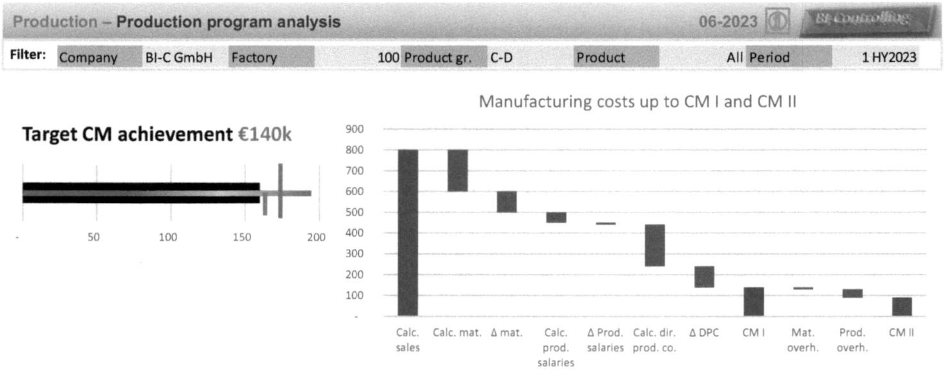

Product hierarchy	Quantity produced	Ending inventory	Calculated sales	Calculated materials	Δ material	Calculated production salaries	Δ Production salaries	Calc. direct production costs	Δ DPC	CM I	CM I %	CM II	Production manufacturing costs	Production costs Inventory
+Total	4,290	1,750	7,550	1,850	520	535	200	1,970	615	1,860	24.6%	1,374	6,177	2,520
+Product group C	2,290	850	3,800	950	360	220	80	1,070	310	810	21.3%	549	3,252	1,207
+Article group C1	1,170	330	1,900	500	200	90	20	520	150	420	22.1%	291	1,609	454
Article C1.1	800	300	800	200	100	50	10	200	100	140	17.5%	90	710	266
Article C1.2	120	30	600	200	50	30	10	200	-	110	18.3%	60	540	135
Article C1.3	250	-	500	100	50	10	-	120	50	170	34.0%	141	359	-
+Article group C2	1,120	520	1,900	450	160	130	60	550	160	390	20.5%	258	1,643	763
Article C2.1	700	400	700	200	10	50	50	200	100	90	12.9%	40	660	377
Article C2.2	120	20	600	100	50	30	-	150	50	220	36.7%	185	415	69
Article C2.3	300	100	600	150	100	50	10	200	10	80	13.3%	33	568	189
+Product group D	2,000	900	3,750	900	160	315	120	900	305	1,050	28.0%	825	2,925	1,316
+Article group D1	1,000	450	1,850	450	60	80	60	550	155	495	26.8%	363	1,488	669
Article D1.1	600	400	600	100	10	10	-	200	100	180	30.0%	135	465	310
Article D1.2	150	50	750	250	-	20	10	200	50	220	29.3%	168	583	194
Article D1.3	250	-	500	100	50	50	50	150	5	95	19.0%	60	440	-
+Article group D2	1,000	450	1,900	450	100	235	60	350	150	555	29.2%	463	1,438	647
Article D2.1	200	50	200	100	-	-	-	50	-	50	25.0%	35	165	41
Article D2.2	200	100	800	150	50	135	50	100	50	265	33.1%	238	563	281
Article D2.3	600	300	900	200	50	100	10	200	100	240	26.7%	190	710	355

Fig. 3.124 Production program analysis

to be adjusted. Furthermore, the contribution margin II shows the contribution margin after full production costs. For further investigations, a jump to the income statement, which has already been presented in the area of the management cockpit, is possible.

For the production program planning, articles with low or negative contribution margins and unfavorable cost structures can thus be identified and removed from the subsequent production program planning. Economic articles, on the other hand, are to be promoted in coordination with sales.

In production program planning, those products and their quantities are planned that are to be produced in the planning period (e.g. year(s), quarter(s), month(s)). The goal is to maximize the profit of a company by taking into account the costs and revenues. The execution of production program planning is only possible, if information about the planned sales quantity, the sales price, the variable production costs per unit, the production time per unit and the available capacities are available.[209] The decisive factor for optimizing the program is, as already mentioned above, the contribution margin

[209] Cf. Weber et al. (Publisher/Setzerei note: 2014, p. 185).

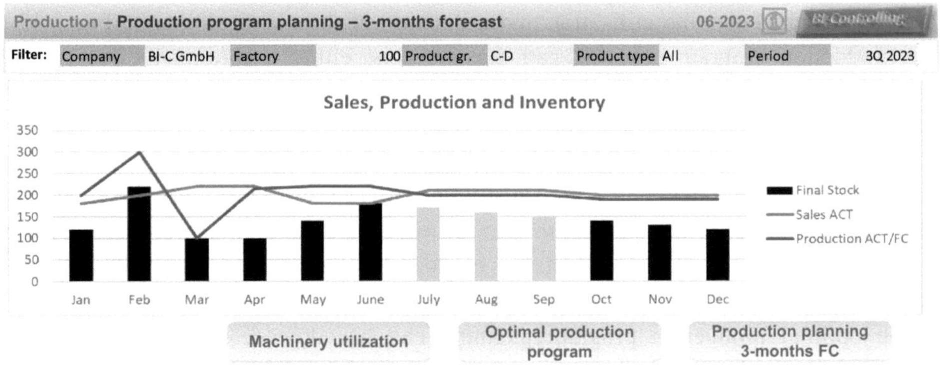

Products	Opening balance inventory	Disposed customer orders	Ø Sales/ Month (6 Mon.)	Planned sales	% Disp. planned cust. order Sales	Open planned sales	Min. quantity	Production demand	Standard lot size	% CM I	Opt. production programme	Jul	Aug	Sep	Planned prod. quantity	FC end. Stock
Article C2.2	100	50	50	100	50.0%	50	50	50	50	36.7%	200	50	0	50	100	100
Article C1.3	100	500	200	600	83.3%	100	100	600	200	34.0%	600	200	200	200	600	100
Article D2.2	100	50	50	100	50.0%	50	100	100	50	33.1%	100	100	50	50	200	200
Article E1.2	200	300	80	300	100.0%	0	50	150	100	32.4%	200	100	0	100	200	100
Article D1.1	200	450	120	480	93.8%	30	100	380	400	30.0%	400	0	400	0	400	120
Article D1.2	250	280	80	300	93.3%	20	50	100	100	29.3%	100	100	0	0	100	50
Article D2.3	500	700	300	1000	70.0%	300	200	700	200	26.7%	750	200	200	400	800	300
Article E1.1	150	250	150	480	52.1%	230	100	430	100	26.5%	450	100	200	200	500	170
Article E1.5	100	50	70	100	50.0%	50	100	100	50	26.4%	100	50	50	100	100	100
Article D2.1	500	650	300	1000	65.0%	350	150	650	100	25.0%	650	300	200	100	650	150
Article E1.4	450	550	280	1000	55.0%	450	250	800	400	21.6%	800	0	400	400	800	250
Article E1.3	100	500	200	550	90.9%	50	150	600	200	19.8%	600	200	200	200	600	150
Article D1.3	150	450	200	600	75.0%	150	150	600	200	19.0%	600	200	200	200	600	150
Article C1.2	200	300	80	300	100.0%	0	50	150	100	18.3%	150	100	100	0	200	100
Article C1.1	300	350	120	480	72.9%	130	100	280	400	17.5%	300	400	0	0	400	220
Article C2.3	600	800	400	1000	80.0%	200	400	800	200	13.3%	800	200	200	200	800	400
Article C2.1	500	650	300	1000	65.0%	350	200	700	400	12.9%	700	400	0	400	800	300

Fig. 3.125 Production program planning—3-month forecast

I per unit after variable production costs. The planning form shown shows a 3-months forecast planning in operational production. controlling, which is based on scheduled and planned sales quantities as well as the current sales prices and cost structures (see Fig. 3.125). The starting point of the forecasts are the current inventory levels, the scheduled customer orders, the average monthly sales quantities of the last 6 months and the planned sales quantities of the period. The open planned sales quantity indicates which production quantities of the planned sales quantities are not covered by already scheduled orders. The production requirement is determined by the inventory level, the sales quantity and the minimum inventory of the item maintained in the inventory management system. The standard lot size should not be undercut when loading. The production planner now has the possibility to sort the products by contribution margin ratios after variable costs and to load those with the highest CM ratios first. As a support, he can use a button to trigger the calculation function to determine the optimal production program. The calculation takes into account the resource bottlenecks required by the articles, e.g. individual plants which have limited capacities available. For this purpose, the times required per capacity are based on the work processes or the calculation values and taken into account. To get a detailed view of the capacity utilization, a further analysis can be accessed which shows the utilization and utilization of the plants similar to Fig. 3.130. The production program planning can thus be done either according to free capacities (only contribution margin per unit orientation) or taking into account capacity bottlenecks. In the operational production program planning with free capacities, only the products that have a positive contribution margin per unit are included in the production program. To calculate the contribution margin per unit, the variable costs are subtracted from the sales price. In the operational production program planning with capacity bottlenecks, the quantities of the products per capacity bottleneck are optimized, so that the sum of all contribution margins for the entire production program is highest.[210]

The optimal production program planning can be adopted, but also individually adjusted in the yellow marked planning fields for the forecast months. If, for example, orders have already been placed in practice, products with a low contribution margin must also be produced. The planning is finalized by the write-back button production planning of the 3-months forecast. The graphics supports the planner and shows him the course of the actual and forecast sales, production and inventory quantities. The three forecast months are highlighted in yellow, the following months are filled by the planning. With the help of the graphics, positive and negative trends can be more easily recognized, and thus, inventory shortages or surpluses can be detected and avoided.

3.9.4.2 Production Result and Costs

A very important controlling instrument for production controlling is the production result and its result components in the form of plant and cost center results. Since production usually does not make sales, the result is traditionally a variance analysis for the costs of the production cost centers condensed to the plant and production result. Here, the actual costs are compared with the target costs. The performance of the production is determined in the allocation of the production output to the cost units. If the actual costs are reduced by the allocated costs, one obtains an under- or overcoverage, which represents the negative or positive production result.

A prerequisite for this is the establishment of a flexible standard costing system, which splits the costs into fixed and variable components. If the standard costs are adjusted to the actual capacity utilization, one speaks of the target costs, which are defined as follows.

$$\text{Target costs} = \text{fixed planned costs} + \text{variable planned costs} \cdot \frac{\text{actual employment}}{\text{planned employment}}$$

The plant-related cost center analysis is similar to the description of the cost center analysis in Sect. 3.9.1.3. Therefore, only the peculiarities of the production cost centers are listed here.

[210] Cf. Weber et al. (2014, p. 186 f.).

The main goal of the target-actual comparison is to uncover inefficiencies of the operational performance process and to initiate appropriate countermeasures by investigating the causes. In order for a meaningful actual cost control to be possible with the help of the planned costs, the pure standard costs must be adjusted to a changed actual employment situation.

The planned direct costs can easily be adjusted to the fluctuations in employment, since they are planned per cost unit and usually belong to the variable costs. Problems in cost planning and adjustment to employment fluctuations are caused by the overhead costs planned per cost center. The flexible standard costing system allows an adjustment of the cost targets to changed cost influencing factors. A prerequisite for this is that, for each overhead cost type of a cost center, a reference variable is determined, to which the cost type is proportional, and that the overhead costs for a certain planned employment can be split per cost type into fixed and variable components. It is then possible to determine target cost functions for each reference variable.

The variance determination in the cost center-related target-actual cost comparison is done per cost type for each cost center by subtracting the determined target costs from the actual costs. If the absolute variance amounts calculated in this way are also related to the corresponding target costs, additional percentage variance indicators can be determined. In order to better track the variance trends, the variance amounts of the period (e.g. month, quarter, half-year or year) are listed for each selected period.

Fixed costs that are not influenced by and independent of the level of activity, such as the depreciation of fixed assets based on the useful life, are allocated according to the principle "target = actual".

If several cost drivers are needed for a causal representation of the quantitative consumption of production factors and thus the cost level in a cost center, there is a heterogeneous cost causation, so that the variable costs of individual cost types can behave proportionally to different cost drivers. Depending on how the actual outputs of the individual cost drivers behave, opposite cost trends can develop, and these can cancel out or neutralize each other in the deviation presentation. In this case, a cost driver-dependent cost type analysis is required within the target-actual comparison.

Due to the rather large number of cost types for a planning and control object such as a cost center, it is also advisable to perform the deviation determination by cost type groups, which can be differentiated again for a deeper deviation analysis.

The standard costing system aims to eliminate certain influences that affect the costs by planning the costs in advance. These cost influences can be caused mainly by:

- fluctuations in the *prices* of the consumption factors,
- fluctuations in the *quantity consumption* of the consumption factors,
- fluctuations in the *degree of utilization.*

In order not to have the cost accounting disturbed by these fluctuations, planned costs (often also called standard costs) are charged instead of actual costs.

Standard costs are not only advance planned costs, but they are also scheduled, i. e. they are incurred with proper and economical execution of the production. They therefore have a target character.

In order to serve operational control, the standard costing system not only determines the difference between planned and actual costs, but also splits the difference into deviations. The following deviations are distinguished:

- **Price deviations** result from the difference between the *planned prices* and *actual prices* at the input of the consumables from the procurement market. They are already separated in advance by so-called price difference accounts and ideally do not end up in the cost center or cost unit accounting since the consumption quantities are not charged with actual prices, but with planned prices.
- **Consumption deviations** arise when the planned and the actual consumed quantities of cost goods do not match. It is calculated as the difference between target costs (planned quantities * planned prices at actual employment) and actual costs (actual quantities * planned prices at actual employment). Negative deviations indicate inefficiencies in consumption, e.g. excess consumption due to scrap or theft.
- **Capacity utilization deviations** can be determined by subtracting the charged standard costs (planned quantity * planned price [at actual employment]) from the target costs (planned quantity * planned price [at actual employment]) at the actual employment level. Negative employment deviations indicate lower utilization in production, something which can occur due to low market demand or loading errors.

A schematic illustration of the deviations in the flexible standard costing system on a full cost basis is shown in Fig. 3.126:

By means of determining the target costs, it is possible within the framework of the variance analysis of a flexible standard costing system based on full and partial costs to identify the part of the cost variance that is due to capacity variances or consumption variances. In the case of applying a pure marginal standard costing system, the presentation of the capacity variance is omitted since only variable costs are directly allocated to the cost objects. In order to avoid misinterpretations when determining the corresponding consumption variances in the target-actual cost comparison, variance causes such as e.g.

- process changes
- batch size changes
- intensity fluctuations
- unplanned process conditions
- changed operating ratios
- performance level fluctuations

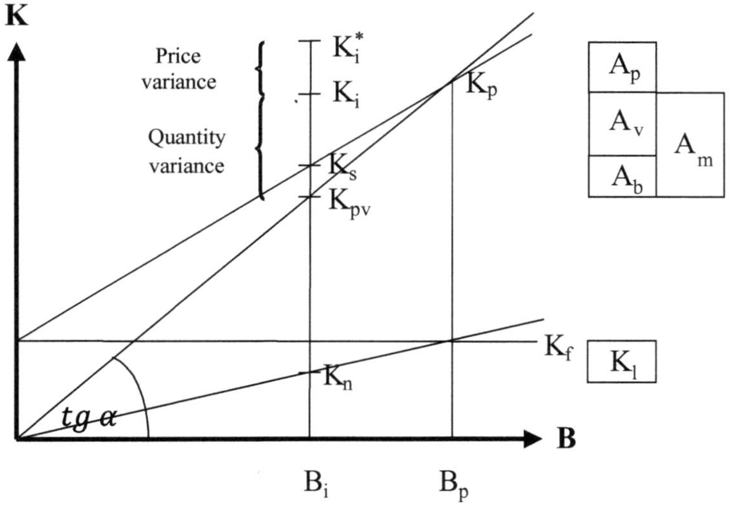

With the following icons:

K:	Costs
Kp:	Total planned costs (planned factor quantities x planned prices) and separated into fixed (Kpf) and variable (Kpv) costs
Ki*:	Actual costs at actual prices (actual quantity x actual prices)
Ki:	Actual costs at planned prices
$tg\ \alpha = \frac{K_P}{B_P}$:	Planned cost rate or clearing rate
B_i	Actual employment
B_p	Planned employment
Kpv:	Planned costs charged for actual employment $\frac{K_p}{B_p} x B_i$
Ks:	Target costs $K_{pf} + \frac{K_{pv}}{B_p} x\ B_i$
Kl:	Idle costs (in actual employment) $K_l = K_f - \frac{K_f}{B_p} x B_i = A_b$
Kn:	Used capacity costs= Kf - Kl
Ab = Ks – Kpv:	Employment variance (= empty costs)
Av = Ki – Ks:	Consumption variance
Am = Av + Ab:	Quantity variance
Ap = Ki* - Ki:	Price variances

Fig. 3.126 Deviations in the flexible standard costing system on a full cost basis

- changed relations between in-house production and external procurement
- technological or organizational changes or
- changed material inputs in the variance calculation

have to be taken into account or eliminated accordingly.

The peak ratio degree of coverage (see Fig. 3.127) indicates how the actual costs are covered by the allocated standard costs in percentage terms. It reflects the achievement of the goal of allocating as many production costs as possible. If the percentage value is below 100%, this is due to the total deviation from employment and consumption deviation. Due to compensation effects between consumption and employment deviations, it makes sense to look at the types of deviations in a differentiated way. Therefore, both

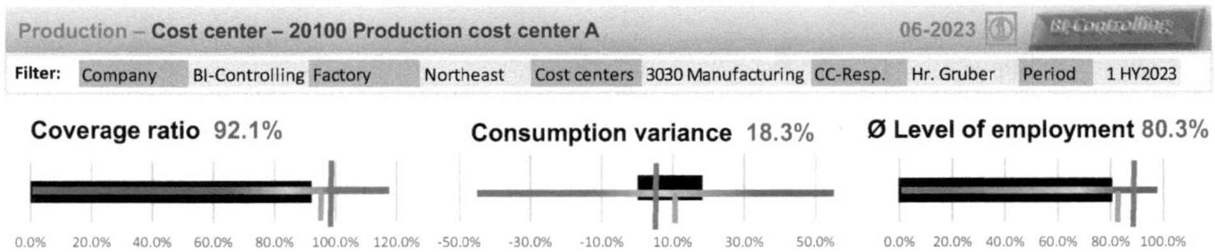

		PL var.	PL fix	PL total	Target	ACT	ΔTA	ΔTA%
Set-up wages	+	25,000	5,000	30,000	25,833	21,530	-4,303	-16.7%
Production wages	+	58,000	12,000	70,000	58,400	61,203	2,803	4.8%
Ancillary personnel costs	+	28,000	8,000	36,000	30,400	32,614	2,214	7.3%
Personnel costs	-	***111,000***	***25,000***	***136,000***	***114,633***	***115,347***	***714***	0.6%
Tool costs	+	5000	2000	7,000	6,000	4678	-1,322	-22.0%
Auxiliary and operating materials	+	5300	1,000	6,300	5,240	4,956	-284	-5.4%
Calcul. depreciation	+	1200	3600	4,800	4,560	5,630	1,070	23.5%
External maintenance	+	1500	4600	6,100	5,800	6,898	1,098	18.9%
Other production costs	+	0	3000	3,000	3,000	3,200	200	6.7%
Total primary costs	-	**124,000**	**39,200**	**163,200**	**139,233**	**140,709**	**1,476**	1.1%
Alocated energy costs		1170	2000	3,170	2,936	3,135	199	6.8%
Alocated room costs		0	800	800	800	800	0	0.0%
Alocated maintenance		500	5000	5,500	5,400	5,421	21	0.4%
Alocated testing costs		1000	6000	7,000	6,800	7,010	210	3.1%
Total secondary costs	-	**2670**	**13800**	**16470**	**15936**	**16366**	**430**	2.7%
TC after internal activity allocation	-	**126,670**	**53,000**	**179,670**	**155,169**	**157,075**	**1,906**	1.2%
Calculated production time (h)		101,670	48,000	149,670	119,736	119,736	0	
Calculated setup time (h)		25,000	5,000	30,000	25,000	25,000	0	
Shortage/surplus	-	**0**	**0**	**0**	10,433	12,339	1,906	18.3%
Measurements	-	**PL var.**	**PL fix**	**PL total**		**ACT**	**ΔPL**	Utilisation rate
Production time (h)				10,000		8,000	-2000.00	80.0%
Set-up time (h)				6,000		5,000	-1000.00	83.3%
Charge rate (€/h)	-							
Production time (h)		10.17	4.80	14.97		16.94	1.98	
Set-up time (h)		4.17	0.83	5.00		4.31	-0.69	
Key figures	-							
Available capacity				12,000		12,000		
Capacity utilisation				83.3%		66.7%		
Downtime costs			0.0%	0	**10,433**	19.7%		
Utilisation costs / Ø employees			100.0%	**53,000**	**42,567**	80.3%		
Coverage				100.0%		92.1%		

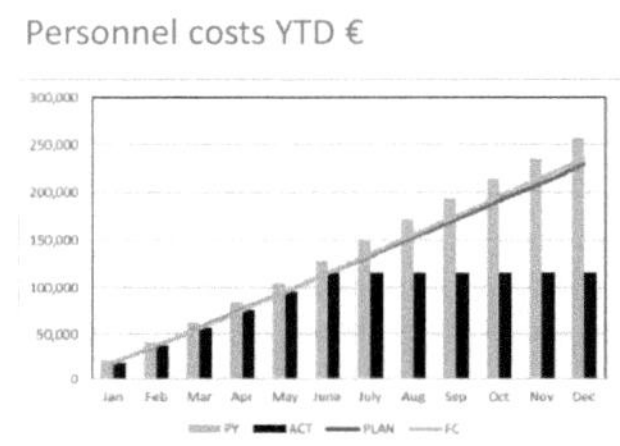

Fig. 3.127 Production result/production cost center analysis

the percentage peak ratios “consumption deviation” and “average employment level” and the absolute consumption and employment deviations can be analyzed in the line under- and overcoverage. The positive or negative consumption deviation per cost type can be quickly recognized by the positive or negative bar chart. Useful and idle costs as well as the capacity utilization can be seen in the lower part of the table. If you subtract the allocated standard costs from the target costs, you get the employment deviation. The employment deviation corresponds to the **downtime costs** and the allocated standard fixed costs to the **useful costs** of the cost center.

For the calculation, the actual prices for the respective reference quantities of the cost centers are shown in addition to the current standard prices. If price deviations in one direction are permanently recognizable, this is an indication to adjust the standard cost rates for the calculation, also called allocation prices or allocation rates, accordingly.

If you determine the capacity utilization from the performance side, you can determine it for each reference quantity, in the example this is done for the setup hours and production hours, differentiated by dividing the actual times by the planned times.

The management can analyze the cost development of individual cost types or the entire cost center in the monthly development and accumulatedly in the graphs. By changing the cost type or cost line, the diagrams change dynamically.

3.9.4.3 Productivity Analysis

The productivity analysis serves in the company to analyze and control the performance. It can be used for various business areas and shows which output was achieved with which input. In profitability management, for example, return on investment was introduced, which shows how much result was achieved with the capital employed (see Sect. 3.9.1.1). In production, productivity according to Erich Gutenberg is defined as the ratio of produced quantity to employed production factors:[211]

$$\textit{Productivity} = \text{Production quantities/Factor input quantities}$$

Through the law of diminishing returns, especially in agricultural production, it is traditionally known that with increased use of production factors (e.g. fertilizer or human labor), the yield is initially increased above average (progressively). If the use of production factors is increased further, the increase is only below average (degressively), and the marginal yield decreases.[212] In industrial production, linearly limitedproduction functions tend to dominate, in which the factor input ratio is constant regardless of the output quantity. The planned input factors per output quantity are defined by the bill of materials and work plans.[213] However, the ratio between input factors and output can deviate. The reasons for this are manifold, e.g.:

- Scrap is produced.
- Different production methods, operating ratios and performance levels lead to different consumption of input factors.

To track down these excess consumptions, it makes sense to determine and analyze the productivity. In practice, however, the key figure listed above quickly reaches its limits as most companies produce a variety of products and use several input factors for most products. Therefore, it makes sense to determine the quantity- or value-based partial productivity of the individual production factors:[214]

labor productivity = *produced quantity* of valued produced quantity/*labor hours*
machine productivity = *produced quantity* of valued produced quantity/*machine*
material productivity = *produced quantity* of valued produced quantity/*material*

In the cockpit example, the partial productivities were determined in terms of value (see Fig. 3.128). The produced good quantities were evaluated with their standard produc-

[211] Cf. Camphausen et al. (2014, p. 95).
[212] Cf. Wöhe (2000, p. 396 ff.).
[213] Cf. Wöhe (2000, p. 407 ff.).
[214] Cf. Camphausen et al. (2014, p. 95 f.).

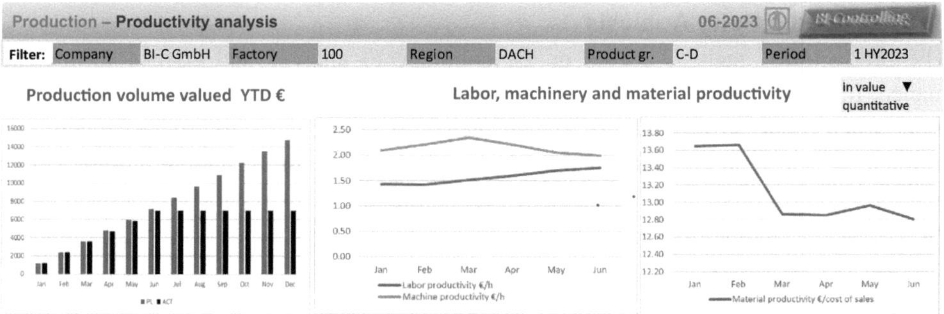

Product hierachy	PL	Prod. quantity	ΔPL%	Good quantity	GQ %	GQ €	Prod. hours	Machinery and setup hours	Cost of sales	Labor productivity €/h	Machinery productivity €/h	Material productivity €/cost of sales
+Total	7,180	Ist 7,070	-1.5%	Ist 6,976	98.7%	73,301 €	46,739	Setup hours 34,158	5,584	1.57	2.15	13.13
+Product group C	4,030	3,855	-4.3%	3,777	98.0%	40,150 €	24,873	19,145	2,806	1.61	2.10	14.31
+Article group C1	2,090	1,945	-6.9%	1,892	97.3%	19,216 €	11,586	6,208	1,428	1.66	3.10	13.46
Article C.1.1	1,170	985	-15.8%	970	98.5%	10,185 €	6125	3215	559	1.66	3.17	18.22
Article C1.2	800	820	2.5%	790	96.3%	7,552 €	4230	2345	634	1.79	3.22	11.91
Article C1.3	120	140	16.7%	132	94.3%	1,478 €	1231	648	235	1.20	2.28	6.29
+Article group C2	1,940	1,910	-1.5%	1,885	98.7%	20,935 €	13,287	12,937	1,378	1.58	1.62	15.19
Article C2.1	1,120	1,050	-6.3%	1,040	99.0%	12,792 €	7231	6984	673	1.77	1.83	19.01
Article C2.2	700	760	8.6%	749	98.6%	7,160 €	3951	3857	469	1.81	1.86	15.27
Article C2.3	120	100	-16.7%	96	96.0%	982 €	2105	2096	236	0.47	0.47	4.16
+Product group D	3,150	3,215	2.1%	3,199	99.5%	33,151 €	21,866	15,013	2,778	1.52	2.21	11.93
+Article group D1	1,750	1,620	-7.4%	1,614	99.6%	21,716 €	12,605	6,363	1,763	1.72	3.41	12.32
Article D1.1	1,000	950	-5.0%	950	100.0%	11,780 €	5678	2654	945	2.07	4.44	12.47
Article D1.2	600	550	-8.3%	545	99.1%	8,480 €	4612	2345	654	1.84	3.62	12.97
Article D1.3	150	120	-20.0%	119	99.2%	1,455 €	2315	1364	164	0.63	1.07	8.87
+Article group D2	1,400	1,595	13.9%	1,585	99.4%	11,436 €	9,261	8,650	1,015	1.23	1.32	11.27
Article D2.1	1,000	1,150	15.0%	1,144	99.5%	8,466 €	5345	4956	498	1.58	1.71	17.00
Article D2.2	200	250	25.0%	247	98.8%	1,373 €	1245	1348	368	1.10	1.02	3.73
Article D2.3	200	195	-2.5%	194	99.5%	1,597 €	2671	2346	149	0.60	0.68	10.72

Fig. 3.128 Productivity

tion costs. This has the advantage that the respective productivities can be compared with each other on condensed analysis levels, such as the product groups. Alternatively, the analysis can be switched to quantity-based partial productivities, something that is suitable for the detail level of individual items and parts. For labor productivity, production hours and machine productivity, the machine and setup hours are used as input factors. High setup hours indicate small lots and thus frequent tool changes at the machines, something which lowers productivity. The basic input size for material productivity is the valued material consumption. In the example, the development of productivities varies. While the machine productivity has fallen since April, the labor productivity has been rising continuously since February. For the material productivity, a clear break in the trend can be seen. The productivities provide initial clues that can be further pursued in the special analyses, e.g. for the equipment, the employees or the material consumption.

The analysis is complemented by the accumulated production quantities. Here, the planned and actually achieved good quantities can be compared in the table below and in the graph above left. The presentation is done as with the productivities also by choice in terms of value or quantity.

3.9.4.4 Equipment Effectiveness

One of the most important indicators in an equipment-intensive production is the overall equipment effectiveness (OEE), as this indicator gives a comprehensive impression of the performance of production equipment.[215] The OEE is an indicator that was developed by Seeichi Nakajima and the Japan Institute of Plant Maintenance as part of the decades-long development of the TPM concept (TPM: Total Productive Maintenance)[216] The OEE is a relative indicator and relates the achieved effectiveness to a perfectly effective machine that operates without interruption, without quality loss, and at the highest speed.[217] The OEE consists of the indicators availability, performance and quality.

$$OEE = \text{Overall equipment effectiveness} = \textit{Availability rate} * \textit{Performance rate} * \textit{Quality rate}$$

To find a cause for the wasted potential of a plant, the OEE should be broken down into its components.

The availability rate relates the planned availability, which was reduced by planned maintenance and downtime, to the actual availability, which was reduced by unplanned downtime, e.g. due to lack of material for production.[218]

$$\text{AR} = \textit{Availability rate} = \textit{actual availability in hours / planned availability in hours}$$

The performance rate considers the actual output against the possible output (normal output), which can be achieved at the planned availability of the plant. For example, 100 pieces per hour can be made according to the work plan, but only 80 pieces are produced. The possible output quantity can be calculated retrogradely based on the standard times in the work plans. As with the REFA method, the standard times should be checked and adjusted from time to time, especially when optimizations lead to an improved production time.[219]

---

[215] See Schnell (2018, p. 60).

[216] See Nakahima (1995) and May and Koch (2008, p. 245).

[217] See Focke and Steinbeck (2018, p. 3).

[218] Cf. Biedermann (2005, p. 36 f.).

[219] Cf. Focke and Steinbeck (2018, p. 9).

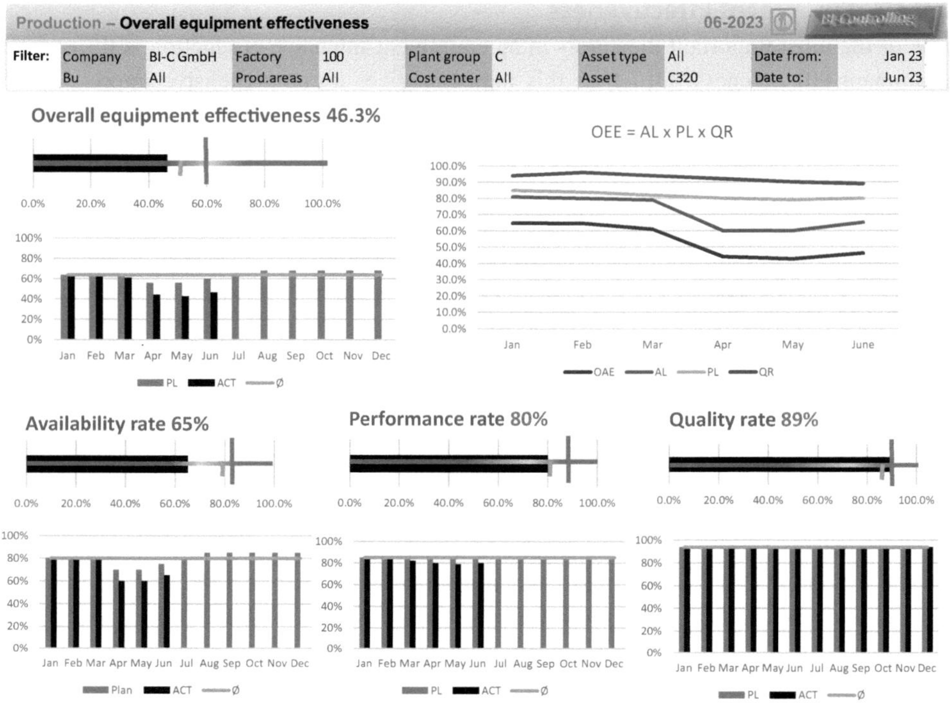

Fig. 3.129 Overall equipment effectiveness

PR = *Performance rate =*
actual output in pieces / possible output in pieces

The quality rate indicates the proportion of flawless products in the total output quantity. Which products do not fall into the category of flawless products must be defined by the company beforehand. These could be functional or performance characteristics that are outside a deviation tolerance, or optical defects. However, only the production quality is considered and not the delivery quality. Thus, it is not taken into account whether these products were improved later on.[220]

QR = Quality rate =
number of flawless products in pieces / actual output in pieces

[220] Cf. Focke and Steinbeck (2018, p. 10).

To give the user of the cockpit a comprehensive overview of the productivity of the plants, OEE analysis looks at not only the peak indicator OEE but also the individual sub-indicators (AR, PR and QR) in more detail (see Fig. 3.129).

These are presented to the user in the report in the form of a bullet chart based on the selected period. As usual, the planned and previous year values serve as comparison values.

In this example, all indicators (OEE, AR, PR and QR) are displayed up to the current month in the actual compared to the plan. By displaying the indicators, first signs of potential for improvement can be shown and thus recommendations for action for measures can be given. With the help of the time slider function, the graphs can be adjusted to any time periods, e.g. cross-year view, monthly view and daily view. This allows for a detailed root cause analysis to improve the equipment effectiveness.

The graphs show the course and special developments of the individual indicators. In the example, it becomes clear that the OEE is mainly influenced by the availability rate in the last months. The performance rate and the quality rate are satisfactory in the example.

3.9.4.5 Machinery Utilization

The evaluation of the machinery utilization serves to control the intensity of use of the production plants and thus indirectly the cost control of the manufacturing. Since the costs of the plants are divided by the number of pieces produced, the unit costs decrease progressively with an increase in utilization.[221]

The machine utilization or the plant utilization rate takes into account both the theoretically possible running performance and the actual performance of the plant.[222] The indicator is calculated as follows:

$$\text{Macthinery utilization rate} = \textit{actual hours} \; * \; 100/\textit{total capacity in hours}$$

To flexibly utilize the capacities, it makes sense to use several suitable capacity limits for the calculation. The **maximum technically possible capacity** indicates the technically and legally maximum possible use of the plant, e.g. 24 hours, 7 days a week and 365 days a year. Depending on the chosen **shift model**, the capacity can only be used in certain time periods, and based on these the utilization can be adjusted. The most important capacity limit is the **planned capacity**. It reflects the planned target capacity utilization. Here, e.g., necessary maintenance times and times for test runs have to be taken into account. Thus, in Fig. 3.130 for plant 2, a utilization of 58.9% of the planned capacity is obtained. The capacity utilization is often visualized by area diagrams, as shown in the example. By using different filter options, the production managers can examine, for example, different plant types, plants and time periods.

---

221 Cf. Gottmann (2019, p. 54).

222 Cf. Sejdic (2019, p. 184).

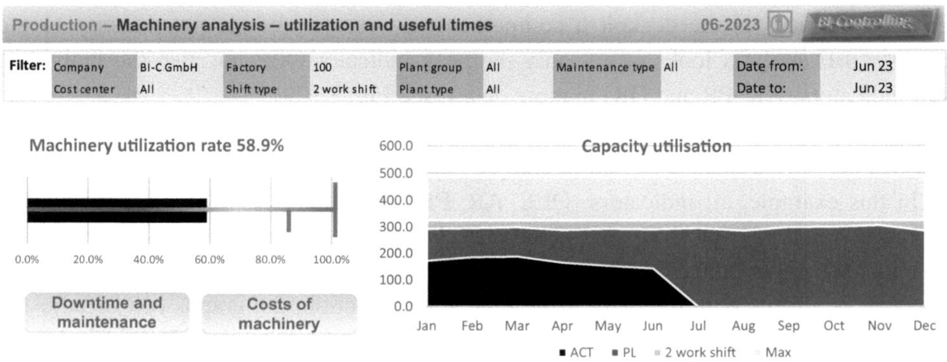

Plant	Max. technical capacity	2 shift operation	Planned capacity	Contract time	Utilization at planned capacity	Utilization at 2 shift operation	Utilization at max. capacity	Production time	Setup time	Setup quota
Total	**45318**	**30212**	**27191**	**20970**	**77.1%**	**69.4%**	**46%**	**16379**	**4591**	21.9%
Plant 1	3120	2080	1872	1374	73.4%	66.1%	44%	1230	144	10.5%
Plant 2	2925	1950	1755	1034	58.9%	53.0%	35%	985	49	4.7%
Plant 3	2730	1820	1638	987	60.3%	54.2%	36%	874	113	11.4%
Plant 4	3081	2054	1849	1647	89.1%	80.2%	53%	1523	124	7.5%
Plant 5	3003	2002	1802	1640	91.0%	81.9%	55%	1020	620	37.8%
Plant 6	2769	1846	1661	1236	74.4%	67.0%	45%	1034	202	16.3%
Plant 7	3120	2080	1872	1436	76.7%	69.0%	46%	987	449	31.3%
Plant 8	2730	1820	1638	1203	73.4%	66.1%	44%	845	358	29.8%
Plant 9	2535	1690	1521	1134	74.6%	67.1%	45%	756	378	33.3%
Plant 10	2340	1560	1404	1200	85.5%	76.9%	51%	987	213	17.8%
Plant 11	2847	1898	1708	1490	87.2%	78.5%	52%	1245	245	16.4%
Plant 12	3042	2028	1825	1643	90.0%	81.0%	54%	1123	520	31.6%
Plant 13	3081	2054	1849	1547	83.7%	75.3%	50%	1164	383	24.8%
Plant 14	2652	1768	1591	1234	77.6%	69.8%	47%	1034	200	16.2%
Plant 15	2535	1690	1521	1140	75.0%	67.5%	45%	874	266	23.3%
Plant 16	2808	1872	1685	1025	60.8%	54.8%	37%	698	327	31.9%

Fig. 3.130 Machinery utilization - useful times

In the overview table below, deviations in utilization can be detected in detail for each filtered plant, supported by the color-coding. If planned order times are analyzed in future time periods, utilization deficits can be identified early. The order time is still differentiated according to production and setup time. A too high setup ratio indicates deficits in the order loading and the sequence planning of the orders.

The button 'Downtimes and maintenance' or the context menu leads to the machinery analysis of downtimes and times for maintenance (see Fig. 3.131). The peak indicator chosen here is the breakdown and maintenance rate. The detailed information is contained in the analyst below in the overview table. Here, the downtime, the downtime rate, the failures and the failure hours are displayed first. Then, actual and planned maintenance hours are listed. For the actual times, for example, repairs, maintenance and inspections are incurred. For a more detailed overview, different maintenance types (e.g. testing, cleaning and measuring times) are shown in a pie chart. If planned maintenance deviates significantly negatively from the actual maintenance hours, this indicates a lack of maintenance work, which may lead to unplanned failures. This is clearly related to the company's maintenance strategy. In addition to the summarized breakdown and maintenance rate, the spare part requirements for the machinery are also displayed in euros. If

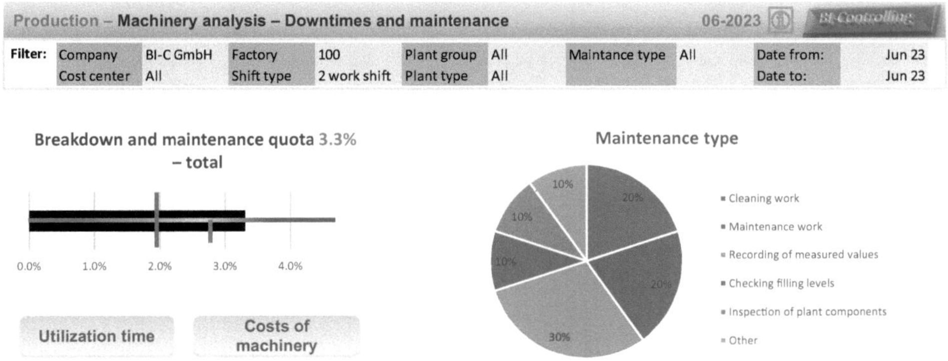

Plant	Downtime 2 shift (h)	Downtime quota	Number of disruptions	Downtime disruption (h)	Repairs (h)	Maintenance (h)	Inspection (h)	Maintenance / Insp. (h)	Inspection. plan (h)	Δ Inspection.	Disruptions / maint. rate	h without employment	Spare parts demand €
Total	**9242**	**34.0%**	**38**	**524**	**120**	**161**	**94**	**255**	**349**	**-94**	3.3%	**8343**	**21,888**
Plant 1	706	37.7%	2	16	8	8	4	12	16	-4	1.9%	670	1,000
Plant 2	916	52.2%	4	48	16	16	10	26	16	10	5.1%	826	2,031
Plant 3	833	50.9%	2	60	8	8	6	14	16	-2	5.0%	751	800
Plant 4	407	22.0%	2	50	8	10	6	16	16	0	4.0%	333	1,131
Plant 5	362	20.1%	1	32	8	12	2	14	8	6	3.0%	308	894
Plant 6	610	36.7%	5	71	8	8	0	8	16	-8	5.2%	523	3,654
Plant 7	644	34.4%	6	110	8	14	2	16	24	-8	7.2%	510	6,451
Plant 8	617	37.7%	0	0	0	6	4	10	16	-6	0.6%	607	0
Plant 9	556	36.6%	0	0	0	14	8	22	16	6	1.4%	534	0
Plant 10	360	25.6%	2	16	8	12	12	24	16	8	3.4%	312	1,009
Plant 11	408	23.9%	4	41	8	7	10	17	16	1	3.9%	342	2,345
Plant 12	385	21.1%	2	16	8	8	8	16	16	0	2.2%	345	564
Plant 13	507	27.4%	2	16	8	10	4	14	16	-2	2.1%	469	761
Plant 14	534	33.6%	2	16	8	14	8	22	16	6	2.9%	488	1,013
Plant 15	550	36.2%	2	16	8	8	6	14	16	-2	2.5%	512	0
Plant 16	847	50.3%	2	16	8	6	4	10	8	2	2.0%	813	235

Fig. 3.131 Machinery utilization - downtimes and maintenance

all downtime and maintenance times are subtracted from the total downtimes, the downtimes without employment remain. The utilization time analysis can be switched back and forth to view the employment times.

To avoid bottlenecks, machines that fail more frequently should be monitored more closely with regard to their maintenance strategy.

A further analysis allows the investment costs to be analyzed at the time of order (see Fig. 3.132). In order to keep the unit costs related to the plant low, a degression effect must be achieved at high utilisation of the plant. They are 39.77 €/h per period for the selected plant 7 in the example and show the highest value of all plants. Furthermore, the report on the analysis of the plant costs can serve as a whole and thus forms an interface to the plant management (asset management) as well as to the investment management. Important key figures of the plants are listed here, such as the acquisition date, the acquisition costs, the useful lives, the accumulated and periodic depreciation and the residual value of the plant, whose values usually come from the fixed asset accounting. Further plant costs must be separately allocated by the accounting. These include accounted maintenance and repair costs as well as operating costs (e.g. energy costs such as electricity, gas etc.) of the plants. Especially for machines with a high capital commitment,

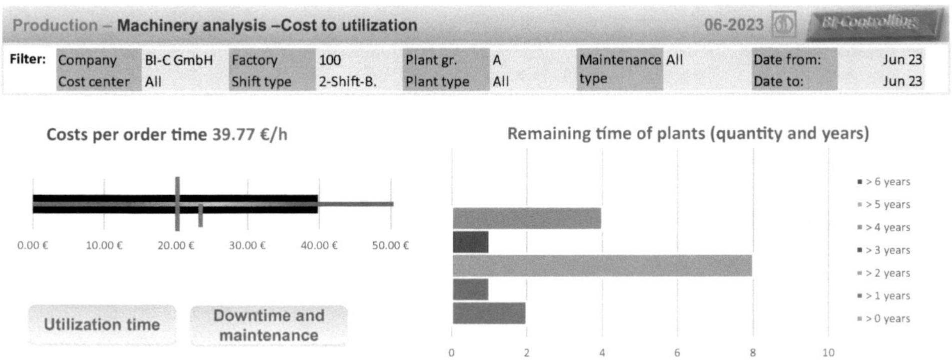

Plant	Acquisition	Acquisition costs	Duration of use	Remaining term	Accumulated depreciation	Residual Value	Depreciation/ Period	Running Costs/Period	Maintenance / Period	Repair cost/ Period	Plant Cost/ Period	Order time	Cost/ order hour	Utilization of planned capacity	Fault and Maintenance Rate
Total		**2,829,846**			**1,707,701**	**1,122,145**	**283,993**	**28,812**	**23,324**	**9,227**	**345,357**	**20970**	16.47	**77.1%**	**3.3%**
Plant 1	01.03.2019	110,599	5	0.7	95,873	14,726	11,060	1,519	3,653	2,031	18,263	1374	13.29	73.4%	1.9%
Plant 2	15.04.2018	200,567	4	-1.2	261,287	- 60,720	25,071	2,507	4,210	1,365	33,153	1034	32.06	58.9%	5.1%
Plant 3	29.09.2020	50,364	5	2.2	27,707	22,657	5,036	504		1,203	6,743	987	6.83	60.3%	5.0%
Plant 4	15.07.2020	340,154	6	3.0	167,747	172,407	28,346	2,835	2,014		33,195	1647	20.15	89.1%	4.0%
Plant 5	20.09.2019	40,648	5	1.2	30,714	9,934	4,065	406	1,035		5,506	1640	3.36	91.0%	3.0%
Plant 6	15.03.2020	86,474	5	1.7	56,954	29,520	8,647	865	1,503		11,015	1236	8.91	74.4%	5.2%
Plant 7	23.05.2022	500,623	5	3.9	110,549	390,074	50,062	5,006	2,045		57,114	1436	39.77	76.7%	7.2%
Plant 8	15.05.2021	103,645	4	1.9	55,088	48,557	12,956	1,296	605		14,856	1203	12.35	73.4%	0.6%
Plant 9	19.06.2017	203,478	6	0.0	204,593	- 1,115	16,957	1,696	3,045		21,697	1134	19.13	74.6%	1.4%
Plant 10	15.03.2020	56,987	5	1.7	37,533	19,454	5,699	570	320	1,365	7,954	1200	6.63	85.5%	3.4%
Plant 11	15.08.2019	76,985	5	1.1	59,690	17,295	7,699	770	1,254		9,722	1490	6.53	87.2%	3.9%
Plant 12	26.02.2020	552,987	5	1.7	369,668	183,319	55,299	5,530	560		61,389	1643	37.36	90.0%	2.2%
Plant 13	30.03.2022	201,547	5	3.7	50,470	151,077	20,155	2,015		2,305	24,475	1547	15.82	83.7%	2.1%
Plant 14	14.06.2019	103,647	5	1.0	83,883	19,764	10,365	1,036	1,520		12,921	1234	10.47	77.6%	2.9%
Plant 15	17.11.2021	102,578	5	3.4	33,162	69,416	10,258	1,026			11,284	1140	9.90	75.0%	2.5%
Plant 16	12.12.2020	98,563	4	1.5	62,783	35,780	12,320	1,232	1,560	958	16,070	1025	15.68	60.8%	2.0%

Fig. 3.132 Machinery utilization - costs to utilization

the utilization is decisive to keep the unit costs low.[223] The residual book value serves as an indicator for the necessity of replacements. In connection with the remaining service life of the plants including the visualization in the bar chart, one can quickly identify those plants that should potentially be replaced in the near future. Depending on the selection of certain plants, cost centres etc., structural investment backlog problems can be identified. Supportive factors here are utilization ratios, disruption and maintenance ratios as well as the concrete downtimes and disruption frequencies of the idle time analysis and the maintenance and repair costs for decision-making.

In order to increase the utilisation, various optimisation measures can be carried out, which can be assigned to the following categories:[224]

Plant-specific measures

- renewal
- automation

[223] Cf. Gottmann (2019, p. 54).

[224] Cf. ORGLINEA (2019, p. 4).

- standardisation
- factory layout
- expansion
- …

Employee-dependent measures

- qualification
- incentives
- shift models
- dimensioning
- plant occupancy
- …

Process-dependent measures

- setup times
- operating resources provision
- inflow and outflow logistics
- preventive maintenance
- …

Control-specific measures

- control room
- sequence planning
- production meetings
- monitoring
- …

Finally, it should be said that machine utilization should not serve its own purpose. According to the principle of lean production, production capacity should be aligned with demand. Ideally, the principle of the **breathing factory** is pursued. This means that the capacity is aligned with the basic utilization. Periodic peak demands are secured by outsourcing or by an extended workbench etc. Otherwise, too large fixed cost potentials arise in the plant area, since the plants are only economically utilized in times of good order conditions. A conditionally possible stock production can support this utilization problem periodically, by producing accordingly in advance. The emergence of too large finished stocks should be avoided, however, as they may not be sold later and cause too high capital tie-up costs. Furthermore, the capacities should not be reduced that far that the external or internal sales are no longer secured. The capacity utilization is therefore always to be managed in the context of the availability of the goods, the delivery reliability to the customer and the capital tie-up costs.[225]

3.9.4.6 Employees

In addition to the repetition factors (materials, purchased parts, etc.) and the equipment, the employees are another important input factor in the production of a company to be taken into account in the cockpit control of production controlling. They form together with the equipment the potential factors which can be used as a resource for the value creation process over a longer period of time.

The most important target and control variables for the production employees are:

- the increase of productivity and performance
- optimal use of employees (productive times) and
- the avoidance of absenteeism and accidents

The labor productivity indicates for the area under investigation how much time the employees have to spend to achieve a quantitative output, e.g. tons per hour:

> Labor productivity [unit/h] = output quantity/input hours

The performance level is especially interesting in the production areas where different performance levels occur, such as in piecework. The performance can be determined by the following indicator:

> Performance level = actual performance/normal performance

If different goods with different values are produced in the considered production area, a value-based labor productivity can also be calculated if, for example, the planned allocated production or manufacturing costs of the products are used as a valuation approach, e.g. production costs per hour.

> *Labor productivity [currency unit/h] =*
> *valued output quantity/input hours*

[225] Cf. Sejdic (2019, p. 184).

The use and utilization of the employees can best be analyzed by means of the time types. Here, productive, indirect and absentee times are distinguished. Productive times include the times for production, which are directly value-adding, such as setup and production times. Indirect times include maintenance, repair, cleaning times, but also training, further education and administrative times (among others, team meetings and project work) can belong to this. The indirect times are also important working times which are used, for example, to maintain or increase the performance or the resources or for dispositional and planning purposes. In some companies, the term "productive" is not popular as it suggests that all other times are not productive. This is of course not correct. One should consider and clarify, when introducing such a term, that indirect times are also value-adding, but not directly at the production process, but rather indirectly.

The utilization rate shows the sum of the actual times (productive production times and indirect times) to the target working time, which is explained further below:

Utilization rate = *actual attendance time/target working time*

The productivity ratio is an important control variable which shows the direct value-adding activity at the production process.

Productivity ratio = *direct productive actual attendance time/target working time*

Absenteeism can be differentiated into planned and unplanned absenteeism. The planned times include vacation times, times for reducing overtime, but also parental leave. Sometimes, educational times are also counted in this category. The unplanned times include maternity leave and sickness absence of the employees. Sickness absence is again divided into paid and unpaid times (with and without continuation of payment). The unplanned absenteeism caused by sick days has a high impact on the capacity and is normally not taken into account in the capacity planning.

The determination of the lost working time can be done directly via the above-mentioned absenteeism or indirectly via the target working time and the following formula:[226]

time absent = *target time present* − *actual time present*

[226] Cf. Gottmann (2019, p. 105).

The absenteeism rate then results as:

*time absent (%) = (target time present − actual time present) * 100/target time present*

The starting point of the target working time is the paid working time according to the employment contract. If the absenteeism is defined more narrowly, the planned absenteeism such as vacation etc. are deducted from it. The absenteeism rate then corresponds to the sickness rate. The actual time present is the recorded time present of the employees.

Absenteeism or sick days can be an indicator of the climate in a company or business unit, so that if absenteeism is steadily increasing, it must be considered that there may be fundamental problems in the daily work routine, but not necessarily.

This would have to be checked. Possible disease epidemics (flu wave etc.) are to be excluded, for example.

Especially in today's time of skills shortage, in which qualified employees are urgently needed on the labor market, the employee satisfaction is an important factor to keep the employees in the company. To reduce the increasing absenteeism, it can therefore be useful to question the work atmosphere and conditions. But also health-promoting programs such as nutrition advice or sports support can reduce absenteeism and increase satisfaction.[227] Another important indicator for securing the human workforce and for motivating occupational safety measures is the accident rate:

Accident rate = Number of accidents per period/Working days of the period

Accidents can also be differentiated according to commuting or direct work accidents and according to their severity, e.g. minor, serious, life-threatening, fatal etc.

Furthermore, the payment of the hours is different depending on the time of deployment. There are shift allowances and Sunday and holiday surcharges which all deviate from the normal payment. These additional costs can represent a significant cost driver, so that they have to be taken into account in the analysis of the employees depending on the industry and company.

The following cockpit for the area Employee Production shows selected key figures from the above list (see Fig. 3.133).

The key figures in production can be filtered for various objects of investigation. The most important hierarchy is the cost center hierarchy, which provides a break down from

[227] Cf. Gottmann (2019, p. 105).

Production – Employees 06-2023

Filter:										
	Company	BI-C GmbH	Factory	100	Cost center	KST-Gr. A	Workplace	All	Period	1 HY2023
	Cost center	All	Employees gr.	All	Professional	All	Qualification	All	Shift	All

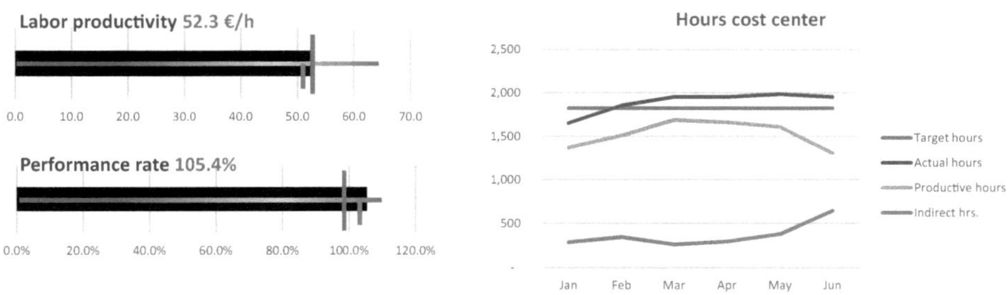

Cost centre hierarchy		Target hrs.	Actual hrs.	Utilization rate	Productive hrs.	Indirect hrs.	Plan prod. Rate	Actual prod. rate	Overtime	Hrs. with surcharge	Absente-eism
Total		**213,109**	**206,997**	**97.1%**	**149,312**	**57,685**	**65.0%**	**72.1%**	**14,161**	**20,243**	**11,960**
Cost centre 1	+	10,927	11,353	103.9%	9,144	2,209	75.0%	80.5%	1,135	2,271	568
Cost centre 2	+	18,164	16,530	91.0%	10,654	5,876	60.0%	64.5%	2,480	1,653	1,157
Cost centre 3	+	20,505	19,875	96.9%	12,568	7,307	60.0%	63.2%	2,584	2,981	1,988
Cost centre 4	-	73,802	72,129	97.7%	51,470	20,659	70.0%	71.4%	3,606	2,885	4,328
Workstation 1	+	15,321	15,564	101.6%	10,535	5,029	60.0%	67.7%	1,632	1,465	1,032
Workstation 2	+	18,430	18,791	102.0%	12,546	6,245	70.0%	66.8%	1,364	1,265	1,452
Workstation 3	-	40,051	37,774	94.3%	28,389	9,385	75.0%	75.2%	567	321	254
Employee A	+	7,605	8,412	110.6%	6,450	1,962	70.0%	76.7%	914	654	751
Employee B	+	16,806	14,320	85.2%	10,685	3,635	70.0%	74.6%	716	1,718	644
Employee C	-	15,640	15,042	96.2%	11,254	3,788	70.0%	74.8%	752	1,805	677
Cost centre 4	+	50,364	45,879	91.1%	35,692	10,187	70.0%	77.8%	2,294	5,505	2,065
Cost centre 5	+	39,347	41,231	104.8%	29,784	11,447	70.0%	72.2%	2,062	4,948	1,855

Fig. 3.133 Production employees

the overall area of production via the plants to the individual cost center groups and cost centers. The employees are often assigned to a fixed main cost center. The hierarchy can be extended by the workstations per cost center. When using jumpers and changing workstations, the use of pool or team cost centers is recommended. The employees can be distinguished by employee groups (e.g. white and blue collar), by professional groups (master, qualified technician or staff of certain disciplines) or by qualifications (trainer certificate, equipment inspection etc.).

The peak indicator labor productivity was presented as a value-based indicator in euros per hour. Detailed information on quantities and production or manufacturing costs can be found in other analyses. The actual and normal performance for determining the performance grades can also be analyzed via detailed reports.

The line graph allows for faster recognition of positive and negative developments in productive and indirect times compared to the target working time. The main table below shows, among other things, the utilization rate, the planned and actually achieved productivity ratio, as well as overtime, surcharge times, and absenteeism for the selected area of investigation. Detailed analyses in the personnel and production area lead to further key figures such as the overtime ratio, the sickness rate, and the accident rate.

3.9.4.7 Inventory Management

The operative inventory planning is carried out in the company's merchandise management system. It is necessary for providing the right resources for the manufacturing of the products. The right input factors in the right quality and quantity, at the right place, at the right time, at the right costs and supplemented by the right information must be made available (7 "R" criteria). The inventory planning ensures an optimal material inventory. It is important not to endanger the performance of the production and the delivery capability in the shipping to the customer by a too low inventory or to reduce the profitability of the company by too high inventories and thus too high storage and capital commitment costs or the risk of losing the goods due to unsalability.[228] The controlling cockpit complements the operative inventory planning by structural warehouse analyses. Exemplary cockpit reports on warehouse analyses in the area of purchasing for the logistics of the incoming goods warehouse are presented in Sect. 3.9.5.5. In the area of production logistics, the required input factors must be provided in sufficient form for each value-added or production stage from semi-finished to finished products with the R criteria listed above. The shipping logistics follows seamlessly here.

The basic information comes from the operating merchandise management system. The starting point for the inventory planning are the types of inventory. The following types of inventory are distinguished:

- **Inventory stock:** Is the actual stock of material, which can be determined by using the inventory method.
- **Warehouse stock:** This is the stock electronically determined in the warehouse according to the merchandise management system at the time of planning. It can coincide with the inventory stock or deviate from it. The basic formula for the warehouse stock determination is:

> Opening stock + incoming goods − outgoing goods = closing stock.

Incoming goods can be created by orders and deliveries, but also by transfers or returns with rework. Withdrawals or outgoing goods occur e.g. by the material consumption in the production order, scrap, transfers etc.

- **Available stock:** This type of stock refers to the inventory plus open orders and minus the number of reserved materials e.g. from production orders.
- **Reserved stock:** This is the stock that is reserved for the execution of customer orders or planned production orders.

---

[228] Cf. Jacob (2012, p. 188).

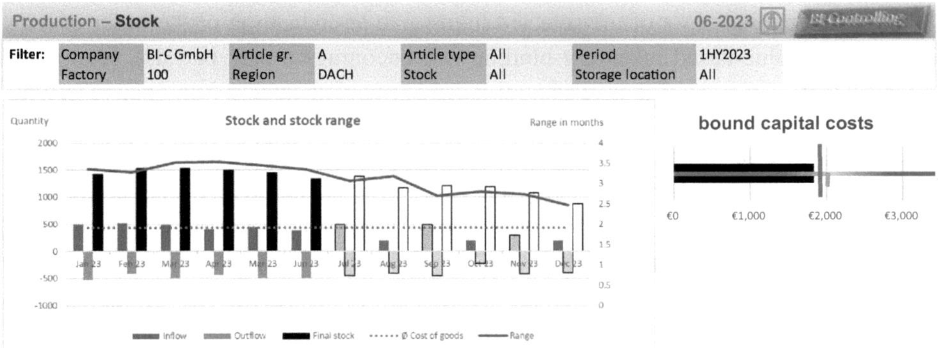

	Stock (pcs.)	Stock €	Ø Stock inventory	Stock turnover	Storage period in days	Avail. st. after open orders	Reserv. stock	St.after reserve	Minimum stock	Maximum stock	number of. prod. orders	Ø number of pieces prod. ord.	Ø Consum-ption piece/day	Ø Delivery times; Ø Lead times	Range in days
+Total	**8,722**	**87,312 €**	**9,300**	**1.14**	**26.25**	**450**	**350**	**8,822**	**5,600**	**10,500**	**70**	**125**	**59.1**	**78.6**	**69.1**
+Article group A	**8,722**	**87,312 €**	**9,300**	**1.14**	**26.25**	**450**	**350**	**8,822**	**5,600**	**10,500**	**70**	**125**	**59.1**	**78.6**	**69.1**
+Article group A1	**4,401**	**47,076 €**	**4,658**	**1.32**	**22.77**	**200**	**150**	**4,451**	**2,900**	**6,000**	**35**	**126**	**34.1**	**67.0**	**62.1**
Article A1.1	1,331	26,620 €	1,392	2.07	14.50	100	50	1,381	900	2,000	10	133	16.0	20.1	63.1
Article A1.1	1,702	13,616 €	1,767	1.00	30.11	0	100	1,602	1,000	2,000	15	113	9.8	89.6	84.5
Article A1.3	1,368	6,840 €	1,500	1.00	30.03	100	0	1,468	1,000	2,000	10	137	8.3	91.2	73.2
+Article group A2	**4,321**	**73,374 €**	**4,642**	**0.97**	**31.00**	**250**	**200**	**4,371**	**2,700**	**4,500**	**35**	**123**	**25.0**	**90.2**	**83.0**
Article A2.1	1,404	7,020 €	1,484	0.94	32.03	0	150	1,254	800	1,500	10	140	7.7	89.3	92.5
Article A2.2	1,267	15,204 €	1,349	0.93	32.38	100	50	1,317	900	1,500	10	127	6.9	92.6	89.8
Article A2.3	1,650	51,150 €	1,809	1.02	29.30	150	0	1,800	1,000	1,500	15	110	10.3	88.6	71.8

	Stock (pcs.)	Stock €	Ø Stock inventory	Stock turnover	Storage period in days	Avail. st. after open orders	Reserv. Stock	St.after reserve	Minimum stock	Maximum stock
+Total	**1,331**	**10,648 €**	**1,392**	**2.07**	**14.50**	**100**	**50**	**1,181**	**900**	**2,000**
+Storage location A	**653**	**5,226 €**	**690**	**2.50**	**11.98**	**60**	**20**	**573**	**500**	**1,000**
Storage bin A1	653	5,226 €	690	2.50	11.98	60	20	573	500	1,000
Storage bin A2	0	0 €	0	0.00	0.00	0	0	0	0	0
+Storage location B	**678**	**5,422 €**	**702**	**1.64**	**18.28**	**40**	**30**	**608**	**400**	**1,000**
Storage bin B1	678	5,422 €	702	1.64	18.28	40	30	608	400	1,000
Storage bin B2	0	0 €	0	0.00	0.00	0	0	0	0	0

Fig. 3.134 Inventory management

- **Maximum stock:** Indicates the amount of materials that may be present in the warehouse at most.
- **Reorder point:** When this stock level is reached, an order or production order is initiated.
- **Minimum or safety stock:** This type of stock represents the amount of materials that the stock must not fall short of. When the safety stock is reached, an order or a production order is generated.

In the listed cockpit for inventory management, a structured analysis of the inventory movements for selected production areas (plants, warehouses, item groups, region etc.) is possible (see Fig. 3.134).

Using the diagram, bottlenecks or excesses of inventory levels in their temporal development can be analyzed for each selected row in the item group hierarchy or in the warehouse hierarchy. This allows the analyst to see inventory levels as well as inflows and outflows in the period under review and in their forecast for future periods. The average material usage serves as a guide for the inventory movements. Another additional line shows the range in days with which the security of the required stocks can be controlled.

The bound capital costs of the selected items give an indication of the order of magnitude of the value-based inventory binding and encourage a safe but also economical inventory level. This key figure can be replaced or supplemented by alternative key figures in inventory management. Important indicators are e.g. the **inventory turnover**, the **inventory range,** but also the quantity- or value-based as well as average **inventory level**. A detailed analysis of these indicators is possible for items and warehouses in the lower tables. Fast and slow movers can be identified by the inventory turnover frequency supported by the miniature bar chart and by the signal color.

As described above, the available stock also shows open orders that are planned for the selected period, but have not been delivered yet. If they are expected to arrive soon, they increase the available stock. Reserved stocks are already unavailable stocks due to customer or production orders. The minimum and maximum stock with the respective signal identifiers indicate whether the stock value has fallen below or exceeded these critical limits. Depending on the value-added stage in production, the raw materials and purchased parts or semi-finished products required for production are analyzed from upstream production stages to the finished products ready for picking and shipping to the customer. Depending on the item type, the number of orders or the number of production orders of the respective factors of use are interesting as they allow to keep an eye on the average quantity of the order or the lot size of the production orders. Based on the average consumption or material usage and the average delivery time for raw materials or the average throughput time of the production orders, the important safety indicator "range in days" can be determined and analyzed. It signals dangers of under-supply, but also risks of over-supply.

In inventory management, structural problems of inventory management as well as ordering in purchasing or loading of production orders (scheduling and capacity planning of the PPS system) in work preparation can thus be detected. The operational inventory management in the merchandise management system provides the data for this, which determine the material consumption. This process is done by documenting and evaluating all material withdrawals that have occurred, e.g. by production orders, by inventory differences, by material overconsumption or by scrap. The subtasks of inventory management include, among others, providing current information on the stocks by type and quantity, carrying out company-ordered and legal inventories, determining gross and net requirements, and providing data that are relevant for processing orders and production orders. Inventory management can be done both in terms of quantity and value.[229]

After determining the material requirements, the order is placed. Depending on the type of demand determination, different ordering systems or scheduling methods are used. The methods include order-driven, plan-driven and consumption-driven scheduling.[230]

---

[229] Cf. Jacob (2012, p. 189 f.).

[230] Cf. Jacob (2012, p. 190 f.).

In order-driven scheduling, the order is placed based on customer orders. Only the material that is needed for processing the orders is ordered.

The plan-driven scheduling requires a highly developed planning system. The advantage of this scheduling method is that the company's capital commitment costs can be minimized by low inventory levels. The scheduling risk depends on the quality of the planning. The prerequisites for this are a functioning production control and a high data quality in the production process, especially well-maintained bills of materials.

In consumption-driven scheduling, the orders are based on the consumption values from the past. The prerequisite for this is a current and correct inventory management. Disadvantages arise here, among other things, from the shift in demand, so that past values no longer match the requirements.

The following basic strategies can be distinguished in inventory planning:[231]

- With **stock procurement**, high inventories are ordered in stock. This leads to high inventory holding costs and is suitable for inventories that can be stored for a long time. The advantage is that a large quantity can be purchased at low purchasing and transport costs.
- With **single procurement**, the inventories are only procured when there is a demand. This type of procurement leads to low inventory holding costs and fits the order-related scheduling.
  - With **synchronous delivery,** also called just-in-time principle, the required materials are delivered at the time of production. In most cases, this results in no or hardly any inventory holding costs, as the materials are ideally delivered only when they are needed for production. A small buffer stock can still be useful to counter the risk of production interruption due to shortages.

3.9.4.8 Dates and Orders

The dispatchment and capacity planning is directly dependent on the order-related quantity planning. Based on the production program planning, a dispatchment can be determined, based on the existing customer orders or stock orders, which result, for example, from a sales forecast, to replenish the existing stock quantities.[232] The sequence planning of the orders in a temporal order and the assignment of the orders to the corresponding plants in their sequence of operations is carried out by the work preparation in the PPS system. With the help of network planning, "the determination of the earliest and latest start and end dates of the production orders to be executed is carried out, in order to ensure that they are completed by the delivery date at the latest".[233] This serves above all to avoid delivery delays and customer dissatisfaction.

[231] Cf. Jacob (2012, p. 190 f.).

[232] Corsten and Gössinger (2016, p. 493).

[233] Taken from Corsten and Gössinger (2016, p. 532).

The adherence to deadlines or deadline reliability is a central goal in production controlling. It is determined by the following key figure:

$$\text{Deadline reliability} = \textit{number of } \text{on-time finished } \textit{orders} \times 100/\textit{number of all orders}$$

In this case, the manufacturing of certain products is considered to be on schedule if the parts required for the production are available at the corresponding production stages or the finished products to the customer at the agreed date. If the order is finished too early or there are delays, this often leads to higher costs for the company. Orders completed too early in the pre-production stages and too long lead times result in additional storage and bound capital costs. Orders completed late result in possible penalty payments agreed in advance with the customer. In addition, not reaching the deadline targets (desired or order date) for the orders leaves a negative impression on the customer, something which damages the image. As part of the planning and control of the production processes, information on the capacity utilization, material backlogs, the scheduling of the orders and the order sequences must therefore be presented transparently in order to reach the best possible lead times and a high degree of adherence to deadlines.[234]

The lead time is the time that elapses between the start of production and the completion of an order. The reduction of the lead time is as important for many companies as the cost reduction, because this has the effect that the products can be produced and delivered faster, which thus leads to the elimination of costs such as e.g. storage costs.

The throughput time is calculated as follows: [235]

$$\textit{throughput time} = \textit{order time} + \textit{transition time}$$

The pure production time is composed as follows:[236]

$$\text{Production time} = \textit{batch size} \times \text{production } \textit{single time} + \textit{setup time}$$

In case of different operating relations of the production staff to the machine, machine and production time are also distinguished.

[234] Cf. Littkemann et al. (2018, p. 192 f.).

[235] Cf. Grabner (2017, p. 132 ff.).

[236] Cf. Grabner (2017, p. 135).

The setup time is the time that is required for a change of order on a plant due to the setup process. For example, a tool is changed or the setting of the plant has to be adjusted. Ideally, an optimal lot size is aimed for to minimize setup and storage costs. The work planning creates production orders for this purpose, which fulfill this criterion as best as possible. However, other goals such as the minimization of lead times and the adherence to deadlines also compete for the capacities, so that the optimal lot sizes cannot always be achieved.

The transition time consists of various transport and wait times:

$$transition\ time\ = transport\ time\ + wait\ time$$

The wait times are divided into process-related wait times, such as e.g. wait times due to an already utilized plant and production-related wait times, such as e.g. ripening times in cheese production.

The transport times result from before or after the respective production process at the plant. In state-of-the-art productions, products are often transported in a flow system via automatic conveyor systems to the next production point, so that in this case essentially the cycle time determines the throughput time from production and transport time. Conveyor systems can also take over storage and waiting functions before the next production stage.

Furthermore, control times can be evaluated separately as specific production time or as additional transition time. In the cockpit example, this time was depicted as part of the production time (see Fig. 3.135).

In the cockpit, the peak indicator of deadline compliance is supported by the average delay of the orders, which in turn is composed of the average deviation of the production time and the average deviation of the transition time from wait time and transport time.

Detailed analyses are possible in the lower table of the production order time analysis. Here, among other things, the throughput time and the individual indicators delay, production time, wait time and transport time and their deviations can be recognized. The miniature bar charts show both negative and positive time deviations, so that compensation effects can also be detected.

The analyst has to specify the analysis spectrum in the production order time analysis using numerous selection filters. This allows the production management to perform top-down analyses from the plant to the individual order. For further order analyses, a jump to the ERP system or PPS system is provided. Ideally, the time analyses are possible across all analysis objects, resulting in a variety of analysis options for the production management, such as specific production areas based on the cost center group or specifically for certain plants and plant types. In addition, special order types such as customer or warehouse orders or only certain orders with the status completed can be analyzed.

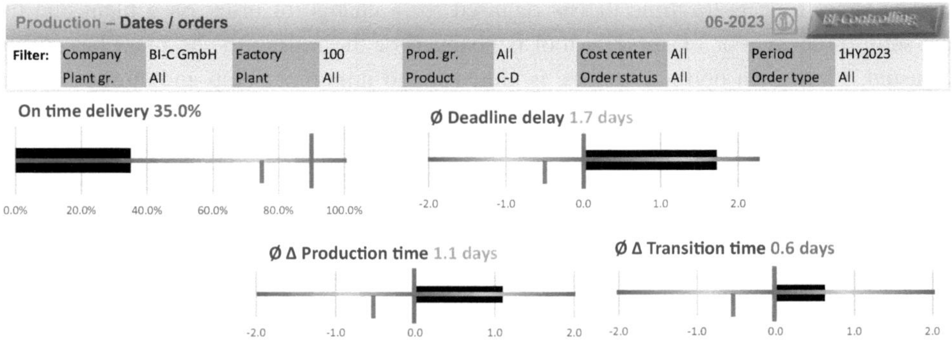

Order	Status	Order start	Desired date	Planned date	Actual date	Lead time	Deadline delay	Produc-tion	Δ Production	Waiting time	Δ Waiting time	Transpor-tation	Δ Trans-portation
Average						12.4	1.7	6.6	1.1	3.8	0.3	2.1	0.3
Order 1	finished	10.01.2023	18.01.2023	20.01.2023	21.01.2023	11.0	1.0	5	0.5	4	0.3	2	0.2
Order 2	finished	16.01.2023	02.02.2023	04.02.2023	04.02.2023	19.0	0.0	8	0.5	7	-0.5	4	0
Order 3	finished	16.02.2023	18.02.2023	20.02.2023	22.02.2023	6.0	2.0	3	1.5	1	0.3	2	0.2
Order 4	finished	18.02.2023	28.02.2023	05.03.2023	10.03.2023	20.0	5.0	9	2.0	8	1	3	2
Order 5	finished	26.02.2023	05.03.2023	15.03.2023	14.03.2023	16.0	-1.0	7	-0.4	5	-0.3	4	-0.3
Order 6	finished	03.03.2023	10.03.2023	15.03.2023	15.03.2023	12.0	0.0	6	0.2	4	-0.1	2	-0.1
Order 7	finished	16.03.2023	18.03.2023	20.03.2023	20.03.2023	4.0	0.0	3	-0.5	0.5	0.3	0.5	0.2
Order 8	finished	17.03.2023	22.03.2023	26.03.2023	26.03.2023	9.0	0.0	6	1.5	2	-1	1	-0.5
Order 9	finished	20.03.2023	28.03.2023	30.03.2023	01.04.2023	12.0	2.0	6	1.0	4	0.8	2	0.2
Order 10	finished	25.03.2023	02.04.2023	08.04.2023	10.04.2023	16.0	2.0	8	1.0	5	0.7	3	0.3
Order 11	finished	02.04.2023	10.04.2023	15.04.2023	20.04.2023	18.0	5.0	10	3.0	5	1	3	1
Order 12	finished	10.04.2023	20.04.2023	25.04.2023	30.04.2023	20.0	5.0	11	2.0	6	2	3	1
Order 13	finished	20.04.2023	22.04.2023	28.04.2023	02.05.2023	12.0	4.0	6	3.0	4	1	2	0
Order 14	finished	30.04.2023	05.05.2023	08.05.2023	08.05.2023	8.0	0.0	6	0.0	2	0	0	0
Order 15	finished	02.05.2023	05.05.2023	10.05.2023	10.05.2023	8.0	0.0	6	0.0	1	0	1	0
Order 16	finished	08.05.2023	12.05.2023	20.05.2023	26.05.2023	18.0	6.0	10	4.0	5	1	3	1
Order 17	finished	16.05.2023	22.05.2023	30.05.2023	30.05.2023	14.0	0.0	8	0.5	4	-0.3	2	-0.2
Order 18	finished	25.05.2023	30.05.2023	02.06.2023	06.06.2023	12.0	4.0	6	3.5	4	0.3	2	0.2
Order 19	in progress	06.06.2023	15.06.2023	28.06.2023		11.0		5	0	5	2	1	1
Order 20	open	29.06.2023	05.07.2023	06.07.2023		0.0							

Fig. 3.135 Deadline compliance

3.9.4.9 Quality

An important objective of production is to comply with internal and/or external predefined quality standards with regard to the goods to be produced. To meet this objective, the cockpit displays the indicator production error rate (PER).

> *production error rate* (%) =
> *faulty* parts in *production* × 100/*total* number of parts

This rate gives a statement about the extent to which the production quantity produced in production meets the relevant quality requirements. It should be noted that the quality of the produced quantities can vary depending on the order type.[237] The number of defective parts or products is set in relation to the total production for the considered production section. Parts that are damaged during transport or storage within the production

[237] Cf. Littkemann et al. (2018, p. 194).

section should be separated as best as possible.[238] A part is counted as defective, if at least one defect is present. Alternatively, one could also count several defects more precisely according to the reasons for the defect. In practice, one would have to design the defect detection more detailed than is often the case. In this case, the indicator Defects per million opportunities (DPMO) is suitable, which is defined as follows:

DPMO = number of total defects × 1,000,000/(number of processed parts × number of defect opportunities per part)

Since the defect rates are often very low and zero-defect tolerance is aimed for from a quality cost perspective, the factor "million" provides a better view of these low defect rates.

Alternatively to the production defect rate and the indicator DMPO, the indicator parts per million (PPM) is also often used, which multiplies defective parts by the factor of one million to the total number of parts.

PPM = defective parts × 1,000,000/total parts

With only one defect opportunity per part, the value of the indicator DPMO corresponds to the value of the indicator PPM. The indicator DPU (defects per unit) rounds off this set of indicators, in which the average number of defects per part is calculated.

DPU = number of total defects × 100/number of total parts

Other important quality variables are the yield of the good quantity. It is calculated as follows:

Yield = number of defect-free parts × 100/number of total parts

If a company-specific high scrap rate is observable during or at the end of the production process, the causes of the quality defects in the company must be remedied. The scrap quantities cause costs until their defect detection, which are the higher the later the poor

[238] Cf Gottmann (2019, p. 100).

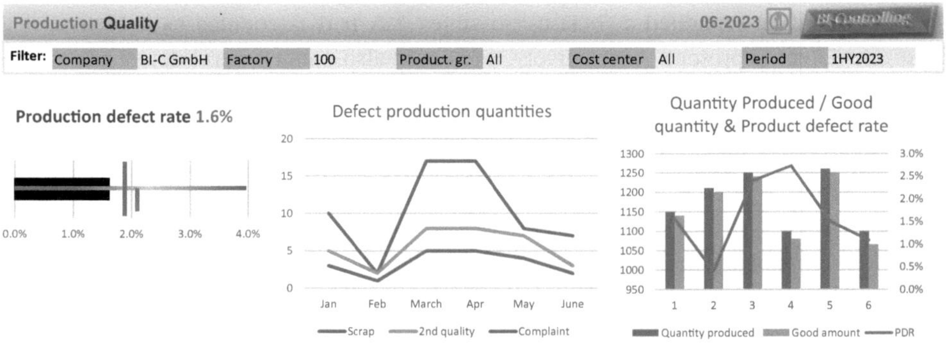

Product hierarchy	PL	Quantity prod.	ΔP%	Good quantity	Scrap	2nd quality	Complaint	Scrap %	2nd quality %	Complaint %	Product defect rate
+Total	**12,520**	**7,070**	-43.5%	**6,976**	**61**	**33**	**20**	0.9%	0.5%	0.3%	1.6%
+Product group C	**7,820**	**3,855**	-50.7%	**3,777**	**54**	**24**	**18**	1.4%	0.6%	0.5%	2.5%
+Article group C1	**1,170**	**1,945**	66.2%	**1,892**	**38**	**15**	**10**	2.0%	0.8%	0.5%	3.2%
Article C1.1	800	985	23.1%	970	10	5	3	1.0%	0.5%	0.3%	1.8%
Article C1.2	120	820	583.3%	790	20	10	5	2.4%	1.2%	0.6%	4.3%
Article C1.3	250	140	-44.0%	132	8	0	2	5.7%	0.0%	1.4%	7.1%
+Article group C2	**6,650**	**1,910**	-71.3%	**1,885**	**16**	**9**	**8**	0.8%	0.5%	0.4%	1.7%
Article C2.1	3,700	1,050	-71.6%	1,040	5	5	1	0.5%	0.5%	0.1%	1.0%
Article C2.2	1,950	760	-61.0%	749	9	2	5	1.2%	0.3%	0.7%	2.1%
Article C2.3	1,000	100	-90.0%	96	2	2	2	2.0%	2.0%	2.0%	6.0%
+Product group D	**4,700**	**3,215**	-31.6%	**3,199**	**7**	**9**	**2**	0.2%	0.3%	0.1%	0.6%
+Article group D1	**1,750**	**1,620**	-7.4%	**1,614**	**0**	**6**	**1**	0.0%	0.4%	0.1%	0.4%
Article D1.1	300	950	216.7%	950	0	0	0	0.0%	0.0%	0.0%	0.0%
Article D1.2	950	550	-42.1%	545	0	5	1	0.0%	0.9%	0.2%	1.1%
Article D1.3	500	120	-76.0%	119	0	1	0	0.0%	0.8%	0.0%	0.8%
+Product group D2	**2,950**	**1,595**	-45.9%	**1,585**	**7**	**3**	**1**	0.4%	0.2%	0.1%	0.7%
Article D2.1	300	1,150	283.3%	1,144	4	2	1	0.3%	0.2%	0.1%	0.6%
Article D2.2	1,750	250	-85.7%	247	3	0	0	1.2%	0.0%	0.0%	1.2%
Article D2.3	900	195	-78.3%	194	0	1	0	0.0%	0.5%	0.0%	0.5%

Fig. 3.136 Quality cockpit - Production defect rate, rejects, 2nd choice and complaints

quality is detected in the value chain.[239] The defective production can be caused by various reasons, such as contamination, start-up problems, incorrect settings in machines and plants, but also by processing the wrong parts or materials due to possible provisioning problems.[240]

Another reason for observing this indicator is the fact that an increasing scrap rate can also indicate an impending machine failure that needs to be fixed. This failure is equivalent to consequential costs, but also to a standstill and a corresponding downtime of the respective machine.

In the following quality cockpit example (see Fig. 3.136), the production defect rate is used as the top indicator. In the data table, planned quantity, produced quantity, good quantities, rejects, second choice products and complaints can be captured at a glance. Another good indicator that could be reported and analyzed here is the rework rate. High deviations of individual items can be analyzed more closely based on the longer bar

[239] Cf. Littkemann et al. (2018, p. 194).

[240] Cf. Gottmann (2019, p. 100).

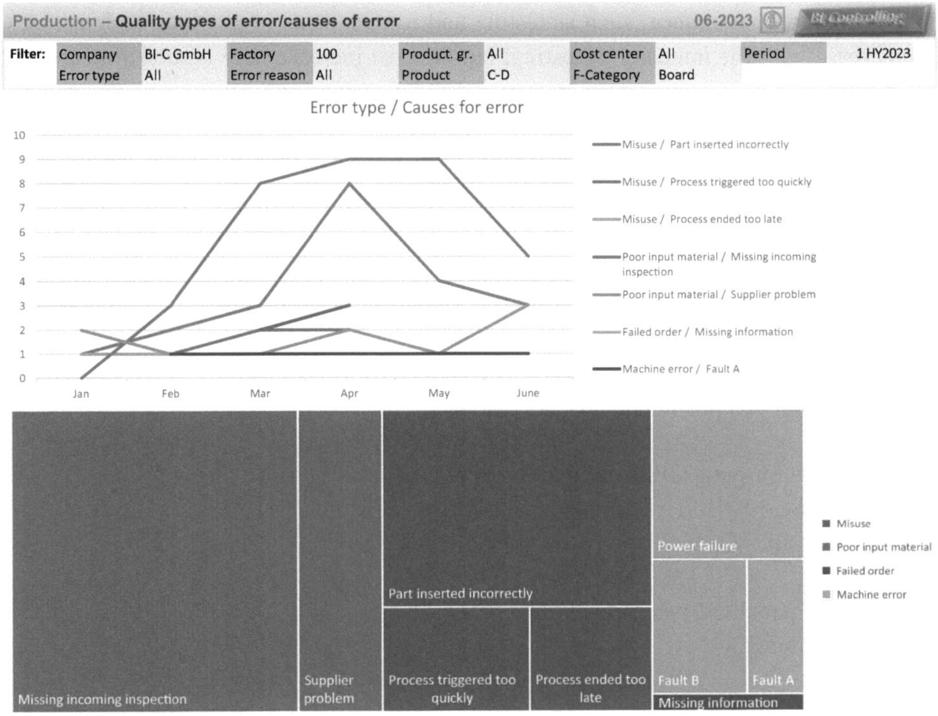

Fig. 3.137 Quality cockpit—error type/causes for error

deviations. The line and bar charts indicate seasonal effects of the selected row along the product hierarchy. High reject rates occurred in March and April. Here, the jump to a detailed analysis of the types and causes of errors is useful (see Fig. 3.137).

Using differentiated selection filters, error categories, error types and error causes can be examined for individual products, but also for entire areas. With the help of the line chart, the trend of errors can be analyzed. In the treemap below, the eye is immediately drawn to the most frequent error types and causes, which need to be eliminated. A failure mode and effect analysis (FMEA) is useful here to identify, assess and correct these quality risks.[241]

If errors are detected late, e.g. in the final assembly or even at the customer, the quality costs associated with them, such as rework costs, return logistics costs and warranty penalties, are very high. It is assumed that there is an increase factor of 10 in quality costs per value-added phase and thus an exponential increase in costs. The costs for qual-

[241] See Horváth and Urban (1990, p. 71).

ity inspection and assurance, such as testing and rework costs in all value-added stages, also increase with the intensity of testing, but the cost increase here is moderate and not exponential. For this reason, many companies today strive for a zero defect rate in connection with a total quality management.[242] In science and in some companies in practice, an analysis of the quality costs is suggested by separating all quality costs of the company in a so-called quality cost accounting[243] into the following components:

Quality costs for quality conformity

- general quality costs for quality assurance such as training, certificates, attestations
- testing costs for the conformity of the examined good parts

Quality costs due to detected quality deviations

- testing costs for parts with quality deviations
- rework costs
- costs for scrap
- returns
- contractual penalties, warranty claims
- damage to image

Since such a separation would entail a considerable amount of posting effort in most cost accounting systems, this breakdown is rarely found in practice. It is especially difficult to compile the cost types from different cost centers in a special calculation.

Therefore, many companies in quality management rather pursue the quality side from the performance or defect side and not from the cost side.

For the quality costs, mainly the costs of the quality assurance units and departments such as testing laboratories etc. are analyzed via the cost center report already described. The costs are usually allocated as a manufacturing overhead surcharge or more modernly via cost drivers of the process costing, e.g. costs per test position.

3.9.5 Cockpit Reports for Purchasing

The purchasing management and responsible employees in purchasing need, like the corporate management and the controlling, information for the successful management of the purchasing area or the procurement. The term procurement is used synonymously with the term purchasing. The most important main targets of purchasing regarding the provision of goods and services for the company can be expressed in the seven "Rs":

---

[242] This is also underlined by the contribution of Wildemann (1992, pp. 761–782).

[243] See Seghezzi and Fries (1995, p. 4).

Providing the

- right product/right goods,
- in the right quantity,
- in the right quality,
- at the right place,
- at the right time,
- with an economically appropriate use of resources (at the right cost)
- and supplemented by the right information on the flow of goods.

The seventh "R" bridges the gap to our modern information and communication society. Here, the logistics objectives are complemented by the **right information on the flow of goods**, as today customers and suppliers ideally should always know where the precisely specified goods are located.

Furthermore, procurement, on the one hand, should be as flexible as possible with regard to performance, quantities, time and place and be able to react to changes in demand. On the other hand, the procurement risk (e.g. dependence on a supplier) should be as low as possible, so that demands can always be met on time.[244]

The success of purchasing has an important leverage effect for the company's success. Compared to sales increases, whose result effect lies in the earned contribution margins after variable costs, the savings in purchasing are usually directly effective on the result.

From a control perspective, it makes sense again to divide the cockpit into the following areas via the menu bar:

- Operational controlling
- Strategic controlling
- Planning (budget/forecast/medium-term planning)

The entry area is the operational purchasing controlling. Here, the most important current information on purchasing can be analyzed with key figures and management reports.

The strategic purchasing controlling comprises the strategic analysis of the purchasing markets, the formation of strategic purchasing objectives as well as the strategic purchasing planning and control. In the strategic analysis, the future perspectives regarding the required products and potential suppliers are examined. From the status analysis, strategic objectives and plans can be derived, which, for example, result in a roadmap planning for purchasing. Here, strategic purchasing projects (e.g. the development of strategic supplier partnerships or the identification of new purchasing needs due to product innovations) are scheduled and controlled.

---

[244] Cf. Klein and Schentler (2016, p. 33 f.).

Fig. 3.138 Purchasing: Start cockpit

In the area of purchasing planning, the entry into the procurement-relevant parts of the budget planning, the forecast or the medium-term planning takes place. A status and tracking report shows the current processing status of the individual planning areas and approval rounds as well as their deadlines. For purchasing, the following planning areas are particularly important:

- Resource planning (e.g. by deriving the quantity requirements from the bill of materials and work plans (compare the MRP runs in Sect. 3.12.3) or the calculation structures as well as other resource planning)
- Planning of incoming freight and other incidental procurement costs
- Purchasing cost planning

In the following, the example focuses on the operative purchasing controlling.

In the start cockpit, the analyst of procurement should get a quick overview of the current situation in purchasing based on key figures and diagrams (see Fig. 3.138). It should cover central questions and important areas of action of purchasing, e.g.:

- Does purchasing achieve the set goals of purchasing planning in terms of purchasing volume and costs? How does the purchasing volume develop?
- How much purchasing volume do our buyers achieve on average?
- How high is the saving potential that was exploited?
- How high are the purchasing costs per order on average?
- How well do our suppliers deliver in terms of quantity, quality and deadlines?
- How fast is the average delivery time?

Important hierarchies in purchasing are especially the product group hierarchy as part of the product hierarchy, the supplier hierarchy and hierarchies or structures of the warehouses, the purchasing organization and the delivery regions.

The key figures in the purchasing area are uniformly structured in accordance with the overall system. For this purpose, speedometers, tiles, column or bar charts are suitable for visualization to reflect the state of the current situation. In the present example, actual values are always displayed with a black bar. The target value is shown by means of a "blue" marker line and the previous year value by a "grey" bar. The threshold ranges can be quickly analyzed using the traffic light color of the orientation line. The number of the achieved actual values is highlighted in color and indicates to what extent the actual value falls below, reaches or exceeds the target value.

The top performance indicator in the operational purchasing cockpit is the performance indicator **purchasing volume per employee**. Based on the number of purchasers, it serves as a benchmark between companies and other purchasing organizations. The development of the top performance indicator **purchasing volume** is shown graphically in an accumulated way, so that the achievement of objectives and, if recorded, a forecast can be quickly analyzed. Alternatively, the current purchasing volume could be, for example, **the weekly or monthly value**, which depends on the temporal control intensity of the purchasing department.

The purchasing success can be measured by the performance indicator **saving quota**. It shows the achieved saving volume of the purchasing department. This performance indicator is not easy to determine in practice, as the achieved prices have to be measured against existing or set price structures. The inventory management system should ideally enable such price comparisons and provide algorithms for the existing prices. As a reference value, for example, the last achieved price or target prices (e.g. market prices) could be maintained. If the performance indicator cannot be measured, it is advisable to move up other performance indicators. For example, the performance indicator bound capital costs can be included; this indicates how much costs are necessary to tie up the inventories. It raises the analyst's awareness to the storage and interest costs of the capital tied up in the warehouse. However, the performance indicator could also be assigned to the logistics area.

The performance indicator **cost/order process** indicates how expensive an order process is on average. It is a performance indicator for measuring the efficiency and measures the ratio between the resources used and the number of order processes. An alternative performance indicator, which measures the economic use and the appropriate deployment of purchasing resources for the purchasing tasks, is the performance indicator **purchasing costs/purchasing volume**.[245]

The **delivery date reliability** indicates in percent how punctually our suppliers have delivered. If the performance indicator is extended by the correct delivery parts (by quality and quantity), it can be gradually expanded to the entire **delivery reliability**. Alternatively or additionally, the performance indicator **incorrect delivery rate** can signal the proportion of incorrectly delivered goods here.

The performance indicator **average delivery time in days** indicates how long the company has to wait for needed goods starting from the order.

A basic selection of detailed analysis reports for purchasing can be accessed via a menu control bar (on the right in the example). The following areas are distinguished:

- purchasing analyses
- buyer analyses
- price/condition analyses
- supplier analyses
- inventory and delivery analyses
- cost analyses
- comments and actions

The areas of cost analysis and comments and actions have already been dealt with in the area of corporate management and are not redundantly listed below.

3.9.5.1 Purchasing Analyses

The purchasing analysis allows to start the examination of the **purchasing volume** and other purchasing-relevant key figures (such as the **purchasing quantity**) along the product group hierarchy down to the individual purchasing item (see Fig. 3.139). The purchasing analysis can be evaluated in various ways using the selection criteria of the menu bar (product group, supplier, purchasing organization, order type, etc.).

The selected key figure in purchasing can be evaluated monthly and accumulatedly in the uniform column scheme. Alternative key figures can be changed via the filter. The growth compared to the previous year and the target achievement compared to the planned value are visually highlighted for the key figure.

The diagrams behave dynamically to the selected rows. Thus, individual items or selected product groups can be analyzed with regard to their monthly development and in their accumulated view for all value types (actual, plan and forecast).

A powerful tool in the context of purchasing analysis is the ABC analysis (see Fig. 3.140).

245 Cf. e.g. Gladen (2014, p. 281).

Material group hierarchy	PY	PL	ACT	ΔPY	ΔPL	ΔPY%	ΩPL%	PY	PL	ACT	ΔPY	ΔPL	ΔPY%	ΩPL%
+Total	**1,080**	**1,188**	**1,225**	**145**	**37**	**13.4%**	**103.1%**	**5,661**	**6,020**	**5,960**	**299**	-60	**5.3%**	**99.0%**
+Product group A	**565**	**618**	**635**	**70**	**17**	**12.4%**	**102.8%**	**3,411**	**3,570**	**3,570**	**159**	**0**	**4.7%**	**100.0%**
+Article group A1	**350**	**395**	**426**	**76**	**31**	**21.7%**	**107.8%**	**2,121**	**2,300**	**2,210**	**89**	-90	**4.2%**	**96.1%**
Article A1.1	200	215	186	-14	-29	-7.0%	86.5%	1,221	1,400	1,350	129	-50	10.6%	96.4%
Article A1.2	90	100	150	60	50	66.7%	150.0%	540	500	560	20	60	3.7%	112.0%
Article A1.3	60	80	90	30	10	50.0%	112.5%	360	400	300	-60	-100	-16.7%	75.0%
+Article group A2	**215**	**223**	**209**	-6	-14	**-2.8%**	**93.7%**	**1,290**	**1,270**	**1,360**	**70**	**90**	**5.4%**	**107.1%**
Article A2.1	30	43	39	9	-4	30.0%	90.7%	240	260	230	-10	-30	-4.2%	88.5%
Article A2.2	100	120	100	0	-20	0.0%	83.3%	600	610	620	20	10	3.3%	101.6%
Artikcl A2.3	85	60	70	-15	10	-17.6%	116.7%	450	400	510	60	110	13.3%	127.5%
+Product group B	**515**	**570**	**590**	**75**	**20**	**14.6%**	**103.5%**	**2,250**	**2,450**	**2,390**	**140**	-60	**6.2%**	**97.6%**
+Article group B1	**250**	**290**	**330**	**80**	**40**	**32.0%**	**113.8%**	**1,110**	**1,200**	**1,180**	**70**	-20	**6.3%**	**98.3%**
Article B1.1	100	110	90	-10	-20	-10.0%	81.8%	610	600	650	40	50	6.6%	108.3%
Article B1.2	90	100	150	60	50	66.7%	150.0%	250	300	210	-40	-90	-16.0%	70.0%
Article B1.3	60	80	90	30	10	50.0%	112.5%	250	300	320	70	20	28.0%	106.7%
+Article group B2	**265**	**280**	**260**	-5	-20	**-1.9%**	**92.9%**	**1,140**	**1,250**	**1,210**	**70**	-40	**6.1%**	**96.8%**
Article B2.1	80	100	90	10	-10	12.5%	90.0%	350	400	390	40	-10	11.4%	97.5%
Article B2.2	100	120	100	0	-20	0.0%	83.3%	590	600	610	20	10	3.4%	101.7%
Article B2.3	85	60	70	-15	10	-17.6%	116.7%	200	250	210	10	-40	5.0%	84.0%

Fig. 3.139 Purchasing: Purchasing analysis

The **dynamic ABC analysis** filters the selected result indicator (e.g. purchasing volume) on several selected result objects. Unlike in the filter area, the importance (ABC classification) of the objects can be used directly visually for the analysis. Due to the limitation of the screen area, four dynamic ABC analysis tables are usually recommended. The most important result objects in the example are the items based on the product group hierarchy and the suppliers (top), followed by the warehouses and the buyers (bottom). The result objects can be exchanged, for example, by the purchasing organization or the region. An alternative would be the placement of corresponding filters in the menu bar. The order of the objects is usually displayed in the ABC logic from high to low. A change is possible via the radio button from fast to slow sellers. In addition to the colour coding based on set limits for the ABC categories, miniature bar charts help to assess the short-term trend of purchases.

The classification of the classes is to be done individually by the company. Usually, the following 3 classes (A, B and C) are formed:

- A-items: already 20–30% of all items account for 70–80% of the total purchasing volume.

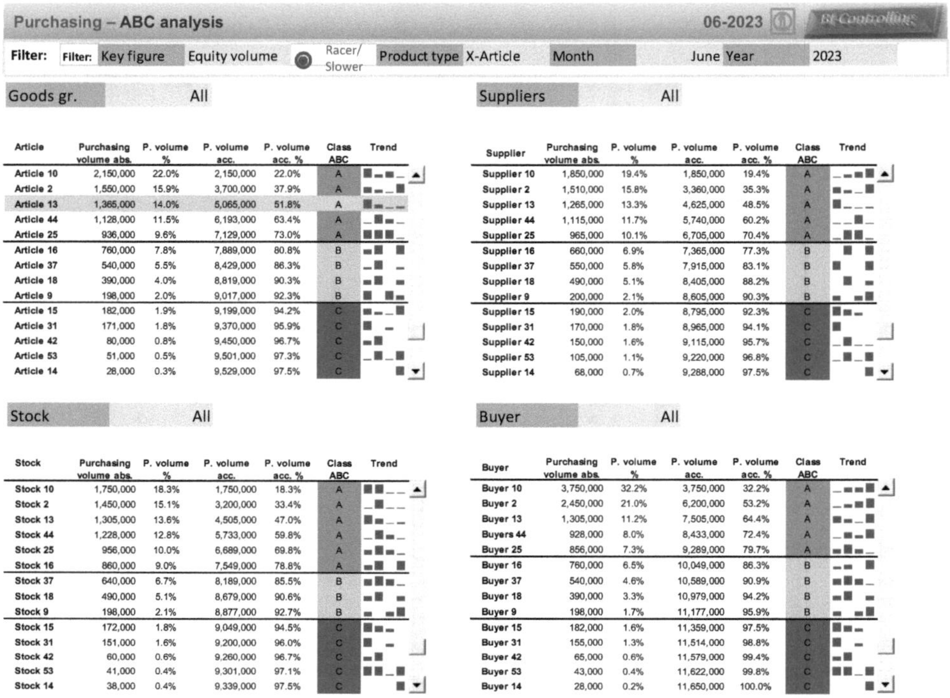

Article	Purchasing volume abs.	P. volume %	P. volume acc.	P. volume acc. %	Class ABC	Trend
Article 10	2,150,000	22.0%	2,150,000	22.0%	A	
Article 2	1,550,000	15.9%	3,700,000	37.9%	A	
Article 13	1,365,000	14.0%	5,065,000	51.8%	A	
Article 44	1,128,000	11.5%	6,193,000	63.4%	A	
Article 25	936,000	9.6%	7,129,000	73.0%	A	
Article 16	760,000	7.8%	7,889,000	80.8%	B	
Article 37	540,000	5.5%	8,429,000	86.3%	B	
Article 18	390,000	4.0%	8,819,000	90.3%	B	
Article 9	198,000	2.0%	9,017,000	92.3%	B	
Article 15	182,000	1.9%	9,199,000	94.2%	C	
Article 31	171,000	1.8%	9,370,000	95.9%	C	
Article 42	80,000	0.8%	9,450,000	96.7%	C	
Article 53	51,000	0.5%	9,501,000	97.3%	C	
Article 14	28,000	0.3%	9,529,000	97.5%	C	

Supplier	Purchasing volume abs.	P. volume %	P. volume acc.	P. volume acc. %	Class ABC	Trend
Supplier 10	1,850,000	19.4%	1,850,000	19.4%	A	
Supplier 2	1,510,000	15.8%	3,360,000	35.3%	A	
Supplier 13	1,265,000	13.3%	4,625,000	48.5%	A	
Supplier 44	1,115,000	11.7%	5,740,000	60.2%	A	
Supplier 25	965,000	10.1%	6,705,000	70.4%	A	
Supplier 16	660,000	6.9%	7,365,000	77.3%	B	
Supplier 37	550,000	5.8%	7,915,000	83.1%	B	
Supplier 18	490,000	5.1%	8,405,000	88.2%	B	
Supplier 9	200,000	2.1%	8,605,000	90.3%	B	
Supplier 15	190,000	2.0%	8,795,000	92.3%	C	
Supplier 31	170,000	1.8%	8,965,000	94.1%	C	
Supplier 42	150,000	1.6%	9,115,000	95.7%	C	
Supplier 53	105,000	1.1%	9,220,000	96.8%	C	
Supplier 14	68,000	0.7%	9,288,000	97.5%	C	

Stock	Purchasing volume abs.	P. volume %	P. volume acc.	P. volume acc. %	Class ABC	Trend
Stock 10	1,750,000	18.3%	1,750,000	18.3%	A	
Stock 2	1,450,000	15.1%	3,200,000	33.4%	A	
Stock 13	1,305,000	13.6%	4,505,000	47.0%	A	
Stock 44	1,228,000	12.8%	5,733,000	59.8%	A	
Stock 25	956,000	10.0%	6,689,000	69.8%	A	
Stock 16	860,000	9.0%	7,549,000	78.8%	A	
Stock 37	640,000	6.7%	8,189,000	85.5%	B	
Stock 18	490,000	5.1%	8,679,000	90.6%	B	
Stock 9	198,000	2.1%	8,877,000	92.7%	B	
Stock 15	172,000	1.8%	9,049,000	94.5%	C	
Stock 31	151,000	1.6%	9,200,000	96.0%	C	
Stock 42	60,000	0.6%	9,260,000	96.7%	C	
Stock 53	41,000	0.4%	9,301,000	97.1%	C	
Stock 14	38,000	0.4%	9,339,000	97.5%	C	

Buyer	Purchasing volume abs.	P. volume %	P. volume acc.	P. volume acc. %	Class ABC	Trend
Buyer 10	3,750,000	32.2%	3,750,000	32.2%	A	
Buyer 2	2,450,000	21.0%	6,200,000	53.2%	A	
Buyer 13	1,305,000	11.2%	7,505,000	64.4%	A	
Buyers 44	928,000	8.0%	8,433,000	72.4%	A	
Buyer 25	856,000	7.3%	9,289,000	79.7%	A	
Buyer 16	760,000	6.5%	10,049,000	86.3%	B	
Buyer 37	540,000	4.6%	10,589,000	90.9%	B	
Buyer 18	390,000	3.3%	10,979,000	94.2%	B	
Buyer 9	198,000	1.7%	11,177,000	95.9%	B	
Buyer 15	182,000	1.6%	11,359,000	97.5%	C	
Buyer 31	155,000	1.3%	11,514,000	98.8%	C	
Buyer 42	65,000	0.6%	11,579,000	99.4%	C	
Buyer 53	43,000	0.4%	11,622,000	99.8%	C	
Buyer 14	28,000	0.2%	11,650,000	100.0%	C	

Fig. 3.140 ABC analysis

- A and B items: approx. 50% of all items already account for up to 90% of the total purchasing volume altogether.
- C-items: the remaining 50% of all items account for only 10% of the total purchasing volume.

If the purchasing and logistics costs are at the centre of the analysis, the order or picking frequency of the items can alternatively be used as a key figure for the ABC analysis instead of the purchasing volume.

The dynamic ABC analysis demonstrates its power for analysis purposes in the selection of specific rows in any ABC analysis table. For example, if you select an item of the product group in the first ABC table, all values of the other ABC tables are restricted to this selected item. If you further limit the view in the buyer ABC table to a specific buyer, you can analyze his or her individual purchases for the selected item by supplier and delivery region in the other two ABC tables. The four combinations in the analysis provide many views on the processing of the purchasing market. Important purchasing-related questions can be easily analyzed, such as:

- From which suppliers and with which buyers is the largest value volume or quantity purchased? Who buys little?

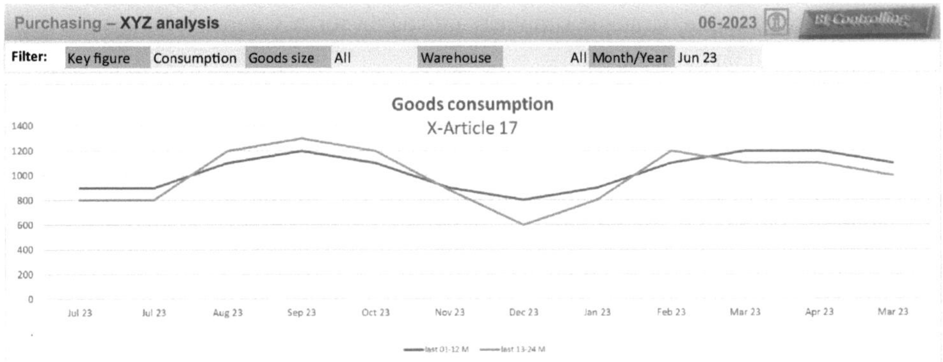

X-Articles

X-Articles	Goods consumption absolute	Consumption of goods %	Mean values of the goods consumption	Variant coefficient
Total	3,964,170	43.4%	330,348	12.6%
Article 2	1,510,000	16.5%	125,833	10.5%
Article 10	850,000	9.3%	70,833	9.8%
Article 41	523,174	5.7%	43,598	21.3%
Article 61	369,857	4.0%	30,821	17.6%
Article 32	236,044	2.6%	19,670	16.8%
Article 91	125,698	1.4%	10,475	12.3%
Article 44	115,000	1.3%	9,583	14.8%
Article 24	103,004	1.1%	8,584	14.8%
Article 56	65,981	0.7%	5,498	22.1%
Article 72	65,412	0.7%	5,451	19.3%
Article 88	34,210	0.4%	2,851	12.6%
Article 38	25,541	0.3%	2,128	19.5%
Article 17	12,400	0.1%	1,033	13.9%

Y-Articles

Y-Articles	Goods consumption absolute	Consumption of goods %	Mean values of the goods consumption	Variant coefficient
Total	3,445,741	37.7%	287,145	21.3%
Article 37	1,550,000	17.0%	129,167	33.4%
Article 16	660,000	7.2%	55,000	30.6%
Article 18	490,000	5.4%	40,833	44.1%
Article 9	200,000	2.2%	16,667	46.8%
Article 11	125,470	1.4%	10,456	30.5%
Article 13	111,045	1.2%	9,254	34.2%
Article 12	105,367	1.2%	8,781	49.5%
Article 23	98,622	1.1%	8,219	43.7%
Article 14	54,230	0.6%	4,519	32.7%
Article 63	23,100	0.3%	1,925	26.8%
Article 25	12,678	0.1%	1,057	28.9%
Article 73	9,865	0.1%	822	31.4%
Article 87	5,364	0.1%	447	44.9%

Z-Articles

Z-Articles	Goods consumption absolute	Consumption of goods %	Mean values of the goods consumption	Variant coefficient
Total	1,729,027	18.9%	144,086	61.4%
Article 31	1,170,000	12.8%	97,500	71.6%
Article 15	190,000	2.1%	15,833	66.4%
Article 42	150,000	1.6%	12,500	81.3%
Article 76	113,987	1.2%	9,499	891.4%
Article 53	105,040	1.1%	8,753	90.4%
Article 19	98,500	1.1%	8,208	91.4%
Article 26	88,066	1.0%	7,339	191.4%
Article 27	76,000	0.8%	6,333	291.4%
Article 39	67,410	0.7%	5,618	491.4%
Article 36	55,000	0.6%	4,583	391.4%
Article 48	12,067	0.1%	1,006	691.4%
Article 47	10,268	0.1%	856	591.4%
Article 59	5,047	0.1%	421	791.4%

Fig. 3.141 Purchasing: XYZ analysis

- Which items are ordered most or least frequently from which suppliers?
- Which warehouses are used most and least intensively for which products by which buyers or with which suppliers?
- Which buyer buys which items most frequently?
- ...

The ABC analysis is usefully complemented by the XYZ analysis in purchasing. The XYZ analysis illustrates the cyclical behavior of the goods consumption (see Fig. 3.141). The following classes are distinguished:

- X: regular, relatively constant goods consumption.
- Y: seasonal, periodically fluctuating goods consumption
- Z: irregular goods consumption

In the strategic purchasing planning, the item class combinations from the ABC and XYZ classification are often used to develop strategic goals, projects or measures for purchasing. For example, strategic suppliers are sought for AX items, whereas CZ items are only purchased on demand.

The XYZ analysis provides the classification of the individual articles e.g. with the help of the coefficient of variation. The coefficient of variation results from the division of the average mean value and the standard deviation of the goods consumption of a considered period and can e.g. be calculated already in the merchandise management system. Alternatively, the merchandise management system only provides the basic data, and the calculation is done in the data warehouse. The following classifications for the XYZ articles are typical:

- X articles: The coefficient of variation is between 0–25%
- Y articles: The coefficient of variation is between 25–50%
- Z articles: The coefficient of variation is greater than 50%

The analyst can dynamically display the goods consumption graphically for a selected article. For this purpose, he or her has e.g. the last 24 periods available. If the curves show similarities, seasonal effects can be detected. The selected article in the example, despite its X assignment (coefficient of variation is 13.9%), has regular consumption around €1,000 in addition to seasonal fluctuations (peaks in September and March/April).

For structural and regional considerations in purchasing, geographical and structural analyses are again suitable (Fig. 3.142). Here, the key figures to be analyzed (especially purchasing volume and quantities) are displayed graphically on a map (e.g. delivery or storage locations) or an object plan (e.g. storage facilities with storage locations).

For the structural analysis, pie charts, donut or sunburst charts are suitable for few analysis objects, as well as bar or column charts or treemap charts for several objects. This allows large and small result object areas such as product groups, supplier groups, order size classes, etc. to be quickly visually identified. The color classifications of the objects should ideally be consistent across the reports. The same colors should not be used for different objects. With many result objects and classification needs, the color assignment can be difficult to maintain throughout the cockpit.

The diagrams react dynamically to the selected filters in the menu bar and to the selected result objects in the diagrams below (e.g. by clicking on a segment in the pie chart or an area/column on the map).

3.9.5.2 Buyer Analyses

The performance of purchasing depends on its employees, especially the buyers. Here, the supply of goods is ensured and savings are achieved which have a direct impact on the increase of the company's profit. The buyer analysis offers, especially in connection with the reports of the purchasing analysis (dynamic ABC and XYZ analysis, geographical and structural analysis), starting points for an improved market development (see Fig. 3.143).

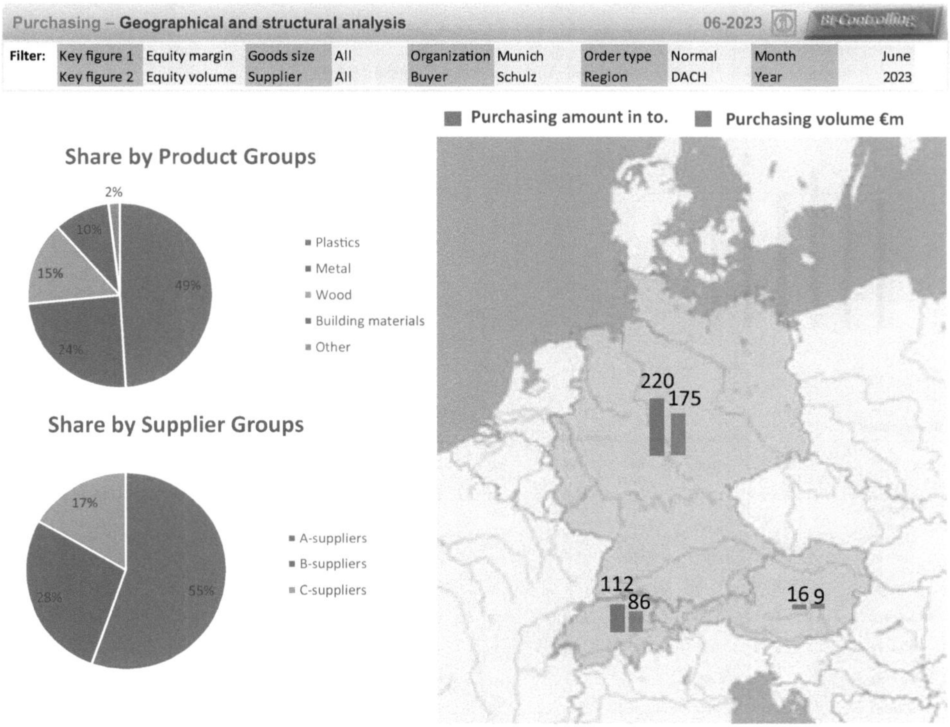

Fig. 3.142 Purchasing: Geographical and structural analyses

The buyer analysis shows the monthly and accumulated purchasing performance of the buyers as well as their development in the current year. Especially the key figures **purchasing volume** or **purchasing quantity** are selected. For this purpose, both the dynamic diagrams already known and the table values "growth compared to the previous year" and the "target achievement compared to the plan" can be used as visual support. The key figures **number of orders** and **average order** allow to assess the order sizes and the order performance. Tendentially, larger order volumes are more favorable in terms of order process costs. However, larger order quantities also may lead to an increased bound capital, and this must be kept in mind through the inventory analysis.

The key figures **saving** or **saving quota**express absolutely or as a percentage the savings achieved in the order/negotiation process. The determination of these key figures is not easy, as mentioned before. For this purpose, comparison prices for the achieved order prices must be determined or recorded. For example, existing achieved prices, target prices or market prices with the best or most cost-effective supplier at the same quality can be used. The saving quota should not be used too dominantly or exclusively in the control. The reliability and the quality of the goods delivery are other important control variables that should also be monitored.

Buyer groups	Trend	PY	PL	ACT	ΔPY	ΔPY%	YTD ΔPL	ΔPL%	Number of Orders	Ø Order	Saving	% Saving
+Total		**5,640**	**5,100**	**5,960**	**320**	**5.7%**	**860**	**116.9%**	97	61	173	3%
+Buyer group A		**3,960**	**3,600**	**4,260**	**300**	**7.6%**	**660**	**118.3%**	45	95	103	2%
Buyer A1		2,820	2,500	2,980	160	5.7%	480	119.2%	20	149	149	5%
Buyer A2		600	700	720	120	20.0%	20	102.9%	15	48	-29	-4%
Buyer A3		540	400	560	20	3.7%	160	140.0%	10	56	-17	-3%
+Buyer group B		**1,680**	**1,500**	**1,700**	**20**	**1.2%**	**200**	**113.3%**	52	33	70	4%
Buyer B1		840	800	850	10	1.2%	50	106.3%	21	40	34	4%
Buyer B2		240	200	230	-10	-4.2%	30	115.0%	16	14	-14	-6%
Buyer B3		600	500	620	20	3.3%	120	124.0%	15	41	50	8%

Fig. 3.143 Purchasing: Buyer analysis

The division of the buyers into a hierarchy allows a graded analysis, in which, for example, buyer groups, teams or the entire purchasing organization can be examined.

With the buyer analysis, the company and purchasing managements have a tool at their disposal that should be used very sensitively. Here, managers receive very personal information which can be misused. For example, if you put the buyers under constant pressure to achieve predefined goals (e.g. saving quotas), fears may arise that they will not be able to withstand the pressure, something which may lead to resignation and demotivation. This does not serve either the employee or the company. Rather, the instrument should be used for individual motivation and support. For example, the following question patterns could be helpful for this:

- Which buyers or purchasing organizations are currently very successful? By asking questions, sensitive market information and market development techniques can be recognized and multiplied.
- Which buyers are successful with which suppliers? This way, better access to the supplier can be identified after consultation with the buyer.

- In which regions is a lot or successful purchasing currently taking place? This way, regional differences and regional peculiarities can be asked.
- How large is the overall target deviation in the purchasing organization? In the team, you can motivate and encourage better to achieve or even exceed goals.
- Which suppliers or order types are doing very well or poorly with the top buyers? This way, the buyers can be alerted to sensitive industries and suppliers.

3.9.5.3 Price/Condition Analyses

Prices and conditions are very important cost drivers in purchasing for both retailers and manufacturers and they can have a significant impact on the company's results. This applies especially to the important trade items or the most important raw or operating materials that are needed in the value creation process. Here, price changes of important basic materials (e.g. oil or metal prices) can indicate important signals for the change of purchasing prices.

Important questions here are for example:

- What are the current prices?
- How will the prices develop?
- How large are the savings of the achieved prices compared to the reference prices, e.g., list prices and market prices?
- How do the prices change depending on the benchmark prices?
- What conditions could be achieved?
- Which suppliers grant the best or the worst conditions?

The purchase price analysis shows the change of the average **net purchase prices** compared to the previous year based on the product group hierarchy (see Fig. 3.144). Since prices usually affect quantity changes, the changes of quantity and purchase volumes are shown in parallel. Significant **price, quantity and volume changes** can be quickly visually captured by the positive and negative deviations of the bar lengths. The net purchase price development is analyzed in the dynamic charts above in the temporal development. Changes to the negotiation basis, to the average market price or a target price such as the price ceiling can be quickly examined by the combined column-line chart representation. Current prices can either be displayed constantly or, if there are too many entries, only when the respective column or line is marked with the mouse pointer in a desktop application or with the finger on the tablet.

The price dependencies of the net purchase price on central raw materials such as oil or metals (in the example above the zinc price LME) can be displayed and compared via benchmark prices. Thus, price changes can be anticipated earlier. In the example above, the fluctuations of the change of the benchmark price are much larger than those of the net purchase price changes. Alternatively, the price-sales function could be included as a graph which shows for which purchase quantities which prices were paid, so that quantity-related price jumps can be analyzed.

The price analysis is refined by the condition analysis (see Fig. 3.145).

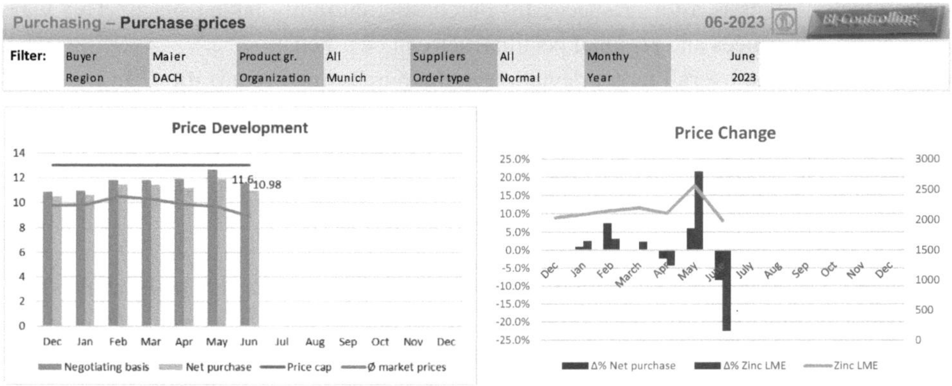

		PY	ACT	ΔPY	YTD ΔPY% Price	ΔPY% Quantities	ΔPY% P. vol.
+Total	**Purchasing volume**	**10,269**	**9,336**	**-933**			**-9.1%**
+Commodity gr. A	Purchasing volume	**4,108**	**3,734**	**-373**			**-9.1%**
Article 1	**Ø Net price**	9.64	10.98	1.34	13.9%		
	Quantities	12	13	1		8.3%	
	Purchasing volume	116	143	27			23.4%
Article 2	**Ø Net price**	10.63	9.64	-0.99	-9.3%		
	Quantities	205	215	10		4.9%	
	Purchasing volume	2,179	2,073	-107			-4.9%
Article 3	**Ø Net price**	120.86	116.85	-4.01	-3.3%		
	Quantities	15	13	-2		-13.3%	
	Purchasing volume	1,813	1,519	-294			-16.2%

Fig. 3.144 Purchasing: Purchase price analysis

The savings achieved by purchasing can be examined in stages by supplier group down to the supplier level with the help of the condition analysis. Starting from the negotiation basis, the **saving quota** shows the savings potential that could be achieved with the supplier. This allows identifying those who, for example, have hardly achieved any price reductions. In combination with the filtering of further attributes, such as the items in the product hierarchy or the order type, suggestions for the purchasing negotiations can be derived. The total saving quota results from the purchasing volume valued at net purchasing prices by granted and realized cash discounts and rebates. The purchasing volume reduced by this is also called purchasing volume valued at net-net purchasing price (here briefly NNPP) in the trade. Cash discounts are revenue reductions granted by the supplier, if the goods are paid by the company within a specified period. This so-called supplier credit is usually very high, so it is advisable, if the liquidity situation allows, to exploit the cash discount income by paying early. Rebates are revenue reductions that are often granted due to negotiated (higher) purchase quantities or other negotiated services.

Bundling effects in purchasing usually offer great advantages in achieving savings in price negotiations, especially the achievement of rebates. However, the bundling and optimization in purchasing and thus the achievement of lower purchasing costs must be

Purchasing – Conditions 06-2023 BI-Controlling

Filter:								
	Buyer	Sippel	Supplier	All	Order type	Normal	Date from:	01.01.2023
	Organization	Munich	Product gr.	All	Region	DACH	Date to:	30.06.2023

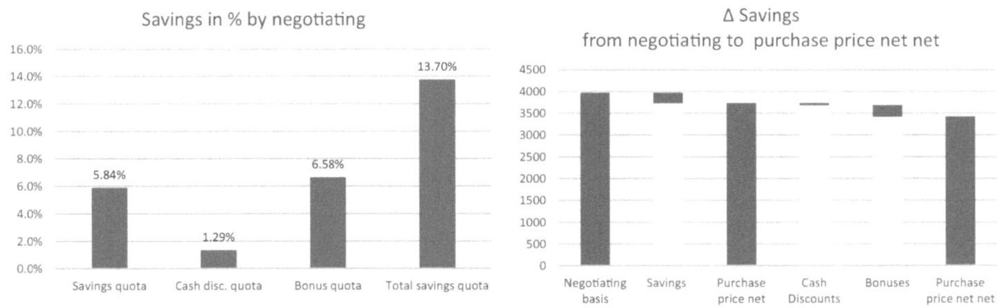

	Negotiating basis	Savings	Savings quota	Purchase price net	Cash Discounts	Bonuses	Purchase price net net	Total savings	Savings quota
+Total	9,869	533	5%	9,336	138	522	8,676	1,192	12%
+Supplier Gr. I	3,966	232	6%	3,734	51	261	3,422	543	14%
Supplier 1	822	75	9%	747	15	37	695	127	15%
Supplier 2	1,970	103	5%	1,867	19	187	1,662	308	16%
Supplier 3	382	28	7%	353	7	18	329	53	14%
Supplier 4	182	5	3%	177	4	9	164	18	10%
Supplier 5	245	7	3%	238	5	7	226	19	8%
Supplier 6	108	3	3%	105	2	3	100	8	8%
Supplier 7	90	4	5%	86	0	0	86	4	5%
Supplier 8	78	2	3%	76	0	0	76	2	3%
Supplier 9	88	3	4%	85	0	0	85	3	4%

Fig. 3.145 Purchasing: Condition analysis

coordinated with the optimization of inventory management and goods provision for production or shipping and the higher inventory and capital commitment costs that arise here. Here, it is useful to jump in the cockpit to the inventory analysis (see Sect. 3.9.5.5).

If savings in purchasing, e.g. rebate rates, are planned with the suppliers, the target achievement should be supplemented in the report. The most important conditions and savings can be visually captured via the line selection over the dynamic diagrams shown above.

3.9.5.4 Supplier Analyses

The most important business partner of the purchasing department is the supplier. In the operating supplier management, the purchasing department wants to analyse the suppliers as comprehensively as possible from different perspectives. The following questions are in the focus:

- With which suppliers are which purchasing volumes and quantities realized for which articles or product groups? Were the purchasing target quantities and volumes achieved?

Purchasing – Suppliers 06-2023 BI Controlling

Filter:								
	Buyer	Maier	Region	DACH	Order type	Normal	Month	June
	Product gr.	All	Organization	All	Supplier size	All	Year	2023

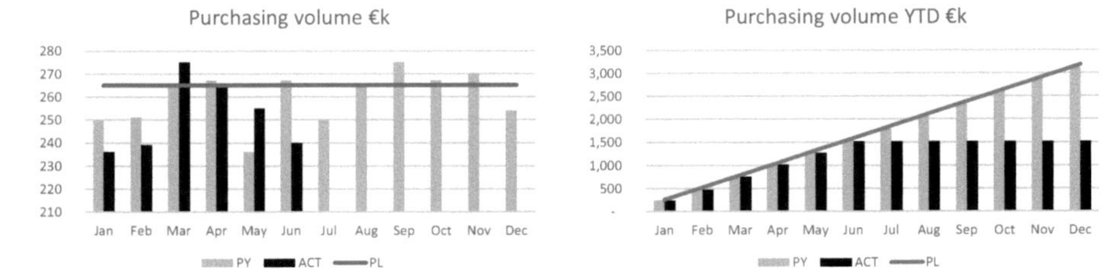

	Purchasing volume PL	Purchasing volume ACT	ΔPL%	Purchasing quantities	Ø Purchase price net	Savings Quota	Delivery reliability	Ø Delivery time	Complaint
+Total	**9,000**	**8,589**	95.4%	**871**	**9.86**	**7%**	**89.0%**	**21.8**	**8**
Supplier 1	1,400	1,340	95.7%	150	8.93	8%	92.0%	23.5	0
Supplier 2	1,590	1,510	95.0%	200	7.55	15%	81.0%	25.3	5
Supplier 3	1,200	1,265	105.4%	96	13.18	12%	95.0%	15.4	1
Supplier 4	600	856	142.7%	65	13.17	5%	94.0%	28.4	0
Supplier 5	2,500	2,364	94.6%	160	14.78	0%	96.0%	30.6	2
Supplier 6	1,710	1,254	73.3%	200	6.27	4%	90.0%	20.5	0

	Purchasing volume PL	Purchasing volume ACT	ΔPL%	Purchasing quantities	Ø Purchase price net	Savings Quota	Delivery reliability	Ø Delivery time	Complaint
+Total	**1,590**	**1,510**	95.0%	**61**	**24.75**	**15%**	**81.0%**	**25.3**	**5**
Article 1	300	302	100.7%	10	30.20	10%	75.0%	20.5	1
Article 2	150	151	100.7%	15	10.07	20%	69.0%	25.9	0
Article 3	490	453	92.4%	9	50.33	25%	56.0%	30.3	3
Article 4	70	76	107.9%	3	25.17	26%	81.0%	32.4	0
Article 5	250	227	90.6%	4	56.63	8%	100.0%	16.5	0
Article 6	330	302	91.5%	20	15.10	7%	91.0%	15.6	1

Fig. 3.146 Purchasing: Supplier analysis

- How high are the average net purchasing prices and saving quotas for different suppliers?
- How satisfied was the company with the punctuality, quantity availability and quality of the delivered goods?

These central questions can be comprehensively answered by the dynamically combinable tables of the supplier analysis in combination with the dynamic diagrams (see Fig. 3.146). In the top table, the suppliers can be examined according to the supplier hierarchy. Here, by restricting the filter area, all articles, product groups and individual articles in the product group hierarchy can be selected. The key figures can be analysed in the column structure. The target achievement of the **purchasing volume** is highlighted in the table with traffic light signal colours. The temporal distribution of the purchases and the accumulated view of the purchases can be read from the dynamic diagrams (top) for each selected row combination of suppliers and articles.

Furthermore, the **purchasing volume quantities** and **average purchase prices** provide indications for the supplier selection and the future negotiation strategy. The length of the miniature bar charts indicate the **saving quota** , **average delivery times** and the **number of complaints**. In addition, the **delivery reliability** is an important key figure for assessing the suppliers with regard to quantity- and deadline-compliant delivered goods. These important key figures form the foundation of a 360-degree analysis in the supplier analysis, which allows further detailed investigations regarding purchasing organization, buyer, order types and regional aspects through the filter restriction.

3.9.5.5 Inventory and Delivery Analyses

One of the most important goals of purchasing is certainly the economic assurance of the availability of the goods necessary for production or sales. To achieve this goal, appropriate inventory and delivery analyses are necessary for controlling purchasing. Central questions are:

- How long will the stocks last to meet the supply of production or shipping?
- How high is the warehouse utilization and the average inventory level?
- How long does the delivery take? How long is the capital tied up in the warehouse?
- How expensive is the capital commitment?
- How high is the average order value, how often are orders placed?

As peak indicators of the inventory analysis (see Fig. 3.147), various sizes are suitable, e.g. the **inventory turnover**, the **inventory range,** but also the **inventory level** and the associated bound capital costs. Since the **bound capital costs** rather represents a comprehensive measure for economic efficiency control, this indicator was highlighted as a single indicator. The other sizes can be analyzed in detail for the product groups and items or the warehouses and storage locations in the dynamic tables and graphs.

Further objects of investigation in this area can be limited by the filters, such as the purchasing organization, the suppliers or the order types. The analysis of the stock of goods in the considered warehouses is possible in terms of quantity and value. As a guide, the **average stock / inventory level** and the **stock / inventory turnover** are displayed. Based on the average stock / inventory level, it can be seen how high the stocks of a selected period were on average. It can be displayed for both quantities and value sizes. The stock / inventory turnover quickly indicates, by the signal color, which items are "fast and slow movers", that is, the storage period in the warehouse until the use or shipment of goods is short or long. Very low inventory turnover frequencies indicate so-called zero movers, which have not been sold for a longer period of time. The **storage period** in days is additionally visualized by the deflection in the miniature bar chart. Zero movers are to be marked especially.

An essential lever in purchasing is the bundling of goods for ordering and transport. This allows for more favorable conditions and lower order and transport costs per unit. On the other hand, there are the storage and interest costs of the tied-up capital. For this reason, it makes sense to consider the number of orders, the average quantity and amount of the order value, and the average consumption. High inventory levels may be justified by strongly bundled orders, if the expected consumption is high and the savings are worthwhile. Excessive order quantities, on the other hand, should be avoided, if the storage and

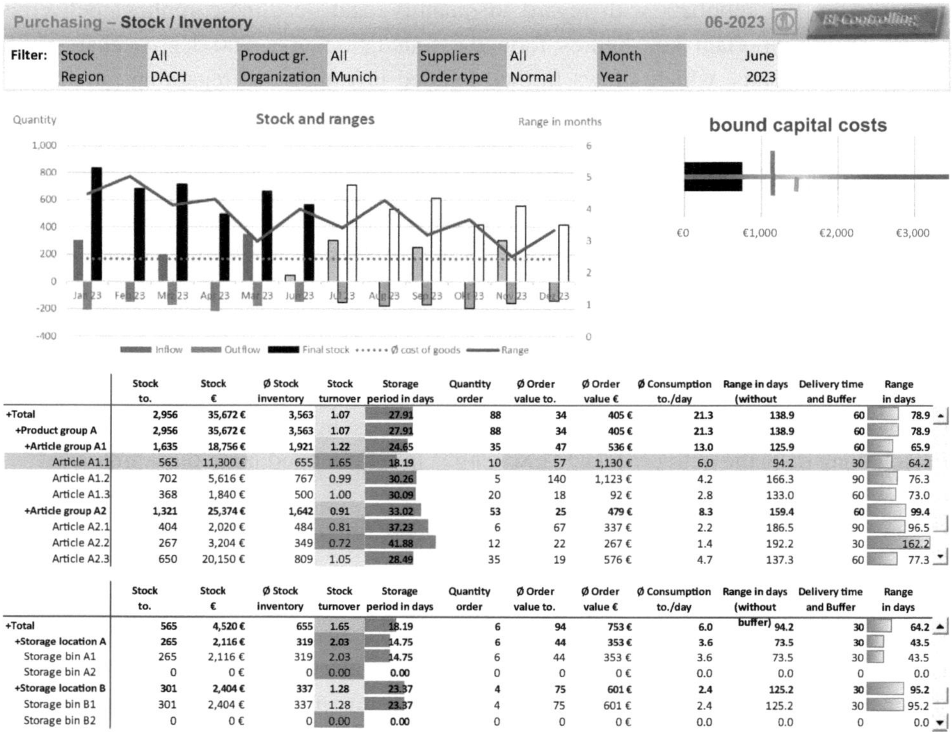

	Stock to.	Stock €	Ø Stock inventory	Stock turnover	Storage period in days	Quantity order	Ø Order value to.	Ø Order value €	Ø Consumption to./day	Range in days (without buffer)	Delivery time and Buffer	Range in days
+Total	2,956	35,672 €	3,563	1.07	27.91	88	34	405 €	21.3	138.9	60	78.9
+Product group A	2,956	35,672 €	3,563	1.07	27.91	88	34	405 €	21.3	138.9	60	78.9
+Article group A1	1,635	18,756 €	1,921	1.22	24.65	35	47	536 €	13.0	125.9	60	65.9
Article A1.1	565	11,300 €	655	1.65	18.19	10	57	1,130 €	6.0	94.2	30	64.2
Article A1.2	702	5,616 €	767	0.99	30.26	5	140	1,123 €	4.2	166.3	90	76.3
Article A1.3	368	1,840 €	500	1.00	30.09	20	18	92 €	2.8	133.0	60	73.0
+Article group A2	1,321	25,374 €	1,642	0.91	33.02	53	25	479 €	8.3	159.4	60	99.4
Article A2.1	404	2,020 €	484	0.81	37.23	6	67	337 €	2.2	186.5	90	96.5
Article A2.2	267	3,204 €	349	0.72	41.88	12	22	267 €	1.4	192.2	30	162.2
Article A2.3	650	20,150 €	809	1.05	28.49	35	19	576 €	4.7	137.3	60	77.3

	Stock to.	Stock €	Ø Stock inventory	Stock turnover	Storage period in days	Quantity order	Ø Order value to.	Ø Order value €	Ø Consumption to./day	Range in days (without buffer)	Delivery time and Buffer	Range in days
+Total	565	4,520 €	655	1.65	18.19	6	94	753 €	6.0	94.2	30	64.2
+Storage location A	265	2,116 €	319	2.03	14.75	6	44	353 €	3.6	73.5	30	43.5
Storage bin A1	265	2,116 €	319	2.03	14.75	6	44	353 €	3.6	73.5	30	43.5
Storage bin A2	0	0 €	0	0.00	0.00	0	0	0 €	0.0	0.0	0	0.0
+Storage location B	301	2,404 €	337	1.28	23.37	4	75	601 €	2.4	125.2	30	95.2
Storage bin B1	301	2,404 €	337	1.28	23.37	4	75	601 €	2.4	125.2	30	95.2
Storage bin B2	0	0 €	0	0.00	0.00	0	0	0 €	0.0	0.0	0	0.0

Fig. 3.147 Purchasing: Stock / Inventory analysis

interest costs outweigh the savings achieved. The order sizes should be appropriate. Outliers up and down can be quickly found and analyzed in detail. To estimate the **range** of the inventory, the **delivery and buffer time** is displayed in addition to the consumption. The range analysis signals dangers of under-supply, but also risks of over-supply.

The important key figures inventory and inventory range can be used in the dynamic diagram for each selected row in the product group hierarchy or the warehouse hierarchy to analyze the temporal development. Important order and consumption cycles can be recognized by the inflows and outflows. The average of the goods consumption is additionally shown in a dotted line. The current inventory level is highlighted in black in the current month. Taking the planning into account, future supply bottlenecks or surpluses can be anticipated, if the line to the inventory range drops or rises too sharply. As a complement to the range analysis, the key figure "failure rate" can also be used; this figure measures the number of orders that cannot be delivered on time due to insufficient inventory levels.

The overall report allows the competing targets in purchasing (especially high availability, low inventory, high throughput) to be viewed holistically. This analysis is interesting in a similar form for the logistics or shipping area. In logistics, capacity utilization analyses and special storage, transport and picking services are also to be included in the focus of control.

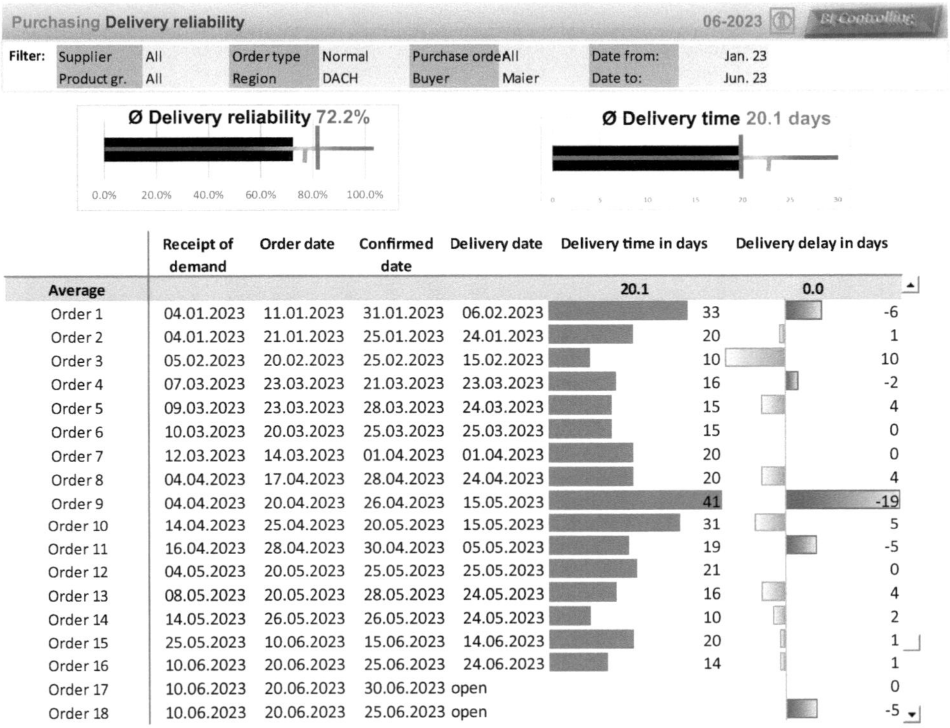

	Receipt of demand	Order date	Confirmed date	Delivery date	Delivery time in days	Delivery delay in days
Average					20.1	0.0
Order 1	04.01.2023	11.01.2023	31.01.2023	06.02.2023	33	-6
Order 2	04.01.2023	21.01.2023	25.01.2023	24.01.2023	20	1
Order 3	05.02.2023	20.02.2023	25.02.2023	15.02.2023	10	10
Order 4	07.03.2023	23.03.2023	21.03.2023	23.03.2023	16	-2
Order 5	09.03.2023	23.03.2023	28.03.2023	24.03.2023	15	4
Order 6	10.03.2023	20.03.2023	25.03.2023	25.03.2023	15	0
Order 7	12.03.2023	14.03.2023	01.04.2023	01.04.2023	20	0
Order 8	04.04.2023	17.04.2023	28.04.2023	24.04.2023	20	4
Order 9	04.04.2023	20.04.2023	26.04.2023	15.05.2023	41	-19
Order 10	14.04.2023	25.04.2023	20.05.2023	15.05.2023	31	5
Order 11	16.04.2023	28.04.2023	30.04.2023	05.05.2023	19	-5
Order 12	04.05.2023	20.05.2023	25.05.2023	25.05.2023	21	0
Order 13	08.05.2023	20.05.2023	28.05.2023	24.05.2023	16	4
Order 14	14.05.2023	26.05.2023	26.05.2023	24.05.2023	10	2
Order 15	25.05.2023	10.06.2023	15.06.2023	14.06.2023	20	1
Order 16	10.06.2023	20.06.2023	25.06.2023	24.06.2023	14	1
Order 17	10.06.2023	20.06.2023	30.06.2023	open		0
Order 18	10.06.2023	20.06.2023	25.06.2023	open		-5

Fig. 3.148 Purchasing: Delivery reliability

Special reports on delivery (see Fig. 3.148) and delivery quality (see Fig. 3.149) deal with the following questions:

- Was the supplier able to meet the promised delivery times?
- What was the proportion of incorrect deliveries?
- How long do the deliveries take?
- How satisfied or dissatisfied is the company with the deliveries? How high are the returns, complaints and credit notes?
- For what reasons is the company dissatisfied with the deliveries?

The delivery reliability can be evaluated in detail for the most important control-relevant objects in purchasing (supplier, item, order types, regions etc.) using the filter criteria. The top indicators are the **delivery reliability** and the **average delivery time** for the selected filter criteria (see Fig. 3.148).

For the individual orders, the demand entry, the order date, the promised delivery date and the delivery date are displayed. The **delivery time in days** and the resulting **delivery delay** can be quickly recognized by the miniature bar charts. The sorting function allows the most important outliers to be summarized.

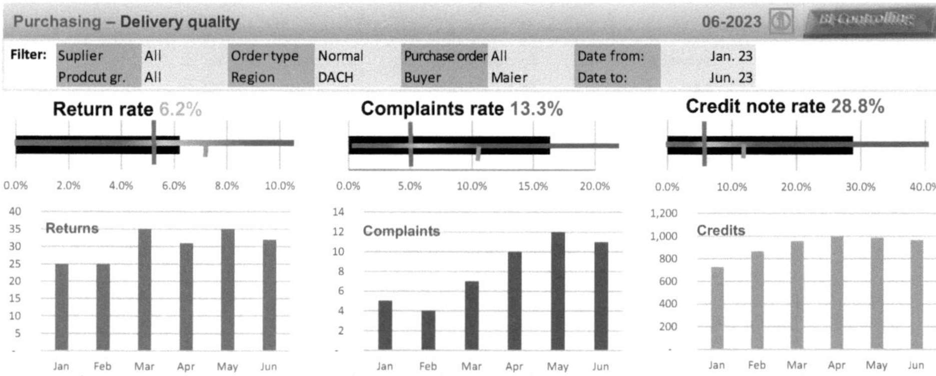

	Quantity			No. of	No. of	Purchase value			No. of
	Incoming goods	Return delivery	Return rate	Reversals goods receipt	Complaints	Incoming invoice	Incoming credit note	Credit note rate	Reversals of incoming invoice
+Total	**2,954**	**183**	**6.2%**	**30**	**49**	**19,119**	**5,497**	**28.8%**	**21**
Supplier 1	180	8	4.4%	4	8	2,632	265	10.1%	4
Supplier 2	190	3	1.6%	-	-	3,654	-	0.0%	-
Supplier 3	260	10	3.8%	1	2	4,632	198	4.3%	2
Supplier 4	150	5	3.3%	3	5	3,489	200	5.7%	1
Supplier 5	210	25	11.9%	5	9	2,398	1,164	48.5%	5
Supplier 6	150	-	0.0%	-	-	2,314	-	0.0%	-
Supplier 7	350	50	14.3%	5	6	5,600	1,364	24.4%	1
Supplier 8	254	30	11.8%	3	4	3,641	1,236	33.9%	5
Supplier 9	100	15	15.0%	4	6	3,500	869	24.8%	2
Supplier 10	150	-	0.0%	-	-	2,364	-	0.0%	-
Supplier 11	150	10	6.7%	2	3	1,367	-	0.0%	-
supplier 12	350	6	1.7%	1	2	5,600	201	3.6%	1
Supplier 13	250	6	2.4%	1	2	3,478	-	0.0%	-
Supplier 14	210	15	7.1%	1	2	2,314	-	0.0%	-

Fig. 3.149 Purchasing: Delivery quality

In addition to **delivery reliability** and **average delivery time**, the **return rate**, the **complaint rate** and the **credit note rate** are listed as key indicators for measuring the quality of delivery (see Fig. 3.149). The return rate (return quota) shows the deliveries whose goods were sent back to the sender due to incorrect deliveries, delays, quality defects and other reasons. It is shown in terms of quantity and number (cancellation of goods receipts). The complaint rate measures the number of complained deliveries to the total number of goods deliveries. It illustrates the dissatisfaction of the company with the goods delivered by the supplier. The credit note rate indicates the percentage of the value-based credit notes measured against the amount of the incoming invoice sum. In addition to the value-based credit note sum, the number of cancelled incoming invoices is listed. This allows to distinguish between complete credit notes and partial credit notes, e.g. for goods of second choice or goods with defects.

Return, complaint and credit note rates can be usefully supplemented by a statistic of the causal reasons in detail.

The dynamic charts react to the row selection in the data table, based on which the development of returns, complaints and credit notes can be examined.

3.9.6 Cockpit Reports for Project Management

The project management in companies is of central importance, especially for those companies whose core performance is based on project production, such as in plant engineering, consulting, software development or civil engineering. In addition, project management also has a high significance in all other economic sectors, such as in trade, industry or banking and insurance, for example, for research and development work as well as for operational and strategic projects. The following reports for the operational project management in single and multi-project controlling are examined in more detail.[246] Important questions that are answered in the operational project controlling are:

- How do my projects look in an overview?
- Which projects achieve which margins? Which projects are loss-making?
- How high are the actual and forecast costs in relation to the delivered or planned performance?
- Which projects are delayed?
- Are project outflows covered by the counter-financing?
- Which opportunities and risks have to be kept in mind in the projects?
- What does the cross-project utilization and productivity of my project resources, such as employees or equipment, look like?

The presented project cockpit has to be adapted to the individual requirements of project management in the industry or company.

In the start cockpit of project controlling, the responsible manager receives key figures that provide an overview of the most important success drivers in the project business (see Fig. 3.150). In the example, the **contribution margin quota** indicates the margin of the project success that is available to cover further overhead costs beyond the direct project costs. The achievement of the planning target can be tracked accumulatedly over time for the project contribution margin by the combined bar and line chart. The aggregated project services are illustrated by the two key figures **productivity** and **schedule delay**. The productivity shows the number of productive project hours based on

[246] The cockpit illustrations in this section are self-created. A detailed explanation of project controlling based on the example shown can be found in the following article: Schön (2017, p. 507–537). The areas "strategic project management" and "planning" are left out as in the previous sections.

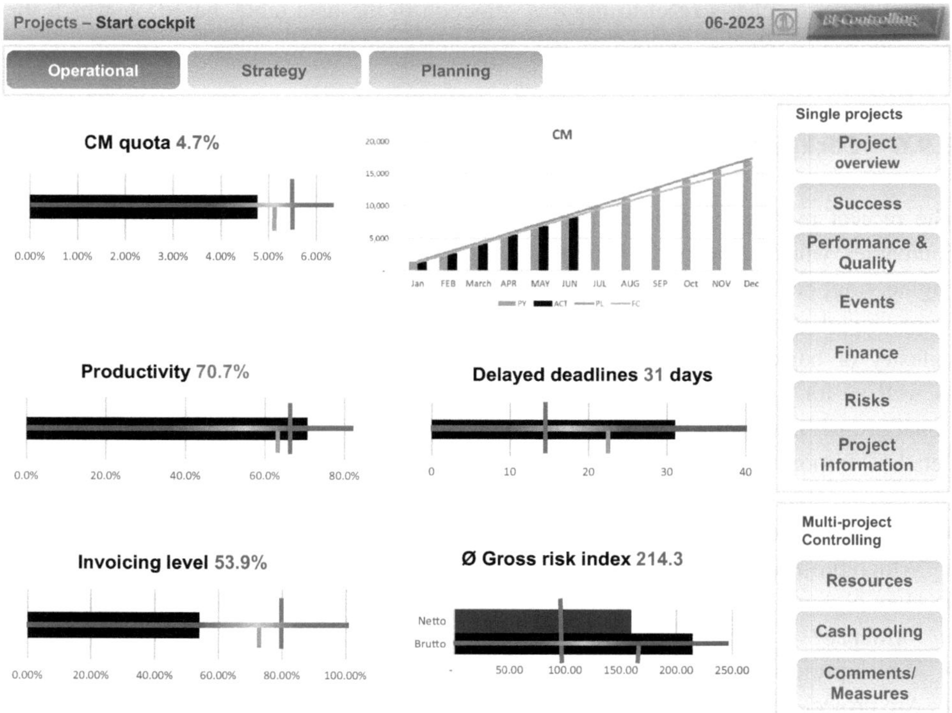

Fig. 3.150 Projects start cockpit

the planned target hours. It thus measures the economic use of resources. Project delays are often an indicator of inefficiency, e.g. in the form of increased project costs and risks, and also cause dissatisfaction among customers. Therefore, the avoidance of delays should have a high priority in the project monitoring. Unbilled project services not only lead to a worse cash flow but can also hide risks, such as customers disputing or delaying payments for services rendered. The **invoicing level** shows, in the ideal case, the proportion of billable services that have already been invoiced. The project business is usually highly complex and contains risks that cannot always be expressed quantitatively in values or quantities. For example, coordination with other trades and companies is necessary or there are problems with the delivery of goods. Furthermore, there can be different interpretations of the services, as the rights and obligations of the business partners are not clear in the contract. These and other project-specific risks are condensed for control in a **risk index**. The net risk index shows to what extent the **gross risk** can be reduced by initiated measures.

The individual reports for operational project management are divided into the following areas:

- Individual project controlling
  - project overview
  - project success
  - project performance & quality
  - project deadlines
  - project finances
  - project risks
  - project information
- Multi-project controlling
  - project resources
  - cash pooling
  - comments/actions

While the reports in individual project controlling cover the most important control aspects of the individual project, multi-project controlling takes over the cross-project control areas such as resource management and cash pooling.

3.9.6.1 Project Overview

For individual project controlling, an overview of all projects is initially useful in which the key figures of the individual projects can be analyzed (see Fig. 3.151). The filter, sorting and traffic light functions help to identify outliers upwards or downwards and to enter the detailed analysis of a project. The menu filters allow restrictions to be made according to the requirements of the analyst, whether, for example, only one division, one branch or one project type should be examined.

The analysis table is compactly tailored to the most important key figures of the individual project control in areas. The order value indicates the monetary importance of a project undertaking. Large projects can thus be quickly identified using the sorting function. In the control area "success", the **forecast of the contribution margin** shows the projected margin over the direct individual project costs. In addition, the **cost performance index (CPI)** indicates the cost deviations (target costs/actual costs) in the project. In the area of "performance", the individual projects can be analyzed with regard to their **schedule performance index (SPI)** and the degree of completion. The performance index shows, based on the **Earned Value Analysis**, to what extent the performance (target costs) lags behind the planned performance.[247] The **degree of completion** shows the

[247] Cf. for the Earned Value Analysis e.g. Fiedler (2016, p. 182–195), Bea et al. (2008, p. 302–310) and Schlick et al. (2016, p. 197 ff.).

Projects – **Project overview** 06-2023 BI-Controlling

Filter:								
	Customer	All	Type	Cologne	Building	All	Month	June
	Branch	All	Project Leader	All	Division	Construction	Year	2023

Project overview	Total	Success		Performance		Deadline	Finance		Risk
	Order value	CM FC	CPI	SPI	Degree of completion	Delay	Invoicing rate	CF FC	Risk index
+P13 Building P1	1,200	-244	0.75	0.64	28.6%	-28	58.3%	-244	86
+P21 Outdoor hall	5,600	361	1.20	1.10	89.0%	-10	62.0%	123	75
+P25 Shopping centre north	4,325	130	0.95	0.94	82.3%	-6	80.4%	10	50
+P 31Administrative building ZAG	2,364	-152	0.86	0.84	94.5%	11	95.6%	260	42
+P52 District sports complex	2,600	347	1.15	1.05	86.4%	0	100.0%	360	32
+P63 West school complex	3,640	211	1.03	1.10	64.2%	-61	43.0%	-132	80
+P78 Nursing home	800	164	0.96	0.86	76.4%	-14	76.0%	-86	65
+P81 Centre for the elderly S1	950	30	0.81	0.92	83.0%	-30	92.0%	53	32
+P83 Senior citizens' centre T1	2,421	-142	0.75	0.75	92.1%	-30	93.1%	250	40
+P84 Centre for the Elderly V7	2,536	347	1.15	1.05	86.4%	10	100.0%	360	66
+P91 Administration centre	3,750	325	0.98	0.86	64.2%	7	86.0%	50	75
+P93 Clinic BTZ 1	852	56	0.96	0.84	75.3%	-6	76.0%	-54	66
+P94 Clinic BTZ 2	952	31	0.80	0.91	84.0%	-5	93.2%	50	53
+P95 Clinic BTZ 3	1,203	-155	0.79	0.82	91.2%	-10	82.3%	-104	33
+P96 Clinic BTZ 4	2,614	101	0.81	0.84	85.0%	14	93.2%	64	64
+P98 Centre for the elderly D	3,210	200	0.78	0.75	84.0%	-67	93.2%	85	98
+P99 Centre for the elderly F	1,346	23	0.79	0.82	84.0%	-103	93.2%	-104	65
+P104 Centre for the elderly TA	2,436	254	1.06	1.00	95.3%	5	93.2%	360	66
+P102 Nursing centre	3,641	325	0.98	0.86	64.2%	7	86.0%	50	75
+P103 Clinic ZA 1	954	56	0.96	0.95	75.3%	-6	76.0%	-54	66
+P106 Clinic ZA 2	864	31	0.80	0.91	84.0%	-5	93.2%	50	53

Fig. 3.151 Project overview

actual performance of the still forecasted total performance (forecast).[248] Different measurement techniques of the degree of completion can be found among others in Schreckeneder.[249] Supported by the length of the miniature bars, the user can quickly see which projects are still far behind their completion. The **delay** supports this view in the area of "schedule" where the forecasted delay days from the original planned completion date are displayed.

In the area of "finance", in addition to the already explained **invoicing level**, the **project cash flow** is listed as the peak indicator. It shows the internal financing power of the project and should be as positive as possible. The final area of risk shows the **risk index** of the individual project, so risk-relevant projects can be identified more quickly.

The project P13 is very critical to evaluate in terms of success in the example shown, so a deeper analysis makes sense (see Fig. 3.152). First, the manager gets an overview of the most important information about the project.

[248] Cf. DIN 66901 – 5.3.18 (2009).

[249] Cf. Schreckeneder (2013, p. 160–161).

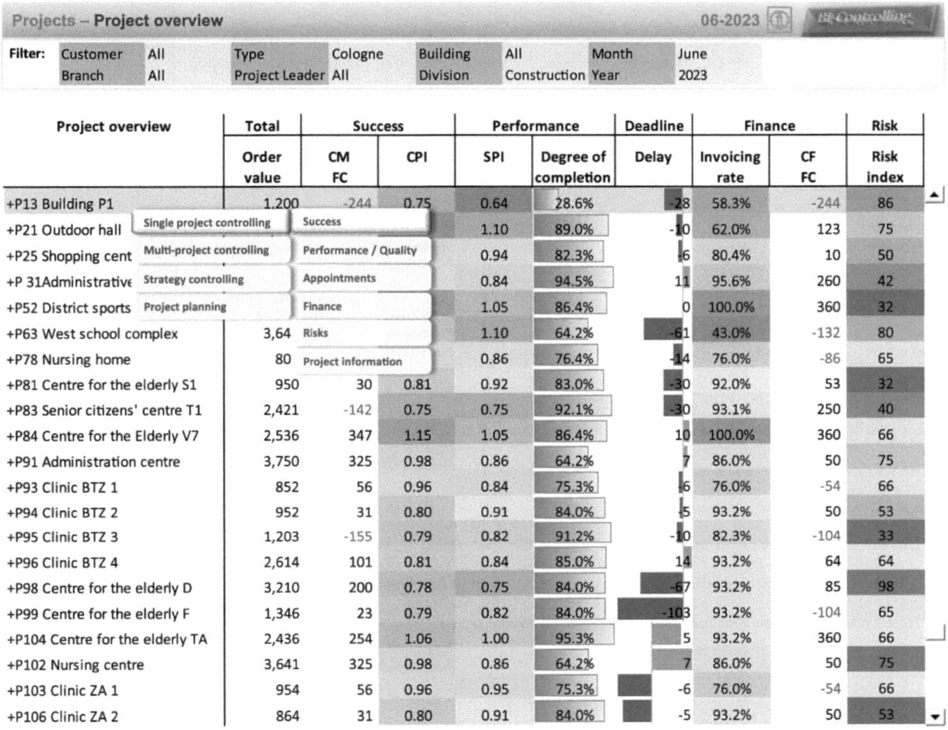

Project overview	Total	Success		Performance		Deadline	Finance		Risk
	Order value	CM FC	CPI	SPI	Degree of completion	Delay	Invoicing rate	CF FC	Risk index
+P13 Building P1	1.200	-244	0.75	0.64	28.6%	-28	58.3%	-244	86
+P21 Outdoor hall				1.10	89.0%	-10	62.0%	123	75
+P25 Shopping cent				0.94	82.3%	-6	80.4%	10	50
+P 31Administrative				0.84	94.5%	11	95.6%	260	42
+P52 District sports				1.05	86.4%	0	100.0%	360	32
+P63 West school complex	3,64			1.10	64.2%	-61	43.0%	-132	80
+P78 Nursing home	80			0.86	76.4%	-14	76.0%	-86	65
+P81 Centre for the elderly S1	950	30	0.81	0.92	83.0%	-30	92.0%	53	32
+P83 Senior citizens' centre T1	2,421	-142	0.75	0.75	92.1%	-30	93.1%	250	40
+P84 Centre for the Elderly V7	2,536	347	1.15	1.05	86.4%	10	100.0%	360	66
+P91 Administration centre	3,750	325	0.98	0.86	64.2%	7	86.0%	50	75
+P93 Clinic BTZ 1	852	56	0.96	0.84	75.3%	-6	76.0%	-54	66
+P94 Clinic BTZ 2	952	31	0.80	0.91	84.0%	-5	93.2%	50	53
+P95 Clinic BTZ 3	1,203	-155	0.79	0.82	91.2%	-10	82.3%	-104	33
+P96 Clinic BTZ 4	2,614	101	0.81	0.84	85.0%	14	93.2%	64	64
+P98 Centre for the elderly D	3,210	200	0.78	0.75	84.0%	-67	93.2%	85	98
+P99 Centre for the elderly F	1,346	23	0.79	0.82	84.0%	-103	93.2%	-104	65
+P104 Centre for the elderly TA	2,436	254	1.06	1.00	95.3%	5	93.2%	360	66
+P102 Nursing centre	3,641	325	0.98	0.86	64.2%	7	86.0%	50	75
+P103 Clinic ZA 1	954	56	0.96	0.95	75.3%	-6	76.0%	-54	66
+P106 Clinic ZA 2	864	31	0.80	0.91	84.0%	-5	93.2%	50	53

Fig. 3.152 Project overview with context menu

The report Project information/comments/measures lists the most important project information in the upper part (see Fig. 3.153). These include the already known peak indicators from the project overview. Below are important information on the master data of the project (order number, customer, organization, project type, etc.) and its contacts, especially the phone number and email address of the project manager. The project status and the most important dates for the project complete this information block.

In the lower part of the report, the comments of the project manager and current measures for the project are shown. Via the already explained connection to the ticket system, tasks can be assigned to responsible persons and tracked with the help of tickets.

3.9.6.2 Project Performance and Success

If the project performance and the project success are to be further considered, the user can jump to the report Success via the context menu (right mouse button). Alternative jump options would be possible via the hidden menu bars in the overall navigation.

The report Project success shows the income statement of the selected project (see Fig. 3.154). The costs, revenues and contribution margins as well as the calculated project result can be analyzed in detail across all value types (plan, actual, forecast). Project

Projects – Project information/comments/measures 06-2023 BI-Controlling

Filter: Project P13 Building P1 Branch Cologne Project manager All Month/Year Jun 23

Project overview	Total	Success		Performance		Deadline	Finance		Risk
	Order value	CM II FC	CPI	SPI	Degree of completion	Delay	Invoicing rate	CF FC	Risk index
+P13 Building P1	1,200	-244	0.75	0.64	28.6%	-28	58.3%	-244	86

Project information

Offer number:	An102-31	**Client:**	4785 Clinic Beta	**Project manager:**	GB	++49-221/1245-6	GB@bi-consult.de
Order number:	Au102-31	**Place:**	45703 Essen	**Site Manager:**	SF	++49-221/1245-21	SF@bi-consult.de
Bidder:	PT	**Country:**	Germany	**Project Controller:**	CF	++49-221/1245-41	CF@bi-consult.de

					Plan	FC	Actual
Company:	BI-Consult	**Industry:**	Health care				
Branch:	Cologne	**Contract type:**	Standard	**Project start:**	02.05.2018	02.05.2018	02.05.2018
Division:	Structural engineering			**Project end:**	27.11.2018	25.12.2018	open
Building type:	Parking garage	**Project status:**	In progress	**Project duration:**	209	237	59

Comment / Measure

The project shows very high time and cost deviations due to incorrect deliveries and a lack of incoming inspections.

There are calculation errors and a lack of supplement management.

The project manager is too inexperienced.

- Check whether recourse can be taken against the supplier. Who: GB, When: Jul.-Aug.
- Carry out incoming goods inspections according to specifications. Who: GB, When: Jul.-Aug.
- Project manager receives qualification measure. Who: GB, When: Sep.

Ticket

T32031
T32032
T32033

Fig. 3.153 Project information/comments/measures

indicators such as the degree of completion and the performance and cost deviation support a quick success analysis.

The project success calculation shows the value creation stages of the project success in the line scheme. The most important starting value here is the **total output**, which results from already invoiced revenues (partial or final invoices) and the change in inventory. Long project durations lead to the fact that performance creation and revenue generation are far apart in the project business. For this reason, partially completed or processed project services have to be determined in the position change in inventory. The **completed-contract method**, which is frequently found in Germany and which evaluates partially completed project services exclusively with the (incomplete) direct production costs (individual costs) incurred for them, is not transparent here. It leads to the fact that revenues are only reported in the period in which the turnover is generated. This is often only the case at the time of the final invoice. Therefore, the **percentage-of-completion method** is preferable for inventory valuation. The performance valuation here includes the full production costs including project overheads and calculated profit statements of the partial services performed, taking into account the degree of completion in the calculation. This allows a periodic profit determination for the project. For the partial

Projects Success 06-2023 BI Controlling

Filter:									
	No.	P13	Branch	Cologne	Building type	HB 21	Status	in progress	Month	June
	Project	Buildung P1	Project manager	Müller	Division	Construction			Year	2023

Project income statement	Order PL	Supple-ments PL	Revised PL	Costs ACT	Residual power	FC	Ω FC %	ΔPL-FC	ΔFC-PL%
Sales	1,000	200	1,200	200	1,000	1,200		0	0.0%
Change in inventories (POC)				143	-143	0		0	
= Total output	**1,000**	**200**	**1,200**	**343**	**857**	**1,200**		**0**	**0.0%**
Cost of sales	200	**100**	300	100	**300**	400	25.0%	100	33.3%
External services	100	**50**	150	50	**150**	200	25.0%	50	33.3%
= Gross profit	**700**	**50**	**750**	**193**	**407**	**600**		-150	-20.0%
Personnel costs	420	**42**	462	210	**462**	672	31.3%	210	45.5%
= CM 1 Project result a. direct costs	**280**	**8**	**288**	-17	-55	-72		-360	-125.0%
Material overhead	6	**3**	9	3	**9**	12	25.0%	**3**	**33.3%**
Production overhead	100	**10**	110	50	**110**	160	31.3%	**50**	**45.5%**
= CM 2 Project result a. overhead	**174**	-5	**169**	-70	-174	-244		-413	-244.4%
Impending losses	0	**0**	0	174	-174	0		**0**	
Accruals	0	**0**	0	0	**0**	0		**0**	
= CM 3 Project result after risks	**174**	-5	**169**	-244	**0**	-244		-413	-244.4%
Calc. interest	0	**0**	0	0	**10**	10		**10**	
= CM 4 after calculated Interest	**174**	-5	**169**	-244	-10	-254		-423	-250.3%
Percentage of completion				**28.6%**	71.4%	100%	**28.6%**		

Earned Value Analysis	PL	Target	Performance	Actual	Costs
Period costs	485	309	-176	413	-104
			SPI		CPI
Index			0.64		0.75

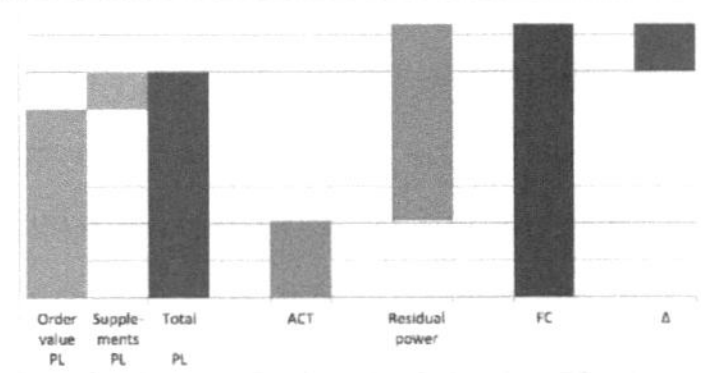

Fig. 3.154 Project success

services, it is important here that they are documented, accepted by the customer, and auditable partial services.

The degree of completion is calculated based on the accumulated actual costs in relation to the projected project costs in the forecast and thus shows the ratio of actual performance and projected total performance.

The costs are deducted step by step from the total output. First, the material costs and the external services are deducted and the **gross profit** is obtained, which in the project business is alternatively called **own performance**[250]. Subtracting direct personnel costs results in **contribution margin** 1 after project individual costs. This contribution margin represents the covering power of the project for further overheads of the project and the company. If the allocable material and production overheads of the project are deducted in the next result stage, contribution margin 2 is obtained, which represents a full cost result of the project before risk positions and further calculated costs. Contribution margin 3 takes these values into account in the form of impending losses, provisions and cal-

[250] Cf. Lutz et al. (2011, p. 63 f.).

culated interest. Contribution margin 4 corresponds to a **project result** in the form of a calculated full cost result. It is an important control variable for the project success both in the offer and in the order phase.

The column scheme of the project result calculation shows, as the starting point of the control, changes and modifications besides the original plan of the order calculation; in doing so, always a current revised total plan value of the project is available for the control in the sense of the change management. In comparison, the accumulated actual costs and the remaining services including obligations in the forecast value of the expected performance have to be compared. With the help of the so-called forecast and the actual costs, the **target achievement** and the **degree of completion** of the project can be calculated.

The biggest problem in controlling a project is the forecast of the remaining services. Here, the project managers are required who have to update or revise the remaining costs of individual items within the framework of the ongoing project planning. For this purpose, appropriate simple planning forms are available, which are maintained either in the pre-system of the project control or in the planning area.

With the help of the waterfall diagram at the bottom right, the change of the project total result due to change orders of the change management as well as the accumulated and forecasted remaining performance to the forecast can be captured faster.

In addition to the project success calculation with degree of completion determination, the determination of project key figures of the **Earned Value Analysis** described in the overview is also suitable. Here, the performance index and the cost index are in focus. If the performance index is significantly below 100% or below the factor 1, the target performance of the project is significantly behind the planned performance. The factor 0.64 indicates that only 64% of each planned euro of the project has been completed by the current month. Whether the services rendered were actually produced economically is shown by the cost index. If the cost index is below 100% or the factor 1, the actual services have been provided at significantly higher actual costs than the estimated target costs. In the example, the cost deviation is currently €104. The cost index of 0.75 in the example shows that only 75 cents of value have been created for one euro of project costs!

If the project control at the highest project level is not sufficient, the success report should be broken down according to the project hierarchy. Here, the subprojects and project work packages are to be analyzed in the same way as the overall project.

The aggregation of the project results can in turn be carried out for further result objects, e.g. branches, divisions and project types.

3.9.6.3 Project Deadlines

An important success factor for a project is the adherence to agreed deadlines (see Fig. 3.155). For this reason, it is advisable to continuously carry out a project schedule control and identify time deviations. The milestones defined in the project planning serve

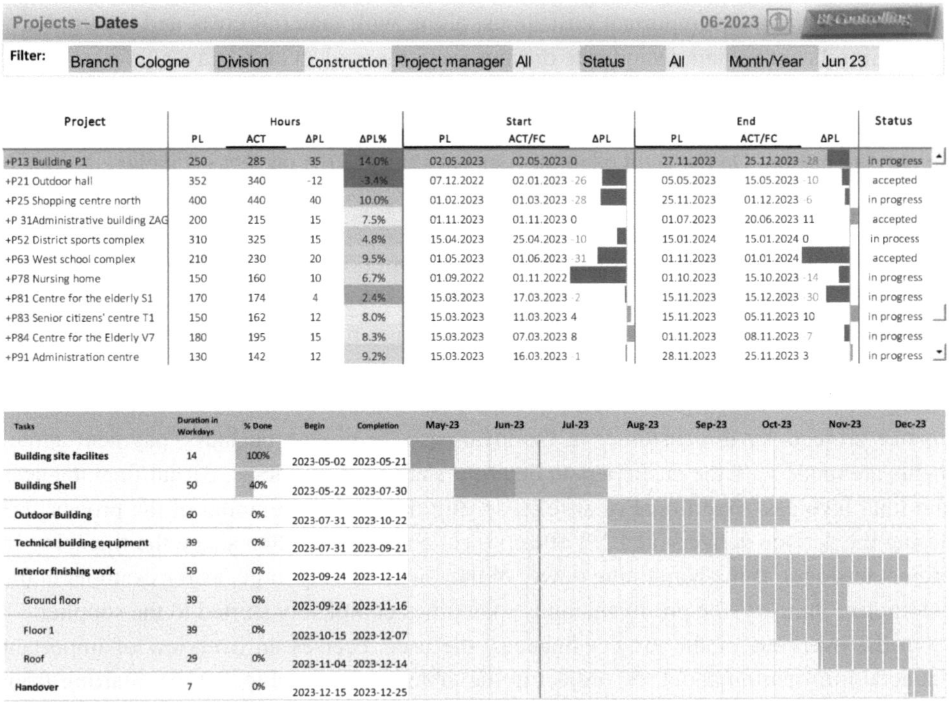

Projects – Dates 06-2023

Filter: Branch Cologne | Division Construction | Project manager All | Status All | Month/Year Jun 23

Project	Hours				Start			End			Status
	PL	ACT	ΔPL	ΔPL%	PL	ACT/FC	ΔPL	PL	ACT/FC	ΔPL	
+P13 Building P1	250	285	35	14.0%	02.05.2023	02.05.2023	0	27.11.2023	25.12.2023	-28	in progress
+P21 Outdoor hall	352	340	-12	-3.4%	07.12.2022	02.01.2023	-26	05.05.2023	15.05.2023	-10	accepted
+P25 Shopping centre north	400	440	40	10.0%	01.02.2023	01.03.2023	-28	25.11.2023	01.12.2023	-6	in progress
+P 31Administrative building ZAG	200	215	15	7.5%	01.11.2023	01.11.2023	0	01.07.2023	20.06.2023	11	accepted
+P52 District sports complex	310	325	15	4.8%	15.04.2023	25.04.2023	-10	15.01.2024	15.01.2024	0	in process
+P63 West school complex	210	230	20	9.5%	01.05.2023	01.06.2023	-31	01.11.2023	01.01.2024		accepted
+P78 Nursing home	150	160	10	6.7%	01.09.2022	01.11.2022		01.10.2023	15.10.2023	-14	in progress
+P81 Centre for the elderly S1	170	174	4	2.4%	15.03.2023	17.03.2023	-2	15.11.2023	15.12.2023	-30	in progress
+P83 Senior citizens' centre T1	150	162	12	8.0%	15.03.2023	11.03.2023	4	15.11.2023	05.11.2023	10	in progress
+P84 Centre for the Elderly V7	180	195	15	8.3%	15.03.2023	07.03.2023	8	01.11.2023	08.11.2023	-7	in progress
+P91 Administration centre	130	142	12	9.2%	15.03.2023	16.03.2023	-1	28.11.2023	25.11.2023	3	in progress

Tasks	Duration in Workdays	% Done	Begin	Completion
Building site facilites	14	100%	2023-05-02	2023-05-21
Building Shell	50	40%	2023-05-22	2023-07-30
Outdoor Building	60	0%	2023-07-31	2023-10-22
Technical building equipment	39	0%	2023-07-31	2023-09-21
Interior finishing work	59	0%	2023-09-24	2023-12-14
Ground floor	39	0%	2023-09-24	2023-11-16
Floor 1	39	0%	2023-10-15	2023-12-07
Roof	29	0%	2023-11-04	2023-12-14
Handover	7	0%	2023-12-15	2023-12-25

Fig. 3.155 Project deadlines

as the most important control points, especially for complex project structure plans. Negative effects of schedule deviations can be, for example, contractual penalties. In addition, delays in the project business cause increased costs in subsequent work packages. For example, materials have to be stored at a cost.

For the selected project, a compact schedule for the most important project phases is displayed in the graphical overview (below) using the bar chart technique. This allows important time information about the project to be analyzed quickly. The project progress can be read from the actual status. The expected project end takes into account necessary shifts of project tasks and thus reveals important deviations at the start and end of individual milestones. A special type of milestone is the quality gate. A quality gate can only be passed, if certain requirement criteria are met.[251]

The table above shows the hours and key dates for the project. The first area shows the deviation of the required hours from the planned hours. The personnel hours are

[251] Cf. Fiedler (2016, p. 90 ff.).

often one of the most important cost items, along with material costs and external services. Significant deviations indicate difficulties in the project and delays.

In the second and third area of the table, the delay of the start or end date of the project is visually highlighted with miniature bar charts. This way, larger outliers can be quickly identified. On the right edge, the current status of the project is displayed.

3.9.6.4 Project Finances

The presentation of the project finances includes a comparison of the customer's payments and the project's payments to third parties as well as payments for the project's own services. The difference between these two values results in the peak indicator "project cash flow". Project cash flow indicates the financing power of the project and should ideally be positive throughout the project process. For this reason, it is advisable to prefinance projects by the customer in the ideal case. In project business, payment agreements are made with the customers to determine advance payments, partial final invoices and final invoices. Traditional rules, e.g., 1/3 are due at the beginning of the project, 1/3 during the service delivery and 1/3 after project acceptance, often shape the financing in project business. If the bargaining power of the clients is very large, as for example in the automotive industry, the pre-financing is sometimes completely shifted to the supplier.

In the overview of the project finances, the user receives an overview of important financial information about the project in the table above (see Fig. 3.156). Starting from the order value, it is shown how high the sum of the advance payments, partial final, and final invoices are per project.

The **invoicing level** shows the ratio of the invoices issued to the order value. Low invoicing levels indicate unnecessary bound capital costs and default risks of services. If the company is able to determine billable service levels for the project separately, the invoicing level can refer to this value.

The sum of the **outstanding payments** as opposed to the **received** payments is an indication of delayed payment behavior and can indicate, for example, disturbances in the project. If the outstanding payments become too large, it is advisable to discuss the reasons with the project manager and the customer. As a visual aid, the payment behavior can be well read from the indicator **payment degree** . As already shown for the company in the receivables management, outstanding payments (overdue open items) are also to be classified by their due dates in the project. Due dates over 30 days and longer should be avoided.

The graphical overview (bar chart) shows the project cash flow for the selected project. At the front, the total project cash flow for the projected project end is displayed. To the right of it, the development of the planned, current and forecasted cash flow over time can be analyzed by the positive and negative fluctuations in the bar chart.

Projects – Finance 06-2023

Filter: Branch Cologne | Division Construction | Project Leader All | Status All | Month/Year Jun 23

Projects	Order value	Down payment	Partial / final invoice	Invoicing level	Received	open	Payment degree	overdue > 10 days	overdue > 30 days	overdue > 50 days
Total	30,186	9,700	16,265	53.9%	8,904	2,494	54.7%	0	200	2,294
+P13 Building P1	1,200	200	400	33.3%	200	200	50.0%		200	
+P21 Outdoor hall	5,600	1,000	3,214	57.4%	2,000	1,214	62.2%			1,214
+P25 Shopping center north	4,325	2,000	2,145	49.6%	2,000	145	93.2%			145
+P31Administrative building ZAG	2,364	1,000	1,523	64.4%	945	578	62.0%			578
+P52 District sports complex	2,600	500	1,234	47.5%	1,000	234	81.0%			234
+P63 West school complex	3,640	1,000	2,045	56.2%	2,045	0	100.0%			0
+P78 Nursing facility	800	200	314	39.3%	314	0	100.0%	0		
+P81 Senior citizens' center S1	950	300	523	55.1%	400	123	76.5%			123
+P83 Center for the elderly T1	2,421	1,000	423	17.5%	400	23	94.6%			23
+P84 Center for the elderly V7	2,536	2,000	2,430	95.8%	2,430	0	100.0%			0
+P91 Administration Center	3,750	500	2,014	53.7%	1,500	514	74.5%			514

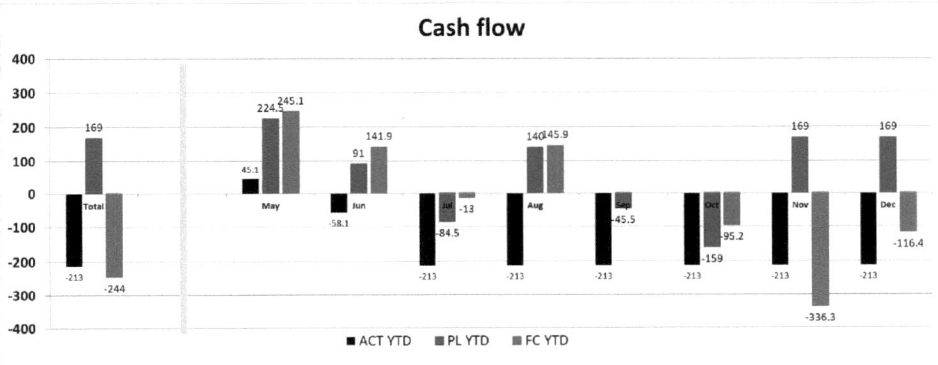

Fig. 3.156 Project finance

3.9.6.5 Project Risks

Projects are often risky. Important reasons for the emergence of risks in project business are shown by the following incomplete list:[252]

- innovation and technology risks regarding the novelty of the project task
- schedule risk (poorly predictable durations)
- calculation and price risks due to high competitive intensity with ruinous price competition
- contract and legal risks due to tightened legal framework conditions
- utilization risks due to cyclical employment
- project complexity risks
- performance risks due to non-transparent and partly unknown services in the offer and acquisition phase

[252] Cf. Schön (2017, p. 507–537).

- country risks
- pre-financing risk of services
- coordination risk between sub- or subcontractors, suppliers, joint venture partners and own central service units

The executives should therefore manage the risks using an appropriate risk management, especially in project business. Here, the risks should be collected (risk identification), analyzed and evaluated (risk analysis and assessment), controlled (risk control) and continuously monitored (risk monitoring and documentation). The latter is done via the project risk report (see Fig. 3.157).

Based on the graphic, the risk portfolio (risk map) can be analyzed with regard to the probability of occurrence and the extent of individual risks. Depending on the setting of the radio button, gross and net risks can be displayed. While the gross risk shows the initial assessment of the risk, the net risk shows the reduced risk assessment that is assumed for the risk after the implementation of risk measures.

The size of the area of the risk network shows the risk assessment of evaluated important risk criteria in the project business. Here, the risk criteria are assessed using a mostly

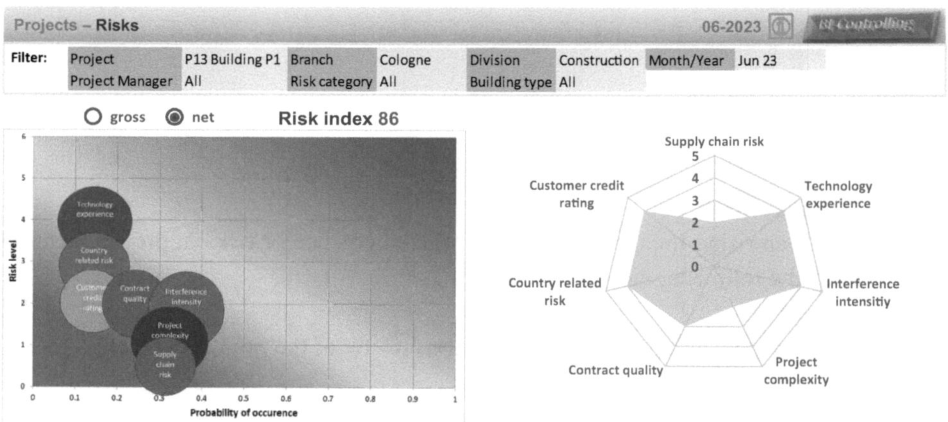

Risk identification				Risk analysis				Risk Monitoring		
Risk Basis	Category of Risk	Risk	Type	Cause-effect relationship	Risk level	Probability of occurence	Risk index	Controlling instrument	Responsibility	Actions
Supply chain risk	Output	Procurement	Quality risk	Poor quality goods have an impact on other trades	1	2	2	Supplier selection and tighter incoming goods control	Head of Goods Receiving	Supplier selection
Technology experience	Output	Sales	Product risk	Competitors' product innovations displace own sales	5	2	10	Adapt marketing strategy, emphasise USP	Marketing Manager	Increase R&D expenditure, customer needs analysis
Interference intensity	Personnel	Process	Experience	Lack of experience leads to project errors	5	5	25	Train internal project procedures	CTO	Training
Project complexity	Output	Organisation	Project manage-ment risk	Slowed information flow influences time of project completion	4	3	12	Regular control of project progress and buffer consumption	Project Manager	Calculate buffer time appro-priately
Contract quality	Contract	Personnel	HR availability	Lack of definition and delimitation of project content increases staff input	5	3	15	Check contracts for legal security	Lawyer	Support contract preparation
Country related risk	Output	Sales	Import risk	Volatile political situation influences own import	5	2	10	Search for alternative suppliers, increase stock and minimum stock levels	Head of SCM	Reduce depen-dence on suppliers, forward contracts
Customer credit rating	Finance	Customer	Payment behaviour	Payment morale affects own liquidity and interest income	4	3	12	Dunning system, factoring	Head of Finance	Tighter dunning system

Fig. 3.157 Project risks

simple ordinal rating, e.g. "1" very low to "5" very high. The larger the marked area of the spider diagram, the greater the overall risk of the project.

Details on the risk assessment, responsibilities and measures to limit the risks can be found in the risk catalog below. The selection of individual risks is dynamically designed. The display of the risks in the risk catalog or in the graphics is adjusted accordingly.

3.9.6.6 Project Resources

The control of project resources involves the identification of deviations regarding the resource utilization in the project process. In most projects, this refers to the work performance of the project staff or the equipment or plant performance that is used for the project. Since project resources are usually used across projects, this is a task of the multi-project controlling and is carried out for the corresponding organizational units (e.g., branches, personnel pool or equipment pool) of the company.

For selected areas, e.g., branches and activity groups, the performed and productive hours for the deployed personnel resources can be analyzed (see Fig. 3.158). The most important key figures for the resource control are the utilization degree and the productivity. The **degree of utilization** shows the proportion of the actual hours performed in comparison to the target hours. Alternatively, extended utilization degrees can also be reported and analyzed for the technical or maximum total capacity. The latter is suitable, for example, for companies with shift models or other working time models. If the utilization is very high, there may be overloads that can only be absorbed with the help of overtime, external staff or other adjustment measures. If the utilization is low, there is a cost-intensive provision of resources, which can be countered, for example, with short-time work, redundancy, redistribution of resources or other measures.

While the actual hours include both productive and indirect hours (e.g. for training, administrative work, quality assurance or cleaning), the key figure **productivity** only shows the number of productive hours as a percentage of the target hours. The mini-bar length allows the performance differences of the branches to be easily compared with each other. The deviation of the actual productivity from the target value can be examined either accumulated in the table or monthly in the temporal development using the diagram. Overtime, hours with surcharges and absences complete the resource analysis in the considered organizations. They provide indications of the workload of the resources and can thus be used to regulate the workload, for example, if resources can be deployed in a supportive way in other overloaded organizational units.

In summary, this section shows for the project controlling how quickly one can analyze steering-relevant information in the project controlling cockpit. Peak indicators help to quickly get an overall overview. With a few clicks, strategic and operational analysis areas for single-project and multi-project controlling can be reached.

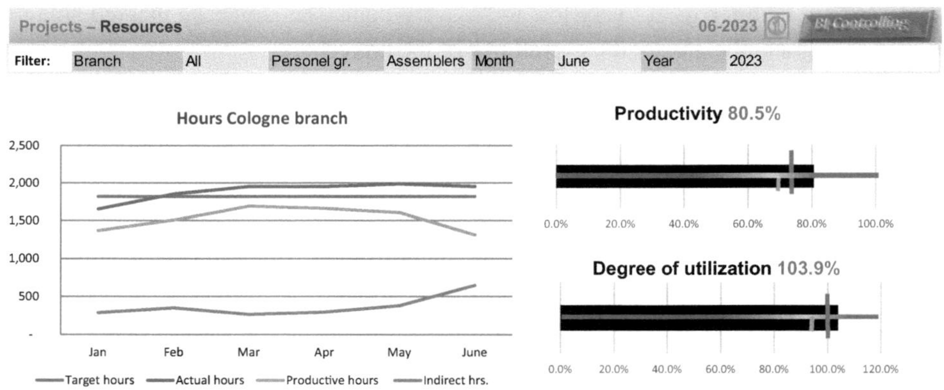

Subsidiaries	Target hrs.	ACT hrs.	Load factor	Productive hrs.	Indirect hrs.	Productivity PL	Productivity ACT	Overtime	Hrs. with Surcharge	Absenteeism
Total	**131,009**	**129,906**	**99.2%**	**91,813**	**38,093**	**65.0%**	**70.7%**	**12,218**	**13,155**	**8,837**
Cologne	10,927	11,352	103.9%	9,144	2,208	75.0%	80.5%	1,135	2,270	568
Hanover	18,164	16,530	91.0%	10,654	5,876	60.0%	64.5%	2,480	1,653	1,157
Berlin	20,505	19,875	96.9%	12,568	7,307	60.0%	63.2%	2,584	2,981	1,988
Munich	14,619	16,531	113.1%	13,547	2,984	70.0%	81.9%	827	827	992
Magdeburg	15,321	15,564	101.6%	10,535	5,029	60.0%	67.7%	1,632	1,465	1,032
Frankfurt a.M.	18,430	18,791	102.0%	12,546	6,245	70.0%	66.8%	1,364	1,265	1,452
Stuttgart	8,632	8,531	98.8%	5,684	2,847	65.0%	66.6%	567	321	254
Hamburg	7,605	8,412	110.6%	6,450	1,962	70.0%	76.7%	914	654	751
Dresden	16,806	14,320	85.2%	10,685	3,635	70.0%	74.6%	716	1,718	644

Fig. 3.158 Project resources

3.10 Planning Forms and Their Peculiarities

In an integrated planning and controlling model, planning and reporting should be aligned. This means that planned values should also be controlled and analyzed in terms of their achievement or failure in reporting. However, the granularity of planned and actual information is usually different. In the actual, all business transactions of the company have to be recorded and documented in different areas and systems and condensed and further processed until reporting. Planning cannot start at this level of detail, but is based on a condensed level. Usually, it takes place at the levels of accounts or revenue and cost types, but sometimes only at the respective group level of the accounts or revenue and cost types of the company. Nevertheless, it is important that the planned results can be compared in reporting at the same value level. For this reason, it makes sense to design the planning forms in accordance with the reporting.

Just like in the reports, filter and selection functions are used in planning forms, which allow a targeted plan entry. Identical to the report design, the basic design of report headers, tables and diagrams up to the comments is also the same.

In contrast to the reports, planning forms have special features that have to be considered in the design:

- **Reading in default values as orientation-or transfer sizes for planning**
  In planning, default values, e.g. old planned values, previous year values etc., serve as orientation sizes for the planner for the new planning round and thus simplify the planning. They can be changed and transferred by the planner into the new plan "one to one", or also by revaluation functions or manual adjustment. The disadvantage of transferring old planning sizes is that the planner may not make an effort to anticipate future plan values and is more concerned with securing budgets from the past or simply adopting target specifications.
- **Direct input option for target values and other planning variables (manual planning)**
  The most important difference to reports is the input option for target values, i.e. the editability of quantity, price and value structures. In addition to the direct input option for target values, planning mainly requires the maintenance of target quantities and prices, but also of percentage values and other conversion and distribution variables for the determination of target values.
  To make the input as comfortable and flexible as possible, the following functions help:
  - absolute value input (e.g. 800) or percentage value input (e.g. +6%)
  - short input, e.g. for 7 million = 7 m or 3000 = 3k,
  - undo function: Undo the last inputs
  - time settings via Time controller (Timeslider) as an alternative to selection
  - distribution aids (e.g. seasonal distribution curves)
  - documentation aids (work instructions for the planning forms)
  - comment functions
- **Reading or calculating target values from upstream planning areas**
  - Similar to the default values, the planning forms of the downstream planning areas already set outgoing planning variables of the upstream planning areas as planning result variables. These have already been fixed there and cannot be changed here, e.g. the depreciation values from the asset accounting for the existing fixed assets.
- **Access and locking functions**
  If several users work on planning content at the same time, access rules and locking functions are also necessary since at one time only one authorized user can enter planning changes for certain selected data areas. Otherwise, synchronization functions of differently changed values are necessary for parallel inputs.
- **Use of planning functions**
  While data is only read and output in reports, planning functions are to be triggered in planning forms or upstream planning areas. The most frequently used planning functions (distributions, extrapolations, forecasts, revaluations, consolidations,

plan evaluations and calculations, simulations and sensitivity analyses) are briefly explained below.

- **Distributions**
  Distributions are made from a total initial value to a number of detail values, such as the annual value to the monthly values. For this purpose, various keys (percentage values, absolute values) can be used as distribution keys. For many planning areas such as sales planning, it is useful if manually planned values are not overwritten by a distribution run, but only those detail values are overwritten that have not yet been manually planned. This distribution is called, among other things, residual value distribution. Various top-down distribution and bottom-up consolidation functions are shown in Fig. 3.159. Here, the plan entries are highlighted in color (e.g. yellow).
  The data entry and the distribution of values to underlying objects is also called splashing in some software applications. It has the character of breaking down higher-level values using distribution algorithms or simple historical distribution parameters, e.g. in customer-oriented sales planning, the distribution to products based on the actual values of the previous year.
  In addition to splashing, there are other support options for data distribution, e.g.:[253]
- **Copy function**: The entered plan values are transferred to other plan input fields.
- **Base value distribution**: The shares of previously maintained base elements are used to distribute the entered plan value.
- **Additive and percentage increases/decreases**: The entered absolute or percentage plan values are added or subtracted.
- **Fixed value distribution**: A fixed input value cannot be overwritten during distribution. The plan value is distributed around the fixed value (last example in Fig. 3.159).

• **Extrapolations/forecasts/predictive analytics**
Automated and fast determination of target values should be possible for some planning items by using various extrapolation functions for reasons of efficiency and planning simplification.[254] The user should be able to choose between different extrapolation methods that are suitable for different purposes. The basis for extrapolations are the actual data. Often, the actual data are extrapolated to the end of the year or the end of the project. The remaining target values of the remaining periods to be

[253] Cf. BARC (2016, p. 5–6).

[254] Cf. Reichmann (2006, p. 130 f.).

Top-Down

	Sales PY	Sales PLAN
Company total	1,000	1,200
Sales region north	250	
Sales region south	300	
Sales region west	200	
Sales region east	250	

	Sales PY	Sales PLAN
Company total	1,000	1,200
Sales region north	250	300
Sales region south	300	360
Sales region west	200	240
Sales region east	250	300

Bottom-Up

	Sales PY	Sales PLAN
Company total	1,000	
Sales Region North	250	250
Sales region south	300	360
Sales Region West	200	240
Sales Region East	250	300

	Sales PY	Sales PLAN
Company total	1,000	1,150
Sales region north	250	250
Sales region south	300	360
Sales region west	200	240
Sales region east	250	300

Counterflow (Top-Down then Bottom-Up)

	Sales PY	Sales PLAN
Company total	1,000	1,200
Sales region north	250	
Sales region south	300	
Sales region west	200	
Distribution Region East	250	

	Sales PY	Sales PLAN
Company total	1,000	1,200
Sales region north	250	300
Sales region south	300	360
Sales region west	200	240
Sales region east	250	300

	Sales PY	Sales PLAN
Company total	1,000	1,140
Sales region north	250	320
Sales region south	300	340
Sales region west	200	200
Sales region east	250	280

Distribution without overwriting fixed values

	Sales PY	Sales PLAN	
Company total	1,000	1200.00	
Sales region north	250	306.25	
Sales region south	300	367.50	
Sales region west	200	220.00	fixed (manually planned)
Sales region east	250	306.25	

Fig. 3.159 Distribution and aggregation functions

planned are either taken from the original plan or planned anew. For this purpose, statistical-mathematical formulas or methods (such as trend exploration) can be used, which are based on the past actual values.[255] A simple form of extrapolation is the trend calculation from the actual data of the past periods. Forecasts are also called Forecasting today. Forecast values can alternatively be performed for a fixed future period (e.g. always 12 months) instead of the end of the year. Then it is a rolling planning. Predictive analytics (predictive models) are statistical forecasting functions that are used to predict future events and developments. The linear regression analysis is a well-known example for this, where the behavior of a dependent variable is calculated with the help of a past data series (one or more independent variables).

- **Revaluations**

  Price or rate adjustments or changes in quantity and value estimates can be taken into account within the planning with the instrument of planning revaluation.

[255] Cf. Warnick (1992, p. 1301). The extrapolation must be done separately for variable and fixed costs, for example, since the fixed costs are independent of the output.

- **Consolidations**
  Consolidations are the aggregation of detailed target values to a higher level, e.g. from individual target values per cost type to a cost type group.
- **Plan evaluations and calculations**
  Plan evaluations can result from various values, e.g. the planned quantity multiplied by the planned price gives the planned value (e.g. revenue), or the discount is determined by percentage evaluation based on the revenue. Other examples are interest calculations or currency conversions. In addition, more complex calculation functions, such as the bill of materials explosion, the planning iteration in the internal cost allocation or the program planning, belong to this. More complex calculation functions are often not performed in the planning form but in upstream ERP systems with special statistics programs (e.g. PASW the successor of SPSS) in the data warehouse or in other systems.
- **Simulations[256]and sensitivity analyses (What-If and How-to-Achieve simulations)**
  In the What-If simulation the impact of changing one or more input factors on one or more output variables is examined. Example question: How does the profit change, if the sales volume increases by 2% and the price decreases by 10%?
  In the sensitivity analysis (How-to-Achieve simulation) it is investigated, starting from one or more output variables, to what extent these can be influenced and achieved by input factors (individually or jointly). Example question: What price change and sales volume change is necessary to achieve a profit of €1 million?

An example of the integration of various input and planning functions with the software product "prevero" is shown in Fig. 3.160.

3.11 Alignment of Planning and Reporting Content

The data for the operational business reflect the economic reality of the company in the actual state. The data can be analyzed down to the level of individual documents. In reporting, this individual document information is already aggregated on condensed levels of the accounting systems on the levels of accounts or cost and revenue types. However, there is the possibility to navigate with drill-down- or drill-through functionality in reporting down to the lowest document level.

For planning, this level of detail down to the document level is not suitable, so planning must take place at a higher level. In planning, the value-creating factors should be in the foreground with regard to their impact on the business result.

[256] Cf. on simulations: Romeike and Spitzner (2013).

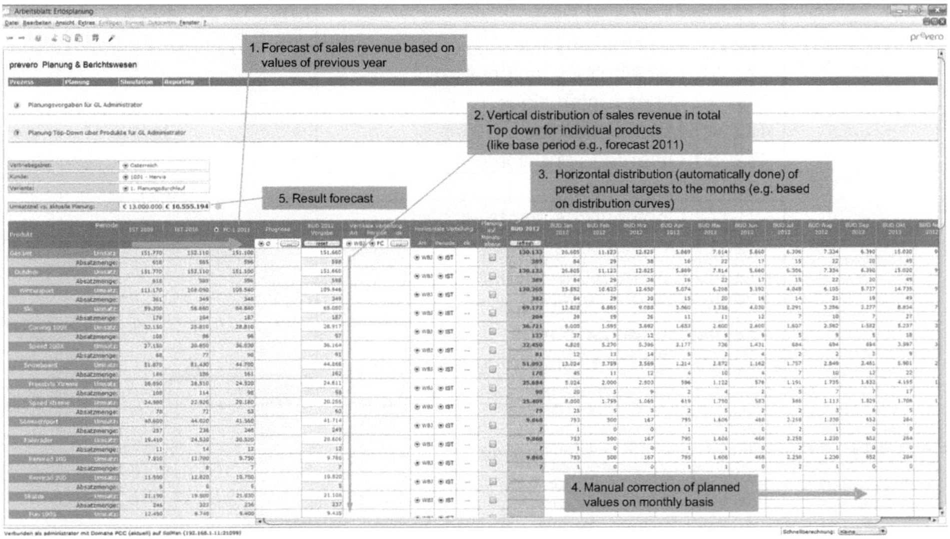

Fig. 3.160 Different input and planning functions using the example of revenue planning with prevero. (Slightly modified from the presentation slides of Hein 2011, p. 42)

For this reason, planning for these important value-creating factors takes place at the detailed level of accounts, cost and revenue types. For other items, it makes sense for economic reasons to plan at a higher level, e.g. on aggregated planning accounts or planning cost and revenue types. Often, the formed groups and hierarchy levels of the hierarchies of accounts or cost and revenue types are suitable for this purpose, as shown in Fig. 3.161 using the example of contributions and fees. Depending on the software product, however, accounts or groups cannot be posted. In this case, the establishment of special aggregated planning accounts or planning cost and revenue types is necessary. In practice, a detailed cost type or account is then often also determined for planning. For transparency reasons, however, the introduction of a planning object (account/cost/revenue type) is recommended (Fig. 3.161).

If a comparison on these compressed objects is not sufficient, the planned values can be distributed to the detail objects, e.g. according to stochastic rules (e.g. as in the previous periods), so that a plan-actual comparison can take place on the detail objects.

Otherwise, the following basic rules apply for the alignment of planning and reporting content:

- The structure and classification of the objects should be identical.
- The selection and filter criteria should be identical.
- The design of the reports and planning forms should be similar in layout.

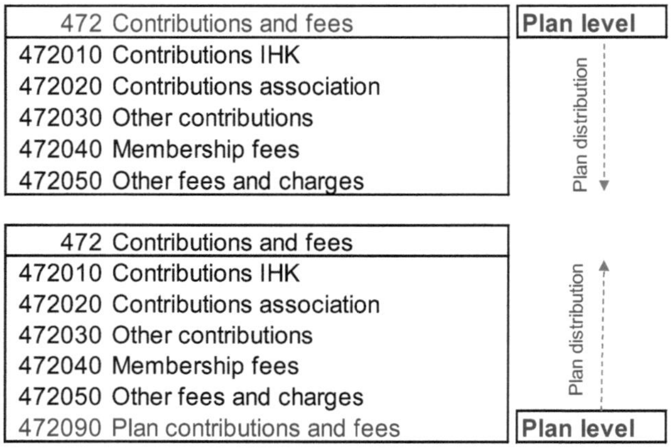

Fig. 3.161 Alignment of planning and reporting content

3.12 Exemplary Planning Areas in the BI-Supported Controlling Cockpit

In the following subchapters, exemplary planning forms and reports for selected planning areas are shown integrated in the BI-supported controlling cockpit:

- Balanced Scorecard in the context of strategic planning
- Sales and revenue planning
- Resource planning
- Purchase price planning
- Personnel, plant and capacity planning

3.12.1 Balanced Scorecard in the Context of Strategic Planning

For the strategic management, the Balanced Scorecard is suitable in terms of content (cf. Sect. 2.2). It is used in addition to the classic instruments of strategic planning, such as the SWOT analysis and the portfolio technique, in strategic planning and serves here especially as a link to the tactical and operational control. The result of the strategy finding is the determination of the strategic targetsobjectives, the risks and the necessary strategic projects. Measured variables and default values are recorded for each target. The Balanced Scorecards can be set up for the entire company or for subareas, e.g. business units or regional areas. A Balanced Scorecard, which was displayed with the

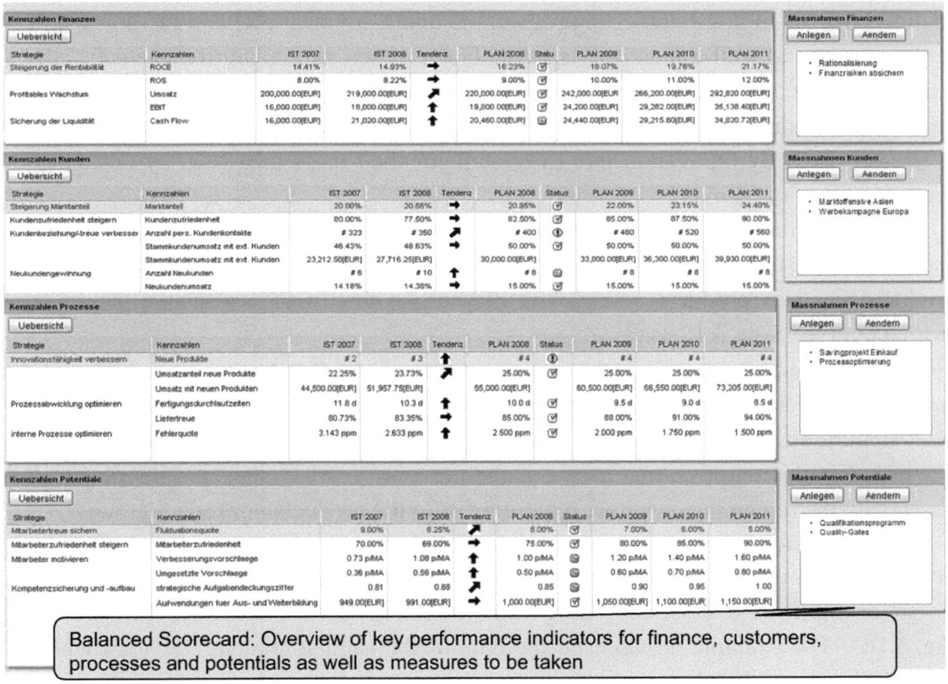

Fig. 3.162 Exemplary Balanced Scorecard planning with SAP NetWeaver technology. (Taken from Schön et al. 2010, p. 25)

SAP NetWeaver technology, is shown in Fig. 3.162. Here, the 4 basic perspectives of the Balanced Scorecard with their strategic targets are shown. For each strategic target, one or more indicators are defined. The indicators are used for planning and controlling the strategic targets. Trend arrows and color status symbols indicate the need for action. Strategic projects and measures are created to achieve the targets.

3.12.2 Sales and Revenue Planning

Sales and revenue planning usually forms the important starting point of operational planning as all other important resource plans (assets, personnel, purchasing, etc.) and result plans depend on it. The approach of sales and revenue planning varies greatly among companies and can therefore only be given as an example. Industry and company specific requirements must be taken into account. These include macroeconomic factors, such as changes in gross domestic product, the unemployment rate and incoming orders in the industry, as well as company specific factors, such as competitive intensity, advertising activities, price developments and existing customer relationships. The quantity-based sales planning of external sales is the basis for the value-based revenue planning. It is carried out by the responsible sales staff for their sales area (e.g. by customer or product groups or regions).

When planning the copious products and different product groups, it is generally necessary to distinguish between products that are already in series production and new products. For new products, so-called reference planning, used in resource planning, is carried out (see Sect. 3.12.3).[257]

The sales and revenue planning process is carried out step by step, e.g. first analytically for the secure order backlog and framework contracts, then for the offers and other sales opportunities. In a further step, the remaining current customer requirements are planned. For this purpose, one uses e.g. the previous year's values and distributes stochastically the remaining customer requirements to the products or product groups.

For the different planning steps, simple extrapolation, revaluation and allocation functions as well as individual planning masks for planning are available.

An example of a detailed rolling sales planning for the sales area is shown in Fig. 3.163. The example is based on the data of the sample company Pantara Holding using Cubeware. The planning is done according to the hierarchy levels of the articles and customers. Optionally, the view of products and customers can be changed. The planning table is divided into three blocks. In the first block, the average values of the last 12, 6 and 3 months are shown as orientation values. The trend is displayed using miniature bar charts. In the second block, the last 5 months are shown for comparison. In the last block, the planned figures are maintained periodically. The maintenance is supported by aggregation and distribution functions, e.g. the splash functionality. Manually planned values are not overwritten by a planning distribution. As an example of taking company-specific influencing factors into account, revenue changes based on already confirmed, still planned advertising campaigns or listings at the retailer can be maintained in the planning lines. When designing the planning form, it is important to include the most important success-driving variables for planning and to support the user as best as possible with distribution and other planning functions.

Planning functions and evaluations can be sensibly selected via product and customer group hierarchies as well as other result object levels, such as plants, sales organizations and company codes. This increases the performance and acceptance of the planning and helps to avoid overlaps.

After the direct sales quantities have been planned, the revenue planning follows. It includes the planning of the sales prices as well as the revenue and cost planning of the sales-related items such as outgoing freight, packaging, licenses and commissions. In

[257] See also Schön and Irmer (2010, p. 49–56).

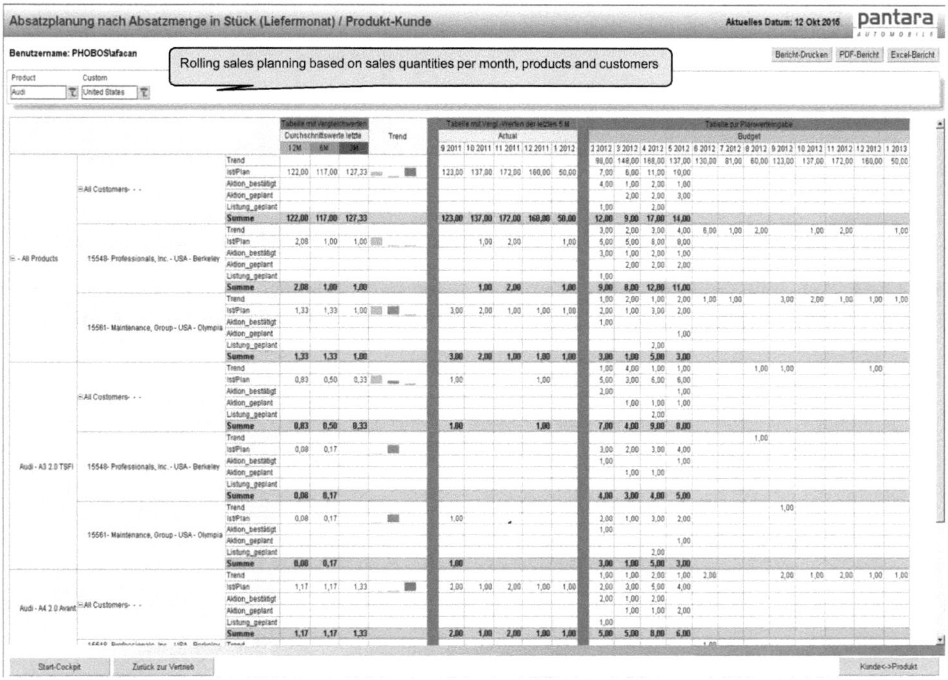

Fig. 3.163 Rolling sales planning with the Cubeware cockpit

order to make the revenue planning as fast and as close to the real valuation conditions as possible, the individual price components are valued with the same pricing procedure that is also used in the operating ERP system. Thus, with proper maintenance of the master data, the valuation can be done automatically and without much effort. The manual maintenance effort in the revenue planning is limited to the planning of expected price changes (e.g. due to savings of an OEM), conditions (bulk discounts, cash discounts, etc.) as well as the adoption of the automatically determined conditions (e.g. scale-dependent gross or net prices as well as outgoing freight, packaging and commissions). The planning is either global, selective by selected groups (see Fig. 3.164) or individual (see Fig. 3.165). It can be planned annually, periodically or with distribution functions in different plan versions. Default or starting values of the planning, such as prices of the ERP systems, are adopted.

For the later integration of the financial planning, the receivables and sales taxes dependent on the revenue planning have to be determined based on payment terms and tax payment dates. These plan data are subsequently required for the sub-plans balance sheet and income statement as well as the cash-flow planning. For the integration into the result planning, the revenues are made available in the provided and enriched result

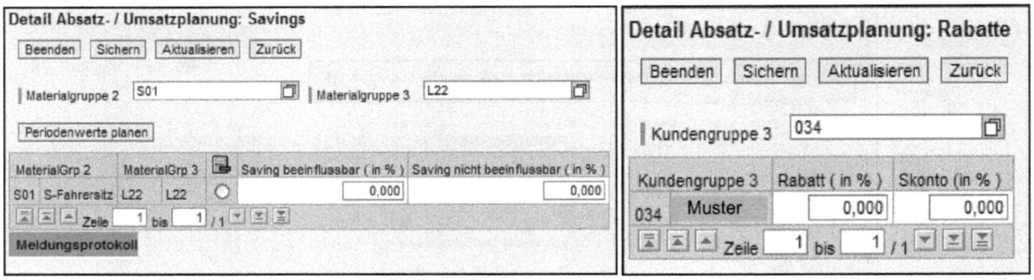

Fig. 3.164 Selective group planning of price changes and conditions (*Source*: Schön and Irmer 2010, p. 52. This is an implementation using the BEx Web Application Designer of SAP AG)

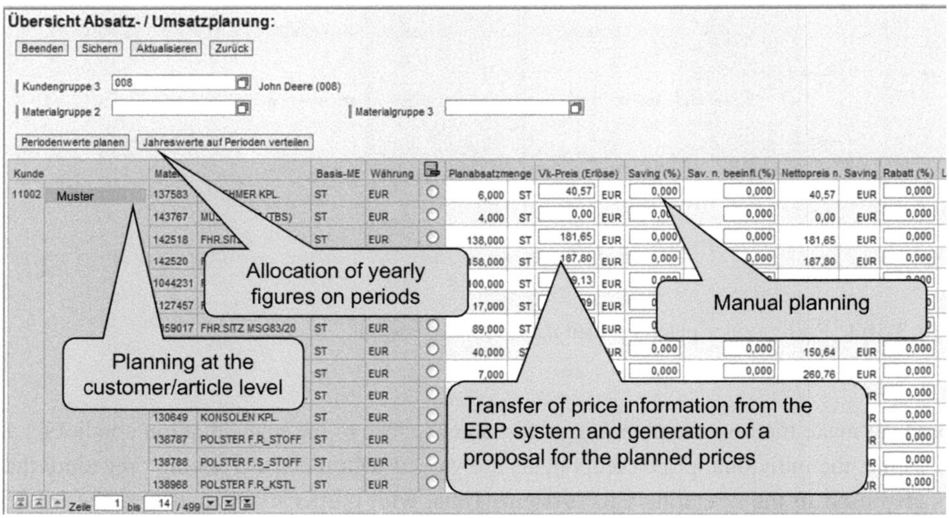

Fig. 3.165 Detailed revenue planning at customer/product level (*Source*: Schön and Irmer 2010, p. 52. This is an implementation using the BEx Web Application Designer of SAP AG)

characteristics (e.g. material group, customer group, business unit, profit center, sales organization, region) for the result planning.

The intercompany revenues (internal revenues) are automatically determined via the resolved IC-sales quantities (from the resource planning) with the help of an iterative valuation per component (see Sect. 3.12.3). To identify the IC-prices, it is necessary to identify the connection of the sending and delivering plant, as well as the group supplier and customer number. For the valuation, in addition to the automatically determined manufacturing costs, IC-surcharges, freight, customs and fees have to be taken into account according to the uniformly regulated price formation guidelines in the group.

To determine the scheduling of the sales quantities in a Europe-wide or worldwide operating company more precisely, both the external and the internal sales quantities are determined taking the route planning into account . This is necessary when long transport distances between the affected plants have to be bridged.

3.12.3 Resource Planning

Resource planning is an important link for the downstream planning areas of purchase price, personnel and capacity planning. It has to be designed individually for each company, just like sales planning. Resource planning comprises the planning of the quantities of raw materials and components as well as the planning of the necessary production services for the planned sales or production quantities. These are, on the one hand, the material requirement quantities for the purchase price planning and, on the other hand, the performance requirement quantities (times per type of service) for the capacity requirement plan for production-related personnel and for the equipment used at the cost centers.[258]

The calculation of the quantity and time frameworks of the sales planning is done with the help of the structures for calculation or the bill of materials and work plans (e.g. from the ERP system with the support of the MRP runs), so that data integration and thus planning reliability are ensured here as well.[259] For new product planning, structures of reference calculations are used. These calculations must correspond to the content structure of a standard calculation.

In contrast to pure cost structure planning, this resource-oriented approach allows a transparent conclusion on resource requirements in purchasing, personnel and equipment (see Sect. 3.12.5). This is a decisive competitive advantage, as one can react much faster and more precisely by targeted resource adjustments due to changing market conditions. This is important for the operational planning, for the forecast and for simulations or scenarios.

As part of the resource planning, it is necessary to recognize the intercompany relationship in order to get a complete view of revenues and costs.

Both the automatic quantity determination (multi-level resolution) and the price determination of the intercompany parts (multi-level aggregation) require an iterative approach for which a powerful functionality must be provided.

With regard to the calculation valuation, various simulation steps are necessary; these steps successively enable planning results at different points in time. By loading currently used, i.e. not yet planned, purchasing conditions and performance rates, e.g. quantity-related effects on the planning result, are visible. By transferring the quantity

[258] Cf. Schön and Irmer (2010, p. 49–56).

[259] Cf. in contrast the process in MRP systems: Steven (2007, p. 235 ff.).

frameworks to the downstream planning areas "material valuation planning and performance requirement planning", the price and rate structures can be further planned and their impact simulated step by step. The feedback of the purchase prices from the material valuation planning and the new rates from the overhead planning complete the planning price information. With the help of the iterative calculation determination and the consideration of the IC-relationships as well as further cost surcharges, the manufacturing costs for the items as well as *for the IC-components* can be determined.

For this planning step, no planning form is required, but a planning function is triggered from the planning process.[260]

3.12.4 Purchase Price Planning

Based on the results from the resource planning, the purchase price planning (material valuation planning) is carried out. The purchase price planning evaluates the resolved components, quantities and raw materials of the resource planning with planned prices in order to determine the planned material consumption.[261]

The planning of the purchase prices is done separately for raw materials and IC-materials. The process of IC-price planning has to be done iteratively, since intercompany relationships have to be resolved step by step, as already described in the context of resource planning (see Sect. 3.12.3). For the planning of the raw materials and parts to be procured, planning templates can be used in which ERP system prices for these materials can be taken over or overwritten as proposal values in the planning dialog.

With regard to the price determination in purchasing, different paths have to be taken; here, the rule is that special and current price information is considered before general price information. The price information can be derived, for example, with the help of the material master records, the order books, the purchase information records, the delivery schedules and the framework contracts of the ERP system. The forms of the purchase planning have to be designed individually according to the requirements of the company.

For the manual material price planning, a suitable interface (e.g. a web-enabled planning interface) should be provided to the planner. The planner can narrow down his planning objects by using various selection options, such as plant, material group, commodity group, valuation class and supplier. For the included individual materials, the price planning is carried out. The planner decides whether the proposed price is transferred to the plan price or whether it is maintained manually.

[260] Cf. Schön and Irmer (2010, issue, p. 49–56).

[261] Cf. Schön and Irmer (2010, p. 49–56).

For price planning, a step-by-step approach is often suitable (see Figs. 3.166 and 3.167), e.g.:

1. Surcharge/discount flat rate for a supplier broken down to plant-material-price combinations
2. Surcharge/discount flat rate for commodity group (commodity) broken down to plant-material-price combinations

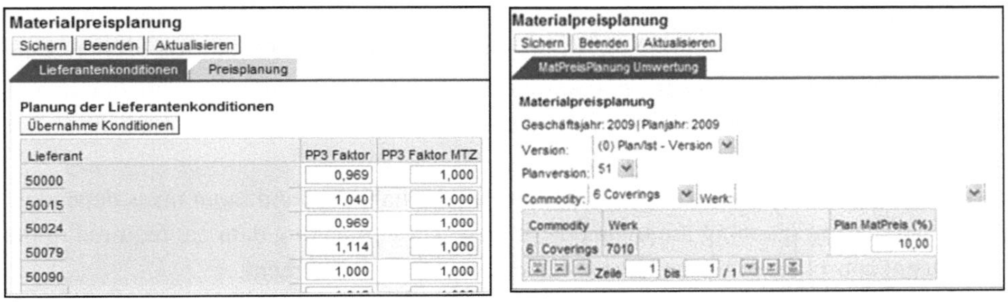

Fig. 3.166 Exemplary surcharges in purchasing price planning (*Source*: Schön and Irmer 2009, p. 296. This is an implementation using the BEx Web Application Designer of SAP AG)

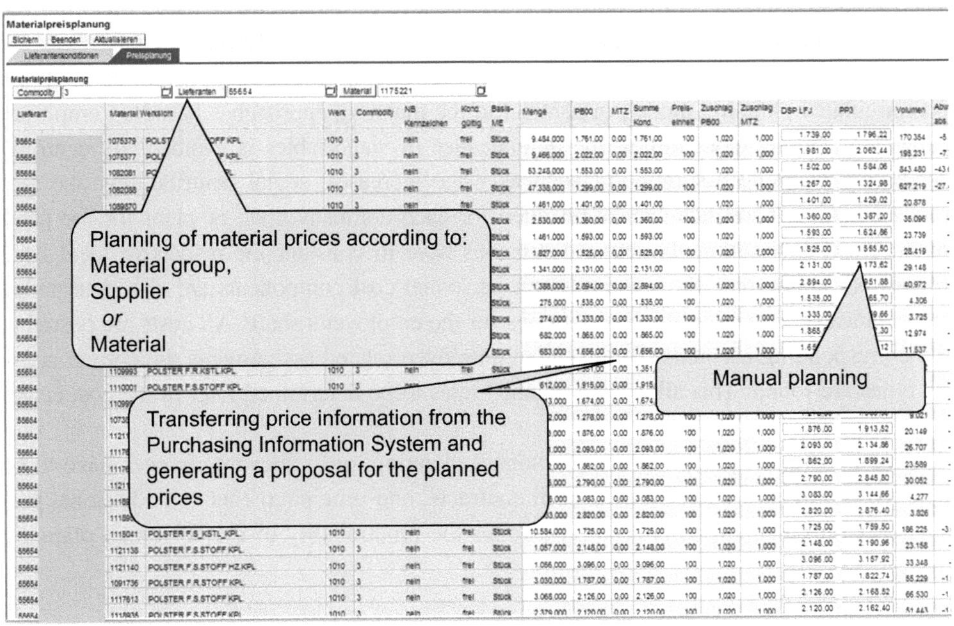

Fig. 3.167 Detail planning template for purchase price planning (*Source*: Schön and Irmer 2010, p. 54. This is an implementation using the BEx Web Application Designer of SAP AG)

3. Fixed plan price for a material in all plants (cross-plant)
4. Fixed plan price for a material in one plant (plant-specific)

When planning the model and the planning templates, the following topics have to be considered in the purchasing area, among others: clear derivation of price and incoterm information, correction of price determination errors, maintenance of master data (valuation classes, quantity units, etc.), currency usage, authorization concept, handling in the planning layouts and the treatment of scale prices and conditions.

With regard to the integration with the activities in the ERP environment (execution of calculations, valuations, price negotiations), the material prices determined within the scope of material price planning are to be automatically retracted into the ERP system. Thus the integration with regard to coordinated planning and control information avoids numerous costly sources of error in the future, such as incorrect calculation bases for offer price determination.

For the later integration of financial planning, the liabilities and input taxes dependent on the purchasing planning are to be determined. These planning data are required in the subsequent sub-plans of the balance sheet and the income statement.

3.12.5 Personnel and Plant Planning

Thepersonnel planning can be done employee-related or anonymously based on employee categories. In the employee-related personnel planning, the planning is based on the actual data which are extracted from the payroll systems at the employee level. With the person-related planning model, in contrast to the planning approaches based on employee categories, among other things, the maintenance of wage tables is avoided. By centrally specifying planning assumptions, such as wage increases, social contributions and factory calendars, these data can be revalued for each company code or plant for the planning period. Internationally acting companies have to consider the respective local legal framework conditions. Provision-relevant personnel cost components are also determined systematically, as well as the contributions for the employer's share. All costs are posted to the corresponding accounts. By means of employee-related assignments the correct activity types are found. This allows the planned rates to be determined later in the cost center accounting.[262]

Based on the employee-related personnel planning, only relevant changes have to be planned now, e.g., transfers, fixed-term contracts, one-time payments and additional personnel requirements. Figure 3.168 illustrates the functionality of the personnel planning

[262] Cf. Schön and Irmer (2007, p. 245–255, 2010, p. 49–56).

with regard to the coordination with the capacity requirements which were determined in the resource planning.

At the level of individual employees, their potential performance is determined and condensed into the performance offer of the cost center. The excess or shortfall in capacity derived from the sales planning is shown in the capacity reconciliation at the cost center level. The capacity is then adjusted within the framework of personnel planning based on the hiring concerning a correspondingly valued excess or shortfall in employees (possibly anonymous: NN = No Name). Further alternatives for capacity adjustment result from changing employee- or cost center-related assumptions, such as performance levels and shift models. Exemplary planning templates or evaluations of personnel cost planning are shown in the two Fig. 3.169 and 3.170.

In personnel planning, both the headcount and the personnel costs have to be planned. For this purpose, the actual data from the payroll system at employee level are extracted as a template. Relevant changes, such as transfers, fixed-term contracts, one-off payments, additional employee requirements and releases, can now be easily added. In addition, general information such as wage increases and social contributions are centrally specified for the personnel cost planning and wage determination of the cost centers. From the personnel planning, the possible performance offer of the cost centers can also be determined and made available in the capacity reconciliation.

As part of the data integration, the plants are retracted as capacities of the cost centers from the ERP system. The presentation of the capacities is in hours per performance type. By multiplying by the production days, the capacity offer per capacity type, e.g. machine, is calculated according to the factory calendar per period.

The capacity requirements determined in resource planning can be adjusted, for example, by justified manual corrections and special additional requirements (see Fig. 3.171). The capacity planning finally balances the capacity supply in machine hours

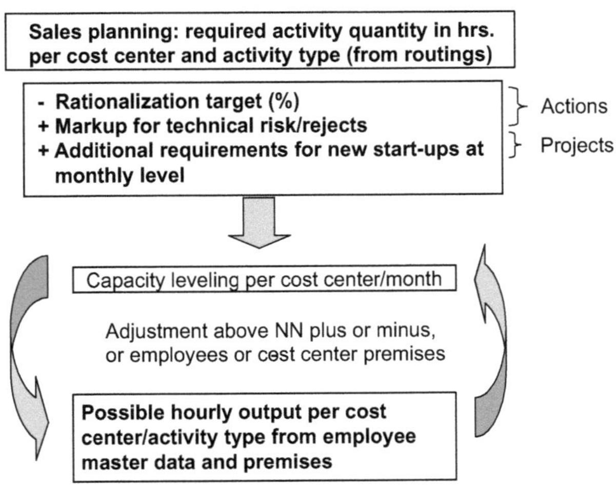

Fig. 3.168 Coordination of personnel planning (Taken from: Schön and Irmer 2009, p. 301)

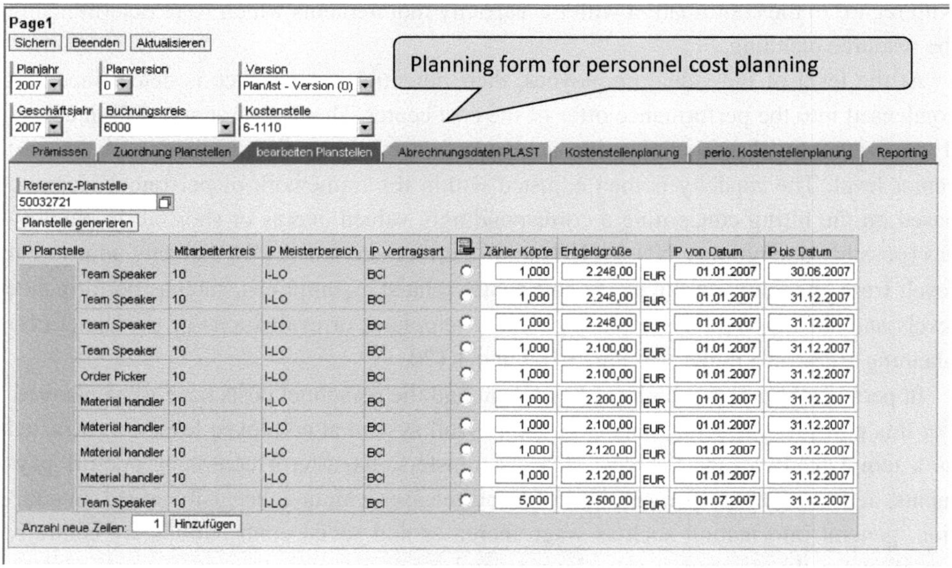

Fig. 3.169 Planning form personnel cost planning (*Source*: Schön and Irmer 2007, p. 253. This is an implementation using the BEx Web Application Designer of SAP AG)

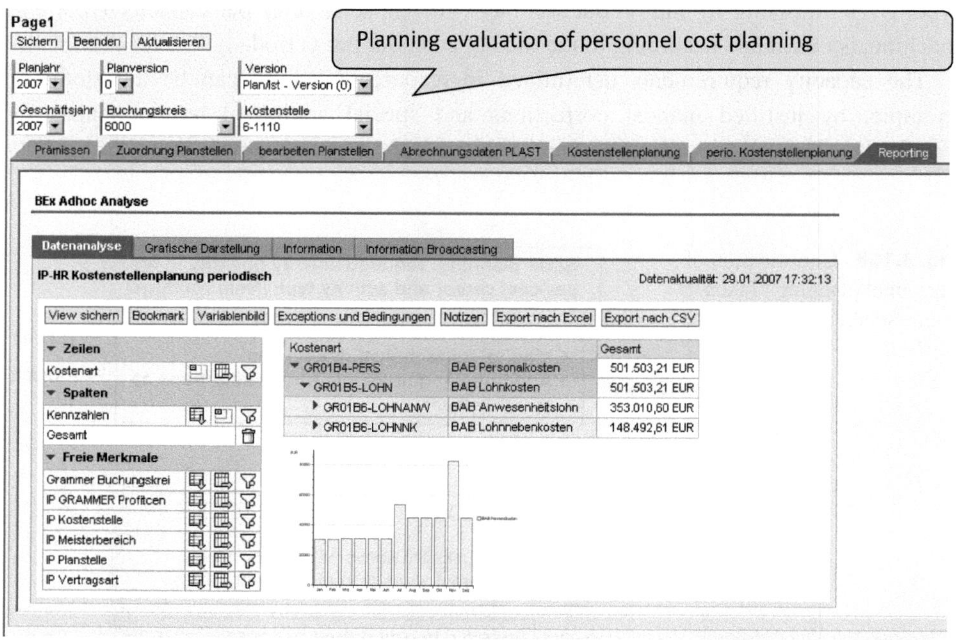

Fig. 3.170 Planning evaluation of personnel cost planning (*Source*: Schön and Irmer 2007, p. 254. This is an implementation using the BEx Web Application Designer of SAP AG)

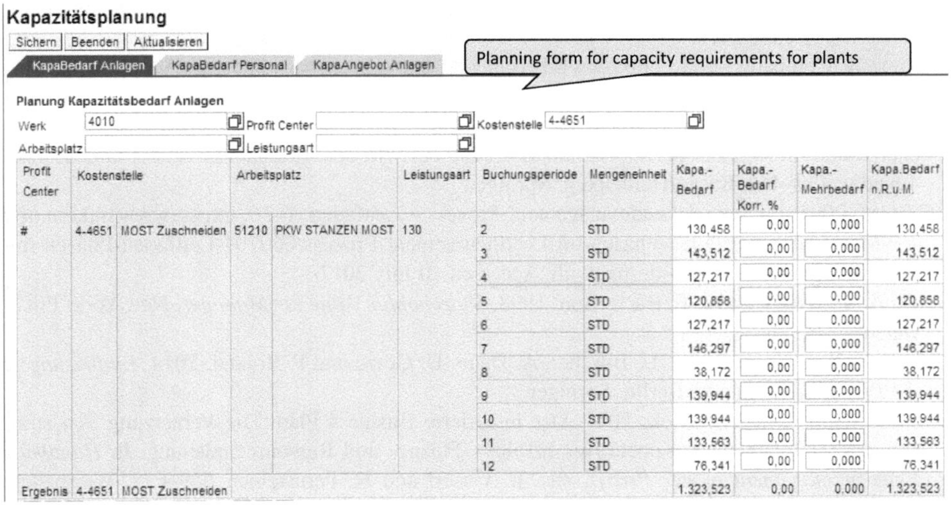

Kapazitätsplanung

Sichern | Beenden | Aktualisieren

KapaBedarf Anlagen | KapaBedarf Personal | KapaAngebot Anlagen

Planning form for capacity requirements for plants

Planung Kapazitätsbedarf Anlagen

Werk 4010 | Profit Center | Kostenstelle 4-4651

Arbeitsplatz | Leistungsart

Profit Center	Kostenstelle		Arbeitsplatz		Leistungsart	Buchungsperiode	Mengeneinheit	Kapa.-Bedarf	Kapa.-Bedarf Korr. %	Kapa.-Mehrbedarf	Kapa.Bedarf n.R.u.M.
#	4-4651	MOST Zuschneiden	51210	PKW STANZEN MOST	130	2	STD	130,458	0,00	0,000	130,458
						3	STD	143,512	0,00	0,000	143,512
						4	STD	127,217	0,00	0,000	127,217
						5	STD	120,858	0,00	0,000	120,858
						6	STD	127,217	0,00	0,000	127,217
						7	STD	146,297	0,00	0,000	146,297
						8	STD	38,172	0,00	0,000	38,172
						9	STD	139,944	0,00	0,000	139,944
						10	STD	139,944	0,00	0,000	139,944
						11	STD	133,563	0,00	0,000	133,563
						12	STD	76,341	0,00	0,000	76,341
Ergebnis	4-4651	MOST Zuschneiden						1.323,523	0,00	0,000	1.323,523

Fig. 3.171 Planning form for capacity requirement plants (*Source*: Schön and Irmer 2010, p. 55. This is an implementation using the BEx Web Application Designer of SAP AG)

with the planned requirements (demand). The adjustment of the plant capacity is done, similar to personnel planning, either by planning an excess or shortfall of facilities or by changing the capacity or cost center premises (e.g. by changing the shift models or a performance adjustment). In addition, economization potentials and additional requirements for new launches can be planned.

References

Al-Laham, A., M. K. Welge and M. Eulerich. 2017. *Strategisches Management: Grundlagen – Prozess – Implementierung*, 7th edn. Wiesbaden: Springer Gabler.

Ansoff, H. I. and E. J. McDonnell. 1990. *Implanting Strategic Management*, 2nd edn. New York: Prentice-Hall.

BARC. 2016. BARC Product Review – Jedox Suite 6, 5–6. http://barc.de/docs/jedox-jedox-suite. Accessed 4 July 2016.

Bauer, H. H. 1995. Marketing-Planung und -Kontrolle. In *Handwörterbuch des Marketing*, eds. B. Tietz, R. Köhler and J. Zentes, 1653–1668. Stuttgart: Schäffer-Poeschel.

Bauer, J. 2017. *Produktionscontrolling und -management mit SAP® ERP*, 5th edn. Wiesbaden: Springer Vieweg.

Bea, F. X. and J. Haas. 2013. *Strategisches Management*, 6th edn. Munich: UVK.

Bea, F. X., S. Scheurer and S. Hesselmann. 2008. *Projektmanagement*. Stuttgart: UVK.

Biedermann, H. 2005. *Instandhaltungscontrolling und -budgetierung im Wandel: Planung, Kontrolle und Koordination der Instandhaltung*. Cologne: TÜV Media GmbH TÜV Rheinland Group.

bime Showcase. 2017. Pie- and donut-chart. https://showcase.bime.io/dashboard/ecommerce. Accessed 23 June 2017.

Bissantz & Company. 2011. http://www.bissantz.de/sparkmaker/beispiele.asp. Accessed 12 May 2011.

Bissantz, N., G. Mertens, G. Butterwegge and V. Christ. 2013. Visualisierung betriebswirtschaftlicher Daten. In *Analytische Informationssysteme, Part III*, eds. P. Chamoni and P. Gluchowski, 4th edn., 439–462. Berlin/Heidelberg: Springer.

BITCOM. 2005. Phasen im Leadmanagement-Prozess – Leitfaden, 1–36. https://www.bitkom.org/NP-Marktdaten/Leitfaden-Phasen-im-Leadmangement-Prozess/060707-Leitfaden-Phasen-im-Leadmanagement-Prozess-dc-final.pdf. Accessed 10 July 2017.

Black, A., P. Wright and J. E. Bachmann. 1998. *Shareholder Value für Manager*. New York: PricewaterhouseCoopers.

Bloech, J., R. Bogaschewsky, U. Buscher, A. Daub, U. Götze and F. Roland. 2014. *Einführung in die Produktion*, 7th edn. Berlin: Springer.

Blumenschien, F. and R. Dick. 2004. Der integrierte Business Plan: Die Vernetzung von strategischer Zielplanung und operativer Erfolgs-, Finanz- und Ressourcenplanung. In *Handbuch Finanzmanagement in der Praxis*, eds. R. Guserl and H. Pernsteiner, 659–678. Wiesbaden: Springer Gabler.

Bruhn, M. 2010. *Marketing*, 10th edn. Wiesbaden: Springer Gabler.

Bühner, R. 2005. *Personalmanagement*, 3rd edn. Munich: Oldenbourg.

Camphausen, B., T. Vollmer, J. Jandt, F. Levin and B. Eichler. 2014. *Grundlagen der Betriebswirtschaftslehre*, 3rd edn. Munich: Oldenbourg.

Coenenberg, A. G. 2009. *Jahresabschluß*, 21st edn. Landsberg/Lech: Moderne Industrie.

Copeland, T., T. Koller and J. Murrin. 1998. *Unternehmenswert – Methoden und Strategien für eine wertorientierte Unternehmensführung*, 2nd edn. Frankfurt a. M.: Campus.

Corsten, H. and Gössinger, R. 2016. *Produktionswirtschaft. Einführung in das industrielle Produktionsmanagement*, 14th edn. Berlin: De Gruyter Oldenbourg.

Cubeware Bildergalerie. 2011. Cubeware Bildergalerie. http://www.cubeware.de/produkte/online-demo-cockpit-v6pro.html. Accessed 20 July 2011.

Denk, R., K. Merkelt-Exner and R. Ruthner. 2008. *Corporate risk management, 124*. Vienna: C.H. Beck.

Deyle, A. 2001. Soll-Ist-Vergleich, Erwartungsrechnung und Führungsstil. In *Controller Praxis – Führung durch Ziele, Planung und Controlling*, Vol. II, 31, 4th edn. Wörthersee-Etterschlag: Verlag für ControllingWissen.

Dillerup, R. and R. Stoi. 2006. *Unternehmensführung*. Munich: Vahlen.

DIN 66901. 2009. https://www.din.de/de/meta/suche/62730!search?query=69901. Accessed 5. July 2017.

Drews, H. and S. Schilling. 2014. Reporting-Regeln situationsgerecht anwenden: Hinweise zum Umgang mit Hicherts Success-Modell. In *Der Controlling-Berater, Reporting und Business Intelligence – Update*, eds. R. Gleich and A. Klein, Vol. 32, 125–144. Freiburg i. Br.: Haufe-Lexware.

Egger, N., J. M. Fiechter, C. Rohlf, J. Rose and S. Weber. 2005. *SAP BW Planung und Simulation*. Bonn: Galileo Press.

Ehrmann, H. 2016. *Marketing-Controlling*. Herne: NBW.

Eichenberg, T., M. Hahmann, O. Hördt, M. Luther and T. Stelzer-Rothe. 2017. *Unternehmensführung: Fallstudien, Klausuren, Übungen und Lösungen*, Lehr- und Klausurenbücher der angewandten Ökonomik, Vol. 5. Berlin: De Gruyter Oldenbourg.

Eisl, C., H. Losbichler and P. Hofer. 2015. *Grundlagen der finanziellen Unternehmensführung*, Vol. IV, 3rd edn. Vienna: Linde.

Fandel, G. 2007. *Produktion I*, 7th edn. Berlin: Springer.
Farris, P., N. Bendle, P. Pfeifer and D. Reibstein. 2007. *Marketing messbar machen – Die 50 wichtigsten Methoden aus dem Marketing die jeder Manager kennen sollte*. Munich: Pearson Studium.
Fiedler, R. 2016. *Controlling von Projekten*, 7th edn. Wiesbaden: Springer.
Fischer, P., S. Kowalski and U. K. Wissmeier. 2005. *Die besten Tools für Marketing und Vertrieb. Professionelle Excel-Lösungen für kleine und mittelständische Unternehmen*. Freiburg i. Br.: Haufe.
Fischer, R. 2003. *Unternehmensplanung mit SAP SEM – Operative und strategische Planung mit SEM-BPS*. Bonn: Galileo Press.
Focke, M. and J. Steinbeck. 2018. *Steigerung der Anlagenproduktivität durch OEE-Management*. Wiesbaden: Springer Gabler.
Friedemann, W. and C. Neumann. 2007. Kundenzufriedenheit und Kundenbindung. In *Wirtschaftspsychologie*, ed. K. Mose. Heidelberg: Springer.
Frontline-Consulting. 2011. Tipps: Gestaltung von Präsentationsmedien. https://www.frontline-consulting.de/tipp-mediengestaltung.pdf. Accessed 19 Apr 2011.
Gehringer, J. and W. J. Michel. 2000. *Frühwarnsystem Balanced Scorecard. Unternehmen zukunftsorientiert steuern; mehr Leistung, mehr Motivation, mehr Gewinn*. Berlin: Metropolitan in Walhalla u. Praetoria.
Georges, P. M. 2000. The Management Cockpit – The human interface for management software: reviewing 50 user sites over 10 years of experience. In *Wirtschaftsinformatik* 42(2):131–136. Gabler. https://link.springer.com/article/10.1007%2FBF03250727. Accessed 14 Sept 2018.
Gerths, H. and R. Hichert. 2011. *Geschäftsdiagramm mit Excel nach den SUCCESS-Regeln gestalten*. Freiburg i. Br.: Haufe-Lexware.
Gladen, W. 2014. *Unternehmensperformance, Controlling mit Kennzahlen*, 6th edn. Wiesbaden: Springer Gabler.
Gleich, R. and A. Klein. 2016. *Der Controlling-Berater*, Vol. 44. Freiburg-Munich: Visualisierung von Managementberichten.
Gleißner, W. 2000. Risikopolitik und Strategische Unternehmensführung. *Der Betrieb* 2000(33):1625–1629.
Gleißner, W. 2011. *Grundlagen des Risikomanagements*. Munich: Vahlen.
Gleißner, W. and F. Romeike. 2005. *Risikomanagement*. Planegg: Haufe.
Godmode-trader. 2011. R. Berteit: SOFTWARE AG wieder im Käufermarkt. http://www.godmode-trader.de/nachricht/SOFTWARE-AG-wieder-im-Kaeufermarkt-Software,a2580092,b1.html. Accessed 5 July 2011.
Göpfert, I. 2006. Berichtswesen. In *Wirtschaftslexikon*, Vol. 2, 692–702. Stuttgart: Haufe.
Gottmann, J. 2016. *Produktionscontrolling*. Wiesbaden: Springer Gabler.
Gottmann, J. 2019. *Produktionscontrolling*, 2nd edn. Wiesbaden: Springer Gabler.
Grabner, T. 2017. *Operations Management*, 3rd edn. Wiesbaden: Springer Gabler.
Growe, A., M. Boehle and D. Schön. 2020. BI-gestütztes Controlling-Cockpit für den Unternehmensbereich IT. *Controlling* 32(3):80–85.
Haman, M. K. 2013. *Rollen & Verantwortlichkeiten in der Produktion*. Books on Demand.
Heigl, A. 1989. *Controlling – Interne Revision*, 2nd edn., 128. Stuttgart/New York: Gustav Fischer.
Hein, A. 2011. *prevero Unternehmen und Lösungen – Unternehmenssteuerung mit prevero*. Vienna: prevero Austria. Lecture script.
Hermann, S. 2015. *Preisheiten – Alles, was Sie über Preise wissen müssen*, 2nd edn. Frankfurt a. M.: Campus.
Herrmann, F. 2010. *Kurzvorträge BWL/VWL*. Wiesbaden: Gabler.

Heuser, R., F. Günther and O. Hatzfeld. 2003. *Integrierte Planung mit SAP – Konzeption, Methodik, Vorgehen*. Bonn: Galileo Press.
Hichert+Partner AG. 2008. Gestalten von Berichten und Präsentationen, 1–17. http://www.hichert.com/de/_media/broschueren/broschuere_2008.pdf. Accessed 15 July 2011.
Hichert+Partner AG. 2011. Management Information Design – Damit Berichte etwas berichten, 1–29. http://www.hichert.com/de/_media/broschueren/handout.pdf. Accessed 15 July 2011.
Hofbauer, G. and C. Hellwig. 2016. *Professionelles Vertriebsmanagement*, 4th edn. Erlangen: Publicis.
Homburg, C. 2017. *Marketingmanagement*, 6th edn. Wiesbaden: Springer Gabler.
Hornung, K., T. Reichmann and M. Diederichs. 1999. Risikomanagement – Teil I: Konzeptionelle Ansätze zur pragmatischen Realisierung gesetzlicher Anforderungen. *ZfC* 11(7):317–325.
Horváth, P. 2008. *Controlling*, 11th edn. Munich: Vahlen.
Horváth, P. and G. Urban. 1990. *Qualitätscontrolling*. Stuttgart: Poeschel.
Hostettler, S. 2000. *Economic Value Added – Darstellung und Anwendung auf Schweizer Aktiengesellschaften*, 4th edn. Bern: Haupt.
Ibarra, A. 2005. *Anforderungen an Führungskräfte in der Produktion*, 73941-Bericht. Konstanz: Christiani.
IIRC. 2013. The International ⟨IR⟩ Framework. http://www.theiirc.org/wp-content/uploads/2013/12/13-12-08-THE-INTERNATIONAL-IR-FRAMEWORK-2-1.pdf. Accessed 16 Mach 2015.
Jacob, M. 2012. *Informationsorientiertes Management*. Wiesbaden: Springer Gabler.
Keist, N.-K., S. Benisch and C. Müller. 2016. Möglichkeiten und Grenzen der plattformübergreifenden App-Entwicklung. In *Mobile Anwendungen in Unternehmen*, eds. T. Barton, C. Müller and C. Seel, 109–119. Wiesbaden: Springer.
Kemper, H. G., H. Baars and W. Mehanna. 2010. *Business Intelligence Grundlagen und Praktische Anwendungen*, 3rd edn. Wiesbaden: Springer Vieweg.
Kilger, W. 1988. *Flexible Plankostenrechnung und Deckungsbeitragsrechnung*, 10th edn. Wiesbaden: Gabler. Revised from K. Vikas.
Kistner, K.-P. and M. Steven. 2001. *Produktionsplanung*, 3rd edn. Heidelberg: Physica.
Klein, A. 2014. *Reporting und Business Intelligence*. Freiburg i. Br.: Haufe-Lexware.
Klein, A. and P. Schentler, eds. 2016. *Einkaufscontrolling: Instrumente und Kennzahlen für einen höheren Wertbeitrag des Einkaufs*. Freiburg i. Br.: Haufe.
Knorren, N. 1998. *Wertorientierte Gestaltung der Unternehmensführung*. Wiesbaden: Springer Gabler.
Köhler, R. 1993. *Beiträge zum Marketing-Management: Planung, Organisation, Controlling*, 3rd edn. Stuttgart: Schäffer-Poeschel.
Kohlhammer, J., U. Proff and A. Wiener. 2013. *Visual business analytics*. Heidelberg: dpunkt.
Kommission der Europäischen Gemeinschaft. 2001. *Grünbuch Europäische Rahmenbedingungen für die soziale Verantwortung der Unternehmen, 7*. Brüssel: European Commission – CSR Green Paper Consultation.
Krause, H.-U. and D. Arora. 2008. *Controlling-Kennzahlen Key Performance Indicators*. Munich: Oldenbourg.
Krüger, J. 2015. *SAP simple finance an introduction*. Bonn: Galileo Press.
Krupp, A. D. 2014. *Unternehmensplanung und Kontrolle*. Norderstedt: Kompakt.
Küpper, H. U. 2004. *Controlling – Konzeption, Aufgaben und Instrumente*, 4th edn. Stuttgart: Schäffer-Poeschel.
Lachnit, L. and S. Müller. 2012. *Unternehmenscontrolling*, 2nd edn. Wiesbaden: Springer Gabler.
Laier, R. 2011. *Value reporting*. Wiesbaden: Springer Gabler.

Linden, M. 2016. *Geschäftsmodellbasierte Unternehmenssteuerung mit Business-Intelligence-Technologien – Unternehmensmodell – Architekturmodell – Datenmodell*. Wiesbaden: Springer Gabler.

Lippold, D. 2017. *Marktorientierte Unternehmensführung und Digitalisierung. Management im digitalen Wandel*. Berlin: De Gruyter Oldenbourg.

Littkemann, J., K. Derfuß and M. Holtrup. 2018. *Unternehmenscontrolling – Praxishandbuch für den Mittelstand – Konzepte, Instrumente, praktische Anwendungen mit durchgängiger Fallstudie*, 2nd edn. Herne.

Lutz, S., C. Back and W. Becker. 2011. *Gabler Kompaktlexikon Modernes Rechnungswesen*, 3rd edn. Wiesbaden: Springer Gabler.

Lux, W. 2006. Performance Management und Controlling Cockpit (2/2). Aufbau eines Cockpits und deren Unterstützung durch Software. *KMU-Magazin* (6):28–32. https://www.researchgate.net/publication/315642760_Performance_Management_and_Controlling_Cockpit_Part_2. Accessed 14 Sept 2018.

May, C. and A. Koch. 2008. Overall Equipment Effectiveness (OEE) Werkzeug zur Produktivitätssteigerung. *ZUb H* 6:245–250.

Mehanna, W. 2016. Digital forecasts. *Business Intelligence Magazine* 2016(1):22–25.

Mehlan, A. 2007. *Praxishilfen Controlling: die besten Controlling-Instrumente mit Excel*. Freiburg i. Br.: Haufe.

Meier, M., W. Sinzig and P. Mertens. 2002. *SAP Strategic Enterprise Management/Business Analytics: Integration von strategischer und operativer Unternehmensführung*. Berlin: Springer.

Mentzel, K. 2015. *Strategisches Management & Controlling*. Dortmund: W3L-Verlag.

Microsoft. 2017a. Sunburst-chart. https://docs.microsoft.com/en-us/sql/reporting-services/report-design/tree-map-and-sunburst-charts-in-reporting-services. Accessed 23 June 2017.

Microsoft. 2017b. Treemap-chart. https://docs.microsoft.com/en-us/sql/reporting-services/report-design/tree-map-and-sunburst-charts-in-reporting-services. Accessed 23 June 2017.

Möller, K., F. Federmann, S. Pieper and M. Knezevic. 2017. Predictive Analytics zur kurzfristigen Umsatzprognose. *Zeitschrift für Controlling* 28(8–9):509–518.

Müller, H. 1995. Grundlagen und praktische Anwendung der Primärkostenrechnung. In *Modernes Kostenmanagement – Grenzplankostenrechnung als Controllinginstrument*, eds. W. Männel and H. Müller, 35–44. Wiesbaden: Springer. Contributions of the Plaut-Gruppe.

Müller, R. M. and H.-J. Lenz. 2013. *Business Intelligence*. Berlin/Heidelberg: Springer Vieweg.

Nakahima, S. 1995. *Management der Produktionseinrichtungen*. Frankfurt a. M.: Campus.

Oehler, K. 2006. *Corporate Performance Management*. Munich/Vienna: Hanser.

Online Chart Makers – amCharts. 2017a. Funnel-chart. https://www.amcharts.com/demos/funnel-chart/. Accessed 23 June 2017.

Online Chart Makers – amCharts. 2017b. Bullet-Charts. https://www.amcharts.com/demos/bullet-chart/. Accessed 23 June 2017.

Opresnik, M. and C. Rennhak. 2012. *Grundlagen der Allgemeinen Betriebswirtschaftslehre*. Wiesbaden.

ORGLINEA. 2019. Die Maschinenauslastung steigern. https://www.orglinea.com/app/download/5804271521/Projektbericht_Maschinenauslastung_2015.pdf.

Panke, B. 2015. Geringe Margen erfordern schnelle Entscheidungen – das Management-Cockpit als Basis. *Getränkefachfrosshandel* (2):11–13. Online-Quelle: http://www.kantiko.com/wp-content/uploads/2015/02/Geringe-Margen-erfordern-schnelle-Entscheidungen-%E2%80%93-das-Management-Cockpit-als-Basis.pdf. Accessed 8 Apr 2018.

Paxmann, S. and G. Fuchs. 2010. *Der unternehmensinterne Businessplan. Neue Geschäftsmöglichkeiten entdecken, präsentieren, durchsetzen*, 2nd edn. Frankfurt a. M.: Campus.

Poensgen, O. H. 1981. Break-Even-Analysis. In *HWR*, ed. E. Kosiol, 2nd edn., 303–313. Stuttgart: C.E. Poeschel.

Porter, M. E. 2013. *Wettbewerbsstrategien. Methoden zur Analyse von Branchen und Konkurrenten*, 2nd edn. Frankfurt a. M.: Campus.
Porter, M. E. 2014. *Wettbewerbsvorteile. Spitzenleistungen erreichen und behaupten*, 8th edn. Frankfurt a. M.: Campus.
PPM Execution. 2017. Bubble-chart. http://ppmexecution.com/portfolio-reports-portfolio-bubble-charts/. Accessed 23 June 2017.
Pradtke GmbH. 2021. Dynamische Vollkräfte-Statistik mit Heartbeat – dem Führungsinformations- und Personal-Controlling-System im Gesundheitswesen.
Pufahl, M. 2015. *Sales performance management*. Wiesbaden: Springer Gabler.
QlikView-Demo Market Share. 2015. Qlik View-Demo Market Share. http://eu-b.demo.qlik.com/QvAJAXZfc/opendoc.htm?document=qvdocs%2FMarketPS.qvw&host=demo11&anonymous=true. Accessed 18 Febr 2015.
react. 2017. Bubble-chart. https://www.npmjs.com/package/react-bubble-chart. Accessed 23 June 2017.
Reichmann, T. 2006. *Controlling mit Kennzahlen und Managementberichten*, 7th edn. Munich: Vahlen.
Reichmann, T. 2011. *Controlling mit Kennzahlen, Die systemgestützte Controlling-Konzeption mit Analyse- und Reportinginstrumente*, 8th edn. Munich: Vahlen.
Reichmann, T., M. Kißler and U. Baumöl. 2017. *Controlling mit Kennzahlen – Die systemgestützte Controlling-Konzeption*, 9th edn. Munich: Vahlen.
Reinecke, S. and S. Janz. 2007. *Marketingcontrolling*. Stuttgart: Kohlhammer.
Reiterer, H., T. M. Mann, G. Mußler and U. Bleimann 2000. Visualisierung von entscheidungsrelevanten Daten für das Management. *HMD, Praxis der Wirtschaftsinformatik* 212(4):71–83.
Romeike, S. and J. Spitzner. 2013. *Von Szenarioanalyse bis Wargaming. Betriebswirtschaftliche Simulationen im Praxiseinsatz*. Weinheim: Wiley.
SAP AG. 2003. BExMap, Beispiel Market Potential USA. Walldorf: SAP AG.
Schlick, C., M. Schenk, D. Spath and W. Ganz. 2016. *Produktivitätsmanagement von Dienstleistungen – Modelle, Methoden und Werkzeuge*. Berlin: Springer Vieweg.
Schmid-Gundram, R. 2014. *Controlling-Praxis im Mittelstand*. Wiesbaden: Springer Gabler.
Schmidt-Volkmar, P. 2008. *Betriebswirtschaftliche Analyse auf Operationalen Daten*. Wiesbaden: Springer Gabler.
Schneider, W. and A. Henning. 2008. *Lexikon Kennzahlen für Marketing und Vertrieb – Das Marketing-Cockpit von A–Z*, 2nd edn. Heidelberg: Springer.
Schnell, H. 2018. Produktionscontrolling: Selbstverständnis, Aufgaben und Instrumente. In *Der Controlling-Berater*, ed. Andreas von Klein, Vol. 4, 1st edn., 21–40. Freiburg i. Br.: Haufe.
Schoeneberg, K.-P. 2014. *Komplexitätsmanagement in Unternehmen*. Wiesbaden: Springer Gabler.
Schön, D. 1999. *Neue Entwicklungen in der DV-gestützten Kosten- und Leistungsplanung*. Frankfurt a. M.: Lang.
Schön, D. 2003. Stichwortgruppe: Projekt-Controlling (ca. 100 Stichwörter). In *Vahlens Großes Controllinglexikon*, eds. P. Horváth and T. Reichmann, 2nd edn. Munich: Vahlen.
Schön, D. 2004a. Moderne EDV-gestützte Planning- und Reportingtools. In *19. Deutscher Controlling Congress, Tagungsband*, ed. T. Reichmann, 287–337. Dortmund: Gesellschaft für Controlling e.V.
Schön, D. 2004b. Moderne Planungskonzepte und Reportingtools. In *19. Deutscher Controlling Congress, Tagungsband*, ed. T. Reichmann, 287–337. Dortmund: Gesellschaft für Controlling e.V.
Schön, D. 2017. Kapitel 10 Projekt-Controlling. In *Controlling mit Kennzahlen*, eds. T. Reichmann, M. Kißler and U. Baumöl, 9th edn., 507–537. Munich: Vahlen.
Schön, D. and K. H. Irmer. 2007. Integrierte Unternehmensplanung bei der Grammer AG. *Controlling* 19(4):245–255.

Schön, D. and K.-H. Irmer. 2009. Integriertes Planungssystem bei der Grammer AG. In *24. Deutscher Controlling Congress, Tagungsband*, ed. T. Reichmann, 279–305. Dortmund: Gesellschaft für Controlling e. V.

Schön, D. and K. H. Irmer. 2010. Effiziente Steuerung mit Forecasting und Integrierter Unternehmensplanung bei der GRAMMER Gruppe. *Controlling* 22:49–56.

Schön, D. and R. Müller. 2010. Mittelstandscontrolling für Inhaber und Manager. In *25. Deutscher Controlling Congress, Tagungsband*, ed. T. Reichmann, 123–165. Dortmund: Gesellschaft für Controlling e. V.

Schön, D., V. Busch and M. Diederichs. 2001. Chancen- und Risikomanagement in Unternehmen mit Projektgeschäft – Transparenz durch ein DV-gestütztes Frühwarninformationssystem. *Controlling* 13(7):379–387.

Schön, D., J. Thünken and A. Johanning. 2010. Strategische Planung mit der Balanced Scorecard im SAP Visual Composer. *Controlling Magazin* 35(5):22–29.

Schön, D., M. J. Reinfelder and M. Drozdzynski. 2011. Modernes Reporting im Mittelstand am Beispiel der Hemmelrath Lackfabrik GmbH. *Controlling* 5(4):230–236.

Schön, D., M. Ständer and M. Liebe. 2013. Unternehmens- und Projektsteuerung mit Hilfe eines BI-gestützten Controlling-Cockpits am Beispiel der Securiton GmbH. *Controlling* 25(4/5):252–260.

Schreckeneder, B. 2013. *Projekt-Controlling – Projekte überwachen, steuern, präsentieren*, 4th edn. Freiburg i. Br.: Haufe.

Schutt, A. 2014. *Das Management Cockpit auf Basis von SAP NetWeaver zur Unterstützung des Performance Measurement; Konzeption und Realisierung*. Hamburg: GRIN-Verlag.

Seghezzi, H. D. and S. Fries. 1995. Zweckmäßige Fehlerkostenrechnung anstatt traditioneller Qualitätskostenrechnung. *KRP* 39:87–91.

Sejdic, G. 2019. *Produktionscontrolling im Kontext von Industrie 4.0*, 1st edn. Baden-Baden: Nomos.

Seppelfricke, P. 2005. *Handbuch Aktien- und Unternehmensbewertung*, 2nd edn. Stuttgart: Schäffer-Poeschel.

Steger, J. 2014. *Kennzahlen und Kennzahlensysteme*. Herne: NBW.

Steuernagel, A. 2017. *Strategische Unternehmenssteuerung im digitalen Zeitalter. Theorien, Methoden und Anwendungsbeispiele*. Wiesbaden: Springer Gabler.

Steven, M. 2007. *Handbuch Produktion*. Stuttgart: Kohlhammer.

Steven, M. 2016. *Produktionscontrolling*, 1st edn. Stuttgart: Kohlhammer.

Sullivan, L. and W. W. LaMorte. 2017. Box-whisker-plot. http://sphweb.bumc.bu.edu/otlt/mph-modules/bs/bs704_summarizingdata/bs704_summarizingdata8.html. Accessed 23 June 2017.

Tremmel, J. 2003. *Handbuch Generationengerechtigkeit, Stiftung für die Rechte zukünftiger Generationen*, 2nd edn. Munich: Ökom-Verlag.

Tufte, E. 2011. The work of Edward Tufte and Graphics Press. http://www.edwardtufte.com/. Accessed 12 May 2011.

Unrein, D. 2016. *Excel im Controlling*. Munich: Vahlen.

Varnholt, N. T. and M. Dagit. 2009. Rolle der BWA im Risk Performance Management. In *Risk Performance Management*, eds. R. M. Hilz-Ward and O. Everling, 191–208. Wiesbaden: Springer Gabler.

Voß, S. and K. Gutenschwager. 2001. *Informationsmanagement*. Berlin/Heidelberg: Springer Gabler.

Warnick, B. 1992. Typische Funktionsumfänge der Standardsoftware zur Kosten- und Leistungsrechnung. In *Handbuch Kostenrechnung*, ed. W. Männel, 1295–1307. Wiesbaden: Gabler.

Weber, J., U. Bramsemann, C. Heineke and B. Hirsch. 2017. *Wertorientierte Unternehmenssteuerung*, 2nd edn. Wiesbaden: Gabler.

Weber, W., R. Kabst and M. Baum. 2014. *Einführung in die Betriebswirtschaftslehre*, 9th edn. Wiesbaden: Springer Gabler.

Weißenberger, B. E. 2014. Integrated Reporting Fragen (und Antworten) aus der Diskussion um die integrierte Rechnungslegung. *Controlling* 26(89):440–446.

Weissman, A., T. Augsten and A. Artmann. 2014. *Das Unternehmenscockpit*, 2nd edn. Wiesbaden: Springer Gabler.

Wildemann, H. 1992. Kosten- und Leistungsbeurteilung von Qualitätssicherungssystemen. *ZfB* 62:761–782.

Wöhe, G. 2000. *Einführung in die allgemeine Betriebswirtschaftslehre*, Munich.

XLSTAT. 2017. Histogramm. https://www.xlstat.com/de/loesungen/eigenschaften/histogramme. Accessed 23 June 2017.

Organization and Processes

4

Contents

4.1 Organizational Integration

For the design of the planning and reporting solution, the organizational anchoring is important. This is mainly due to the different information needs at all levels of management responsibility in the company. The information requirements are very different, starting from the service-providing units to the top management. Regarding the organizational integration, the following topics are examined:

- Company size
- Corporate connections
- Organizational structure
- Leadership style

D. Schön, *Planning and Reporting in BI-supported Controlling*,
https://doi.org/10.1007/978-3-658-41044-5_4

- Involved
  - recipients
  - coordinates
  - creator/Sender

4.1.1 Company Size

The company size has a significant influence on the scope and depth of the planning and reporting of the companies. Generally, the larger a company, the more complex the control requirements for planning and reporting. In this chapter, therefore, the classification of company sizes is worked out. Subsequently, differences and similarities for medium-sized and large companies for planning and reporting are worked out for selected criteria.

A uniform definition and classification of size classes for companies, e.g. in small, medium and large companies in Germany, cannot be found. Leading institutes in this field use similar, but not identical definitions. The most important criteria for differentiation are the number of employees, the annual turnover and the balance sheet total.

The EU Commission, for example, has been using the thresholds for the SME (small and medium-sized enterprises) sector since 01.01.2005 as shown in Table 4.1.

The Institute for SME Research Bonn has been defining the SME sector since 01.01.2002 as shown in Table 4.2.

The limits shown for the SMEs should not represent a quantitative wall for the design of planning and reporting. In business practice, it is shown that also larger companies, as shown in the classes, feel they belong to the medium-sized class. Here, the number of employees can also go far beyond 1,000.

Table 4.1 Thresholds for the SME sector according to the EU Commission (Table contents taken from EU Commission 2003)

Company size	Number of employees	and	Turnover €/year	or	Balance sheet total €/year
Micro	Up to 9		Up to 2 million		Up to 2 million
Small	Up to 49		Up to 10 million		Up to 10 million
Medium	Up to 249		Up to 50 million		Up to 43 million

Table 4.2 Threshold values for the SME sector according to IfM Bonn (Table contents taken from IfM Bonn 2006)

Company size	Number of employees	and	Turnover €/year
Small	Up to 9		Up to less than 1 million
Medium	Up to 499		Up to less than 50 million
(SME) together	Under 500		Under 50 million

Table 4.3 Size class divisions of medium-sized enterprises according to empirical studies (Table contents taken from: Ossadnik et al. 2010, p. 11)

Size class according to number of employees		Kosmider (1993)	Legenhausen (1998)	Dintner/ Schorcht (1999)	Kappler/ Scheytt (2000)	Zimmermann (2001)	Ossadnik/ Barklage/van Lengerich (2003)
Small	Up to	100	50	49	49	25	100
Medium	From To	101 500	51 200	50 499	50 250	26 250	101 200
Large	From To	500 1000	201 500	–	–	251 1000	201 500
Very large	From To	–	–	–	–	1001 2000	–

Table 4.4 SME definition of EKAM (Table contents taken from Becker and Ulrich 2009, p. 3)

Company size	Employees	Annual turnover
Micro-enterprises	Up to approx. 30	Up to approx. 6 million EUR
Small enterprises	Up to approx. 300	Up to approx. 60 million EUR
Medium-sized enterprises	Up to approx. 3000	Up to approx. 600 million EUR
Large enterprises	Over 3000	Over 600 million EUR

Other studies on controlling in medium-sized enterprises by Kosmider, Legenhausen, Dintner/Schrocht, Kappler, Schreytt, Zimmermann and Ossandnik/Barklage/van Lengerich have further summarized the size classes here using the example of the number of employees.[1] The delimitation criterion of these studies was the company size, whereby the range was from 25 to 2,000 employees (see Table 4.3).

The discussions about the numbers to be determined were considerably influenced by a study by Simon. In an investigation, he determined new threshold values on the basis of quantitative and qualitative criteria. According to his study, there are companies that can be attributed to the medium-sized sector even though their numerical values are much larger than the threshold values mentioned above.[2] Following this path, the European Competence Centre for Applied SME Research (EKAM) at the University of Bamberg also sets the threshold values much higher (see Table 4.4).[3] The most important

[1] Cf. Ossadnik et al. (2010).

[2] Cf. for the study by Simon (1992) the contribution by: Becker et al. (2009, p. 258).

[3] Cf. Becker and Ulrich (2011, p. 29). The predecessor institution of EKAM was the Deloitte SME Institute (DMI).

qualitative criteria that the *EKAM* uses to define SMEs are "owner-managed companies" and "family businesses".[4]

Since the quantitative variables are not sufficient as delimitation criteria for the planning and reporting according to company size, further qualitative criteria should be worked out. Here, in contrast to broad-based studies on the delimitation of SMEs and large enterprises, such as by Pfohl or Rheinard et al.,[5] the requirements for planning and reporting are in the foreground.

As a first point of reference for finding criteria, a representative empirical study serves done by the author and Diamant Software GmbH from April 2010. Here, with the help of a cross-sector questionnaire survey, the controlling in medium-sized enterprises was analyzed.[6] With a response rate of approx. 5%, 190 of the approx. 4,000 requested companies listed important deficiencies and obstacles to the design of efficient planning and reporting solutions:

- Lack of staff and time
- Lack of a holistic controlling for all functional or decision-making areas:
  - The focus of the reporting system is often only on the central evaluations of the accounting system (business economic analysis, cost center and income statement) and the financial accounting (liquidity development).
  - There are no segment- and cost unit-based income statements.
  - Reports for other functional areas such as procurement, sales, production etc. are often completely missing.
- Inconsistent reports and planning forms often prepared with Excel.
- Deficiencies in the content design of the reports and planning forms.
- Deficiencies in the processes for report design. The data preparation is error-prone and time-consuming, partly manual.
- The data integration of further systems missing (e. g. inventory management, time recording, PPC, liquidity, logistics etc.); there are heterogeneous software landscapes.
- IT support not consistently exploited.

Furthermore, the study showed that especially owner-managed companies only desire limited full result transparency for the departmental/area management levels as well as group and team management levels in the reporting. For these management levels, only partial transparency is sought for the selected areas. The full result transparency is reserved for the company management. Therefore, it is important for the reporting and

[4] Cf. Becker et al. (2016, p. 6).

[5] Cf. Pfohl (2006, p. 331–355) and Rheinhardt et al. (2007).

[6] Cf. Schön and Müller (2010, p. 123–165).

planning to provide efficient authorization systems and to build user group-oriented reporting and planning concepts.[7]

If one summarizes further results of the study, further requirements for planning and reporting in medium-sized enterprises can be identified. For planning and reporting, an integration of management-relevant data from various upstream systems with automatic data transfer should be available.[8] In addition to the target group-oriented preparation of the reports and planning results[9], a stronger strategic orientation of planning and reporting is demanded. Moreover, planning and reporting should be anchored in the process and organizational structure as well as in the management culture of the enterprise.[10] This is independent of the nationality of the company as the internal planning and reporting system is of high importance in international controlling.[11]

If one considers the difference between large enterprises on the one hand and the smaller and medium-sized enterprises as medium-sized enterprises on the other, the following delimitation criteria can be found (see Table 4.5).

4.1.1.1 Ownership and Corporate Governance

The group of addressees in medium-sized enterprises consists of a small circle of owners (often family members) and not of a multitude of known and anonymous investors, unlike in large corporations. The owners derive their current income in the form of management salaries or profit distributions from the company, therefore they need information about the earnings situation and especially about the earnings potentials and the liquidity strength of the company. In this respect, the preservation of the source of income is of the highest priority. Therefore, the need for information and planning extends to both the recognition of positive and negative developments in the company, as these support or endanger their income stream in the long term.[12] Ownership, management and economic security form a unity here.

While in large corporations there is usually a separation of owners and management according to the principal-agent theory, SMEs are often directly managed by owners or families. An owner-managed company or a family-run company is characterized by the fact that one or more natural persons are active in the management and they together or their families hold at least 50% of the shares in the company.[13] Owners can also have a significant influence on the management activity through the advisory board or the supervisory board. Due to the concentration of management by owners and families, the

[7] Cf. Schön and Müller (2010, p. 142–143).

[8] Cf. Reichmann (2008, p. 690; 2011, p. 7).

[9] Cf. Horváth (2008, p. 37).

[10] Cf. Weber et al. (2008, p. 27).

[11] Cf. Hoffjan (2008, p. 657).

[12] Cf. Janssen (2008, p. 89 ff.).

[13] Cf. Wallau (2008, p. 9 f.).

Table 4.5 Delimitation criteria of large enterprises and SMEs with regard to planning and reporting

Delimitation criterion	Large enterprise	SME/medium-sized enterprise
Ownership and management	Ownership is rather distributed among many shareholders. The large enterprises are usually management-led (principal-agent relationship)	Ownership is concentrated among few shareholders. The enterprise is often owner- or family-led, but sometimes also management-led
Strategy alignment	Rather higher	Rather lower, but with an increasing upward tendency
Operational action and decision-making paths	More complex hierarchies require longer decision-making paths with approval processes	Flexible and manageable business processes; simple decision-making paths, flat hierarchies
Limited human resources	Human resources are available. Methodological and instrumental know-how available	Human resources are severely limited. Methodological and instrumental know-how partly insufficient
Planning and reporting intensity	Rather higher	Lower, increasing upward tendency
IT-supported controlling systems	Rather integrated system landscapes and interlinked ERP systems, reporting and planning are increasingly carried out with database or data warehouse-based solutions, but partly also with MS Excel	Heterogeneous software landscape, partly lacking data integration, planning and reporting often done with MS Excel
Capital and liquidity dependence	Access to capital market larger, liquidity dependence rather lower	Access to capital market more limited, liquidity dependence rather higher
Ownership structures	Usually many national and international corporate connections	Fewer national and international corporate connections, but increasing for larger SMEs

decision-making paths are much shorter, and necessary adjustments are implemented more quickly. Planning and reporting systems must therefore also adapt quickly to these changes in the company.

Management-led companies, in the medium-sized sector as well as large companies, are characterized by the fact that persons are employed for the management who are not or only to a lesser extent directly involved in the company. In management-led companies, the shareholders do not exercise a direct control and steering function on the company. This is done by the management. In management-led companies, the managers are connected to the interests of the shareholders, whose return targets tend to be more short-term. Since they are not committed to the family bond and sustainability of generations, managers are more likely to make a change in the strategy and in the business field orientation. In addition, they pursue their own personal, success-related and financial goals which are particularly highlighted in the principal-agent theory.[14] To achieve their internal goals, the managers (agents) have to prove their successful action to the owners (principal) and thus present it transparently to the outside world.

It can therefore be assumed that management-led companies rely more on planning and reporting content and controlling instruments due to the presentation requirements. The owners, on the other hand, depend on this information to control their goal achievement.

Since both owner- or family-led and management-led companies can be found in the medium-sized sector and large companies, both management directions must be taken into account for planning and reporting.

4.1.1.2 Strategy Anchoring

Surprisingly, only a certain proportion of SMEs have a "written" or "communicated" corporate strategy, which can guide the reporting and planning. Thus, the first problem for controlling is to provide an internal reporting and planning for the management that can be aligned with a strategy. An empirical study by Rautenstrauch in 2005 showed that only 45% of medium-sized enterprises have a strategic guiding principle or a vision.[15] 12% of the companies here even had no explicitly formulated strategy. The number of companies with a corporate strategy was measured here at 58%, whereas only 33% and 14% had further business unit or functional area strategies.

Results of another empirical study by Feldbauer-Durstmüller show a continuously increasing use of strategic planning tools compared to older findings by Niedermayr and Wimmer.[16] It becomes clear that, in contrast to large enterprises, medium-sized enterprises lag behind in strategy anchoring and impact in the company, but are gaining in importance. However, for the alignment of planning and reporting to strategic goals, much greater obstacles are to be expected in medium-sized enterprises than in large enterprises. This is especially true if the strategic goals are not explicitly formulated.

In summary, it can be said that planning and reporting for SMEs as well as large enterprises must first be adapted to the strategic goals. If these are not or only insufficiently defined, this must be done or improved (see Sect. 3.1). In addition, the business model and its value-creating factors should have a significant influence on planning and reporting (see Sect. 3.2).

---

14 Cf. Andreae (2007, p. 23).

15 Cf. Rautenstrauch (2005, p. 1–17).

16 Cf. Feldbauer-Durstmüller and Wimmer (2008, p. 31–52).

4.1.1.3 Operational Action and Decision-Making Paths

In contrast to large enterprises, medium-sized enterprises are characterized by the flexibility and manageability of their business processes. In addition, there are flatter hierarchies and approval paths in SMEs than in large enterprises, and thus decisions can be made and implemented more quickly. Direct communication channels and the proximity of employees to decision-makers support this. In contrast to large enterprises, he requirements for reporting and planning are lower for medium-sized enterprises. Reporting and planning can be set up and used more flexibly and without great approval hurdles. While in large enterprises decisions are only made and secured on the basis of extensive reporting and planning results, reporting and planning in SMEs can be used more selectively, quickly and flexibly for decision-making.

4.1.1.4 Limited Human Resources

Compared to large enterprises, SMEs have fewer human resources available for reporting and planning. Planning and reporting in the company is mainly supported by controlling. However, the controlling tasks in SMEs are carried out by only a few employees, often even by only one employee. Often there is no separate controlling department or position, but the controlling tasks are covered as activities in the area of accounting and finance.[17] Competing activities such as the preparation of monthly and annual financial statements in accounting and ongoing booking activities significantly reduce the potential time for controlling tasks, such as management reporting and planning. But also in the other departments, the staffing situation regarding reporting and planning is relatively predictable. The conclusion for small and medium-sized enterprises from this is that the systems for planning and reporting have to be developed and maintained with these limited resources. Any knowledge gaps have to be closed.

Another problem arises from the lack of technical know-how. Modern reporting and planning tools are not or only insufficiently known in small and medium-sized enterprises.

In large enterprises, the personnel base for planning and reporting tasks is significantly higher. In coordination with the management and the departments, mainly the central and decentralized controlling are responsible for the planning and the reporting (see Sect. 4.1.5). Content and technical know-how is more often available, but this does not mean that there can be no know-how deficits in large enterprises.

4.1.1.5 Planning and Reporting Intensity

Also with regard to the criterion of planning and reporting intensity, SMEs lag behind large enterprises. This is not surprising, as the complexity of a company tends to increase with its size and thus, as a rule, the need for control and planning also becomes greater. In large enterprises, it can generally be assumed that the content and processes of planning and reporting are much more structured, detailed and extensive than in SMEs.

---

[17] Cf. Schön and Diamant Software GmbH (2009, p. 4).

However, it is encouraging that the proportion of more intensively planning SMEs is increasing. According to the empirical study by Schön and Diamant Software GmbH in 2009, already about 33.4% of SMEs have fully implemented an integrated planning for cost- and success-relevant areas in the company, while another 29.7% of the companies have done so at least partially.[18] Nevertheless, one will also find SMEs that have not or only very rudimentarily carried out planning so far.

In SMEs, it is noticeable that the reporting is often historically grown and that there is no basic reporting concept in the sense of a holistic controlling for all functional or decision-making areas. The focus of reporting in SMEs is often only on the central evaluations of accounting (e.g. sales statistics, business economic analysis, cost center and income statements as well as liquidity developments). There are no segment- and cost unit-based income statements. Reports for other functional areas such as sales, production and technology are often completely missing.

With regard to the planning and reporting intensity, there is a need for improvement both conceptually and technically. This applies especially to SMEs, but even in large enterprises there are deficits (e.g. proliferation of reports) in reporting and planning. Optimization potentials can be identified in all company sizes.

4.1.1.6 IT-Supported Controlling Systems

In contrast to medium-sized enterprises, larger companies often have a more homogeneous software landscape with a comprehensively developed (central) ERP system. The controlling functionality of the ERP system is extended by reporting and planning systems based on business intelligence and data warehouse (cf. Sect. 5.5 and 5.6). However, this does not mean that there cannot be heterogeneous software landscapes in larger companies. Especially in decentralized units, the functional departments often use their own IT systems that are not or only partially integrated into the overall IT landscape.

In SMEs, heterogeneous software landscapes are more common. Software for human resources and accounting as well as inventory management systems are often from different software vendors and have to be connected via interfaces. Often, manual data transfer paths can also be found. The planning calculations and controlling evaluations can only be done in parts with the reporting functions of the accounting system. For the final execution, the spreadsheet program MS Excel is often used (cf. Sect. 5.4.3). Modern data warehouse solutions with reporting and planning cockpits are not as common as in larger companies.[19] This applies especially to SMEs with less than 100 employees. Here, a size threshold for the acceptance and use of business intelligence could be identified. For larger companies, the use of business intelligence solutions is steadily increasing.[20] Likewise, larger companies are trendsetters for new technological topics such as

[18] Cf. Schön and Diamant Software GmbH (2009).

[19] Cf. Schön (2011).

[20] Cf. Schön (2011, p. 10).

AI-based and big data-based analytics (cf. Sect. 5.7 and 5.8) as well as mobile BI (cf. Sect. 5.10).

With regard to IT support, such improvement potentials should be pursued that are independent of the company size. Depending on the current status of the IT landscape of the company, the systems for planning and reporting should be appropriately and sustainably developed further. Important topics here are the integration of the upstream systems (inventory management, recording systems, etc.) as well as the content design of the commercial systems and especially the expansion of the profit and cost center accounting. Today, an important link in the information chain are reporting and planning tools with modern data warehouse or business intelligence technology. Further important trends such as AI, big data and mobile BI can be found in Chap. 5.

4.1.1.7 Capital and Liquidity Dependence

Compared to large enterprises, the access to the capital market is limited for SMEs. While large enterprises have access to the organized capital market and other forms of financing, the SMEs use bank loans and increasingly also institutional investors such as private equity firms or silent partners in addition to the shareholder contributions.

Due to the limited access to the capital market and the limited financial possibilities, the planning of the liquidity needs and the reporting of the liquidity development have a high importance for SMEs. For large enterprises, the planning and reporting of the liquidity and the capital needs have always played an important role. Increased credit collateral requirements of the banks due to Basel II and III require further qualitative information about the performance drivers (e.g. about processes, employees and customers) of the enterprises in addition to the financial ratios of the traditional annual financial statement analysis.[21] Due to the rating requirements for the enterprises, the planning and reporting of all important risks and opportunities within the framework of risk management and risk controlling become more important.

4.1.1.8 Ownership Structures

Enterprises in which strangers or other enterprises have a larger share than 25% of the voting rights or ownership interests are no longer economically independent. If the share is between 25% and 50%, they are called partner enterprises. If it is higher, they are called affiliated enterprises.[22] Large enterprises usually have group structures that form a network of subgroups and individual companies nationally and internationally. In this network of corporate connections, however, many SMEs are involved, and these have to meet the reporting requirements of the superior group units. Depending on the type of group (financial holding, strategic holding, management holding), the requirements for reporting and planning are higher or lower (see following Sect. 4.1.2). The influence of

---

[21] Cf. Schulte-Mattler and Manns (2010, p. 83–126).

[22] Cf. Lerch (2008, p. 43 ff.).

the superior group company is significantly higher in the case of the management holding than in the case of the financial holding. It is possible, therefore, that the influence on the involved (medium-sized) enterprise is rather low and thus the scope for action can be mostly exercised freely.[23] Although the independence as a criterion of a medium-sized enterprise is lost for the affiliated SMEs, many of these affiliated companies still have medium-sized structures, such as the limited personnel resources, the heterogeneous software landscapes and the low strategy anchoring.

Unlike the smaller enterprises, one finds increasingly more companies among the larger medium-sized enterprises that act as a corporate group with several smaller individual companies in the market. The requirements for planning and reporting in terms of investment controlling increase here, too.

4.1.2 Corporate Connections

Of course the requirements of groups are more relevant for larger enterprises. But many larger medium-sized enterprises are organized as a corporate group, too, and have the same problems in mapping their interconnections in planning and reporting as group enterprises (see Sect. 4.1.1.8).

Depending on the corporate connection, there are specific requirements for reporting and planning, which are the basis for the consolidation but also for the management of the corporate group.

Some the tasks of consolidation are, among others:[24]

- capital consolidation
- consolidation of receivables and liabilities
- consolidation of revenues and expenses
- correction and closing entries

Additional tasks of controlling a corporate group are, among others:

- cash pooling
- participation reporting
- corporate valuation

The control of the corporate group is determined by the type of holding. It decides on the level of detail of the information to be provided in the participation reporting and in the participation planning and intervenes more or less strongly in the strategic and

[23] Cf. Reinemann (1999, p. 661–662).

[24] Cf. Coenenberg (2000, p. 567 ff.).

operational control of the subgroups, strategic business units and individual companies. The types of groups are distinguished here as financial holding, strategic holding and management holding.

A **financial holding**considers the participations largely from the perspective of a financial investment and thus does not directly intervene in the operational control of the participations, but rather carries out a value-oriented control, in which the achievement of monetary key figures is in the foreground. Key figures for value-oriented corporate management are, for example, the Economic Value Added (EVA),[25] the Discounted Cash Flow (DCF) or the Cash Flow Return On Investment (CFROI).[26] Planning and reporting can thus be kept lean in the relationship of the holding to the participations and focus on the value-oriented control variables.

In the case of a **strategic holding**there is greater influence of the group on the participations. In addition to the value-oriented control, the strategic control in the group network becomes particularly important. The strategic orientation of the overall group is broken down to the subgroups and the individual companies. In addition to financial variables, key figures of other strategic target areas, such as sales, production, processes and resources, are used for strategic control. The planning and reporting of a "strategic holding" to its participations must therefore be extended to include the tracking of the strategic key figures.

A **management holding**extends the control to the operational level. For example, there is a group-wide pricing strategy or bundling effects are exploited in purchasing. In addition to the strategic key figures, operational key figures are also used for control up to the detail level of the business units and functional areas. The form of management holding thus has the greatest influence on planning and reporting.

Fig. 4.1 summarizes the three basic forms of holding and their influence on planning and reporting.

In order to meet the consolidation requirements in planning and reporting, the intercompany relationships and the strategic business units must be defined. Ideally, all forecast and booking records should be identifiable according to the intercompany relationships, but also according to the assignments of the strategic business units (SBU). Furthermore, the intercompany relationships and business units must also be identified and analyzed at higher levels of aggregation. This is often not so easy because the information about the intercompany relationship or the assignment to the strategic business unit is missing. For this reason, intercompany rules and assignment rules must be applied to these values, which result from a matrix of the IC relationships and SBU assignments. How difficult this can be for groups with many individual companies is shown in

---

[25]The Economic Value Added (EVA) is a registered trademark of Stern Stewart & Co. management consultancy.

[26]Cf. Stiefl and von Westerholt (2007), Hüllmann (2003), Coenenberg and Salfeld (2003), Rappaport (1986) or Copeland et al. (2002).

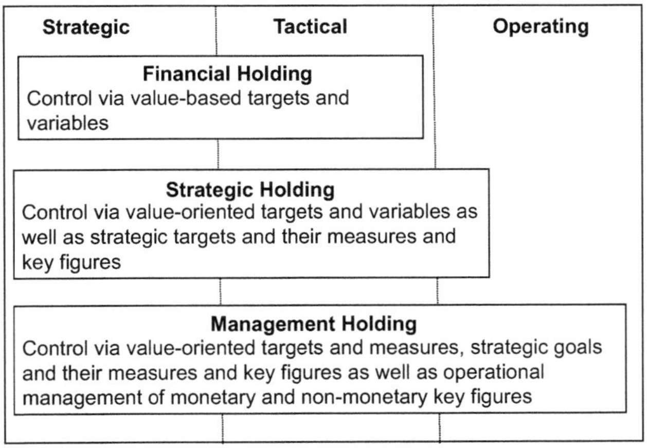

Fig. 4.1 Influence of the type of holding on planning and reporting (Slightly modified compared to: Weber et al. 2008, p. 25)

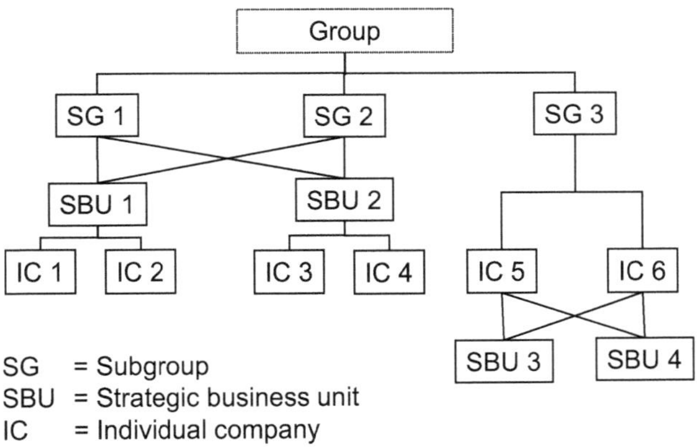

Fig. 4.2 Possible group interrelations with regard to SBU and company relations

Fig. 4.2. The subgroups 1 and 2 have different strategic business units, which are covered by different individual companies, while subgroup 3 consists of several individual companies, which can include several SBU.

In corporate practice, there are all possible group interrelations. Subgroups comprise, for example, several strategic business units. Individual companies cannot always be clearly assigned to the strategic business units. This is most difficult for companies referred to in practice as "zebra" companies. As shown in the example, companies 5 and 6 comprise several strategic business units. In contrast to the pure intercompany relationship, the SBU must now also be identified for consolidation.

Another basic prerequisite for planning and reporting with intercompany relationships is the use of consistent structures, especially master data and dimension values. Ideally, a

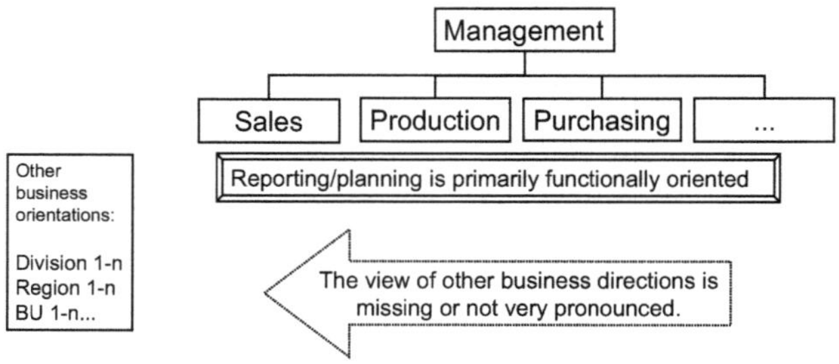

Fig. 4.3 Functional organization

uniform group chart of accounts, a coordinated group-wide cost center and carrier plan, but also a standardization of the item and material master data is necessary. As already shown in the planning area of sales and revenue planning as well as resource planning (see Sect. 3.12.2 f.), internal revenues, e.g. by triggering the bill of materials or calculation structures, have to be identified in the group. Here it must be ensured that the components used can be identified as IC parts. In addition, the prices for IC settlements have to be calculated and used for valuation.

4.1.3 Organizational Structure

The basic forms of organizational structures (functional, divisional, regional, matrix) have an important influence on the design of planning and reporting.

In a **functionalorganization**, planning and reporting are aligned with each functional area from the top management (see Fig. 4.3). The large functional areas of sales, purchasing, production and also the other areas such as logistics, research & development etc. require reports and key figures relevant for control.

Reports and plans that coordinate and reconcile the cross-functional value creation are often missing. Since, for example, a product manager, a division or business unit manager is missing, the integrative, interlocking element is lacking. In addition, the view of other business orientations is missing or less pronounced. This is important, however, as changes in the business areas, regions, etc. have an impact on the functional areas, for example, when sales volumes of certain products decline.

In a **divisional organization**, the focus of planning and reporting from the top management is on the individual divisions, business fields or business units (see Fig. 4.4). Here, the interlocking of cross-divisional functional and central areas and the regional orientation are neglected. The control of the resources and capacities used across divisions is rather neglected, too.

In a **regional organization** (no illustration) the planning and reporting is aligned with the regional units, such as branches and offices. Similar to the division-oriented organization, the integration of the cross-functional and central areas is neglected.

A major advantage of a one-dimensional alignment (primarily functional, regional or divisional) is the low coordination effort. In a multidimensional alignment, the interface

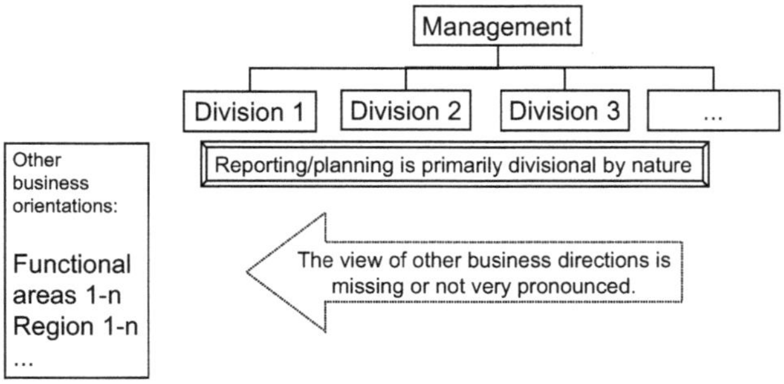

Fig. 4.4 Divisional organization

and coordination requirements for reporting and planning are significantly higher. A disadvantage of a one-dimensional alignment is mainly the lack of or reduced consideration of other business orientations, such as product groups, customer groups or regions.[27]

Due to the mentioned shortcomings of the purely functional, divisional or regional oriented companies, a matrix organization with divisional/regional or functional dominance is becoming more and more prevalent in practice.[28]

In a **matrix organization**a corresponding multidimensional alignment of reporting and planning is required for all perspectives (see Fig. 4.5). Product managers receive a division-based income statement. The sales manager needs the overall view to be able to steer his resources as best as possible across divisions.

In doing so, the coordinating and coordinating element gains a higher value in both planning and reporting. Modern IT systems such as data warehouse and BI-based controlling systems support the multidimensionality in the analysis much better than traditional planning and reporting solutions (see Chap. 5).

The design and maintenance of the planning and reporting systems is ideally centrally organized (usually controlling assumes the professional responsibility in cooperation with the departments and IT the technical responsibility). The report creation should be as automated as possible on a standard data set provided.

The report analysis and commentary is prepared with the support of the responsible controlling, depending on the organizational form, centrally or decentrally by the functional area. Controlling supports the responsible managers, makes suggestions for measures, and helps with decision- making. The responsibility for the analysis, commentary and control is carried by the responsible managers in their functional area.

The controlling organization as the decisive technical authority for planning and reporting usually follows the management organization of the enterprise.[29] Controlling is often functionally and hierarchically subordinate to the commercial management of

[27] Cf. Gleich and Temmel (2008, p. 84–87).

[28] Cf. Freese (2005, p. 445).

[29] Cf. Weber et al. (2001, p. 8).

Matrix	*Divisional/Regional*			
	Division 1	Division 2	...	Division n
	Office 1 ..n	Office 1 ..n	...	Office 1..n
Functional Sales				
Purchasing				
...				

Fig. 4.5 Matrix organization

the company. While in larger companies there are often central and decentralized controlling departments in the organizational structure, in medium-sized businesses there is often only a staff unit that centrally supports the reporting and planning functions of the company as a service area. In smaller medium-sized companies, the commercial service areas of finance, accounting and controlling are also often combined under one department, so that the controlling function is established as a subtask of the operational accounting. In these cases, there is no separate controlling department in the organizational chart. For this reason, the human resources for controlling tasks are scarce. There is a lack of technical know-how and IT support.

Due to the many communicative and coordinative tasks involved in setting up a planning and control system with a data warehouse and the complexity of business intelligence projects, it is increasingly recommended by business informatics to establish a BI Governance Committee as a steering and decision-making body for business intelligence and a separate BI Competence Center (BICC) as a separate BI-organization in the company. The BICC is supposed to act as an interface between IT, controlling, business units and management regarding business intelligence.[30] This is, in my opinion, an ideal-typical conception and would unnecessarily increase the personnel demand and the organizational structure. Ideally, in the sense of "**simultaneous engineering**", a BI team with representatives from the areas of controlling, IT and the business units should take over these tasks without having to anchor a new institution in the organization. A representative of controlling is suitable for leading such a BI team, as the controlling institution already takes over the coordinative and communicative tasks for the planning and control system in the company and supports the business units and the management regarding their content-related requirement profiles. The IT is an indispensable part of this team, especially to solve the technical issues.

[30] Cf. Kemper et al. (2010, p. 189 ff.).

4.1.4 Leadership Style

In Sect. 4.1.1 "Company size" it was already shown that owner- or family-run companies tend to build up a less extensive reporting and planning system than management-run companies. This is mainly due to the shorter decision-making paths in owner- and family-run companies and the concentration of leadership in the hands of a few people. In management-run companies, reporting and planning have a higher value as the organizational structures and decision-making paths are often more complex, and reporting, among other things, also serves to document and secure the achievement of the managers' goals vis-à-vis the owners.

Furthermore, there are different leadership styles in both owner- and management-run companies, which have an influence on reporting and planning.

In the MIND studies, which were scientifically accompanied by the IfM Bonn, specific stylistic leadership types for the SME sector were worked out in 2004.[31] Based on 1.6 million owners and managing partners, personality profiles of the executives could be analyzed. As a result of the study, four basic types were worked out: the strategist, the doer, the pragmatist and the patriarch.[32]

The **strategist**is characterized by a good education, by professional management and his will to implement his own ideas, e.g. to occupy promising market niches. He is more often found in large companies and in larger medium-sized companies, in which several business areas exist and in which he can implement his innovation ideas. For the analysis and control of his strategic measures, he tends to need a pronounced reporting and a detailed planning with both operational and strategic orientation.

The **doer**is distinguished by a good education, usually in technical professions or in another field of expertise. He sees his strengths in the direct implementation of his ideas, which tend to be not secured by detailed plans and reports.

The **pragmatist**is more likely to be found in small businesses, e.g. in the crafts sector. He is concerned with handling his existing business solidly and less characterized by implementing new business ideas. Reporting and planning tasks therefore also tend to focus on the representation of the central contents of the business and are not overly pronounced.

The **patriarch**is often found in owner- or family-run businesses, where this classic entrepreneur is characterized by the fact that he makes his decisions essentially alone. He has often built up the business himself or taken it over from the previous generation, so he knows the environment and the business processes very well. As a result, he is not dependent on a pronounced reporting system and a detailed planning, but he determines himself the design and the level of detail of the planning and the reporting as it is useful for him.

---

[31] The acronym MIND stands for the SME sector in Germany.

[32] Cf. IfM (2006, p. 58); Schweinsberg (2006 p. 64).

Another distinction of the leadership style can be made in connection with the involvement of other executives in the control process of the company. Here, the authoritarian and the cooperative leadership styles can be distinguished.

The **authoritarian leadership style**tends to set goals and guidelines for the executing areas and responsible persons. Decisions are made and communicated alone. The achievement of goals is regularly checked, and sanctions are imposed in case of non-achievement. In this case, the planning is done top-down, whereby the decentralized areas have to accept the planning specifications. The reporting is more focused on the control function.

The **cooperative leadership style**is characterized by the fact that decentralized areas and decentralized responsible people are involved in the control process. The planning is done in the counter-current procedure, so that top-down specifications can be concretized and possibly changed by bottom-up planning. The reporting serves both the analysis, control and monitoring of the jointly agreed goals.

4.1.5 Participants

The participants in the planning and reporting are distinguished, on the one hand, according to the addressees or recipients and the senders or creators and coordinators, on the other, whose different positions and connections are in the foreground in the next two chapters. In IT, adequate permissions or role profiles have to be formed and assigned to all participants (see Sect. 5.5.3).

4.1.5.1 Addressees/Recipients

For therecipients of planning results and reports, **external addressees** (e.g. shareholders, customers, suppliers, creditors, legislators, stock exchange supervisors) and **internal addressees** (e.g. management, controlling, executives, department heads, project managers) can be distinguished. They are usually also responsible for the content of the plans and reports or derive their control needs from them. Due to the acceptance and better control possibilities with the information, all addressees should be consistently considered in the information needs analysis (see Sect. 4.2.1.2). In addition to the information needs analysis in the introduction phase, the recipients are mainly involved in the following processes: testing the IT solutions in the phase of IT implementation (see Sect. 4.2.1.6), information analysis and control (see Sect. 4.2.3.4), planning execution and approval (see Sect. 4.2.2.3 and 4.2.2.4) and quality assurance of the information needs (see Sect. 4.2.4.1).

The addressees are not alone and isolated, but there is a **partnership** (e.g. equal department heads) and often also a **hierarchical relationship** between the addressees in which detail and overall responsibility meet. Especially among the internal addressees, the hierarchical relationship ranges from the top management level to the lowest leader-

Fig. 4.6 Hierarchical relationship in the company

ship levels in the company. In Fig. 4.6 this is shown by the example of the branch and department management.

There is a kind of contractual relationship between the client and the contractor, here only between the sender (responsible and operating instance) and the recipient (control and steering instance). In the hierarchy of the leadership levels, the agreed contents are created and analyzed within the framework of planning and reporting in order to derive control-relevant measures and decisions from them. Here, the contents are presented more and more compressed on each higher leadership level in order to ensure a control-relevant overview of the entire area of responsibility. This does not mean, however, that detail information is lost or inaccessible. If necessary, a top-down analysis can be performed down to the most detailed information.

The divergence between control-relevant result objects and the reporting or planning responsible can lead to inconsistencies in the control. Four cases can be differentiated:[33]

Full agreement (ideal case): The control-relevant result object has a clear planning or reporting responsible (example: the product manager is responsible for the product groups assigned to him).

Partial agreement: Only a part of the control-relevant result objects has a clear planning or reporting responsible (example: The product manager is responsible for the product groups assigned to him, which are sold in different regions, but the regions in which the products are distributed are not managed by him or by a regional manager). In this case, either the unassigned result object is assigned to one or more responsible people. In the latter case, the different product managers can control the result object region via the evaluation view region.

No agreement: There is no agreement between the control-relevant result object and the reporting responsible (example: The material procurement process has no specially assigned responsible; due to the functional organizational orientation, the purchasing, quality assurance and incoming goods departments are involved in the process). Here too, it makes sense to assign the unassigned result object to one or more responsible persons.

[33] Cf. Gräf and Nase (2008, p. 47).

Multiple responsibility: The control-relevant result object has several planning and reporting people responsible (example: In a matrix organization, product and service managers have the result object product in their joint responsibility). Here it makes sense to delimit the assigned result object so that the responsible areas sales and product result as well as result from service and maintenance are separated.

4.1.5.2 Coordinators

An uncoordinated reporting and an uncoordinated planning are characterized by the fact that both the decentralized and central units create their reports and plans independently of each other and forward them to the upstream management levels, e.g. division management, functional area management or the top management. This procedure is not recommended as it is prone to errors and often requires extensive rework.

For this reason, it is advisable to use the controlling department as a coordinating institution for planning and reporting. The controlling department should accompany all process steps during the introduction, the continuous implementation as well as the maintenance and quality assurance (see Sect. 4.2). In larger medium-sized companies and large corporations, the management of the decentralized areas coordinates with their decentralized controlling. Furthermore, there is a coordination of the decentralized controlling areas with the central controlling. The central controlling has the uniform design and specification function of central standards for planning and reporting (e.g. uniform layout specifications, key figure calculations and uniform planning and reporting contents). The decentralized controlling fulfills the planning and reporting requirements of the decentralized management within the framework of the central standards. If there are individual decentralized requirements, they are to be supplemented area-specifically without violating the standard. The top management thus gets a central institution as a contact person for the coordination and reconciliation of the data (see Fig. 4.7).[34]

The controlling department does not take over the responsibility for the planning data or the realized reporting results. These remain with the decentralized and central managers in their respective functional areas. The controlling department supports the managers in planning, reporting, analysis and commentary and thus helps as a business partner and consultant in supporting decisions.[35]

In smaller (medium-sized) companies, the task of controlling is often performed by only one person or a few people (see Sect. 4.1.1.4). In this case, the coordination is done by the *controlling*-executing unit.

An important role in the technical provision of the professional information requirements is played by the IT-department. The controlling department coordinates the content and professional requirements and communicates them to the IT which is responsible for the technical implementation and maintenance. Technical queries, e.g.

[34] Cf. Weber et al. (2008, p. 50).

[35] Cf. Weber et al. (2006, p. 44).

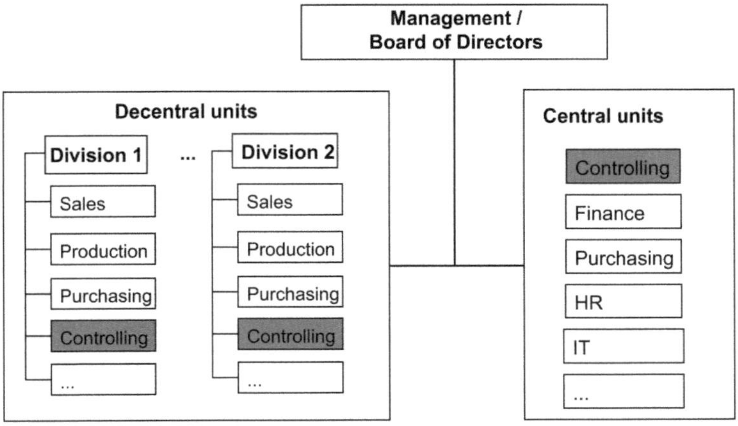

Fig. 4.7 Controlling in larger (medium-sized) companies

regarding the data source, the interface connection or the desired reporting implementation, are solved in dialogue between IT, controlling and functional departments. Due to the important coordination function of controlling, extended knowledge of the applied information technologies proves to be helpful. The knowledge spectrum of the controller expands in this direction.

Basic database knowledge, simple programming tasks, the customization and application of the software products used in the area of traditional ERP systems and modern business intelligence technologies (see Chap. 5) expand the knowledge spectrum of today's controllers. For technically demanding tasks, the own IT department or external experts are available.

4.1.5.3 Creator/Sender

The authority over the contents of the planning and reports to be created lies primarily with the executives and management as the recipients of the reports. The creation of the planning and reports can be done by the executive himself, but is usually often done decentrally by employees of the specialist department, by a decentralized or a central controlling. In some cases, IT staff are deployed for report creation. As already discussed in the previous Sect. 4.1.5.2, controlling is mainly responsible for the coordination of planning and reporting in the company.

Ideally, the task of conception and further development of planning and both standard and analysis reporting is taken over by controlling. For the implementation of the planning areas and forms as well as the reporting system, technical know-how is usually required, and therefore, support by IT staff is inevitable. The proportion of the professional and controlling-specific creation depends largely on the user-friendliness of the IT solutions used and the available IT know-how of the staff. The direct consideration of professional aspects is advantageous in the creation by the specialist department and controlling. Disadvantages are the lack of knowledge regarding the available data and their structures as well as the low know-how regarding the application of information technology, e.g. the report generators and design tools.

4.1.5.4 Self-Service BI

While, for example, simple special and ad hoc reports can be created more or less quickly by staff of the specialist department and the controlling department with spreadsheet and office programs without deeper IT knowledge, the establishment of a standard and analysis reporting with this know-how is only insufficiently possible. Therefore, modern IT tools are expected to be usable by simple specialist users without extensive IT prior knowledge and training. With the discussion about the use of modern data warehouse and BI-supported controlling systems, this topic has been addressed in recent years under the keyword "Self-Service BI". The term **Self-Service BI** is intended to illustrate that specialist users can create analyses and planning independently with suitable software without recourse to the know-how of IT departments. According to the worldwide BARC study "BI Survey 14", 50% of companies want to give their employees more freedom in reporting through self-service BI in the specialist area.[36] In the BI trend monitor 2017 of the Business Application Research Center, self-service BI is the second most important trend after data discovery.[37] Self-service BI aims primarily at the software providers in the BI environment developing easy-to-use user interfaces and data integration possibilities with higher processing speed.[38] This enables the staff of the specialist department to create independent analyses such as ad hoc reports, dash boards etc. independently. In the specialist departments and especially in controlling, the proportion of so-called BI power users is growing, who, in contrast to rigid standard reporting, receive more advanced analysis and forecasting functions, for example, to extend data models and expand or modify reports. With the support of in-memory technology, data discovery or visual discovery tools can also be used, which simplify data connection, data transformation and modeling to a limited extent (see Sect. 5.7.2.1). In addition, in-memory technology achieves a higher processing speed in the analysis preparation. Collaborative tools help teams to work more effectively together, for example by administering planning results and analyses via workflows and without media breaks. In addition to desktop installations, device-independent mobile solutions should also be used for independent use of the services.[39] With the help of the concept of "sandboxing", selected data areas should be made available to the specialist users for free use in order to create individual analyses for the area of responsibility. These have to be checked for their value in a later evaluation process before they are transferred to the regular operation of the control.[40]

[36] Cf. Bange (2014, p. 28–29).

[37] Cf. BARC (2017).

[38] Cf. Taschner (2013, p. 163 f.).

[39] Cf. Dittmar and Schmidt (2014, p. 24–27).

[40] Gluchowski (2014, p. 240).

The advantage of shifting the work from the IT department to the business department is that the analysis results are created in a targeted way and used more intensively. The controlling department, as the central responsible unit for reporting and planning, still has to ensure that there is no proliferation of evaluation techniques without restricting the creative use of data. Especially in the cooperation across departments and hierarchies, standard reports remain indispensable, but not inviolable. By involving the business department more strongly through self-service BI, a continuous optimization of reporting and planning for the entire company should be pursued. Since only selected, mostly tech-savvy employees of the business department qualify as power users, they should be identified as key users and integrated into a working group led by the controlling department, including IT staff. This working group usually emerges as the nucleus for the continuous development of BI-supported controlling in the company. Important topics such as data quality, life cycle of report and data usage, improvement of planning forms and reports, documentation and metadata management are discussed here.

In connection with the IT support by Big Data and Predictive Analytics, the deployment of a so-called **DataScientist** is also demanded. He is supposed to close the gap between IT and business users, a gap which arises from the data pattern recognition (data mining/predictive analytics) and the data storage technologies (RDBMS, data warehouse and especially big data). According to Davenport, he is well versed in common programming languages (hacker), has scientific curiosity (scientist), works with large quantitative data sets (quantitative analyst) and is simultaneously a business expert and communicative in his ability to grasp, moderate and look for technical implementation solutions for the decision problems in the company from a data processing perspective (trust adviser).[41] Especially for data analysis and forecasting, the data scientist masters practical methods and theoretical approaches from information technology, mathematics and especially statistics. It is therefore a personnel profile that best meets the requirements for data management in terms of business administration and IT. This profile has very strong overlaps with controlling and the IT staff and is rarely found in one person in the labor market. For this reason, the bundling of the required personal competencies in most companies is better covered by a team of controlling, IT and business department. As already mentioned in Sect. 4.1.3, "Organizational structure", I recommend the formation of a BI team instead of a separate BI competence center organization. The requirements that are placed on the data scientist are taken over by the **BI team**. A central function in the BI team is taken by the controller, who is primarily responsible for the information supply of the management from a professional point of view. As already mentioned in Sect. 4.1.5.2, the knowledge area in controlling expands in the direction of IT support. The controller brings along his business know-how as well as basic computer science skills for the administration of the application programs and sometimes even for the databases. In addition, there will be one or more people in the BI team from the

---

[41] Davenport (2014, p. 88 ff.).

IT department who support the functions of the data scientist from a technical point of view, such as the identification and linking of data sources as well as the structuring and modeling of data storage. They also need to have knowledge of the technical software and hardware products, and their performance and scalability.[42] Furthermore, experts are sought who can select and model the right algorithms and tools for data forecasting and analysis.

4.2 Processes

From a process-oriented perspective, various tasks arise for planning and reporting. Here, the introduction process, the cyclical planning or reporting process and the quality assurance process have to be distinguished (see Fig. 4.8).

The **introduction process** is required when setting up a planning and reporting system for the first time in a newly founded company, for example, or when a company intends to completely or extensively redesign the existing planning and reporting structure due to major deficiencies. Since the introduction process for planning and reporting systems is quite similar, the following chapters only highlight planning and reporting tasks where there are significant differences.

The **cyclical execution process** includes the actual execution of the planning and the regular reporting process in the company. Since the cyclical process of planning and reporting has clear differences, the presentation is done in separate sections.

The **quality assurance process** serves to continuously improve and adapt the content-related, organizational, process-related and IT-related requirements in planning and reporting. The quality assurance process for planning and reporting systems is similar in many parts, so the presentation is done together again and only contents are highlighted where there are significant differences.

4.2.1 Introduction Process

The introduction process is to be divided into different phases due to its complexity in the form of a project. A possible phase division is shown in Fig. 4.9.

4.2.1.1 Definition of the Framework Conditions

When a company intends to introduce a new reporting or planning system, one should keep an eye on the timing of the productive start. For reporting projects it is advisable to put the new reporting system into operation at the change of a fiscal year and to replace old systems. This is especially advantageous because the annual closing bookings in the

[42] Cf. Horváth and Aschenbrücker (2014, p. 47–62).

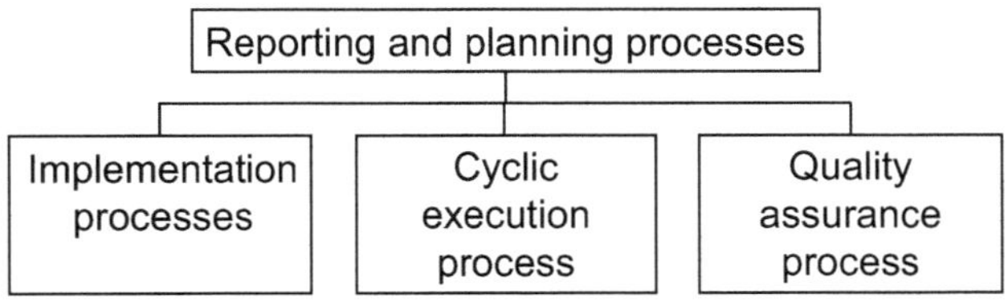

Fig. 4.8 Processes of planning and reporting

Fig. 4.9 Introduction process for planning and reporting

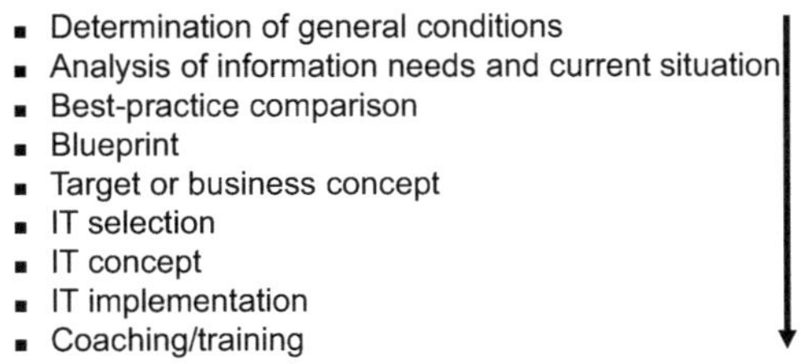

old system can be completed and the data transfer (e.g. balance transfers) can take place at the start of the new fiscal year. All bookings of the new fiscal year are then made in the new system.

For planning projects the introduction project should be completed by the start date of the budget planning (usually mid/end of the 3rd quarter of the fiscal year) in order to be able to start with the new planning system. For the introduction project, a corresponding project duration with buffer times should be planned.

At the beginning of the project, the following framework parameters should be defined as part of the project preparation.

- Project management
  Ideally, the project manager should be someone who is familiar with the topic of reporting and planning as broadly as possible and who has strong communication skills to coordinate the many project participants. A management team is advisable for larger projects, in which specific areas of responsibility are divided (e.g. IT and controlling). A project manager (spokesperson) should be determined in the management team. Due to his coordinative tasks, the controller is especially suitable for this.
- Steering committee
  The steering committee serves as a control body for the acceptance of the goals defined in the project plan for the individual project steps and the overall result. It is often shaped by the management, but can also consist of or be supplemented by other persons. Without a steering committee, there is a risk of missing the target and of inefficient execution of the project.
- Project team
  In addition to the project manager or the project management, the project team includes the main person responsible for the reporting or planning. This task usually

falls to Controlling. Furthermore, important report/planning recipients of the functional areas should be won over for the project team because they are essential for the determination of information needs in the project team. The larger the number of report recipients is, the more advisable it is to bundle report recipients in suitable working groups, whose spokesperson is then a member of the core team. The members of the working group then belong to the extended project team and are consulted as needed in the project.

Due to the often lacking cross-cutting know-how about modern planning and reporting systems and their design, it is advisable to involve external consultants in the process and in the project team.

- Project plan
  The following sub-items have to be defined and possibly adjusted during the project course in the project plan:
  - work packages and project steps (tasks) with milestones
  - deadlines
  - responsibilities
  - resources
  - finances

Furthermore, the management has to be won over as a **project promoter**. Without the promoter support from the management, the power may be lacking in difficult project phases to implement uncomfortable project steps that entail changes.

4.2.1.2 Information Needs and "as-is" Analysis

The acceptance of new reporting and planning solutions can only be achieved in the company if the affected employees and managers are involved in the design process and are picked up at the current "as-is" state. For this purpose, an as-is analysis is suitable, which examines the strengths and weaknesses of the previous system in all facets already mentioned (content, organization, processes and IT).

The information needs analysis should identify for the users and decision-makers which information is useful and decision-relevant for the operational control. The goal of the information needs and "as-is" analysis is to obtain a strength and weakness profile for the analysis fields, based on which improvement suggestions and recommendations for action can be developed for the new development or further development of the planning and reporting system. As far as comparison possibilities to best-practice solutions are available, it is advisable to use them for the analysis and design.

After the strength and weakness profile and the best-practice analysis, a blueprint for the future planning and reporting system and concrete implementation suggestions in the form of project steps have to be developed.

Information needs analysis and content status assessment

The **first step** of the content status assessment and the information needs analysis deals with the alignment of the planning and reporting content with the **strategy** and the **business model of the company** (see Sect. 3.1 and 3.2). However, in medium-sized enterprises, a formulated strategy is not always available as already shown in the introduction (see Sect. 4.1.1). If this is the case, it is recommended to develop a strategy orientation with the management, based on the vision and guiding principles of the enterprise, which ideally is also broken down to the individual business units and functional areas. For the formed strategic goals, corresponding control-relevant indicators should be developed.

Unfortunately, there are also situations in business practice where such a preliminary strategy formation project is not possible and thus no strategy is available as a basis for planning and reporting. In this case, one should gain clarity about the existing business model with its value-creating factors and align or further develop the planning and reporting accordingly. This way, at least the operational control variables of the company are identified. For the representation of planning and reporting processes, it is useful to develop use cases, which can illustrate and improve the input and planning steps.[43]

Once the strategic and operational information needs have been derived from the strategy and the business model of the company (deductive information needs analysis), the planning and information needs of the individual areas and report recipients are to be recorded and further developed (inductive information needs analysis). For this purpose, it is advisable to carry out the recipient-oriented information needs analysis in the **second step**. First, all relevant planning and reporting areas as well as the most important participants are identified (see Sect. 4.1.5). This is done in connection with the analysis of the organizational structure in the company, in which the most important senders and receivers in planning and reporting are determined and defined for the project team and their workshops.

In order to make the workshops as successful as possible, it is advisable to carry out an analysis of the existing planning and reporting documents in the context of a status assessment beforehand. Checklists and questionnaires can help to identify planning deficits and information gaps in a structured way, as well as to determine planning and information needs. The basic W-questions for Reporting and planning have to be answered, whereby the "Why?" question regarding the reporting purpose is the most important question:[44]

- Why is planning and reporting done? Planning and reporting purpose
- For whom is planning and reporting done? Planning and reporting recipients
- Who is doing the planning and reporting? Planning and reporting creators

[43] See Schmidt (2020, p. 12–13).

[44] Extended from Koch (1994, p. 59), Blohm (1974, p. 13 f.) and Ziegenbein (2012, p. 589).

- What is planned and reported? Planning and reporting objects and content
- How is planning and reporting done? Design forms, presentation media, transmission channels, processes and software used for planning and reporting
- When is planning and reporting done and for what period? Dates and periods of planning and reporting
- Where is planning and reporting done? Localization of planning and reporting
- With what is planning and reporting done? Systems and technical instruments of planning and reporting

For the status analysis, concise documents, which roughly show which reports and key information are currently produced and used by which person or user group, should be created in advance. In addition, the involved IT systems and data sources have to be analyzed.[45] By comparing the status analysis and the information needs analysis, overlaps and gaps in the information supply can be detected and eliminated. Reports and information have to be prioritized according to their importance and gradually specified in the content target or functional concept (see Sect. 4.2.1.3).

Experts, e.g. external consultants, can contribute with best-practice comparisons from other companies and projects to generate new ideas for the future design of the planning and reporting solution.

In addition to classical value-oriented variables such as sales, costs or returns, important qualitative factors and performance variables for the central and decentralized business areas must also be worked out. All indicators must be examined with regard to their **plannability** and their **information content**. Further important criteria are:

- strategic relevance and control relevance
- scope and appropriateness
- measurability and influenceability
- comprehensibility and presentation possibility
- timeliness and temporal availability
- quality and efficiency

A tendency toward more and more information requests must be prevented when defining the planning and information requirements. The statement that applies here is: "Less is often more". The planning and data collection as well as their benefit must be in a reasonable economic relation to each other.

The results of the information requirements analysis are to be summarized as a basis for the blueprint and the target concept.

[45] Cf. Strauch and Winter (2002, p. 359–378).

Organizational analysis

The aim of the organizational analysis is to determine which planning and reporting recipients need which information for which management areas and tasks. Planning and reporting follow the organizational structure of the company and serve all decision-makers, starting from the top management to the lower management levels.

In addition to the internal addressees, external addressees (such as banks, private equity companies, parent companies, etc.) must also be taken into account; they usually participate to a limited extent in the results of the planning and reporting.

For the reporting, at least one reporting organizational matrix should be set up to assign the reports of the individual reporting areas to the recipients such as the management circles or management bodies (see Table 4.6). Further information, such as the reporting intervals, presentation techniques, user roles, dates, locations, persons or groups, can be added accordingly.

For the planning, the planning forms, planning versions can be assigned to planning areas. Furthermore, it must be determined who is responsible for the planning execution and the approval (see Table 4.7; planning organization matrix).

Table 4.6 Example of a reporting organizational matrix

Reporting areas	Reports	Reporting interval	Management circle
Top management	Balance sheet	Quarter	Management
	P&L	Quarter	Management
	Income statement	Quarter	Management
	Sales statistics	Week	Management
	Product income statement	Month	Management
	...	...	...
Sales	Sales statistics	Week	Sales Management
	Product profit and loss statement	Month	Sales Management
	Sales statistics	Week	Sales Management
	...	...	...
Production	Capacity statistics	Month	Technical Management
	Product success calculation	Month	Technical Management
	Hourly statistics	Week	Technical Management
	Maitenance statistics	Quarter	Technical Management
...	...	...	...

Table 4.7 Example of a planning organization matrix

Planning areas	Planning forms	Planning versions	Planning execution	Approval
Strategic planning	SWOT	Long-term planning	Management	Management
	Portfolio	Long-term planning	Management	Management
	BSC	Budget, medium-term, long-term planning	Management	Management
	…	…	…	…
Sales	Sales planning	Budget, medium-term planning	Sales management	Management
	Revenue planning	Budget, medium-term planning	Sales management	Management
	Product result planning	Budget, medium-term planning	Sales management	Management
	Cost center planning	Budget, medium-term planning	Sales outlet management	Sales management
	…	…	…	…
Production	Resource planning	Budget, medium-term planning	Technical management	Management
	Staff planning	Budget, medium-term planning	Technical management	Management
	Plant planning	Budget, medium-term planning	Plant operator	Technical management
…	…	…	…	…

Typical problems that are discovered in the organizational analysis include:

- The committee formation does not match the organizational chart of the company
- There are overlaps in responsibilities
- There is no assignment of the addressees, planning areas and reports to the committees
- The reporting intervals are not clear
- The user roles in the IT systems do not match the organizational matrix

Should inconsistencies and deficiencies be detected, they should already be recorded as a recommendation for the organizational improvement in the blueprint, which is used as the basis for the target concept.

The structure of the reports according to reporting areas and report recipient groups or the planning areas and their responsible persons help to define the teams for the working groups doing the information needs analysis and content status assessment and to develop the later target or functional concept (see Sect. 4.2.1.3). Furthermore, the reports form the basis for the development of the role and authorization concept (see Sect. 5.5.3).

Process analysis

In the process analysis, those cyclically recurring planning and reporting processes are examined that are described in detail in Sect. 4.2.2 and 4.2.3. The cyclical reporting process includes the sub-processes of report creation, analysis preparation, report provision, and report analysis and control.

The planning comprises the strategic planning, the medium-term planning and the subareas of budget planning and forecasting (intra-year and rolling forecasts). The respective planning processes differ in the sub-processes of planning preparation, execution, coordination and approval.

The process analysis records the status of the processes, identifies deficiencies and shows improvement possibilities.

IT analysis

The inventory of the IT lists all IT components involved in the reporting process, such as hardware and software. The levels of data supply and provision (interfaces, ETL or ELT process), data processing and storage, and data analysis possibilities are examined. Important criteria for evaluation are:

- user-friendliness (creation and analysis)
- layout design possibilities
- content-related/professional design possibilities
- flexibility (creation and analysis)
- data quality
- effort
- speed/performance
- data storage and processing
- data integration of upstream information systems
- further processing of the information in downstream systems
- planning and analysis functionality
- training sessions
- support assistance
- evaluation criteria for the software manufacturer (creditworthiness, security, independence, etc.)

The IT analysis records the status of data processing, identifies deficiencies, and shows improvement possibilities. For missing and inadequate IT components, a selection process for the procurement and implementation of suitable new IT components must be initiated.

Strengths and weaknesses profile and best-practice analysis
The results of the "as-is" analysis are summarized in a **strengths and weaknesses profile** according to the analysis areas (content, processes, organization and IT), so that a basis for decision-making for the targeted improvement in planning and reporting is possible. Recommendations for action and project proposals focus especially on the biggest deficiencies and weaknesses.

Typical weaknesses that are uncovered in the context of a reporting analysis are:

- There is historically grown reporting from individual reports of various source systems, which are processed with e.g. Excel.
- There is a lack of a reporting concept.
- Data from the merchandise management and other upstream systems are not connected with the commercial data in the reporting system.
- The data content and quality needs to be significantly improved and expanded in some areas.
- Changes of historically grown structures (e.g. cost carrier, product, project and cost center hierarchies), previous contents (report contents and key figures) and processes (change of the booking logic for certain transaction types) become necessary.
- There is a lack of process coordination and organizational anchoring.
- …

Typical weaknesses that are uncovered in the context of a planning analysis are:

- There is a lack of a strategy reference in the planning.
- The planning inputs in the planning forms (e.g. in MS Excel) are cumbersome and error-prone.
- There is a lack of a planning concept.
- There is a lack of process coordination and organizational anchoring.
- There is a lack of alignment of the planned values between the planning areas.
- Planning contents are unstructured, and planning forms exist in different variations.
- Subsequent changes can only be done with difficulty in the entire planning.
- The integration of the plan data from other source systems is only done manually or is not available.
- …

In addition, there are further general problem areas such as e.g.:

- Low human resources in the involved functional areas with competing activities such as monthly and annual financial statements (accounting/controlling) or hardware, software and IT architecture changes (IT)
- Heterogeneous software landscapes
- …

The strength and weakness profile can be very well illustrated by using a scoring or utility analysis (see Sect. 3.7.10).

The **best-practice analysis** complements the strength and weakness profile in a meaningful way. For this purpose, implementation and usage experiences from other companies must be available; their requirements must be comparable to the own planning and reporting system and whose solutions are considered as best or desirable. By comparing with the "best", the decision makers get further ideas for improving their own planning and reporting solution (see Fig. 4.10).

Blueprint with project proposals

As the last phase of the "as-is" analysis, a blueprint with project proposals is created to prepare the decision-making for the improvement of the planning and reporting. The blueprint shows, from a bird's eye view, the most important parameters and design factors for the future solution. The rough structure and scope as well as the main contents and basic design suggestions are to be listed and serve the subsequent target concept for the detailed design.

For the blueprint, graphical representations are suitable and provided with short comments. The blueprint concludes with concrete project proposals for the conception and implementation of the new planning and reporting solution. Ideally, several alternatives

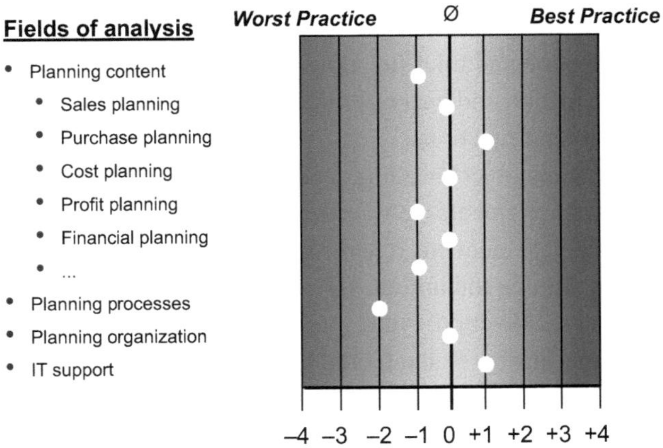

Fig. 4.10 Scoring analysis (here the planning area)

with their advantages and disadvantages are to be presented. Preferred solutions and recommendations are to be expressed.

The conclusion of this phase is the decision for the desired planning and reporting solution, whose next phase is the target concept.

4.2.1.3 Target or Functional Concept

Based on the results of the "as-is" analysis and the blueprint, the detailed requirements for the future planning and reporting system are developed in the target or functional concept.

For planning and reporting, the focus is on the content and professional design of the individual reports and planning forms, their structure, the design of the processes, and the organizational anchoring including an authorization concept. The area of data processing is excluded here as the results of the target concept form the basis for the later IT concept (see the following Sect. 4.2.1.4).

As a tool for the target concept, a detailed documentation is suitable and should later also be available as meta-documentation in the software program for further development and maintenance.

An overview of the reports and planning areas should provide the following information:

- name of the report/planning area
- purpose and significance of the report/planning area
- user or user groups of the report/planning area
- example/draft of the report/planning form
- notes on the data sources and systems for the reports or planning areas which are to be specified in the subsequent IT concept

When designing the individual reports and planning forms, the following criteria, among others, are to be defined in a documentation (see in detail on the design possibilities and recommendations Sect. 3.8):

- head information (headings, responsibilities, logos etc.)
- selection and filter functions
- navigation and analysis functions
- planning functions
- dynamic object headings and texts that change according to object selection
- tabular elements (rows/columns)
- graphical elements (charts, traffic lights etc.)
- key figures (definition, formula, unit, value types, target size, data source, application areas, collection cycle, meaning) and their object assignment according to dimensions, attributes and hierarchies
- value types (actual, plan, forecast etc.)
- further requirements such as comment functions

Table 4.8 Key figure sheet (Cf. for the design of key figure sheets e.g. Schulte 2002, p. 170 and Eisele and Doyé 2010, p. 384)

Key figure: Asset coverage (%)
Definition: The asset coverage indicates to what extent the fixed assets are covered by long-term capital (equity + long-term debt)
Formula: $\text{Asset coverage (\%)} = \frac{\text{equity + long-term debt}}{\text{fixed assets}} \times 100$
Unit: percentage value
Target size: 110–150%
Value types: Actual, Plan, Forecast
Base data and data source: Balance sheet ratios: equity, long-term debt and fixed assets (from financial accounting)
Object assignments: Company
Areas of application: Key balance sheet figures
Recipient/role assignment: Management, leadership level
Survey cycle: Quarter, year
Meaning: In the context of monitoring the long-term financial structure, the equity-to-fixed-assets ratio plays an important role. The principle underlying this ratio is the maturity-matched financing of the long-term tied assets to be controlled by the equity-to-fixed-assets ratio and has the function of revealing long-term structural imbalances. Long-term assets should also be financed long-term (golden balance sheet rule)! Therefore, the equity-to-fixed-assets ratio should be clearly above 100% (target e.g. 150%). The further the equity-to-fixed-assets ratio is above 100%, the more the current assets are also financed by long-term capital and thus a higher financial stability of the company is given. If the fixed assets are e.g. partly financed short-term, the company could get into payment difficulties when short-term liabilities are due since the current assets are not sufficient and the fixed assets are not quickly liquidable.

Reports, planning forms and their information are to be prioritized, coordinated and specified according to their importance. An example of the documentation of key figures in a key figure sheet is shown in Table 4.8.

4.2.1.4 IT Concept

The IT-concept translates the content requirements of the target or functional concept into technical specifications for the implementation in an existing or new IT system landscape. It can also be used for the process of IT selection as a requirement profile.

This includes the following tasks:

- determination of the technical data sources
- integration of the upstream and downstream IT systems
- data storage and data modeling
- data supply concept including ETL process (extraction, transformation and loading processes, cf. Sect. 5.5.2) or ELT process (extraction, loading and transformation processes, cf. Sect. 5.7)

- query, function and report design

For a company-wide reporting and an integrated planning, the source data are located in different pre-systems and, parts, have to be additionally entered manually. The IT landscape often consists of several heterogeneous systems, in which, for example, the data of the human resources, the inventory management and the accounting, but also of external sources, such as the internet, have to be meaningfully combined.

First, the technical data sources in the respective pre-systems have to be clearly and completely determined in all value expressions (e.g. actual, plan and forecast). If there are any inconsistencies (e.g. who is the customer, the invoice or the goods recipient?) or missing data, they have to be identified and clarified.

As part of the data supply concept, the information flows have to be defined. Here, the interfaces of the upstream systems to the internal and external accounting should be examined closely. All accounting objects (e.g. cost units, cost centers, projects) that are necessary from the conception have to be either transferred with the master and transaction data of the inventory management to the cost accounting or be determined from derivation rules (e.g. the industry can be identified as an attribute from the customer). Often, interfaces have to be newly implemented, adapted or extended. Data from the pre-systems that are not yet harmonized have to be harmonized in the transformation process.

Cost accounting is a central system for management reporting, in which the value flow of the company is mapped, among other things, for the short-term income statement. Cost accounting must be designed in such a way that the costs and services are allocated to the result objects in accordance with the cause. Unfortunately, there are often deficiencies in this area in medium-sized businesses that should be remedied beforehand. Some typical problems are, among others:

- There is no cost unit accounting; the calculation can only be analyzed in the merchandise management system.
- There is only a cost center-based income statement for selected result objects; for example, evaluations for differentiated cost units are missing.
- The structures of cost accounting are too rough for control, and result object structures may even be missing.
- There is no planning or the planning is only rudimentary and not integrated.
- There is no performance measurement system established.
- The overhead cost allocations are based on historically or politically determined values and assumptions.
- Material purchases are booked directly as expenses, instead of booking them through inventory and material consumption.
- …

Depending on the selected planning and reporting tool, the data bases have to be defined and the data model has to be designed. In the IT concept, the data supply and maintenance processes have to be determined. In the case of a data warehouse-supported reporting solution with an analysis cockpit, for example, the data supply of the data warehouse and the data modeling for the analysis evaluations have to be defined within the scope of the IT concept. The data supply involves the ETL process, starting from the extraction of data from the source systems, through the transformation, to the loading processes into the data warehouse (see Sect. 5.5.2). An important task here is the harmonization of data and the assignment of namespaces for the technical objects and contents to be used.

The data modeling prepares the data structures and contents for an optimal and performant analysis and planning by defining so-called data targets (e.g. a "multi-cube" or an "operational data store object").

In the last step of the IT concept, the queries for the planning forms and reporting are to be described on the basis of the content-related target concept and the developed data model. Furthermore, the navigation, planning and other functions as well as the layout design of the reports and planning forms are to be technically documented (cf. Sect. 3.3, 3.4, 3.5, 3.6, 3.7, 3.8, 3.9, 3.10, 3.11 and 3.12). In addition, cross-cutting tasks, such as the creation of the authorization concept, are required (cf. Sect. 5.5.3).

4.2.1.5 IT Selection

If the IT analysis reveals deficiencies or missing IT components, an IT selection process must be carried out, whose steps are only listed here in a bullet point list:

- creation of functional and technical specifications
- vendor inquiry
- preselection
- presentations of selected vendors
- test model implementations (prototyping)
- evaluation and decision
- implementation
- training and introduction
- ongoing support

Central studies on planning and reporting tools, such as the analyses of InfoSoft AG or the BARC studies, provide a fairly good overview of the software market situation and can be used as an inspiration in the preselection phase. However, individual criteria of the company, and especially the context of the IT landscape and the requirements of the departments, make a separate consideration and analysis of the software packages in a selection decision later inevitable. Software experts such as consultants are to be used effectively to achieve an economically and technically good solution. Here, the content still determines the technology. Technical features and possibilities only define the framework.

Regardless of whether a purchase decision or the self-production of a suitable IT-supported system takes place, it is advisable to create a technical specification which contains all the above-mentioned requirements for a software program regarding planning and reporting. It can either be used as a checklist for checking the performance of different software programs from different external manufacturers or as a programming specification for the own or external software production. A technical specification is the more valuable the more concrete the requirements and their weighting (e.g. "Must, Can, Nice to have") are formulated. In addition to the business requirements for the IT system, general questions about the software program (product features, installation numbers, user-friendliness, integration capability, database and hardware technology etc.) and questions about the manufacturer (company characteristics, pricing, references, hotline, service, training, maintenance etc.) are to be listed.[46]

Top IT solutions that are offered on the market can be assigned to two approaches: "best-of-suite" and "best-of-breed".

The "best-of-suite" solution is an integrated platform as a complete package from one manufacturer. The individual modules are coordinated with each other.[47] The "best-of-breed" approach integrates the best software or hardware components and services in a solution even if they come from different manufacturers. The individual components can thus exploit their individual strengths in the overall solution.[48]

4.2.1.6 IT Implementation

The last step of the reporting introduction comprises the IT implementation. The results of the IT concept form the basis of the IT implementation. The mostly content-related requirement profiles of the IT concept are also supplemented by technical and organizational framework parameters in the sense of a specification sheet or a requirement specification which form the basis for the software development.

The responsible development activities are usually carried out by IT staff, purchased external developers from consulting, and software companies or Poweruser from the BI team, e.g. from Controlling. The individual steps are:

- Ensuring the data supply from the source systems
- Integration of the upstream systems
  - necessary adjustments in the upstream systems
  - interface creation or adaptation
- Setup and maintenance of the data storage, e.g. of the data warehouse up to the data targets
- Development of the analysis queries and functions

46 Cf. Schön and Krause (1997, p. 57 f.).

47 Bitkom (2013, p. 30).

48 Cf. Martin (2013, p. 139).

- Creation of the individual planning and reporting layouts
- Implementation of the cross-cutting tasks
  - groupings
  - portal integration
  - authorization system
  - sizing/performance
  - …
- Test and acceptance of the IT implementation based on the established requirement profiles in the IT concept (quality assurance of the IT implementation)

Since new insights leading to change requirements arise throughout the entire phase of introducing planning and reporting, a **prototyping** is recommended for the introduction in practice. Here, the individual process steps are not to be carried out to the last detail, but to a level that can be further processed by the next stage. In this way, parts of the planning model or the reporting can also be created step by step.

Disadvantages of prototyping include:

- Due to later concept changes, parts of the development work in pronounced prototypes are no longer usable. The time and effort for the final development are incurred again.
- If the prototypes are designed too simply, it is difficult to recognize deficits and potentials of the systems.

Advantages of prototyping include:

- The users receive project results faster and can better identify the need for changes.
- The future users and maintainers of the systems gain higher learning effects regarding the possibilities and limitations of the available systems, something which leads to a better design of the overall system.
- Change requests can be quickly incorporated into the respective process steps.

Especially the advantage that the users are involved in the development phase at an early stage through prototyping and can better initiate their requirements through the concretization of the prototype often speaks in favour of prototyping.

In larger data warehouse and BI projects, long development times for new or changed requirements are often complained about. Since classical development processes can only handle change requests e.g. in a waterfall model with difficulty, new project approaches in software development are preferred. Under the keyword "**agile BI**" the development model "**Scrum**" is discussed,[49] which establishes mixed develop-

[49] Cf. Gluchowski (2017, p. 9–10).

ment teams similar to "simultaneous engineering" that deliver development results in short intervals of a few weeks in so-called "sprints".[50] The word scrum is used e.g. in rugby where it describes a method of restarting the play after a minor infringement.[51]

Before the start of the sprint, the content-related product requirements are recorded in a product backlog and converted into technical backlog tasks. The product owner is responsible for the product requirements and for the success of the development project.[52] He coordinates requests and product characteristics with the stakeholders of the product. Daily meetings take place in the scrum process to inform the members of the development team about the development step and, if necessary, to agree upon adjustments (new "backlog tasks" in the sense of change orders). Within the sprint, however, no change requests that affect the goal of the sprint are allowed. At the end of the sprint, the results are presented and compared with the requirements. New requirements are recorded by the product backlog, and a new sprint can be initiated.

As in simultaneous engineering, the scrum process could be improved by allowing the development team to incorporate change requests in the development process during the sprint, even if this extends the sprint goal. However, they should only be included if the changes are appropriate. In this case, costly coordination, loops and subsequent changes in the slightly adapted development process could be reduced.

A third role, the so-called scrum master, is responsible for the framework conditions in the scrum development model. He sets up rules for the success of the scrum process and takes care of disturbances in the ongoing process.[53]

4.2.1.7 Coaching/Training/Quality Assurance

In order to achieve a high acceptance for the new planning and reporting solution, the main supervisors and the most important users of the system should be involved in the introduction process, in addition to the information needs assessment and the creation and determination of the target concept.

For the supervisors (e.g. BI team, Controlling and IT) of the systems, a coaching approach is recommended. The coaching approach envisages that the future supervisors of the planning and reporting system are involved in all steps of the introduction phase, to learn content-related and technical requirements and implementation steps. The know-how of the experts from external consulting and software companies is thus gradually transferred to the supervisors in the company. Professional and technical documentation also helps to secure the knowledge for the further development of the system in the long term.

---

[50] Cf. Dräther et al. (2013, p. 125 ff.).

[51] Cf. Foegen (2014, p. 50–51).

[52] Cf. Schwaber and Sutherland (2013, p. 13 ff.).

[53] Cf. Pichler (2009, p. 20–23).

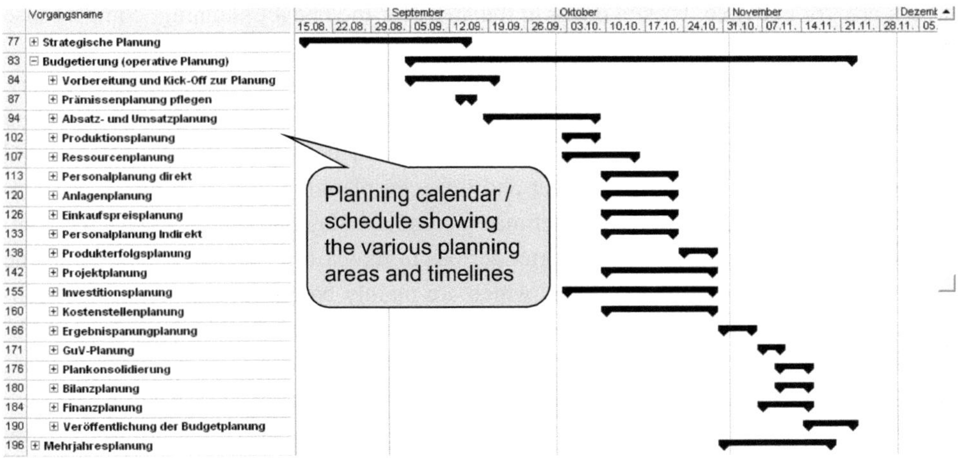

Fig. 4.11 Planning calendar (The planning calendar was created with MS Project.)

After the IT implementation, the planning and reporting addressees have to be trained and instructed regarding the contents of the planning and reporting system and its use. This can be designed, for example, in the form of core training and extended training rounds up to individual training. The involved employees should acquire the knowledge of how the reporting system in the company is handled and used.

The introduction process ends with the handover and acceptance of the system. The cyclical planning and reporting processes follow and recur in the course of the fiscal years (see Sect. 4.2.2 and 4.2.3).

As a quality assurance measure of the development process, continuous quality assurance measures for the further development of the planning and reporting solution in the company should be established. The ongoing quality assurance represents a separate process, which is described in Sect. 4.2.4.

4.2.2 Cyclical Planning Process

The planning process is divided into planning areas in business practice; these areas have to be passed through successively in order to obtain an overall plan. Due to the complex interrelationships, simultaneous planning is practically not feasible. Fig. 4.11 shows a possible planning calendar, in which the planning areas are listed.[54]

The starting point of the planning is, as already explained in Sect. 3.1.1, the strategic planning. It provides the planning premises and the strategic planning cornerstones,

[54] An integrated reporting and planning calendar can be found in Fig. 3.11. A simple process flow description can be found, for example, in Weber et al. (2009, p. 30).

which are, for example, broken down to the areas for the budget planning, compare also Sect. 4.2.2.1 below.

The operational budget planning is aligned with the bottleneck area of the company, which in most cases is the sales area. From there, the other sub-areas follow according to the dependencies in the value creation, e.g. the production and resource planning, and then extend to the profit, balance sheet and financial planning. While in the past operational planning cycles lasted 3 to sometimes 6 months, today one tries to reduce the planning duration to 4 weeks and even shorter periods in forecasting. This is also possible due to the BI-supported planning systems, which are mainly based on data warehouse technology (see Sect. 5.5 and 5.6). Furthermore, it can be assumed that the in-memory technology will also contribute to the shortening of the planning cycles in the medium term, when e.g. complex planning models and algorithms are calculated (see Sect. 5.8.2.1).

The medium-term planning (multi-year planning, e.g. 1-3 years) often follows directly after the strategic planning in practice, and this has the disadvantage that it is not well coordinated with the budget planning. If the budget planning, in which the coming fiscal year is planned in detail, follows only after a few months, then the first year of the medium-term planning becomes obsolete. Therefore, it makes sense, in my opinion, to carry out and align the medium-term planning directly after the budget planning. The first year of the medium-term planning then corresponds to the results of the budget planning.

Planning is a convergent approach, in which the planning results are gradually refined and coordinated.

Using scenario planning, optimistic, pessimistic and realistic planning alternatives (best, worst, middle case) can also be created; these show alternative future perspectives and serve as orientation and determination of a final plan version. The planning of a worst and best case variant is also called bandwidth planning. In connection with premise planning (cf. Sect. 3.1.2) or value driver-oriented planning (cf. Sect. 3.2) and modern data warehouse and BI technology, it can be carried out in a reasonable time if one focuses on a few central premises or driver variables.

The planning areas are primarily processed successively due to their dependencies (e.g. sales, production and resource planning). If there are no or only conditional dependencies between the planning areas, they can also be carried out overlapping or in parallel (e.g. personnel and purchasing planning). However, in the planning process there are also planning loops, where a return to a previous planning area is necessary and the reprocessing of this and the subsequent planning areas has to be done. For example, an important customer with a very high sales share drops out and the sales plan has to be revised. This affects many downstream planning areas up to the result and financial planning. Such planning loops are often initiated by:

- New events and insights from the environment and the operation which became known only at a later point in time
- Faulty planning results which were uncovered by checks in downstream planning areas

- Coordination between the responsible management areas which leads to improved the planning results in the approval process

While the operational planning is usually carried out in the third quarter of the fiscal year, the strategic planning often takes place in the first quarter. To derive the top-down targets, it would be more advantageous if the individual planning components moved closer together or were processed continuously in the sense of rolling planning. Rolling planning is also referred to as rolling forecast. In forecasting, two variants are basically to be distinguished:

- The forecast is based on the current actual values for a fixed point in time, i.e. usually until the end of the fiscal year (year-to-end).[55]
- The forecast is carried out on a rolling basis for a fixed period of time, e.g. for 12 months in advance (see Fig. 4.12).

The forecasting is shown in the illustration of the reporting calendar (see Fig. 4.15) because it is carried out in the fiscal year with the quarterly or semi-annual financial statements or other intervals of the reporting. This way, the reporting system is supplemented with the updated forecasts based on the current actual values. Due to the effort required, monthly forecasting rounds are rather rare in practice. Here, the forecasts take place rather two to three times a year, e.g. quarterly, whereby the last quarter coincides with the budget planning of the new year.

All planning areas certainly have their specialties and planning approaches. Nevertheless, cyclical planning process steps can be determined for all planning areas, something which can be seen from Fig. 4.13.

Before the individual cyclical planning steps are described, the general basic orientation of the planning is explained first, whereby a distinction is made between top-down and bottom-up planning as well as the counter-current method.

4.2.2.1 Basic Orientation of Planning (Top-Down, Bottom-Up or Counter-Current Method)

When deciding who starts with the planning and what coordination is required, three basic orientations can be distinguished (see Fig. 4.14):[56]

- top-down
- bottom-up or
- Mutual (counter-current method)

[55] Cf. Becker et al. (2013, p. 131).

[56] Cf. Wahls (1993, p. 83 ff.) and Schlegel (1996, p. 66).

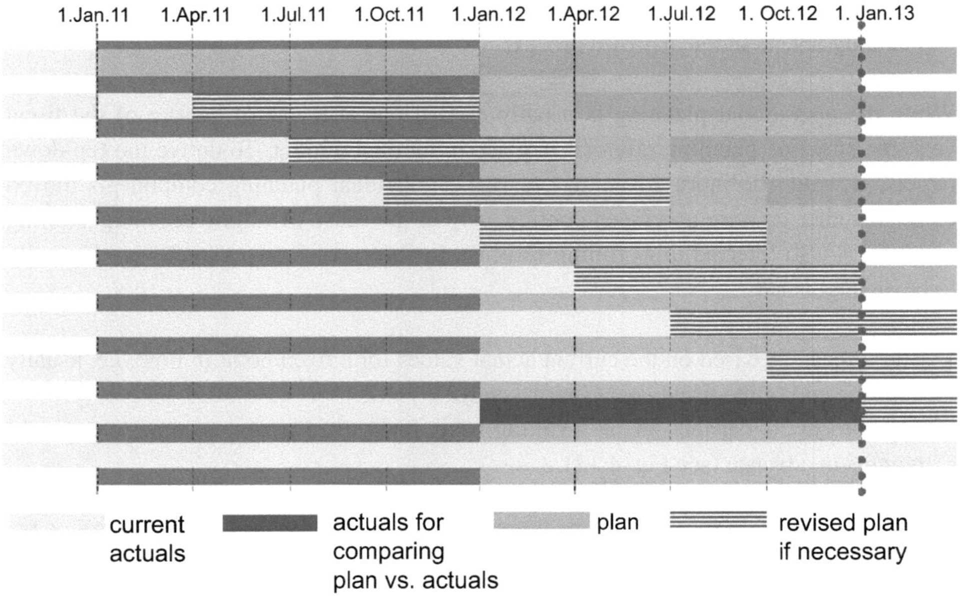

Fig. 4.12 Rolling planning/rolling forecast

Fig. 4.13 Cyclical planning process

Cyclical planning process:

- Planning preparation
  - Schedule
  - System settings
  - Procurement and transformation of standard and orientation values for planning
  - Providing of planning forms and planning functions
- Planning execution
  - Execution of automatic planning functions and manual planning entries
  - Plausibility check and verification of planning results
  - Commenting on planning
- Reconciliation and approval of planning
  - Presentation, discussion and coordination of planning contents
  - Maintenance and correction of changes
  - Approval of planning
  - Fixing the planning

In a pure top-down approach, the planning specifications are broken down from the top management level to the responsible result areas. The specifications are thus fixed and are not changed by the decentralized managers. The specifications serve as a control and management instrument for compliance with or deviation from set goals. The strategy orientation and alignment with the management's control needs speak for the top-

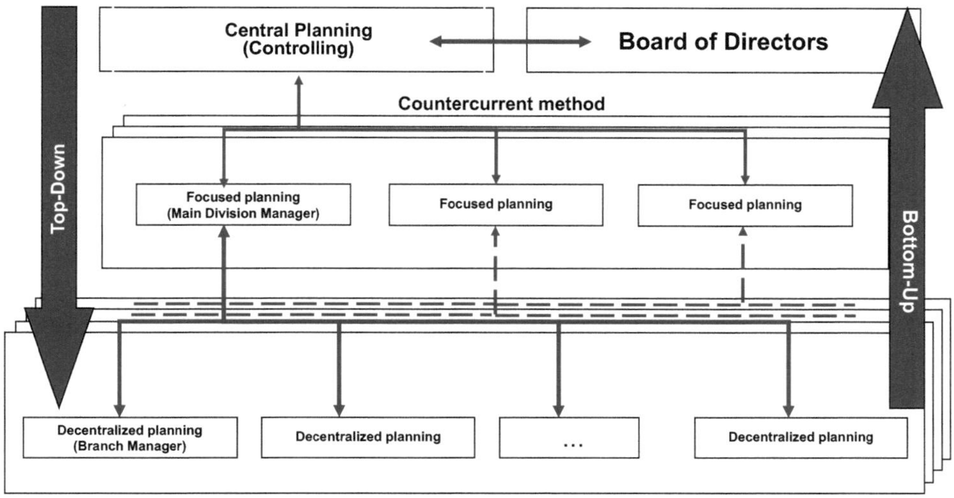

Fig. 4.14 Basic planning orientation

down planning approach.[57] The advantage of the top-down approach is also the fast and economically favorable implementation. The disadvantage is the non-inclusion of the decentralized decision-makers, which means that important detailed information from the decentralized areas for planning and coordination is not taken into account. Due to the lack of participation, the acceptance of the planning by the employees is lower, and the basis for control and the incentive to achieve the goals is missing.

In the bottom-up methodthe planning is carried out starting from the decentralized managers. The planning results are aggregated and coordinated step by step from the respective corporate levels from bottom to top up to the overall company. If the planning results do not meet the expectations of the top management, a revision of the planning is necessary. The advantage of the bottom-up approach is the proximity to the data generation and delivery and the operational business processes through the participation of the decentralized decision-makers. The main disadvantage is the long process and the complex revision processes that are usually required. In addition, the bottom-up planning can lead to plan values that do not meet the management requirements, for example, if revenues are budgeted too low and costs too high and thus the tension level is too low.

The counter-current method is a combination of both top-down and bottom-up planning. The starting point is the integration of the strategic goals and the management requirements into planning guidelines for the most important key figures and result objects, which are then broken down top-down to the other result objects of the lower responsibility areas. This should be done with the distribution algorithms of the software

---

[57] Cf. Weber et al. (2008, p. 22).

solutions used, which can be manually adjusted if necessary. This procedure is fast and economical. In the counter-current method, these top-down guidelines are then concretized and designed by the bottom-up planning. The target values of the top-down planning are used as orientation and coordination parameters, but this does not mean that the targets cannot be discarded. This is a successive procedure, which should be done by a communicative process between the involved recipients and creators. Only in this way misalignments and design deficits can be avoided. The revision steps can be reduced to a reasonable level. The tension level and the coordinative alignment of the plan values are significantly more appropriate than in the top-down or bottom-up planning.

The fixation of the budget planning is finally the last step to a coordinated planning between the management levels and used for the control of the company.

The mutual alignment of the planning in the counter-current method with broken-down top-down targets thus contributes to the best solution.

4.2.2.2 Planning Preparation

The planning preparation includes the scheduling, the preparation of the system settings in the software, the procurement and transformation of the target and orientation values, the provision of the individual planning forms as well as additional planning functions for the planners.

In the scheduling, based on the kick-off date, all planning processes of the individual sub-areas with their start and end dates, the dependent planning steps, and the responsible employees should be listed. In addition, dates for the coordination workshops and other planning rounds should be communicated early. For this purpose, it is advisable to use communication software such as office products (e.g. Lotus Notes, MS Outlook) or communication platforms (forums, chats etc.).

In the software applications for planning, the plan versions and other central parameters for the planning year are to be set at the beginning of the planning round.

For the individual planning areas, it must be ensured that their pre-modeled target and orientation values in the concept are provided in the planning forms. While some values, e.g. the old plan values of the previous year, can be taken from the historical data sources, some values depend on upstream planning areas, such as the premise planning for the personnel cost planning or the sales planning for the production planning. A monitoring of the planning statuses of the individual areas and their release should ideally indicate to the planner of the planning area whether all data and prerequisites for starting the planning area are available.

The data and form provision is otherwise subject to the same or similar requirements as those that arise in reporting when creating reports (see Sect. 4.2.3.1). In addition to providing the planning forms, planning functions are also to be initiated in the planning, such as allocations, distributions and revaluations, whose results are displayed in planning logs and planning reports.

4.2.2.3 Planning Execution

When all systems and information for planning are available, the planning is carried out. Here, either the planning functions are initiated or the plan values are entered manually in the planning forms. Once the planning is completed, the planning results are to be checked. As a plausibility aid, instruments such as reconciliation reports, error logs etc. can also be used.

Finally, the planning is to be supplemented with targeted comments that make the further processing in the subsequent planning areas more transparent. In addition, the comments help with the analysis, coordination and approval of the planning results in the respective management bodies.

4.2.2.4 Coordination and Approval of the Planning

The planning results of most planning areas require coordination and approval by the responsible persons of the respective management levels and, if necessary, adjacent planning areas.

These processes are to be marked as milestones in the planning calendar.

In certain steering committees of the company, the planning results are usually presented, discussed, and finally coordinated. If the expectations of the respective management levels and planning areas differ significantly, corrections and changes in the planning are usually necessary. Therefore, the results of these workshops should be fixed in writing before the maintenance and correction of the changes follow. If consensus planning results have been achieved, the planning is approved. The planning status of the planning area is frozen (fixed) and released for further planning steps. When all planning areas have been completed, the integrated overall company plan with all sub-planning areas is fixed. The final approval is given by the management or the board of directors and is finally decided by the shareholders' meeting or the supervisory board and the general meeting.

The plans are finally released for disclosure and reporting.

4.2.3 Cyclical Reporting Process

The reporting in the company is characterized by the monthly, quarterly and annual financial statements in the fiscal year. They form a main task for the controlling and the accounting department. The timing of the reporting intervals can be taken from the following reporting calendar (see Fig. 4.15). Depending on the planning system, forecasting is also integrated in the reporting process (see Sect. 4.2.2). As a subject, forecasting belongs to planning and is carried out, for example, quarterly.

The processes can be broken down into further detail steps. An example of the monthly closing is shown in Fig. 4.16, where the reports have already been sent after 4 days and analyzed and used for decisions after 10 days.

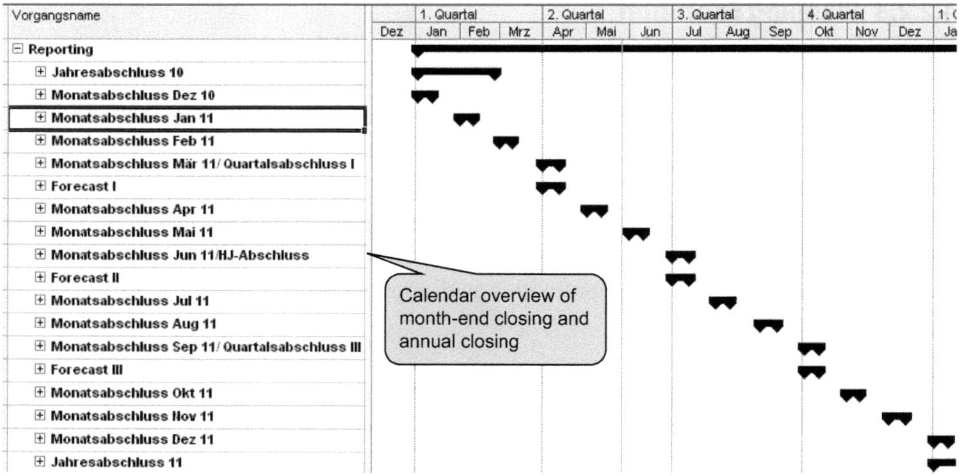

Fig. 4.15 Reporting calendar (The reporting calendar was created with MS Project.)

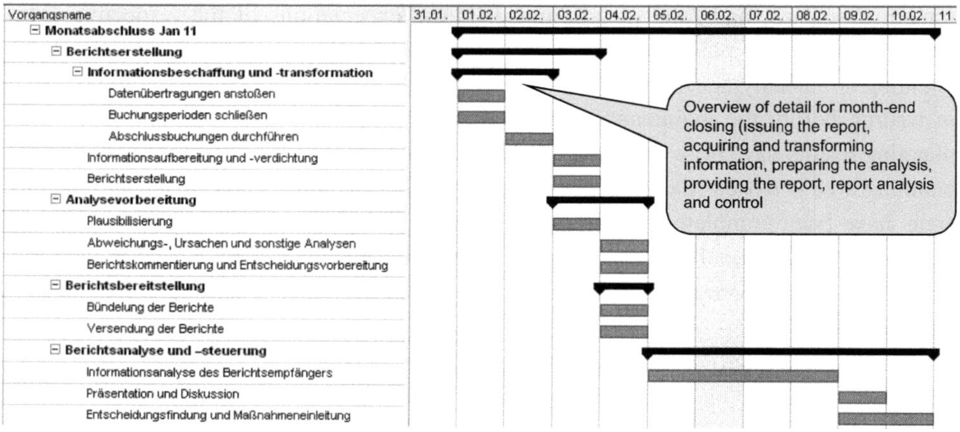

Fig. 4.16 Reporting calendar monthly closing (The reporting calendar (monthly closing) was created with MS Project.)

The cyclical reporting process (see Fig. 4.17) comprises the process steps that are discussed in detail in the following chapters.

4.2.3.1 Reporting

The cyclical reporting process starts with the reporting. It includes the processes of information acquisition, processing and consolidation as well as the reporting in particular.[58]

---

[58] Cf. on the reporting process among others Sanders and Schafft (2014, p. 297 f.).

Fig. 4.17 Cyclical reporting process

Cyclical reporting process

- Reporting
  - Information acquisition and transformation
  - Information preparation and summarization
  - Report generation
- Analysis preparation
  - Plausibility check
  - Deviation, cause and other analyses
  - Report commenting and decision preparation
- Report preparation
  - Bundling reports
  - Dispatching reports
- Information analysis and control
  - Information analysis by the report recipient
  - Presentation and discussion
  - Decision-making and action initiation

The process of reporting is essentially based on the results of the information needs analysis (see Sect. 4.2.1.2) and the reporting requirements of the report recipients; these have already been defined in the target and IT concept (see Sect. 4.2.1.3 and 4.2.1.4) and implemented in the IT implementation (see Sect. 4.2.1.6). The process of cyclical reporting has a high degree of standardization,[59] therefore, it is advisable to support it centrally. From a professional point of view, controlling as a central institution is the ideal choice. In case of difficult technical questions and tasks, support is provided by the IT department and experts.

Information acquisition and transformation

For the cyclical reporting, the required data must first be collected. Here, manual and automatic data capture as well as the location of data capture have to be distinguished. Manual data capture can take place either in the reporting unit or in other parts of the company in the functional areas and then be queried centrally. In automatic data capture, the required data are transferred via interfaces or other system accesses from the upstream systems directly to the downstream systems. This can happen continuously at regular intervals or on request. Due to the high workload and error-proneness, manual data capture should be avoided wherever it is economical to replace it by automatic data capture. The input of data still has to be done manually, at best directly in the IT system. However, technical aids, such as barcode scanners, RFID or QR code readers, can help with the capture.[60] In the age of increasing digitalization, more and more source data are also generated by the systems themselves, e.g., by measuring stations and machine

[59] Cf. Gleich and Temmel (2008, p. 78).

[60] RFID stands for radio-frequency identification. Here, objects that are marked with a transponder can be identified by means of electromagnetic waves. QR code stands for Quick-Response-Code. It is a two-dimensional code that is used to identify objects and associated information.

evaluations. Further data input is received by the systems through user inputs, e.g. on the websites of the company.

Since the data in the upstream systems are not always consistent, they sometimes have to be adjusted during data collection. As part of the transformation, the data can be adjusted in a way that, for example, different designations are standardized (harmonized) or missing information such as abbreviations or long text descriptions are enriched.

For most billing systems (e.g. payroll and financial accounting), booking months are locked and closed for the monthly, quarterly, and annual financial statements, so that changes to the actual bookings are no longer possible unless the booking months are retroactively reopened; this should not be the case but is often encountered in practice. The initiation of audit and reconciliation activities, the execution of closing bookings, and the document archiving complement these routine closing activities in the billing systems (e.g. in accounting and the ERP systems).

For the supply of a data warehouse with data, software tools for data collection and transformation are used, for example, that support the so-called ETL process (extraction, transformation, and loading process) (see Sect. 5.5.2). Here, the manufacturers usually use powerful own or external applications that fill up their data base. Compared to individually written interfaces, these are more flexible, usually faster, and cheaper due to the standardization. The programs for application integration are responsible for the possibly necessary export in addition to the import function, if data have to be played back into the upstream system or downstream systems need to receive data.

The speed of information acquisition and transformation is one of the essential criteria for the users. As an example of data and application integration, Hyperion Application Link is shown here (see Fig. 4.18) with which a high data integration can be achieved with existing transaction systems (e.g. ERP systems such as SAP), existing databases, but also simple Excel tables with a data warehouse.

Information processing and consolidation

Since the data in the phase of data collection are not always complete, harmonized, and in compatible structure and quality, the data have to be prepared for the consolidation in a further step. Here, it is important to structure and store the data as optimally as possible for the actual reporting.

Depending on the use of the reporting technology, this phase has a different role. For example, if the reports are prepared very simply with MS Excel, many time-consuming (often manual) tasks, such as checking, harmonizing, aggregating and consolidating, are involved. If a data warehouse is used, the data administration ensures that the data preparation is done automatically as much as possible by harmonization rules into so-called data targets. The data targets are, for example, multidimensional cubes or flat tables (Operational Data Store Objects), based on which the report queries (queries) and report design are done (see Sect. 5.5).

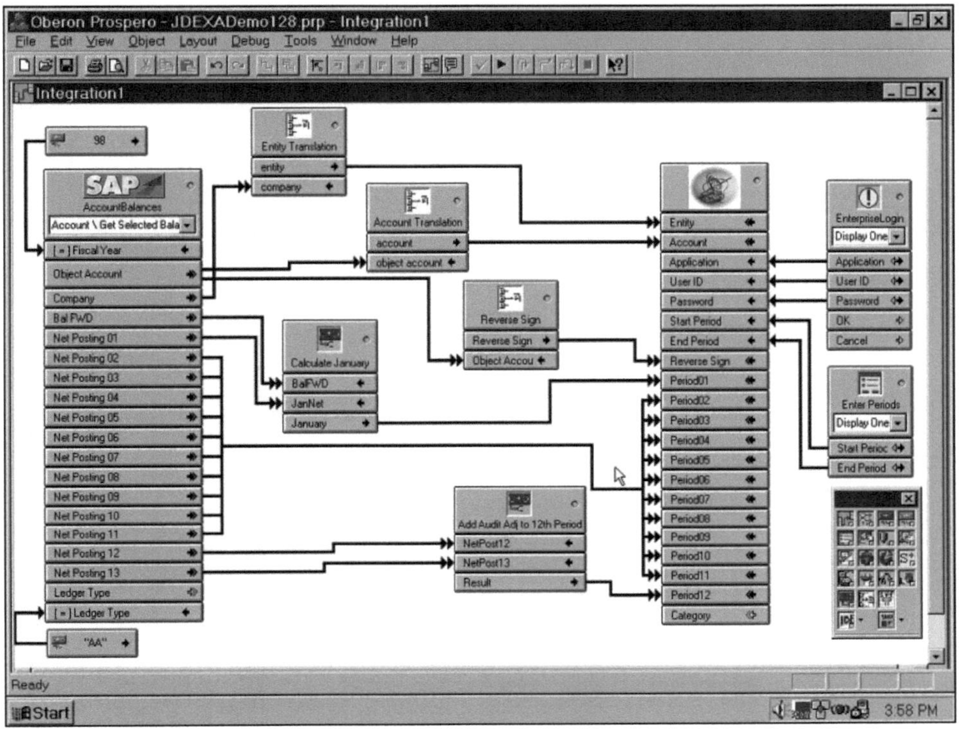

Fig. 4.18 Hyperion Application Link (Taken from Schön 2004, p. 323)

Report creation/design

The report creation supports standard, analysis, ad-hoc and exception reporting. Important basic forms of reporting, layout tips and report design objects (comments, tables and graphics) have been described in detail in Sect. 3.6, 3.7 and 3.8. The phase of report creation can be distinguished according to the initial setup, the design and generation of the report, and the cyclical retrieval of already generated reports. If powerful report generators are applied to, for example, existing relational databases or data warehouse solutions, the initial report creation is time-consuming. Later, the retrieval of the reports by the push of a button then takes only a few seconds. If reports are created with simple spreadsheet programs, then the initial design is relatively easy, but has to be filled and adapted again and again with regular new requests.

Due to the high standardization possibilities, it makes sense to use powerful report generators. Modern frontend solutions connect directly to the data warehouse and enable multidimensional data analyses, which can be easily created and modified for ad-hoc queries. In the market, three directions emerge for the data warehouse-supported frontend solutions, which are analyzed in more detail in Sect. 5.5.5.2:

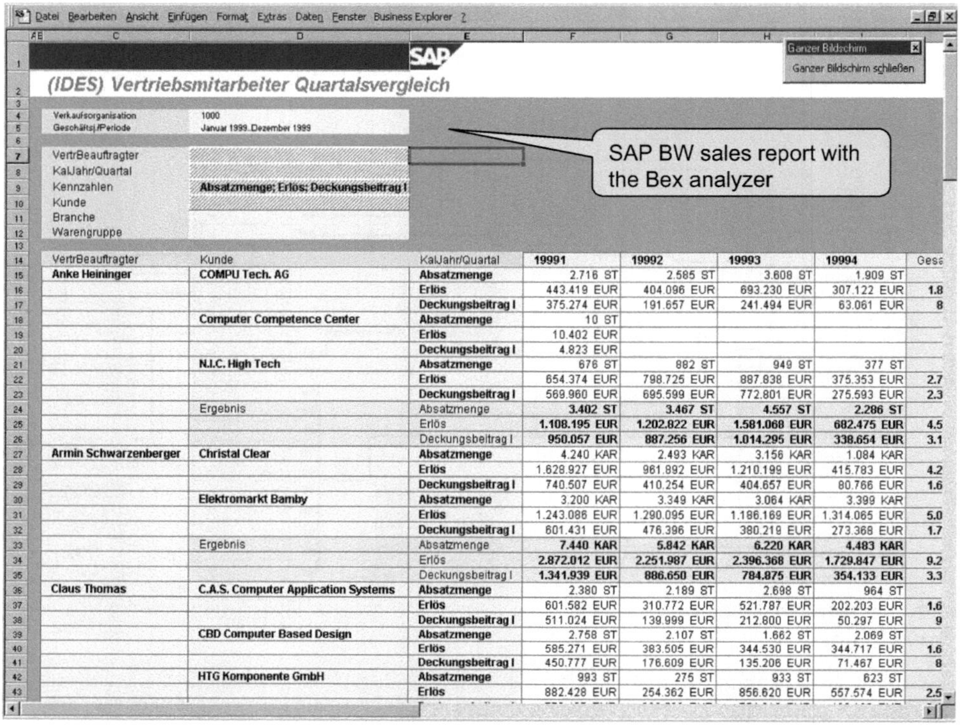

VertrBeauftragter	Kunde	KalJahr/Quartal	19991	19992	19993	19994	Gesa
Anke Heininger	COMPU Tech. AG	Absatzmenge	2.716 ST	2.585 ST	3.608 ST	1.909 ST	
		Erlös	443.419 EUR	404.096 EUR	693.230 EUR	307.122 EUR	1.8
		Deckungsbeitrag I	375.274 EUR	191.657 EUR	241.494 EUR	63.061 EUR	8
	Computer Competence Center	Absatzmenge	10 ST				
		Erlös	10.402 EUR				
		Deckungsbeitrag I	4.823 EUR				
	N.I.C. High Tech	Absatzmenge	676 ST	882 ST	949 ST	377 ST	
		Erlös	654.374 EUR	798.725 EUR	887.838 EUR	375.353 EUR	2.7
		Deckungsbeitrag I	569.960 EUR	695.599 EUR	772.801 EUR	275.593 EUR	2.3
	Ergebnis	Absatzmenge	3.402 ST	3.467 ST	4.557 ST	2.286 ST	
		Erlös	1.108.195 EUR	1.202.822 EUR	1.581.068 EUR	682.475 EUR	4.5
		Deckungsbeitrag I	950.057 EUR	887.256 EUR	1.014.295 EUR	338.654 EUR	3.1
Armin Schwarzenberger	Christal Clear	Absatzmenge	4.240 KAR	2.493 KAR	3.156 KAR	1.084 KAR	
		Erlös	1.628.927 EUR	961.892 EUR	1.210.199 EUR	415.783 EUR	4.2
		Deckungsbeitrag I	740.507 EUR	410.254 EUR	404.657 EUR	80.766 EUR	1.6
	Elektromarkt Bamby	Absatzmenge	3.200 KAR	3.349 KAR	3.064 KAR	3.399 KAR	
		Erlös	1.243.086 EUR	1.290.095 EUR	1.186.169 EUR	1.314.065 EUR	5.0
		Deckungsbeitrag I	601.431 EUR	476.396 EUR	380.219 EUR	273.368 EUR	1.7
	Ergebnis	Absatzmenge	7.440 KAR	5.842 KAR	6.220 KAR	4.483 KAR	
		Erlös	2.872.012 EUR	2.251.987 EUR	2.396.368 EUR	1.729.847 EUR	9.2
		Deckungsbeitrag I	1.341.939 EUR	886.650 EUR	784.875 EUR	354.133 EUR	3.3
Claus Thomas	C.A.S. Computer Application Systems	Absatzmenge	2.380 ST	2.189 ST	2.698 ST	964 ST	
		Erlös	601.582 EUR	310.772 EUR	521.787 EUR	202.203 EUR	1.6
		Deckungsbeitrag I	511.024 EUR	139.999 EUR	212.800 EUR	50.297 EUR	9
	CBD Computer Based Design	Absatzmenge	2.758 ST	2.107 ST	1.662 ST	2.069 ST	
		Erlös	585.271 EUR	383.505 EUR	344.530 EUR	344.717 EUR	1.6
		Deckungsbeitrag I	450.777 EUR	176.609 EUR	135.206 EUR	71.467 EUR	8
	HTG Komponente GmbH	Absatzmenge	993 ST	275 ST	933 ST	623 ST	
		Erlös	882.428 EUR	254.362 EUR	856.620 EUR	557.574 EUR	2.5

Fig. 4.19 SAP BW report with the Bex analyzer (MS Excel integration) (Source: SAP BW report with the Bex analyzer (MS Excel integration) based on data from an SAP-IDES system. Taken from Schön 2004, p. 331)

- Excel-integrated reporting solution
- Special proprietary report generators with graphical user interfaces
- Web-based reporting solutions

For the Excel-integrated reporting solutions, Excel is used as an add-in solution, where the additional functions mainly support the data selection and provision from the data warehouse, but the design functions can be used in Excel (see Fig. 4.19).

Among the software providers, it is noticeable that different strategies are offered for different report forms (standard, analysis, ad-hoc, and exception reporting). While one group of providers uses different tools for different report forms on the same data basis, there are other manufacturers who use only one application for this purpose.

For example, for standard reporting, higher requirements are placed on a print version, which can be better met by special report generators with a pixel-accurate graphical user interface than by web solutions.

A major advantage of web reporting is especially that no client software has to be installed on the desktop anymore. A conventional browser is sufficient for the report

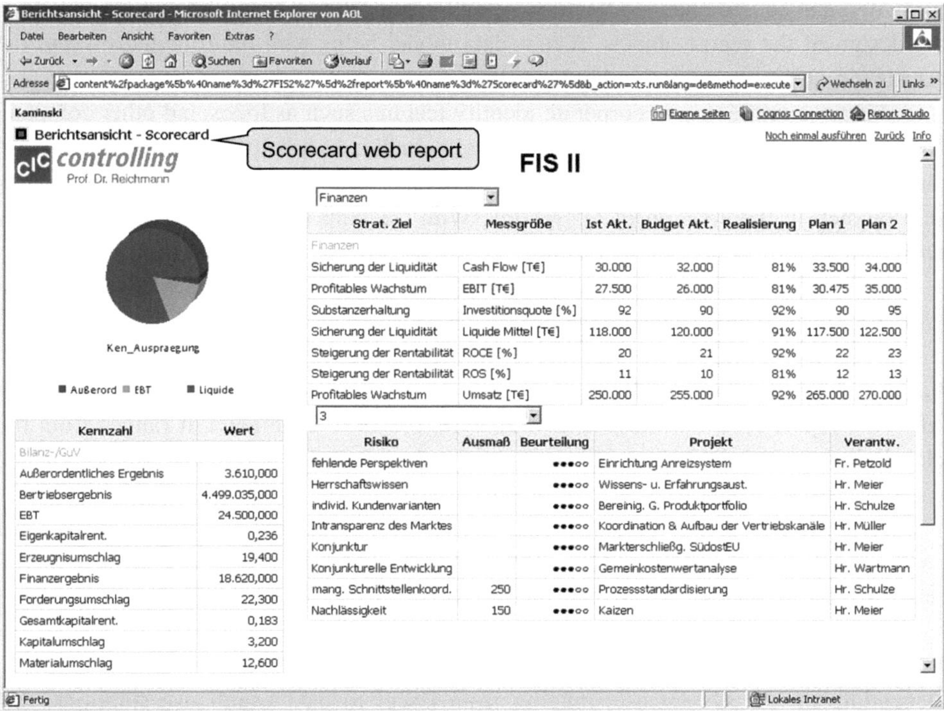

Fig. 4.20 Web report example with Cognos ReportNet (Taken from Schön 2004, p. 324)

presentation and can be accessed from any place in the world where an internet connection is possible (see Fig. 4.20). If, in addition to desktop reporting, mobile reporting on different devices is required, the reports should be developed in such a way that an automatic adaptation of the report contents and functions to the mobile device and its size takes place (see Sect. 5.10.5).

When creating reports, it is not mandatory to have an understanding of database structures and knowledge of programming languages. The reports can be created with modern report generators without the IT department and without IT specialists and consultants. The data connection and analysis is supported by exploratory analysis tools of data mining and data discovery or visual discovery tools, compare also the remarks on data mining in Sect. 5.5.5.3 and on self-service BI in Sect. 4.1.5.4.

The providers distinguish between the report development level and the report analysis level. Important for the report development is the design freedom in the presentation of the report contents. Report templates help in the design of the reports. They contain information about the design and structure of the available information and usable report objects. Frames (report sections) can be used to separate data contents from each other in a report (e.g. on one page). With the drag & drop functionality, so-called report objects (e.g. tables, selection filters, graphics, menu and buttons) are individually positioned, formatted and can be changed flexibly. Selection and filter criteria are to be selected for the design of the report objects. A metadata model helps with multilingualism and the translation of the reports into other languages.

Arbitrary comments and corporate identity features such as logos and other company layouts or product images can be inserted individually. As a help for the report development, so-called "**wizards**" (report creation assistants) can be used; these guide the user step by step in the creation of his reports. With the help of a preview image (**thumbnails**), the report created with the help of the wizard can be viewed in advance.

For the possibilities of report development, compare also the explanations in Sect. 5.5.5.2 and here especially the different user interfaces.

4.2.3.2 Analysis preparation

After the report creation, the phase of analysis preparation follows. It ranges from the plausibility check of the created reports to the deviation and root cause analysis to the commentary of the reports.

This phase is usually carried out centrally by the Controlling function. However, the analyses and commentary tasks also require communication with the subject matter experts or the management and leadership staff.

Plausibility check of data quality

Depending on the system use, ranging from Excel to data warehouse solutions, this phase gets a different weight. By increasing the automation of data transfer, plausibility check processes can be optimized. In the case of very many manual transfer paths, this process step is usually very time-consuming.

The aim of the plausibility check is to detect and correct inconsistencies, errors and quality defects of the data in the created reports (see Sect. 4.2.4.4). Data quality is ultimately the most frequently mentioned problem area of surveys on reporting and their systems, e.g. in the topic area of business intelligence.[61]

The data quality starts already with the **master data** and the **accounting rules**, which have to be standardized and structured for the entire company. Moreover, it is advisable to perform regular **validations** and to establish **standardized creation and maintenance processes of master data** and to support them with certain analysis methods (e.g. **syntax analysis,completeness analysis**) and **synchronization functions.**[62]

Data losses between the systems can be checked, for example, by control sums. **Error logs** of the interfaces show e.g. faulty data transfers or the incompleteness of the data sets. **Reconciliation reportsand reconciliation calculations** help to check the data com-

[61] See e.g. Friedrich (2008, p. 10-11), Finucane and Mack (2011, p. 8-11) and Schön (2011, p. 1-47).

[62] There are already special software solutions for data quality checking (see e.g. Atacama 2011).

pleteness. Unassigned key figures indicate missing information enrichment. Last but not least, **experience** in controlling and in other areas is important in order to find possible faulty spots in the evaluations.

All the found error sources should be corrected at their source. Subsequent corrections in the information process, e.g. by data changes or correction lines, are hardly traceable later.

In practice, one can see that a continuous quality assurance in reporting can achieve an improvement of data quality with decreasing error rates.

Variance, cause and other analyses

A main task of the controller is the function of information provision and decision preparation. He takes on the role of a "consultant" or "sparring partner" for the management. Important foundations are created in the phase of variance and cause analysis.

The variance analysis shows the degree of achievement with regard to the set goals by comparing the actual or forecast values with the planned values.

Best-practice-benchmarks show the analysts improvement potentials for different result objects (departments, processes etc.). The significance of the result objects in relation to the totality of the result objects can be carried out, among other things, with the ABC analysis.

Time series analyses show trends and special drops or peaks of the values considered per result object.

Details on the aforementioned and other reporting and analysis forms can be found in Sect. 3.7.

The controller is responsible for highlighting the most important findings from this "creative" analysis work. Ideally, the cause and effect relationships as well as regularities and special effects should be identified. The controller needs good commercial knowledge for the analysis preparation, knowledge which he has built up in theory and verified and expanded in the company practice. Furthermore, controlling needs an entrepreneurial spirit: controllers must enjoy finding improvement potentials for the company and helping the management to implement them.

Report commenting and decision preparation

In the phase of report commenting, the findings of the preparatory analysis phase are documented. This should be done in a condensed form, e.g. in bullet points and short sentences. In thecommenting, centrally summarized results and individual comments on individual report objects and values can be distinguished (for the design of comments see Sect. 3.8.5). In addition, bookmarks help to find important content.

Based on this, decision needs of the management and leadership staff as well as suggestions for measures should be worked out. These last two steps are often cut short in practice, when the management reserves the sole authority over the interpretation and the decision, or when the possibility of interpretation in controlling is not or only limited possible due to the specific know-how of the specialist department.

If there is a good information and communication culture in the company, ideally the original result reports should be supplemented by corresponding comments and decision needs and suggestions at the end of this phase.

4.2.3.3 Report Provision

After the phase of analysis preparation, report delivery takes place. Report delivery covers the topics of output format, report bundling and the type of report distribution.

The report delivery needs are to be determined with the report recipients in the reporting introduction or reporting quality assurance phase. The execution of report delivery is centrally controlled by Controlling, which uses the technical possibilities of IT (job runs, monitoring etc.).

Exemplary output formats are among others:

- print reports (paper, pdf.)
- graphical output in a presentation program (e.g. PowerPoint, *.pptx)
- graphical output in a spreadsheet program (e.g. Excel, *.xlsx)
- graphical output in a special reporting program
- graphical output in a web browser (*.html)

For the differentiated output formats and media compare Sect. 5.5.5.7.

Bundling of reports

For the bundling of reports, different bundling forms are available depending on the output format, which have been explained in more detail in Sect. 3.3:

- folders, tables of contents, tabs and report folders
- special file directory structures and document folders only accessible for assigned report recipients
- reports and comments bundled in the storyboard
- structuring and navigation options of the reporting programs
- structuring and access options of portal solutions availble via the internet/intranet

Shipping and delivery of reports

Regarding the shipping, transmission (**Broadcasting**) or delivery of the reports, the following ways are common:

- manual shipping and delivery
- electronic delivery
  - by mail
  - by access to a document directory
  - by program access

- by online link
- by web access, e.g. via an intranet-based portal solution
- to mobile devices such as smartphones

For the sake of form, it should be mentioned here that information can also be conveyed verbally. This is also advisable for very important and current information. However, for reporting and planning itself, the delivery is done in paper or electronic form.

The recipients of the reports are to be determined in advance and ideally stored in the systems including their permissions (see Sect. 4.1.5.1 and 5.5.3).[63]

For the technical distribution possibilities of reports and planning forms, compare Sect. 5.5.5.6. The following ways can be distinguished for the type of dispatch:

- manual initiation of report dispatch
- cyclical initiation of report dispatch
- exemption reporting triggering report dispatch when tolerance limits are reached

The modern report transmission and provision is done either on individual, provider-specific, graphical interfaces or on standard web browser interfaces (see Fig. 4.21). In particular, web-based reporting has advantages regarding the distribution of reports as users without installed special applications on the local computer can analyze and navigate the reports.

The distribution function of reports is controlled by the authorization concepts and access rights of the users. For a clear management of the sometimes different report recipients, it is advisable to create so-called report folders, which are assigned to user groups or individual persons.

A further distinction is to be made between push and pull reporting. In **push reporting**, the reports are made available to the recipients according to predefined criteria (e.g. periodically), which can lead to information overload. In **pull reporting**, the recipient requests his reports individually, using, among other things, the help of special search mechanisms. Pure pull reporting, however, entails the risk that users do not receive important information. For these reasons, a combination of push and pull reporting is preferable, in which, for example, a potential report recipient is notified by e-mail or in the user area of the portal about new reports with a short content description (metadata of the report). He can then decide for himself whether he wants to select the report or not.

There is also a need to use the report contents in other downstream systems. For this purpose, the reporting tools provide export functions that serve numerous data formats (e.g. "*.csv" for spreadsheet programs).

[63] See Knöll et al. (2006, p. 200).

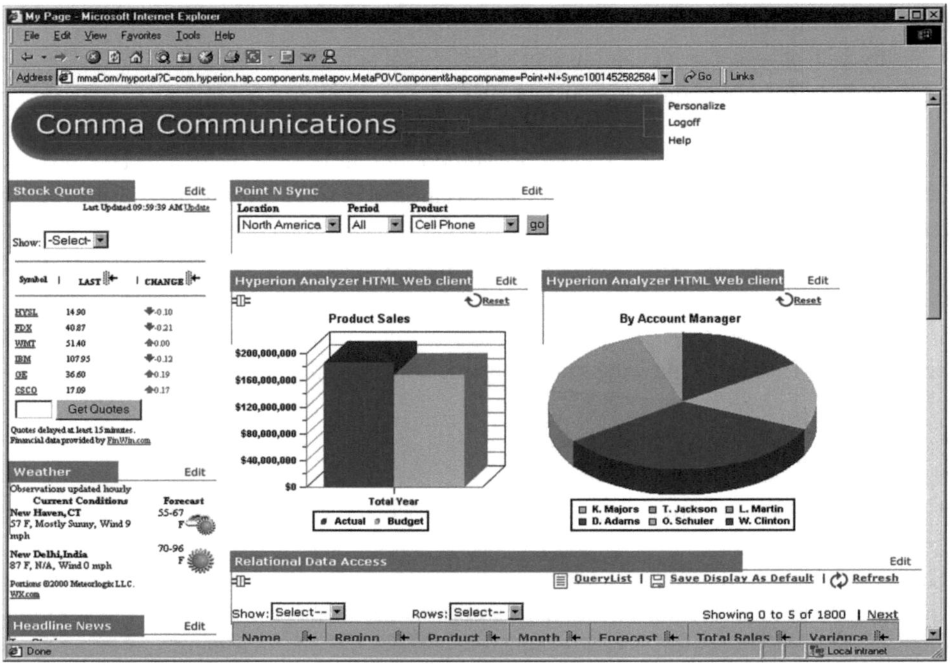

Fig. 4.21 Web-based central portal applications (Hyperion Central) (Taken from Schön 2004, p. 330)

4.2.3.4 Information Analysis and Control

The most important phase of the reporting process is the phase of information analysis and control. It includes the information analysis of the report recipient, the report presentation and discussion of the respective management circles, as well as the decision making and action derivation achieved here.

The information analysis and control is carried out by the management and leadership staff of the respective management level and is supported by the controlling. The communication between the management levels and Controlling is of high importance here. It serves to increase knowledge and create transparency among all participants.

Information analysis of the report recipient

After receiving the report information provided, the report recipient has to work out a picture of the situation of his or her area of responsibility beforehand. The various forms of analysis (see Sect. 3.7), such as deviation analyses, trend analyses or ABC analyses, help to identify strengths and weaknesses or sources of success and loss of the result objects. They build awareness of developments and possibly reveal opportunities and risks. They encourage to uncover further causes and interrelationships and to use them for control. The controller is available as a contact person for questions, advice and decision support, a person who has already supported the decision makers with important hints, comments and suggestions in the analysis preparation.

While the information supply usually takes place bottom-up, i.e. from unconsolidated basic data to information of the highest level of consolidation, the analysis paths have to be designed top-down. An analysis path is an individual or standardized search for causes that can be held responsible for an effect. The localization of the problem-relevant individual information usually takes place on a downstream level.

The following errors should be avoided in the information analysis:

- closed-mindedness toward changes (operational blindness)
- concentration on one or a few aspects
- excessive control
- individual blame assignment
- concentration on the confirmation of made forecasts and assumptions

Instead, the information analysis should be carried out as openly, multifaceted and objectively as possible.

The information intake is ideally carried out digitally in dialogue with the shown business applications in practice. It is a process between the human being and the computer. Printed reports should be avoided more and more.

The most important analysis functions of the software application are the drill-down and roll-up functions, drill-through, slice and dice functions, and pivoting and drill-across functions. With the drill-down and roll-up function, the report contents can be resolved from the condensed state to the data source information and vice versa. With the drill-through function, one has the possibility to branch to the data source in the upstream system. With the OLAP-specific slice and dice function, selected dimensional views (slices and cubes) from the multidimensional data warehouse or data mart can be displayed at high speed. With pivoting and drill-across functions, the selected dimensions (such as customer and product view or time) can be flexibly exchanged (see the section on pivot tables in Sect. 3.8.3). This enables multidimensional data analysis (see Fig. 4.22) at acceptable response times.

In addition, interactive navigation points can be used to branch to other report sources and documents or web pages (via web link connection).

The comments, decision needs and suggestions created by the controlling department help the analyst to quickly absorb important (pre-analyzed) insights and consult with others in order to possibly annotate the reports with their own comments. For the search of important contents, bookmarks can be inserted.

Report presentation and discussion

A further important step of the report control is the discussion of the report contents in the designated organizational leadership circle or committee.

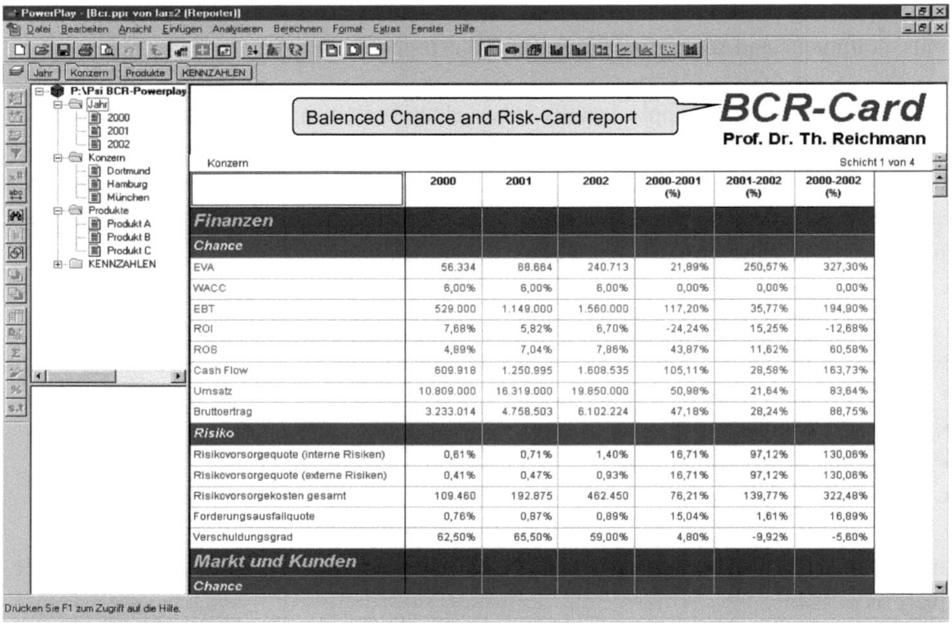

Fig. 4.22 Multidimensional analysis with Powerplay by Cognos (Taken from Schön 2004, p. 332)

An intensive reception, joint analysis, discussion and use of the reports for the control can be achieved in advance by an appealing presentation of the reports.

Good presentations and discussion opportunities for information analysis are characterized by the following points:

- summary of the main results
- commentary on important findings
- predefined analysis paths, starting from condensed information to the important detail information
- provision of further information sources on demand up to the level of detail of the upstream systems (back office)
- formulation of decision needs and proposals
- documentation of results and possibly decisions of the meetings

For the discussion and discussion of report contents in management bodies, classical presentation techniques, such as projector presentations and printed report templates are suitable, but also modern collaboration and groupware functions for scheduling and inviting to meetings, setting up chat rooms, video conferences and other discussion forums. In this way, employees in virtual teams can be brought together worldwide across time zones and geographical conditions in so-called Collaboration Rooms.[64] More and more companies are using these virtual communication possibilities. Independent of the BI software market, special software providers have established themselves here. In addition, mobile reporting is gaining more importance. Users use their mobile devices at any time and at any place (see Sect. 5.7).

Decision making and action initiation
The decision-making and action initiation can, as outlined above, be the result of the management bodies in the context of the report presentation and discussion, but it can also be made in time-shifted management rounds, especially when decisions have to be reconsidered. Often, several decision rounds are necessary to determine appropriate measures or countermeasures or escalation measures.

The measures taken and the projects initiated should be concretized and controlled with regard to the following points:

- Define objectives (e.g. by using performance and success indicators)
- Assign responsibilities
- Allocate resources
- Set time schedules
- Provide financial resources
- If necessary, assess risks

4.2.4 Quality Assurance Process

The quality assurance process comprises the content-related, organizational, process-related and IT-relevant requirements for the entire planning and reporting. The subtasks of the quality assurance process are shown in Fig. 4.23.

4.2.4.1 Information Needs Analysis and Changes
Changes of the environment and changes in the strategic orientation of a company require the review of the information needs at certain intervals. Once a year or at longer

Fig. 4.23 Quality assurance process

Quality Assurance Process:
- Information needs analysis and changes
- Process quality review and changes
- Organizational quality review and changes
- Data quality review and control
- Information system quality review and control

[64] Cf. Knöll et al. (2006, p. 201).

intervals, the information needs and planning contents must also be checked (see Sect. 4.2.1.2). For this purpose, Controlling offers itself as a central coordinating and executing unit in connection with the management levels and specialist departments. To obtain new impulses, the involvement of external consultants is useful.

Once the new or modified planning and information needs have been recorded, the necessary changes must be made and maintained in the systems. In addition, if necessary, transition rules must be created, for example, if organizational units have been merged or split.

For the administration of the reports, the maintenance of metadata is recommended, based on which the employees of IT, Controlling and the specialist departments as well as other recipients and senders of reports can obtain an overview of the reporting stock, their contents, meaning and significance.

As a medium, (IT-supported) reporting manuals and reporting guidelines are suitable in which all requirements for reporting are described in detail. These include descriptions of the development and presentation of reports as well as explanations of the preparation and evaluation of individual key figures.

Furthermore, the technical documentation (see Sect. 4.2.1.4) should also be continuously maintained, starting from the content-related target or functional concept (see Sect. 4.2.1.3) in order to maintain the currency of the data sources, the data modeling and the data supply. Ideally, there would also be automatic documentations that record important changes in the systems, for example, when new data sources have been connected.

4.2.4.2 Process Quality Check and Changes

In the context of the process quality check, the analysis of the planning and reporting process with regard to quality, speed and efficiency is carried out.

For the planning and reporting processes, it must be taken into account that the path from creation to coordination, analysis and control is associated with a "time lag", which results from the time span, for example, between s pozting a deviation in the management information system and recording the source data in upstream systems. To reduce this "time lag" and increase the speed of information supply or planning is one of the biggest challenges in terms of process optimization.

The following tools are suitable for the analysis and documentation of the planning and reporting processes:

- graphical workflows
- process descriptions
- work instructions

If economically viable improvement opportunities arise here over time, they should be implemented, for example, automatic instead of manual data transfers or the introduction of reconciliation reports to check the data quality.

In the reporting processes, the demand to deliver a financial statement or a report quickly (**fast close**) is of high importance. In the business practice, financial statements are unfortunately delivered only several weeks or even months later, so that the timeliness of the information and the resulting control possibility comes too late and may already be obsolete. The goal is to deliver the financial statement report as soon as possible after the end of the period, for example, for the monthly report at the latest by the 10th working day after the end of month, or even sooner, for example, after the 3rd or 4th working day.

Generally speaking, the timing and the cycle of the standard analysis reports have to be checked and determined. Here, the following reporting intervals can be generally distinguished:

- daily reports
- weekly reports
- monthly reports
- quarterly reports
- semi-annual reports
- annual reports
- multi-year reports

The reporting intervals have to be coordinated with the management bodies and analysis possibilities of the report recipients. Important key figures of the company, such as order intake, sales, incoming payments and production figures are to be reported in the short term (weekly to daily). With the modern IT support in today's work environment, the entry report is especially important (see Sect. 3.9). The executive wants to view the most important key figures of his area at a glance and act immediately if necessary for critical parameters. He needs the data in a timely manner (real time) and independent of location (possibly mobile).

The core management reporting will continue to take place monthly, as many result sizes and costs, such as personnel costs, are only available after the booking deadline on a monthly basis. Parameters that do not change significantly in the short term, such as the employee statistics, can be viewed in longer cycles. Of course, the legally mandatory reports, such as quarterly reports and the annual financial statement report, have to be taken into account when scheduling.

For the cyclical report generation, so-called scheduler and monitoring functions are offered, with which a periodic standard reporting system can be electronically scheduled and monitored.

Planning processes are also often excessively long in the business practice and can reach 6 months or more in the budget planning, for example. In this time, the framework conditions have already changed again, so that time-consuming correction planning is necessary. Therefore, the goal here is also to make the planning processes leaner (for example, under 3 months for the budget planning and a few weeks for the forecast) and better coordinated, so fewer corrections and follow-up processes arise.

4.2.4.3 Organizational Quality Control and Changes

Central for the organizational quality control is the assignment of the responsible report recipients, who have to be updated according to organizational changes or personnel changes. The authorization concept and the organizational matrix for planning and reporting as well as their technical implementation, which should be used in addition to the existing organizational and responsibility rules, help with this (see Table 4.6 in Sect. 4.2.1.2).

The authorization concept (see Sect. 5.5.3) assigns different rights and duties for reporting and planning to different persons and groups (reading, selecting, writing, changing, deleting etc.). From the user point of view, simple **users** with reading rights can be distinguished from power users with development rights.

Based on the report assignment concept, individual reports or report folders (briefing books) are assigned to the report recipients individually or role-based.

4.2.4.4 Data Quality Control and Management

The data analyses can only be as good as the input data. The slogan "garbage-in—garbage-out" illustrates this very well and is often used to draw attention to this basic problem of data quality. In addition to the input problem, data quality problems arise in the entire data processing process from the creation or generation to the transformation and analysis of the data.[65]Data quality is a multidimensional measure for the purpose-bound suitability of the use of data and can change over time due to changing needs.[66] The keywords "fitness for use" or "suitability for purpose" are often used for data quality.[67]

The following data quality criteria are generally relevant for IT and especially for business intelligence and big data:[68]

- completeness
- content accuracy/correctness
- relevance
- comprehensibility
- traceability
- technical accuracy/correctness
- consistency
- unambiguity
- timeliness
- non-redundancy

[65] See Naumann (2007, p. 29).

[66] See Würthele (2003, p. 21) and Apel et al. (2015, p. 7).

[67] See Gebauer and Windheuser (2015, p. 88).

[68] Cf. Apel et al. (2015, p. 5).

No data may be missing in the selected area, they must be **complete**. The **content accuracy or correctness of data** means that the data basically correspond to reality. The **relevance of data** refers to the comparison of information supply and demand. Superfluous information should be avoided.[69] The information is to be prepared **comprehensibly** for the recipient. For example, unclear terms and abbreviations should be avoided. The reports are to be designed user-friendly. The relevance and comprehensibility of the report contents should already be considered in the design phase of the reporting (see Sect. 4.2.3).

The **traceability** of data concerns the traceability of data processing from the analysis to the data source. This characteristic also includes the reliability of the data, whether, for example, external data comes from a trustworthy correct source. The **technical accuracy** means that the data is in the correct field format, e.g. with the necessary decimal places and thousand separator. **Consistent**data have no logical contradictions with each other. In relational databases, the following examples of data integrity are meant here. Each object must be uniquely identified by an identification key (entity integrity). An object without a key is not correct. A referential and logical consistency exists, for example, in the connection of objects. (*Example: A customer order is assigned to a customer. The customer cannot be deleted as long as the customer order exists in the data stock. An order for a customer can only be created if the customer has already been created.*) A range integrity exists if a value of an attribute is supposed to lie in a certain value range, e.g. for numerical code ranges for customer orders. **Unique data** should, for example, not have any semantic differences, e.g. different spellings of the same object. The **timeliness of the data** guarantees that the analyses are not based on outdated data.

When considering and evaluating the data quality criteria, all criteria must always be considered in combination. If, for example, the data of a report is complete, but no longer up-to-date due to generous loading cycles, the report will be regarded as qualitatively insufficient in its entirety.[70]

The criterion **redundancy-free data** was deliberately mentioned here as the last criterion. In general, data should not be entered or stored in the data stock twice or redundantly. However, redundancy is preferred in the context of data storage in the data warehouse and especially for distributed data storage in the context of big data. Here, copies (replications) are deliberately created and stored to achieve a better evaluation quality. Only the double entry of values is still not desired.

The criteria for data quality shown here generally apply to all data. In the business world, master and transaction data are distinguished. **Master data** are data that usually remain constant over a longer period of time and describe business objects (e.g. employees, products, customers and suppliers) on which the respective business processes are

[69] Cf. Hinrichs (2002, p. 30 f.) and Zeh (2013, p. 43 f.).

[70] Cf. Leser and Naumann (2007, p. 354 f.).

based.[71]**Transaction data** are business process-related data that arise anew with each business transaction, e.g. with invoices, orders, orders and deliveries, and usually have to be linked with parts of the existing master data.[72]

The causes for poor data quality are manifold and are categorized here as follows:[73]

- Data collection/generation
  - no resources for input available (no inputs)
  - user errors/input errors/lack of know-how for input (wrong inputs, incomplete inputs, inconsistent and redundant inputs)
  - poor input masks
  - errors in the generation algorithm
  - inconsistent data corrections
- Data definition
  - lack of standards for master data maintenance (e.g. missing master data management)
  - missing content and technical rules for the input and maintenance of transaction data (e.g. booking rules, semantic specifications, field formats, number ranges and object determination)
  - missing rules for data sovereignty (responsibility) and data maintenance
- Data processes
  - interface errors, missing extraction, transformation and loading processes
  - media breaks (different software with and without interruption of electronic processing)
  - termination and errors in the extraction, transformation and loading processes
  - missing verification, reconciliation and quality assurance processes
- Data usage
  - wrong interpretation of the provided data
  - too few resources for data analysis
  - lack of know-how for data analysis
  - poor preparation of the reports
- Software and databases
  - heterogeneous software systems
  - heterogeneous databases
  - heterogeneous field formats
- Data decay[74]
  - data latency

[71] Cf. Hildebrand (2015, p. 301).

[72] Cf. Schemm (2009, p. 20).

[73] Cf. Apel et al. (2015, p. 28).

[74] Based on Kemper et al. (2010, p. 9).

    - information updated too late
    - long loading times
    - interruption of loading processes
- analysis latency (too late analysis of the data)
- decision latency (too late decision with the help of the data)
- implementation latency (too late implementation of the decision)

The effects of poor data quality range from the lack of use, to the data quality costs incurred, to the violation of legal requirements. The lack of problem awareness and the cross-functional complexity of data maintenance result in companies investing hardly any efforts in improving data quality. Often, only striking and mostly negative experiences, which caused considerable monetary damage, release financial resources for projects to optimize data quality.

Data quality costs are distinguished into costs that are caused by poor data quality and costs that are needed to improve and ensure an adequate data quality. The costs caused by poor data quality are, for example, revenue losses, inefficiencies, wrong decisions and image losses that resulted from the poor data quality.[75] The costs to improve data quality are differentiated into prevention, detection and correction costs.[76] Similar to the general quality of products and services, a zero-defect quality is also preferred for data quality as data quality errors in early phases of the information processing process multiply by a factor in downstream phases, e.g. master data errors in units lead to negotiation errors and wrong purchases.

In addition to the data quality costs, the legal regulations play a significant role due to the increasing number of provisions. This is especially true for personal data. The data protection law aims to ensure the right to informational self-determination of a natural person.[77] Important regulations in this regard are related to the General Data Protection Regulation (GDPR), which was already presented by the EU Commission in 2012 as part of an EU data protection reform. The regulations of the GDPR do not necessarily have to be implemented into national law. They apply directly from May 25, 2018 to all member states of the EU. The GDPR is published in the Official Journal of the European Union and entered into force on May 25, 2016.[78] The GDPR replaces the EU Data Protection Directive of 1995 and the respective national laws on data protection. Thus, companies, but also other institutions, authorities, associations etc. with their domicile in the European Union face partly completely new data protection obligations while existing ones have been tightened.[79] In Germany, the new Federal Data Protection Act

[75] Cf. Apel et al. (2015, p. 41).

[76] Cf. Won and Choi (2003, p. 73 ff.).

[77] Cf. Albrecht et al. (2013, p. 506).

[78] Cf. DSGVO.expert (2021).

[79] Cf. Voigt and von dem Bussche (2018, p. 3).

(BDSG) came into force simultaneously in May 2018; this act is aligned with the GDPR and addresses the specific concretizations (so-called opening clauses) in the topics. The GDPR is based on the following principles according to Article 5:[80]

- lawfulness, good faith/fairness, transparency
- purpose limitation
- data minimization
- accuracy
- storage limitation
- integrity and confidentiality

The first principle of lawfulness and processing in good faith aims at ensuring that personal data are processed only in a lawful manner and in accordance with the principle of good faith. The term good faith is only used in the German version. In the English version, the term "fairness" can be found.[81] Lawfulness is given according to Art. 6 para. 1 GDPR if either the consent of the data subject is given or this use is based on a legal provision. To achieve this lawfulness of data processing, Art. 6 para. 1 letters a-f GDPR lists the following six conditions, of which at least one must be met for the permissible data processing:[82]

- consent
- contract
- legal obligation
- protection of vital interests
- performance of a task in the public interest
- balancing of interests

For controlling tasks, consent and contractual provisions are particularly important as the other conditions, such as a legally mandatory provision, the protection of vital interests or the performance of a task in the public interest, are less likely to apply here. Conditions for consent are regulated in Art. 7 GDPR, e.g. the controller of the data processing must be able to prove that consent exists.[83] In Art. 6 para. 1 letter b GDPR, the contractual regulation is listed as a condition. The background is that certain data, such as the name and address of the customer, are necessary for the performance of a contract (e.g. order) or for the implementation of a pre-contractual measure (e.g. offer). Transparency about data processing is another key requirement according to Art. 5 GDPR. They are

[80] Cf. Kühling et al. (2018, p. 153).

[81] Cf. Plath (2016, Art. 5 margin note 4).

[82] Cf. Kühling et al. (2018, p. 155).

[83] Cf. Albrecht and Jotzo (2017, p. 52).

specified in the corresponding recitals 39 and 58 of the GDPR. Data subjects should receive and understand all information about the processing of their personal data transparently. The essential rights of data subjects for data processing are regulated in Articles 12–20 GDPR:[84]

- Right to information (Art. 13 and 14 GDPR)
- Right of access (Art. 15 GDPR)
- Right to rectification (Art. 16 GDPR)
- Right to erasure (Art. 17 GDPR)
- Right to restriction of processing (Art. 18 GDPR)
- Obligation to notify (Art. 19 GDPR)
- Right to data portability (Art. 20 GDPR)

The purpose limitation is mentioned as another principle in Art. 5 para. 1 letter b GDPR. Data may only be collected for specified, explicit and legitimate purposes. Further processing for other purposes is not permitted. The principle of purpose limitation is followed by the principle of data minimization. The collection of personal data must be limited to what is necessary for the purposes of processing.[85] An exception arises, for example, if the data come from publicly accessible sources and a notification is disproportionate due to the high number of data subjects. This exception is particularly interesting for big data applications in which social media profiles are analyzed. The often personal data are extracted from the source system and imported into the organization's own data warehouse. The user of the social media platform is not notified of this process.[86]

Furthermore, the processed personal data must comply with the principle of accuracy. They must be factually correct and, if necessary, updated. All data that are incorrect for the purposes of processing must be deleted without delay.[87]

The principle of storage limitation entails the fact that personal data must be stored in a form that allows the identification of the data subjects only as long as it is necessary for the purpose limitation.[88] In recital 39, it is specified that the storage period must be limited to the absolutely necessary minimum. For this purpose, deadlines for the deletion or regular review of personal data are among the options.

As the last principle, integrity and confidentiality are listed. Personal data should be processed in a way that ensures adequate data security. This includes, among other things, protection against unauthorized or unlawful processing, manipulation and against

[84] Cf. Bäcker (2018, Art. 13 GDPR margin note 1) and Kühling et al. (2018, p. 264).

[85] Cf. Albrecht and Jotzo (2017, p. 52).

[86] Cf. Bartsch et al. (2015, p. 194 f.).

[87] Cf. Voigt and von dem Bussche (2018, p. 117) as well as Frenzel (2021, Art. 5 margin note 39).

[88] Cf. Frenzel (2021, Art. 5 margin note 39).

accidental loss, destruction or damage by appropriate technical or organizational measures (short TOM). In this context, the terms data protection and information security must be distinguished. The implementation of data protection is a strict legal requirement. The implementation of information security is a choice, where different concepts can be used.[89] Based on Article 32 of the GDPR, companies must take appropriate TOM to ensure an adequate level of protection. This may include, among other things, encryption, aggregation, anonymization or pseudonymization of personal data. With the help of encryption, the data are protected, for example, by passwords from unauthorized access. The publication of personal data is generally prohibited, but this can be circumvented by aggregation, anonymization or pseudonymization.[90] In aggregation, the data are no longer available at the personal data level. With the help of anonymization, identification features of personal data are deleted and possibly replaced by arbitrary strings. In this case, the personal reference can no longer be established. Pseudonymization means sequences of numbers or characters that replace the personal data. In contrast to anonymization, the identification features are not deleted, but replaced by pseudonyms.[91] The personal reference can only be restored with the help of additional information. The principle of integrity and confidentiality can also be supported by a rights and roles concept, as described in Sect. 4.2.4 "Quality assurance process", to ensure an adequate level of data protection. A rights and roles concept as the basis for an authorization system serves to ensure that only authorized persons have access to the data to which they have an access right. By assigning roles, the rights do not have to be assigned individually to each user, but to a role. The users, in turn, are assigned to a role. This makes it easier to maintain access to data and to track and check data processing operations more clearly. When distributing rights and roles, the principle of minimization of permissions, the so-called "need-to-know principle", must always be followed.

The TOM are intended to ensure not only the integrity and confidentiality, but also the availability and resilience of the systems and services. If the availability is not given, appropriate measures must be taken to restore the availability and access to the data. Companies are obliged to regularly review, evaluate and assess the effectiveness of all these TOM. In doing so, the various requirements for the data protection organization, such as the establishment of a data protection officer and the creation of an efficient data quality management, must be complied with.[92]

In order to sustainably increase the data quality and actively influence the effects of poor data quality, a data quality management should be implemented that continuously checks and corrects the errors and creates sustainable standards for optimizing the data

---

[89] Cf. Rehm (2019).

[90] Cf. Albrecht et al. (2013, p. 508 f.) and Bartsch et al. (2015, p. 191 f.).

[91] Cf. Bitkom (2013, p. 27).

[92] Cf. Voigt and von dem Bussche (2018, p. 119).

quality. Although retrospective error corrections are five to ten times more expensive than preventive measures, a problem awareness for sustainable data quality management is only slowly emerging. As part of the quality planning, it must be determined which quality standards and levels should be adhered to in the company with regard to data quality. The quality control must finally continuously monitor the following tasks with appropriate quality measures:

- Ongoing testing and control of the defined quality standards and levels in all systems, e.g. with the help of:
  - **samples**, **data validation** and **data evaluation** by employees
  - **key figures (systems)** and reports to measure the data quality starting from the data element, over the rules to be followed, up to the formation of a data quality index[93]
  - **data profiling**(tools to check the data for consistency, integrity and faulty values)
  - introduction of a **BI-Monitoring** (BI on BI), where the BI system itself checks its performance based on key figures (see Sect. 5.5.5.6).
- Continuous pointwise and systematic error correction and prevention, among others with:
  - **manual error correction** from the source system to the data target
  - **data cleansing** (tools for automatic data cleansing, where e.g. data is cleaned of duplicates. The term data cleaning is also used synonymously)
  - capture and input aids such as data scanners, transparent input masks, training on data entry etc.
- Systematic instruments for data quality assurance:
  - Creation of a regulatory framework (**data governance** and an appropriate data protection organization (including the establishment of a data protection officer).
  - Creation or further development of standards and guidelines for **data maintenance**, **data management** and **data quality monitoring processes**. This includes the important principle of avoiding possible sources of error at their origin (usually the capture of data) and not only at later control gates. The rule that applies is: “**first time right**”.[94]
  - Optimization of the **data architecture.**
  - Determination of the **master data leading systems**.
  - Definition of **responsibilities** (see below).

Three groups are distinguished for data responsibility:[95]

[93] See also Otto and Legner (2016, p. 552).

[94] Cf. Apel et al. (2015, p. 12).

[95] See also Otto and Legner (2016, p. 553).

- The **data owner** owns the responsibility for the quality of his data area, e.g. the procurement manager for the master data of the purchasing materials.
- The **data user** uses the data and sets the requirements for the data quality for his application area. He supports the data quality management in checking the data and reports anomalies in his application area.
- The **data steward** defines the quality of the data to be maintained across the board and sets rules and a framework for compliance. He initiates the selection and use of tools and instruments to improve the data quality. It is advisable to appoint a data quality officer in the company. Alternatively, the controlling institution can act as a coordinating body here.

Ideally, the errors should be tackled at the root or source ("first time right"), whereby the data-leading source systems are usually affected. The quality management includes, in addition to the planning and control of the data quality, cross-cutting management tasks such as the definition of the quality organization, including the determination of the responsibilities and the communication of quality objectives. In view of the many different systems and data that a company uses today, data quality management reaches its limits. While the quality of the data of the own systems (ERP, CRM etc.) can still be relatively well monitored and improved by the established instruments, this possibility decreases significantly for external data such as community data and big data. Here, control gates help from a certain level to subject the captured data to an appropriate data quality assurance.

4.2.4.5 Information System Quality Testing and Control

From the view of the software manufacturer, regular quality assurance should be carried out within the framework of the **support and maintenance contracts** to guarantee a constant further development of the programs to the latest technical standard and the latest content-related progress. The phase of the new implementation concludes again with tests and acceptances of the respective innovations as a quality-assuring measure (see also Sect. 4.2.1.6). Especially through the implementation of new content and functions, the experience from corporate practice shows again and again that proven functions suddenly no longer work. Therefore, the tests and acceptances should not only focus on the "new" content and functions, but also pay attention to the usability of "old" functionalities.

Errors should be quickly fixed, and new requirements should be recorded and tracked in product management via the hotline and other communication channels.

At the company's end, defects should be continuously queried, reported and recorded by the users and by the centrally coordinating body, e.g. in Controlling. Due to many technical inquiries, the IT department is particularly brought in here.

At regular intervals, e.g. once a year, a review of the IT systems used should be carried out with regard to their quality. Here, possibilities for improvement, e.g. by integrating the systems through defect elimination or by expanding new modules, should be shown. Short-term and operational measures are to be agreed with the IT department and included in the project plan of the company and the affected departments (IT, controlling, etc.).

In order to take up new impulses and advances of the software manufacturers and to check their suitability for use in the company, information about them should be obtained regularly. As sources, professional conferences, congresses, trade fairs (e.g. Tech Summits), internet research, information events of the software manufacturers, professional journals, but also the exchange with other companies, associations and social networks can be considered.

Important quality criteria for the software are:[96]

- **Performance and efficiency**: The software should show an appropriate level of performance in relation to the resources used under defined conditions.
- **Reliability**: The software should guarantee a specified level of performance when used under normal conditions.
- **Usability**: The software should be easily understood and used by the user.
- **Security**: Information and data should be protected so that unauthorized users cannot read or manipulate them.
- **Maintainability**: The software should be maintainable and changeable.
- **Portability**: The software should be usable independently of location.
- **Evolvability**: The software should be adaptable to changed requirements and technology.

If the existing IT landscape no longer meets the content requirements, a change in the entire IT strategy should be considered, in which, for example, complete ERP systems up to the evaluation tools are questioned. The cycle of an IT strategy with the replacement of central software and hardware components covers a longer period of e.g. 5–10 years.

If individual software and hardware components or the entire IT landscape are to be replaced, an IT selection process follows (see Sect. 4.2.1.5).

References

Albrecht, J., et al. 2013. Anwendung. In *Data-Warehouse-Systeme: Architektur, Entwicklung, Anwendung*, Hrsg. A. Bauer and H. Günzel, 4th edn. 373–608. Heidelberg: dpunkt.

Albrecht, J. P. and F. Jotzo. 2017. *Das neue Datenschutzrecht der EU*. Baden-Baden: Nomos.

Andreae, C. 2007. *Familienunternehmen und Publikumsgesellschaft*. Wiesbaden: Springer.

Apel, D., et al. 2015. *Datenqualität erfolgreich steuern*, 3rd edn. Heidelberg: dpunkt.

[96] Cf. Balzert (2011, p. 109–113).

Atacama. 2011. Produkte. http://www.ataccama.de/produkte/dq-analyzer/dqa-uberblick.html. Accessed 22 July 2011.

Bäcker, M. 2018. Art. 13 DSGVO Randnotiz 1. In *Datenschutz-Grundverordnung/BDSG,* Hrsg. J. Kühling and B. Buchner, 2nd edn. Munich: Beck.

Balzert, H. 2011. *Lehrbuch der Softwaretechnik: Entwurf, Implementierung, Installation und Betrieb*, 3rd edn. Heidelberg: Spektrum Akademischer.

Bange, C. 2014. Business Intelligence im Self-Service hat Konjunktur. *IS-Report* 18(8): 28–29.

BARC. 2017. BI-trend monitor 2017. http://barc.de/trend-monitor?s=s_9_550. Accessed 12 June 2017.

Bartsch, M., et al. 2015. Recht. In *Praxishandbuch Big Data: Wirtschaft – Recht – Technik*, ed. J. Dorschel, 167–251. Wiesbaden: Springer Gabler.

Becker, A., J. Leyk and L. Riemer. 2013. Dynamische Unternehmenssteuerung am Beispiel von Beyer MaterialScience. In *Moderne Instrumente der Planung und Budgetierung*, eds. R. Gleich, S. Gänßlein, M. Kappes, U. Kraus, J. Leyk and M. Tschandl, 121–142. Munich: Haufe.

Becker, W. and P. Ulrich. 2009. Mittelstand, KMU und Familienunternehmen in der Betriebswirtschaftslehre. *WiSt* 38(1): 2–7.

Becker, W. and P. Ulrich. 2011. *Mittelstandsforschung in Deutschland: Begriffe, Relevanz und Konsequenzen*. Stuttgart: Kohlhammer.

Becker, W., M. Staffel and P. Ulrich. 2009. Wissensmanagement als Instrument der strategischen Führung im Mittelstand – Konzepte, Modifikationen und Empfehlungen. In *Management-Instrumente in kleinen und mittleren Unternehmen: Jahrbuch der KMU-Forschung und -Praxis 2009*, ed. J. Meyer, 257–282. Lohmar: Eul-Verlag.

Becker, W., P. Ulrich and T. Botzkowski. 2016. *Data Analytics im Mittelstand*. Wiesbaden: Springer Gabler.

Bitkom. 2013. Management von Big-Data-Projekten: Leitfaden. Berlin. https://www.bitkom.org/noindex/Publikationen/2013/Leitfaden/Managementvon-Big-Data-Projekten/130618-Management-von-Big-Data-Projekten.pdf. Accessed 15 Jan. 2017.

Blohm, H. 1974. *Die Gestaltung des betrieblichen Berichtswesens als Problem der Leitungsorganisation*, 2nd edn. Herne: NWB.

Coenenberg, A. G. 2000. *Jahresabschluss- und Jahresabschlussanalyse*, 17th edn. Stuttgart: Moderne Industrie.

Coenenberg, A. G. and R. Salfeld. 2003. *Wertorientierte Unternehmensführung*. Stuttgart: Moderne Industrie.

Copeland, T., T. Koller and J. Murrin. 2002. *Unternehmenswert – Methoden und Strategien für eine wertorientierte Unternehmensführung*, 3rd edn. Frankfurt a. M.: Campus.

Davenport, T. H. 2014. *Big data at work: Dispelling the myths, uncovering the opportunities*. Boston: Harvard Business School.

Dittmar, C. and T. Schmidt. 2014. Self-Service BI benötigt eine ganz spezielle Governance. *IS-Report* 18(8): 24–27.

Dräther, R., H. Koschek and C. Sahling. 2013. *Scrum – kurz & gut*. Cologne: O'Reilly.

DSGVO.expert. 2021. *Expertenwissen zur Europäischen-Datenschutzgrundverordnung*. https://dsgvo.expert/die-datenschutzgrundverordnung/entstehungsgeschichte/. Accessed 15 Febr. 2021.

Eisele, D. and T. Doyé. 2010. *Praxisorientierte Personalwirtschaftslehre. Wertschöpfungskette Personal*, 7th edn. Stuttgart: Kohlhammer.

EU Kommission. 2003. Commission Recommendation of 6 May 2003 concerning the definition of micro, small and medium-sized enterprises (notified under document number K(2003) 1422) (2003/361/EC), which since 1 January 2005, replaces the Recommendation (96/280/EC) in force until then. Official Journal of the European Union, Brüssel.

Feldbauer-Durstmüller, B. and B. Wimmer. 2008. Familienunternehmen und Controlling – Ergebnisse der empirischen Studie: Familienunternehmen in Oberösterreich. In *Familienunternehmen*, eds. B. Feldbauer-Durstmüller, H. Pernsteiner, R. Rohatschek and M. Tumpel, 31–52. Vienna: Linde.

Finucane, B. and M. Mack. 2011. Business-Intelligence-Software boomt in Deutschland. In *is report, Informationsplattform für Business Applications, BARC-Guide Business Intelligence*, 8–11. Dachau.

Foegen, M. 2014. *Der Ultimative Scrum Guide 2.0*. Darmstadt: wibas GmbH.

Freese, E. 2005. *Grundlagen der Organisation*, 9th edn. Wiesbaden: Springer Gabler.

Frenzel, E. M. 2021. Datenschutz-Grundverordnung. In *Beck'sche Kompakt-Kommentare*, eds. B. P. Paal and D. A. Pauly, 3rd edn. Munich: C.H. Beck.

Friedrich, D. 2008. Business Intelligence etabliert sich im Mittelstand, BARC-Guide Business Intelligence, S. 10–11. http://www.barc.de/fileadmin/Fachartikel/10-11_FB_BI-Guide.pdf. Accessed 19 Febr. 2011.

Gebauer, M. and U. Windheuser. 2015. Strukturierte Datenanalyse, Profiling und Geschäftsregeln. In *Daten- und Informationsqualität*, eds. K. Hildebrand et al., 3rd edn., 87–100. Wiesbaden: Springer Vieweg.

Gleich, R. and P. Temmel. 2008. Die Rolle der Organisation des Controllings im Management Reporting. In *Management-Reporting – Grundlagen, Praxis und Perspektiven*, eds. P. Horváth, R. Gleich and U. Michel, 63–91. Munich: Haufe.

Gluchowski, P. 2014. Aktuelle Trends in der Business Intelligence. *Controlling* 26:235–243.

Gluchowski, P. 2017. Die Big-Data-Diskussion verhilft Business Intelligence zum Höhenflug. *isreport* 2017(Special edition August): 7–10.

Gräf, J. and D. Nase. 2008. Objekte des Management Reportings im Überblick. In *Management-Reporting – Grundlagen, Praxis und Perspektiven*, eds. P. Horváth, R. Gleich and U. Michel, 43–61. Munich: Haufe.

Hildebrand, K. 2015. *Master Data Life Cycle – Stammdatenprozess am Beispiel Materialstamm in SAP*. Wiesbaden: Vieweg+Teubner.

Hinrichs, H. 2002. Datenqualitätsmanagement in Data Warehouse-Systemen. Diss. Oldenburg.

Hoffjan, A. 2008. Comparative management accounting. *Controlling* 20:655–660.

Horváth, P. 2008. Grundlagen des Management-Reportings. In *Management-Reporting – Grundlagen, Praxis und Perspektiven*, eds. P. Horváht, R. Gleich and U. Michel, 15–42. Munich: Haufe.

Horváth, P. and A. Aschenbrücker. 2014. Data Scientist: Konkurrenz oder Katalysator für den Controller? In *Der Controlling-Berater – Controlling und Big Data*, eds. R. Gleich and A. Klein, Vol. 35, 47–62. Freiburg: Haufe Verlag.

Hüllmann, U. 2003. *Wertorientiertes Controlling für eine Management-Holding*. Munich: Vahlen.

IfM. 2006. MIND 2006. Der Mittelstand in Deutschland: Das Rückrad der Wirtschaft – Kleine Unternehmen und Dienstleister prägen den Mittelstand – Mittelständler blicken optimistisch in die Zukunft. http://www.impulse.de/unternehmen/:Zahlen-und-Fakten%2D%2DMIND-Downloads/265749.html. Accessed 5 Nov. 2009.

Janssen, J. 2008. Rechnungslegung im Mittelstand, Diss. TU Dortmund.

Kemper, H. G., H. Baars and W. Mehanna. 2010. *Business Intelligence – Grundlagen und praktische Anwendungen*, 3rd edn. Wiesbaden: Springer Vieweg.

Knöll, H. D., et al. 2006. *Unternehmensführung mit SAP BI*. Wiesbaden: Springer Vieweg.

Koch, R. 1994. *Betriebliches Berichtswesen als Informations- und Steuerungsinstrument*. Frankfurt a. M.: Peter Lang.

Kühling, J., M. Klar and F. Sackmann. 2018. *Datenschutzrecht*, 4th edn. Heidelberg: C. F. Müller.

Lerch, V. 2008. *Konzept einer Modellfabrik für integrierte Business Intelligence im Mittelstand – Theorie und Anwendung bei einem mittelständischen Unternehmen der Fertigungsindustrie.* Inauguraldissertation, Universität Mannheim.

Leser, U. and F. Naumann. 2007. *Informationsintegration.* Heidelberg: dpunkt.

Martin, W. 2013. *Business Intelligence trifft Business Process Management und Big Data.* http://www.wolfgang-martin-team.net/BI-BPM-SOA_dt.php. Accessed 14 May 2016.

Naumann, F. 2007. Datenqualität. *Informatik-Spektrum* 30(1): 29.

Ossadnik, W., D. Barklage and E. van Lengerich. 2010. *Controlling mittelständischer Unternehmen.* Berlin: Physica.

Otto, B. and C. Legner. 2016. Datenqualitätsmanagement für den Industriebetrieb. *Controlling* 28(10): 550–557.

Pfohl, H. C. 2006. Abgrenzung der Klein- und Mittelbetriebe von Großbetrieben. In *Betriebswirtschaftslehre der Mittel- und Kleinbetriebe*, eds. H. C. Pfohl, 4th edn., 331–355. Berlin: Erich Schmidt.

Pichler, R. 2009. *Scrum – Agiles Projektmanagement erfolgreich einsetzen*, 20–23. Heidelberg: dpunkt.

Plath, K.-U. 2016. *BDSG/DSGVO. Kommentar zum BDSG und zur DSGVO sowie den Datenschutzbestimmungen des TMG und TKG*, 2nd edn. Cologne: Otto Schmidt.

Rappaport, A. 1986. *Creating shareholder value – The new standard for business performance.* New York: Free Press.

Rautenstrauch, T. 2005. Balanced Scorecard in mittelständischen Unternehmen: Empirische Ergebnisse und Implikationen. In *Einsatz von Controllinginstrumenten im Mittelstand*, Konferenz Mittelstandscontrolling 2005 – TU Kaiserslautern, ed. V. Lingnau, 1–17. Lohmar: Josef Eul-Verlag.

Rehm, S.-M. 2019. *Was unterscheidet Datenschutz von Informationssicherheit.* https://www.haufe.de/compliance/management-praxis/was-unterscheidet-datenschutz-von-informationssicherheit_230130_482568.html. Accessed 18 May 2021.

Reichmann, T. 2008. Die systemgestützte Controlling-Konzeption in Theorie und Praxis. *Controlling* 20:689–700.

Reichmann, T. 2011. *Controlling mit Kennzahlen. Die systemgestützte Controlling-Konzeption mit Analyse- und Reportinginstrumenten*, 8th edn. Munich: Vahlen.

Reinemann, H. 1999. Was ist Mittelstand? Zur Definition der kleinen und mittleren Unternehmen. *WiSt* 28(12): 661–662.

Rheinhardt, R., D. Kilian, B. Kirschner and W. Moriel. 2007. *Wettbewerbsfähigkeit kleiner und mittlerer Unternehmen in Österreich, eine empirische Studie.* Innsbruck: Studia.

Sanders, D. and N. Schafft. 2014. Implementierung eines globalen Berichtswesens. In *Controlling in der Konsumgüterindustrie. Innovative Ansätze und Praxisbeispiele*, eds. M. Buttkus and R. Eberenz, 293–312. Wiesbaden: Springer Gabler.

Schemm, J. W. 2009. *Zwischenbetriebliches Stammdatenmanagement.* Berlin/Heidelberg: Springer.

Schlegel, H. B. 1996. *Computergestützte Unternehmensplanung und -kontrolle.* Munich: Vahlen.

Schmidt, R. 2020. Die drei Erfolgsfaktoren für eine erfolgreiche Data-, BI- & Analytics-Strategie. In BARC Guide Data, BI & Analytics 2020, 12–13. Würzburg: BARC GmbH.

Schön, D. 2004. Moderne Planungskonzepte und Reportingtools. In *19. Deutscher Controlling Congress, Tagungsband*, ed. T. Reichmann, 287–337. Dortmund: Gesellschaft für Controlling e.V.

Schön, D. 2011. Ergebnisse zur empirischen Untersuchung – Business Intelligence für Reporting und Planung im Mittelstand, 1–47. http://www.fh-dortmund.de/de/studi/fb/9/personen/lehr/schdie/103020100000206873.php. Accessed 1 June 2011.

Schön, D. and Diamant Software. 2009. *Moderne Anforderungen und Trends im Finanz- und Rechnungswesen sowie im Controlling – Ergebnisse der empirischen Studie*. Dortmund/Bielfeld: Diamant Software GmbH.

Schön, D. and H. Krause. 1997. DV-gestütztes Krankenhaus-Controlling mit Hilfe von Standard-Software. DV-gestützte und controllinggerechte Kosten- und Leistungsrechnung mit Hilfe von Standard-Software. In *Controlling im Krankenhaus. Ein Handbuch für alle Führungskräfte im Krankenhaus*, ed. E. Hauke, 1–70. Vienna: Ueberreuter. Supplementary delivery 3, Section 9.

Schön, D. and R. Müller. 2010. Mittelstandscontrolling für Inhaber und Manager. In *25. Deutscher Controlling Congress, Tagungsband*, ed. T. Reichmann, 123–165. Dortmund: Gesellschaft für Controlling e.V.

Schulte, C. 2002. *Personal-Controlling mit Kennzahlen*, 2nd edn. Munich: Vahlen.

Schulte-Mattler, H. and T. Manns. 2010. Bedeutung des regulatorischen und ökonomischen Eigenkapitals für das Risikomanagement der Banken. In *Risikomanagement und Frühwarnverfahren*, eds. U. Bantleon and A. Becker, 83–126. Stuttgart: Erich Schmidt.

Schwaber, K. and J. Sutherland. 2013. *Software in 30 days*. Heidelberg: Published Online.

Schweinsberg, K., et al. 2006. Persönlichkeiten – Was macht den Mittelständler aus? In *Praxishandbuch des Mittelstands – Leitfaden für das Management Mittelständischer Unternehmen*, ed. W. Krüger, 63–70. Wiesbaden: Gabler.

Simon, H. 1992. Lessons from Germany's midsize giants. *Harvard Business Review* 70:115–123.

Stiefl, J. and K. von Westerholt. 2007. *Wertorientiertes Management*. Munich: Oldenbourg.

Strauch, B. and R. Winter. 2002. Vorgehensmodell für die Informationsbedarfsanalyse im Data Warehousing. In *Vom Data Warehouse zum Corporate Knowledge Center*, eds. E. Maur and R. Winter, 359–378. Heidelberg: Springer.

Taschner, A. 2013. *Management Reporting. Erfolgsfaktor internes Berichtswesen*. Wiesbaden: Springer Gabler.

Voigt, P. and A. von dem Bussche. 2018. *EU-Datenschutz-Grundverordnung (DSGVO)*, 3. Berlin: Springer.

Wahls, J. 1993. *Unternehmensplanung mit Excel*. Munich: Vahlen.

Wallau, F. 2008. *Institut für Mittelstandsforschung Bonn: Das familiengeführte Unternehmen, Expertenforum Familienunternehmen*. Cologne: Institut für Mittelstandsforschung Bonn.

Weber, J., C. Hunold, C. Prenzler and S. Thust. 2001. *Controllerorganisation in deutschen Unternehmen*, series of publications Advanced Controlling, Vol. 18, 8. Vallendar: Wiley.

Weber, J., B. Hirsch, R. Rambusch, H. Schlüter, F. Sill and A. Spatz. 2006. *Controlling 2006 – Stand und Perspektiven*. Valendar: Wiley.

Weber, J., R. Malz and T. Lührmann. 2008. Excellence im Management-Reporting. In *series of publications Advanced Controlling*, ed. J. Weber, Vol. 22, 25, 50, 62. Weinheim: Wiley.

Weber, J., P. Nevries, D. Breiter, et al. 2009. *Operative Planung*, series of publications Advanced Controlling, Vol. 71, 30. Vallendar: Wiley.

Won, K. and B. Choi. 2003. Towards quantify data quality costs. *Journal of Objects Technology* 2(4): 69–76.

Würthele, V. G. 2003. Datenqualitätsmetrik für Informationsprozesse. Diss. Zürich.

Zeh, T. 2013. Datenqualität. In *Data-Warehouse-Systeme: Architektur, Entwicklung, Anwendung*, eds. A. Bauer and H. Günzel, 4th edn. 49–52. Heidelberg: dpunkt.

Ziegenbein, K. 2012. Controlling. In *Kompendium der praktischen Betriebswirtschaft*, ed. K. Ollfert, 10th edn. Herne: NWB.

IT Support

5

Contents

D. Schön, *Planning and Reporting in BI-supported Controlling*,
https://doi.org/10.1007/978-3-658-41044-5_5

5.1 Operational IT Systems

For business management, and especially for planning and reporting, an effective information supply is needed. Because of the large amounts of data to be processed in planning and reporting, as well as the complexity of the analysis and planning processes, it makes sense to map planning and reporting with the help of IT systems.

The IT-supported implementation of planning and reporting is a human-machine system in which the capabilities of the analyst or planner and the computational possibilities of the computer become an integrated part of the analysis or planning process in an online dialogue.[1] The computer is characterized mainly by its high processing speed and its storage capacity of large amounts of data, while the human abilities are more of a creative and analytical nature.

Following the multidimensional controlling concept of Reichmann, the level for planning and reporting is placed above the recording, disposition and accounting systems.[2] The most important internal information base for controlling is the operational internal and external accounting. For a comprehensive planning and a holistic reporting, many other sources of the recording, disposition and accounting system levels are also needed, such as the payroll for the personnel cost planning or the inventory management for the planning of the logistics and production capacities (see Fig. 5.1). In addition to these internal sources, external information sources are increasingly used to obtain current information, such as exchange rates and price indices for materials. With big data, further data sources for information supply in management become more important, such as web logs from the internet for sales or RFID sensor data in trade and production.

[1] Cf. Mertens and Griese (1988, p. 5).

[2] Cf. Reichmann (2011, p. 18).

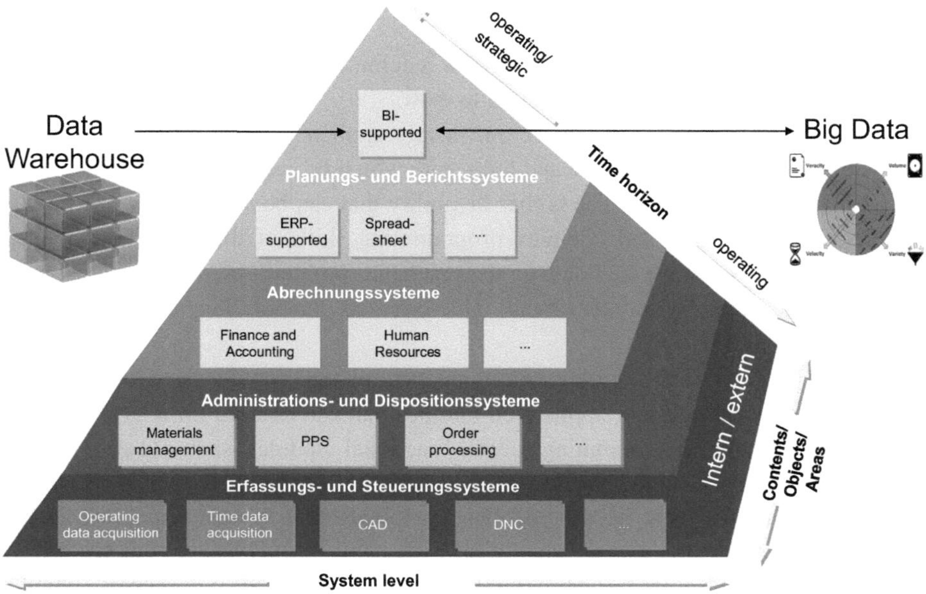

Fig. 5.1 Dimensional levels of operational IT-systems

While relational database systems are usually used for the classical reporting and planning functions of the ERP systems or for the special reporting and planning systems, a uniform database in the form of a data warehouse is introduced as a redundant data source for a traditional business intelligence-supported controlling system. With the core data warehouse, one wants to create a so-called single point of truth (SPoT) as an information source that is centrally used for control purposes. With the introduction of big data technology, this objective is diluted. As will be described in detail in Sects. 5.5, 5.6, 5.7, 5.8 and 5.9, traditional data warehouse and big data technology complement each other to a logical single point of truth, in which the optimal technology for the respective information processing for control support is used.

Since an examination of individual solutions for planning and reporting is not possible due to their diversity, only potential basic orientations will be shown here.

For the respective tasks of planning and reporting, either individual or standard software solutions can be used in principle.[3] Custom solutions are characterized by their individual (own or external) programming, so that especially special requirements of the companies can be taken into account. Standard software is defined as an application software that is offered by software manufacturers for an anonymous market to solve e.g.

[3] Cf. Reichmann (2006, p. 662).

business problems. Standard software is classified into special systems, application languages and standard software families.[4]

Special systems are written specifically for a delimited task area.[5] They are characterized by a high functional and technical know-how.

The language-oriented systems are universally applicable programs that provide the user with a "toolbox" of functions and tools with which he can develop a system to solve his specific problem himself. The higher flexibility that can be achieved by this requires, however, partly detailed and complex programming skills of the user. The language-oriented standard software includes, on the one hand, problem-oriented higher programming languages and, on the other hand, table-oriented planning languages, such as e.g. spreadsheet programs.

Standard software families are generally understood as comprehensive integrated software solutions in the sense of application families that can be used for a larger e.g. business application area. By combining several functional modules, such as sales, materials management, production and finance, a standard software family is created. Many ERP systems are classified here because of their combination of several functional modules (see Sect. 5.4.2). The modular structure allows the compilation of numerous specific systems that have the advantage of being based on a uniform database; thus cross-functional functions can be better and more easily coordinated and aligned.[6] However, since many companies have special requirements for the mapping of their business processes, the standard software has to be adapted to them. This adaptation is done in the standard software programs in specially designated functional areas. This adaptation process is also called parameterizing or customizing.[7]

Table 5.1 lists the advantages and disadvantages of the custom and standard software.

Nowadays, the advantages of standard software systems often outweigh, especially when it comes to more complex tasks of mapping business processes. The problem here is rather to find out, from the multitude of offered programs, the optimal program or the optimal program combination for one's own company. In addition, standard programs should be designed as open systems, so that adaptations to company-specific needs can be mapped with the help of the tools provided and integrated into the existing or to be selected standard system. Individual solutions are only an option if special industry- or company-specific requirements are not met by the standard applications.

Except for some well-structured reports, such as the legally required balance sheet or the income statement, there are no limits to creativity and design for other reports. Therefore, both ERP and special reporting and planning software providers offer tools in their systems that can be used individually for a flexible design of reports and planning

[4] Cf. Scheer (1990, p. 139).

[5] See Weber and Strüngmann (1997, pp. 30–36).

[6] Cf. Scheer (1990, pp. 37, 142 and 153).

[7] Cf. Hesseler and Görtz (2007, pp. 7 f. and 15 ff.).

Table 5.1 Advantages and disadvantages of individually developed and standardized software systems

	Disadvantages	Advantages
Custom software	• Personnel- and capital-intensive (often own programming capacities are lacking) • No guarantee of successful development • Integration capability to downstream and upstream systems must be ensured by oneself • Complexity of requirements usually not manageable and maintainable by oneself • Dependence on individual programmers or programming knowledge	• Flexible further development possible (if corresponding programming resources are available) • Independence from software manufacturers • Individual special requirements can be taken into account
Standard software	• Dependence on software manufacturers • Individual special requirements that go beyond customizing can only be considered for open systems	• Years of experience with various industry requirements can be used • Certain company requirements can be covered by customizing for the intended standard • Integration flexibility to other systems • Less personnel- and capital-intensive • Further development usually ensured by the software manufacturer

forms. For orientation, the manufacturers often provide sample reports and templates for reporting and planning; these can be used e.g. as a copy template.

Before examining the individual basic orientations of potential software solutions, a brief description of the historical development of planning and reporting systems and, in another chapter, a brief overview of the necessary hardware components for planning and reporting will be given first.

5.2 History

IT-supported operational reporting and planning has existed as long as there have been computers. In the early days of data processing, the main focus for operational use was on operational reporting in the form of reports and lists. Since there is already a lot of information in the literature[8] on the historical development of IT-supported reporting and planning solutions, only a brief outline of the history will be shown here.

[8] Cf. on the historical development of reporting and planning solutions, among others, Laudon and Laudon (1988), Gluchowski et al. (2008, p. 55 ff.), Oppelt (1995), Schinzer (1996), Mertens and Griese (2002), Oehler (2006) and Chamoni and Gluchowski (2004, p. 5 ff.).

Application systems, which have the task of supporting specialists and managers through information and communication systems, are summarized under the umbrella term management support systems (MSS).

Under the term management information systems (MIS), the first MSS were developed in the 1960s and created the possibility of information generation by simple reports from the operational data.

In the 1970s, the term decision support system (DSS) (see also Sect. 5.5.5.3) emerged. In contrast to MIS, DSS no longer focused on the pure data supply of specialists and managers, but on the effective support in the planning and decision-making process.

After that, the term executive information system (EIS) emerged in the USA in the mid-1980s. In Germany, the alternative terms management information system (LIS), chief information system (CIS) or board information system (BIS) were used. Due to the technological progress, it became possible to obtain new forms of presentation and information access. These management information systems usually had a user-friendly interface and could be operated without much IT knowledge. They gave users for the first time comfortable evaluation possibilities regarding important operational information.

Since the different terms of MSS are often applied arbitrarily in practice, a precise distinction is difficult. The MSS approaches all failed in the past, among other things, due to the following missing prerequisites:

- powerful and cost-effective processors
- fast, cost-effective and sufficient data storage
- graphical user interfaces
- large databases through integrated operational systems
- fast and comprehensive communication technologies.

With technological progress, the third generation of decision-oriented information systems, which are based on the concepts of data warehousing, finally began to spread in the 1990s. While the predecessor systems, e.g. the ERP systems, are based on On-Line Transaction Processing (OLTP), the data warehouse technology uses the so-called On-Line Analytical Processing (OLAP), which is explained in detail in Sect. 5.5.

The data warehouse technology forms the technical basis for business intelligence and BI-supported controlling. These terms are defined in Sect. 5.6.

Section 5.7 examines the possibilities of using artificial intelligence for controlling with the help of modern planning and reporting systems. Among other things, expert systems, data mining and predictive analytics are presented as pragmatic solutions of a specific (weak) artificial intelligence. With the internet and further technological developments, such as RFID technology with its near-field communication (NFC), more and more data could be generated which can be used in management in addition to the internal company information. For the processing of such huge and mostly unstructured data volumes, big data technology was developed, at the end of the first decade in the new millennium, to an extent that it became interesting for operational use in planning and reporting. Technical foundations, application areas and integration aspects of big data to the traditional data warehouse are explained in Sect. 5.8. They form the basis for exploratory business intelligence and business analytics.

The handling of exploratory data volumes and the desire for the greatest possible availability of information led, in the new millennium, to the further development of distributing data in a cloud for end devices. The various possibilities of cloud computing, as well as their advantages and disadvantages, are presented in Sect. 5.9.

With the technical development of mobile devices such as smartphones and tablets in the last 20 years, the mobile requirements for planning and reporting also increase. The development and technical peculiarities of mobile computing are dealt with in the final Sect. 5.10.

5.3 Hardware and Network

Reporting and planning systems require, like other software programs, coordinated hardware conditions for optimal program use. For this reason, it is advisable to implement the hardware components, operating systems and database systems recommended by the manufacturers. As a rule, several options are offered for the selection of hardware components, operating systems and database systems which are available for the application. Since the recommendations vary greatly from software application to software application, a detailed list is omitted here.

The performance of the servers on which the application and database application run should be calculated and sized sufficiently large by means of a so-called "**Sizing**" of the manufacturer. This is intended to ensure that extensions in reporting and planning are possible without costly retrofitting and that the software's response time (the so-called "**Performance**") is acceptable. For the sizing, various usage parameters, such as the number of users and the expected data sets, are used as the basis for calculation. The procedure is again very different from manufacturer to manufacturer.

Reporting and planning systems should be available in the company's network (e.g. Local Area Network or Wide Area Network) and beyond, like other programs. Here, distributed computer concepts are possible, such as the Client-Server-Model, in which application and database management can be located on different servers and the usage level (client) is set up on the personal computer. In web-based applications, the client communicates via so-called web services with the applications, which can be located on one or distributed on several web servers, a structure which corresponds to a service-oriented architecture. An economic advantage of web-based systems is that no additional client software has to be installed on the desktop and thus less license costs are incurred. In addition, the World Wide Web is available in large parts of the world.

Personal computers, desktops, laptops/notebooks and mobile devices of different sizes (tablet-PC, smartphones etc.) are possible end devices for reporting and planning applications (see also Sects. 5.5.5.7 and 5.8). From the input point of view, these devices can either be operated with keyboard and mouse or via a touchscreen.

Important peripheral devices for reporting and planning are the screens of the PCs and other output devices such as data projector, TV sets and interactive whiteboards for meetings, chats and conferences, as well as printers for paper reports. For larger presentations, TV screens or a datawall for visualization are also suitable.

In addition to numerous internal data sources and applications, the networking of various servers with each other allows the use of numerous information sources from external servers. This results in a cross-company integration of heterogeneous data sources and applications that are partly available worldwide and can usually be accessed via the internet or the company's intranet. With regard to the storage and use of data via the internet, the development of cloud technology is also to be observed (see Sect. 5.10). If it succeeds in ensuring the supply of data securely and stably, then this is an interesting option for the availability of data and applications.

5.4 Software Solutions for Reporting and Planning

For the IT-based mapping of planning and reporting in the company, there are currently four potential basic orientations:

- planning and reporting functions of ERP systems
- spreadsheet programs
- specialized planning and reporting systems based on relational database technology
- BI-based systems with data warehouse and big data technology

The examination of the different basic orientations of planning and reporting systems should be based on various criteria of requirements, something which will be established in the following chapter.

In contrast to the often separate presentation of planning and reporting solutions in practice, the focus here should be on the interlocking of reporting and planning. Often the statements apply to both areas of responsibility because of their many similarities, so an integrated view is sensible. However, special features of the respective area of responsibility of planning and reporting will be highlighted (see also Sect. 3.10).

5.4.1 Criteria of Requirements

There are numerous criteria of requirements for modern and powerful reporting and planning solutions.[9] Some requirements are difficult to compare with regard to the basic

[9] Cf. Wild (1981, p. 36 ff.) or Dahnken et al. (2003, p. 52 ff.).

orientations for reporting and planning systems, so they should not be in the focus of further consideration. These include areas of requirements such as:

- internationality
- currency capability
- authorization systems
- security
- transferability to another technical environment
- training
- maintenance/Service
- multiuser capability
- …

Also excluded are provider-related criteria, such as costs, support, availability, consulting, reputation, creditworthiness and network.[10]

For a directional comparison of the basic orientations listed for reporting and planning systems, the following requirement criteria, which are shaped by content and data, should be used to later analyze the advantages and disadvantages of data warehouse and BI-supported systems compared to ERP systems and spreadsheet programs based on them:[11]

Data connection to upstream and downstream systems
The integration of data sources from different upstream and downstream systems for the reporting and planning system must be ensured. Furthermore, the different internal and external data sources of a company require comfortable import and export interfaces.

Data modeling, harmonization and quality
Through the integration of different data sources for the reporting and planning system, the data must be harmonized as they may be present in different forms and descriptions in the various source systems. If no explicit data model is used to which the data must be adapted, a separate data modeling is also necessary. In addition, new master and transaction data may have to be created for planning. The data quality is one, if not the most important criterion for the acceptance of a planning and reporting system. Due to the

[10] Cf. for further software selection criteria e.g. Becker et al. (2011, p. 16).

[11] The requirement criteria listed supplement the quality characteristics according to ISO 9126 (version until 2005) and the successor standard ISO/IEC 25000. However, they are not defined from a technical point of view (e.g. security and portability) or manufacturer's point of view (e.g. maintenance and service). They rather serve the analysis and comparability of different software directions for planning and reporting solutions.

integration of different data sources and the enrichment of the data stock, inconsistencies occur again and again which are to be avoided.

Speed

The report creation and display as well as the process of planning with the input and determination of the planning results should be easy and fast to perform. Here, special attention should be paid to the calculation speed and the runtime behavior (performance). The system must be available to the user at all times and be usable from the (possibly flexible) workplace immediately and without delay.

Economic efficiency

The benefit of reporting and planning (especially the analysis, design and control functions) must be in an economic relation to the costs incurred for it (e.g. software acquisition, implementation, execution costs).

User-friendliness

Regardless of the functionality of an IT program, the system acceptance, and thus the acceptance of a reporting and planning system, depends to a large extent on the user-friendliness, the ergonomics and the operating comfort. From the user's point of view, well-arranged, clearly and uniformly structured graphical screen interfaces and easy-to-use input options are indispensable for working.

In addition to the keyboard and the function key control, graphical elements such as buttons, selection boxes, tabs or menu bars can be used today per mouse click to execute certain commands and functions or to access reports and planning forms.

The structured screen layout has prevailed in the design of the screen interface, in which certain screen areas are reserved for toolbars, function bars, dialog inputs, messages and news, and other tools. In addition, more and more system-guided interaction techniques are being used, whereby the user receives hints and assistance for the function performed; he is guided step by step through the program.

Meaningful, easy-to-understand documentation and online-help systems complete user-friendly systems.

Analysis and planning functionality

The users need know-how of business models and methods for reporting and planning. For example, the salesperson needs regression analyses regarding the sales development, while in cost accounting a contribution margin calculation must be mapped. In strategic planning, portfolio analyses are used, in purchasing ABC analyses. While standardized key figure and reporting systems are provided for reporting, flexible mathematical and statistical calculation operations, for example in the form of a formula generator, are useful for mapping differentiated planning structures. The variety of reporting and planning functions (e.g. drill-down, aggregations, extrapolations, consolidation, scenario calculations, simulations) which the systems should provide can only be indicated here (see in particular Sects. 3.7 and 3.8).

Since different industries differ mainly in their product range, the value creation process and the production factors used, individual analysis and planning functions may have to be mapped for the industries.

Flexibility and design options

The reporting and planning structures and the associated objects (e.g. areas, products, markets, companies and SBUs) must be constantly updated and adapted to the changed situations over time.

The selection and creation of model and method banks as well as the data structuring should be designed as flexibly as possible. The data basis must be easy to handle, adaptable and analyzable according to different criteria. Layouts are to be developed for the reports and the planning forms, which can be flexibly adapted.

Documentation support

Due to the often complex reporting and planning results, a differentiated, partly object-related documentation support (comments, texts, images, etc.) is useful for better transparency within the planning and control process. The documentation support should assist various aspects, such as the documentation of specific framework parameters and specifications or the commenting of important findings.

Organization and process support

The process and the organization for reporting and planning should be supported by the system. Here, coordinative tasks such as scheduling, distribution of data, central and decentralized coordination of functions, and other assistance are in the foreground.

Not all of the listed requirement criteria have to be fully relevant for a planning and reporting system. Nevertheless, the majority of the established requirement criteria represent a necessity or at least a desirable characteristic for a modern and powerful reporting and planning system.

The efficiency of a planning and analysis system also depends strongly on the acceptance and trust of the affected employees. However, a "good" planning and analysis system can only be a system that is tailored to the needs of the users. Reporting and planning are a matter of trust for all involved parties. Intensity and quality must therefore be in a balanced ratio to each other. Too detailed and differentiated reporting and planning systems hinder the workflow, demotivate people, and lead to losses of flexibility. Since there may be reservations about computer-based systems and some calculation processes and functions can only be traced with difficulty, a "system trust" must be built up especially for IT-supported reporting and planning systems. The more complex the systems are in their structure and functionality, the more trust must be placed in the system. The output of an IT-supported system appears to the user as a result from a "black box" due to the opaque information processes. Here, there is a risk of system credulity, that e.g. IT-supported planning functions are simply used and calculated planning results are adopted without reflection. The IT components and functions, however, only represent instruments or tools and thus serve as aids for determining planning values or reporting.

In addition, a meaningful support of reporting and planning by electronic data processing can only be ensured if appropriately trained personnel are available who are familiar with the existing IT tools. Often, the possibilities of the existing systems cannot be optimally exploited for the company, because the know-how of the employees is not sufficiently expanded, so that not the electronic data processing, but the respective employees form the bottleneck in the actual planning preparation and information evaluation.

Therefore, the systems should be user-friendly, and the users should be sufficiently trained in handling the system. Online help, integrated application examples, comfortable screen interfaces, and other functions help the users to get to know the systems faster and to use them better and more effectively (see also the trend towards self-service BI in Sect. 4.1.5.4).

In summary, reporting and planning systems are required that create information transparency and system trust and support the reporting and planning process in a meaningful and efficient way.

5.4.2 Reporting and Planning Functions of the ERP Systems

A definition for ERP systems is hard to find, to create or to delimit. ERP stands for "Enterprise Resource Planning". It is a term of business informatics and is used in practice by many software manufacturers of standard business applications to describe or classify their software product portfolio for enterprise software. Since ERP systems can include software components for different functional areas of a company, their range of software solutions is very large.[12] It ranges from very powerful ERP systems, such as SAP, Oracle and Infor, which cover a large spectrum of logistical, commercial and other accounting functions, to smaller medium-sized solutions, which focus, for example, only on parts of the logistical or commercial tasks. Often these products are also developed and sold modularly, so that the functional scope can be gradually expanded by implementing additional modules and possibly add-on products. Also special industry developments, e.g. for the construction industry, trade and other economic sectors, can be found.[13]

The probably most widely used ERP system in the business environment worldwide is the formerly called "SAP R/3" system of SAP AG, which was later renamed "SAP ERP Central Components" and, strategically, has been intended to be replaced by SAP S4/Hana for several years. Not meant here are, for example, BI modules of the extended

[12] Cf. Hesseler and Görtz (2007, p. 2 f.).

[13] Cf. Hesseler and Görtz (2007, pp. 17 f., 24).

SAP NetWeaver technology and other acquired BI components of SAP, such as SAP BusinessObjects. The SAP Hana technology will be discussed in Sect. 5.8.3.

The theoretical goal of comprehensive ERP systems is to map (largely) all business processes and resources in a consistently integrated system; these are optimally connected to each other via a common data set.

Frequently encountered functional areas of an ERP standard software are the logistical tasks (sales control, production, planning and control systems, inventory management, research and development etc.) as well as the commercial and accounting tasks (human resources, finance and accounting, asset management, consolidation etc.).

If the use of planning and control is extended beyond the company boundaries, ERP systems can today be supported especially by the rapid spread of internet technology by systems for Supply Chain Management (SCM systems), Customer Relationship Management (CRM) or Advanced Planning and Scheduling (APS systems), which are not further considered here.

In the following, important aspects regarding the above-mentioned requirement criteria for planning and reporting with ERP systems are presented.

Data connection to upstream and downstream systems

While the planning and reporting systems of the ERP systems usually access the existing databases of the integrated ERP system components directly, the other system forms often depend on the data provision of the upstream systems via interface. This is a clear advantage of the ERP systems. However, the integration between the subcomponents also has its limits. For example, isolated personnel costs from human resources or the calculated costs from fixed asset accounting can be made available for cost accounting. Nevertheless, an integrated corporate planning across all subcomponents is very difficult to represent. Like the planning, the reporting in ERP systems is also oriented towards the subcomponents (human resources, inventory management, accounting, etc.). An integration of reporting contents or planning contents of different subcomponents is not carried out with the ERP system, but rather with downstream data warehouse or BI-supported systems. If only a partial planning area is addressed, the bidirectional Excel integration is often used, whereby the default values for Excel are provided by the ERP system and the values planned in Excel are reloaded into the ERP system.

Data modeling, harmonization and quality

ERP systems have a very complex relational data model in the standard version, which can be extended by the user or the software manufacturer with the disadvantage of possibly losing the update and release capability for these further developments. If additional information for planning and reporting from other sources is needed or external data has to be harmonized, it can be very complicated and costly to implement these changes. The data quality is a great advantage for the integrated ERP systems, referring to the internal data. If external data is added, the same data quality problems arise that also occur with the relational or multidimensional database-based systems.

Speed

The advantage of ERP systems lies in the processing of mass data in the form of simply structured document records. Regarding the reporting, very long runtimes occur for very complex queries in the relational data stock, for example, when the segment result calculation for the entire company is triggered. Here, OLAP-based data warehouse applications have a considerable speed advantage.

This also applies to the consolidation of the planning results in the context of planning. However, computationally intensive planning functions such as bill of material explosion, allocation or planned pay-scale determination are already available in the ERP system. Therefore, a change in the execution of the planning task between ERP systems and data warehouse systems is useful for suitable functions.

Economic efficiency

ERP systems are primarily purchased for planning and mapping the various business processes and resources in the company. They are therefore to be seen as a larger item in terms of software expenses, an item which is justified for this operational task. If only the planning and reporting tasks are considered, the economic efficiency is only given if the company can cope with the standard functions. Due to the higher requirements of a holistic reporting and an integrated corporate planning, these standard functions are often not sufficient, and a further development is usually too expensive.

Usability

With regard to usability for planning and reporting tasks, ERP systems often have disadvantages compared to data warehouse and BI-supported systems. The interfaces usually do not correspond to the current technical possibilities. Switching between different views is sometimes only possible by complicated navigation in the system. Help systems are sometimes very complicated, confusing, cryptic or incomplete.

Analysis and planning functionality

The methods and models as well as the resulting analysis and planning functions and techniques of the ERP modules often meet the business requirements of the companies to a large extent. These range from consistent data storage to the mapping of reporting content to the flexible use of planning dialogs (e.g. distributions, planning variants, planning price iterations and planning simulations).

The reporting and planning functions of the ERP systems are also characterized by a certain complexity and thus intransparency with regard to the dialogs to be performed. Only a specialist and system-familiar employee will be able to exploit all the functions that make sense for him. This complexity could be reduced, for example, by discarding the existing field descriptions and dialog masks on partly hieroglyphic and incomprehensible abbreviations and allowing the user to adapt his report or his planning mask to his needs without great programming and customizing knowledge. Planning functions that are not needed should be hidden due to the great lack of clarity of the overall system and reactivated later if necessary.

The delivered and partly parameterizable standard functions are not always sufficient to map all desired company-specific requirements. If such deficits exist, they can only be remedied by costly additional developments or modifications.[14] Here, open systems are recommended which provide the possibility of further development from the outset.

For the integration of individual industry specifications and company requirements as well as special programs in the overall system, e.g. the system "SAP ERP Central Components" offers the possibility to change or add functionalities that are not covered with the help of the ABAP/4 (Advanced Business Application Language) Development Workbench. It is also designed as an open system that enables the interaction and portability of applications, data and user interfaces by taking into account international standards for interfaces, services and data formats (e.g. TCP/IP: Transmission Control Protocol/ Internet Protocol; SQL: Structured Query Language; ODBC: Open Database Connection; OLE: Object Linking and Embedding).[15]

For reporting and planning tasks, however, the trend shows a move away from ERP-supported to data warehouse-supported or BI-supported reporting and planning systems.

Flexibility and design options

With regard to the layout of the structures and master data, the ERP systems are very powerful and can be flexibly changed. With regard to the report design and planning functions, the ERP systems are rather rigid. Row and column information, for example, can only be changed with some effort. Changing views by result objects, time dimensions, etc. are cumbersome to use in the analysis (see Figs. 5.2 and 5.3). Planning forms are relatively fixed and have little design potential (see Fig. 5.4).

Documentation support

ERP systems often provide only little documentation support for comments and other background material (images, texts, drawings, etc.) within their planning and reporting functions. Comments usually have to be stored very individually in the respective sub-applications. Aggregation and level-based commenting are usually not possible.

Organization and process support

A further organization and process support for reporting and planning is usually not available in ERP systems. The bundling of reports, for example, in a cockpit or the mapping of the planning and reporting processes as a control function in the system is often

[14] Cf. Hesseler (2009, p. 52).

[15] Cf. Buck-Emden (1995, p. 29).

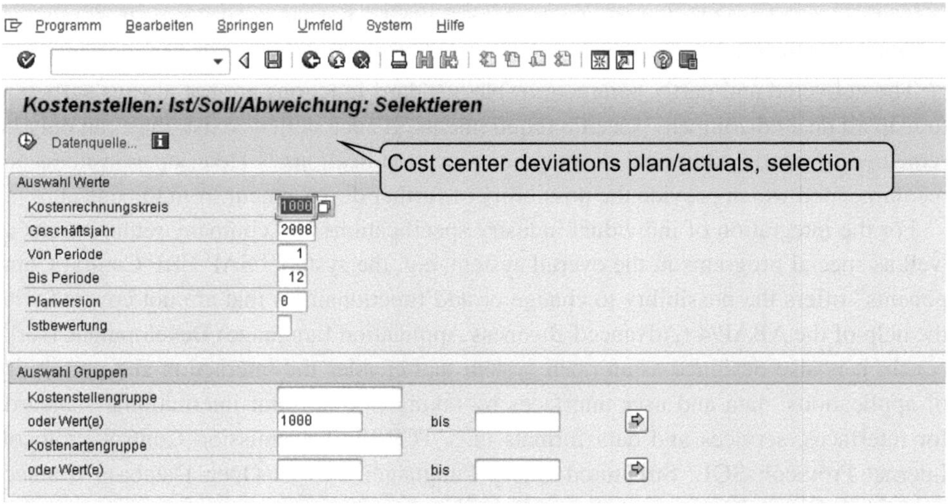

Fig. 5.2 Standard report SAP ERP CC (selection mask) (*Source*: SAP AG (2011) ERP CC System - Standard Report, Selection Mask)

only found in data warehouse or BI-supported systems. If organization and process-supporting functions are to be created for reporting and planning, these are usually complex extensions and additional developments.

In summary, the reporting and planning solutions of the ERP systems are only suitable if one can limit oneself to the respective sub-areas. If holistic and integrated reporting and planning solutions are to be pursued, data warehouse and BI-based systems are to be preferred. Most ERP providers supplement their product range in this direction, so that the ERP data provisioning in these systems is already preconfigured.

For larger business analysis and planning calculations, such as the internal cost allocation and the explosion of the bill of materials and work plans, it is often not worth replicating these more complex planning functions. Here, an interaction with the ERP system is recommended.

5.4.3 Spreadsheet Programs

In the area of reporting and planning, spreadsheet programs, such as Lotus 1-2-3, Numbers and especially MS Excel, are of great importance.

Spreadsheet programs can be described as flexible information technology, as table-driven toolboxes that can also be used quickly and comfortably by non-IT specialists for individual problem solutions. The main focus of these mostly PC-based spreadsheet programs is on the processing of calculation, planning, simulation, optimization and

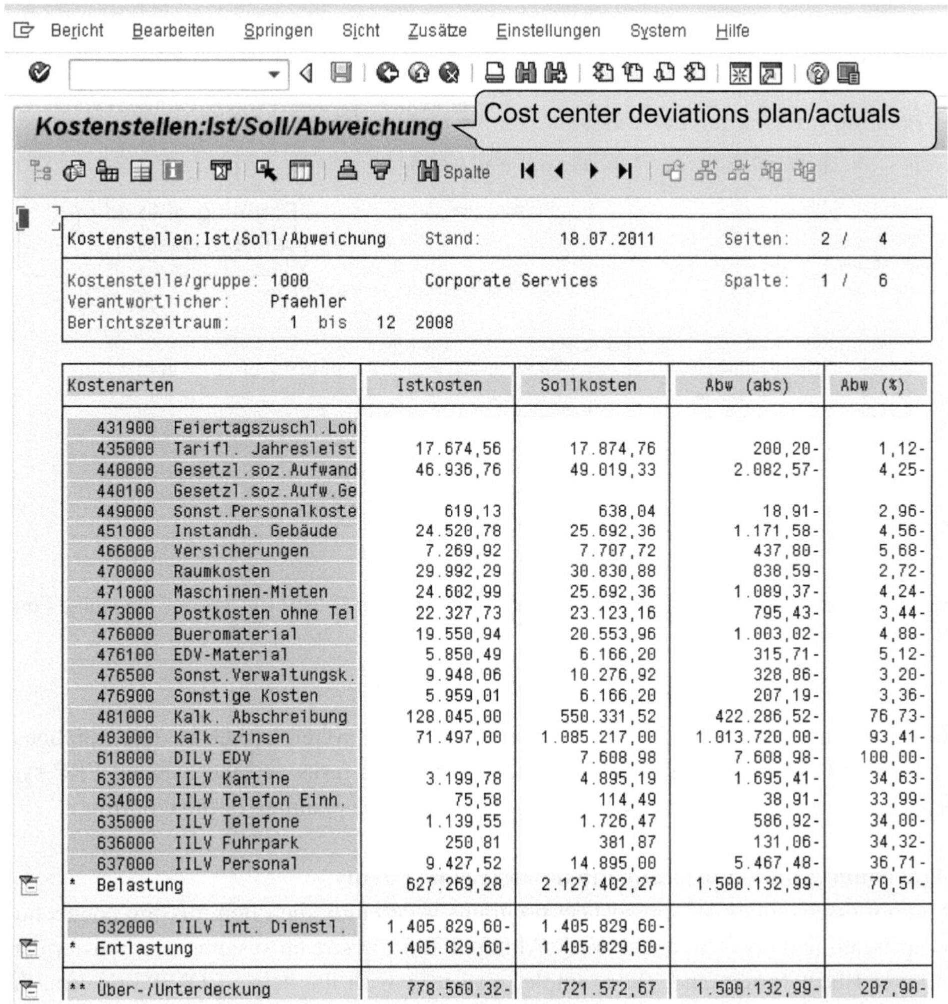

Kostenstellen:Ist/Soll/Abweichung Stand: 18.07.2011 Seiten: 2 / 4

Kostenstelle/gruppe: 1000 Corporate Services Spalte: 1 / 6
Verantwortlicher: Pfaehler
Berichtszeitraum: 1 bis 12 2008

Kostenarten	Istkosten	Sollkosten	Abw (abs)	Abw (%)
431900 Feiertagszuschl.Loh				
435000 Tarifl. Jahresleist	17.674,56	17.874,76	200,20-	1,12-
440000 Gesetzl.soz.Aufwand	46.936,76	49.019,33	2.082,57-	4,25-
440100 Gesetzl.soz.Aufw.Ge				
449000 Sonst.Personalkoste	619,13	638,04	18,91-	2,96-
451000 Instandh. Gebäude	24.520,78	25.692,36	1.171,58-	4,56-
466000 Versicherungen	7.269,92	7.707,72	437,80-	5,68-
470000 Raumkosten	29.992,29	30.830,88	838,59-	2,72-
471000 Maschinen-Mieten	24.602,99	25.692,36	1.089,37-	4,24-
473000 Postkosten ohne Tel	22.327,73	23.123,16	795,43-	3,44-
476000 Bueromaterial	19.550,94	20.553,96	1.003,02-	4,88-
476100 EDV-Material	5.850,49	6.166,20	315,71-	5,12-
476500 Sonst.Verwaltungsk.	9.948,06	10.276,92	328,86-	3,20-
476900 Sonstige Kosten	5.959,01	6.166,20	207,19-	3,36-
481000 Kalk. Abschreibung	128.045,00	550.331,52	422.286,52-	76,73-
483000 Kalk. Zinsen	71.497,00	1.085.217,00	1.013.720,00-	93,41-
618000 DILV EDV		7.608,98	7.608,98-	100,00-
633000 IILV Kantine	3.199,78	4.895,19	1.695,41-	34,63-
634000 IILV Telefon Einh.	75,58	114,49	38,91-	33,99-
635000 IILV Telefone	1.139,55	1.726,47	586,92-	34,00-
636000 IILV Fuhrpark	250,81	381,87	131,06-	34,32-
637000 IILV Personal	9.427,52	14.895,00	5.467,48-	36,71-
* Belastung	627.269,28	2.127.402,27	1.500.132,99-	70,51-
632000 IILV Int. Dienstl.	1.405.829,60-	1.405.829,60-		
* Entlastung	1.405.829,60-	1.405.829,60-		
** Über-/Unterdeckung	778.560,32-	721.572,67	1.500.132,99-	207,90-

Fig. 5.3 Standard report SAP ERP CC (detail view) (*Source*: SAP AG (2011) ERP CC System - Standard Report from an IDES System)

analysis tasks or other application areas that are usually limited to a small amount of data and solve relatively well-structured and closed decision problems.

Especially the layout design of the reporting system and the representation of the planning, such as the preparation of overhead cost plans, sales and revenue plans as well as profit and finance plans, are often carried out in business practice with spreadsheet programs although these are data-intensive and complex application areas. This is often

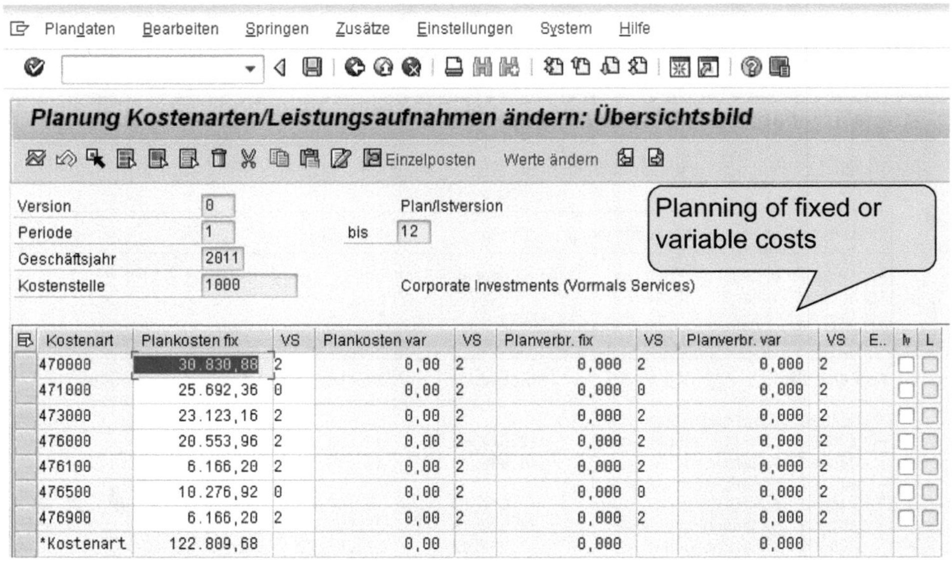

Fig. 5.4 Planning form SAP ERP CC (*Source*: SAP AG (2011) ERP CC System - Planning Form in an IDES System)

due to the fact that in companies there are no powerful systems (such as data warehouse or BI-based reporting and planning systems) for these tasks and the existing ERP systems offer no or only partially powerful solutions for the planning or the reporting.

Data connection to upstream and downstream systems

A major disadvantage of spreadsheet programs is certainly the often missing connection to upstream and downstream sources. Although most upstream systems can transfer data in spreadsheet format and, for example, read it in with the help of ODBC drivers, the merging and checking of the data often has to be done manually in the spreadsheet program. In practice, one still finds frequently the complete manual transfer of data from upstream systems where transmission errors are not uncommon.

In connection with ERP systems and data warehouse or BI-based systems, however, a trend can be seen that these systems increasingly offer bidirectional interfaces to MS Excel, so that data can be relatively comfortably loaded and saved back from the relational or OLAP-based multidimensional data storage. MS Excel is used as an (alternative) add-in application for the planning or reporting interface in such systems.

Data modeling, harmonization and quality

Spreadsheet programs do not have a database-supported data storage, such as relational or OLAP-based multidimensional database systems. The data is entered directly into the cells provided, whereby the developer has complete freedom in the data creation.

Data harmonization in the form of data comparison and adjustment from different data sources is not provided as standard and is often done manually. Due to the weak data integration, often data redundancy and thus errors occur if the same values are, for example, maintained differently at different places of the tables.

If no overview is given when creating data, e.g. in the sense of the IPO principle (input, processing and output), one quickly loses the overview in the tables, and errors and data inconsistencies creep in.

A disadvantage is also the limited data storage capacity in individual tables that reaches its limits with large amounts of data.

With regard to data quality, spreadsheet programs perform worst due to the problems already mentioned regarding data inconsistency and redundancy.

Speed

For smaller reporting and planning applications, where data storage is done directly in Excel, the performance is sufficient. This falls behind the other systems (ERP systems, relational database-based systems, data warehouse or BI-based systems) with increasing application size.

Economic efficiency

The acquisition costs and training costs for spreadsheet programs are by far the lowest compared to the other systems. However, this advantage can quickly turn into a disadvantage as planning and reporting require more personnel effort with increasing complexity, something which is mainly due to the transfer, coordination, maintenance and verification work.

User-friendliness

Due to the high level of familiarity and ease of use, spreadsheet programs are considered to be very user-friendly. This is also a central reason why other planning and reporting systems have partially integrated MS Excel as an interface into their program as an add-in. With regard to data administration and maintenance, however, this user-friendliness does not apply.

Analysis and planning functionality

Spreadsheet programs provide the user with an empty, electronic worksheet (spreadsheet) at the beginning; this sheet is structured like a matrix and divided into rows and columns. Texts, numbers or calculation instructions can be entered into the individual fields (cells). By connecting the calculation fields with formulas, spreadsheet programs are well suited for programming simple calculations. One enters the changed input data and immediately sees the results of the current calculation on the screen. In addition, numerous mathematical and statistical formula collections are available for the calculation operations of individual table fields and areas. With the help of additional macro programming (e.g. with Visual Basic for Applications), additional functions can be created for reporting and planning. However, this requires corresponding VBA programming skills.

In addition to the basic functions of a spreadsheet (cell formatting, formula control, etc.), the programs have powerful, flexible functions such as:

- graphic options (3D graphics, chart manager, etc.)
- scenario and simulation functions as well as options for target value and threshold calculations
- pivot tables to help perform multidimensional data analyses of larger data sets. The analysis is done in a pivot window in which page, row, column and value filters can be used flexibly. The new program version "PowerPivot" from Microsoft, for example, extends the previous pivot functionality in MS Excel.
- possibilities for comfortable programming with the help of powerful, cross-application programming languages as well as simple macro programming
- automatic data exchange with various databases as well as direct database access via ODBC interface, database-like functions for data management as well as multidimensional database evaluations with the help of pivot tables
- numerous simplification functions such as automatic sorting processes, consolidation and query calculations such as the creation of subtotals

Predefined business planning functions such as distributions, consolidations, etc. are not available in the standard and have to be created with a lot of effort. For complex reporting and planning systems, the users and developers of such spreadsheet programs unfortunately lose track of the formulas and calculation steps that have been created, and thus the data quality and consistency can no longer be guaranteed.

Flexibility and design possibilities

For report design, numerous chart functions and formatting options are available in addition to the table functions; this is why spreadsheet programs are also often used for preparing the layout of the reporting system. Spreadsheet programs are also often used for the individual design of the planning forms. Here, the individual use of the tables and formulas is in the foreground with which simple extrapolations, distributions, plan entries, simulations and scenario calculations can be carried out quickly. With ease, presentation-ready tables and graphics can be created and easily transferred to PC-based word processing systems as well as presentation and graphics programs.

An example of a cost and performance planning of a project manufacturer is shown in Fig. 5.5.

Documentation support

In the spreadsheet programs, cell-related comments can be stored which are displayed by markings. If too many cell comments are stored, the commenting often becomes unclear, and important entries are difficult to distinguish from unimportant entries.

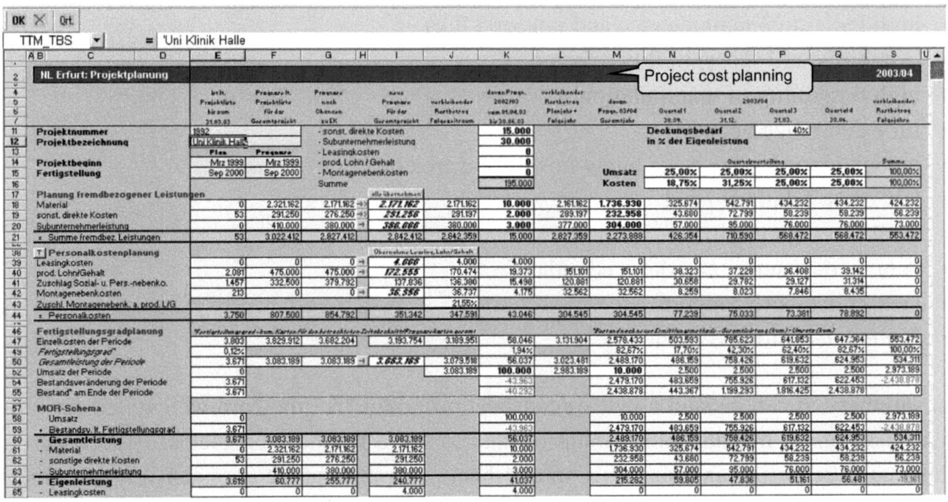

Fig. 5.5 Planning of a project manufacturer with MS Excel (Taken from Schön 2004, p. 572)

In addition to the cell comments, any text information can be added during page layout. An integration of other objects such as images, links, drawings and texts is also possible. However, there is no aggregation function and gradation of comments.

Organizational and process support

There is no pre-installed organizational and process support. However, suitable objects, such as work instructions and process descriptions, can be integrated into the tables. Simple navigation aids for reporting or planning can also be designed using additional macro commands.

Despite the great flexibility, spreadsheet programs are less suitable for complex problems and tasks. This is mainly due to the functional and data-related limitation, which means that computationally intensive functions, such as internal cost allocation and mass data processing in billing runs, are difficult to implement. Furthermore, the initial great flexibility and user-friendliness of the programs when setting up more complex calculations prove to be a programming trap. For example, when consolidations are set up, care must be taken when aggregating the data that later changes to the evaluations and reporting schemes in the row and column scheme are cumbersome to maintain table by table. Formulaic links of cell references and implemented calculation algorithms between several files or worksheets quickly become unclear and can only be adapted with difficulty when changes are made. Inconsistent data storage is the result, whereby errors that have occurred are difficult to locate. It becomes apparent that more complex, self-programmed spreadsheet programs can hardly be understood by third parties.

Important disadvantages of spreadsheet programs are summarized as follows:

- high individual maintenance and upkeep effort
- instability and susceptibility to errors
- high possibility of manipulation
- little centralized control
- model dependence on the know-how of the creator
- low consolidation speed
- weak data integration,
- data inconsistency and data redundancy
- lack of database support
- limited data storage capability.

In summary, it can be said about spreadsheet programs that they are well suited for limited tasks and a manageable amount of data to solve business problems. Especially for the area of reporting and planning, spreadsheet programs can be used sensibly as numerous application examples from business practice show. This is mainly due to the fact that in smaller and medium-sized companies only limited resources of material and personnel are available, so they can use only simple, mostly self-designed spreadsheets for planning calculations, reports and comments.

For larger integrated planning models and more demanding reporting solutions, however, spreadsheet programs reach their limits and should be better replaced by data warehouse or BI-supported reporting and planning systems. With some providers, the interface of MS Excel can be used as an add-in for the analysis and input interface.

5.4.4 Special Software (Based on Relational Database Technology)

Due to the shortcomings of the reporting and planning solution of the ERP systems and spreadsheet programs, some companies have decided to use system solutions based on relational database technology. The spectrum ranges from MS Access as a relational database management system (RDBMS) with simpler planning and reporting functions to special solutions.

Data connection to upstream and downstream systems

Unlike the ERP systems, special reporting and planning systems based on relational database technology depend on the data integration of the upstream systems. The source data is read in via interfaces, e.g. with the help of ODBC drivers or other standards. However, merging and checking the data often has to be done in the interface. Faulty data transfers and data inconsistencies are not always avoidable.

Data modeling, harmonization and quality

Compared to spreadsheet programs, RDBMS-based programs have the advantage of a systematic data storage. An advantage over ERP systems, which usually also rely on relational databases, is the easy possibility of changing the data structures for planning and reporting since here usually no dependencies to other program functions exist and the complexity is lower. With individual developments, which were mostly developed for operational tasks as well as reporting and planning, the same disadvantage exists as with ERP systems. The effort of checking and harmonizing data and the error-proneness are not to be underestimated when integrating many different sources of data.

Regarding data quality, RDBM systems face the same problems as ERP systems and multidimensional data warehouse or BI-based systems due to the integration of data from upstream systems. These are, for example, the inconsistencies of the source data and errors in the loading and transformation processes.

Speed

The time behavior (performance) of special RDBMS-based programs is significantly better than spreadsheet systems in terms of reporting and planning. Due to their specialization, they are also ahead of ERP systems. In contrast to data warehouse-based OLAP systems, the performance is significantly worse. The reasons for this lie in the different data storage and query possibilities that result from this as will be shown in detail in Sect. 5.5.4.1.

Economic efficiency

Special RDBMS-based reporting and planning programs are to be seen as additional acquisition and implementation costs compared to the standard functions of the ERP system, which is primarily purchased for handling the operational business processes. They are usually higher than the costs of spreadsheet programs since the special and often individual developments are much more complex. Even compared to the modern data warehouse and BI-based systems, there are no advantages here, but possibly rather economic disadvantages in development, setup and maintenance.

User-friendliness

Basically, the user-friendliness depends on the use of the technical possibilities and can therefore vary greatly for RDBM systems due to their age. Because of their specialization, RDBM systems should have advantages over ERP systems for reporting and planning. However, this is not always the case with some individual developments since here usually the content and functionality and not the interface or the help systems are in the foreground. Here, the user-friendliness often leaves much to be desired.

If, for example, one compares MS Access with MS Excel, then here the interfaces and help functions provided by the manufacturer Microsoft are technologically and ergonomically on the same level. From the user's point of view, however, MS Excel is more familiar, something which is also due to the high penetration of MS Excel compared to MS Access.

Analysis and planning functionality

With regard to the analysis and planning functionality, a distinction must be made between standard and individual solutions for RDBM systems.

With individual solutions, e.g. also with MS Access, reporting and planning functions, as well as with spreadsheet programs, have to be built up completely. As with spreadsheet programs, mathematical formulas, macro programming with VBA, etc., are available as tools. However, prefabricated business planning functions such as distributions, revaluations, etc. do not exist in the standard, but have to be developed. A report generator is available for the reports. Here, too, the report templates have to be created first. Overall, the analysis and planning functionality of these solutions is usually limited (see Fig. 5.6).

With special RDBMS standard solutions from software manufacturers for reporting and planning, the delivery of numerous business analysis and planning functions as well as templates is included in the package. Here, the functionality is significantly larger and only needs to be checked for the coverage of the company requirements. However, due to the technical development, it can be assumed that the modern data warehouse and BI-based systems will surpass the RDBM systems in terms of both tools and standard templates.

Flexibility and design options

The creation and maintenance of the structures and master data is flexible in most RDBM systems. Again, the individual and standard solutions have to be differentiated. With the individual solutions, the reports and planning functions have to be designed first, and this is relatively flexible and possible depending on the available tool with the form and report generators. The comfort and the results are usually significantly lower than the special standardized RDBM systems, which provide more features and possibilities here. Individual advantages compared to the ERP systems are here in the design with regard to the available planning or report generators. After the first development,

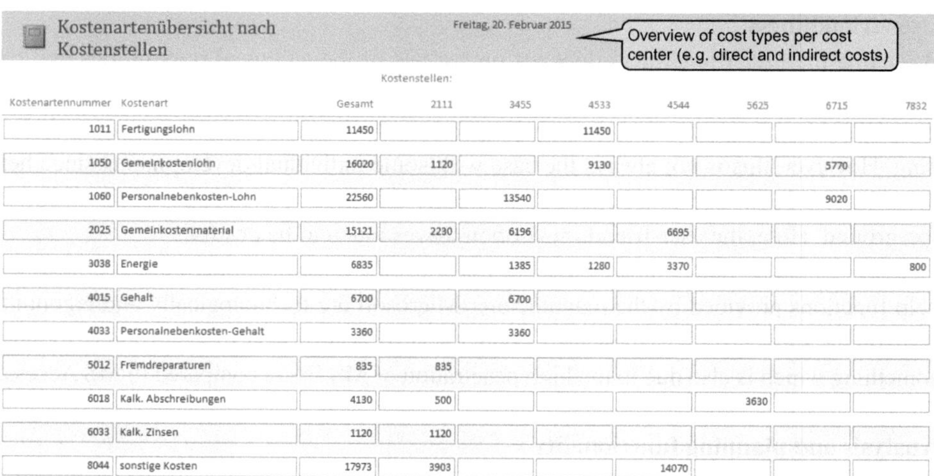

Kostenartennummer	Kostenart	Gesamt	2111	3455	4533	4544	5625	6715	7832
1011	Fertigungslohn	11450			11450				
1050	Gemeinkostenlohn	16020	1120		9130			5770	
1060	Personalnebenkosten-Lohn	22560		13540				9020	
2025	Gemeinkostenmaterial	15121	2230	6196		6695			
3038	Energie	6835		1385	1280	3370			800
4015	Gehalt	6700		6700					
4033	Personalnebenkosten-Gehalt	3360		3360					
5012	Fremdreparaturen	835	835						
6018	Kalk. Abschreibungen	4130	500				3630		
6033	Kalk. Zinsen	1120	1120						
8044	sonstige Kosten	17973	3903			14070			

Fig. 5.6 Reporting with MS Access (Self-created report with MS Access)

changes are possible, but once created, the planning and reporting structures and their functions are often not changed so much as the adjustments are again associated with development effort. An advantage of the special RDBM systems is the use of planning and reporting templates as well as standardized functions, thus that part of the new development is eliminated here.

The design options in reporting and planning will tend to approach those of the modern data warehouse and BI-based systems with special standardized reporting and planning systems. However, the multidimensional analysis possibilities are limited here due to the poorer performance and the more complex relational data modeling and querying required for this.

Documentation support

The documentation support also depends very much on the respective RDBM system. MS Access does not provide a separate (cell-related) comment function compared to MS Excel. Comments have to be developed here in the data model and in the reports and forms. Objects such as images and drawings from other applications can also be linked and updated via OLE/DDE links. However, an aggregation function and staggering of comments does not exist either. Other RDBMS applications offer more (typical for standard developments) or less (typical for individual development) functions for documentation support.

Organizational and process support

The organizational and process support also depends very much on the respective RDBM system.

MS Access, like MS Excel, has no pre-installed organizational and process support. However, simple work instructions and process descriptions can be integrated into the reports and forms. Additional macro commands can also be used to create navigation aids for reporting or planning. For the individual developments and special reporting and planning systems, it depends on whether the manufacturer has provided suitable functions for this.

The software programs specially developed for reporting and planning are characterized primarily by the fact that they provide planning and report templates as well as reporting and planning functions methodically in a structured framework, compared to the spreadsheet programs and ERP systems. Complex planning relationships, such as the consolidation of planning results or planning projections and simulations, can be quickly and comfortably mapped in a database-supported manner, despite constantly changing corporate and cost structures.

The limits of these specially developed software systems, however, also lie in the programmed and thus pre-structured reporting and planning possibilities. Although these are relatively large, there is no standardized reporting and planning tool that covers all conceivable model requirements. This is mainly due to the diversity of the service spectrum and the structure of the companies, considering, for example, different industries. These statements also apply to the modern data warehouse and BI-supported systems insofar as they provide standard templates.

When new investments are made in reporting and planning systems, the trend is not to use RDBMS due to the poorer query speed (performance), but rather to use the modern systems with data warehouse or OLAP technology. In summary, RDBM systems can therefore be described as obsolete models for reporting and planning.

5.4.5 BI-Supported Systems

Business-intelligence supported planning and reporting systems are briefly characterized by the fact that they have a different technological basis. The most important technological changes result from the multidimensional data storage and analysis (OLAP technology) in the data warehouse, from the big-data-technology for handling large and unstructured data volumes, from the web technology as an application interface, and from the app technology especially for mobile but also stationary application software (see Sect. 5.10.3). A more comprehensive preparation of the basics of data warehousing and traditional business intelligence is provided in the two subsequent Sects. 5.5 and 5.6. The extension of traditional BI to exploratory BI by artificial intelligence and big-data technology follows concisely in Sect. 5.7 and in Sect. 5.8. Since modern BI-supported systems for reporting and planning are established, the following chapters will dive deeper into these new solutions.

The performance of the BI-supported reporting and planning systems has increased considerably in recent years. BI providers, such as SAP (BW/SAP BusinessObjects, Lumira, SAP Analytic Cloud etc.), Oracle (Oracle Hyperion), IBM (IBM Cognos), Corporate Planning (CP-Suite), Microsoft (Microsoft Reporting Services), Power BI, SAS-Software (SAS Institute), Infor (dynamic/Enterprise Performance Management), MicroStrategy (MicroStrategy Business Intelligence), Cubeware (C8 Cockpit), Unit4 prevero (prevero Planning and Reporting), Targit (Morton Systems), QlikView and Qlik Sense (QlikTech), Jedox (Jedox AG), Tableau (Tableau Software), Board (Board), evidanza (evidanza MaC), LucaNet (LucaNet), to name just a few, have developed solutions that are user-friendly and include various BI-supported controlling functions. Some of them, unfortunately, focus only on reporting or planning and neglect an integrated solution. In addition to the proprietary software providers, open source products such as Jaspersoft can also be found as a technical alternative on the BI market.

In order to carry out the direct comparison of the BI-supported planning and reporting systems with the requirement criteria for the other system orientations (ERP systems, spreadsheet programs, special RDBM systems), these should be listed in advance.

Data connection to upstream and downstream systems

The integration of the BI-supported systems with other upstream systems is carried out via powerful applications that provide any data and comparative values from ERP and transaction systems, spreadsheet programs and other databases.[16] In the so-called ETL process, the data from the relevant source systems are extracted (**E**xtraction), possibly transformed and enriched (**T**ransformation) and loaded into the data warehouse (**L**oading) (see also Sect. 5.5.2). In the ELT process of a big data application, the order of the last two process steps is reversed (see Sect. 5.8). In both cases, this means a redundant data storage. The effort of a redundant data storage is justified today as the analysis performance is significantly increased by the new technologies.

Data modeling, harmonization and quality

The core of a data warehouse-based reporting and planning solution is the multidimensional OLAP data model, which is responsible for the performant planning and evaluation of the data sets (see also Sect. 5.5.4.1). This has to be developed, if there are no prefabricated models, at the beginning of a BI implementation project from the requirement profiles of the users. An example of a toolset for data modeling of a star schema using Microsoft Visual Studio is shown in Fig. 5.7. The data harmonization takes place in the ETL process. The data quality is strongly dependent on the ETL process and the data modeling of the respective application. When using big-data technology, many structured and unstructured data are loaded via the ELT process. With the help of the map-reduce method, the distribution and merging of the data takes place (see also the details on big data in Sect. 5.8). Apart from inconsistencies in source data, there are problems with the loading processes (e.g. interruption of transmissions) and the transformation processes (e.g. conversion errors). These can contain error sources like interfaces, which have to be avoided. As tools for improving the data quality for DWH solutions, tools for checking the data for consistency, integrity and erroneous values in the upstream systems (data profiling) and tools for automatic data cleansing are distinguished, in which, for example, data are cleaned of duplicates or data are sensibly merged and enriched.[17]

Speed

Due to the OLAP-based data storage designed for extensive, multidimensional and complex data analysis, traditional BI systems already have a very large advantage in terms of the speed of analysis and planning, whereby the user also receives timely response times when filtering and changing dimensions, unlike with relational database technology (see Sect. 5.5.4.1). With the addition of big-data technology, the exploratory BI systems receive another boost in speed. The data processing is done with the help of in-memory technology directly in the main memory. Time-consuming disk accesses are eliminated by this technology (see Sect. 5.8.2.1).

[16] Cf. Bange (2013, pp. 134–135).

[17] Cf. Bange (2013, pp. 98–126).

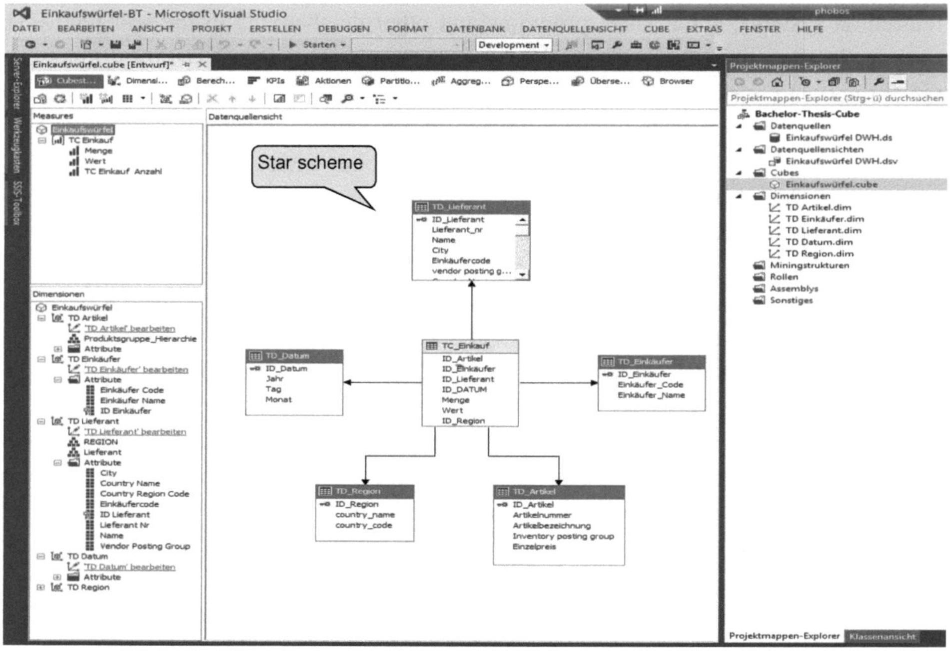

Fig. 5.7 Star schema for purchasing with Microsoft Visual Studio

Economic efficiency

The cost-benefit ratio for the introduction of BI systems is critically discussed in both directions in practice. A clear statement is not yet obtainable by surveys.

The Business Intelligence Study 2010 by the university TU Chemnitz stated as a result that the BI introduction brings advantages for the companies, but the economic efficiency is not measurable.[18]

On the one hand, there are the high introduction costs, which are less associated with the acquisition costs of the software, but rather with the associated introduction projects as they bind own and external resources more intensively and over longer periods of time. As cost drivers, for example, missing and changing requirements for the data model and the evaluations, as well as overdimensioned data warehouse modeling, which is difficult to coordinate, are mentioned. On the other hand, there are the not to be underestimated information advantages for the control of the enterprise, which quickly justify the amortization of the investment with correct decisions.

[18] See Pütter (2011) for this. The study was conducted at the Chair of Business Informatics II of Prof. Dr. Peter Gluchowski in cooperation with the consulting firm *Conunit* (Frankfurt a. M.).

Due to the establishment and technical development of the BI software, it can be assumed that the costs will decrease and the benefits will improve.

User-friendliness

BI-supported reporting and planning systems tend to have comfortable user interfaces on the latest technical level. Their navigation, filtering, selection and sorting functions are intuitive to use, and help systems can be accessed online; thus a high user-friendliness can be assumed. The familiarization with the systems is usually quick. An example of a flexible multidimensional planning form is shown in Fig. 5.8.

Many BI frontend applications have a web interface in addition to their own client interface, so the application can be started, reports analyzed and planning carried out via a simple browser access. The interfaces of some providers are based only on web technology, so the separate installation of an application interface on the client is no longer necessary (see Sect. 5.5.5.2).

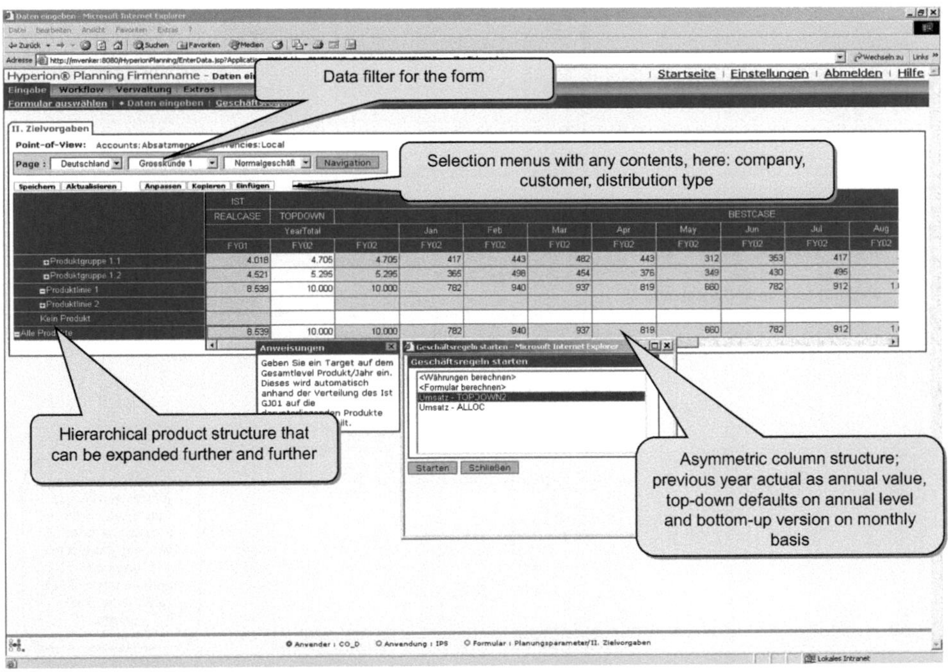

Fig. 5.8 Hyperion Planning: Multidimensional planning forms (Taken from Schön 2004, p. 310)

Through the app technology, the users of different mobile and stationary devices have the opportunity to create their own flat design by individually designing and using tiles with app functionality (see Sect. 5.10.3).

Analysis and planning functionality

In the BI-supported systems for planning and reporting, two orientations can be seen regarding the analysis and planning functionality.

Some providers only deliver basic tools to the company users with models and methods (such as mathematical functions, distribution alternatives, different chart types, design features for report formatting, etc.), which can be used for the development of an individual planning and reporting system (see Fig. 5.9). Based on these basic tools and reference projects, the providers usually also deliver templates and samples for a planning and reporting system, which, however, are usually only used for orientation and as copy templates.

Another part of the providers, on the other hand, deliver important business-oriented planning and reporting systems for subtasks of planning and controlling a company. Complete solutions are not yet to be found. Important subareas are:

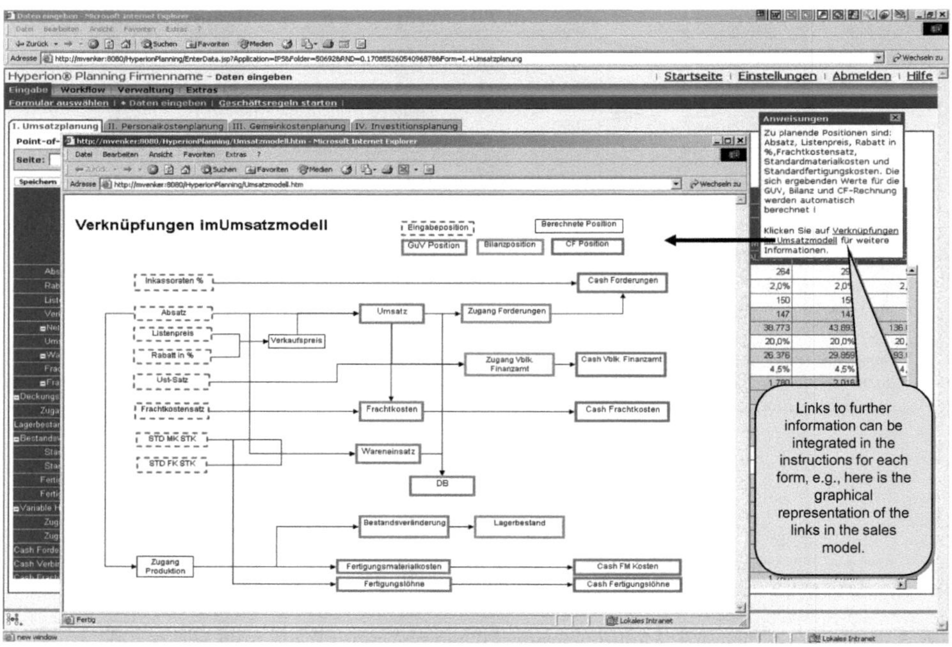

Fig. 5.9 Hyperion Planning: Individual planning models (business rules) (Taken from Schön 2004, p. 574)

- consolidation
- income, balance sheet and financial planning
- Balanced Scorecard
- risk management
- performance measurement systems (e.g. value-oriented management)

In these cases, the standardizations are usually very advanced, and special individual changes are not or only difficult to accommodate (see Sect. 5.5.6).

The model building of the data warehouse or BI-supported reporting and planning tools is not unlimited. If one takes, for example, computationally and data intensive evaluation and planning processes such as the internal cost allocation, the variant configuration or the bill of materials explosion, then it is necessary to check whether these tasks are not better solved in the existing ERP system. Here, a combination between the ERP and the specific BI-supported system can be useful in which the systems exchange necessary data with each other.

Flexibility and design possibilities

Individual reports and planning forms can be easily designed with the help of powerful tools. Even easier is a so-called wizard functionality, which supports the creation by a menu guidance. Input fields, calculation fields and other contents of the planning forms can be separated and protected from unauthorized access in any color. The planning can start at the consolidation level or at the detail level, depending on the request. Reports can be optimized with dynamic chart types and colour coding, to name just a few features.

One of the biggest advantages of BI-supported reporting and planning tools is the enormous flexibility to manage structures and evaluate contents in a differentiated way. This applies to all relevant dimensions such as the time dimensions or the object dimensions for customers, products etc. The planning and evaluations can be consolidated bottom-up or drilled down according to the company orientation.

Evaluation and planning results can be reproduced in any formats and layouts. Besides the system-specific formats, formats such as *.pdf, *.xlsx or *.html are popular. Figure 5.10 shows an evaluation with a web frontend.

Many BI-frontend applications have a web interface in addition to their own client interface, so the application can be started, reports analyzed and planning carried out via a simple browser access. The interfaces of some providers are based only on web technology.

Documentation support

BI-supported systems support reporting and planning by numerous commenting and documentation functions. Comments can not only be created object-related, but also for higher levels and managed in a staggered way (see Fig. 5.11). Documents of different

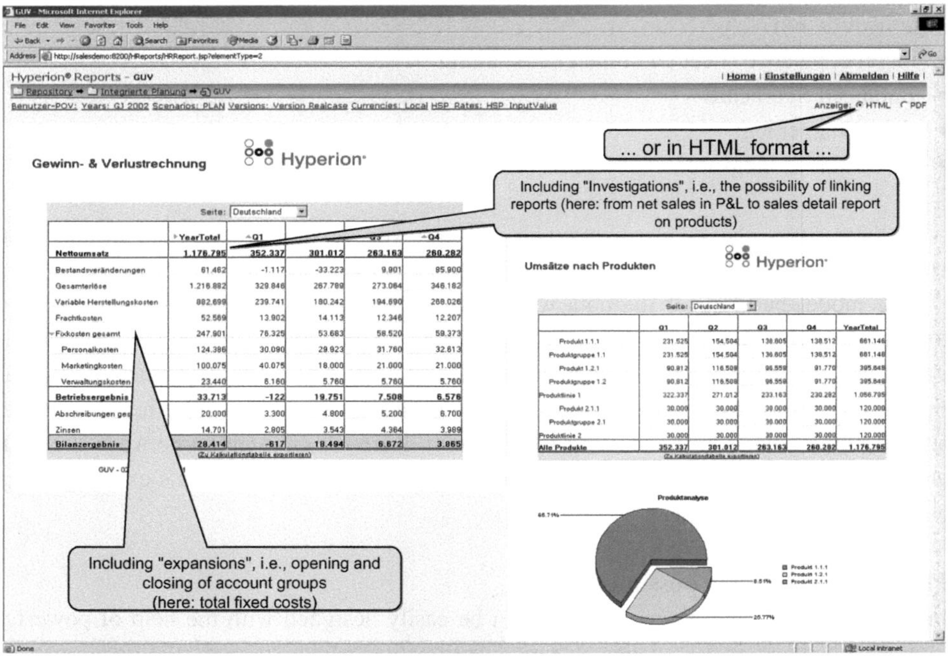

Fig. 5.10 Hyperion Planning: Evaluations (Taken from Schön 2004, p. 576)

types (images, diagrams etc.) can also be managed in the systems and integrated into the reporting and planning interfaces.

Organisational and process support

Many BI systems also offer support for administrative, coordination and process-supporting reporting and planning tasks. For example, Hyperion Planning allows you to design the planning process with the help of a workflow and to forward planning information to planning departments in a system-supported way (see Fig. 5.12). Here, there is usually an integration with typical office applications such as email, chats, calendar management, room reservations and other communication possibilities.

In summary, it can be stated that BI-supported systems will prevail in the medium to long term for more extensive planning and reporting solutions. The benefits of flexible evaluations and their timely availability provide important decision bases for companies, representing competitive advantages over companies that do not have such information. Only for small planning and reporting requirements it is reasonable to resort to spreadsheet programs or simple RDBM systems.

The main advantages of BI-supported controlling systems are:[19]

---

[19] Cf. Schön and Pook (2015, p. 13), Weber (2013, p. 219) and Schön et al. (2013, p. 258).

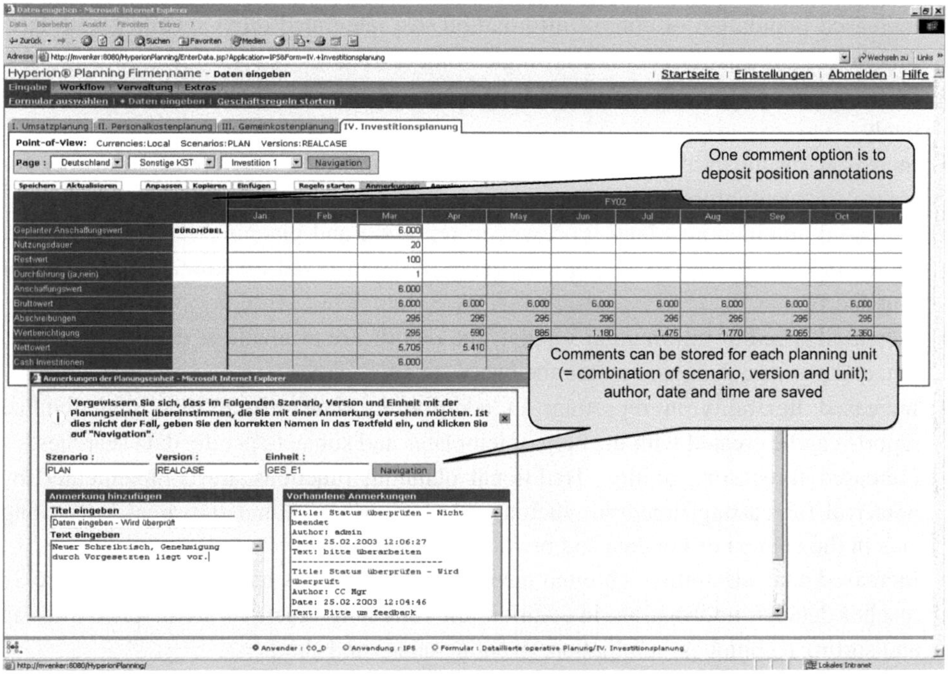

Fig. 5.11 Hyperion Planning: Comment function (Taken from Schön 2004, p. 308)

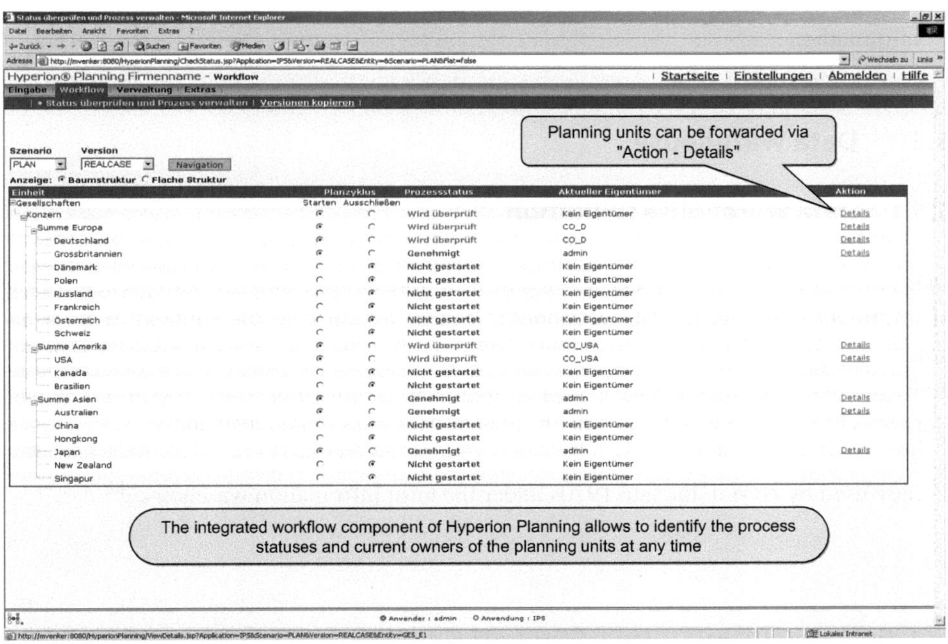

Fig. 5.12 Hyperion Planning: Workflow component (Taken from Schön 2004, p. 575)

- Increased population of control information: An integrated cross-sectional data base for control information is created (Single Point of Truth).
- Increased integration: Reporting and planning are considered and connected integratively.
- Increased sustainability of information: The uniform data base creates a history for control-relevant information, which can also be used for analyses in later periods.
- Increased automation: Many processes in reporting and planning can be automated, something which leads to a significant relief of the involved functional areas, controlling and IT.
- Increased speed of information supply: The reports are available at the push of a button, and pre-calculations increase the query speed.
- Increased flexibility in reporting: In addition to standardized reports, individual reports can be created with the help of templates and subject-specific data searches.
- Increased forecasting ability: Traditional planning functions are complemented by powerful forecasting functions, such as splash distribution and statistical forecasting aids in the context of big data and predictive analytics.
- Increased analysis ability: Through navigation and analysis paths, the analyst quickly reaches detailed information. In addition, the following functions are helpful: filtering and sorting techniques, summation, dynamic tables and charts.
- Increased data quality: Among other things, this is achieved through transformation and cleansing processes.
- Higher data protection and security: Reports can be automatically made available to the right recipients. Unauthorized persons do not receive access rights to confidential information.

5.5 Data Warehouse

5.5.1 Data Warehouse Definition

Due to the connection between the IT-technical data storage and the entrepreneurial use of information, the term “data warehouse” can be assigned to the subfield of information management of business informatics. The previous definitions of the term data warehouse are not uniform, but vary in scope. Mostly, the distinction from other forms of data storage is missing, to which most definitions of databases would also apply.

The term data warehouse was first used by Devlin/Murphy in 1988. The idea was already used by *IBM* in the late 1970s under the term information warehouse.[20]

[20] Cf. Goecken (2006, pp. 11, 15 f.), Gluchowski et al. (2008, pp. 55–88), Chamoni et al. (2010, p. 6 ff.) and Bauer and Günzel (2013, p. 11).

W.H. Bill Inmon, who is often referred to as the father of data warehousing, coined the following definition: "A data warehouse is a subject-oriented, integrated, nonvolatile, time-variant collection of data in support of management's decision".[21]

The four main characteristics are in need of explanation and will be discussed below:

1. **subject-oriented**: A subject or topic orientation for specific facts of the company characterizes a data warehouse, but also other database-supported information systems. If this means rather the **separation** of decision-relevant data from the diverse available operational, rather business process-oriented and other data of a company for planning, analysis and control of the management levels in a prepared new data set, then this is characteristic for the data warehouse definition (separation of dispositional data from the upstream operational source systems).
2. **integrated**: A data warehouse is characterized in particular by the integration of diverse data from different source systems. This can also be the case for other database-supported information systems as a physical centralization of the data in a single data pool, but it is especially standardization and integration of external and internal data and their logical connection which characterize a data warehouse. The integration is characterized by consistent data storage, uniform formats, structures and semantics.[22]
3. **nonvolatile**: In the operational systems, which take over the execution of the value-added processes in the company, the data are characterized by constant change, e.g. in the form of new records. Due to the large amounts of data, the history of these data is only kept for a certain period of time and then archived. The data warehouse is characterized by the fact that the data are used for dispositional control purposes over a longer period of time. The requirement to not change the data once read in and prepared in the data warehouse over time and to consider them as a **permanent collection** can only mean a longer period of time than in the operational systems, since also in data warehouse systems the consideration arises at some point to archive data. A part of the data, e.g. the planned data, change over time and are updated by new forecasts.
4. **time-variant**: Data warehouse solutions consider, in contrast to operational OLTP systems (online transaction processing)[23], rather period-related data with a long-term development than point-in-time data. New requirements of active data warehousing or real time data warehouse require, due to their business model (e.g. trade contracts with constantly varying stock market prices), also current changing data, so that no clear distinction is possible here.

[21] Cf. Behme (1996, p. 31) and Inmon, B.: Definition of a data warehouse. URL: www.billinmon.com (Accessed on 31.07.2002).

[22] Cf. Mucksch and Behme (2000, p. 11 f.) and Hahne (2005, p. 8).

[23] Cf. the explanations on the terms OLAP and OLTP in Sect. 5.5.4.

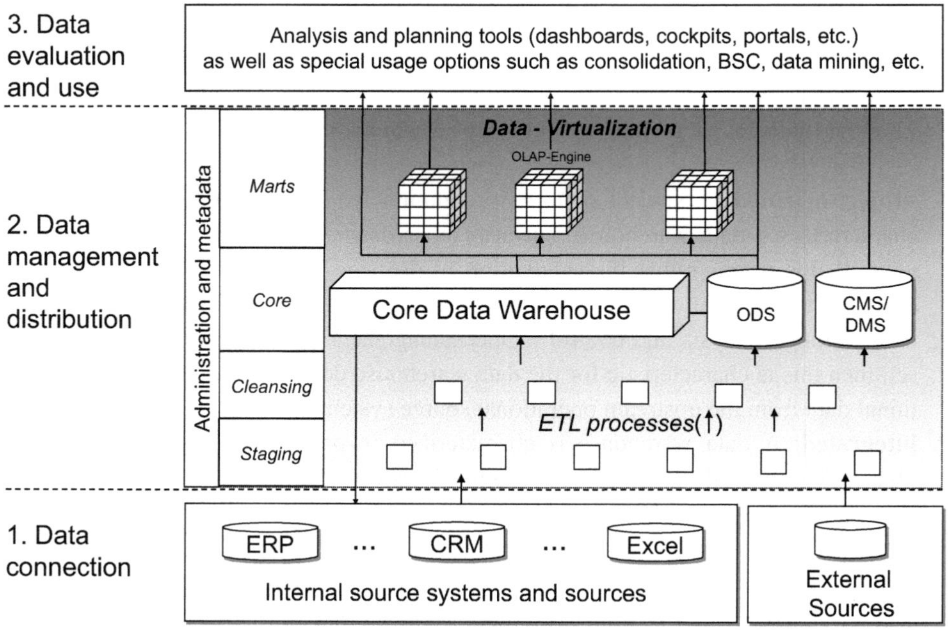

Fig. 5.13 Three levels of the data warehouse architecture

Other authors like Bauer and Günzel choose a very broad definition, which contributes little to a distinction, e.g.: "A data warehouse is a physical database that enables an integrated view of any data." They emphasize here especially the analysis purpose of a comprehensive database.[24] The planning is not emphasized here.

Some authors extend the data warehouse definition in the narrower sense (the pure data collection and its management) by aspects such as loading processes, extraction and transformation of external data, and analysis and presentation with the help of appropriate tools.[25] Figure 5.13 illustrates the three most frequently mentioned levels of the data warehouse architecture.[26]

The three levels can be described as follows:

[24] Cf. Bauer and Günzel (2013, p. 6). Kapp and Kusterer refer the data collection in the data warehouse only to strategically relevant information (Knapp and Kusterer 1996, p. 219 ff.).

[25] Cf. Schinzer et al. (2000, p. 15).

[26] Alternatively, one can also find 5-level representations of the architecture of data warehouse systems, which separately show the data sources, the ETL process, the data management, the data provision for the evaluations via the OLAP server or the OLAP engine, and the presentation level. Cf. Goecken (2006, p. 27). However, since the data sources themselves do not belong to the data warehouse, but only the data connection, and the OLAP engine is a technical system component of the data distribution, the representation with 3 levels is preferred here.

In the **first level**, the **data connection** takes place in which the data from the internal and external sources are obtained and further processed with the first steps of the so-called ETL process. The abbreviation ETL processes stands for extraction, transformation, and loading processes, which are described in detail in the following Sect. 5.5.2. Possible sources for the data connection are mainly internal sources (e.g. ERP systems, accounting and human resources management) but also external sources (e.g. information from the internet or from associations). The range of data sources to be processed in the first level ranges from standard software solutions to individual developments with different databases to individual files in various formats. The *.csv format (comma-separated values) is especially popular as with this, Excel files can be converted. The ETL process, strictly speaking, begins in the level of data connection and ends in the data targets of the 2nd level of data management and distribution (e.g. in the core data warehouse or in the data marts).

In the **second level** of the data warehouse architecture, it is ensured that all data in the system are stored and managed. Through the transformation and especially through the data harmonization, aggregation and enrichment, central tasks of data management have already been taken over. Furthermore, in this level, the further distribution of the data to different data targets takes place. Often, smaller **data marts** for specific tasks are driven from a **core data warehouse** (**Core Data Warehouse**). In this data separation, the core data sets are separated and distributed and prepared for different task areas. This most widely used DWH architecture is also called hub-and-spoke architecture.[27] Compared to the simple stove-pipe approach, which connects the source systems directly with the topic-related data marts, the hub-and-spoke architecture has the advantage of central storage in the core data warehouse. This allows a redundancy-free maintenance and storage of the data.[28]

The second level of the data warehouse architecture is often divided into four areas:

- Staging Area
- Cleansing Area
- Core DWH Area
- Data Marts Area

In the **Staging Area** data from the various source systems can be collected independently of requirements and topic-related in a certain cycle (e.g. daily, weekly, monthly etc.).[29] They are stored until they are passed on to the higher levels. After the transfer, the data in

[27] Cf. Sinz and Ulbrich vom Ende (2010, p. 190 f.).

[28] Cf. Hahne (2016, p. 150 ff.).

[29] Cf. Heuer et al. (2001, p. 469).

the staging area can be deleted. The staging area is an auxiliary area that can be used as a data buffer.

In the **Cleansing Area** the data are cleaned and pre-structured as part of the ETL processes.[30] For example, faulty data are filtered out, corrected or supplemented by so-called Singletons (Default values). In this area, various changes can also be carried out (e.g. currency conversions).

The **Core DWH Area** is the core component of the data warehouse architecture. It represents the central data collection within the data warehouse which is filled from the various internal and external sources via the ETL process.[31] The core data warehouse can comprise database sizes of several terabytes to petabytes. There, the data is stored over a longer period of time (several years). The core data warehouse forms the center of the enterprise information and is also referred to as the single point of truth (SPoT). It is divided into topic-oriented subject areas. If the resilience of the core data warehouse reaches its limits, e.g. in the existence of heterogeneous business fields, the implementation of several core data warehouses is advisable.[32]

The core data warehouse is usually not used directly for analysis for performance reasons. After further transformation processes, the core data warehouse provides (redundantly maintained) data for different evaluation purposes in so-called data marts for different user groups. This area forms the **Data Marts Area**. The term "data mart" refers to functionally limited data warehouse areas or task-related subsets of an existing data warehouse.[33] One reason for the establishment of several smaller data marts as opposed to a large core data warehouse is that the performance of a data warehouse is limited by its size, and this makes a section of a data warehouse more performant for data analysis and use.

The core data warehouse and the data marts as well as all other objects are managed with the help of the **administration** and its functions. The functions include, among others, data modeling and maintenance, the control of read and write accesses (scheduling) and the monitoring of data flows.

Metadata help to describe the meaning and the properties of all objects and functions that are used in the information system, here specifically in the data warehouse. They serve the transparency regarding the analysis, development, maintenance, administration, quality assurance and use of all objects and functions of the information system.[34] For example, the ETL processes from the data sources to the data target are graphically visualized, and their peculiarities are described. But also explanations of business indicators

---

[30] Cf. Jordan and Schnider (2011, p. 7).

[31] Cf. Manhart (2011b).

[32] Cf. Kemper et al. (2010, p. 23) as well as Sinz and Ulbrich vom Ende (2010, p. 188).

[33] Cf. Martin and von Maur (1997, p. 105).

[34] Cf. Vaduva and Vetterli (2001, p. 273).

or of technical mapping tables used can be maintained. Since usually many master data and objects are created in new name spaces in the data models, it makes sense to structure them well and store the documentation in the area of metadata.

The integration of the core data warehouse and the data marts in connection with the administration and the corresponding metadata as well as possibly further objects (e.g. ODS) takes place in a special database management system, which is installed on an OLAP server.

There are two basic types of OLAP-servers that can be distinguished by the way they store multidimensional data. In physical multidimensional data storage, the data are stored multidimensionally in a MDDBMS (Multidimensional Database Management System). In virtual multidimensional data storage, the connection of a special modeling technique within a RDBMS (Relational Database Management System) with an OLAP engine is used. In the first case, one also speaks of Multidimensional OLAP (= MOLAP) and in the second case of Relational OLAP (ROLAP). Mixed forms are also possible, which are called Hybrid OLAP (= HOLAP). See also the explanations in Sect. 5.5.4 (OLAP).

In addition to the core data warehouse and the data marts, the data are also managed and provided for analysis purposes in Operational Data Stores (ODS) and Content Management Systems (CMS) or Document Management Systems (DMS).

The integration of the **Operational Data Store** into a data warehouse is additionally used when operational data from the transactional systems are to be provided for analysis. This involves, for example, data that have individual document character as they already exist in the ERP systems or other OLTP systems.[35] In contrast to the multidimensional data storage in the data warehouse, the data in the ODS are stored in flat tables as in the relational database systems. This form of data storage has some advantages, such as efficient mass data storage, scalability and easy maintenance of the relations.[36] Therefore, it makes sense to set up the data storage in a mixed way (relational and multidimensional) for the corresponding use.

The ODS objects are also used as an extension or precursor of the data warehouse to fill the core data warehouse or the data marts with a suitable ETL process. The data originating from the operational source systems are first temporarily stored in the ODS in highly detailed form, then transformed and cleaned, and finally loaded into the intended data targets.[37]

In contrast to the data warehouse, an ODS thus often does not contain historical, aggregated and flexibly evaluable data, but represents a transaction-oriented data storage that holds detailed data at document level, time-related for certain operational facts and

---

[35] See the explanations of the terms OLAP and OLTP in Sect. 5.5.4.

[36] See Farkisch (2011, p. 27).

[37] See Navrade (2008, p. 20).

regularly updated. The data do not have to be stored for a longer period of time, but can be deleted or archived at certain intervals.[38]

The **content or document management systems** represent, like the ODS objects, an extension of data storage.[39] CMS and DMS are used where many unstructured and non-numerical data are needed for analysis, planning and control in addition to the structured data, such as images, sketches, photos, individual text documents, video or audio sequences. They serve the capture, maintenance, versioning, summarization, quality assurance, workflow and access control, selection, information retrieval and provision of these otherwise very unstructured documents in different media formats within the company.

In DMS, structured but mainly unstructured (e.g. scanned) documents are electronically stored, managed and made available for various purposes. A CMS is an application that allows digital content to be created, stored, managed and published on a platform, e.g. via a website. CMS are characterized by the fact that different file formats (e.g. data, texts, images, audios and videos) can be stored in them. In CMS, a strict and consistent separation of data by content, structure and layout is of particular importance.

In connection with the core data warehouse or the data marts, CMS and DMS serve as source systems, just like ODS objects, when, for example, an image of the product is to be integrated via mash-ups for planning and reporting. Furthermore, the descriptions and metadata for the data warehouse itself, e.g. data flow diagrams and descriptions, can be stored and used as graphic or text documents in the CMS or DMS. In addition to the physical storage of data in the listed objects (core data warehouse, data marts, ODS and CMS/DMS objects) in the data management and distribution layer, a virtualization of the data structures is also possible without storage (data virtualization). They are shown in the figures by the color gradient background. Here, only views on data from the corresponding data sources of the 1st or 2nd data level are modeled in the form of rules. In a virtual query of the data, the data are processed and made available according to the stored rules (on the fly). This virtual way of data query also uses big data technology when it accesses the primary data sources (see Sect. 5.8.2). The construction of such virtual data structures is comparable in terms of modeling effort to the construction of physical data structures. Modeling tools help to build real as well as virtual data structures.

In the **third level** of the data warehouse architecture, the data provided can be accessed using different analysis, presentation tools but also with planning applications, which are presented in Sects. 5.5.5 and 5.5.6.[40] These include:

---

[38] Operational data stores (ODS) are often used to map the requirements of a real-time data warehouse and an active data warehouse (see below), as here operational and constantly updated data are used for business process control.

[39] Cf. Kemper et al. (2010, pp. 12, 141 ff.).

[40] Cf. Goecken (2006, p. 26 ff.).

- Analysis and planning tools
  - free data research
  - OLAP-based report generators
    - proprietary report generators with vendor-specific graphical user interfaces
    - internet-based report generators with web interfaces
    - spreadsheet-based reporting systems with Excel add-in as interface
- Model-based analysis systems
  - EUS/DSS
  - EXP
  - data mining, text mining, web mining and predictive analytics
- Further usage possibilities
  - consolidation
  - Balanced Scorecard
  - risk management
  - performance measurement systems (e.g. for value-oriented corporate management)
  - etc.

In summary, the term data warehouse should be defined as follows:

> A data warehouse is understood to be an integrated central data collection from different data sources that is provided for various, mostly forecasting management tasks in the company, among other things for data analysis and management decisions, but also for planning tasks. The architecture of a data warehouse comprises different levels: data connection, data management and data distribution with the required ETL processes as well as data evaluation and use.[41]

When implementing a data warehouse, different data warehouse types can be distinguished:[42]

- Classic data warehouse
  A classic data warehouse is characterized by the fact that the data is loaded into the data warehouse and processed there at cyclical intervals, such as daily, with the help of the ETL process. Analysis and planning processes finally build on the available data stock. A backward integration of the data into the operational systems is not intended.

---

[41] Extended data warehouse definition of the author based on Mucksch and Behme (2000, p. 6) and Gabriel et al. (2000, p. 76).

[42] Cf. Kemper et al. (2010, pp. 92–96).

- Closed-loop data warehouse
  In addition to the characteristics of a classic data warehouse, a closed-loop data warehouse also provides for the backward integration of the data into the operational systems. This has the advantage that data and results that have been generated in the data warehouse, such as plan data and aggregated values, are also available to the operational systems for analysis and further processing. For example, for the planning of internal services, the primary costs and services of the data warehouse-supported planning are played back to the operational cost accounting system to perform the planning of internal services here and to pass the results back to the data warehouse. Another example of the backward integration of data is the use of sales data from a CRM data warehouse in the operational sales system to support customer processing or the playback of the sales quantities planned in the data warehouse for bill of materials explosion (MRP runs) as part of the operational resource planning.
- Real-time data warehouse
  A real-time data warehouse differs from the classic data warehouse by constantly filling the data warehouse via the loading processes in real time. For example, the current prices of securities are required in securities trading.
- Active data warehouse
  In an active data warehouse, the data of the data warehouse is used for operational business processes. For example, this is how airlines coordinate flight connections and their influences in case of changes (flight bans, cancellations, etc.).

Planning and reporting processes mainly require classic data warehouse types. However, for subareas, such as for the internal service allocation or the securities trading, other data warehouse types, especially the closed-loop and the real-time data warehouse, can also be used.

Focusing on the requirements for reporting and planning, the following tasks have to be fulfilled by the data warehouse which cannot or cannot be done in a timely and economical manner with the operational transaction-driven systems:

- provision and consolidation of condensed information from different data sources to a historically comprehensive data base
- possibility of flexible information composition with regard to multidimensional evaluations and planning
- timely availability of the required complex information queries for analysis and planning

Decision makers, planning managers and executives need, in summary, for reporting and planning an information store that

- relieves the operational systems
- is temporally decoupled from the operational systems

- holds not only detailed data but also condensed data
- holds data and functions optimized for analysis and planning purposes
- standardizes the data from different source systems
- enables the quality assurance of the data from the source systems
- holds information permanently and does not overwrite it *and*
- prepares information in a user-friendly way.

5.5.2 ETL Process

The **ETL process** is defined in the data supply concept for the implementation of the data delivery in the data warehouse and consists of three steps:

In the **first ETL process step**, the **extraction**, the relevant data formats and structures of the source systems are identified and stored as an extract in a work area of the data warehouse system.[43]

The external and internal data sources that are provided to the data warehouse have data that needs to be merged. For these integration tasks, the **transformation**, the **second ETL process step,** is available. The data formats and structures that are located in the work area are adapted to the structures and formats of the target database in the data warehouse system and converted accordingly. This can be done partially automatically or semi-automatically. However, some adjustments or corrections can only be identified and performed manually.

In the transformation process, the tasks can be summarized as follows:[44]

- iltering, elimination, harmonization and supplementation of data
- enrichment with missing hierarchies, dimension values and key figures

The first transformation process step of filtering, elimination, harmonization and supplementation, includes the following tasks:[45]

- cleaning of irrelevant data, e.g. duplicates and unimportant, unnecessary additional fields
- cleaning of syntactic defects, e.g. different formats and different data lengths
- standardization of differently used keys, dimensions, numbers, groupings etc.

[43] Alternatively, there are also virtual data connections without intermediate storage, which access the data of the source systems directly. This approach is less common in practice. With the use of in-memory technology (see Sect. 5.8.2.1), some software vendors increasingly use the direct virtual data access to primary sources without building an intermediate data storage.

[44] Cf. Kemper et al. (2010, pp. 26–38).

[45] Cf. Oehler (2000, p. 21 f.).

- adjustment or compression of data values (e.g. not every document, but only the account sum per month is transferred)
- conversion of units of measurement (e.g. miles to kilometers)
- semantic standardizations (e.g. homonyms are to be separated and synonyms are to be merged)
- addition of missing additional data (e.g. the company number or the country name is missing in a data source)
- topic-related grouping of data according to business contents (e.g. clear customer definition: invoice recipient or delivery recipient).

In the second transformation process step, the data is enriched with missing hierarchies, dimension values and key figures. This allows for aggregation of values according to additional hierarchies (e.g. parallel customer hierarchies) or new dimension values (e.g. industry affiliation) that were not previously available in the source systems. In addition, new previously missing business key figures can be evaluated.

In the **third ETL process step**, the source data are finally transferred to the target database using the extraction and transformation rules with the **loading process** to be physically stored there in most cases.[46] From a temporal point of view, the following loading processes can be distinguished:

- **synchronous** with every change in the source system (real-time data warehousing)
- **asynchronous** at fixed times:
  - periodically (e.g. daily at a fixed time)
  - event-driven (e.g. when an event occurs, such as reaching a data volume)
  - manually (e.g. on request of the user).

The loading processes are either complete (**full upload**) or only related to the changed values (**delta upload**).[47]

Due to the strain on the computing power for large data volumes, loading processes are often initiated in the idle phases of the operation (at night or on weekends). The time of the update represents a snapshot of the business events at that time.

By using appropriate intermediate consolidations of the values on certain aggregation levels, the response times of the evaluations can be significantly improved, but this requires data to be additionally stored.

Figure 5.14 shows the ETL process in a simple representation.

---

[46] Besides the real physical storage in target databases, other forms of virtual storage are also possible, in which only the data structures but not the data contents are stored in the data warehouse system, but these access the source system directly when requested.

[47] Cf. Müller and Keller (2015, pp. 394–395).

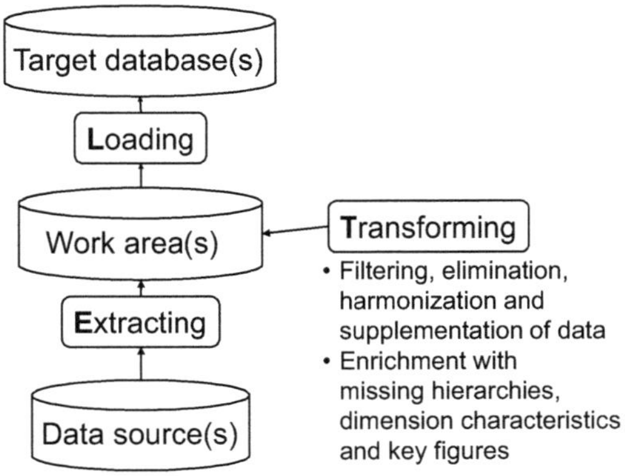

Fig. 5.14 Simple representation of the ETL process

Transformations that serve to standardize the data structure are also known in the software industry as data migration. Data quality is of particular attention in the migration of data. It is not uncommon for insufficient data quality to exist in the source systems, where e.g. redundant, outdated or erroneous values are stored. These can be cleaned up in the transformation process if they are detected in time (see Sect. 4.2.4.4). It should be mentioned here at the end that poor data quality cannot only come from the data sources, but also arise during the ETL process if e.g. errors are built into the transformation rules.[48]

5.5.3 Authorization System and Access Control

For the IT-supported implementation of various reporting and planning tasks of different individuals, departments and hierarchy levels, a tailored data protection concept with the assignment of access rights to different data sets and functions is necessary. User or authorization concepts are very important for the IT-supported reporting and planning process because different employees of different business areas have access possibilities to sensitive data that need to be restricted. For example, personnel data may only be viewed and edited by a selected group of people. Furthermore, access rights can also be used to hide unnecessary overloaded information for the users.

An authorization concept is supposed to minimize a bundle of risks, such as the loss, theft, deliberate manipulation or accidental damage of data. Examples of such risks are:[49]

48 See Apel et al. (2009, p. 67).

49 Cf. IBM Consulting Services (2003, p. 32).

- financial losses due to error, mistake and negligence
- wrong management decisions based on unreliable data
- industrial espionage
- costly error detection

From these hazards, the objectives of establishing an authorization concept are derived:[50]

- Legal and internal requirements for information security and information transparency must be taken into account.
- A possible optimal allocation and restriction of data access should be achieved.
- Data should be protected from accidental destruction and manipulation.
- Data should be protected from intentional manipulation and intentional misuse.
- The confidentiality of sensitive information must be ensured.
- Effort and benefit must be in a reasonable economic relation.

An authorization concept must meet certain requirements. These requirements can either be internal or external to the authorization concept. Internal requirements are imposed on the authorization concept at the levels of individual employees, departments or the entire company directly (e.g. management, controlling, administration, staff in sales and purchasing). These can vary significantly from each other due to the different tasks and needs.

External requirements are imposed on the authorization concept by persons or institutions outside the company. Customers, suppliers and also authorities will, for example, make demands on their contacts in the company regarding their ability to provide information. Furthermore, there are direct requirements for the authorization concept resulting from European and country-specific standardizations, such as the General Data Protection Regulation (GDPR), the German Federal Data Protection Act (BDSG) and the State Data Protection Acts of the different federal states in Germany. Especially the innovations brought about by the GDPR pose a great challenge for the companies, as was already shown in Sect. 4.2.4.4. Likewise, certain legal regulations of the different sectors must be taken into account (e.g. for banks, insurance companies and energy suppliers). The increasing networking of companies with each other leads to the fact that external persons will have more access to the information platforms of companies in the future, e.g. via the internet. Here, customer, supplier and authority relationships such as B2B (business-to-business), B2C (business-to-consumer) and business-to-government (B2G) are particularly important.

[50] Cf. IBM Consulting Services (2003, p. 32).

From the objectives for an authorization concept, the following basic rules result in practice:[51]

- Function orientation
  The granted authorizations should be created based on existing tasks and positions. In doing so, the authorizations should not be tailored to specific persons at first, but contain standardized authorization profiles for a group of persons.
- Consideration of necessity
  Employees should receive the authorizations that are necessary for their tasks. Critical authorizations should not be granted. Uncritical authorizations can be granted as far as possible.
- Level of detail and complexity
  It should be carefully examined what level of detail is needed for granting the authorizations. A too high level of detail and complexity impairs the clarity and maintenance.
- Critical authorizations
  Critical authorizations (e.g. approvals, releases and deletions) should only be available to selected employees based on their position and their professional competence.
- Involvement of the departments
  When developing the authorization concept, qualified employees of the respective departments must be involved. Their expertise on the necessary processes and the associated and required authorizations should be included in an early stage of development.
- Separation of functions
  Especially important processes should be implemented in the authorization concept according to the four-eyes principle. In doing so, these processes must be initiated or countersigned by a second person.
- Test
  Before putting the authorizations into operation, they should be sufficiently tested by employees of the respective departments. However, this proves to be rather difficult in practice during ongoing operation. Therefore, in practice, new authorizations are put into operation gradually.
- Documentation in the system
  The created authorizations must be documented in a comprehensible way for further development.

The access rights in ERP and reporting systems, which are based on relational database technology, are usually regulated separately in the individual systems and must also be

[51] Cf. IBM Consulting Services (2003, pp. 154–155).

maintained separately. This also applies to the access of Excel files, which are often controlled by access controls on document directories.

Data warehouse solutions have the advantage of providing the authorization accesses for all their planning and reporting solutions as an integrated part of the data storage.

Due to the greater flexibility in the development and maintenance of authorization concepts, role-based authorization access controls have prevailed which establish the following rules:

- formation of different collective or user roles
- determination of activities and objects per collective or user role
- assignment of the users to the collective or user roles.

Business tasks are bundled in collective roles. The individual collective roles in turn consist of the respective necessary detail tasks (single roles) for carrying out the duties. A single role comprises those activities and actions that are required to perform a single business task. For example, the individual tasks for buyers, such as "create order", "maintain order", "display and delete order", are summarized in the collective role "process order". Activities in planning are, for example, the display of the plan, the maintenance and deletion of the plan values, and the execution of planning functions.

Furthermore, for the authorization concept, one needs an assignment of the objects to the organizational structure of the company, in which the object values (e.g. cost centers, cost objects, company codes, and companies) are assigned to the organizational groups (e.g. sales, product management, and purchasing). Such organizational assignments can in turn be assigned to a collective role or single role. For reporting and planning, it is especially important that people only receive the data set for which they have access authorization. This also applies to combinations of objects, e.g. cost centers and cost objects.

Finally, the defined collective or user roles are assigned to individual employees or user groups; this has the advantage that authorization changes, e.g. assigning or removing detail tasks or organizational changes, apply to all assigned users with a single change in the single or collective roles.[52]

For the role grouping, the functional areas (sales, purchasing, personnel, etc.) as well as the hierarchy levels (top management, middle management, etc.) are often suitable.

From the perspective of the users for reporting and planning, the following or similar user groups are also often found:

- Administrators (maintenance and administration of the data warehouse solution, usually technical support, no functional area)
- Power Users (reports and planning forms can be created and analyzed in all functional scope)

[52] Cf. Ruprecht (2003, p. 126).

- Simple Users (reports and planning masks can be used and analyzed, analysis and planning functions can be applied directly)
- Simple Analysts (reports and planning results can only be read and analyzed).

5.5.4 OLAP

The abbreviation OLAP stands for the term On-Line Analytical Processing and was first published in 1993 by E.F. Codd, S.B. Codd and C.T. Salley. They examined in this publication to what extent a relational database with an SQL query language is suitable for creating multidimensional data analyses.[53] Instead of the multidimensional data analysis, they recommended calling this analysis process as OLAP. It stands in contrast to the term On-Line Transaction Processing (OLTP), which stands for the operational data processing processes of most business application systems (recording, administration and distribution as well as billing systems) with relational database technology.[54] Characteristic for OLTP are large operational data volumes, which are constantly enriched by new booking documents. Older data are archived earlier in OLTP systems. OLTP queries on data stocks are not as complex as OLAP queries and usually not multidimensional or only to a limited extent.

Codd developed twelve evaluation rules, which should guarantee the OLAP capability when introducing information systems:[55]

1. Multidimensional data view
2. Transparency
3. Access to heterogeneous data sources
4. Stable response times
5. Client-server architecture
6. Generic dimensionality
7. Management of sparse matrices
8. Multi-user capability
9. Cross-dimensional operations
10. Intuitive data manipulation
11. Flexible reporting
12. Unlimited dimension and aggregation levels.

Since the keywords do not directly reveal the meaning, brief explanations should help classify these rules better:[56]

[53] Cf. Bauer and Günzel (2013, p. 44).

[54] Cf. Chamoni et al. (2010, p. 164).

[55] Cf. Chamoni (1997, p. 294) and Codd et al. (1993).

[56] Cf. e.g. Düsing and Heidsieck (2009, p. 108) and Oehler (2000).

1. **Multidimensional data view**: In contrast to flat two-dimensional data structures, the multidimensional view of an analyst should be reflected in multidimensional structures. In this regard, different dimensions and hierarchies should be used. As a symbol for this multidimensional view of the data, data Cubes are often used.
2. **Transparency**: When using OLAP, the end user should not be overwhelmed with system-specific details. A consistent view of the analysis data must be ensured.
3. **Access to heterogeneous data sets**: The OLAP data model draws its data from various heterogeneous data sources.
4. **Stable response times**: Even with complex multidimensional queries, good response times in the second range should be achieved.
5. **Client-server architecture**: The separation of database and application servers, but especially of the clients, is a central feature of the client-server architecture. An OLAP server should be able to integrate different clients with minimal effort.
6. **Generic dimensionality**: The dimensions are equal, that is, there should be only one logical structure for all dimensions. If a dimension is extended by additional functions, these must also be provided for the other dimensions.
7. **Management of sparse matrices**: OLAP tools have the task of automatically adapting the physical schema of the model to the given dimensionality and the distribution of each specific model. This is necessary because some key figures have only few relationships to dimensions and thus many zero entries occur in the matrices.
8. **Multi-user support**: As part of the multi-user support, multiple users can access the same data model at the same time. For this, the integrity of the data base and the data security must be ensured.
9. **Cross-dimensional operations**: The OLAP tool should derive all calculations itself when navigating through the aggregation levels. In addition, the user has the possibility to define his own calculations within a dimension and across different dimensions, something which is important for calculated key figures.
10. **Intuitive data manipulation**: This rule requires that the user interface should be ergonomic and intuitively learnable to ensure that data analyses such as navigating in the data cube are easy to perform.
11. **Flexible reporting**: The evaluation possibilities should match the multidimensional evaluation requirements of the user. An appealing visual presentation and assistance in the multidimensional data selection should be granted. The arrangement of the data should not be restricted by the system.
12. **Unlimited dimension and aggregation function**: The OLAP data model and the tools used should be able to support many dimensions. Due to the rapidly increasing data volume with a high dimensionality, about 10–20 dimensions are sufficient for many business models. The aggregate functions within the dimensions are flexible and applicable without restrictions.

The rules of Codd were reduced in 1995 by Pendse and Creeth to five core contents, which are known under the short form "FASMI". FASMI stands for the initial letters of the term "Fast Analysis of Shared, Multidimensional Information":[57]

- **Fast**: An OLAP system should answer regular queries within a few seconds, complex ones in 20 s.
- **Analysis**: The system should provide an intuitive analysis for arbitrarily complex calculations.
- **Shared**: There is an effective access control, so as to work with multiple users simultaneously.
- **Multidimensional**: At the core there is a multidimensional view of the data, independent of the database structure used.
- **Information**: It should also be possible to scale the application for larger data volumes, so that the information queries are not limited.

Without going into more detail on these Codd and FASMI rules, the difference between OLAP data modeling and storage and conventional relational database structures will be shown in the following two chapters, as well as possible OLAP storage concepts that are important for reporting and planning.

5.5.4.1 OLAP Data Modeling and Storage

OLAP-based data modeling and storage distinguishes data warehouse systems from previous relational database-based application systems, such as most ERP systems. While the latter usually rely on normalized table structures in the 3rd normal form, which are queried using relational database languages of the 4th generation such as SQL (Structured Query Language), a data warehouse that is built on a Relational Database Management System (RDBMS) uses denormalized table structures usually based on the basic form of a star schema or in extended forms such as the snowflake or galaxy schema.

The difference between normalized table structures and denormalized tables in the star schema is illustrated by the following short example. A relational table relationship in the 3rd normal form is shown (see Fig. 5.15), where the primary keys are marked with an underline.

The separation of the tables in the 3rd normal form aims at a redundancy-free data storage without repeating groups. The disadvantage for analytical evaluations is the increased complexity of the queries based on them.

The query "Which customer buys the product 'data medium'?" is formulated with SQL as follows:

[57] Cf. Pends and Creeth (1995).

Order header table

Order number	Customer No	Date
4812	K398	03.01.2011
4918	K007	04.02.2011

Customer table

Customer No	Customer Description
K398	Müller
K007	Meyer

Order item table

Order number	Order item	Product No	Quantity
4812	10	77001	3
4812	20	66666	7
4918	10	77001	2
4918	20	44444	10
4918	30	66666	30

Product table

Product No	Product Description	Price
77001	Manual	34.00
66666	data medium	5.00
44444	Pencil	2.00

Fig. 5.15 Normalized table structure in the 3rd normal form

▶ Query
Select Customer description name
From customer table, Product table
Order header table, order item table
Where customer table.CustomerNo = order header table.CustomerNo AND
Order item table.Ordernumber =
Order header table.Ordernumber AND
Product table.ProductNo =
Order item table.ProductNo AND
ProductDescription = 'data medium';

The query "How much revenue is generated with the product 'data medium' and customer 398?" is formulated with SQL as follows, for example:

▶ Query
SELECT order header table.[customerNo], Product table.[ProductDescription], Sum([Quantity]*[Price]) AS Revenue
FROM order header table INNER JOIN (Product table INNER JOIN order item table ON Product table.[ProductNo] = order item table.[ProductNo]) ON order header table.[ProductNo] = order item table.[Ordernumber]
GROUP BY order header table.[customerNo], Product table.[ProductDescription]
HAVING (((order header table.[customerNo]) = 'K398') AND ((Product table.[ProductDescription]) = 'data medium'));

It becomes clear that the more complex the division of the tables and their relationships are, the more time-consuming the queries become.

In the **star schema**, dimension tables are arranged around a central fact table (see Figs. 5.7 and 5.16). The name derives from the star-shaped arrangement of the dimensions (spikes of the star) around the central fact table. The dimension tables are not linked to each other. This is therefore a denormalization of parts of the "normalized" data model since the transitivity law of the 3rd normal form is not observed.

Dimension Product

Product No	Product Description
77001	Manual
66666	data medium
44444	Pencil

Dimension Order

Order number	Order item
4812	10
4812	20
4918	10
4918	20
4918	30

Fact table

Order number	Order item	Product No	Customer No	Date	Quantity	Price	Revenue
4812	10	77001	K398	03.01.2011	3	34.00	102
4812	20	66666	K398	03.01.2011	7	5.00	35
4918	10	77001	K007	04.02.2011	2	34.00	68
4918	20	44444	K007	04.02.2011	10	2.00	20
4918	30	66666	K007	04.02.2011	30	5.00	150
Summe					**52**	**80**	**375**

Dimension Customer

Customer No	Customer Description
K398	Müller
K007	Meyer

Dimension Time

Date	Year	Month	day
20110102	2011	1	2
20110204	2011	2	4

Fig. 5.16 Star schema with fact and dimension tables

The **fact table** contains only key figures (e.g. values and quantities), whose content description is done via the dimensions (e.g. customer, product, orders and time). The fact represents the smallest unit of information of a data warehouse and the basis for all analyses. Synonyms for the term fact are e.g. measure, key figure and metric. The fact is especially needed to be able to sum and count in an analysis. Aggregation rules can be used to determine how the individual values or quantities of a fact are counted, for example as sum, average, minimum or maximum.[58]

The columns of the dimension tables consist of the primary keys (e.g. customer number) and further attributes (e.g. customer name) of the underlying entities. In the **dimension tables**, all descriptive fields are defined that have something to do with the dimension in terms of content. For example, a dimension *customer* can contain the fields industry, region, etc. as further descriptive attributes. The primary key of the fact table results from the primary keys of the dimension tables.

The evaluations can be made relatively easily by selecting the dimension values of the fact table. If you are looking for the customer revenues for the product "manual", as above, you select the choice according to this product (see Fig. 5.17).

If the query is extended by the time, e.g. February, the selection is extended (see Fig. 5.18).

This representation shows the advantage of this data storage. The queries are easy to execute by selecting the dimensions or the attached attributes. Both the individual records and the sums can be determined quickly.

The representation of the query is usually not done with flat tables as in the above representation. It is done with pivot tables because of the multidimensionality; there rows, columns and page values can be chosen based on the data set which the result area refers to. Figure 5.19 shows the implementation of a pivot table with MS Excel.

[58] See Humm and Wietek (2005, p. 5).

Order number	Order item	Product No	Customer No	Date	Quantity	Price	Revenue
4812	10	77001	K398	03.01.2011	3	34.00	102
4918	10	77001	K007	04.02.2011	2	34.00	68
Total					**5**	**68**	**170**

Fig. 5.17 Table query for the dimension product

Order number	Order item	Product No	Customer No	Date	Quantity	Price	Revenue
4918	10	77001	K007	04.02.2011	2	34.00	68
Total					**2**	**34**	**68**

Fig. 5.18 Table query for the dimensions product and time

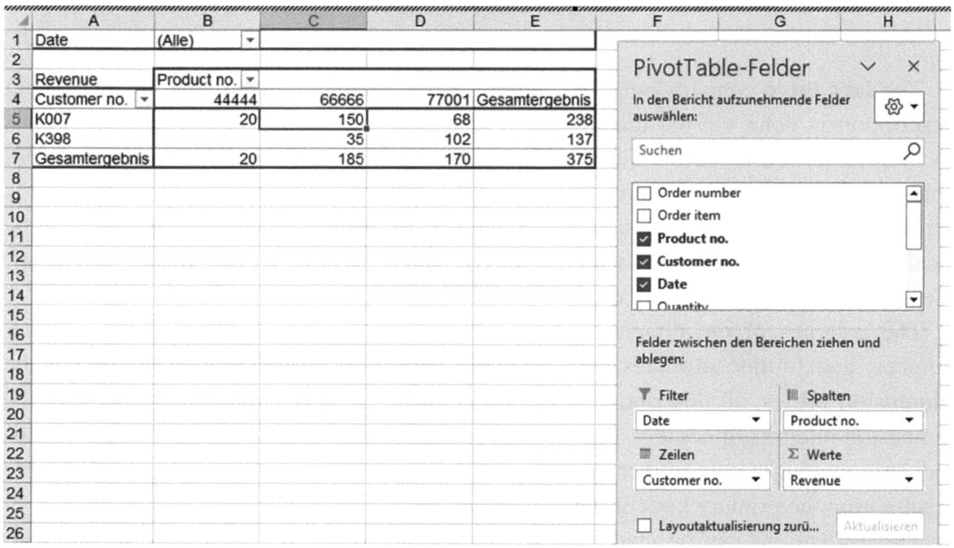

Fig. 5.19 Pivot table with MS Excel

The possibilities of data analysis in pivot tables on multidimensional data sets were already described in Sect. 3.8.3 with its functions rotation/pivoting, slice and dice, roll-up and drill-down, drill-through and drill-across. Another example of mapping the star schema with the help of the data source view workspace in Microsoft Visual Studio (Microsoft SQL Server) is shown in Fig. 5.7.[59] It can be used as a starting point for the data modeling of the exemplary BI-supported controlling cockpit for the purchasing area (see Sect. 3.9.3).

From the perspective of data modeling, the original approach of the star schema contains some deficiencies, which most software vendors try to address. This results in a

[59] Cf. Caesar and Friebel (2011, p. 548).

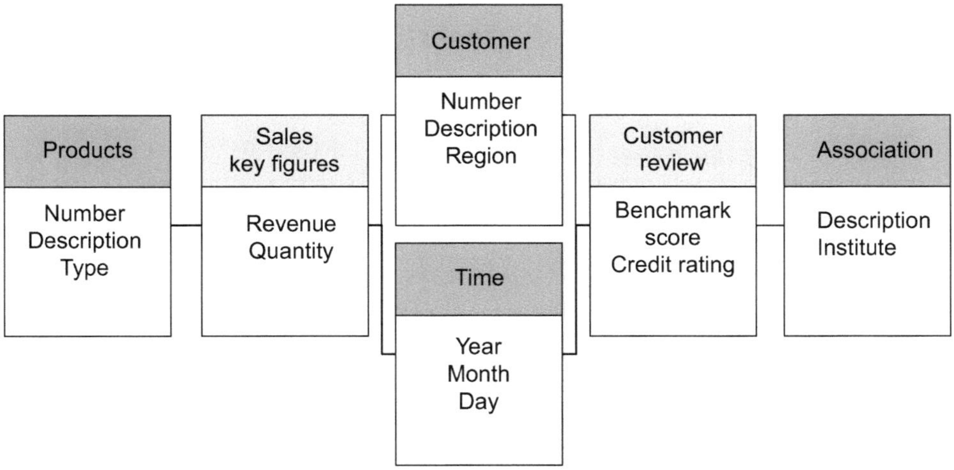

Fig. 5.20 Galaxy schema (Own example. Cf. also Mohr 2006, p. 97)

so-called **extendedstar schema** that supports, among other things, the following additional points in data modeling:[60]

- Multilingualism of attributes
- Hierarchy mapping of master data (independent of attributes)
- Time dependency of master data
- Alphanumeric foreign keys are replaced by so-called surrogate IDs. The reason for the numerical re-encryption of the dimensions with ascending integer numbers (4-byte integer) is that the search for texts and other keys takes longer than with the newly encrypted integer numbers.

If it applies to a use case that several facts can be described by exactly the same dimensions, then the star schema is sufficient for modeling. However, business practice is usually more complex as there are very many facts with very different dimensions. Then it is better, among other things, for performance reasons, to split the key figures of the fact table into different fact tables with only the necessary dimension connections. Several cubes are created side by side, in which the different fact tables are then only partially linked to the same dimension tables. The resulting schema is called a multi-fact table schema or **galaxy schema**.[61]

An example of a small galaxy schema is shown in Fig. 5.20.

[60] Cf. Mohr (2006, p. 93 ff.). For example, SAP AG supplements its star schema for SAP BW with the points listed.

[61] Cf. Bauer and Günzel (2013, p. 204 f.).

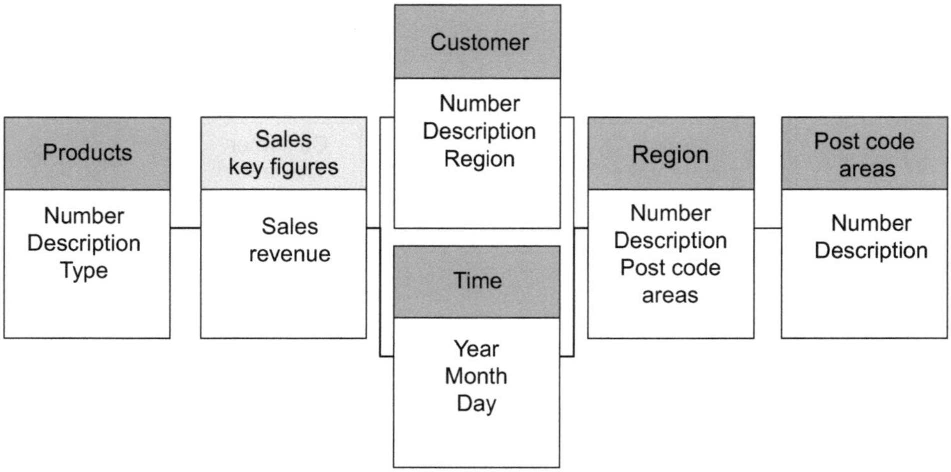

Fig. 5.21 Snowflake schema (Own example. Cf. also Mohr 2006, p. 98)

The fact tables *sales key figures* and *customer rating* are the central fact tables of the galaxy. The dimensions *customers* and *time* are dimensions that are used by both fact tables. The other dimensions are only used by one fact table each.

To prevent the data sets (attributes, hierarchy information) in the dimension tables from becoming too large, one can transform a star schema into a **snowflake schema** in contrast to the denormalization discussed above,[62] in which a normalization of the possibly too large dimension tables is performed again. Unlike the star schema, the attributes and hierarchy information are no longer in one dimension table but distributed among several interconnected tables (see Fig. 5.21). The representation of this data modeling resembles a snowflake, hence the name snowflake schema.[63] Figure 5.21 shows a very simplified example of a snowflake schema.

The dimension "*customers*" was further normalized into the tables "*region*" and "*postal code areas*". The advantage of this structure lies in certain storage space savings and possibly shorter access times, which are however opposed by a higher complexity, especially when the end users have to navigate through the snowflake structure.[64]

When building the multidimensional data model, the advantages and disadvantages of the **account or key figure model** have to be considered early. A change of the data model leads to conversion work, which should be avoided as much as possible. The key

[62] Cf. Azevedo et al. (2005, p. 46).

[63] Cf. Azevedo et al. (2005, p. 52 f.).

[64] Cf. Behme et al. (2000, p. 229).

Dimension tables:

Product ID	Prod. descr.
3312	Ring
4512	Chain

Customer ID	Cust. descr.
471	MTS GmbH
236	SOT AG

Time ID	Year
2018	2018
2019	2019

Fact table:

Dim. ID	Dim. ID	Dim. ID	Key figures	Key figures	Key figures, calculated	Key figures	Key figures, calculated
Product	Customer	Time	Gross sales	Discounts	Net sales	Cash discounts	Net net sales
3312	471	2018	1000	-100	900	-18	882
4512	236	2018	800	-60	740	-15	725
3312	471	2019	950	-50	900	-18	882
4512	471	2018	1250	0	1250	-25	1225

Fig. 5.22 Key figure model

figure model consists of dimensions and key figures as already mentioned for the star schema. All quantities and values such as number, revenues, costs, prices etc. are created as key figures in the fact table. In principle, only basic key figures are needed, which are supplied in the data supply process (ETL process). Key figures that can be calculated from other basic quantities can be additionally created as calculated key figures or calculated elements. The calculated key figures are determined when building the data cube. The following example shows a very small data model for a revenue analysis.

In the **key figure model** only the dimensions with their identification numbers (ID) (customer, product and time) are created. The key figures are the gross sales, the discounts and rebates stored and filled by the data supply process (see Fig. 5.22). The identification numbers of all dimension tables form the primary key of the records of the fact table. The calculated key figures "net sales" and "net-net sales" are created as calculated elements with formulas and calculated when building the data cube. The "net sales" result from the difference between "gross sales" and "discounts". The "net-net sales" frequently used in trade result from the difference between "net sales" and "cash discounts". Technically, "discounts" and "cash discounts" are delivered in our model with negative values in the data supply process, so sums are formed for the calculated key figures to display the signs of the data correctly.

In the analysis frontend, a report with the revenue scheme can now be created by assigning the individual key figures and calculated key figures to a row. If the key figure "discount" is differentiated in the model into three key figures "customer, promotional and special discount", more columns are necessary in the model. If several key figures are assigned to a row, they can be drilled down in the report (see Fig. 5.23).

However, the report scheme has to be set up in the standard and the dynamics are limited to the multiple assignments. The main advantages and disadvantages of the key figure model compared to the account model are:

Report line	*Key figure assignment*
Gross sales	Gross sales
Discounts	Discounts
Net sales	Net sales
Cash discounts	Cash discounts
Net net sales	Net net sales

Report line		*Key figure assignment*
Gross sales		Gross sales
Discounts	+	Customer discounts
		Promotional discounts
		Special discounts
Net sales		Net sales
Cash discounts		Cash discounts
Net net sales		Net net sales

Fig. 5.23 Key figure assignment to report rows

Advantages

- The data volume is lower.
- The data provision of the cube is faster.
- The data delivery is easier for central source tables (ready interface tables with many key figures).

Disadvantages

- Hierarchies cannot be used in reporting.
- Report schemas with fixed aggregation levels must be defined using calculated key figures.
- The data supply is more difficult for heterogeneous source tables (e.g. from different systems).
- The inclusion of new key figures (including calculated key figures) is more difficult. They have to be updated in the data model and in the reporting system.
- The historization (time-dependent storage) of the data, e.g. in case of changed report schemas, is more difficult.

In the **account model**, not so many key figures are inserted in the fact table, often only a few such as quantity and amount. The number of columns is reduced by this. In our example, instead of the key figures "gross sales, discounts and cash discounts", the key figure "amount" is inserted. For this, a dimension for the values is introduced, which in our example is called "dimension accounts". The name account model fits well for business models, as e.g. costs and revenues in accounting or inventory movements in the inventory management system are identified by accounts. These accounts often serve as basic units for the account model. They often form the smallest unit to which value units can be assigned. The fact table gets more entries, the table gets longer and the data volume larger.

General advantages and disadvantages of the account model compared to the key figure model are:

Advantages

- Hierarchies can be used in reporting.
- Report schemas with aggregation levels do not have to be defined using calculated key figures.
- The data supply is easier for heterogeneous source tables (e.g. from different systems).
- The inclusion of new value types (accounts including nodes) is much easier.
- The historization of the data, e.g. in case of changed report schemas, is easier.

Disadvantages

- The amount of data is larger.
- The data provision by the cube is slower.
- The data delivery is more difficult for central source tables (ready-made interface table with preconfigured key figures).

For the use of the hierarchy in reporting, two basic ways can be taken: creating an account dimension either with or without a parent-child hierarchy.

When using the account dimension with a parent-child hierarchy, the hierarchy is uniquely derived from the attribute "predecessor" (see Fig. 5.24). For each account, a unique identification number (ID) is used. Calculations of the result schemes, i.e. levels

Dimension tables:

Product ID	Prod. descr.
3312	Ring
4512	Chain

Customer ID	Cust. descr.
471	MTS GmbH
236	SOT AG

Time ID	Year
2018	2018
2019	2019

Account dimension with parent-child hierarchy:

Account ID	Value	Predecessor	Sort code
100	Net net sales	ZERO	1
8090	Cash discounts	100	2
200	Net sales	100	3
8050	Discount	200	4
8000	Gross sales	200	5

Fact table:

Dim. ID Product	Dim. ID Customer	Dim. ID Time	Dim. ID Account	Key figures Value
3312	471	2018	8000	1000
4512	236	2018	8000	800
3312	471	2019	8000	950
4512	471	2018	8000	1250
3312	471	2018	8050	-100
4512	236	2018	8050	-60
3312	471	2019	8050	-50
4512	471	2018	8050	0
3312	471	2018	8090	-18
4512	236	2018	8090	-15
3312	471	2019	8090	-18
4512	471	2018	8090	-25

Fig. 5.24 Account model with parent-child hierarchy

of final and intermediate results, also receive identification numbers. If the numbers are not sorted, a desired sorting can be taken into account with the help of the sorting code. In the shown example, this affects the lines "discount" and "gross sales", whose account numbers are not numerically sorted in the sense of the revenue scheme.

Advantages and disadvantages of the account model with parent-child hierarchy are:

Advantages

- Levels are easily assigned via parent-child hierarchy (predecessor).
- The table maintenance is easier.

Disadvantages

- Only one hierarchy per dimension can be represented.
- Queries are more difficult to generate.

To be evaluated neutrally is the fact that the recording of values is possible both on accounts and on nodes of the hierarchy. This can be desired or undesired depending on the reporting and planning model!

If the account model without a parent-child hierarchy is used, the account dimension is mapped differently (see Fig. 5.25). Only accounts are mapped as account IDs. The

Dimension tables:

Product ID	Prod. descr.
3312	Ring
4512	Chain

Customer ID	Cust. descr.
471	MTS GmbH
236	SOT AG

Time ID	Year
2018	2018
2019	2019

Fact table:

Dim. ID Product	Dim. ID Customer	Dim. ID Time	Dim. ID Account	Key figures Value
3312	471	2018	8000	1000
4512	236	2018	8000	800
3312	471	2019	8000	950
4512	471	2018	8000	1250
3312	471	2018	8050	-100
4512	236	2018	8050	-60
3312	471	2019	8050	-50
4512	471	2018	8050	0
3312	471	2018	8090	-18
4512	236	2018	8090	-15
3312	471	2019	8090	-18
4512	471	2018	8090	-25

Account dimension without parent-child hierarchy:

Konten-ID	Value	Sort code 3	Level 2	Sort code 2	Level 1	Sort code 1
8050	Discounts	2	Net sales	2	Net net sales	1
8000	Discounts	3	Net sales	2	Net net sales	1
8090	Cash discounts	1	Cash discounts	1	Net net sales	1

Fig. 5.25 Account model without parent-child hierarchy

hierarchy levels of the result schema are defined by attributes (in the example level 1 and 2) and the display order by sorting codes per level. This allows the hierarchy structure to be displayed.

The result schema is created in reverse order as a hierarchy. The net-net sales form the base value of level 1, which results from the net sales and the cash discount. The net sales are formed in the 2nd level from the gross sales and the discounts.

If only one value (in the example cash discounts) is assigned to a level, a content-identical report line is created (marked in yellow in Fig. 5.26). This can be hidden in the hierarchy structure. The following line schema can be displayed step-oriented in the account model for the dynamic revenue analysis (see Fig. 5.27). The consolidation levels can be expanded or collapsed from left to right using the plus sign in the drill-down.

Advantages and disadvantages of the account model without parent-child hierarchy are:

Advantages

- Multiple hierarchies are possible per dimension.
- Queries are easier to generate.

Disadvantages

- The levels of the hierarchies must be predetermined.
- The table maintenance is more complex.

To be evaluated neutrally is the fact that the recording of values in the account model without parent-child hierarchy is only possible on the accounts of the hierarchy. This can be desired or undesired depending on the reporting and planning model!

Fig. 5.26 Hierarchy structure without parent-child hierarchy

Hierarchy structure

Level 1	Net net sales
Level 2	Cash discounts
Value	Cash discounts
Level 2	Net sales
Value	Discounts
Value	Gross sales

Net net sales	+	**Net net sales**	-	**Net net sales**	-
		Cash discounts		Cash discounts	
		Net sales	+	**Net sales**	-
				Discount	
				Gross sales	

Fig. 5.27 Step-oriented revenue schema

Calculations for the use of another key figure in the reporting system can alternatively also be generated in the analysis frontend itself. However, they are calculated when the report is built; this can lead to longer and possibly user-unfriendly call times of the report if there are too many and complex calculations in the report and large amounts of data. Therefore, it is generally recommended considering calculations already in the data model. Alternatively, key figures that need to be calculated can already be delivered as pre-calculated key figures (e.g. via the ETL process), so that no calculations are necessary when building the data cube or the report. In this case, the amount of data to be delivered for the cube increases.

5.5.4.2 OLAP Storage Concepts

For the physical storage of multidimensional cubes, a distinction is made between relational and multidimensional database technologies and hybrid forms. Accordingly, the following terms are used:

- ROLAP (Relational OLAP)
- MOLAP (Multidimensional OLAP) and
- HOLAP (Hybrid OLAP).

ROLAP stands for the storage of the star schema or modified models in the form of flat (two-dimensional) relational database tables. The development of multidimensional views by using two-dimensional relational tables is realized by logical data modeling, which was already shown schematically above for the star schema and the modified forms (galaxy and snowflake schemas). The transformation of multidimensional queries into relational queries requires the use of certain query tools, such as an OLAP engine. The OLAP engine is a query tool whose special ability is to quickly aggregate and evaluate relational data multidimensionally.

In contrast to ROLAP, multidimensional OLAP (MOLAP) stores the data in multidimensional databases with multidimensional cell structures (array), which has the advantage of faster response and query times. By storing in multidimensional cell structures, a physical multidimensionality is present. MOLAP uses the sparse matrix technique to retrieve the data from the multidimensional arrays in the form of dimensional data cubes. Because the cells have a direct addressing, no extensive calculations are required for queries. Compressions of values on various dimensions can be quickly determined. The advantage of fast response and query times is offset by the disadvantage of the increasing data volume. This grows exponentially with the number of dimensions and the depth of the hierarchies. Another disadvantage is the low flexibility of a multidimensional database as only a limited number of dimensions can be maintained and evaluated due to manufacturer constraints.

Another difference to ROLAP is the pre-calculation of compressed values over the dimensions and hierarchies in MOLAP which can achieve a significantly improved performance. With a very high number of dimensions and a very large data volume, this pre-calculation can take a long time, so a temporal outsourcing in off-peak business hours (weekend/night) is advisable.

Goeckel summarizes the advantages and disadvantages of both technologies in a table (see Table 5.2).

To cope with the advantages and disadvantages, some manufacturers also offer hybrid solutions (HOLAP), in which the detailed data is stored in a relational database and the frequently used data is pre-calculated and stored multidimensionally in a compressed form. For these highly compressed data areas, MOLAP techniques are used, which are characterized by low data volume and a manageable number of users. If the user wants to drill down into detailed data areas, he automatically switches to the relational data storage.

5.5.5 Analysis and Planning Tools

Frontend tools or analysis-oriented application systems are applications that have the task of creating the presentation and the report display with the help of tabular, textual

Table 5.2 Advantages and disadvantages of ROLAP and MOLAP (Slightly adapted from: Goecken 2006, p. 42)

ROLAP	
Advantages	**Disadvantages**
• Flexible number of dimensions • Standardized query language with SQL • Know-how of this technology usually available • No pre-calculations required • Large data volumes can be managed • Based on mature, robust technology available in most companies	• No explicit support for multidimensionality • Cannot use sparse matrix technique to query data from a multidimensional database • Performance disadvantages and higher response times
MOLAP	
Advantages	**Disadvantages**

• Explicit support for multidimensionality
• Uses sparse matrix technique to query data from multidimensional database
• Performance advantages for queries and shorter response times
	• Know-how for managing multidimensional databases may not be available
	• Insufficient standardization and openness as essentially proprietary technology is used
	• Possibly limited scalability
	  • Data volume may be limited
	  • Number of dimensions may be limited
	• Lack of robustness as technology is relatively new
	• Extensive pre-calculations, therefore possibly problems with the currency of the data

```
SELECT Customer.CNo, Item.ARTNo, Order.PLIKZ, Customer.Name, Order.CHIER, Item.Color, Sum(Order.[Bottle Sales]) AS [Sum of Bottle Sales], Sum(Order.[Packaging Sales]) AS [Sum of Packaging Sales], Sum(Order.Discount) AS [Sum of Discount], Sum(Order.Discount) AS [Sum of Discount], Sum(Order. Bonus) AS [Sum of Bonus], Sum(Order.Freight) AS [Sum of Freight], Sum(Order.Group Bonus) AS [Sum of Group Bonus], Sum(Order.Cost of Goods Sold) AS [Sum of Cost of Goods Sold], Sum(Order.Commission) AS [Sum of Commission], Sum(Order.Special Bonus) AS [Sum of Special Bonus].FROM Customers INNER JOIN (Color INNER JOIN (Item INNER JOIN Order ON Item.ARTNo = Order.ARTNo) ON Color.ColorNo = Item.Color) ON Customers.CNo = Order.CNo GROUP BY Customer.CNo, Item.ARTNo, Order.PLICZ, Customer.Name, Order.CHIER, Item.Color; SELECT Customer.CNo, Item.ARTNo, Order.PLIKZ, Customer.Name, Order.CHIER, Item.Color, Sum(Order.[Bottle Sales]) AS [Sum of Bottle Sales], Sum(Order.[Packaging Sales]) AS [Sum of Packaging Sales], Sum(Order.Discount) AS [Sum of Discount], Sum(Order.Discount) AS [Sum of Discount], Sum(Order. Bonus) AS [Sum of Bonus], Sum(Order.Freight) AS [Sum of Freight], Sum(Order.Group Bonus) AS [Sum of Group Bonus], Sum(Order.Cost of Goods Sold) AS [Sum of Cost of Goods Sold], Sum(Order.Commission) AS [Sum of Commission], Sum(Order.Special Bonus) AS [Sum of Special Bonus].FROM Customers INNER JOIN (Color INNER JOIN (Item INNER JOIN Order ON Item.ARTNo = Order.ARTNo) ON Color.ColorNo = Item.Color) ON Customers.CNo = Order.CNo GROUP BY Customer.CNo, Item.ARTNo, Order.PLICZ, Customer.Name, Order.CHIER, Item.Color;
```

Fig. 5.28 Example of free data search (e.g. select statement)

and especially graphical representations in a user-friendly and decision-oriented way and enabling analysis and planning functions. In the data warehouse architecture, they usually access the prepared data of the 2nd level (data management and distribution), rarely directly the data of the 1st level (source data systems). If the analysis and planning tools are integrated directly into an application, one also speaks of "Embedded BI".

Analysis tools can be differentiated according to free database queries and query generators, OLAP-based analysis systems and model-based analysis and planning tools.

5.5.5.1 Free Data Searches and Query Generators

Under free data searches (Enterprise Search) one understands the free data search or data selection from an available data stock. The data is queried directly from the database, the data warehouse, the data marts or the ODS using data manipulation languages (DML), programming commands or database languages. For relational database-based systems, SQL (Structured Query Language) has established itself, a programming query language that, in addition to data manipulation, also includes the function of data definition (DDL = data definition language).[65] A simple data search with a select statement is shown in Fig. 5.28. In newer versions of the SQL standard, new navigation and editing functions were added (ISO 2008), so SQL can also be used for multidimensional data structures.[66] As a query language for multidimensional database structures, the query language MDX (Multidimensional Expressions) of the Microsoft Corporation has also established itself as an industry standard among other manufacturers.

For the business user without or with few IT language skills, this form of data search is too technical and not user-friendly because he has to have knowledge of the data storage in addition to the programming.

That is why many software packages offer, in addition to free data research, query generators, as one knows them e.g. from the program MS Access (cf. Fig. 5.29), with which one can create SQL queries via a graphical user interface. These query generators

[65] Cf. Elmasri and Navathe (2007, p. 37 f.).

[66] Cf. Kemper et al. (2010, p. 97).

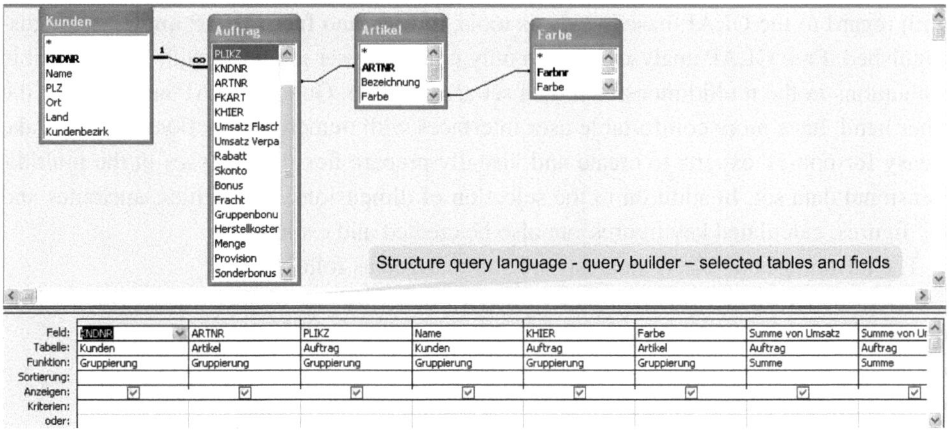

Fig. 5.29 SQL query generator (MS Access) - Query by Example

usually have corresponding functions, such as the use of field selection boxes, the visual design of queries via table relations, and the use of numerous selection criteria (among others, greater, smaller, equal or the Boolean operators). The presentation of the results in the language SQL always takes place in list form of a simple table.

The language MDX has similar components to SQL. Its syntax is based on that of SQL. In contrast to SQL, however, the results of an MDX query are displayed in a cross table. For their presentation, restrictions and formatting must be specified in the query.[67] The biggest difference between the two languages lies in the underlying database. While SQL is optimized for queries in relational databases, MDX is designed for queries from multidimensional databases.

5.5.5.2 OLAP-Based Analysis Systems and Their Interfaces

OLAP-based analysis systems rely on the multidimensional data storage that was already explained in Sect. 5.5.4.1. The physical data storage can take place in both relational and multidimensional databases, which differ in performance and handling in the modeling and the construction of the selections (cf. Sect. 5.5.4.2).

The possibilities of flexible evaluation of the multidimensional data set in data cubes were already presented in Sect. 3.8.3 (pivot tables), to which we refer here:

- rotation/pivoting
- slice and dice
- roll-up and drill-down
- drill-through and drill-across
- split and merge.

[67] Cf. Oehler (2006, p. 93).

With regard to the OLAP-based analysis tools, **guided** and **free OLAP analyses** are distinguished. Free OLAP analyses contain only a limited user guidance with which flexible evaluations in the multidimensional data set are possible. Guided OLAP analyses, on the other hand, have more comfortable user interfaces with numerous functions, which make it easy for non-IT experts to create and visually prepare flexible analyses in the multidimensional data set. In addition to the selection of dimensions, hierarchies, attributes and key figures, calculated key figures can also be created and evaluated.

The advantages of OLAP analysis are summarized as follows:

- easy data query (without IT expertise)
- fast query (even with large amounts of data)
- flexibility for ad-hoc queries by changing the views.

As analysis interfaces, three types can be distinguished for OLAP-based analyses, which are used in different frontend systems:

a) Proprietary report generators with manufacturer-specific graphical user interface
b) Internet-based report generators with web interface
c) Spreadsheet-based report generators with Excel add-in as interface.

a) Proprietary report generators with manufacturer-specific graphical user interface

The proprietary report generators with manufacturer-specific graphical user interfaces (here based on the relational and multidimensional data sets) are the successors of the classic reporting systems, based on the purely relational data sets, as they were used in the formerly designated reporting systems of the management levels:[68]

- MIS (Management Information Systems)
- EIS (Executive Information Systems)
- LIS (Management Information Systems)
- etc.

The providers distinguish between the **report development level** and the **report analysis level** in reporting.

An important aspect of report development is the design freedom for presenting the report contents. For creating reports, no extensive knowledge of database structures and programming languages is required; however, this would be helpful. The reports can usu-

[68] See also the explanations of the historical development of Management Support Systems (MSS) in Sect. 5.2.

ally be created technically without the IT department and without IT specialists and consultants. Nevertheless, some support is usually advisable.

The proprietary report generators have the following functions for the **development level** (development suites, reporting or report editors):

- placement of all objects available in the system, especially the key figures, dimensions, attributes and hierarchies, in the report to be created or maintained
- definition of navigation, selection and grouping aids for use in the report
- placement of design elements (buttons, tabs, text inputs, selection boxes, etc.) in the reports to be designed
- chart creation and integration
- layout functions (placement of report elements such as tables, charts, logos, etc.)
- use of further integrated functions, such as the free data research and query generators for the report design
- use of a macro programming language for developing individual reporting functions such as navigation, triggering of print jobs, etc.
- organizational and process support aids: calendar functions, email generation, etc.
- partially proposal-based user-guided report creation via so-called wizard-functions; these query step by step elements and properties of the future report in the user guidance and thus develop a report template (preview image = thumbnail), which the user can then further develop individually.
- predefined standard report templates, so-called templates, help with the design of individual reports
- with the help of document management functions, storing any "back-office" documents behind the reports (e.g. images, contracts, etc.).

A development suite for report generation like this is shown in Fig. 5.30 using the example of the company *Arcplan* with the product InSight.

Data contents can be separated from each other on a page in a report using frames. With the drag & drop functionality, the report objects (e.g. tables, selection filters, graphics, menu and button fields) can be individually positioned, formatted and changed flexibly. Arbitrary comments and corporate identity features, such as logos and other company layouts, can be inserted individually.

In tool templates, individual presentation forms and graphics can also be selected and designed, e.g. traffic light functions, tables, bubble charts, scoring analyses, point, line and circle diagrams and other presentation forms that have already been introduced in Sect. 3.8.4.

The result can finally be viewed by the user in the reporting analysis view (see Fig. 5.31).

The proprietary report generators have the following functions at the analysis level; these can be easily operated by mouse, keyboard control or touchscreen:

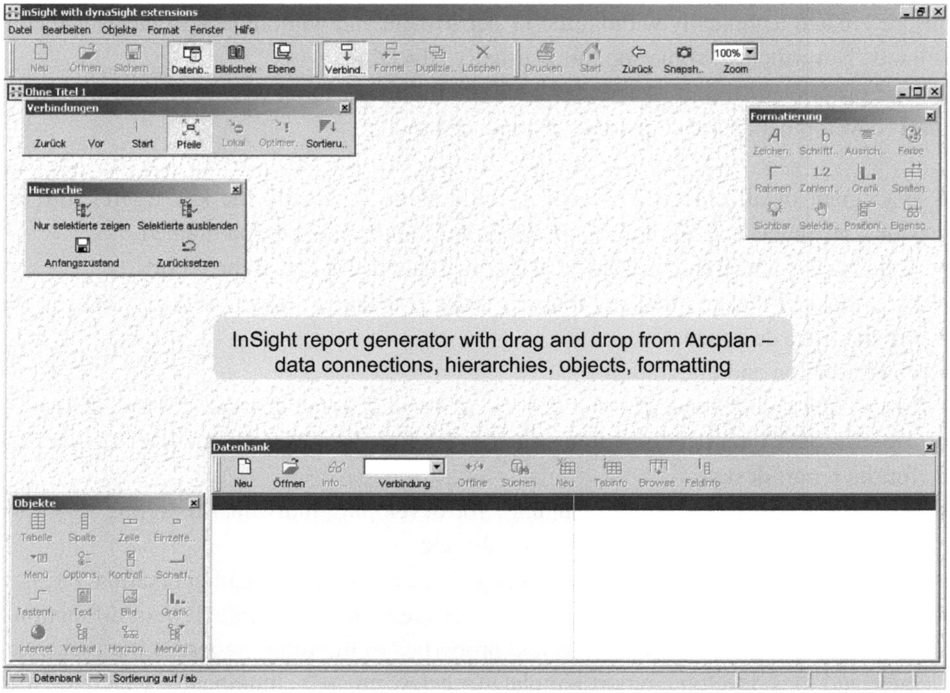

Fig. 5.30 InSight report generator with drag & drop from Arcplan (Taken from Schön 2004, p. 325)

- selection of all possible combinations of key figures with regard to dimensions, attributes and hierarchies
- use of navigation and selection aids
- use of design elements (buttons, selection boxes, etc.)
- dynamic table and chart adaptation, e.g. by selecting specific objects
- triggering of additional report functions (e.g. projections, consolidations, drill-through, drill-down, etc.)
- output and forwarding functions
- use of information or help systems.

The interfaces of the report generators are also called **cockpits** or **dashboards** by the software manufacturers, especially when only the most important data are summarized under one interface for the top management level. Here, both proprietary and internet-based interfaces can be found. Examples for cockpit or dashboard solutions are shown in Fig. 5.32 and 5.33.

Modeled report links allow later analysis to jump to linked reports at the touch of a button.

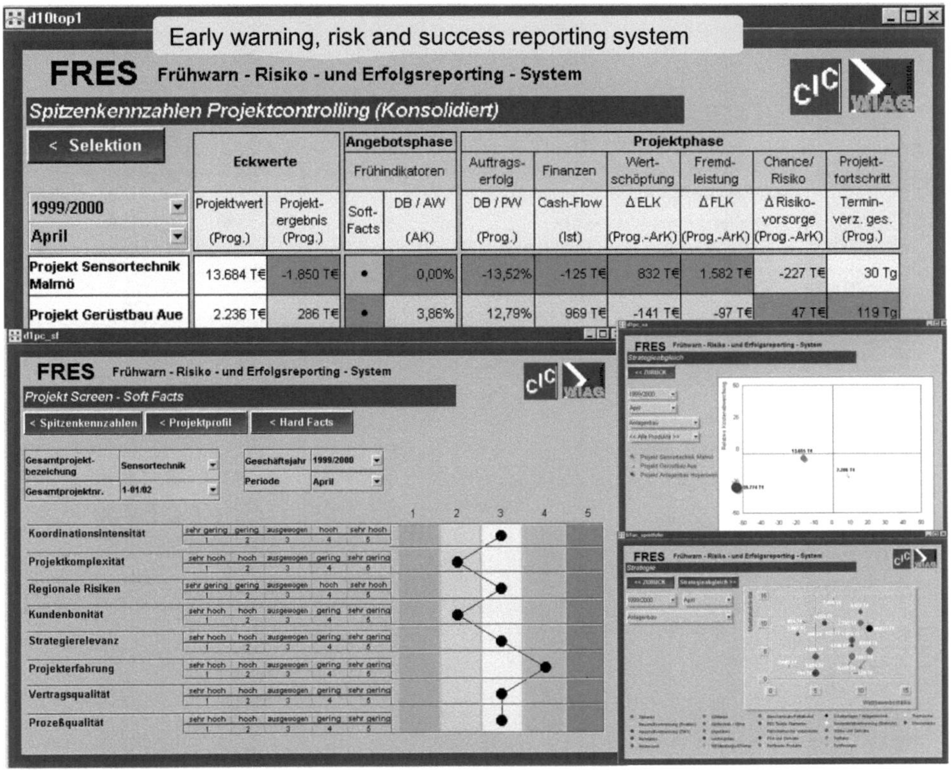

Fig. 5.31 Reporting analysis with traffic light control, utility value analyses and portfolio technique (with DynaSight from Arcplan) (Taken from Schön 2004, p. 326)

In addition, methods for individual design are provided to enable special business analyses with their peculiarities. Examples of this are the ABC analysis and the portfolio analysis.

b) Internet-based report generators with web interface

The internet-based report generators with web interfaces (as cockpit or dashboard solution) differ from the proprietary report generators with manufacturer-specific interface primarily only by the web interface, something which brings the following advantages:

- use of the available and familiar web browser interface
- easy to learn, easy to administer
- no additional installation of an additional graphical user interface

The internet-based report generators use the widespread internet technologies with the common exchange and transmission protocols. The common HTML format is often used.

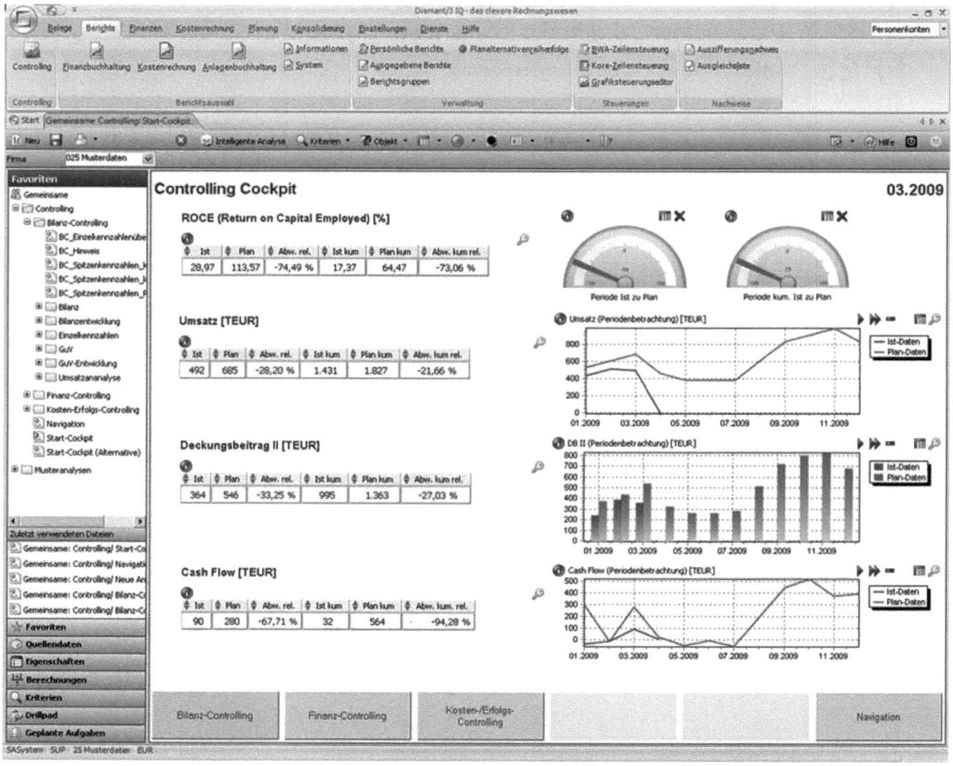

Fig. 5.32 Cockpit/dashboard example (Diamant®/3 IQ) *Source* Schön and Müller (2010, p. 155). Cf. also URL: http://diamant-software.de/291.html (accessed on 12.09.2011)

Other formats such as PDF, CSV, XML and XBRL etc. are also possible. In principle, it is also true for internet-based report generators that no deeper understanding of database structures and no knowledge of programming is required. The reports can usually be created technically without the IT department and without IT specialists and consultants. Nevertheless, some support is usually advisable. The internet-based report generators generally have the same functions, tools and possibilities at the development and analysis level that have already been listed for the proprietary report generators. The only difference is that they are geared to internet technology. Therefore, they are not listed here and reference is made to the previous chapter. In practice, some software solutions from manufacturers who offer both interfaces show that the functionality when using the proprietary interfaces is usually somewhat more extensive than that with the web interface.

c) Spreadsheet-based reporting systems with Excel add-in as interface

As a reporting interface, MS-Excel still has a high share compared to the proprietary and internet-based interfaces and as an output format, which is mainly due to the following reasons:

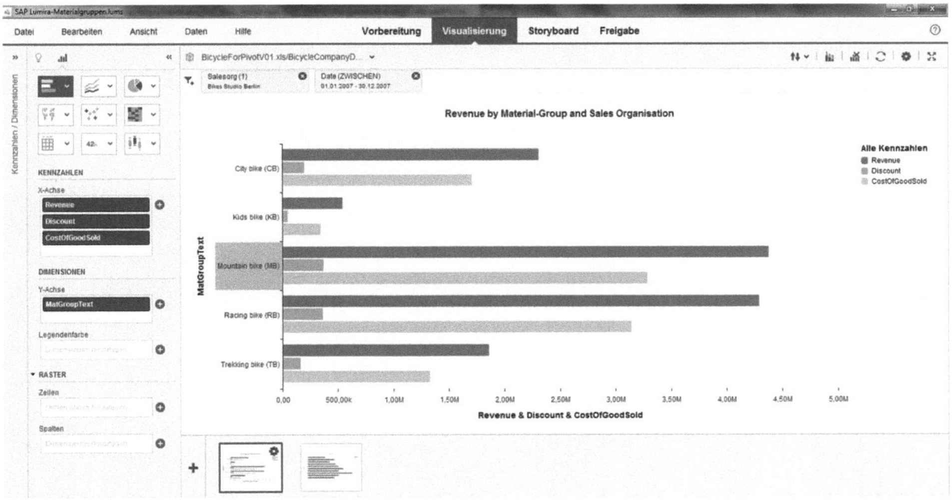

Fig. 5.33 Cockpit/dashboard example (SAP Lumira) (*Source*: SAP Lumira with sample data)

- popularity of MS Excel
- flexibility in use and design
- easy to learn and administer
- low costs

Excel is therefore used by many software manufacturers as an integrated add-in interface for reporting and planning. Here, an integration between the data storage of the data warehouse and the Excel interface as a reporting and input medium is created. As a result, some of the analysis and input values in the Excel table are permanently wired to the data warehouse. To edit (filter, aggregate, etc.) these data, special additional functions are offered as Excel add-ins, which in turn offer many report development and analysis possibilities.

In addition, the usual Excel functions can be applied in the other (non-integrated) parts of the Excel table, which should not be used in the integrated parts of the table.

It is expected that web interfaces and proprietary application interfaces will catch up with Excel-integrated interfaces in reporting. Excel will have a value as an (integrated) interface. The users may switch between application interface types.[69]

The Excel add-in-based report generators have almost the same functions, tools and possibilities in the **development and analysis level** as those already listed for the proprietary and web-based report generators. However, the following points are **not** or only limitedly covered:

- organizational and process support aids
- wizard functions
- predefined standard report templates/templates

[69] See also the survey results of the study by Schön (2011, p. 31).

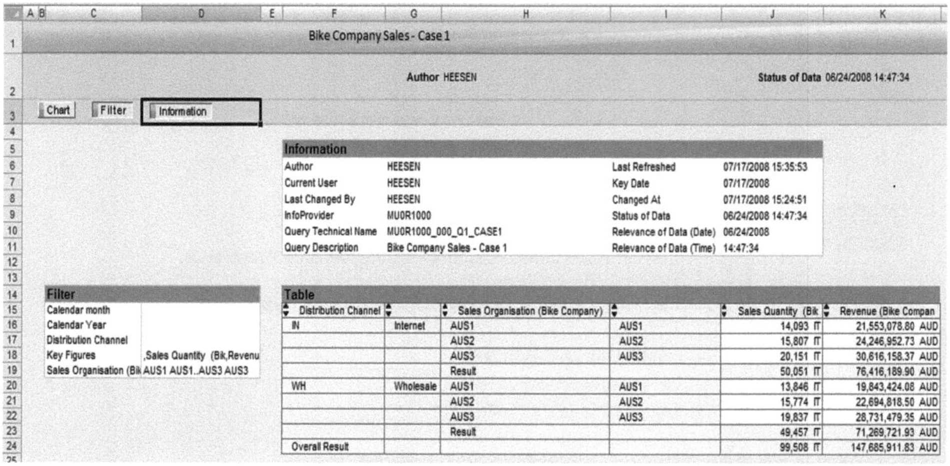

Fig. 5.34 Excel add-in as an interface (here SAP BEx Analyzer) (*Source*: SAP AG (2008), SAP BEx Analyzer)

- use of further integrated functions, such as free data research and query generators

A disadvantage is also the possibility of manipulation and error-proneness of values in the table areas that are not connected to the central data source of the data warehouse, but were additionally entered individually into the Excel tables.

An example of the Excel add-in-based interface is shown in Fig. 5.34.

The basic advantages and disadvantages of the respective surfaces are listed in Table 5.3.

Table 5.3 Comparison of the surfaces

Surface type	Advantages	Disadvantages
Proprietary, manufacturer-specific graphical user interface	• Very high functionality and design possibilities	• Higher learning effort • Own graphical user interface must be installed • Higher time expenditure for the development of the reports
Web-based interface	• High functionality and design possibilities • Only browser installation necessary • Quick familiarization by report recipients • Low costs	• Higher time expenditure for the development of the reports • Functional scope partly lower than with the proprietary user interfaces
Excel add-in as interface	• Flexible design possibilities • High degree of familiarity • Quick familiarization time by report creators and recipients • Low costs	• Limited functionality • Error-proneness and manipulation possibilities in the non-integrated table areas

5.5.5.3 Model-Based Analysis and Forecasting Tools

For reporting and planning, the model-based analysis and forecasting tools are used for special analysis and planning needs. In contrast to free data queries or OLAP analyses, which only require aggregations and minor calculations such as additional key figure calculations, model-based analysis and forecasting tools require more extensive algorithmic or rule-based calculations.

Examples of such systems are those that use operations research, decision support systems, expert systems, data mining and predictive analytics. Systems that apply operations research usually have a problem definition and look for a mathematical model that provides an optimal or near-optimal solution. They are used for well-structured and semi-structured problems. Decision support systems (DSS) provide solution approaches for problems with unknown, semi-structured and unstructured decision situations.[70] This is a classic model-oriented analysis system that provides decision support for complex situations and even when there is insufficient information available for the analysis problem. For this purpose, it requires various components or requirements, which are listed below as follows:

- data base, often data marts or flat files
- model and method databases:
  - method databases that contain standard and special algorithms, especially heuristic, statistical, financial mathematical and prognostic methods, in which one can also store other, partly self-developed methods.
  - model databases, in which a set of methods are summarized.
- application support and dialogue guidance: these help people via the user interface to the computer to use the available method and model databases on the existing data stock comfortably and to evaluate the results.

Smaller DSS solutions can already be implemented with spreadsheet programs or simple database management systems as they already have the three components mentioned above.

Further systems or systematic applications of models and methods, such as expert systems, data mining and predictive analytics, are explained in Sect. 5.7 "Artificial Intelligence".

5.5.5.4 Special Features of Planning Tools

Planning tools,with which simple planning solutions for selected functional areas, but also complex, integrated corporate planning models can be mapped, are based on the

[70] Cf. Turban et al. (2004, p. 103). See also the explanations in Sect. 5.2.

same technical prerequisites of a data warehouse (ETL processes, OLAP etc.), as they are also used for reporting solutions, which were explained in the previous chapters. However, there are special features that specifically concern the planning tools, which are highlighted below.

The software providers offer planning solutions that range from planning platforms to standard solutions.[71] Planning platforms represent an open development environment that provides methods and models as tools for developing a customized planning solution. In addition, simple planning templates can be used and adapted flexibly.[72]

Standard solutions, on the other hand, are planning solutions that are preconceived in terms of business management with many standard templates, e.g. for planning forms and functions, that are standardized and can only be adapted to a limited extent by the company. There are also hybrid forms, where standard templates exist for some areas, and the development environment with methods and models as tools is used for other areas.

Furthermore, the software providers differ in those who understand the planning solutions as part of a whole product range, and those who offer the planning solution as a special program detached from a product range. In the first case, these are, for example, ERP system providers who offer production and logistics as well as accounting and personnel management for the operational business process, and also provide executive systems with planning and reporting components. In the second case, there are other software houses that specialize entirely in the development of tools and templates for planning. These systems have accordingly a higher integration effort when it comes to connecting data sources or passing on information. The planning systems can be designed as platform solutions, as standard solutions or as hybrid forms in both cases.

In terms of **data storage**, the data (Cubes) used in reporting cannot simply be used for planning purposes as well since, from a technical point of view, reporting cubes are only optimized for read accesses and for a comparatively small number of simultaneous accesses.

When generating planning data, however, new data sets are created, and the performance for write accesses becomes more important; for this purpose, e.g., a separate cube or a special partition must be created that are specialized for planning purposes.[73] The multi-user capability must be ensured for both planning and reporting.

The SAP solutions SEM BPS or SAP BI IP, for example, provide a base cube with read-only rights (for reporting purposes) into which data is loaded via the ETL process. In addition, a transactional cube with write and read rights for planning is required, which is now called a real-time capable InfoCube. This cube, which is still called transactional here, must be able to separate, for example, different actions:

---

[71] Cf. Dahnken et al. (2004, p. 55 ff.).

[72] Cf. Meier et al. (2003, p. 90 ff.).

[73] Cf. help texts of the SAP portal (2010).

- transactional cube can be loaded with data → no planning allowed
- transactional cube can be planned → no data loading allowed

In order to avoid switching between planning and data loading for transactional cubes in the SAP solutions mentioned, it is advisable here to create a **virtual connection** of the base and transactional cubes with the help of a so-called **multicube** with which planning and reporting purposes can be fulfilled simultaneously.

This allows to circumvent the restriction that the actual data, which are often required as a reference for planning, cannot be loaded directly into a transactional cube from upstream sources.[74] In this case, three objects are needed for data storage and maintenance:

- A conventional (non-transactional) **base cube**, which is filled with data to be loaded via the modeled ETL process.
- A **transactional cube**, in which planning data are generated with reference to the loaded data of the conventional cube.
- A multicube (virtual layer), which virtually connects the base and transactional cubes to support both planning and reporting purposes.

Due to the write accesses of different planners, planning systems need special service functions that enable parallel work. Objects that are currently being planned by one planner cannot be planned by others at the same time. Moreover, loading and writing processes must not interfere with each other. Here, locking concepts and functions that steer this task help.

Filling a transactional cube can be done via three ways: via manual planning in the planning form, via the application of planning functions, or via loading data from upstream sources.

Another special feature of planning tools are the methods and models for representing the planning. In a **planning environment**, all planning objects and planning functions can be administered. It thus forms the central work environment for planning. For a planning system solution, the following objects and functions are needed:

- Planning areas
- Planning functions
- Planning forms
- Planning process support

Planning areas comprise self-contained planning areas that deal with a part of the overall planning. It makes sense to assign, for example, sales, financial and cost center plan-

---

[74] Cf. help texts of the SAP portal (2009).

ning etc. to their own planning areas. Also, the subdivision of these areas into smaller planning areas is useful depending on the complexity of the planning model.

An important planning task for complex, integrated corporate planning models is the **integration and coordination of the sub-plans**. For this purpose, the planning systems also offer solutions to consolidate and compare data. In addition, they support coordination and approval processes.

A planning area in the SAP planning system SEM BPS or SAP BI IP contains, for example, a data basis in which the plan data including master data (attributes and variables, such as planning periods) are provided, and it includes additional settings that allow the manual or machine processing of the plan data. Several planning areas can also be virtually combined into a multi-planning area.[75]

Below the multi- or planning areas, there are often further administrative levels such as planning levels or planning packages, in which, among other things, the actual data can be loaded as default values, and the characteristics, characteristic values and key figures to be planned can be filtered or selected. Furthermore, planning functions can be made available here.

Planning functions change transaction data to generate plan data. When defining a planning function, it is determined, among other things, which values should be used as base size and as calculation size to determine the planning values as target sizes. Typical planning functions are (see also Sect. 3.10):

- copy
- distributions, e.g.
  - top-down with or without fixing non-changeable values
  - residual value distributions
  - reference distributions
  - seasonal distributions
  - key distributions
  - …
- aggregation
- consolidation
- reconciliation functions
- undoing data changes (rollback function)
- revaluations
- conversions of units and currencies
- extrapolation and forecast functions
- simulations and sensitivity analyses (among others what-if- and how-to-achieve-simulation)

[75] Cf. help texts of the SAP portal (2009), URL: http://help.sap.com/saphelp_sem60/helpdata/de/05/242537cedf2056e10000009b38f936/frameset.htm (accessed on 09.02.09).

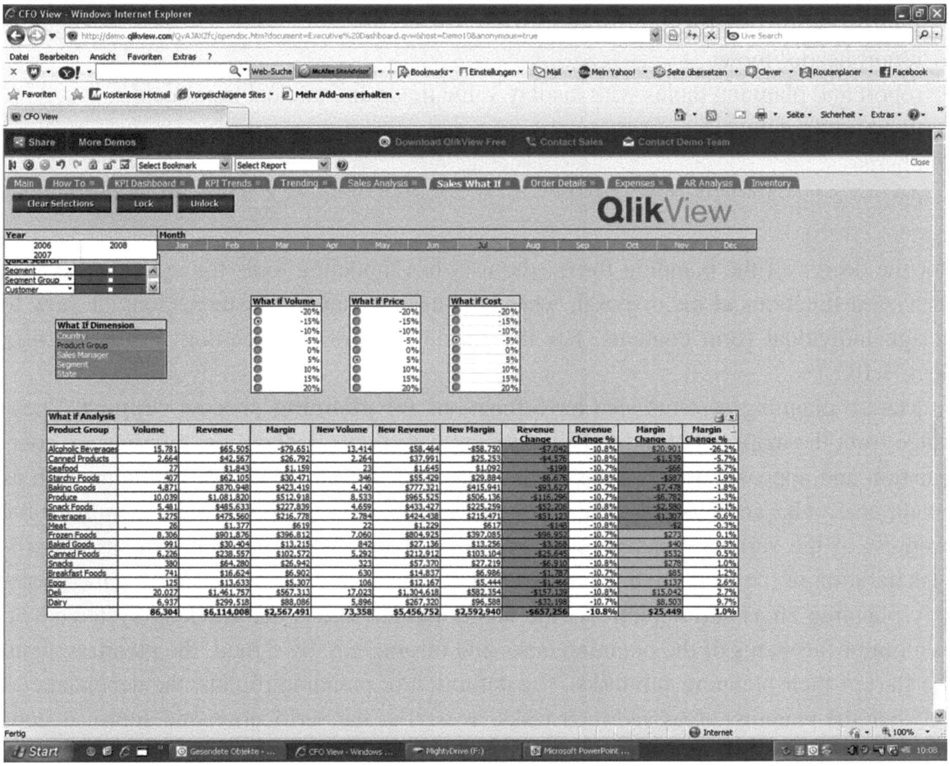

Fig. 5.35 What-If-Simulation (Example QlikView) (*Source*: QlikView-Demo Sales Analysis (2011))

- mathematical and statistical formulas
- transfer
- deletion
- manual entry/change

For many planning functions, a result is usually expected immediately that can be triggered and analyzed "on the fly". An example of a What-If simulation, which instantly calculates the change of the input factors on the outcome, is shown exemplarily in Fig. 5.35.

A **planning form** is basically a form for capturing planning data and the corresponding comments. The user is given the possibility to plan new data or to use loaded or calculated values for reporting and analysis purposes.[76] Maintenance functions allow the layout of the planning form to be defined in all facets:

[76] Cf. Egger et al. (2005, p. 163 ff.).

- filter/selection criteria
- report header/footer
- report and planning tables with display value fields and manual input fields
- planning functions
- charts
- etc.

For the design of the planning forms, the user has modeling tools for individual design or wizard functions at his disposal, where the user is guided by suggestions on how to design individual form contents. For the special features of planning forms, see also Sect. 3.10.

Mature planning systems also have functions for **planning process support**. These range from the mapping of individual planning workflows, scheduling, monitoring, coordination and approval of sub-plans to the integration of collaboration-functions such as email, chat, wiki and blogs. So-called **status and tracking systems** help to monitor the progress of the various planning tasks in the company. These systems support the process iteration of the respective planning tasks and help the planners to process the necessary planning areas and objects. As the name implies, status and tracking systems help to monitor the status of the planning tasks and inform, e.g. by e-mail, the planners about the start of their planning sub-tasks. The definition of planning rounds, the dependencies between the planning areas and their tasks, as well as the verification and approval steps and the associated releases are also supported.[77]

As planning interfaces, the user interfaces already introduced in Sect. 5.5.5.2 are possible:

- MS Excel as add-in solution
- Proprietary graphical user interface
- Web interface

5.5.5.5 Portals and Embedded BI

In **portals**, various analysis systems and information (e.g. based on the data warehouse but also on other systems) are connected by a uniform user interface and usually made accessible on the web. For a better handling and more comfortable operation, the portal is personalized according to the user, which means that the interface and use of his or her required analysis and planning tools etc. can be individually adapted and optimized for the end user. For example, it is possible to create and manage one's own folder structures for reporting.[78] On an aggregated level, role-based portal alignments for specific user groups are also useful.

[77] Cf. Knöll et al. (2006, pp. 212–215).

[78] Cf. Egger et al. (2009, p. 101 f.).

In contrast to standard software reporting solutions, which allow the creation of reports and plans only from a specific workstation or another device, portals can integrate different applications sensibly using internet technology. Examples of such portals are e.g. SAP Business Objects with Info View from SAP AG or the SharePoint portal from Microsoft. All information and applications can be displayed and used in a web browser, without the need to call up another program separately. Therefore, the single sign-on, the one-time login for various programs and tasks, is an important advantage of **portal or SharePoint** solutions. Only the individual or role-specific content and applications are made available via the login check of the authorization profile. Generally publicly accessible and useful information, such as current company announcements, business calendar, stock prices etc., enrich the information possibilities of the portals.

Another advantage of portal and SharePoint solutions is the interaction with other users. For example, chats and video conferences for presentations are enabled or reports and documents can be worked on together. The corresponding software applications for this can be started via the portal.

The following types of portals can be distinguished:

- public portals (catalog, search, news services, social networks etc.)
- personal portals (personal web pages)
- enterprise portals (e.g. company intranet)

To enable simultaneous access to multiple connected applications, portals use so-called portlets. Each portlet forms an independent window for the associated application within a report page. These portlets can be arranged individually on the screen and minimized by the user. Portals are generally based on internet technology today and are therefore largely opened in web browsers.[79]

Reporting and planning are integrated within the enterprise portals. They can be accessed worldwide from any device, either workstation or mobile devices (cell phones, PDAs, smartphones etc.) via the internet and the secured portal access.

Due to the important decision support function for the management, reporting and planning are important components for portal solutions and can be usefully complemented by the communication function and other information areas, as shown in Fig. 5.36. In addition to the reporting and planning applications, collaboration and office functionalities, such as calendar, email services, search engines and chats etc. can also be integrated. The joint work on projects and tasks is another functional area that can be usefully supported by portals, especially when many different people are involved. Here, individual project tasks, processes and requests can be planned, assigned, controlled and coordinated (project, hotline, task and workflow management).

[79] Cf. Gluchowski (2010, p. 278).

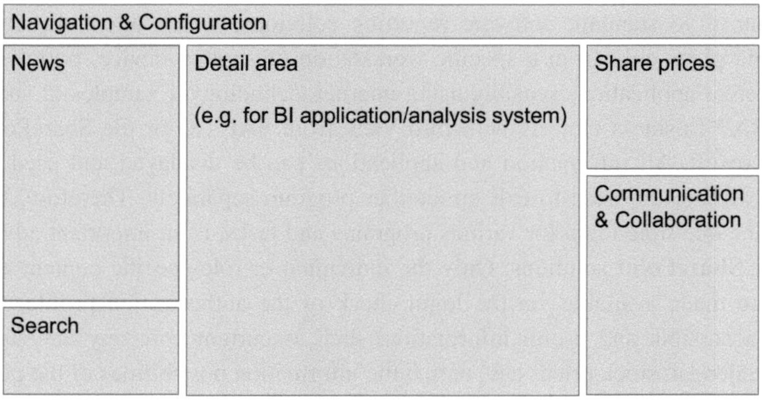

Fig. 5.36 Schematic structure of a portal (Taken and shortened by the border from: Kemper et al. 2010, p. 158)

Under the term **Embedded BI** established ERP system providers try to extend their solutions by embedded BI functions. The ERP system's own functions for reporting and planning are thus usefully supplemented or partly even completely replaced. Here, for example, individual BI objects such as reports or complete dashboards can be integrated.[80]

5.5.5.6 BI Monitoring, BI on BI and Reporting 2.0

The bigger a BI system becomes, the more difficult it is to keep an overview of the entire system. Nevertheless, it is important for companies to know how effectively their BI systems are used. **BI Monitoring** or **BI on BI** refers to the ability to monitor and control the BI system itself and its performance based on key indicators. For example, the number of data accesses, the queried reports or the number of errors in the data loading processes should be analyzed. The goal is to gain insights into the performance, data quality and usage rate of the BI system. This allows measures to be taken that continuously improve the BI system, so that it can be used more effectively in the company.[81] From the perspective of Forrester Research4 experts, the market for BI monitoring solutions is not very mature yet as even companies that already offer BI monitoring usually focus on the presentation of quantifiable indicators such as performance and rarely draw attention to how the systems are used in the company.[82] An example of a BI monitoring shows the main dashboard of Qlik View (Fig. 5.37).

[80] Cf. Search Business Analytics (2016).

[81] Cf. Lixenfeld (2015, p. 24).

[82] Cf. Forrester Research (2016). Forrester Research is a US-based market research company for the information technology sector.

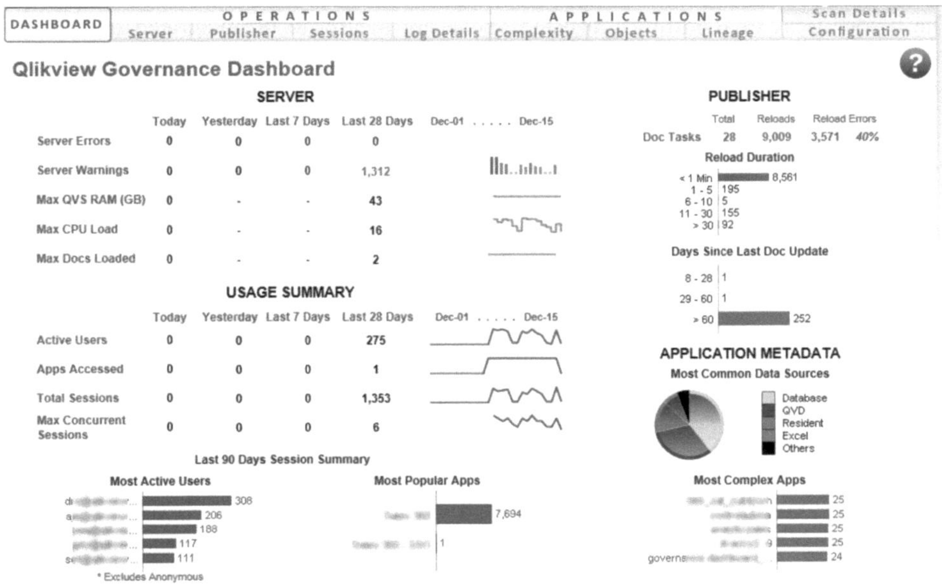

Fig. 5.37 Main dashboard of the Qlik View Governance Dashboard (*Source*: QlikTech (2017))

Reporting 2.0 refers to a form of reporting that emerged in alignment with the presentation of the use and evaluation of profiles in social networks or on the Web 2.0. This is about reporting on the usage behavior and the evaluation of the reporting itself. For example, users are analyzed in the portal or in the BI system with regard to reporting. The following list shows possible evaluation areas:[83]

- A history reporting shows which reports the user himself (and which users in general) have visited in the past period and which user last analyzed the report (last user).
- A ticket reporting shows the activities (to-dos) for users that need to be done. They were sent, for example, by a push report from a sender or due to a reached tolerance value.
- A rating reporting shows reports in terms of a scoring. Evaluation can be done, for example, by giving "likes", grades or other evaluation methods.
- A user reporting shows all users who have analyzed the same report or a similar report profile as the user himself.
- A tags reporting contains keywords and markings that are stored in the reporting system and can be evaluated in terms of frequency and usage.

[83] Cf. Gluchowski (2010, p. 278).

The report areas of reporting 2.0 listed above are ideally to be integrated into the BI monitoring. They complement the performance analyses in the BI monitoring with regard to usage and evaluation.

5.5.5.7 Distribution Possibilities

In this section, the distribution possibilities for planning and reporting are briefly outlined:[84]

For the **distribution**, the following ways are basically available:

a) Use of the reporting and planning functions within the analysis tool:
   - creation of one-level or hierarchical folder structures, lists, tabs or report folders (Briefing Books), e.g. for persons, roles and areas,
     in which the reports are stored in a format or
     in which direct access to the reports and planning forms in the respective interface (e.g. proprietary, web or Excel add-in) is possible.
   - creation of navigation reports and menu controls, e.g. for persons, roles and areas, from which the reports can be retrieved in an output format or
     in which direct access to the reports and planning forms in the respective interface (e.g. proprietary, web, Excel add-in) is possible.

b) Distribution of reports and plans outside the analysis tool in the following forms:
   - various output formats with email (manually or possibly supported by the analysis tool)
   - various output formats in simple document directories
   - various output formats in content/document management systems
   - various output formats to special reporting and planning servers via managed query environment
   - various output formats to portal solutions
   - various output formats for specific communication services (chat, video conferences, etc.)
   - online links
   - direct integration of the analysis and planning systems in portal solutions.

Due to the overload of the central servers regarding simultaneously accessed reports with many queries, a different approach is pursued under the keyword "**Managed Query Environment**" (MQE). In the MQE, a separate reporting server is used in the 3-tier architecture between the database and the client application level. Its tasks are to prepare and enrich the raw data from the database for the reports and to perform the pre-formatting for the reports. The reports are stored and managed as copies on the reporting server

[84] Cf. among others Kemper et al. (2010, pp. 148–153).

(content descriptions, output formats, etc.) and retrieved and evaluated by the end user. The goal of MQE is to relieve the database and application servers.[85]

By pre-calculating the reports, the frontend computers are also freed from complex calculations during the report creation. However, the disadvantage here is that the currency of the values then depends on the loading and preparation time of the reporting server. For this reason, complex calculation runs for report creation on the reporting server, which are resource- and time-consuming, are shifted to low-load computer times. Here, scheduling and monitoring functions help.

Advantages also result from the significantly lower network load as a pre-formatted report contains much less data volume than the raw data base on which the report is based.

5.5.5.8 Output Formats and Media

As output formats for planning and reporting, the following formats are mainly available:

- Document formats
  - web formats (HTML etc.)
  - print formats (*.pdf etc.)
  - text formats (*.docx, *.txt etc.)
  - presentation formats (among others *.pptx from MS Powerpoint)
  - graphic formats (*.jpg, *.bmp etc.)
  - video and audio formats (*.mp3, *.avi, *.wmv etc.)
  - e-books (*.epub [open standard for electronic publication])
  - …
- Machine-readable data formats and standard formats for specific application components
  - CSV files (Comma Separated Values)
  - spreadsheet format (among others xlsx from Microsoft)
  - proprietary format of the analysis tool
  - XML-format (Extensible Markup Language)
  - XBRL-format (Extensible Business Reporting Language)
  - PMML (Predictive Model Mining Language) for data Mining models
  - …

Reports and planning forms are often converted into different output formats for further processing for printing, for downstream systems or for presentation.

Pure document formats are well suited for paper-based reporting and sending electronic mails or storing fixed reports in portal and web solutions or in special document directories, content or document management systems up to own reporting servers.

[85] Cf. on MQE among others Manhart (2011a).

Presentation output formats are well suited for presenting the report contents e.g. in management circles, assemblies or other meetings. In contrast to the fixed representation of report contents as in print, the analysts can partly even interactively navigate in the data set in the presentations, without having to branch directly into the system. This is also useful for storyboarding where a summarized report is enabled by using various analyses and comments. This allows for explanations and statements on what happened, reasons for forecasts as well as recommendations and measures for the future to be embedded in an overall context (storytelling).

Machine-readable data formats and standard formats for special application components are used for further processing of reports and plans in downstream systems. Often MS Excel is used as a spreadsheet system, in which however the maintenance of the plans and reports is partly very complex and time-consuming. Less common is the export function for other systems (e.g. to an ERP system or to a database management system such as MS Access).

The software applications for reporting and planning usually have their own (proprietary) graphical user interface and proprietary output formats with which the reports are provided. These can be accessed directly via the reporting software by IT access. In management circles and other meetings, the reporting software can be used in analysis meetings in office, conference or presentation rooms with projector, screens or other technical display options. The reports and the selected data set can be analyzed flexibly. The disadvantage is that the software and the graphical user interface must be provided on the computer or via the computer access, something which is usually associated with license costs.

Since the introduction of the World Wide Web, the internet- and intranet-capable web formats such as HTML are indispensable for the analysis of reports. Some software vendors rely only on the web technology for the output format of the reports and even for the entire application. If the PC or another device on which the analysis is accessed has a browser, the report can be viewed independently of the reporting software. The advantage here is that no license and no additional installation costs for the software on the respective devices for the analysis are incurred. For many internationally active companies, it is also advantageous that reports and plans can be accessed from anywhere in the world where there is an internet access. In addition, the output function should be possible on any stationary and mobile devices with the reports adapting to the size and the chosen resolution of the devices (see Sect. 5.10.5).

For the bundling of reporting, planning, but also many other applications, web-enabled portal solutions are increasingly used. They have the advantage that the whole information diversity of the company for all its external and internal users and business partners can be designed and used under one interface.

For the exchange of business data, other languages such as XML or further developments such as XBRL are also increasingly used,[86] which is an international standard

[86] Cf. Pastwa (2010, p. 11 f.).

among others for the exchange of an electronic balance sheet (e.g. E-Tax Balance Sheet)).[87]

From the user side, the following output media can be differentiated:

- Paper printout
  - in various sizes (DIN A4 etc.)
  - coloring (black and white/colorful)
  - binding (spiral, folder etc.)
- Screens and presentation media
  - in various sizes (e.g. 15-inch, 17-inch, 19-inch etc.) and resolutions
  - output medium: flat screen, interactive whiteboards, television, projector, display etc.
  - location-bound devices (PC, network terminal etc.)
  - mobile devices (laptop, notebook, smartpad, smartphone, ebook reader etc.)

When designing the reports and planning forms, different output media have to be considered, which mainly influence the page view and size.

Print reports

Print reports are usually created in Germany in the common A4 format (in the USA in the letter format) landscape or portrait. If print reports are to be generated in accordance with screen presentations, the landscape format is recommended as otherwise the conversion of the screen format output would be larger. The modern reporting design programs today often convert from the screen reports to printable presentation reports and scale the formats from the screen resolution to the print format size.

In the case of reports sorted by objects, such as the cost center report by areas and individual cost centers, it is important to include the general report header information, so that the reports can be clearly assigned later.

In print reports, the color design is still important. If cost-saving black-and-white printers are used, the layout and especially the design of the diagrams must be taken into account, so that the readability of the report is not impaired or even made impossible in some places due to the lack of color. In the case of color-intensive management reports, color printing is preferable due to the higher recognition of the screen report and the better readability. Nevertheless, it can make sense to use the cost-saving black-and-white version for simple report forms.

[87] Cf. Manhart (2011a), Krudewig (2012, p. 29) and Feindt (2014, p. 53 ff.).

Due to the many modern electronic presentation media, print reports will lose importance. Cost savings can be generated here. Many companies and organizations are striving for the paperless office.

Screens and presentation media

Screen reports and other presentation media such as projectors, large screens or large monitors are often used for management reporting, regardless of the output format of the software. Advantages of this presentation are the flexibility in navigating the reporting system, and the color perception.

Depending on the type and media, the screens and other presentation media have different sizes and resolutions. These are partly adjusted by the scaling of the reporting software. However, if the screen size is too small in its resolution, parts of the report information are not visible at first glance and have to be reached by scrolling the screen to the right and left or up and down. To achieve this, some reporting tools help with the report creation by adjusting the report objects to the available screen size. This adjustment is also called "responsive design" (see also Sects. 5.10.3 and 5.10.5.2).

The screen size becomes even more decisive when the management reporting is to be displayed on smartphones. Here, the size of the 3.5″ screen, such as on an iPhone, is not sufficient to reproduce the information of a 19″ monitor. Here, the information division and compression is indispensable.

The currently most extensive information provision for the user takes place directly to his mobile device. Smartphones, smartpads and other mobile devices are already so powerful that they can display detailed report contents. They range from accessing simple PDF documents to web-based reporting systems to direct access to the respective applications. The advantage is the constant availability of information at all mobile locations in the world. The disadvantage is the small resolutions, which are not suitable for more complex reporting material as well as for planning tasks (see also Sect. 5.10).

In general, there is a great technical, but also content-related challenge for reporting and planning in the coordination of the report sizes of print and screen reports. The user should find it easy to switch between the media and to orient himself quickly under the respective new interface.

With regard to the technology, screen size and screen resolution have to be considered. While the screen size (shown in inches) describes the diagonal of the screen area, the screen resolution indicates the dot density per column and row, which can be used to determine the number of pixels of the image (see Fig. 5.38).

For mobile reporting, the following points are important due to the smaller screen resolution:[88]

[88] For the technical design and application possibilities of mobile devices, see also Sect. 5.10.3.

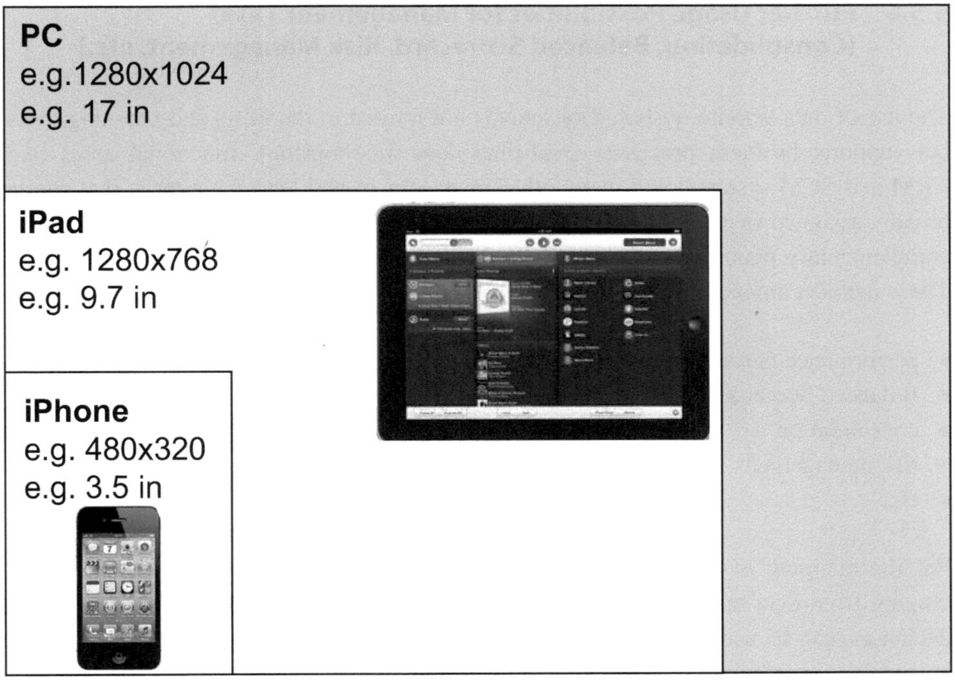

Fig. 5.38 Screen resolution and Screen size. (The brands, images and symbols of the iPhone and iPad are the exclusive property and trademarks of Apple Inc.)

- PDF documents in e-book format with adjusted font size
- Page-by-page navigation instead of active navigation in the detailed report, e.g. with icons
- hiding of unnecessary functions
- reduction of content to core information for One-page reporting
- graduated navigation from core to detailed information
- simple highlights such as with color coding (traffic light colors) or with conditional formatting without overloading with details
- tabs for browsing instead of complex pull-down menus
- context menus of the PC application are not possible on mobile devices.

The disadvantage of the small screen size can be reduced, if possible, by combining with other devices (e.g. TV sets and screens) if the mobile devices are connected to these devices. With the Airplay functionality of Apple, this is possible wirelessly, for example, but again requires compatible devices that can be connected, for example, via a Bluetooth interface.

5.5.6 Further Usage Possibilities for Management Tasks (Consolidation, Balanced Scorecard, Risk Management, etc.)

The use of data warehouse-based systems is not limited to reporting and planning tasks, but supports business processes (real-time data warehousing), functional areas (e.g. CRM and SCM systems) and many other management and leadership tasks that can be broadly assigned to planning and reporting tasks. They can be mapped as independent or complementary planning and reporting areas in the data warehouse and in the frontend. These include, among others:

- performance measurement systems e.g. for value-oriented corporate management
- Balanced Scorecard
- consolidation
- risk management
- etc.

For the mapping of these task areas, the tools described for planning and reporting, ranging from data modeling to data analysis as well as the existing method and model databases, can be used for an individual custom development. This applies especially to business tasks that require many planning and reporting functions, such as the Balanced Scorecard and the performance measurement systems.

Performance measurement systems can be found, among others, in classic concepts such as the RL performance measurement system, the Du-Pont or ROI performance measurement system or in new concepts of "value-oriented corporate management" (Value Based Management); they can be easily created in tabular and graphical form with the reporting and planning tools listed. Modern indicators such as the Discounted Cash Flow (DCF), the Economic Value Added (EVA) or the Cash Flow Return on Investment (CFROI) can be mapped just as easily as traditional indicators. Even the presentation of the value driver trees can be realized graphically, as shown in Fig. 5.39. For special functions, such as the simulation of the indicator development in the value driver trees, extended standard solutions have also been established.

An example of the implementation of the **Balanced Scorecard** using existing methods and model databases was already shown in Sect. 3.12.1 using the SAP NetWeaver technology as an example. The advantage of such individual solutions is the flexible structure and implementation of the individual requirements that a company places on a business method such as the Balanced Scorecard. The main disadvantage is the higher effort for the conception and implementation. If the Balanced Scorecard methodology is expanded beyond the presentation of the pure key figure table with method-specific functions, standard solutions can be considered. Such a standard solution for the Balanced Scorecard includes the following functions in addition to the presentation of the key figure table:

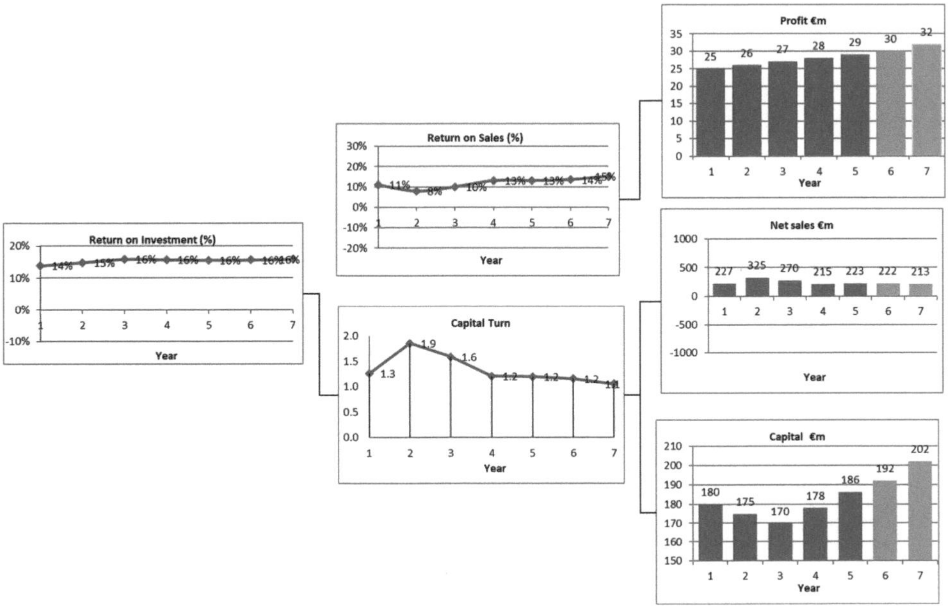

Fig. 5.39 Exemplary Du-Pont key figures tree

- cascading and alignment of the Balanced Scorecard from the central corporate level to area scorecards (e.g. strategic business units and functional areas)
- documentation and communication functions for strategy development
- integration of strategic instruments such as GAP, SWOT and portfolio analysis
- project management functions for the control of strategic activities/measures
- representation and use of feedback diagrams for the presentation and planning of cause-effect relationships

An example of an analysis of the causes and effects is shown in the following Fig. 5.40.

In the areas of **consolidation** and risk management, the number of method-specific functions is even more extensive as many tasks go beyond the pure reporting and planning functions.

In **consolidation**, these are mainly the following tasks:

- mapping the structure of the companies to be consolidated (e.g. full/partial consolidation)
- recording, monitoring, validating, and preparing the data of the companies to be consolidated
- legal group consolidation with capital, debt, expense and income consolidation as well as elimination of intercompany results

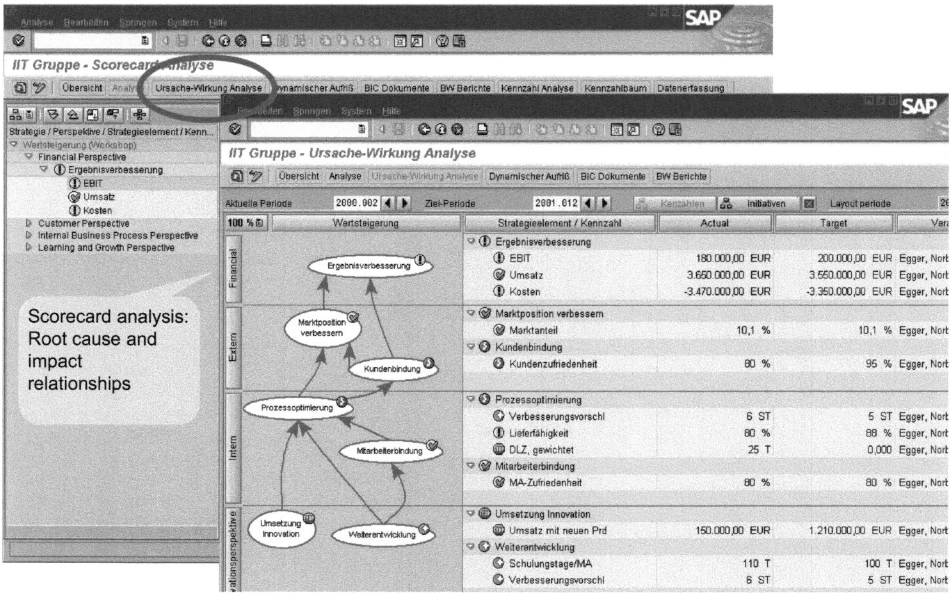

Fig. 5.40 Scorecard analysis with SAP SEM (*Source*: SAP AG (2003), SAP SEM, Scorecard analysis and cause-effect analysis)

- additional accrual and closing entries
- currency conversions
- management consolidation (resolution of intercompany business also for divisions and other result object groups (SGE, business units etc.)
- reporting (balance sheet, income statement, cash flow statement, segment reporting, notes, management report etc.)

In **risk management**, the following tasks are mainly to be mapped:

- risk identification
- risk assessment
- risk reporting
- risk control with project-based measure tracking
- support of the risk management organization (structural and procedural structures)

Due to the many management tasks that go beyond pure planning and reporting functions, standard solutions can be found especially for the above-mentioned management methods or instruments (consolidation, key figure systems, Balanced Scorecard and risk management), which can be based on both relational database technology and OLAP or data warehouse-based systems. Further fields of activity and management instruments, such as benchmarking, life cycle and conjoint analysis, can also be supported by data warehouse technology, to name just a few more possibilities of use.

5.6 Business Intelligence and BI-Supported Controlling

For the mapping of modern reporting and planning solutions, there has been a trend for over 10 years to use systems that are based on data warehouse technology and are described under the keyword "business intelligence" (short BI).[89]

It is undisputed that business intelligence is a very popular term that is used both by science and by software manufacturers and companies in practice. If you search for this term on the internet, there are overwhelming amounts of hits.

However, there is no generally accepted definition for the term business intelligence.[90] The literal translation, for example, with "business intelligence" leads even to the wrong track. The word "intelligence" should rather stand for information service or further for the transformation of information into knowledge.[91] In accordance with the knowledge pyramid, information is linked with experiences, e.g. business insights, and thus results in knowledge.[92]

Especially the distinction from other terms such as Controlling and data warehouse are unclear. It is clear that this term is attributed to the scientific field of business informatics. Mertens, for example, has worked out seven variants from the multitude of definitions:[93]

1. BI as a continuation of data and information processing: information processing for the management
2. BI as a filter in the information flood: information logistics
3. BI = MIS (Management Information System), but especially fast/flexible evaluation
4. BI as an early warning system
5. BI = data warehouse
6. BI as information and knowledge storage
7. BI as a process: symptom detection → diagnosis → therapy → prognosis → therapy control

---

[89] Cf. Winterstein and Leitner (1998, p. 34), Kemper et al. (2010, p. 10) and Chamoni and Gluchowski (2004, p. 119).

[90] Cf. among others the definitions of Schrödel, King and the authors mentioned in the further course of this chapter: Schrödl (2009, p. 9) and King (2014, p. 37).

[91] Cf. Hanning (2008, p. 77).

[92] Cf. Taschner (2013, pp. 9–11).

[93] Cf. Mertens (2002, p. 4).

Most of these variants, however, show that the definitions often delimit themselves by the systems used. The business management methods and control tasks are not considered here at all.

Interesting is variant 5, in which BI is equated with data warehousing. If one looks at the BI framework that Kemper (see Fig. 5.41) has developed, the comparison with the three described data warehouse levels (see Sect. 5.5.1) becomes quickly visible.

The levels of data connection and data provision are shown in the lower part of the graphic, whereby it remains unclear whether the extraction of the data connection is part of the BI or not. Due to the integration of the ETL process, the extraction is, in my opinion, to be taken into account. The data analysis and use is divided into the levels information generation (analysis systems) and distribution (distribution) as well as the information access, here connected with the portal integration.

Interestingly, under analysis systems, concept-oriented systems such as the Balanced Scorecard, the planning, the consolidation and the value-oriented management are also included, which, in my opinion, are to be classified as business methods or instruments for management support. So here too, if not visible at first glance, a connection between business methods and IT is established.

Gluchowski tries to describe BI in the context between the process phase of data provision and analysis as well as the focal points technology and application (see Fig. 5.42).

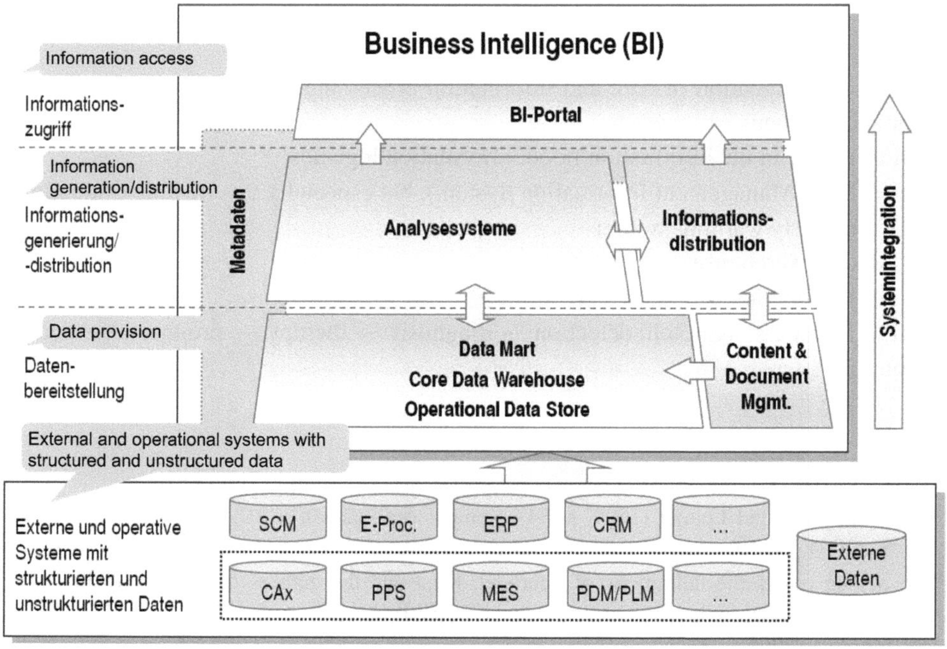

Fig. 5.41 BI framework (*Source*: Kemper et al. 2010, p. 11)

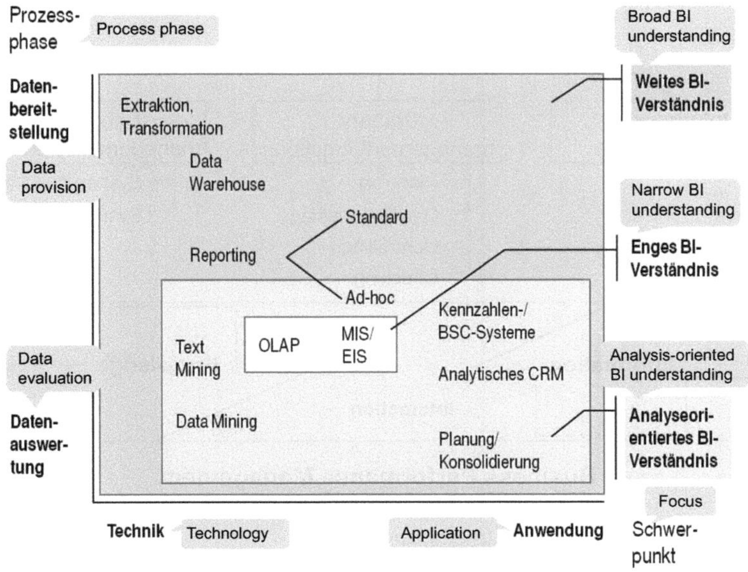

Fig. 5.42 Different facets of BI (Adopted by Kemper, modified from Gluchowski 2001, p. 7. See Kemper et al. 2010, p. 4)

To emphasize is that this classification tries to connect the technical system levels and the (professional) application with each other. Unclear, however, is why planning is rather assigned to the application and reporting than to the technology. Also the vague technical assignments are hard to understand. Unclear is e.g. why planning is more assigned to data analysis than reporting. This is not comprehensible. Therefore, the demarcation must be done more clearly.

If one looks back at the originator(s) of the term BI, one recognizes here first also that the definition focuses on the information system side and less on the professional, business application of the management methods: the *Gartner Group* understands business intelligence as the creative, intelligent use of company-wide available knowledge. Howard Dresner of the Gartner Group coined this term in 1989 for the first time as a transformation process of data into information and, by means of exploration, into knowledge.[94]

He also coined the more advanced term **Business Performance Management (BPM)**. "Performance management" refers to the management of an organization that deals with the performance control (employees, teams, departments, processes, etc.)

[94] See Behme and Mucksch (1997, p. 15).

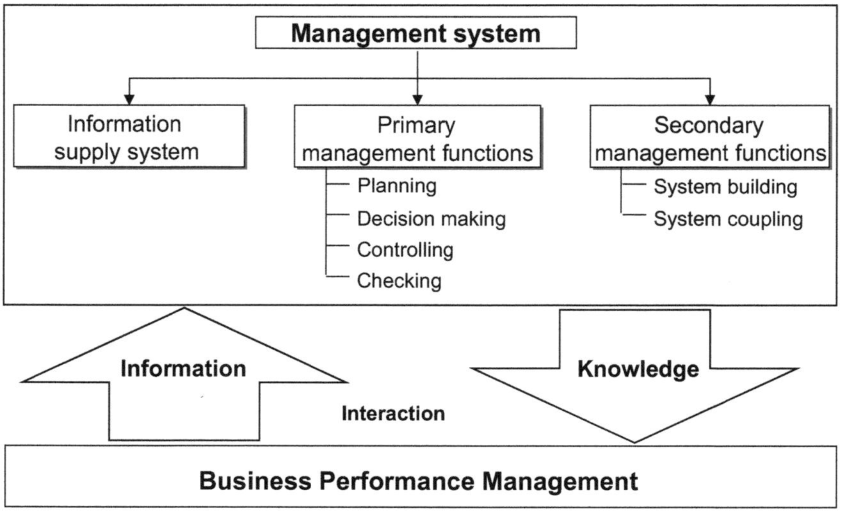

Fig. 5.43 Business Performance Management

in the company.[95] To control and monitor the performance, target specifications, result controls, performance evaluations and the remuneration of the performance, especially of the employees and managers in the company, are used.[96] The measurement of the performance is done by a mix of quantitative and qualitative measures. For performance control, instruments such as the Balanced Scorecard are proposed. The data for measurement and the generated information for performance control are usually provided via the data warehouse levels.

The term business performance management thus represents a further development of business intelligence.[97] BPM comprises business and IT-related methods, tools and processes to improve the performance and profitability of companies.

While business intelligence is rather characterized by data and information processing, business performance management summarizes the interactions between the information supply system and the primary and secondary management functions (see Fig. 5.43). BPM thus integrates information technology and business methods for managing and controlling a company.

The term BI has established itself more strongly in practice than the term BPM and today, according to a broader definition, interlinks information technology and business

[95] Cf. Bange et al. (2009, p. 7).

[96] Cf. Jetter (2004, p. 33).

[97] Cf. Gleich (2001). Alternatively to the term Business Performance Management, the term Corporate Performance Management (CPM) is also used.

management support. Engels' statement regarding BI underlines this connection: "The main objective of business intelligence is the targeted preparation of data and information to improve business decisions."[98]

The demarcation from controlling is also difficult to draw. If controlling is usually associated with the task of supporting the decision-making of managers in the company and with the tasks of planning, analysis, management support, control and their coordination,[99] then support with IT-based systems is indispensable. Business intelligence is therefore defined here as follows:

Business intelligence is the integration of professional management methods, IT procedures and analytical processes that cover both the preparation and provision of data as well as the discovery of relevant relationships and the communication of the gained insights for decision support for the management and, for this purpose, use powerful IT such as data warehouse and big data technology.[100]

An essential distinguishing feature here is the reference to the IT support of controlling-relevant tasks, which is why the term "**BI-supported controlling**" makes this integrative fact between business methods and technical implementation clearer, as the following definition describes:

BI-supported controlling is the goal-oriented decision support for the executives in the company and helps the management with the tasks of planning, analysis, control and their coordination; it uses powerful information technologies in the context of business intelligence such as e.g. data warehouse and big data technology.

BI-supported controlling serves to improve the decision bases for the management, creates the necessary transparency for corporate actions and shows relationships between previously isolated information.[101] From this point of view, BI-supported controlling is a clarification of the supporting character of business intelligence for controlling, which links the use of new information technologies, especially the data warehouse and big

[98] Cf. Engels (2015, p. 15).

[99] Cf. Horváth (2008, p. 125) and Reichmann (2006, p. 13).

[100] Business intelligence definition by Prof. Dr. Dietmar Schön in the field of controlling at the University of Applied Sciences and Arts Dortmund, July 2017.

[101] Cf. Schrödl (2009, p. 13 f.).

data technology, with the professional application methods of controlling for the management. This corresponds to the broad definition of business intelligence.

The overall context of narrow and broad definition for business intelligence can be seen in Fig. 5.44. The term IT-supported controlling is broader than BI-supported controlling. It covers all information technologies and does not limit itself to business intelligence technologies. In IT-supported controlling, the basic technologies e.g. of the ERP systems for business processing and control are also included.

If one only considers the evaluation and usage level of data e.g. via frontend systems, report generators, cockpits, portals etc., then this is a very narrow BI definition. Often the term business intelligence is also only associated with these frontend solutions. In this case, one also speaks of business intelligence tools.

The definition can now be gradually expanded on the side of the IT view or the business view.

Possible operational areas of application are:

- reporting in all functional areas and management levels
- planning, budgeting and forecasting systems
- consolidation systems
- customer relationship management
- supply chain management
- risk management
- stakeholder information systems
- Balanced Scorecard systems

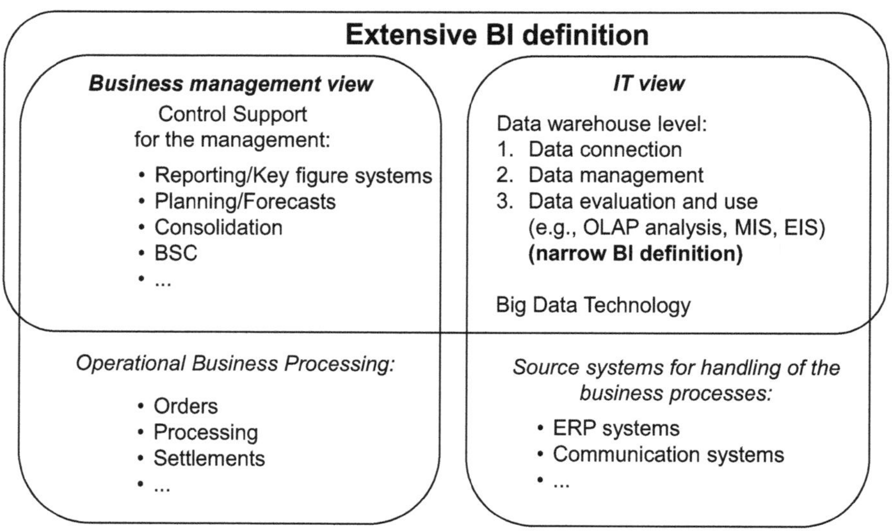

Fig. 5.44 Narrow and broad BI definition

- performance measurement systems, e.g. for value-oriented corporate management
- benchmarking
- …

Figure 5.44 illustrates the relationship between the narrow and broad interpretation of the business intelligence definition.

From an IT perspective, transactional source systems, such as ERP and CRM systems, are delimited. The remaining DWH and big data technologies, including their frontend solutions, are left. In addition, the business management support of the management without the operational business processing is added. For this reason, the broad definition of business intelligence corresponds to the BI-supported controlling. The potential users of transactional systems differ from those of a data warehouse and big data. While the transactional systems are mainly used by the executing level (e.g. clerks), the analytical systems such as the data warehouse are used by controllers, managers or analysts of the functional areas.[102]

Kemper illustrates in Fig. 5.45 the use of BI application systems for controlling activities at all management levels. BI takes on the role of the management system and supports the application of business functions such as planning, controlling and monitoring.

Here too, the bracket between the IT and business perspectives becomes clear, so that in my opinion the term business intelligence increasingly follows the broad definition. This is also shown by his definition of BI: BI "refers to an integrated, company-specific, IT-based overall approach to operational decision support."[103] However, the term IT-based overall approach does not specify the technology used and could thus also include ERP systems. Here, a clear reference to the BI technologies such as data warehouse and big data technology would be necessary.

5.7 Artificial Intelligence, Expert Systems, Data Mining, RPA and Chat Bots

5.7.1 General and Specific (Weak) Artificial Intelligence

Artificial Intelligence (AI) is a subfield of computer science that deals with explor the automation of mechanisms of human and other intelligent behavior (intelligence).[104] For this purpose, information technology, computers, robotics, statistics, machine learning,

---

[102] Cf. Bange et al. (2013, p. 9 ff.).

[103] Cf. Kemper et al. (2010, p. 9).

[104] On the term AI, see Ertel (2016, p. 1ff.) and Lubos (2020, pp. 45–49). On AI in accounting and controlling, compare e.g. Wullenkord (2018, pp. 113–129).

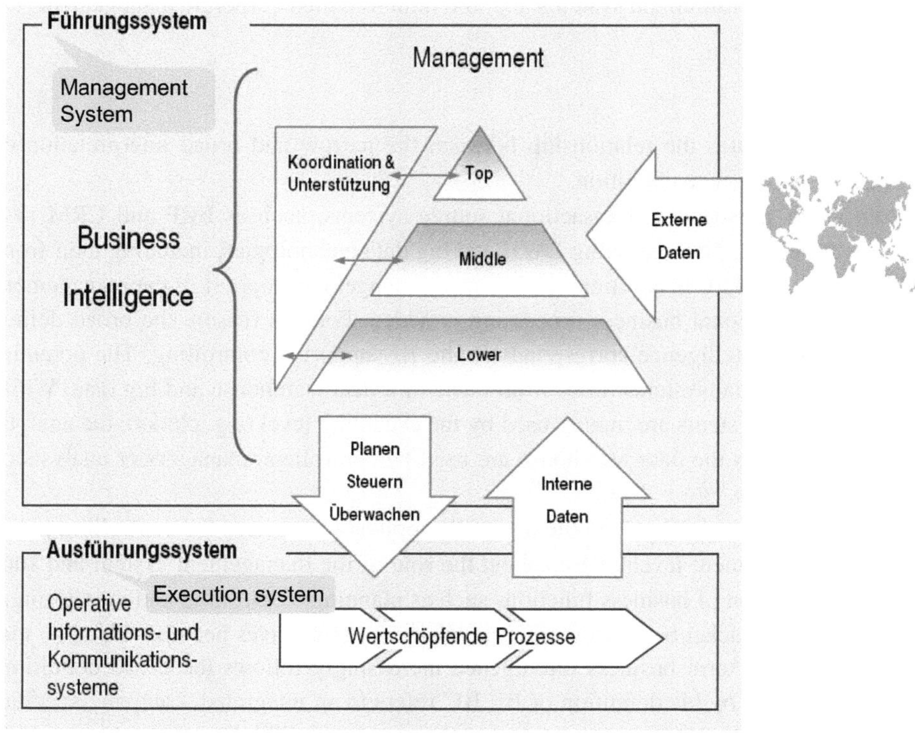

Fig. 5.45 Fields of application of BI application systems (Cf. Kemper et al. 2010, p. 9. The BI use in controlling, especially in reporting and planning, is emphasized by Seufert and Oehler, among others (Seufert and Oehler 2009))

deep learning, etc. are used to strive for intelligent behavior such as perception, language ability, learning ability, judgment ability, adaptability, autonomy, reflection ability, decision and control ability, etc. in the development with AI.

Numerous examples of autonomous and adaptive systems with some artificial intelligence have more or less already arrived in everyday life or will be established in the next few years:

- autonomous driving, delivery robots, drones and autonomous ships
- navigation systems, early warning systems, traffic control systems
- personalized content and responses (online advertising, social media information, music and streaming offers, chatbot responses, etc.)
- image and face recognition
- voice assistants, software, game, work and care robots
- …

Artificial autonomy is the ability to perform tasks in a more or less complex environment without permanent guidance by a human. The adaptability of artificial intelligence consists in learning from experience and thereby continuously improving performance. The experiences can come from humans, from various data analyses of existing systems (e.g. machines, devices and computers) and from own experiences with systems based on artificial intelligence.

The term "intelligence" is both part of the word pair "Business Intelligence" and "Artificial Intelligence". It is difficult to classify and define. What is intelligence and when is someone intelligent? Unlike humans, for example, a language robot can speak many languages and has a much larger vocabulary. Yet humans may be better able to interpret ironic wording.

We have a similar problem when defining quality: What is quality? Quality is often defined indirectly: The customer determines by his or her demands whether a product meets his or her quality perception. But that does not mean that the optimal quality is achieved.

If we also define intelligence indirectly, something is intelligent if it meets the expectations of the observer. In the case of business intelligence, this is the shareholders' expectation for successful management or leadership of the company. In the case of artificial intelligence, it is the imitation of a certain behavior (e.g. talking, analyzing, driving, forecasting, grasping, etc.) as well as the combination of different skills. Whether, for example, the shortest route in navigation is always the best depends on the observer and...

- their objectives (e.g. low fuel consumption, low time, low effort, low costs, scenic route, etc.),
- the conditions (e.g. available technology),
- a special event (e.g. traffic jam),
- the time of day (within or outside of rush hour) and
- other factors.

Artificial intelligence is often distinguished into strong and weak AI or general and narrow AI. Strong or general AI is equipped with a "mind" and "consciousness" (like humans). This means that strong or general AI is aware of itself and can make decisions and actions based on its assessment (intelligence). The environment and the decision options are complex and diverse. Such strong or general AI has not been produced or available on Earth in the last 50 years and is not likely to be available in the foreseeable future. It is only staged in science fiction films based on human imagination.[105]

Weak or narrow AI is a limited but pragmatic intelligence that we have known for years, namely systems that exhibit intelligent behavior for more or less complex tasks,

[105] Cf. University of Helsinki and Reaktor (2020).

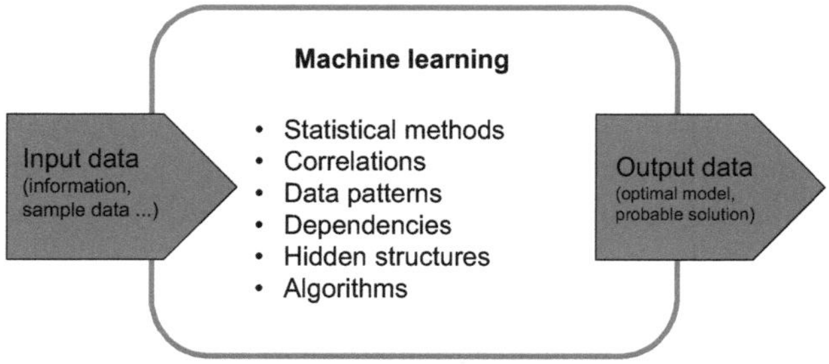

Fig. 5.46 Machine learning

decision situations and conditions, although they are "only" IT-supported computers. The examples listed above are always weak or narrow AI. As with autonomous driving, navigation systems, chatbots, etc., these are limited complex environments in which the advantages of data storage, data processing, the computational models and algorithms used in the IT solutions are exploited. Unlike classical IT solutions without AI, such as simple search functions, weak or narrow AI solutions are able to imitate certain behaviors flexibly, and autonomously decide, learn and adapt.

5.7.2 Machine Learning, Deep Learning and Neural Networks

Narrow AI solutions often rely on "machine learning", "deep learning" and neural networks, so these terms are briefly explained.

Machine Learning is about systems that improve their performance in a specific task through more experience, especially through collected data and used statistical methods, without necessarily being completely programmed through. The focus of machine learning is on the development of computer programs that learn independently by accessing data and thus generate results.[106] A schematic relationship is shown in Fig. 5.46. In order for the programs to make decisions, a human being must train the algorithm, for example. By providing training data, the algorithm can recognize relationships and patterns and learn from the data (step by step), which is why this process can also be called model training.[107] Here, the programs use self-learning algorithms. An algorithm is initially a clear sequence of instructions for solving a problem or several classified problems. Algorithms are usually programmed using software, but it could also be instructions from humans or the combination of humans and computers. Essentially,

[106] Cf. Frochte (2019, p. 13) and Portal (2020, p. 69).

[107] Cf. Wuttke (2020).

however, the algorithms in the respective syntax of the computer programs are meant, which consist of finitely many, well-defined clear process steps and generate a specific output based on a specific input.[108] Examples are search algorithms for search engines and shortest-path algorithms for navigation systems. In problem solving, a specific input is transformed into a specific output. Self-learning algorithms collect more data and thus insights about an intelligent or learning agent during the application. An intelligent or learning agent is, mathematically speaking, a function that maps a feature vector to a discrete class value or, more generally, to a real number. This function changes during the learning phase under the influence of the training data. It is not programmed.[109]

For larger machine learning projects, data science platforms are used that provide special components to support the machine learning process. An important component of a data science platform is a feature store. It is used as a central storage that manages data and features that are specially prepared for the data science process, i.e. special program modules (e.g. for the application of methods, models and algorithms). Other components of the platforms include management, analysis and monitoring components. If the above-mentioned data science platforms are extended in the sense of **auto-machine learning** this means a further automation of the machine learning process steps. Here, the intervention of the human (especially the data science specialists) regarding the data preprocessing, architecture and model selection is replaced by the system. This brings time and cost savings for these projects. For the human as an AI developer, the incoming processes such as the goal definition, the design of the data sets as well as the outgoing processes such as the interpretation of the model results remain due to his or her domain expertise.[110]

The most important functions or learning methods in machine learning, which are also used to a large extent in data mining and predictive analytics, can be described as follows:[111]

- Classification
  In order to obtain a better assignment, the data is divided into classes. A well-known example for the assignment in classes are the damage classes of the insurance companies, in which the customers of the respective insurance are classified. For example, when placing online advertising, the nearest neighbor classifier is used to search for the person or group that is closest to the customer based on his or her characteristics.

---

[108] Cf. Rogers (1967, p. 2) and Leiserson et al. (2010, p. 5 f.).

[109] Cf. Ertel (2016, p. 194).

[110] Cf. Wuttke (2020).

[111] Cf. among others Schrödl (2009, p. 26 f.), Gluchowski et al. (2008, p. 196 ff.), Chamoni et al. (2010, pp. 329–356), Cleve and Lämmel (2014, pp. 57–192), Alpar and Niedereichholz (2000, p. 11), Küsters (2001, pp. 95–130), Gabriel et al. (2009, pp. 144–276), Runkler (2010, p. 96), Petersohn (2005, pp. 73–255), University of Helsinki and Reaktor (2020) and Ertel (2016, p. 194).

- Clustering
  This part deals with the grouping of data according to certain characteristics. For example, the aggregation of similar customers is counted. This aggregation is based on selected criteria. In contrast to classification, the clustering groups are not known in advance but represent the result of the clustering procedure.
- Principal component analysis
  The principal component analysis aims to describe objects more simply by examining their main factors.
- Factor analysis
  A factor analysis is a statistical method in which the characteristics are related to each other and groups of characteristics are formed that show a correlation.
- Dependency discovery
  This process reveals the relationship patterns of the feature levels among each other. The dependency discovery is used, for example, for the shopping basket analysis in retail.
- Deviation discovery
  This examines whether features are present that differ greatly from others. With this method, one can, for example, examine productivity fluctuations in a fixed period of time.
- Discriminant analysis
  The discriminant analysis represents a method that detects significant differences of features and that enables the assignment of a new object to existing classes.
- Contingency analysis
  The contingency analysis analyzes the dependency or independence of two or more nominal-scaled variables.[112]
- Variance analysis
  The variance analysis examines whether significant differences exist for a metric variable depending on certain groupings according to one or more nominal variables, e.g. whether there are significant differences in the buying behavior of male or female consumers for a product group.
- Decision trees
  A decision tree can be imagined as a tree, where the branches represent the question and the twigs the different answer options. Decision trees serve to represent the decision paths that lead to a certain statement.
- Attribute weighting
  The attribute weighting methods are used to determine the determining factors for a result, e.g. a decision process. The preference order of the attributes is done by weighting.
- Regression analysis

[112] Cf. Hammann and Erichson (2006, p. 322 ff.).

The regression analysis explores the relationship between a dependent variable and one or more independent variables. A dependent continuous variable is explained by one or more independent variables, e.g. the sales forecasts in the automotive industry by the gross domestic product and other factors. This is a linear regression analysis, which allows predictions in classes or values. With the help of logistic regression, it is also possible to classify the uncertainty of the prediction by probabilities.

- Time series analysis
  The time series analysis deals with time-dependent forecasts of the development of observed variables. It is a special form of regression analysis.
- Association analysis
  With the help of an association analysis, rules are defined that describe relationships between occurring elements of data sets of a data stock. Rules are established that, for example, assert that if element A occurs, then element B also occurs.
- Monte Carlo simulation
  The Monte Carlo simulation is a statistical simulation method that uses random numbers to approximate a target variable. For example, the Monte Carlo simulation is used for the aggregated risk planning of the value at risk in the financial and insurance sector.
- Support vector machine
  The support vector machine is a mathematical method of pattern recognition that is implemented in computer programs. Here, the method divides a set of objects into classes in two steps. It is a promising approach that provides a solution for linear and non-linear models.
- The Bayes classifier
  The (naive) Bayes classifier is a technique of machine learning that can be used to classify objects into two or more classes. The classifier learns with the help of training data in which the correct classes are given. It is especially used when the probabilities of the classes based on a series of observations are to be determined in the analysis. The model assumes that the variables are conditionally independent depending on the class. The basis for this method is Bayes theorem (Bayes' rule) known from probability theory.

The following types can be distinguished in machine learning (ML):[113]

- **Supervised machine learning**: In supervised learning, one uses known training data to recognize patterns and relationships from them. Here, the connection to a target variable is always learned, and the machine tries to predict it correctly.
- **Unsupervised machine learning**: In unsupervised machine learning, the algorithms do not receive targeted training data but training data from which they are supposed to

[113] Cf. Scheuer (2020, p. 20 f.), Berry et al. (2018, p. 4), Buxmann and Schmidt (2019, p. 10 f.) and Folkers (2019, p. 3 ff.).

discover interesting, hidden groups and patterns independently. The target variable is thus not known. The results have to be evaluated by the expert, e.g. the data scientist, regarding to what extent they are useful for specific applications, e.g. in the business area.
- **Semi-supervised machine learning**: Semi-supervised machine learning is a hybrid form of the first two ML types, where only a small amount of data with known target variable is required for the learning process and a large amount of data, where this target variable is not yet present, can be used. The advantage is that one can start with a smaller amount of known data.
- **Reinforcement learning**: In reinforcement learning, the algorithms interact with the environment and are evaluated by a cost function or reward system to gradually learn a way to solve the problem independently; the reward is positive or negative. Alternatively, one also speaks of a positive or negative feedback process. The advantage is that the algorithms do not need any training data in advance.

Especially the data pattern recognition, as it is presented in Sect. 5.7.4 using the example of data mining and predictive analytics, uses the methods and techniques of machine learning to obtain better results for the user's objective. They use the functions and learning methods listed above in a task-related way, for example, for targeted online advertising or for responses to a chatbot request.

Similar to the CRISP data mining process, there is a process for machine learning with the following steps:[114]

- problem description and goal definition of the use case
- data acquisition, transformation and ideally feature extraction from a feature store
- learning phase, in which the machine learning algorithm is trained
- interpretation of the results
- productive use.

Many of the AI problems, especially in connection with robotics, can be best solved by machine learning, which makes machine learning a central branch of AI for robotics. Robotics, as the name suggests, deals with the creation of robots. Unlike simple machines, these are to be built and programmed to function autonomously and adaptively in complex and realistic scenarios. Since, for example, industrial robots and care robots have to learn and master many different skills, such as touching, recognizing and grasping, the interaction of the skills in almost all areas of AI poses a great challenge for robotics.[115] In Sect. 5.7.5 it is shown, in particular for software robots using the example of (simple and intelligent Robotic Process Automation (RPA)), to what extent these robots can be used for tasks of planning and reporting.

---

[114] Cf. Wuttke (2020).

[115] Cf. University of Helsinki and Reaktor (2020).

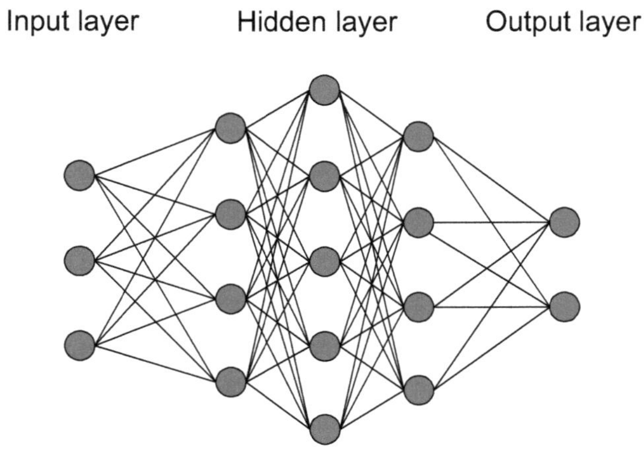

Fig. 5.47 Artificial neural networks

Deep learning defines certain methods of machine learning, in which the different layers of simple processing units are connected to a network. It is a subfield of machine learning. After entering the data into the input layer of the system, they pass through all layers one after the other, similar to how a visual observation is transferred from the eye of the human being via the retina into the brain, in order to recognize, for example, an object.[116] In deep learning, artificial neural networks are built and used for machine learning. The depth of the network allows it to learn more complex structures without processing unrealistically large amounts of data. One of the most widely used neural models is the so-called backpropagation algorithm.[117] Figure 5.47 shows the structure of a neural network, from the input layer via hidden layers to the output layer.

Artificial neural networks consist of neurons that are connected to each other by weighted links.[118] Neural networks are used when the large number of data objects makes the pattern sought difficult to find. Artificial neural networks resemble the human brain (real biological neural network). They are able to generate solutions and make decisions independently through experience (e.g. deviations from the target) and through newly acquired knowledge.[119] Powerful neural networks consist of a multi-layered or richly layered network architecture. In the input layers, the system receives the input signals (input data, etc.), which are passed on to further, partly hidden layers and processed there. These in turn output signals to further layers. Finally, the final output layer delivers the output signals of the entire network.

[116] Cf. University of Helsinki and Reaktor (2020).

[117] Cf. Ertel (2016, p. 291).

[118] Cf. Mertens and Barbian (2019, p. 11).

[119] Cf. Alexander et al. (2018, pp. 11–19).

Neurons of the first hidden layer, for example, break down the input data (e.g. images) into components (image fragments). The neurons of the next layer recognize structures (eyes, ears, etc.) in the data in the image fragments. In the next layer, the components are analyzed for similarities with comparison data (e.g. different animals). A result with a relative probability is transferred to the output layer (which animal it could be). The examples here come from the beginnings of image recognition. But examples can also be given for the economic sector. For example, the probability of insolvency of a company is predicted. Input data would be e.g. monetary key data (balance sheet, income statement, etc.) and non-monetary key data (e.g. age of management, industry, etc.), which are processed and classified in the layers. Using comparison values of historical insolvencies, probabilities of insolvency can be output.

Some neural networks already work with up to 1000 layers to further improve simultaneous interpreting, speech recognition, gesture control, face recognition, autonomous driving and human-robot cooperation.[120]

Due to the complexity and obscurity of the network architecture, one also speaks of **deep learning** that can learn complex structures due to the network architecture without being dependent on unrealistically large amounts of data. Typically, non-linear activation functions and learning rules such as the backpropagation algorithm are used in the multilayered networks.

With the help of the support of **Generative Adversarial Networks (GAN)**, differently trained artificial neural networks are used in a competition against each other so that they can learn and continuously develop through the output signals of the other network.

5.7.3 Expert Systems (XPS)

Expert systems (XPS) are needed where decision makers need support due to the problem complexity and the amount of data material required for the solution. For example, when checking the creditworthiness of a customer, information and indicators for the credit check are analyzed, which are time-consuming to obtain and provide. A credit inquiry at an institution accesses such a special analysis tool. This is a model-oriented analysis system (see Sect. 5.5.5), which is continuously improved by expert knowledge. Due to the faster data processing in the model-oriented analysis system, the system's decision proposals become better than those of individual experts themselves in the long run. Special expert knowledge and problem-solving mechanisms developed over many years are incorporated and made usable in expert systems.[121]

[120] Cf. Winter (2018, p. 66).

[121] Cf. Gabriel (2010) http://www.oldenbourg.de:8080/wi-enzyklopaedie/lexikon/ (Accessed on 15.01.2011).

Expert systems can be attributed to the early research area of artificial intelligence. As early as the 1960s and 1970s, medical expert systems were developed to improve diagnosis and therapy decisions.[122]

Rule-based expert systems are characterized by the fact that knowledge from different sources is collected in them. Conclusions are derived from the knowledge depending on predefined rules. These rules are implemented in the form of if-then relationships.[123]

Expert systems are described by the following three basic components:

1. In the **knowledge base**, all the knowledge is collected and stored. The knowledge comes in the form of factual knowledge, as in classical databases and books, but also in the form of experiential knowledge, which has been built up by experts over the years. Especially the representation of expert knowledge in these systems is very complex and requires teams of computer scientists and experts from the respective fields, who work together and feed in the knowledge.
2. The **rule base** defines the scope of action of the expert system.
3. The **rule interpreter** draws conclusions based on the fact base and the predefined rules.[124]

These types of systems are rarely referred to as expert systems today with the advent of artificial neural networks and machine learning. Expert systems cannot improve themselves as in machine learning. Nevertheless, expert systems have a high significance for artificial intelligence. AI systems can build on the expert systems and learn independently and expand the human horizon of knowledge.[125]

5.7.4 Data Mining, Text Mining, Web Mining and Predictive Analytics

The systematic application of statistical methods and models to large data sets, as they are explained in more detail for example in Sect. 5.8 Big Data, can be summarized under the term data mining. Data mining pursues the goal of identifying insights, such as data patterns, from data and making them usable for decision purposes. The "buzzword" data mining stands for data extraction (mining or digging of data). By linking information, **data mining** provides new insights for the analyst.[126] Here, the data are systematically

---

[122] Cf. Shortliffe (1976): *Computer-Based Medical Consultations: MYCIN*. Elsevier, New York 1976.

[123] Cf. Sperner (2020).

[124] Cf. Gentsch (2018, pp. 32–34).

[125] Cf. Sperner (2020).

[126] Cf. Bissantz et al. (2001, pp. 130–131) or Determann and Rey (1999, p. 143).

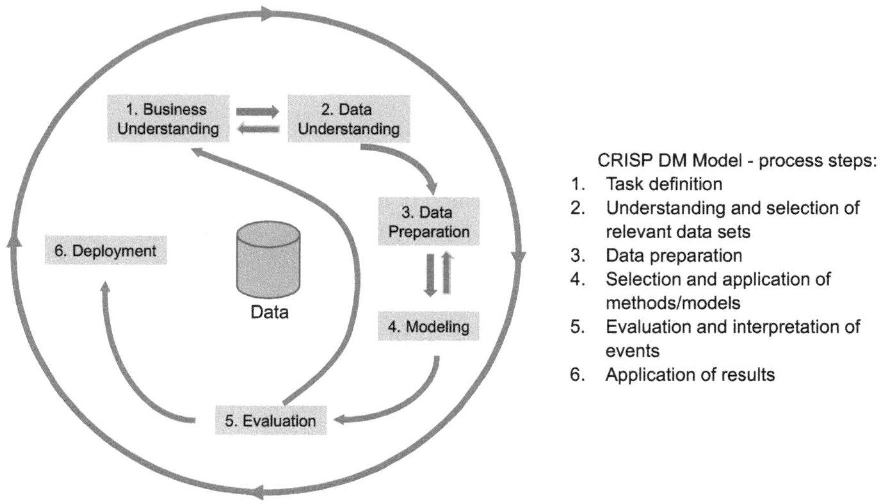

Fig. 5.48 CRISP data mining model—process steps. (Slightly modified representation based on Shearer 2020, pp. 13–22)

queried for unknown relationships. If a specific relationship is found with a statistical probability, data mining delivers this relationship. For example, specific customer profiles are generated for sales and used for advertising and selling a product. Data mining is thus understood as a process that recognizes meaningful relationships as well as patterns and trends by searching a large data base using pattern recognition (statistical and mathematical methods). The data mining process is also discussed under the term **Knowledge Discovery in Databases** (KDD), which characterizes **data selection, processing, transformation, data mining and interpretation** as process phases around data mining.[127] At the beginning of the millennium, a uniform standard for data mining processes was created with the CRISP data mining model, based on an EU-funded project with company participation; its aim was to provide a standardized process flow of data mining for companies that is independent of industry, software and application. The CRISPdata mining model distinguishes the following phases, which are arranged iteratively in a circular model (see Fig. 5.48):[128]

CRISP data mining model—process steps

Predictive analytics models are models that can derive meaningful insights (e.g. data pattern recognition) and predict future outcomes based on data from various sources.[129] The

[127] Cf. Kononenko and Kukar (2007).

[128] Cf. Shearer (2020, pp. 13–22).

[129] Cf. e.g. Siegel (2013).

areas of application in companies cover the entire area of planning already shown, ranging from corporate and financial planning to functional area planning of procurement, production and sales, as well as cross-sectional functions such as human resources and quality management. The data pattern recognition serves for example the recognition of purchase patterns, behavior patterns and process patterns.

The accuracy of forecasts and their models increases with the amount of qualitative data available. Therefore, the use of big data technologies (see Sect. 5.8) plays a particularly important role in the context of predictive analytics.

There is hardly any difference between the term "predictive analytics" and "data mining", so both terms are discussed together here. As already stated above, data mining is the process of gaining knowledge with the help of available data.[130] Descriptive, predictive and prescriptive data mining were already distinguished in data mining. The goal of descriptive data mining is to recognize data patterns, while predictive data mining determines forecasts and predictions for a target value. Prescriptive analysis is the derivation of optimized decisions and actions based on different expectation scenarios of several predictive analytics models.[131] While descriptive methods try to make normative statements and formulate general results based on empirical studies, a prescription first forms a hypothesis, which is then validated by empirical studies or experiments.

With regard to the distinction from data mining, predictive analytics is particularly highlighted by the use of modern BI and big data technologies. A fundamental difference from data mining is not recognizable.[132]

Predictive analytics is usually sketched with the following 3 to 5 phases, whereby in the three-phase description, the first phase is often not mentioned, and phases 2 and 3 are summarized.[133] Thus, the phases correspond to those in data mining.

1. **Identifying phase**: What should be investigated? What hypothesis has been formulated and should be tested? What goal should be pursued or achieved? Formulation of the question on the subject of investigation, e.g. as a problem description, goal or hypothesis.
2. **Descriptive phase:** What has happened, what is happening right now (monitoring)? Collection and description of the data for pattern recognition.
3. **Diagnostic phase**: Why did it happen? Data analysis and determination of the causes with regard to the detected symptoms.

[130] See Petersohn (2005, pp. 10–11) and Cleve and Lämmel (2014, p. 38).

[131] See Feindt and Grüßling (2014, p. 181 f.).

[132] See Felden (2010, pp. 307–328).

[133] Cf. Burow et al. (2014, pp. 13–20), Bitkom (2014, pp. 21–24, 45–47) and Schubert (2013).

4. **Predictive phase**: What relationships can be recognized? What will happen in the future? Model development for data pattern recognition and forecasting. After defining the model to be applied, irrelevant data are cleaned, and required data are selected and transformed accordingly. With the help of model development, the data can be evaluated, patterns recognized and forecasts made. In addition, the models have to be checked and optimized for application in this phase.
5. **Prescriptive phase**: What do I have to do to achieve the desired goal? Ideally, automatic recommendations for action (e.g. exploiting trends, avoiding risks) should be derived from the insights of the data patterns and forecasts. For example, automatic dispositions to optimize inventory or the initiation of maintenance orders when detecting anomalies in devices are meant here.

The methods and algorithms selected and used within the models are interdisciplinary and come from scientific disciplines such as mathematics, statistics and computer science as well as biology and physics. The process steps listed have great similarities with the procedure for early warning information or detection systems.[134]

In contrast to data mining and predictive analytics, the evaluation and the derivation of measures in early warning information systems are supported by expert knowledge (cf. Sect. 5.7.3) and not primarily by IT models. In data mining and predictive analytics, IT-based statistical methods are usually used, in which models and methods are predefined by the developer. Here, the spectrum of the statistical methods already listed above ranges from cluster analysis to Bayes' theorem (from the a-priori chance to probability ratios to the naive Bayes classifier).[135]

Furthermore, there are self-learning algorithms that independently develop models from the data set used and derive suitable methods. These are summarized under the term "machine learning" as already mentioned in the previous Sect. 5.7.2.[136] Data mining and predictive analytics use these static methods and learning methods.[137]

The creation of a data mining or predictive analysis model requires extensive domain and technical knowledge. The implementation is essentially based on four pillars:[138]

[134] Cf. Gehra (2005, p. 22 f.), Krystek and Moldenhauer (2007, p. 124) and Hammer (1998, p. 252 ff.).

[135] Cf. University of Helsinki and Reaktor (2020).

[136] Cf. Baars (2016, p. 175).

[137] Cf. among others Schrödl (2009, p. 26 f.), Gluchowski et al. (2008, p. 196 ff.), Chamoni et al. (2010, pp. 329–356), Cleve and Lämmel (2014, pp. 57–192), Alpar and Niedereichholz (2000, p. 11),Küsters (2001, pp. 95–130), Gabriel et al. (2009, pp. 144–276), Runkler (2010, p. 96) and Petersohn (2005, pp. 73–255).

[138] Cf. Schrödl (2009, p. 28 f.).

1. Data basis: The quality and availability of data are an important factor for the overall project success. A map to provide an overview of customer needs of the end customers (Customer Design Map) requires at least five data sources, for example:[139]
   a) Product information from the internal merchandise management systems
   b) Click data and search queries on the websites
   c) Explicit data with ratings
   d) Data of the purchase context
   e) Data of the social context
2. Clarity of the question: The question to be investigated must be clearly formulated. Otherwise, there may be a deviation from the target result.
3. Quality and selection of the method: To obtain a result, the methods of analysis must match the question and the data.
4. Interpretation of the results: It is particularly important to emphasize that concrete actions are derived from the results achieved, thereby creating added value for the relevant business process.

To better interpret the results, new and familiar visualization techniques are available, which among other things show the frequencies, relationships and connections that were identified in the context of data pattern recognition. The graphical analysis and prediction of such data is referred to as visual analytics and discussed in connection with **data discovery** and **data visualization**.[140] Figure 5.49 shows examples of these new chart types from top left to bottom right: bubble chart calender charts, collapsible treeview, choropleth map, collapsible intended treeview, flare chart, sunburst chart, Word-Cloud, zoomable treemap, chord chart, hierarchy bar, global chart.[141] Other chart types are e.g. the heatmaps (thermal images), dendrograms (hierarchical tree diagrams), network diagrams (visualization of connections) and Sankey diagrams (flow diagram with different quantity-proportional input and output arrows). Some of the charts that are frequently used for business questions, such as bubble, sunburstand treemap charts, were already explained in detail in Sect. 3.8.4. Due to their good visualization, these chart types are often used in BI systems that focus on areas such as big data analytics, data mining, and predictive analytics. The visualization function is also supported by data discovery tools, which can easily link and analyze previously tedious to integrate third-party data with the data of the data warehouse via ad hoc connection.

Some BI software providers offer their own tools for visualization, data mining or predictive analytics. Others use an open source programming language like *R* for statisti-

[139] Cf. Weigend (2017, p. 16).

[140] Cf. Ruf and Schwab (2016, pp. 495–501) and BARC (2017).

[141] Taken from: Freiknecht (2014, p. 345).

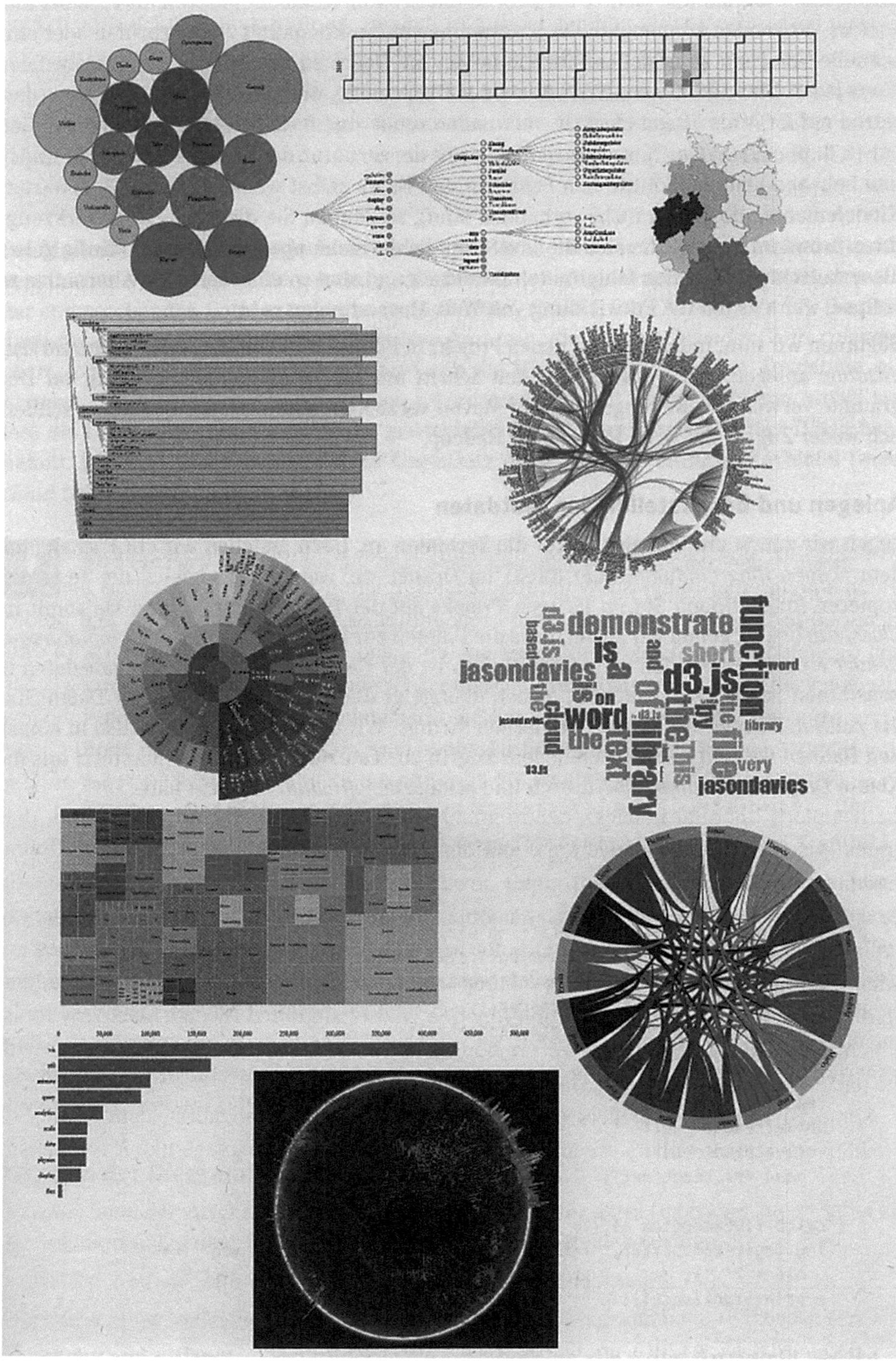

Fig. 5.49 Charts for big data data visualization

cal calculations, data mining, predictive analytics and the creation of graphics.[142] *R* is freely available and can be used for statistical data analysis and data mining.[143] In the development environment RStudio *R* offers among others the following functions: analysis of variance, factor analysis, contingency analysis (cross or contingency table), Lorenz curve and normal distribution. Thus, the user is given many possibilities for statistical test procedures, time series and cluster analyses, regression and association analyses, text mining as well as linear and non-linear modeling. In recent years, the programming language *Python* has gained popularity. *Python* offers in particular a framework for the topic of deep learning. As open source analytics platforms, the tools *KNIME* and *Rapidminer* are also worth mentioning for data mining and predictive analytics.[144] Fields of application of predictive analytics or data mining in connection with big data largely correspond to those listed in Sect. 5.8 in Table 5.5.

Related to data mining and predictive analytics is also the method of case-based reasoning (case-based inference). Case-based reasoning is a machine learning technique that uses analogies to find solutions for problems. Here, similar historical problems and their solutions are searched for new problems and applied either directly or adapted.[145] Practical applications can be found in customer service, e.g. as helpdesk systems.

Text mining defines a mostly automated process of knowledge discovery in textually represented data, which should enable an effective and efficient use of available document collections.[146] In analogy to data mining, relationships between texts should be recognized. The term text mining was introduced by Ronen Feldman and Ido Dagan as "Knowledge Discovery from Text" (KDT).[147] From the analyzed text documents, essential insights (hypotheses, relationships, etc.) should be determined with the help of statistical and linguistic means of the text mining software. As a form of visualization, the word cloud is often used in text mining (see Fig. 5.49 in the middle right).

Related to text mining is audio mining, in which audio recordings or other audio files are analyzed for their patterns and relationships. If the documents are extended to include video and image files, multimedia mining is also discussed.[148] If the aim is to go beyond the internet or web information, one speaks of **web mining**.

Due to the new possibilities that the internet offers in the field of information retrieval, it is obvious to import data and information from the internet into a data warehouse. The

---

[142] For example QlikTech (2016) and Jedox (2016).

[143] Cf. R (2016).

[144] Cf. Wuttke (2020).

[145] Cf. Richter (2003, pp. 407–430).

[146] Cf. Mehler and Wolf (2005, p. 2).

[147] Cf. Hotho et al. (2005, pp. 19–62).

[148] Cf. Mertens (2002, pp. 17–19), URL: http://www.wi1-mertens.wiso.uni-erlangen.de/veroeffentlichungen/download/Business_Intelligence-ein_Ueberblick_Arbeitspapier_der_Universitaet_Erlangen-Nuernberg.zip, (accessed on 23.07.2011).

WWW is the most important source of information, and it is relatively easy to extract the existing data from the structured websites with HTML or XML format and import them into the data warehouse. This variant of the data warehouse concept is called **web farming**, **web casting** or **web mining**.[149] The main areas of knowledge discovery are the content, the structure of the web data and the behavior of the users. The terms **web scraping**, **web crawling**, **web data extraction** or **screen scraping** refer to the automated reading of information from websites. This technique is a basic technology for collecting data for web mining and beyond. The information is not only images and texts, but it can also be URLs or dynamic content such as from a CMS, PHP and JavaScript.

Systems such as DynaSight from Arcplan or Decision from Comshare integrate, for example, data sources from the intranet and the internet, whereby web-casting applications check and update the internet content for changes.[150] So-called internet agents, such as BullsEye, SearchPad and InfoMagnet, have a similar purpose and are equipped with meta-search engines with data acquisition functions.[151] Web farming or internet agents are at the beginning of their career in terms of their use and design. Especially for strategic planning, they can play a central role with functions of net echoing, net monitoring and net scanning. Changing market conditions can be recognized faster through the information from the internet. RSS channels provide users with short information blocks, similar to a news ticker, that consist of, for example, a headline, a brief summary of the text of a website or a blog, and the corresponding link to the website. This way, the user can supply himself with more targeted news.[152]

5.7.5 Robotic Process Automation (RPA)

The starting point for the development of Robotic Process Automation (RPA) was the observation that, despite the use of information technology (such as ERP, CRM and SCM systems), costly manual activities are still necessary to operate the systems.[153]

The digital transformation of software-supported processes within the context of the advancing digitization in companies receives a new IT approach through robot-controlled process automation, which is also discussed for accounting and controlling processes, such as planning and reporting.[154] RPA accesses existing applications without changing them and executes well-structured processes automatically. Thus, the

[149] Cf. Behme and Mucksch (1997, p. 150) and Schinzer et al. (1999, pp. 284, 314 f.).

[150] See Leßweng (2004, p. 43).

[151] Cf. Leßweng (2004, pp. 41–49).

[152] Really Simple Syndications (RSS) is a family of formats for the simple and structured publication of changes on internet pages.

[153] Cf. Barton et al. (2018, p. 116).

[154] Cf. Langmann (2019b).

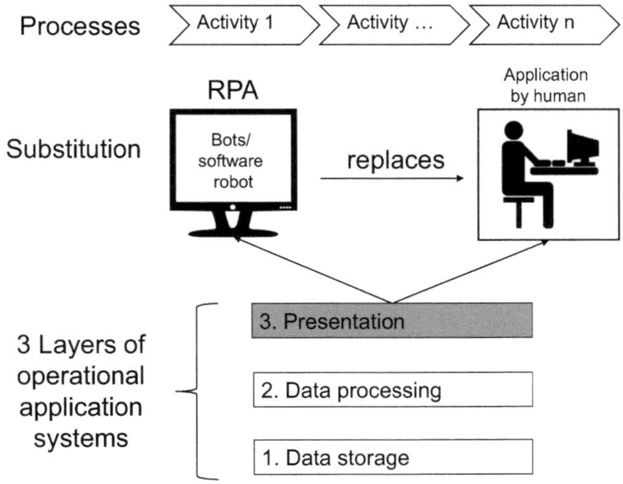

Fig. 5.50 Architecture of RPA (Figure similar to Barton et al. 2018, p. 117)

work effort by the employee is replaced by the RPA software deployment. Ideally, RPA handles the communication between multiple software programs or digital systems. The main advantages are therefore potentials for resource saving and error reduction of human work performance as well as the subjective liberation from time-consuming routine tasks.[155]

RPA is a relatively young field of IT. It is a software for process automation, in which processes from the front and back office are automated by so-called software robots (bots). The software robots take over the tasks of the user and imitate the human operation of computers at the user interface. These bots work directly with the existing user programs, so no new or additional programming or creation of new interfaces is necessary.[156]

Architecture of RPA

Unlike classical process optimization approaches by IT solutions, RPA only accesses the required software applications at the surface of the various software programs (presentation layer, see Fig. 5.50). Classical IT process optimization approaches use, for example, new or modified program modules, routines or queries as well as interface programs, which are used in the first two layers (data storage and processing) of the affected operational application systems.

Whether **RPA systems**, classical **IT process optimization approaches** such as interfaces or the **human** as process user should be used is a pure question of economic efficiency. Here, all costs and benefits have to be weighed against each other. The

[155] Cf. Dinnessen and Halfmann (2018).

[156] Cf. Smeets et al. (2019, p. 8).

application possibilities of RPA systems can be limited by the frequency of cases and the degree of case differentiation. For processes with high frequencies of cases and a low degree of case variability, traditional process automation, such as program modules, interfaces, etc. may be more suitable.[157] In addition, RPA systems are unsuitable for heterogeneous processes that occur rarely and are very different to handle. Here, the human has an advantage in process execution due to his cognitive abilities. For the introduction of RPA, simple, well-structured and frequently recurring processes from the operational area are particularly well suited. For example, a price maintenance can be mentioned, where prices from a standardized, externally provided Excel file are transferred to a software program (e.g. CRM system) weekly. Instead of the price identification and entry done by the human, the RPA takes over this previously manual process. This can reduce the number of errors in highly repetitive and monotonous tasks.

The following procedure is suggested for identifying and implementing processes with RPA systems:[158]

1. Identification of suitable processes and goal setting
2. Detailed process analysis
3. Optimization and standardization of the processes
4. Final process description
5. Implementation in the RPA software
6. Process monitoring and compliance
7. Expansion

After the first rough identification of suitable processes, all components of the process (e.g. logging in to the system, opening the program, positioning the cursor in the correct field, **copying** the field, opening the further program, executing the required menu function, entering the value, saving the value, closing all programs) must be clearly defined, optimized and standardized if necessary, and described in the process analysis for the RPA. Only then are the individual components replicated in the RPA software. The providers of RPA systems lure companies and organizations with the promise that the implementation of the application is simple and quick. However, it must be considered that various tasks await the organization before and after the actual implementation of the software. The preliminary steps (1–4) are conceptual work as they are also applied in process optimization projects. The process monitoring (number of processes, time required per process, etc.) as a subsequent task can be well supported by the RPA software as it saves all important process information when performing. If certain rules

[157] Cf. Smeets et al. (2019, p. 10).

[158] Cf. Martens (2019).

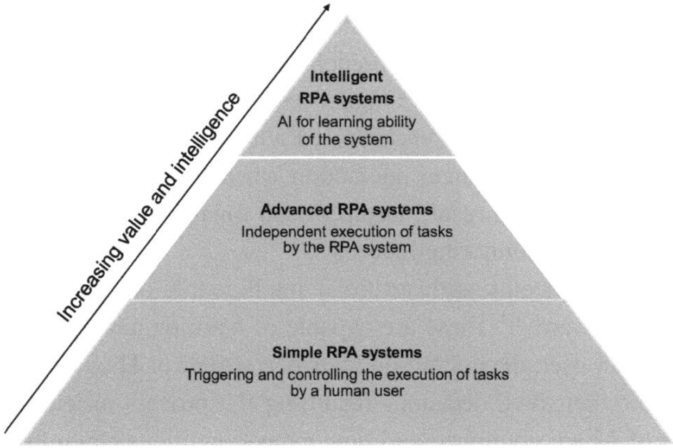

Fig. 5.51 Types of RPA solutions (Slightly modified from Sellmair et al. 2019)

(e.g. legal rules, company instructions, etc.) are to be complied with, this can be built in as a check step in the RPA. The error rate of the processes can be reduced, and the compliance can be automatically maintained.[159] If the later data situation of the processes does not meet the requirements for performance or compliance, these RPA processes can be separated and evaluated. Identified problems can be eliminated and resolved in the expansion phase.

Unlike human users, the software robots do not know any rest periods. They can work continuously 24 hours a day, even on weekends or holidays, which represents a great advantage from a purely economic point of view.

Three types of RPA solutions are distinguished (see Fig. 5.51). These differ essentially in their autonomy and learning ability:

Types of RPA solutions

Simple and enhanced RPA systems operate according to fixed workflows and rules. They cannot, unlike humans, exercise discretion and are not capable of learning. Therefore, ambiguities or gaps in the process lead at best to interruptions or even to erroneous processing by the bots. The processes must then be checked and adjusted by human users if necessary. With simple RPA systems, the start of a process is triggered by a human user. Due to the human involvement, they are also called supervised robots (attended robots). An example of this is Robotic Desktop Automation (RDA), where a macro installed on a desktop is used, which first records the inputs of a user during the process and then executes them identically.[160]

---

[159] Cf. Tripathi (2018, p. 12).

[160] See Smeets et al. (2019, p. 9).

With advanced and unsupervised RPA systems (unattended robots), the process is started by so-called triggers. A trigger requires a predefined event that triggers the process.[161] Such an RPA platform is usually located on a higher-level server or in a cloud environment and has standardized workflows that are synchronized with specific commands.[162] While supervised (attended) robots require human interaction when the process requires it, e.g. at the start or at the verification of the result, unsupervised (unattended) robots are monitored by central control centers with monitoring functions.[163]

By combining RPA systems with artificial intelligence, one obtains intelligent RPA systems, which are still rare.[164] These are capable of learning autonomously and independently from the human user, depending on their performance. They can adapt to changing conditions and make their own decisions regarding the process execution. Depending on which techniques of AI are combined and how many conditions can influence the process, large amounts of data are needed to map the processes with intelligent RPA systems. These data volumes and the high implementation effort of intelligent RPA systems are a major disadvantage and have to be economically justified.[165] In addition, the fields of action in planning and reporting should be examined more closely. The activities in planning often require the assessment of the planning manager. The report analysis and derivation of recommendations for action for the management decisions are currently hardly automatable and too complex. Here, cognitive, but also emotional and social skills of the human are needed.[166] Activities such as report creation, financial statements, report distribution and cost allocation etc. are already automated by the accounting and controlling software and thus covered.[167] In the BI-supported controlling cockpit, everything is best automated except for the manual analysis and planning function. The use of simple or advanced RPA systems is suitable for the following tasks in accounting and controlling: tasks of auditing, confirming and archiving data, master data maintenance etc.[168] Report creation examples with the spreadsheet program Excel, which are based on different data sources, are often used as a good example for RPA.[169] With reasonable data integration in the data warehouse, the report creation should be better done by a powerful reporting software, something which has been presented in this book with numerous BI frontend solutions.

The selection of an RPA system should always depend on the respective process. For simple processes, the effort of implementing an intelligent RPA system is usually

---

161 See Sellmair et al. (2019).

162 See Smeets et al. (2019, p. 9).

163 See Langmann and Turi (2020, p. 12).

164 Cf. Barton et al. (2018, p. 120).

165 Cf. ibid.

166 Cf. Allweyer (2016, p. 35).

167 Cf. among others Alexander et al. (2018, pp. 11–19) from Bearingpoint and Deloitte (2019).

168 Cf. Obermaier (2019, p. 692).

169 Cf. Peper (2018, p. 27 ff.).

not worthwhile. Here, traditional IT process optimization methods can be used. On the other hand, simple and advanced RPA systems are unsuitable for very complex and constantly changing processes. RPA is suitable for frequently recurring, standardizable and rule-based processes, which are performed by human users in a labor-intensive, monotonous and cross-software manner on several different software interfaces. These processes should be less complex and should have as few exceptions from the rule application as possible. Furthermore, few manual interventions and decisions of the human should be required.[170]

If many manual steps of complete process chains (end-to-end processes) are to be taken over by software robots, it must be taken into account that each of the bots used must have its own user identification. Just like the human user, each bot user must log in to the individual systems in order to access them. The limit on the number of robots that can be used is therefore the number of client requests that can be processed on the server side, which in turn must be taken into account when calculating the license costs.[171]

5.7.6 Chatbots/Conversational UX

Chatbots are computer-based dialogue systems that communicate with humans in a chat, and some of them are based on artificial intelligence. These chatbots are used, for example, audibly via voice assistants or textually in the form of a chat interface via a corresponding website. A well-known example of a voice assistant, including private use, are Amazon's Alexa or Apple's Siri.[172] The representation of an online chatbot is an animated or static avatar.[173] The term conversational user experience (Conversational UX) is also used as a collective term for chatbots, messenger bots, voice bots and assistants. More and more companies are using simple or AI-based systems for voice, messaging or text communication on their websites and online shops.

The primary tasks of a chatbot are to recognize, analyze, process and answer the inputs, questions or requests of the respective user in real time. The processing process described here takes place with the help of keywords, parameters and defined rules. Furthermore, the bot researches in the company's internal database or on the internet. The text or voice inputs of the user can be, for example, in the form of a question or in the form of a request to perform a specific action.[174] If necessary, the chatbot also asks follow-up questions if the information is too vague or not defined.

---

[170] Cf. Langmann (2019b).

[171] Cf. Barton et al. (2018, p. 117).

[172] Cf. Friedl (2019, p. 35).

[173] An avatar is an artificial person or graphic character that is assigned to a web user in the virtual world, for example in a computer game or as here in a chat bot.

[174] Cf. Schonschek and Haas (2020, p. 1 ff.).

Using a chatbot and the feedback of the users leads to the fact that, if the bot is based on artificial intelligence, it continuously develops or optimizes itself by means of machine learning. This results in more complex queries being handled more competently.

Chatbots can also act instead of or as a customer service for a company. Certain inquiries or queries from customers can, for example, be processed online by a chatbot and do not require any employee contact. This can be more cost-effective for the respective company, but in any case guarantees availability around the clock.[175] The following types of chatbots can be distinguished:

- **Rule-based chatbots** do not have artificial intelligence. They are not capable of solving problems independently and have no learning ability. They are usually used for answering questions. The biggest disadvantage of rule-based chatbots is based on the already described characteristics, from which the limited processing of user inputs results. Unknown questions cannot be answered and would have to be reprogrammed.[176] The programming has an elementary influence on the performance of the chatbot, as new rules have to be manually stored in case of processing errors and thus create an effort. The advantages of a rule-based chatbot are the easy implementation and the already stored answers of the programmer, which exclude unwanted expressions.[177] A specific form of rule-based chatbots are non-intelligent **FAQs** (Frequently Asked Questions), which answer frequently asked questions with fixed rule-based answers.
- **Intelligent chatbots** are based on artificial intelligence. They are therefore capable of learning and more competent than rule-based chatbots.[178] Intelligent chatbots try to capture the human behavior in requests in all its facets and answer them as best as possible. This includes not only pure speech recognition but also e.g. the recognition of moods and background information. Tasks that such a chatbot can take over in the company for employees and customers are diverse, e.g. managing appointments, answering inquiries, preparing information for customers and employees, recommending products or services, and managing bookings and cancellations. The following types of intelligent chatbots can be distinguished:[179]
  - **Intelligent FAQ bots** have a matching system that compares certain inputs, e.g. keywords, with existing and learned information in the system or in the database. Through feedback and learning ability, the intelligent FAQ bot adapts over time.

[175] Cf Hundertmark (2020, p. 1 f.).

[176] Cf. Mori et al. (2017, p. 395 et seq.).

[177] Cf. McTear et al. (2016, p. 125 et seq.).

[178] Cf. Kumar and Tiwari (2017, p. 60).

[179] Cf. Stephan (2020, p. 1 f.).

The advantage for companies is the already described possibility of automated customer service.

- **Scriptbots** are known for performing standardized processes that are characterized by a high degree of interaction with predefined dialogue trees. They ask the customer, for example, for necessary information within the scope of an ordering process. Call center agents or web forms can be replaced by the scriptbot as it can perform processes faster.
- **Natural language understanding chatbots** have natural language processing and a context memory. This type of chatbots can not only analyze texts but also recognize moods and relate statements. An example of a context memory is when a user of a mobile device asks for the shortest route from A to B. The bot recognizes the potential departure location and the destination. It calculates the arrival time and can e.g. suggest a booking for a train ticket as an alternative to using the car. Information is therefore not queried one after the other as with a scriptbot. The recognition of mood and language processing becomes evident when irritated users are forwarded to an employee in the call center.
- **Virtual agents** are the most complex chatbots and have the largest range of functions. Through flexible dialogue structures and the strongest memory, they can handle complex requests such as "the fastest way to work". For this, the location of the workplace and the location information on the phone are accessed. The context memory becomes evident by the fact that the bot retains and can handle the old information even when leaving or switching topics.

In the field of conversational UX, for example, SAP AG uses both classical technologies of computational linguistics and new deep learning methods to process large unstructured text contents as well as speech. Traditionally, the complex resolution of ambiguities in a natural language is simplified by deep learning and contextualization. Key functions include the recognition of intents (intent matching) and conversation topics (entity recognition). The intents of the users behind a statement are checked and classified according to a predefined list of possible intents, often according to the specified backend functionalities. Basic knowledge, such as domain-specific documents or semantically equivalent training inputs, help the deep learning algorithm to create a reliable intent matching model. The recognition of conversation topics helps to identify or parameterize statements more precisely. For example, a product name or a delivery date as well as quantities and units can be extracted from a natural user input to enable more accurate processing.[39]

The possibilities of using conversational UX or chatbots in business practice are diverse. For operational processes, e.g. from purchasing and sales but also examples from the human resources area, e.g. in application and hiring processes, good examples can already be found in practice. The use of chatbots in controlling and especially for the task of planning and reporting are still rare. However, they are increasingly discussed in

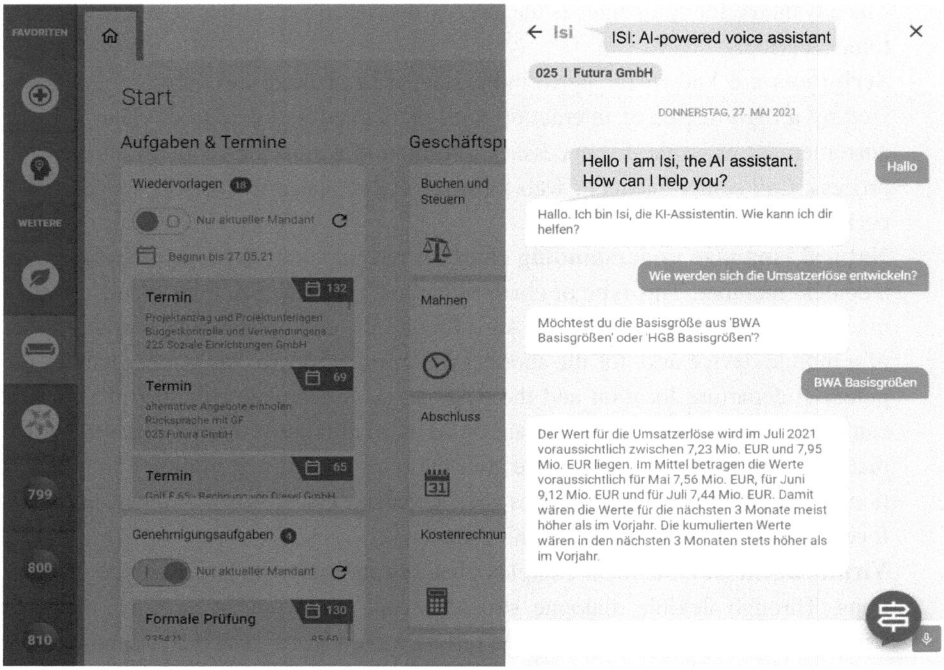

Fig. 5.52 ISI: Voice assistant in the Diamant software Diamant/4. (*Source*: Diamant Software 2021)

the course of digitalization.[180] Even if there are hardly any examples of chatbots implemented in controlling practice at present, one can well imagine that chatbots have a good potential for planning and analysis tasks and are especially used as communication assistants in the digital development speed of the next years. For simple inquiries about available reports and planning areas, the use of chatbots is already very conceivable. For this purpose, the context information (e.g. formulas and meaning of a key figure) of the available contents of the planning and reporting tools would have to be used. Suggestions for further information on the report or planning area could be linked and thus addressed directly by the user as long as he has the necessary access rights. An example of such a voice assistant in the Diamant software Diamant/4 is shown in Fig. 5.52.

The same applies to the analysis and interpretation of key figures, table contents and graphics. Standard interpretation possibilities and standard action recommendations could be provided rule-based. For inquiries with simple analysis and forecast complexity, intelligent chatbots could prepare answer suggestions, thus supporting the analysis and forecast task. For example, when estimating the seasonal development of the sales quan-

[180] Cf. Friedl (2019, p. 35 f.).

tity, the sales quantity of the last months and the comparison period of the previous year could be displayed. Furthermore, a planning form could be pointed out which forecasts the sales quantities based on the historical data and the guide price development. Alternatively, this planning form could also be executed directly, and only the results could be played back. This is then a mixed application of human-chatbot and IT-system coupling.

For complex analysis and forecasting requests from management with different objectives, a system with intelligent chatbots quickly reaches its limits. The manifold cause-and-effect relationships of business issues are too complex for the current AI. In practice, therefore, simple rule-based and intelligent chatbots will prevail, whose horizon focuses on clearly defined tasks. Management and other users can direct numerous such simple queries directly to the digital assistant and thus simplify the task of the controlling function. Simple questions such as where is the report, how does the planning form work, what does the indicator mean, how does the revenue develop, etc. are answered by the chatbot in connection with the existing IT systems for planning and reporting. The time gained by this can be directed by controlling and management to more demanding tasks, such as the analysis and commentary of the monthly result or the quarterly planning.[181] However, it will take some time before such links between chatbots and IT systems are available in a marketable version. The classification of further possible uses of chatbots in planning and reporting is made in connection with other IT solutions in the following Sect. 5.7.7.

5.7.7 Use of AI, RPA and Chatbots in Planning and Reporting

Planning and reporting has existed since the beginnings of information technology and received a significant boost through business intelligence and the use of data warehouse technology (see Sects. 5.4, 5.5 and 5.6). Due to the increased discussion of artificial intelligence in connection with machine learning, deep learning, data mining, predictive analytics and RPA, the question arises to what extent these technological developments support or even completely take over the planning and reporting tasks in the controlling process of a company from managers, controllers and other employees. The latter can be excluded due to the non-existent, strong or general AI to date (see Sect. 5.7.1). However, weak and narrow AI provides good approaches to complement or support the previous tasks in planning and reporting. This applies especially to clearly defined tasks for which a large amount of data is available and usable.

Similar to the analogy to the maturity level in the development of "autonomous driving" in the automobile, it is conceivable that systems with artificial intelligence could also support controlling, at least in the form of a co-pilot. Corresponding planning and

[181] Cf. Friedl (2019, p. 35 ff.).

reporting processes could be carried out semi-autonomously or possibly in larger parts also completely autonomously.[182]

This chapter therefore aims to show the possibilities and limitations of AI use in comparison to previously used IT solutions for the main tasks of planning and reporting within the framework of controlling. It will analyze which tasks can be performed by humans, by traditional or new IT solutions in a overall comparison to the other possible solutions. Table 5.4 provides an overview.[183]

The most important tasks in planning and reporting are to be examined in more detail here:

- **Strategic planning**

In many companies, strategic controlling is carried out with little IT support. Often, strategic analyses and plans are supported with simple office software such as MS Excel, MS PowerPoint and MS Word. For portfolios, Balanced Scorecards and other strategic instruments, special software solutions with relational database technology may be used for presentation and planning.

A powerful alternative to these simple office and special IT solutions can be BI-based solutions. Section 3.9.2 shows an example of a BI-based controlling cockpit for strategic controlling. Here, the management acts as an expert in the strategic planning process and is supported by traditional BI technology. Exploratory data analyses with AI for specially delimited strategic analyses can, however, be integrated into the BI-based system. With the help of AI, the process of strategic analysis and planning can be supported selectively if the company succeeds in using suitable data from internal and external sources. Suitable analysis examples can be found for many strategic tasks. With regard to product program planning, trends in research needs due to changes in customer needs can be checked. For the competitive analysis, e.g. patent applications and new company formations can be systematically analyzed.[184] The diversity of the strategic analysis and planning tasks with regard to environmental, corporate, competitive analysis etc. as they are listed in the example of Sect. 3.9.2 cannot be fully supported by AI-based systems, but only selectively. As mentioned, AI can be used exploratively for suitable large amounts of data and thus complement the traditional instruments sensibly. Gradually, it can succeed in improving strategic analyses by recognizing strategically relevant relationships, deriving better forecasts and suitable metrics and adapting business models.

[182] Cf. Sauer and Sturm (2019, p. 35).

[183] Own compilation: Due to the many row and column information, the table was divided for better readability.

[184] Cf. Gentsch, Peter (2019): Artificial Intelligence for Sales, Marketing and Service, 2nd ed., Springer Gabler Verlag. p. 71.

Table 5.4 Assignment of planning and reporting tasks from human to AI—Part 1

Tasks (Planning and Reporting)	Human (e.g. Manager/Controller)	Office Software (e.g. MS Excel)	ERP Systems	BI-based Systems
Strategic Planning				
Use of strategic analysis and planning tools such as SWOT and portfolio analysis	The management is comprehensively responsible for the strategic analysis and planning. They can use strategic tools such as the SWOT analysis etc. with different IT solutions (Excel, special systems etc.) and data sources (e.g. network, associations, trade fairs). IThey can also consult other experts (controlling, executives, external) besides their own expert knowledge	With the help of office software, simple strategic analysis and planning tools, such as the SWOT analysis, portfolios and trend analysis, are implemented on a case-by-case basis. In addition, the tools can be used for organizational activities and documentation tasks	Strategic tools are usually not implemented in the ERP systems. ERP systems support the strategic analysis and planning with basic operational and multi-year data of the company	Building and using tools for strategic analysis and planning (e.g. SWOT analysis, portfolios, Balanced Scorecard) are possible in BI-based systems. Here, basic data from various internal and external sources are connected
Identification of relevant strategic indicators/key figures	The management's experience helps to identify strategic indicators. They can use the IT solutions and other data sources (e.g. network, associations, trade fairs) listed for this purpose	Can be used for documentation tasks, e.g. for the structuring and assignment of strategic indicators. With the use of common mathematical and statistical functions, dependency analyses of strategic indicators on target variables can also be supported by calculation. However, the amount of data required for the analysis is limited	Supports the search for indicators with basic operational data of the company in the ERP system	Can be used for analysis and documentation tasks to identify strategic indicators. With the use of mathematical and statistical functions and models, critical success factors, e.g. by means of dependency analyses, can be determined and analysed. The potential amount of data is extended by the data of the data warehouse. Extensions by big data analytics are possible

(continued)

Table 5.4 (continued)

Tasks (Planning and Reporting)	Human (e.g. Manager/Controller)	Office Software (e.g. MS Excel)	ERP Systems	BI-based Systems
Forecasts of relevant strategic indicators/key figures	The management can form expert forecasts from their own experience. The management is supported by other experts such as controlling, executives, and external consultants. They can use the IT solutions and other data sources (e.g. network, associations, trade fairs) listed for this purpose	Supports the occasional use for various strategic forecasts with the help of common mathematical and statistical functions based on smaller amounts of data	Supports other systems with basic operational data of the company for the forecasts from the ERP system	Extended integrative planning models across several planning areas are possible. Supports the use for various strategic forecasts with the help of common mathematical and statistical functions and models based on larger amounts of data in the DWH. Extensions by big data analytics are possible
Development of strategic objectives	Management task with support from executives, controlling and possibly external parties	Documentation	–	Documentation
Derivation of strategic measures/projects	Management task with support from executives, controlling and possibly external parties	Documentation	Strategic projects can be created as cost collectors, e.g. order	In addition to the documentation, standard strategies for certain data constellations of the WH can be suggested
Review and adjustment of the business model and the organization to the strategic planning	Management task with support from executives, controlling and possibly external parties	Documentation	Adjustment of the business processes and organizational structure in the ERP structure data	Adjustment of the business processes and organizational structure in the BI structure data

(continued)

Table 5.4 (continued)

Tasks (Planning and Reporting)	Human (e.g. Manager/Controller)	Office Software (e.g. MS Excel)	ERP Systems	BI-based Systems
Economic efficiency calculations of strategic projects	Management task with support from executives, controlling and possibly external parties	Documentation, spreadsheet as calculation basis for monetary and non-monetary evaluation	Supports the evaluation with basic operational ERP data of the company, e.g. costs	Documentation and calculation methods for specific planning areas for monetary and non-monetary criteria for the economic efficiency calculations using the DWH data
Development of vision, mission, guiding principles	Management task with support from executives, controlling and possibly external parties	Documentation	–	Documentation
Alignment of the strategy with stakeholders	Management task with support from executives, controlling and possibly external parties	Documentation	–	Documentation and communication via the cockpit
Communication of the strategy	Management task with support from executives and controlling	Documentation	–	Documentation and communication via the cockpit
Monitoring of the strategy planning and implementation	Management task with support from executives, controlling and possibly external parties. They can use the IT solutions and other data sources (e.g. network, associations, trade fairs) listed for this purpose	Documentation	Supported by basic operational ERP data of the company, which are used, for example, to achieve the targets for strategic indicators and measures	Monitoring of the strategy planning and implementation, e.g. for objectives, key figures, projects with BSC management can be displayed via the BI frontend analyses

Table 5.4 Assignment of planning and reporting tasks from human to AI—Part 2

Tasks (Planning and Reporting)	Big Data Analytics, Data Mining, Predictive Analytics, AI	RPA	Chatbots/Conversational UX	Collaboration Tools (e. g. Online Meetings, Task Management)
Strategic Planning				
Use of strategic analysis and planning tools such as e.g. SWOT and portfolio analysis	Building and using tools for strategic analysis and planning (e.g. SWOT, portfolio etc.) are not provided in big data based systems. However, the punctual analysis of important (critical) success factors using additional (large) amounts of data (big data) from different internal and external sources can be very useful as a complement for strategic planning	–	–	–
Identification of relevant strategic indicators/key figures	By using mathematical/statistical functions and models up to AI-supported models, success factors can be determined and analyzed for definable tasks based on (larger) amounts of data (big data)	–	Through the question-answer functions of AI-supported chatbots, results on potential critical success factors can be output which were identified with the help of available data sources using model-based and AI-supported systems (see big data analytics)	–
Forecasts of relevant strategic indicators/key figures	By using mathematical/statistical functions and models up to AI-supported models, strategic forecasts can be determined for definable tasks based on (larger) amounts of data (big data)	–	Through the question-answer functions of AI-supported chatbots, forecast results can be output which were identified with the help of available data sources using model-based and AI-supported systems (see big data analytics)	–

(continued)

Table 5.4 (continued)

Tasks (Planning and Reporting)	Big Data Analytics, Data Mining, Predictive Analytics, AI	RPA	Chatbots/Conversational UX	Collaboration Tools (e. g. Online Meetings, Task Management)
Development of strategic goals	–	–	–	Can be used as an alternative medium for face-to-face meetings (= online communication), in which strategic goals are discussed and developed
Derivation of strategic measures/ projects	For clearly defined tasks, standard strategies for certain data constellations (big data) can be suggested	–	Through the question-answer functions of AI-supported chatbots, standard strategies for clearly defined tasks can be output which were identified with the help of available data sources using model-based and AI-supported systems (see big data analytics)	Online communication supplemented by task management systems through which corresponding project tasks can be supported
Examination and adaptation of the business model and the organization to the strategic planning	–	–	–	–
Economic calculations of strategic projects	–	–	–	–
Development of vision, mission, guiding principles	–	–	–	Online communication
Alignment of strategy with stakeholders	–	–	–	Online communication

(continued)

Table 5.4 (continued)

Tasks (Planning and Reporting)	Big Data Analytics, Data Mining, Predictive Analytics, AI	RPA	Chatbots/Conversational UX	Collaboration Tools (e. g. Online Meetings, Task Management)
Communication of strategy	–	–	Recipients of the strategic planning can be informed about the strategic plans (e.g. goal, guiding principles, vision, mission and strategic measures) via chat bots (e.g. via FAQ-based chatbots)	Online communication
Monitoring of strategy planning and implementation	–	–	Recipients of strategic planning can be informed about deviations of strategic plans via chat-bots (e.g. via FAQ-based chatbots)	–

Table 5.4 Assignment of planning and reporting tasks from human to AI—Part 3

Tasks (Planning and Reporting)	Human (e.g. manager/controller)	Office software (e.g. MS Excel)	ERP Systems	BI-based Systems
Operational planning/forecasting (prediction)				
Design/develop planning	Management task with support from controlling and possibly external parties. Basis: human experience/expert knowledge, e.g. about business processes and strategic orientation	Supports the occasional use of simple planning models, e.g. pure result planning of the BEA items. Support of documentation tasks	Clearly delimited subareas of the ERP-related planning areas, e.g. cost center planning, production program planning can be supported with conventional planning functions of the ERP system (in mostly rigid) predefined dialogs	With the help of DWH and BI-supported software, comprehensive integrated planning models with various planning areas can be developed individually for the company. For this purpose, there are development tools with which, for example, planning inputs and functions can be implemented in planning forms
Identification of planning value drivers (recognition of cause-effect relationships)	Management task with support from controlling and possibly external parties. Basis: expert knowledge	Supports the identification of the planning value drivers possibly with simple additional information (e.g. tables with external data)	Supports other systems in the identification of the planning value drivers with the data of the ERP system	Provides the possibility to develop statistical planning models for different planning areas, which support the identification of value drivers based on larger data sets. Big data and predictive analytics models can be integrated in the BI system
Design simple planning functions (distribution, revaluation, allocations etc.)	Management and controlling decide which planning functions are best suited for the planning and control system. If necessary, support from external sources	Simple planning functions can be mapped with spreadsheet programs	The ERP software offers standardized selectable planning functions for its delimited planning areas, e.g. allocation in cost center planning	Provides the possibility to develop individually needed planning functions for different planning areas, which are based on larger data sets in the DWH. It also offers the possibility to link the data from different planning areas in an integrated corporate overall planning system

(continued)

Table 5.4 (continued)

Tasks (Planning and Reporting)	Human (e.g. manager/controller)	Office software (e.g. MS Excel)	ERP Systems	BI-based Systems
Use more complex forecasting methods (statistical models e.g. for simulation, sensitivity analysis and scenarios)	Management task with support from executives, controlling and possibly external sources. Basis: expert experience	Provides simple mathematical and statistical methods that can be used for small planning models for delimited planning areas	Provides simple mathematical and statistical methods that can be used for small planning models of specific delimited planning areas within the ERP system	Provides mathematical and statistical methods available in the system that can be used for small to large planning models for various planning areas based on the data in the DWH. Simulations, sensitivity analysis and scenarios are possible not only for delimited planning areas, but for the entire integrated corporate overall planning model
Data acquisition for orientation and target values	Provides ideas for potential data sources based on human experience	Spreadsheet programs can be used as simple data sources. Data as orientation and target values often have to be inserted manually or via interface	Provide larger structured data sets from the individual ERP modules. Data as orientation and target values can be drawn from the data of the ERP system in the limited planning area	Depend on upstream data sources such as ERP systems and other third-party data sources. The data loaded into the DWH via ETL processes can be used as orientation and target values in various planning areas of the integrated corporate overall planning
Planning model development	Provides ideas for potential usable planning models for different planning areas and their integration. Decides on the use of the models	Supports the punctual use of simple planning models, e.g. pure result planning of the BEA items. Also supports the documentation, conception and design in the model development	The planning models available in the ERP software are parameterizable, but essentially standardized to the predefined planning area and its functions	Complex overall models for integrated corporate planning with individually required planning areas can be built with development tools

(continued)

Table 5.4 (continued)

Tasks (Planning and Reporting)	Human (e.g. manager/controller)	Office software (e.g. MS Excel)	ERP Systems	BI-based Systems
Develop planning areas and forms	Provides ideas for potential usable planning forms for different planning areas and their integration. Decides on the use of the planning forms and areas	Supports the use of simple planning forms for single planning areas with possibly simple integration of few areas. Complex integrated overall models are not possible. Support of conception and documentation tasks	Contains mostly standardized planning forms for certain delimited planning areas of the ERP software, which can only be slightly customized	Complex overall models for integrated corporate planning with individually required planning areas and individual planning forms can be built
Planning coordination	The management sets the framework conditions and target values for the planning. Controlling coordinates the overall planning. Creates planning calendars and provides/trains tools for planning. If necessary, support is provided by other internal (e.g. IT) or external persons	Can be used as documentation and for designing coordinative tasks, e.g. planning calendar, status control, responsibilities etc.	Provides (rarely) for its own planning areas coordination-supporting functions, such as a status tracking system, if necessary	In the BI-supported planning system, individual planning areas for planning the dates and responsibilities can be built e.g. in a status tracking system for coordinative planning tasks

Table 5.4 Assignment of planning and reporting tasks from human to AI—Part 4

Tasks (Planning and Reporting)	Big Data Analytics, Data Mining, Predictive Analytics, AI	RPA	Chatbots/Conversational UX	Collaboration Tools (e.g. Online Meetings, Task Management)
Operational Planning/ Forecasting (Prediction)				
Design/develop planning	They are not suitable for building comprehensive planning models with integrated planning areas With the help of big data based models, data mining or predictive analytics models can be built for limited planning tasks, for example for sales forecasts based on identified value drivers	–	–	Online communication
Identification of planning value drivers (detection of cause-effect relationships)	Provides the possibility to develop statistical planning models for different but delimited planning areas to support the identification of value drivers based on big data. Besides simple statistical models, complex statistical models as well as machine learning methods up to deep learning models within the framework of artificial intelligence can be developed	–	–	–
Design simple planning functions (distribution, revaluation, allocations etc.)	–	–	–	–
Use more complex forecasting methods (statistical models e.g. for simulation, sensitivity analyses and scenarios)	Provides statistical methods available in the system, which can be used for complex planning models based on big data for specially delimited planning areas. With AI technology, simple or learning algorithms can be generated for delimited forecasting tasks (e.g. predictive maintenance, and sales forecasts). Simulation, sensitivity and scenario analyses are only possible for delimited planning areas, but not for the integrated overall corporate planning	–	AI-supported chatbots can output forecast results, which are generated in interaction with model- and AI-supported systems, via question-answer functions	–

(continued)

Table 5.4 (continued)

Tasks (Planning and Reporting)	Big Data Analytics, Data Mining, Predictive Analytics, AI	RPA	Chatbots/Conversational UX	Collaboration Tools (e.g. Online Meetings, Task Management)
Data acquisition for orientation and target values	Orientation and target values for input planning forms are not provided. These IT solutions are designed specifically for forecasting, simulation and sensitivity analysis tasks, which are based on simple to complex statistical models as well as machine learning methods up to deep learning models in the context of artificial intelligence They use upstream internal data sources such as ERP systems and DWH as well as external big data sources	–	–	–
Planning model development	With AI technology, only suitable forecasting models can be developed for limited planning tasks (e.g. predictive maintenance), but not for more complex business planning areas of an integrated overall corporate planning	–	–	Online communication
Develop planning areas and forms	–	–	–	Online communication
Planning coordination	–	–	–	Online communication complemented by task management systems, which can support corresponding project tasks

Table 5.4 Assignment of planning and reporting tasks from human to AI—Part 5

Tasks (Planning and Reporting)	Human (e.g. Manager/Controller)	Office software (e.g. MS Excel)	ERP Systems	BI-based Systems
Communication before, during and after planning	All participants such as management, controlling, employees and external parties are involved in the communication before, during and after planning according to their role	Office documents can be used as tools in the communication for planning in various ways	–	–
Perform planning/forecasting	Manual plan data entries, but also the triggering of simple planning functions and complex planning procedures with the different software solutions used (see following columns) are carried out by the planner	Use of simple planning forms, functions and models	Use of the special ERP planning forms, functions and models	Use of the planning forms, functions and models of the planning areas stored in the BI system up to an integrated overall corporate planning
Calculation as a special task		The calculation can be supported by spreadsheet programs for simple requirements and small data volumes	The calculation is usually done in the ERP system, e.g. derived from bill of materials and work plan items valued with prices, plus surcharges for overhead costs etc.	For special calculation requirements, individual calculation areas can be created. Usually, the results of the calculation are taken over from the upstream ERP systems
Plan result preparation/ consolidation	Provides concept proposals for plan result preparation, e.g. row and column structure of the result presentation, graphics etc. If no automation is possible in data preparation, he/she may help manually with data preparation	Can be supported for simple planning models. The more extensive the data has to be condensed and consolidated, the more likely one reaches the technical limits here	Supports the plan result preparation, condensation and possibly the consolidation (consolidation modules in external accounting) for delimited planning areas of the ERP system	Supports the plan result preparation and condensation for the developed planning areas of the BI-supported software in the integrated corporate overall planning model. For special consolidation tasks, special BI-supported software components take over the consolidation tasks, such as internal sales, debt and capital consolidation

Table 5.4 (continued)

Tasks (Planning and Reporting)	Human (e.g. Manager/Controller)	Office software (e.g. MS Excel)	ERP Systems	BI-based Systems
Analysis and interpretation of the plan results	Management task with support from controlling and possibly external parties. For this purpose, the planning reports of the IT solutions used are utilized	Provides analyzable planning reports for smaller planning models that can be used	Provides analyzable planning reports for the standardized planning areas in the ERP system that can be used	Provides analyzable planning reports for the different planning areas, e.g. as part of an integrated corporate planning, that can be used for analysis and interpretation
Communication of planning results for different audiences (presentation, meetings)	Management task with support from controlling and possibly external parties. For this purpose, planning systems are used up to collaboration tools. The planning results are presented with the IT solutions used, directly (e.g. with reports and storyboards) or further processed with office products	Planning results are often compiled with office products for presentation purposes	–	Planning results are compiled in individual reports or with storyboard or similar functions for presentation purposes
Preparation of decision proposals	Provides concept proposals for the preparation of the decision proposal of planning results, e.g. row and column structure of the result presentation, graphics, texts etc.	Decision proposals are often compiled with office products in the context of approval processes	–	Decision proposals are compiled with storyboard or similar functions in the context of approval processes
Approval processes	Management task on the different organizational decision levels with support from controlling and possibly external parties. Use of the decision proposals	Documentation in the approval process	–	Documentation in the approval process
Plan approval	Management task	–	–	–

(continued)

Table 5.4 (continued)

Tasks (Planning and Reporting)	Human (e.g. Manager/Controller)	Office software (e.g. MS Excel)	ERP Systems	BI-based Systems
Transfer of the approved planning to the accounting/ DWH	Controlling and accounting staff initiate the transfer processes for plan data acquisition at best. At worst, manual maintenance of the data	Data must be loaded into the accounting/DWH via interface	Data must be loaded into the accounting/DWH via interface	Data must be loaded into the accounting via BI-directional interface
Superordinate planning and analysis tasks				
Planning and reporting distribution	Manual or digital distribution of the created reports, e.g. via print and postal delivery, via file systems, via email or with the help of other distribution functions of the used IT systems	Files with different office formats (pdf etc.) can be distributed e.g. via mail or file storage systems	Possibly offers functions for automatic distribution of own reports and plans. Users usually access the reports directly in the ERP system	Offers functions for the automatic distribution of own reports and plans. The main goal, however, is to let the different users access the BI frontend systems directly
Authorization and access control	Provides specifications for the authorization system. The individual users are allowed to act in the different systems depending on the stored access permissions. Errors in data access can be manifold (password, settings etc.). Critical is especially data misuse (e.g. manipulation, spying)	Has only very simple file, sheet and cell protection functions	Usually has more or less powerful functions for setting permissions and access control for the ERP data and functions	Usually has more or less powerful functions for setting permissions and access control for the data in the DWH and the analyses and planning forms as well as analysis and planning areas of the BI frontend

Table 5.4 Assignment of planning and reporting tasks from human to AI—Part 6

Tasks (Planning and Reporting)	Big Data Analytics, Data Mining, Predictive Analytics, AI	RPA	Chatbots/Conversational UX	Collaboration Tools (e.g. Online Meetingd, Task Management)
Communication before, during and after planning	–	–	Chatbots can be built and used as a simple or learning FAQ for general questions about planning (e.g. explanations of the planning calendar and the status of planning). Thus, they relieve the controlling function in communication. However, they do not help with planning itself	Online communication
Perform planning/forecasting	Use of the specific, available for delimited planning areas, big data based planning models. In addition to simple statistical models, complex statistical models as well as machine learning methods up to deep learning models within the framework of artificial intelligence can be used	RPA can be used for uniform recurring planning, input and data transmission processes with different software programs and data sources if economically justified and this task cannot be performed directly with the automation functions of the planning software	–	–
Calculation as a special task	–	RPA can be used for uniform recurring calculation processes with different software programs and data sources if economically justified As a rule, it makes sense to perform this task directly with the automation functions of the specific calculation software	–	–

(continued)

Table 5.4 (continued)

Tasks (Planning and Reporting)	Big Data Analytics, Data Mining, Predictive Analytics, AI	RPA	Chatbots/Conversational UX	Collaboration Tools (e.g. Online Meetingd, Task Management)
Planning result preparation/consolidation	–	RPA can be used for uniform recurring planning result preparation and consolidation tasks with different software programs and data sources if economically justified and this task cannot be performed directly with the automation functions of the other IT systems	–	–
Analysis and interpretation of the planning results	Can be used for delimited planning areas of the big data based models for analyses of the input and output values of the planning model. In addition to simple statistical models, complex statistical models as well as machine learning methods up to deep learning models within the framework of artificial intelligence can be used	–	Through question-answer functions, model- and AI-based chatbots learn to develop answer patterns for common questions and provide assistance for the analysis and interpretation of the results of the planning systems	Online communication
Communication of planning results for different audiences (presentation, meetings)	Forecast results are provided in individual reports. The compilation and preparation of the results for presentations is usually done via other IT solutions (including Office, BI, AI-supported chatbots)	–	Using the question-answer functions of model- or AI-based chatbots, results can be attached to existing storyboards and reports or newly compiled	Online communication

(continued)

Table 5.4 (continued)

Tasks (Planning and Reporting)	Big Data Analytics, Data Mining, Predictive Analytics, AI	RPA	Chatbots/Conversational UX	Collaboration Tools (e.g. Online Meetingd, Task Management)
Preparation of decision proposals	–	–	–	–
Approval processes	–	–	–	Online communication
Plan approval	–	–	–	Online communication
Transfer of the approved planning to the accounting/ DWH	–	–	–	–
Superordinate planning and analysis tasks				
Planning and reporting distribution	Provides, if applicable, functions for automatic distribution of own reports and plans. Users usually access the reports of the big data based systems directly	Distribution of the reports and plans created in other software tools via RPA processes are possible via various ways (e.g. email) and are intended to economically replace manual distribution processes	Through the question-answer functionality of the AI-based chatbots, individual information and report requests can be supplied	In the collaboration tools, files with reports can be provided for the meetings
Authorization and access control	Usually has more or less powerful functions for setting permissions and controlling access to the data and functions of the specific big data based tools	RPA have their own licensed access when accessing the other software systems, so the access control can be mapped in compliance with the rules	Intelligent AI-based chatbots can control information sharing in accordance with existing authorization systems and access compliance	A conditional rough authorization and access control to information in the meeting is done via the invitation function

Table 5.4 Assignment of planning and reporting tasks from human to AI—Part 7

Tasks (Planning and Reporting)	Human (e.g. Manager/Controller)	Office software (e.g. MS Excel)	ERP Systems	BI-based Systems
Analysis and Reporting				
Determination of information needs and design/ conception of the analyses and reporting	The management knows or develops requirements regarding their information needs. The controlling supports the information needs analysis and makes suggestions especially for the standard reporting. The orientation is based on the strategic goal alignment and the value-creating factors of the business model. Suggestions are obtained from external experts, benchmarks and reporting standards, among others	Office software can be used as a tool for designing and conceiving the reporting and individual analyses as a documentation tool	–	With the help of tracking systems, the usage behavior of the users can be examined, so frequently or rarely used reports, filter settings, dropouts etc. can be recognized. They thus provide suggestions for a new reporting system
Data acquisition	Management, controlling and IT staff as well as possibly other external and internal staff have knowledge about the data origin of required information from the IT systems. In addition, manual data entries in the systems are required if no automatic data transfers via interfaces etc. are set up or if this information is not available in any data source so far	Spreadsheet data can be data source, but also data target for analyses. Problems with data acquisition are, among others, the susceptibility to errors, the manipulability and the limited performance in maintenance	ERP data can be data source, but also data target for analyses in the ERP system. ERP systems as well as other source systems such as CRM systems, inventory management systems, payroll systems etc. usually provide data via interfaces for downstream analysis and reporting systems	BI systems need data from the upstream source systems (ERP systems, CRM systems, inventory management systems, payroll systems, flat files from spreadsheet programs etc.) that are read into the DWH by ETL processes

(continued)

Table 5.4 (continued)

Tasks (Planning and Reporting)	Human (e.g. Manager/Controller)	Office software (e.g. MS Excel)	ERP Systems	BI-based Systems
Data/information preparation and management	Controlling and other internal (e.g. IT and other specialist functions) and external employees are often still too much involved in data preparation and manual correction processes when working with office products. In the ERP systems, special software solutions, BI and big data based systems, many parameterization activities and other setting tasks (e.g. maintenance of the cost center hierarchy, maintenance of the row and column structures for reports) have to be performed and the interface processes have to be monitored	Office products allow a very flexible, but also limited possibility to prepare and manage data mostly by manual input. Disadvantages are e.g. the error-proneness, the manipulability, the limitation to a small data space and the poor performance with larger data volumes	The data preparation and management is done in the databases and standardized settings and functions of the ERP system, which are done by human parameterization. Processes that cannot be mapped in the standard can possibly be mapped by individual adjustments	The data preparation and management is usually done in the data warehouse or less often by virtual data integration from the source systems. The DWH is usually developed and built (e.g. by the IT or the controlling with self-service BI). Standardized settings and functions of the DWH can be used for data preparation and management. Individual processes can be mapped
Report preparation/development	If the company uses office products for reporting, the human has to manually build or adapt reports and pay attention to structural changes. This is often done for each reporting request. When using other systems (ERP, BI-based systems etc.), the reports are only built by the human in the development phase and adapted when the information needs change	Very individual and flexible in layout and calculation functions, e.g. via cell-based formulas. Disadvantages are mainly time-consuming rework for structural changes as well as the error-proneness and manipulation of the data	Standardized report editors with limited functionality are often available. The report generators quickly reach the limits of layout and flexible individual analysis requirements More standardized list reports are generated	BI frontend tools usually have very powerful tools for developing individual flexible analysis requirements, especially for visualization and dynamic selection adjustments

(continued)

Table 5.4 (continued)

Tasks (Planning and Reporting)	Human (e.g. Manager/Controller)	Office software (e.g. MS Excel)	ERP Systems	BI-based Systems
Report call	Report call, for example, by clicking in the program in the corresponding menu or in office products by opening the file. With AI-supported chatbots, voice control is also possible	Report call by opening the file in the respective program	Report call by clicking in the program in the corresponding menu	Report call by clicking in the BI frontend in the corresponding menu, for example, via dashboard tile or via directory structures
Analysis and interpretation (cause-effect relationships, deviations, anomalies, etc.)	Management, controlling and other external and internal employees analyze cause-and-effect relationships, deviations and anomalies in the reports and use these insights for future actions and recommendations. They range from professional, experiential to expert knowledge	Office products, especially spreadsheet programs, can display cause-and-effect relationships, deviations, anomalies, etc. in limited smaller-scale analyses, which are based on the (smaller) data volumes of the office products	ERP systems offer limited analysis capabilities for their task area via their own reporting, which are based on the structured data of the ERP system	BI-supported systems provide analysis and interpretation capabilities for more extensive task fields of corporate management via the reporting cockpit, which are based on the mostly structured data of the DWH
Determine, evaluate, prioritize and decide on recommendations and measures	The management determines, prioritizes and decides on appropriate recommendations and measures. ontrolling as well as other external and internal employees support the decision-making process and develop suggestions for action	Can be used as a tool for documenting recommendations and measures	ERP systems provide limited analysis capabilities for their task area through their own reporting, which can be used as a data basis for corresponding recommendations and measures	Special BI-supported analysis and planning areas can be created for the collection, evaluation and prioritization of recommendations and measures within the framework of the overall model. BI-supported systems usually have extended commenting functions

(continued)

Table 5.4 (continued)

Tasks (Planning and Reporting)	Human (e.g. Manager/Controller)	Office software (e.g. MS Excel)	ERP Systems	BI-based Systems
Communication of the results of the analyses and reporting for different addressees (presentations, meetings)	Management task with support from controlling and possibly external parties. For this purpose, collaboration tools are used. The planning results are presented directly (e.g. with storyboards) or further processed with office products using the IT solutions used (from Excel, ERP, special software to BI and big data supported systems)	Analysis results are often compiled with office products for presentation purposes	–	Analysis results are compiled with storyboard or similar functions for presentation purposes, which are often offered in the BI-supported systems

Table 5.4 Assignment of planning and reporting tasks from human to AI—Part 8

Tasks (Planning and Reporting)	Big Data Analytics, Data Mining, Predictive Analytics, AI	RPA	Chatbots/Conversational UX	Collaboration Tools (e.g. Online Meetings, Task Management)
Analysis and Reporting				
Determination of information needs and design/conception of the analyses and the reporting	With the help of the eye-tracking method the reading behavior of the report analyst can be examined, so frequently used information or information that is not used at all can be recognized. They thus provide suggestions for a new report design or report system	–	With the help of model- and AI-based chatbots, suggestions for the reporting can be obtained, by recognizing, for example, frequently requested reports, key figures, forecasting models, visualization techniques, etc.	Can be used as an alternative medium for face-to-face meetings to determine information needs and design analyses (= online communication)
Data acquisition	Require structured and unstructured data from upstream source systems (DWH, ERP systems, CRM systems, inventory management systems, flat files from spreadsheet programs, internet files, web logs, social media, image, sound and video data, etc.) for their own models for result processing, which are processed by ETL and ELT processes. They do not acquire data for general business reporting	Data acquisition can alternatively be done by software robots for manual data maintenance if it is a similar, repetitive activity, where, for example, always the same structured data from different source systems (software programs) are pulled	–	–

(continued)

Table 5.4 (continued)

Tasks (Planning and Reporting)	Big Data Analytics, Data Mining, Predictive Analytics, AI	RPA	Chatbots/Conversational UX	Collaboration Tools (e.g. Online Meetings, Task Management)
Data/information preparation and management	Data preparation and management is usually done in special tools for big data-based systems (e.g. "R", SPSS Modeler, Rapid Miner or others). These tools are characterized by the fact that they provide many stochastic models for solving a problem up to building neural networks for deep learning. However, these are limited analysis areas that are used for data mining or predictive analytics purposes (e.g. cluster analysis, forecasts, simulations). The settings and functions of the programs, especially the data selection and test runs, are often done by human parameterization. An integrated and comprehensive enterprise reporting system is not supported here	Can be done as an alternative for manual data preparation processes by software robots if it is a similar, repetitive activity, where, for example, always the same structured data from different source systems (software programs) are pulled	–	–

(continued)

Table 5.4 (continued)

Tasks (Planning and Reporting)	Big Data Analytics, Data Mining, Predictive Analytics, AI	RPA	Chatbots/Conversational UX	Collaboration Tools (e.g. Online Meetings, Task Management)
Report preparation/development	Provide reports to present results of the used (statistical) models or use frontend tools, e.g. of the BI systems, to display results for the applied statistical models, e.g. with the help of sunburst, treemap or heatmap graphics	Is usually suitable when the reporting is mostly developed and prepared manually and can be taken over by RPA. This is the case when it is about similar, repetitive tasks, where e.g. always the same structured data from different source systems (software programs) are pulled	Language-based chatbots can provide reports independently based on existing data spaces and development toolsets for limited selective information requests, e.g. for time intervals, key figures, result objects, visualization types etc.	–
Report call	Report call by click in the program in the corresponding menu. With AI-supported chatbots also possible by voice control	Report calls in the other software tools possible via RPA processes	Chatbots can e.g. provide links to the report call by voice control based on specific requests of the human or open the report directly. Besides single calls, e.g. overviews and further information such as common navigation paths can be obtained	–
Analysis and interpretation (cause-and-effect relationships, deviations, anomalies, etc.)	Analyses and interpretations are possible in clearly defined tasks with large data sets (big data) with the help of statistical models up to the application of AI, e.g. for the identification of important influencing criteria, anomalies and peculiarities in large data sets	–	AI-based chatbots learn to develop and improve answer patterns to common questions through question-answer feeding, which can provide assistance for analysis and interpretation	Online communication

(continued)

Table 5.4 (continued)

Tasks (Planning and Reporting)	Big Data Analytics, Data Mining, Predictive Analytics, AI	RPA	Chatbots/Conversational UX	Collaboration Tools (e.g. Online Meetings, Task Management)
Determine, evaluate, prioritize and decide on recommendations for action and measures	Provide for delimited tasks by supporting the forecast with simulations and scenario analyses indications of suitable change variables (value drivers), which can be used for the derivation of recommendations for action and measures	–	AI-based chatbots can collect and process data through question-answer functions, which can learn and provide answers for limited statements and recommendations for action as well as for simple questions and tasks as long as they are relatively well structured	–
Communication of the results of the analyses and reporting for different audiences (presentations, meetings)	Results of AI-based analyses and forecasts can be evaluated alone or appended to existing BI-based storyboards for presentation purposes	–	Through the question-answer functions of AI-based chatbots, results, e.g. model- or AI-based system, can be appended or newly compiled to existing storyboards and reports	Online communication

The definition of corporate guiding principles, the development of a strategic vision, the derivation of strategic objectives and the decisions on strategic measures remain, however, the responsibility of management.

- **Operational planning** and **Forecasting**

One of the main tasks of controlling is the determination of budget and forecast values for the key success factors of a company such as sales volumes, revenue or profit. This task takes up a significant part of the controlling work in the year. For the basics of planning and especially the planning areas, see Chaps. 2 and 3. Planning and forecasting form the basis for the variance analysis, one of the most important tasks in controlling, in order to take timely countermeasures in case of negative deviations.

The methods of machine learning and deep learning are already capable of creating good forecasts for focused and delimited tasks in terms of their ability (see Sect. 5.7.2). Research has shown for some applications that forecasts based on artificial intelligence are often better than those created by humans. Statistical methods such as linear regression, where predictive values are derived from many influencing variables, are used. But also more complex methods such as deep learning are useful in certain situations for the forecasting task when suitable large amounts of data in the sense of big data (see Sect. 5.8) can be evaluated.[185]

In operational planning and forecasting, many companies still use Excel solutions or planning solutions of the ERP or accounting software (see Sect. 5.4). In addition, more and more companies also use BI-supported solutions as listed in Sect. 5.4.5, for an integrated corporate planning.

In terms of content, for example, statistical forecast values are automatically determined based on past-oriented values or an analytical planning is carried out by the planning-responsible experts (e.g. cost center manager, key account manager, division manager, etc.), who enter manual planning values in the planning forms of the respective IT solution. Based on the data series of individual key figures, trends can be identified, and simple extrapolations or more complex static methods such as exponential smoothing can be used for forecasting models. In addition, numerous IT functions, such as monthly distributions or revaluations with value drivers, can be used to facilitate the planning work (see Sects. 3.10, 3.11 and 3.12). There is therefore already a partially automated planning in predefined planning structures, which try to map the future development of the business in a business-economic framework with result objects such as business areas and contents such as an income statement.

AI solutions can support and improve certain planning processes in parts. However, they cannot replace the existing planning solutions, as they are not able to model and change planning structures of a future business independently in the foreseeable future.

[185] See Friedl (2019, p. 36).

This shall be examined more closely for value driver planning with simulations, sensitivity and scenario analyses:

- **Value driver planning with simulations or sensitivity analyses**

Simulations help to make transparent the cause-and-effect relationships of operationally relevant factors on business objectives. This way, critical situations can be identified early and supportive or countermeasures can be initiated.[186] If one finds the right cause-and-effect relationships and incorporates them into the planning models, the forecasts created with them can be more accurate and efficient.

In **simulations**, one or more of the input parameters are varied, and their impact on the target variables is examined. "What-if questions" are answered. In **sensitivity analyses**, it is examined how strongly target variables react to the change of certain input parameters. In the search, the input variables are changed to find out which combinations lead to the desired result. The following question is answered: "What has to change in order to achieve a target value?" Traditional IT solutions have been using simulation models and sensitivity analyses for a long time. Here, the company-specific input parameters and the target variables are predefined by the business experts of the company or by the software developers for general systems. Here, existing expert models use numerous established success and cost drivers (such as prices, quantities, discounts, variable and fixed costs, etc.) as input parameters to determine established business target variables (such as profit, return, cash flow, etc.).

AI solutions, data mining and predictive analytics can primarily help to discover new input parameters for the models that can be applied to the business target variables. For this purpose, suitable data spaces have to be identified and analyzed for individual planning areas. If suitable new parameters have been discovered, they should be integrated into the existing planning models. BI-supported controlling systems are suitable for this purpose. For example, the sales planning for certain business areas is to be improved and automated as best as possible with the help of AI, data mining or predictive analytics. If, for example, dependencies of the quantity planning on the weather are found in the data space to be analyzed, the forecast for the next week could be significantly improved. The annual planning, on the other hand, could not. If dependencies of placed marketing actions and price developments of lead prices on the result are found, the planning of the actions in the model of the operational budget planning and the forecast has to be integrated. The lead prices have to be made available in the model as input parameters, and the forecasts of the price changes can be anticipated with statistical models for the respective planning period. The marketing actions planned manually by the sales department in the past can now be classified and statistically evaluated with the help of IT. The model enriched with the AI variables should finally deliver more accurate forecasts than

[186] Cf. Gleich and Tschandl (2018, p. 189).

the conventional model. The range goes from simple independent variables to the modeling of correlation between many model variables. For example, in the simple model, a correlation between currency rate and price could affect the quantity change.

The integration of AI solutions, data mining and predictive analytics can thus make value driver models such as simulations and sensitivity analyses more robust.[187] There remains the possibility to identify or verify individual causal relationships with regard to their relevance to reality based on empirical observations with the help of suitable data spaces. For this purpose, classical statistics offer various possibilities (see Sect. 5.7.2). They can be applied with or without the use of AI. AI is suitable for large amounts of data in the sense of big data. In addition, AI-based data mining or predictive analytics solutions can generate plausible input data and well-founded "if-then" rules that previously had to be created manually.[188]

This also coincides with a study by the management consultancy Spitzner Consulting, in which three essential criteria were identified that, from the participants' point of view, make a good simulation. They named the traceability of the model assumptions, the transparency of the model structure and the selection of relevant parameters.[189] In principle, there is also the possibility to replace a simulation model with a machine learning method and do an own simulation instead of only determining the input variables as a support function. For example, the ML model could use the data from sales to evaluate the influencing factors and drivers for the customer's purchase decision, such as granted discounts, the number of advertisements placed or the visit frequency of the field staff. On this basis, a "what-if" analysis for the target variables could then be carried out via the essential input variables. Such a model takeover has the advantage that the AI-supported model automatically delivers results and sales staff can additionally consult a chat function to obtain, for example, the answer to the question of which discount level is ideally suited for a customer group to achieve an optimal sales volume through quantity increases. Unfortunately, this approach would not be transferable to many tasks and decision problems in a company as those usually involve more complex economic interrelationships. This means intra- and extra-company causal chains that controlling tries to analyse. In a simulation that consists purely of machine learning, certain causal relationships would remain opaque.[190] The example of the discount decision alone has manifold consequences. Short-term sales volume increases lead to saturation effects later on. In addition, for example, production capacities with their utilization rate and the liquidity effect as well as the profit effect have to be considered, to name just a few interrelationships. Management and controlling have the claim to a holistic view of economic

[188] Cf. Oehler (2020, p. 27).

[189] Cf. Spitzner and Schneider (2015, p. 5).

[190] Cf. Oehler (2020, p. 28).

[187] Cf. Oehler (2020, p. 23 ff.).

interrelationships. The models required for this are far away from the AI technology that is possible today. With regard to the ML models, it is important to know that the result of machine learning is based, for example, on correlations, i.e. statistical relationships. Intermediate results in the models are not output and cannot be used for other planning areas and tasks. In addition, there may be spurious causality, which is misinterpreted by the system.[191] This is one of the reasons why human-made hypotheses and interpretations are still necessary to derive causal relationships from them. Only these enable management to make well-founded decisions for steering the company.[192]

A traditional value driver planning and simulation model (cf. e.g. Figs. 2.3 or 3.11 in the preceding Chaps. 2 and 3) not only provides the simulation result, but also supports the model of integrated business planning by determining intermediate variables to identify causal relationships in other planning areas. For example, in sales planning, these would be the sales volume for inventory and production planning or the revenue for operational cash flow planning. Key figures that refer to intermediate variables can no longer be derived in an ML simulation. Due to the lack of integration in the overall view, there would be a risk of making even wrong decisions; for example, granting too high discounts does not only lead to short-term sales increases, but has a negative impact on later purchase decisions, which among other things lead to capacity fluctuations in seasonal production. At this point, there would be a risk of misinterpretation in an ML simulation due to the overlapping effects of the various influencing factors on different economic goals. A traditional simulation, on the other hand, can make the various effects clear in the models of integrated business planning and is transparently comprehensible for humans. The quality also depends here on the chosen integrated model. With AI technology, it remains to be seen to what extent the models can be economically created. In addition, the singular models aligned to specific goals would have to be further split up in order to be able to use intermediate results for other planning areas. The set of required split-up ML models would then have to be linked together in a similar way as in integrated business planning in order to provide an adequate planning and decision quality as in traditional integrated business planning models. Ultimately, it is a question of cost and benefit which systems will prevail.

- **Scenario analyses**

A well-known problem in the management of a company lies in the fact that future-oriented decisions are associated with uncertainties. An alternative to simulations and sensitivity analyses are **scenario analyses**, in which alternative states of a future are predicted in the form of different scenarios. It is thus an intensification of a simulation for several future states.

---

[191] Cf. Oehler (2020, p. 30).

[192] Cf. Davenport and Kirby (2016, p. 21 ff.).

Result – **Scenario planning** 06-2023 BI-Controlling

Filter: Company BI-C GmbH | Business Unit BU HB1 | Product gr. All | Month / Year Jun 23
Region All | Customer gr. All

	Worst	Jan	Feb	Mar	Apr	May	Jun	Jul	Aug	Sep	Oct	Nov	Dec
Price changes SPr	-5.0%	-5.0%	-5.0%	-5.0%	-5.0%	-5.0%	-5.0%	-5.0%	-5.0%	-5.0%	-5.0%	-5.0%	-5.0%
Price changes PPr	10.5%	5.0%	6.0%	7.0%	8.0%	9.0%	10.0%	11.0%	12.0%	13.0%	14.0%	15.0%	16.0%
Wage increase	3.0%	0.0%	0.0%	0.0%	0.0%	0.0%	0.0%	6.0%	6.0%	6.0%	6.0%	6.0%	6.0%
	Mittel												
Price changes SPr	0.0%	0.0%	0.0%	0.0%	0.0%	0.0%	0.0%	0.0%	0.0%	0.0%	0.0%	0.0%	0.0%
Price changes PPr	0.0%	0.0%	0.0%	0.0%	0.0%	0.0%	0.0%	0.0%	0.0%	0.0%	0.0%	0.0%	0.0%
Wage increase	1.0%							2.0%	2.0%	2.0%	2.0%	2.0%	2.0%
	Best												
Price changes SPr	5.0%	5.0%	5.0%	5.0%	5.0%	5.0%	5.0%	5.0%	5.0%	5.0%	5.0%	5.0%	5.0%
Price changes PPr	-3.3%	-0.5%	-1.0%	-1.5%	-2.0%	-2.5%	-3.0%	-3.5%	-4.0%	-4.5%	-5.0%	-5.5%	-6.0%
Wage increase	0.0%	0.0%	0.0%	0.0%	0.0%	0.0%	0.0%	0.0%	0.0%	0.0%	0.0%	0.0%	0.0%

Fig. 5.53 Scenario planning—input parameters

As a consequence, a company must weigh up expected opportunities and risks. A first possibility of dealing with opportunities and risks is to consider all possible future scenarios. At this point, however, the company will encounter the problem that the selection of conceivable future scenarios is infinitely large and a selection of a few possible scenarios is predominantly arbitrary in nature.[193]

Typically, companies like to apply simple scenario alternatives in practice, such as best- and worst-case scenarios. Here, the input variables are changed depending on the scenario with regard to the target variables. In practice, scenario analyses are not so common for operational planning and forecasting in the level of detail. This is because planning effort is incurred with each scenario to be planned. Thus, many companies limit themselves to planning scenarios only for higher-level planning models (result planning with few value drivers). As with the value driver-based simulations and sensitivity analyses, scenarios are then carried out in business practice with the common IT solutions such as spreadsheet programs or more advanced with special planning software or BI-supported solutions, in which the input parameters are modeled with regard to the target variables. The simplicity or complexity here depends on the underlying model. An example of the parameter input and result impact for scenario planning is shown in Figs. 5.53 and 5.54.

AI solutions, data mining and predictive analytics can, as already described above for the value driver-based simulations and sensitivity analyses, primarily help to discover new input parameters for the models that can be applied to the business target variables. They can help to find the better model parameters by detecting relationships and patterns as well as to determine the model cause-effect chains for the result output, e.g. in the form of "if-then" rules, when using a large amount of data and applying statistical methods and algorithms. However, the results cannot be blindly adopted. Some relationships

[193] Cf. Geißner and Wolfrum (2015, p. 243).

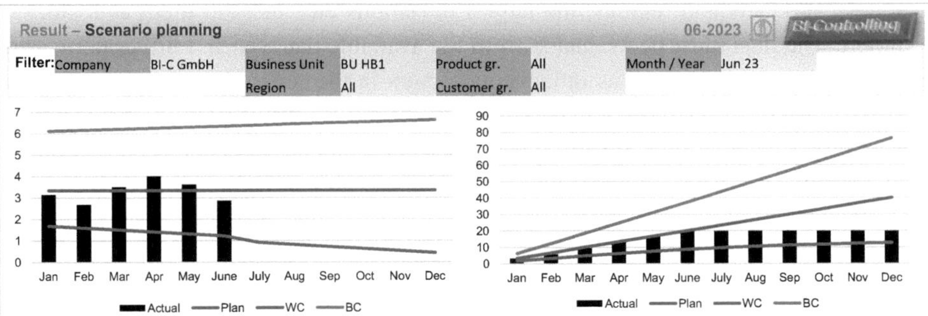

		Worst Case	PLAN	Best Case	ΔWC	ΔBC	ΔWC%	ΔBC%	Plan % Total output
Net	+	133	140	147	-7	7	-5.3%	4.8%	98.6%
Change in stocks	+	2	2	2	0	0	0.0%	0.0%	1.4%
Overall	+	135	142	149	-7	7	-5.2%	4.7%	100.0%
Discounts	+	4	4	4	0	0	0.0%	0.0%	2.8%
Bonuses	+	14	14	14	0	0	0.0%	0.0%	9.9%
Outbound freight	+	5	5	5	0	0	0.0%	0.0%	3.5%
Net net sales	-	112	119	126	-7	7	-6.3%	5.6%	83.8%
Materials	+	39	35	34	4	-1	9.5%	-3.4%	24.6%
External services	+	24	22	21	2	-1	9.5%	-3.4%	15.5%
Gross profit	+	49	62	71	-13	9	-26.5%	12.5%	43.7%
Var. Costs	+	22	22	22	0	0	2.0%	-1.0%	15.5%
CM 1	-	27	40	49	-13	9	-50.5%	18.5%	28.2%
Material overhead costs	+	3	3	3	0	0	0.0%	0.0%	2.1%
Production overhead	+	10	10	10	0	0	0.0%	0.0%	7.0%
CM 2 n. HK	-	14	27	36	-13	9	-98.9%	25.2%	19.0%
Administrative overheads	+	2	2	2	0	0	0.0%	0.0%	1.4%
Sales overheads	+	5	5	5	0	0	0.0%	0.0%	3.5%
Calculated Result	-	7	20	29	-13	9	-204.2%	31.2%	14.1%
Delta. BU overheads	+	0	0	0	0	0	0.0%	0.0%	0.0%
CM Business Unit	-	7	20	29	-13	9	-204.2%	31.2%	14.1%
CM 1 %GL		19.7%	28.2%	32.9%					
CM BU %GL		4.9%	14.1%	19.5%					

Fig. 5.54 Scenario planning - results

are statistically significant, others are rather vague in nature. In principle, however, AI can find new previously unknown input variables that complement or replace existing or unusable variables.[194]

AI-based chatbots in combination with predictive analytics are useful in both planning and report analysis as they are able to provide relevant correlations, trends and other data patterns or background data for forecasting or report analysis. The features "Smart Insights" and "Smart Discovery" are, for example, innovative augmented analytics features of SAP Analytics Cloud; these features are based on built-in techniques for machine learning and enable analyses of deeper influencing factors as well as forecasts with simulation and scenario orientation.[195] For the analysis, statistically significant value deviations (e.g. outliers, maximum and minimum values), value distributions, trends and relevant relationships between data can be displayed. Based on a key figure value, the analyst of a sales value, for example, can display the trend, the most important

[195] Bliznak, Karol (2020, pp. 160–167).

[194] Cf. Oehler (2020, p. 27).

customers, products and regions, the largest deviations etc. and add them as a story to the report. The data must be evaluable from the available data space. Furthermore, critical value drivers can be used as influencing factors in the sense of cause and effect for forecasts. "What-if" analyses are possible by calculating the impact of the identified influencing parameters (e.g. price increases and the number of customer visits) on the target variable (e.g. sales). Conversely, scenario analyses can be supported which combine different constellations of input variables for achieving desired target values. For example, scenario variants for different constellations of changes in prices, the number of customer visits or the number of marketing actions are to be determined, so that a desired target sales is achieved. The possibility of these analysis and forecasting functions depends significantly on the extent to which available data spaces can be provided which can be used to identify the dependencies of the influencing factors on their target variables by means of suitable algorithms and models. Finally, good forecasting models have to be developed, which can also calculate realistic forecast values. Highly developed systems should provide assistance here and suggest or automatically generate models that deliver the best results for the task.

- **Analyses and reporting**

The analysis and reporting tasks are carried out IT-supported in most companies. Here, the spectrum ranges from the simple use of office software with a spreadsheet program like MS Excel to the reporting functions of the ERP software in use to special software solutions with relational database technology to the use of BI-supported IT solutions for analysis and reporting (see Chap. 3 especially Sects. 3.9 and 5.4).

For special analyses, exploratory data analyses with AI can be performed separately or integrated in interaction with a BI-supported system. They are especially useful when new previously unknown data relationships and new criteria can be recognized based on suitably large amounts of data from internal and external sources. If this succeeds, planning and forecasting (see above) as well as reports for controlling the business model can be adapted to the new insights. In this context, one also speaks of "augmented analytics" (enhanced or AI-supported analysis).[196] The results of an AI-supported data analysis itself can be presented with the help of simple standardized reports for the result presentation of the models used. Alternatively, they use suitable frontend tools, such as those of the BI systems, for the visualization of their own model results in order to display the results of the statistical models used transparently with special visualization graphics, such as sunburst, treemap and heatmap graphics.

AI-based systems are not suitable for direct data management and the creation of business reporting systems. They are not designed for this purpose. However, by using AI, existing planning and reporting systems can be improved. As already shown, the

[196] Bliznak, Karol (2020, p. 156).

AI-based models can identify relationships and dependencies of relevant input and output variables of the business model which can be used for planning and reporting. Therefore, some BI providers have recently been striving to integrate AI-based software components into their systems, by integrating established systems such as "R" as an open source solution, for example.

AI-based systems require structured and unstructured data from the upstream source systems (DWH, ERP systems, CRM systems, inventory management systems, flat files from spreadsheet programs, internet files, web logs, social media, image, sound and video data, etc.) for their own models for result preparation, which are processed, for example, via ELT processes (see Sect. 5.8.3).

The reporting process itself can be indirectly improved by using AI to analyze the behavior of the report creator in data collection and processing or the reading behavior of the report recipient. Based on the creation or reading behavior, the AI can, for example, recognize which data are important to the report creator or the analyst. The user behavior of a report consumer is, for example, examined by means of an **eye-tracking method**. By analyzing the reading behavior of the recipient, for example, frequently used key figures or reports are identified, and this allows conclusions about the individual and common information needs.[197] The findings can be used to, for example, discover reports and key figures that were used very frequently or not at all and then are possibly not relevant. However, these analysis results should not automatically lead to favoring or excluding. It could be, for example, that reports and key figures that were not used are actually important, but the users were not properly trained and thus did not communicate those for the purposes of corporate management.

Also, the positioning of the information in the report layout or the use of visualization techniques, selection filters and navigation paths can be adequately examined. However, it should be critically questioned whether the most frequently used forms of analysis are always the best ones. The so-called swarm behavior (see below) could be supported here. The individuality in report design would be lost.

On the other hand, the eye-tracking analyses could adapt the report structures more individually, so that in the future each analyst can access his or her individual information according to his or her needs and preferences much faster. However, the uniformity and the common use of central information in the organization are to be criticized here. The systems would have to be intelligent enough to weigh individual and common use and to consider individual and common control requirements as best as possible.

Furthermore, the problem remains that only existing information may be evaluated. New information and impulses would not be identified by the eye-tracking method. How, for example, should good indicators and reports from other companies or from science

[197] Cf. Werner, Roland (2021): PwC: Reporting 5.0. URL: https://www.pwc.de/de/im-fokus/finance-transformation/future-ofsteering/reporting-5-0-ki-revolutioniert-das-reporting.html from 20.03.2021.

that are not yet known be discovered in order to improve one's own reporting? A local eye-tracking in a closed facility is stalled at the existing analyzed data stock and knowledge. Reporting, however, is also always stimulated and improved by the impulses of executives, of employees inside and of experts from outside, e.g. via benchmarks. This could only be solved by a general or global AI, whose technical feasibility is currently not realistic as stated at the beginning of the chapter. Therefore, it is recommended to use these new AI-supported analysis techniques mainly for clearly defined tasks. They can provide good impulses for improving reporting by analyzing the reading behavior of the existing knowledge through the eye-tracking method. In this context, the basic problems in reporting are currently mainly supported:

- Which data are relevant for the area and corporate management? (reports, report objects such as graphics, tables, filters, indicators, report positions, size, etc.) For this purpose, the indicators determined by the eye-tracking method, e.g. fixation points, gaze shifts and dwell time in the report, must be evaluated in a targeted manner.
- Which software and hardware is needed for eye-tracking?
- What know-how is required for the application, e.g. know-how of the operation and the statistical or AI-supported models?
- How should the results of the eye-tracking method be handled (automatic, semi-automatic or manual interventions in the change of reporting)?

Chatbots can support this learning process in a meaningful way by allowing analysts to ask questions about the reports. Answers are found and provided by machine or deep learning. In addition, improvement suggestions for report creation can also be generated and used through this, for example, by evaluating which information was requested most frequently. A new function "Search to Insight" in the SAP Analytics Cloud, for example, provides such a query interface that delivers corresponding information as an answer to natural language questions for individual information requests. Here, selective restrictions such as time intervals, result objects, indicators and presentation forms such as graphic types can be entered and later refined and extended and delivered as output.[198] For this, language algorithms have to correctly interpret the query, and then check the requirement based on existing data spaces and trigger corresponding reporting development toolsets that initiate a simple report creation. This support will gain popularity due to the simple search field question or possibly later even audio language input. However, this technology is still in its infancy. Language inputs are, for example, only possible in English. However, it can be assumed in the future that this technique will be strongly pushed and widely spread. The search field query or language input has the advantage of quickly accessing information compared to the conventional analysis paths via menu navigation by tile, icon, tree structure, etc. The knowledge of existing reports remains

[198] Bliznak, Karol (2020, pp. 158–159).

additionally advantageous, however, as standard reports usually have a higher information density and are, among other things, better suited for common control coordination. If the analyst knows the report names, he can of course also display a standard report directly via the language assistant. For a better search of further interesting information, overviews of existing report structures can also be shown; these can also be accessed.

The AI-supported technology with chatbots and predictive analytics also enables the support of report commentary, e.g. in the form of a storyboard, based on analytically determined data. For this purpose, the additional insights gained, are appended e.g. to the existing reporting.

More difficult, but certainly conceivable, is the use of algorithms for automatically deriving recommendations for action for management or for making decisions by AI.[199] The AI trend for planning and reporting will have to deal increasingly with intelligent systems and the associated data availability, quality and processing in the coming years. Here, the AI capabilities are to be used to increase the meaningfulness of analytics in planning and reporting. The increasingly precise processing of structured and especially unstructured data is part of this, as well as, for example, an AI-supported suggestion system for meaningful data combinations or suitable visualization.[200] If the report recipient has questions about important data and key figures, he asks a chatbot, which answers in real time based on a question-answer tree and displays the answer to the key figure in graphical form, for example. Here, too, the system learns with an increasing number of questions.[201]

If the companies increasingly use collaboration tools as well as stationary and mobile devices for planning, organizing and conducting meetings for steering the company, the circle of digital use of information is complete. Printing and travel costs can be saved as the information is available digitally and location-independent. In meetings, notes and bookmarks can be inserted directly into the report. Minutes can be attached to the reports or in the storyboard. Measures can be planned and initiated via integrated task management functions. Distribution and approval processes are thus easy to carry out.

- **Superordinate planning and analysis tasks**

Similar to classic reporting tools, BI and also AI-supported analysis systems must comply with data protection rights (such as the GDPR), something which is covered in most cases by authorization systems and access rights. The challenge is especially with the

---

[199] Cf. Schneider Steffen (2021) Reporting Pulse Check.—What topics move the experts from Controlling & Finance? URL: https://www.haufe.de/controlling/controllerpraxis/reportingpulse-check-das-bewegt-finanzexperten_112_471720.html from 24.03.2021.

[200] 126 Cf. Schmitz, Robert (2018). BI Scout: BI Trends 2019: Integrating Artificial Intelligence in Analytics Scenarios. URL: https://www.bi-scout.com/bi-trends-2019-integration-vonkuenstlicher-intelligenz-in-analytics-szenarien from 21.03.2021.

[201] Cf. Werner Roland (2018) PwC Germany: Reporting 5.0. URL: https://www.pwc.de/de/pressemitteilungen/2018/pwclauncht-reporting-5-0.html from 18.03.2021.

data use of AI-supported systems as the correct use must be clarified already at the access to the data source.

The RPA solutions require their own licensed access to other software systems, so that access control can be mapped in accordance with the rules. The advantage of technical solutions in general is that rule-compliant data accesses can be better adhered to than with manual processes.

Through the question-answer functionality of the AI-supported chatbots, individual information and reporting requests of the users can be supplied. This extends the previous possibilities of push or pull supply of the users compared to the previous possibilities of the existing systems with simple report distribution functions via file directories and email or directly via the menu navigation considerably.

Specific analytical problems of AI-supported analyses will be shown at the end. They should be considered in the AI projects and their results:

- **Filter bubbles, echo chambers, fake news, troll factories:** The phenomenon of filter bubbles and echo chambers is discussed in the context of the use of data in the social media domain. This means that the social media user only receives selective, user-specific information through AI-supported algorithms and thus, due to the data isolation in a cluster, opposite viewpoints are not shown at all.[202] Filter bubbles are also an ideal breeding ground when placing fake news. If information is deliberately manipulated, e.g. by so-called troll factories, the false news can be easily multiplied.
- **Swarm intelligence:** In a similar direction as filter bubbles, the term swarm intelligence can be understood as collective intelligence or group intelligence in which groups of individuals can make intelligent decisions through collaboration, regardless of the intelligence of the individual members. In the context of distributed artificial intelligence, complex networked software agent systems try to solve certain tasks in the swarm, following the model of state-forming insects or bird flocks.

5.7.8 Challenges and Recommendations for Action for the Use of AI, RPA and Chatbots in Planning and Reporting

As shown above for the tasks of planning and reporting, the use of AI, RPA and chatbots in planning and reporting offers many opportunities and potential benefits. Finally, the challenges of the new technologies for the analysis and planning tasks in controlling shall be briefly summarized:

[202] Cf. Pariser (2011) Eli Pariser: *The Filter Bubble: What the Internet Is Hiding from You.* Penguin Press, New York, 2011.

For the use of AI technology in planning and reporting, the following points need to be solved, with first recommendations for action highlighted in italics:

- **Finding or training suitable IT experts for business intelligence, AI, RPA and chatbots with a personal requirement profile toward data scientist.** Here, the skills that are particularly important are those to identify suitable data spaces and to use appropriate statistical models and algorithms for the management-relevant business tasks. Of particular importance here is the further development of the executives and employees in the company in the affected areas (e.g. IT, controlling and other functional areas). The *recommendations for action range from targeted recruiting to further education to the establishment of strategic partnerships with technology companies or universities.*
- **The determination of the status quo and the maturity level of one's own control situation with regard to the tasks in planning and reporting in IT-supported controlling.** *For this purpose, it is advisable to seek external support from technology-oriented consulting firms or other suitable experts. When determining the status quo and the maturity level, the weaknesses of the existing IT landscape and the software solutions used must be taken into account. For example, if the data quality of the existing source systems is insufficient or important business and control processes are not well supported, these fundamental weaknesses must be addressed. In the first maturity level with new technology, those subtasks should be identified that can be used economically for appropriate IT support (BI, AI, RPA, chatbots, etc.). In the next maturity level, cross-functional processes can already be supported by automation and, if necessary, simple decisions can be made semi-autonomously. A maturity level with fully automated process support and continuous autonomous decision support is currently not realistic due to the lack of performance of the available AI technology. For this stage to be reached, purely technically speaking, the supply of relevant data for complex relationships would have to function almost in real time.*[203]
- **The identification and technical use of suitable large amounts of data for the planning and analysis tasks listed.** The data quality and the availability of a solid information base form the essential basis for the use of a modern technical solution for each identified application area. Therefore, the company must ensure effective data processing and management. *The recommendations for action are far-reaching. They range from avoiding data entry errors in the existing internal source systems to building data catalogs in the data warehouse to using suitable new external data sources in the sense of big data.*

---

[203] Cf. BitKom (2017): Understanding artificial intelligence as automation of decision making. URL: https://www.bitkom.org/sites/default/files/file/import/Bitkom-Leitfaden-KI-verstehen-als-Automation-des-Entscheidens-2-Mai-2017.pdf from 24.03.2021.

- **The investment and implementation of suitable hardware and software in the IT landscape of the company to be developed.** Companies need new resources for the new IT solutions, such as databases, computers, platforms, networks, software for BI, AI, RPA, etc. and have to integrate or replace them in parts in the existing IT structure. *The IT projects should be economically evaluated and prioritized by an investment analysis. Both quantitative and qualitative criteria should be considered. For the implementation, a tight project management is necessary as was presented, for example, in* Sect. 4.2.1.
- **The integration into the processes and organizational structure of the company as well as the creation of acceptance among the managers and employees.** There is the fundamental problem that results of AI systems are no longer comprehensible for humans, e.g. due to the depth of the neural networks and applied models and algorithms.[204] With the help of artificial intelligence, processes can be optimized and (partially) automated. Therefore, many employees are afraid of change and job loss. New requirements from the technology to the employees require new skills. *A successful IT development and implementation must therefore be accompanied by the right communication and involvement of all stakeholders, among other things with appropriate training programs.*

5.8 Big Data

The internet and the social media platforms are moving more and more into the center of society. In the World Wide Web, countless data are uploaded, stored or shared every day, consciously or unconsciously. This applies not only to the private sector, but also to business life and the public administration.

The information systems used in the companies, public institutions and other organizations are becoming more and more numerous. As a result, more and more data have to be processed, as already shown intensively in the topic area of data warehouse. The companies and institutions have recognized not only to process the internal data from the daily business processes, but also to use the potential of the data from the internet, the social networks and other data sources, for example, to better adapt to the future customer needs, avoid risks and differentiate themselves from the competition. Blog entries, web logs and many other data can provide valuable information here. Furthermore, our mobile devices such as the cell phone and notepad provide many unstructured information such as photos, videos, audios and navigation data. The data volume is also enriched by the many sensory data such as motion and temperature data from components, machines, cars and buildings and other infrastructure facilities (e.g. power lines).[205]

[204] Cf. Davenport and Kirby (2016, p. 21 ff.).

[205] Cf. Seufert (2014, p. 25).

These data are only valuable, however, if the information that allows the right conclusions to be drawn and decisions to be made is identified from the amount of internal and external data.

The amount of data is not only growing steadily, but rather exploding exponentially. This is evidenced, among other things, by a study by IDC (International Data Corporation) in 2011, which found that the volume of data worldwide more than doubles every two years.[206] In 2011, the global volume of data amounted to a total of 1.8 zettabytes (1.8 trillion gigabytes). A follow-up study by IDC in 2012 found that the volume of data increased by 0.9 to 2.7 zettabytes in 2012.[207] According to another study by IDC, the global annual generated data volume will increase from 33 zettabytes in 2018 to 175 zettabytes in 2025.[208] Data-intensive companies today already reach a volume of one petabyte (= 10^{15} bytes = 1000 terabytes) or more. Relational databases and also the data warehouse-based OLAP databases shown above reach their limits when processing such large and largely unstructured data volumes and can no longer handle them efficiently. The main disadvantages of these systems, besides the inadequate processing of unstructured data, are mainly the lack of scalability of data storage (across computer boundaries) and the consistency assurance of data, which requires a structured structure and a distribution of data within a mostly physical relational or multidimensional data model (see Sect. 5.5). This is where big data comes in.

Unstructured data should be analyzed, stored and made available to various users in a cost-effective, location-independent and high-speed manner. Big data provides tools and techniques for capturing, storing and analyzing such large unstructured, structured and semi-structured (mixed) data volumes in a technically feasible and economically attractive way. Big data is intended to provide new insights from information use (e.g. deviations, correlations or trend detection) to improve decision making at all management levels, thus contributing to the value creation of the company.

Frequently mentioned application fields of big data solutions and their value creation are summarized in Table 5.5.[209]

The biggest problems for companies in introducing and applying the big data technology are the lack of technical and professional know-how as well as the high costs for the development of the data bases and the IT solutions for the so-called advanced analytics. Furthermore, companies as well as society in general see special dangers concerning data protection and data security.[210]

---

206 Cf. IDC (2011).

207 Cf. Gesellschaft für Informatik et al. (2013).

208 Reinsel et al. (2018, p. 3).

209 Similar examples can be found in different sources, among others Dorschel (2015, p. 109) and Bitkom (2013).

210 Cf. e.g. BARC (2014, pp. 23–24) and Dorschel et al. (2015, p. 2) as well as Institute for Business Intelligence (2013).

Table 5.5 Big Data: Fields of application and value creation

Fields of Application	Value Creation through Big Data
Marketing/Sales/ E-Commerce/Trade	Better alignment with customer needs, e.g. based on trends, which are determined with the help of data from online forums, social networks or other online platforms, based on clickstreams, log data from websites or other entries. In addition, evaluation of further market and customer-related data from internal and external sources. This results, among other things, in:
	• Targeted product advertising, offers (without scattering losses) and product advice for a specific customer clientele (customer segmentation)
	• Customer needs analysis for targeted new product development or for the development of new business models
	• Improvement of the response time to current market events, e.g. with the help of sentiment and trend analyses as well as point-of-sale data
	• Acquisition and retention of customers through knowledge advantage as well as individual offers and support (e.g. better accessibility)
	• Targeted up-selling and cross-selling or bundled sales
	• Identification of purchase cycles and trends
	• Improved sales/revenue planning as well as market/competition analysis
	• Improved sales forecasts and return forecast
	• (Partial) automation of the scheduling
	• Better coordination of the field and office staff on the buying behavior or intentions of the customers to move
	• Identification of customers with the highest value/potential and better customer segmentation
	• Individual pricing and conditions for individual customers and customer segments
	• Pattern recognition in the area of customer complaints and warranty analysis
	• Churn management support to prevent customer movement

(continued)

Table 5.5 (continued)

Fields of Application	Value Creation through Big Data
Production/Procurement/Inventory and Asset Management/Logistics	Machine learning over networked sensor data, intelligent components (= intelligent devices, i.e. devices with internal computing power such as computers, mobile phones, cars but also kitchen appliances and airplanes) and RFID scans (Internet of Things) and Smart Factory (Industry 4.0): Networking of products, plants, buildings, logistics via intelligent objects, which are connected in an internet of data and control themselves. This results in, among other things:
	• Event-driven and proactive identification of defects and monitoring of machines, devices, components, IT systems for cost-effective maintenance and repair as well as the provision of spare parts and the reduction of downtimes
	• Improved forecasts of raw material price developments
	• Quality assurance of production processes, detection of disturbances and errors; avoidance of failures
	• Improved production, plant and demand planning and control
	• Quality assurance of delivered products and plants
	• Performance and efficiency improvements through optimal resource use (learning systems with sensor data)
	• Production process optimization and assortment optimization
	• Improvement of efficiency in the use of service units
	• Improvement of energy efficiency
	• Optimization of transport and storage processes (driving speed, distances, inventory levels etc.)
	• Improvement of road guidance systems, high-bay storage systems, chaotic storage systems
	• Improvement of logistics processes to optimize logistic objectives such as delivery reliability, inventory optimization and delivery capability
Research & Development	Evaluation of market-related, technical and other internal and external data. This results in, among other things:
	• Generation of new product ideas through trend/market analysis (e.g. patent analysis) as well as the identification of customer needs
	• Shortening of development times by faster consideration and implementation of customer needs
	• Improved analysis of measurement data as well as improved testing of new products/processes

(continued)

Table 5.5 (continued)

Fields of Application	Value Creation through Big Data
Management in general	Evaluation of internal and external data related to market, environment, context and company. This results in, among other things:
	• Improved management decisions by monitoring and analysis of business-relevant events (market, competition, environment …)
	• Process and organizational improvement
	• Improved participation and overall corporate planning and forecasting
	• Shorter response times to events and faster decision preparation
Risk management	Improved risk prediction and analysis. This includes, among others:
	• Damage analysis and assessment
	• Risk and vulnerability analyses
	• Risk index determination for segmenting customers and suppliers with regard to contracts, credit financing, etc.
	• Fraud prevention and detection
	• Early warnings regarding new risks
Finance/Accounting/Controlling	Analysis of detailed data from the areas of finance, accounting and controlling as well as external data. This includes, among others:
	• Improved forecast of costs and success as well as financial and capital requirements and asset development
	• Credit rating for contract release, pay determination, insolvency prediction
	• Better assessment of and response to payment behavior
	• Simulation, prediction, scenarios of key figures
IT	Evaluation of log files and other IT data. This results in, among other things:
	• Improved search for IT problems, e.g. security gaps and performance problems
	• Improved IT usage and performance optimization
	• Predictive IT maintenance and repair

(continued)

Table 5.5 (continued)

Fields of Application	Value Creation through Big Data
Human resources	Evaluation of internal and external data relevant to personnel. This results in, among other things:
	• Improvement of personnel selection by targeted search using personnel criteria
	• Deepening of company knowledge by enriching external data (knowledge gain)
Banks/Insurance companies	Internal and external data in the banking and insurance business. This results in, among other things:
	• Improved interpretation of the investor mood on the stock market and its significance for share price development by sentiment analysis
	• Credit rating to determine loan granting, overdraft limits and interest rates
	• Damage and risk analysis and forecasts
	• Fraud detection, e.g. credit card fraud
	• Structure optimization for charges and premiums
	• Preventive churn management
	• Improved customer loyalty
	• Decision support for portfolio changes and portfolio optimization
Energy supply	Data from energy suppliers. This enables, among other things:
	• Improved forecast of energy demand
	• Detection of customers willing to switch for new customer acquisition or customer retention
	• Structure optimization of rates and prices
	• Better use and utilization of resources (network, personnel …)
Healthcare	Data from medical devices, as well as diagnostics and treatment (e.g. on the electronic health card) and many other data from the healthcare sector. This enables, among other things:
	• Better predictions of life expectancies
	• Improvement of drug treatment
	• Better predictions of the onset and course of diseases

(continued)

Table 5.5 (continued)

Fields of Application	Value Creation through Big Data
Telecommunications	Data from the telecommunications industry. This enables, among other things:
	• Identification of customers willing to switch for customer acquisition or retention
	• Structure optimization of rates and prices
	• Better use and utilization of resources (network, staff …)
Public administration/Politics	Internal and external data from public administration and society in general. This enables, among other things:
	• Better predictions of social data: population development, crime rates etc.
	• Pattern recognition of fraudulent actions (e.g. tax evasion)
	• Analysis and integration of investigative data by the police
	• Improvements in opinion research and election forecasts
	• Better use of geographical and GPS data
Environment	Environment-related data. This includes, among other things:
	• Better forecasts and control options regarding emission loads
	• Improved weather and climate forecasts

5.8.1 The 3-V Model or 4-V Model

The young research field of big data has not yet produced a generally accepted definition for big data.[211] In practice and in studies, explanations often occur that understand big data as the processing of large amounts of data, e.g. of social media and internet data, and use better techniques for data management, such as distributed data storage and real-time data analysis, for this purpose.[212] In most cases, IT and domain experts agree that big data can only be spoken of if at least one criterion of the 3- or 4-V model (Volume, Variety and Volatility/Velocity or Veracity) is met.[213]

The 3-V model consists of the terms: Volume (the amount of data), Variety (the data diversity or complexity) and Volatility or Velocity (the inconstancy or speed of the data or data change). The 4-V model extends the 3-V model by the term Veracity (accuracy or credibility) of the data (see Fig. 5.55).[214]

- Volume (amount of data)
  The volume describes the amounts of data that are available to companies internally and externally, data which are too large to cope with using the traditional data storage methods, such as relational databases. Big data allows data to be stored and processed on a scale that far exceeds the previously known limits.[215] A study by IDC, for example, found that sensor data account for about 2% of the total data volume worldwide and that this share will rise to 10% worldwide by 2020 and to 14% in Germany.[216] The more data a company extracts, the more likely the relationships between them will become apparent and decisions can be made more precisely based on verified data.
- Variety (data diversity/complexity)
  The characteristic variety describes the data diversity and complexity of data. The data to be stored usually do not have the same structure. Structured data and unstructured data are distinguished. Structured data are, for example, data from a relational table with name, age or income of an employee.[217] Examples of unstructured data include machine-generated data such as sensor data and web logs, as well as images, audio and video files, whose data type is different.[218] There are also mixed forms,

[211] Cf. Finlay (2014, p. 13).

[212] Cf. IBM Institute for Business Value and Säid Business School (2012).

[213] Cf. Gartner (2015) and Brücher (2013, p. 41 ff.).

[214] Cf. Schroeck et al. (2015, p. 3 f.).

[215] Cf. Finlay (2014, p. 13) and Gesellschaft für Informatik et al. (2013).

[216] Cf. TECChannel (2014).

[217] Cf. Finlay (2014, p. 13).

[218] Cf. Sack (2013).

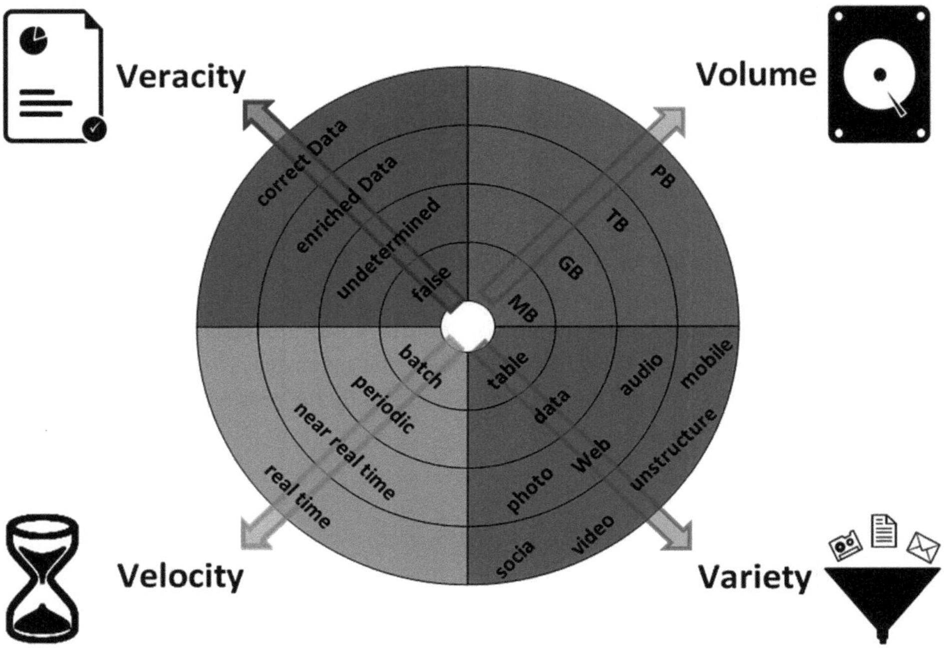

Fig. 5.55 4-V model. (Own representation based on: Klein et al. 2013)

so-called semi-structured data, which can be found, for example, in an e-mail. An e-mail contains structured data in the header, such as the sender and the date of sending, but often contains unstructured data in the text field, such as the text entry with attachments and images. When using structured and unstructured data, one also speaks of polystructured data.

- Velocity (data speed) and Volatility (data inconstancy)
  The third characteristic deals with the speed (velocity) and volatility of the data. Data, for example on the internet and in social networks, are constantly supplemented or updated, which means that in addition to the processing amount, the data processing speed and the updating are of great relevance. Data should be able to be evaluated as close to real time as possible. Data changes have to be traceable, old entries have to be replaced or supplemented by new entries.
  Data that are relatively constant or not constant have to be distinguished. Relatively constant data include, among others, names of customers, gender or nationality. Inconstant data are data that change (frequently), such as the price and quantity of an order from a supplier or the speed of a machine.[219]

[219] Cf. Finlay (2014, p. 13).

- Veracity (accuracy/authenticity of data)
  IBM introduced the accuracy and authenticity (veracity) of data as a fourth criterion.[220] Since the data come from different sources and have different structures, the data must be harmonized and aligned. False data are to be deleted, uncertain data are to be supplemented by transformation, so that the information obtained does not contradict. The data quality and security of the data must be considered when processing and analyzing the data (see Sect. 4.2.4.4). Examples of false data are e.g. translation errors or misleading advertising.[221] Inconsistent data result e.g. from the use of old sources, such as an old delivery address or phone number.

Since all criteria of the 3- or 4-V model do not work alone in practice, but usually together, the thesis can be put forward that big data solutions are interesting for companies and institutions when all 4 criteria apply to the data to be used. In more recent times, a fifth V for defining big data is listed. The value of the data should serve as an additional criterion. Only useful and valuable data should be used for a specific big data application.[222] In my opinion, data-driven applications generally require beneficial data. Therefore, this criterion does not distinguish itself from other non-big data applications, and neither does the criterion of veracity. However, unlike this criterion, many non-beneficial data, which should be correct in principle, are also processed in big data applications. Often, it is important in big data and AI-based applications to identify and separate beneficial and non-beneficial data from the large amount of data, and then to further analyze beneficial data.

Big data will initially offer interesting extension possibilities in data processing for larger companies, but later also for medium-sized businesses.

5.8.2 Technical Foundations for Big Data

To understand big data, four technical foundations are explained first: the in-memory technique, NoSQL databases, Hadoop as a Java framework and analytical databases.

5.8.2.1 In-Memory Technology

With the help of in-memory technology, data is stored on the main memory and not on hard disks as usual. This allows faster access and data analysis to be performed more quickly.[223] Queries from RDBMS on hard disks are in the range of milliseconds, access times of in-memory database queries are in the range of nanoseconds. The increased

---

220 Cf. IBM Institute for Business Value and Säid Business School (2012, p. 4).

221 Cf. Freiknecht (2014, p. 13).

222 Cf. Kreutzer and Sirrenberg (2019, pp. 78–80).

223 Cf. e.g. Gluchowski and Chamoni (2016, p. 189).

use of main memory is made possible by the enormous price decline. Furthermore, the development of 64-bit systems allows a more effective use, which means that modern servers now have several terabytes of main memory.[224] A particularly important factor for the advancement of big data applications, e.g. in the form of artificial intelligence, is the great progress in the computing power of processors. To process large amounts of data, a strong computing power is needed. A direct driver for big data applications and AI are the graphics processing units (GPU). These processors were originally developed for the calculation of complex mathematical functions, to make the graphics in computer games appear more realistic.[225]

However, storing data in main memory also entails the great risk that, for example, in the event of a system failure due to a power outage, all data in main memory can be lost.[226] For this reason, it makes sense to store all important data on the persistent hard disk or other storage devices as a matter of course. A persistent data storage has the advantage that the data is still stored after a power loss.

In-memory concepts distinguish between pure in-memory systems and hybrid in-memory systems. Pure in-memory systems are systems where all data is stored on the main memory and hard disks are used as persistent storage for security. This allows faster access to the data in main memory. However, it must be ensured that enough memory is available for changes. Furthermore, for pure in-memory systems it has to be distinguished whether the data is used on a single server or on distributed servers. In distributed in-memory systems, the distribution is done via in-memory data grids. This allows both the entire use of the main memory of a server (scale-up) and the possibility to store its data on additional servers or main memories (scale-out). The grid ensures that it is possible to access different servers.

In hybrid in-memory systems, only part of the data is stored on the main memory and the rest on the hard disk. These hybrid systems are also called temperature model as the relevant data in main memory is called hot data and the less relevant data is called cold data. This has the advantage that only the required hot data has to be provided in main memory in order to be able to process and evaluate the data faster.[227]

5.8.2.2 NoSQL Databases

NoSQL-databases have to be distinguished from relational databases and OLAP-databases (see Sect. 5.5.4). NoSQL does not stand for "no SQL", but for "not only SQL".[228] NoSQL is not meant to replace traditional SQL-databases in enterprises, but to comple-

---

[224] Cf. Brenckmann and Pöhling (2012).

[225] Cf. Gentsch (2018, pp. 21–23) and Bauer (2020).

[226] Cf. Schmitz (2015, p. 236).

[227] Cf. Bitkom (2014, pp. 21–24, 45–47).

[228] Cf. Walker-Morgan (2010).

ment them sensibly. NoSQL can also be understood as a union of all non-relational databases that are able to process large amounts of unstructured and structured data.

NoSQL-databases are characterized by the fact that they can process and manage unstructured data such as data from social networks (images, texts, videos etc.) better. They ensure the storage of large amounts of data, while doing without absolute data consistency and the structuring of data as in SQL- or OLAP-databases.[229] The NoSQL-database structures distribute their computing power and storage capacities on a system network, which can also be operated with less powerful and cheaper hardware.[230]

If one tries to define NoSQL, then NoSQL-databases are mainly characterized by the following properties: They are non-relational, schema-free and open for changes, the data can be easily distributed, the data are horizontally scalable and replications of data are easily possible.[231]

To store and analyze the data, they do not always have to be related. Therefore, the data are stored without a fixed schema and structure.

NoSQL-databases do not comply with the ACID-properties of relational databases, which are of great relevance for security-critical data such as account movements and invoice items. ACID stands for: Atomicity, Consistency, Isolation and Durability.[232] For the example of a transfer as a financial transaction, this means that a transaction consists of several commands executed one after the other or in parallel (payment entry, verification, booking) and it is only considered completed if all commands are successfully completed (atomic transactions). Consistent data only exist if data integrity exists, for example, the account specified in a payment transaction must exist. Furthermore, several transactions (transfers) must not be executed in parallel, but only one after the other (isolation). The deposits and withdrawals from the transfer are permanently stored in the system (durability) and can also be traced in the account after a system crash.

For unstructured data, such as in social networks, not the consistency, but the short response time and the fault tolerance as well as the availability of the data are in the foreground. Therefore, the requirements of the ACID-principle have to be relaxed. With the help of the Brewer-theorem or the **CAP-theorem** (Consistency, Availability, Partition tolerance) the relation of these three properties is described:[233]

Consistency: After completing a transaction, the database must be left in a consistent state. This means that if a node within a cluster is changed, all other nodes that have read or write access to this node must be informed (updated), something which affects the response time of the system when there are more than a thousand nodes.

---

229 Cf. Sack (2013).

230 Fasel and Meier (2016, p. 6 f.).

231 NoSQL Databases (http://nosql-database.org/. Accessed on 15.12.2014).

232 Cf. Warner (2007, pp. 480–485).

233 Cf. Edlich et al. (2010, pp. 31–33).

Availability: The response times of a system should be acceptable and the system available.
Partition tolerance: The failure of a node in the system (cluster) should have no impact on the entire system. The system must therefore be able to continue working as if the affected, failed node did not exist. This fault tolerance is only possible with data replication.[234]

In practice, it is easy to see that it is difficult to meet all three requirements simultaneously in distributed systems with large amounts of data. For example, if availability and fault tolerance are considered important for a database (social networks), the theorem states that the requirement for consistency should be relaxed. Depending on the application area of a database, databases can be distinguished that only meet two of the three properties of the CAP theorem. The following three types of database systems can be distinguished:

CP systems completely forego availability, but guarantee fault tolerance and consistency. They are suitable, for example, for distributed banking applications (ATMs and central financial system), where data consistency (correct account movements) is the highest priority and system availability is not prioritized.
CA systems support the properties of consistency and availability of data, but dispense with fault tolerance of nodes in the system. This combination cannot be applied to distributed systems as the system would no longer function properly in the event of any node failure. They are therefore used for classic relational database systems that do not work with distributed systems and nodes, but with one system. Examples are inventory management and ERP systems.
AP systems do without the complete consistency of data and put availability and fault tolerance in the foreground. This means that not all nodes in the system always see the same value of an attribute. There can be significant delays in updating the data until data consistency is restored. These systems are mostly used for DNS (Domain Name System) or cloud computing, where it can take hours or days for a change to become visible. They are suitable, for example, for social networks, where availability is more important than data consistency.

Figure 5.56 shows examples of selected databases for different system types in the assignment of the CAP theorem.

Since strict consistency, as described in the ACID properties, would be difficult to implement in distributed systems and a decision for a system selection in terms of the CAP theorem always brings disadvantages by neglecting a property, alternative solutions were sought. For this reason, BASE (Basically, Available, Soft state, Eventually consist-

[234] Fasel and Meier (2016, p. 12).

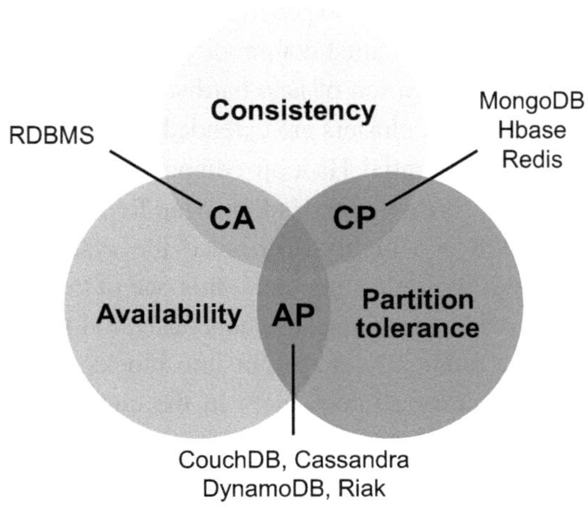

Fig. 5.56 CAP theorem. (Based on NN/w3resource 2015)

ent) was developed, which is used in most NoSQL databases. This is an advancement that tries to overcome the disadvantages of the ACID properties as well as the weaknesses of the CAP theorem. In this model, availability with acceptable response times is paramount. Fault tolerance must be guaranteed. Consistency is secondary, as it is assumed that it will be achieved over time.

An important basic technology of NoSQL databases is therefore Multiversion Concurrency Control (MVCC). MVCC was developed to allow multiple users to make changes to data sets at the same time without locking them as in relational databases and thus only allowing sequential work. This is achieved by creating a new version every time a record is inserted, deleted or modified. Each version contains a timestamp or a sequential transaction number that refers to the previous version. Since multiple versions of the same record are stored, this leads to a higher storage requirement. To prevent this from getting out of hand, old versions are deleted at regular intervals. If different attributes (e.g. the billing and delivery address) of the record are changed by different users at the same time, synchronization is required. If there are conflicts, i.e. an attribute (e.g. the name) has been changed differently by two or more users, the users receive a conflict message that needs to be resolved.

5.8.2.3 Hadoop

Hadoop is a free Java-based open source software. It was originally developed at the company Yahoo and was later transferred to the Apache Foundation and further developed by an open source community.[235] It is a Java framework that can store data distributed on many computers within a system and can process the data in parallel. The framework is designed for processing large amounts of unstructured and structured data.

---

[235] Fasel and Meier (2016, p. 124).

It does not rely on expensive hardware, but on inexpensive systems (e.g. normal PCs), which are also called commodity hardware. Against this background, the scale-up effect (i.e. the acquisition of new hardware) already described is not used, but rather the existing servers or clusters are extended by additional nodes. This is then the scale-out effect.

Three essential Hadoop components are distinguished: the HDFS (Hadoop Distributed File System), the Hadoop Map Reduce and YARN.[236]

HDFS is a highly distributed file system for storing large amounts of data on the nodes of multiple computers. It is one of the core elements of Hadoop, which allows data to be read and written simultaneously, unlike relational databases. To store large amounts of data efficiently, it is split into blocks using the map-reduce method and distributed evenly across all computers in the cluster. In this context, the term cluster computingis used. To be distinguished from cluster computing is the term grid computing where many loosely connected computers over a network provide unused capacity for applications with high resource requirements.[237] Both cluster and grid computing are technologies that can be used for big data analysis.

Hadoop Map Reduce is a programming framework for distributed processing of data according to the two-phase processing by map and reduce functions. It was developed by Google for indexing websites.[238] The method uses the distributed storage of data on multiple computers and runs in parallel on these computers. To make large calculations faster, they are split across multiple computers (map function) and after the calculations, the framework ensures the merging of the partial results (reduce function). With the map-reduce method, the fault tolerance in a cluster is very high, as it contains many computers and the blocks are sent as duplicates to individual computers. If a node fails, another one that has the copy can take over the work. Thus, there is no delay or interruption of the process. A disadvantage compared to RDBMS is that the definition and execution of queries in a map-reduce system takes significantly more time.

YARN (Yet Another Resource Negotiator) is "a programming model that allows any application to run distributed on a Hadoop cluster and to reserve a precise amount of resources for these executions".[239] YARN manages all the resources available in the cluster, such as CPU, RAM and storage, in the sense of an intermediate layer. With the help of YARN, in-memory processing systems such as Spark and real time streaming frameworks such as STORM can be directly connected to HDFS. Apache Spark is also used for processing and storing very large amounts of data and can be considered as a successor of Hadoop. The technology is also free to use and based on a hardware cluster. In addition, Spark can perform real-time data analysis even with very large amounts of data, compared to Hadoop. This is because Spark, unlike Hadoop, which is based on the

236 Cf. Freiknecht (2014, p. 20).

237 Cf. Luber and Litzel (2017).

238 Cf. Rouse (2014).

239 Cf. Freiknecht (2014, p. 20) and Bitkom (2014, p. 39).

map-reduce method, relies on in-memory data processing. The data is stored and processed in the memory of the cluster nodes. In addition to data from the Hadoop Distributed File System, data from relational databases such as Hive and NoSQL databases can also be processed. Overall, Spark can handle much more complex tasks than Hadoop due to a machine learning algorithm.[240] With STORM, data streams can be analyzed in real time (real time) as they occur, and do not have to be stored in a database first.[241] Although HDFS, Map Reduce and YARN can also be used independently of each other, Hadoop unfolds its strengths through their integrative use.[242]

The use of Hadoop has implications for the 3- or 4-V model. Hadoop technologies affect the data volume, data velocity, data variety and data accuracy.

Hadoop is also used as a transformation platform in the context of the ELT process. In contrast to the ETL process, the phases in the **ELT process** are changed in order. [243] After extraction, the loading and transformation processes swap places. Unlike the ETL process, in the ELT process all data from the various source systems are extracted, cleaned and then stored, regardless of whether they are relevant for the result or not. The transformation then converts the unstructured data into a uniform format. It only takes place when the data are specifically relevant for a query and are requested. This process consumes more storage space due to the amount of stored data that may never be queried.

Since many relational database systems are still in use in many companies, there is also the possibility to export or import data from relational databases to Hadoop. For this purpose, Sqoop can be used, for example. [244] Interesting for data warehouse solutions is the export of the results of the Hadoop processing, for example, to data marts, while the raw data exist in the Hadoop system. The advantage of this approach lies in the performant use of SQL or MSQL as a query language, especially for "refined structured" data in the DWH. Although Hadoop offers a corresponding database with SQL-like language, Hive, this is not nearly as powerful as the relational or multidimensional databases of the DWH solutions. Therefore, it makes sense to combine DWH and Hadoop data storage in a meaningful way. But there are also reverse ways possible by sending large historical data sets of the DWH, which are often also called legacy data, to the data storage of the Big Data technology and analyzing them from there with the advantages of the in-memory technique.

---

[240] Cf. Big Data Blog (2015).

[241] Cf. Kaufmann (2014, p. 369).

[242] Cf. Freiknecht (2014, p. 20).

[243] Data Academy and Davenport (2008).

[244] Cf. Wartala (2012, pp. 180–183).

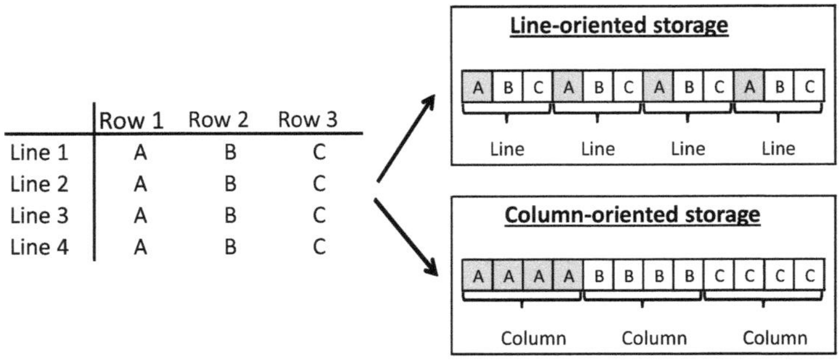

Fig. 5.57 Row- or column-oriented storage (Cf. Plattner 2011, p. 37)

5.8.2.4 Analytical Databases

Databases that process huge amounts of data and are oriented towards analytical applications are also called **analytical databases**. They are characterized by a combination of column orientation, data compression, in-memory technology and parallel storage and processing. They also achieve high query frequencies in multi-user operation. The development of analytical databases reflects advances in database technology that have the following features:[245]

- in-memory storage (see Sect. 5.8.2.1)
- different storage forms (especially column-oriented but also row-oriented and hybrid data storage)
- massive parallel processing of data
- high compression of data for holding "in memory" or for reducing disk accesses and processing compressed data without decompression

The difference between row- and column-oriented storage concepts is illustrated in Fig. 5.57.

In row-based storage, the data is stored consecutively in a row (tuple). Additional rows can be easily added or updated. This form of data storage is unsuitable for analytical purposes because all, and thus also unnecessary data, is read. The data of a row is not stored in one, but distributed across different databases and is loaded specifically for the query, something which makes the query take longer. This method of storage is useful when many columns and only individual rows are needed for queries.[246]

In the context of column-oriented data processing, the columns are stored one after the other in blocks and only the required columns are read for queries. This form of stor-

[245] Cf. Müller (2014, p. 450) and Alexander and Grosser (2017).

[246] Cf. Berg and Silvia (2013, p. 41).

age is advantageous when many rows and only a few columns are used for queries. It is suitable for evaluations that often apply aggregation functions to individual columns. If data records change frequently, column-oriented storage is not very efficient because for the write process, the correct column and then the correct row must be identified first.[247]

In the massive parallel processing of data, conventional databases are used, which are preconfigured and implemented on multiple specialized hardware systems (appliances). These can be used simultaneously. Using special algorithms and access methods, the accesses and calculations are synchronized and performed in parallel on the distributed systems using data clusters. Massively parallel processing databases are also known by the abbreviation MPP.[248] In data compression, among other things, materialized views (MatViews) are used. MatViews contain copies of data that have been pre-aggregated, calculated, recoded or renamed, for example. Analyses access the MatViews instead of the base data to increase the query speed since calculations and aggregations are omitted, for example. This also reduces the accesses to the hard disk or the compressed data are processed directly in the main memory. Another possibility to achieve a reduction of disk accesses is a better organization of the physical data on the data carriers.[249]

In the further development of database technologies, the performance in terms of data query speed must be kept in mind. Currently, there is much to suggest that big data will not simply replace the traditional relational database systems and the OLAP-based data warehouse systems, but rather that a powerful integration of different analytical platforms will emerge while maintaining high data quality.[250] The advantages and disadvantages of the respective data storage must be weighed accordingly and taken into account in the context of an economic analysis for the IT strategy and investment decision. Big data is not a guarantee for competitiveness and success, but is subject to the same economic considerations of costs and benefits in the context of IT investments.[251]

Examples from business practice show that both ERP systems and data warehouse solutions are used with big data technology and thus the data storage is handled with big data technology.[252] However, there are also examples and opinions that use big data technology as a complement to the existing systems. In this case, hybrid technology approaches are also spoken of.[253] The hybrid technology is also supported by the "Best of Breed" approach, in which the best technology for the respective application purpose should be used.[254]

---

[247] Cf. Berg and Silvia (2013, p. 41).

[248] Cf. Alexander and Grosser (2017) and Intelligence.de (2017).

[249] Cf. Intelligence.de (2017).

[250] Cf. BARC (2014, pp. 23–24).

[251] Cf. Baumöl and Berlitz (2014, p. 169).

[252] For example, SAP BW on Hana at the shoe company Reno, Schäfer (2014).

[253] Cf. Welker (2015).

[254] Frietsch (2016, p. 169 f.).

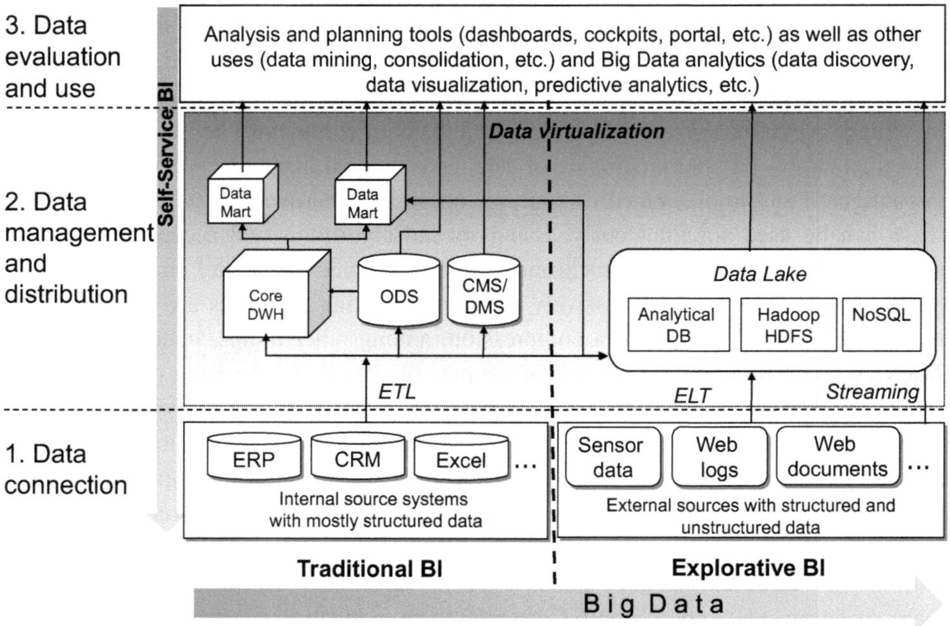

Fig. 5.58 Traditional and explorative BI architecture

5.8.3 Explorative Business Intelligence and Business Analytics

If one extends the traditional business intelligence approach (Traditional BI) with a classic DWH architecture, then extension possibilities arise in the context of big data to an explorative business intelligence approach which is shown in Fig. 5.58. The term business intelligence is supplemented by the word "explorative". Explorative here means exploring and researching many mostly unstructured data, of which only little knowledge and connections for the company could be used so far. The explorative BI extends the traditional BI with big data analytics (among others in connection with predictive and prescriptive analytics). In particular, the mathematical-statistical methods used, which are needed for predictive and prescriptive forecasts, support this choice of terms.

In the **first level**, the **data connection** takes place, which is enriched in exploratory BI with numerous mostly external sources with structured and unstructured data (RFID sensor data, web logs, etc.). The data is transferred in the staging area via ELT processes to the data storage of the big data technologies described above (NoSQL databases, Hadoop/HDFS or analytical databases). This data collection within the framework of big data technology is summarized under the term "**data lake**".[255] In the data lake, all data

[255] Frietsch (2016, p. 171) and Seiter (2017, p. 83).

is stored in its original form regardless of source and structure and only prepared when needed. This principle is called "schema on read".[256] In the context of traditional BI, the mostly structured data from the internal source systems is provided for the different data targets (core data warehouse, data mart or operational data store) of the **second level** via the ETL process already described in Sect. 5.5.2.[257] Here, the principle of "schema on write" applies. Data is defined according to a developed data model (e.g. star schema) before the data from the source systems are extracted, transformed, mapped and loaded.[258] The content or document management systems (CMS/DMS) used in traditional BI for unstructured data are, if they are already used, continued to be used. Alternatively, the data storage can be replaced by the data storage of exploratory BI.

Both in traditional and exploratory BI, the data management and distribution of data takes place in the **second level** via the mapping of physical or virtual data targets (data virtualization). It may make sense to make the data from the data lake (Hadoop and analytical data storage) also available for the data targets of traditional BI in the data warehouse, in order to use here the advantages of the powerful SQL or MDX queries for the third level of data analysis. By distributing, connecting and caching the data, relational or multidimensional queries can be performed efficiently and promptly. Conversely, it may also make sense to make large amounts of data from the traditional data targets available in the area of data storage of exploratory BI in order to use here the advantages of speed (e.g. in-memory technology) and availability (e.g. parallel processing).

In the **third level** of data analysis and use, the analysis and planning systems already explained are used which can process the corresponding data of the data targets of traditional or exploratory BI. In particular, by extending with big data technologies, the use of predictive and prescriptive analytics will gain more weight here than in traditional BI (among others with data mining), as larger amounts of data can be processed faster. The analysis of large amounts of data as well as the manifestations of data mining in the sense of predictive and prescriptive analysis with big data technology is today also summarized under the term advanced analytics[259] or **big data analytics**[260]. It should also be mentioned the possibility, in the sense of an operational real-time BI, to access the data of the source systems directly via the datalake. This process is also called streaming, or Complex Event Processing (CEP) and the analyses streaming analytics[261]. Examples of applications in practice are online shops and securities trading, both of which process and evaluate changes in the source data immediately. Many more examples that are

---

[256] Cf. Ballhorn (2017).

[257] The differentiated presentation of the levels Staging, Cleansing, Core DWH and Data Marts was omitted here. See Fig. 5.13.

[258] Cf. Ballhorn (2017).

[259] Cf. Gluchowski (2016, p. 277).

[260] Cf. Chamoni and Gluchowski (2017, p. 9) and Felden (2017, pp. 1–8).

[261] Cf. Hortonworks (2013, p. 4).

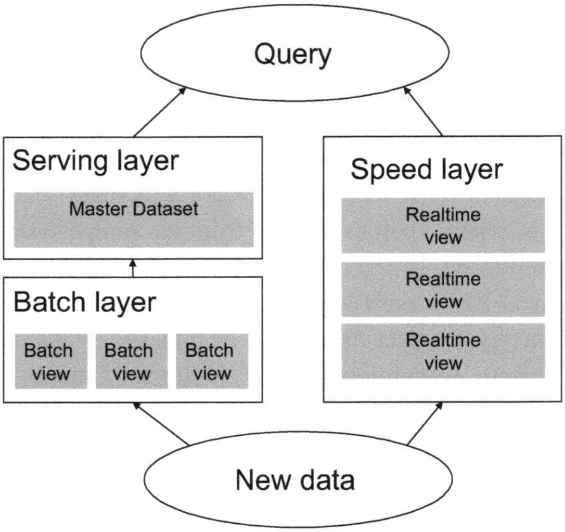

Fig. 5.59 λ-architecture for real-time data processing

suitable for the use of streaming analytics can be found in Table 5.5 at the beginning of Sect. 5.8. For streaming, the so-called **λ-architecture** (lambda architecture) is used, where the symbol λ stands for the Greek letter lambda. It essentially consists of three components: the batch layer, the serving layer and the speed layer (see Fig. 5.59).[262] Here, new data is loaded in parallel into the batch and speed layers. For the batch layer, a very large storage capacity is necessary as here the new data is stored permanently in an immutable form. Since the analysis time of the batch layer is very high, the analysis results are made available to the serving layer when the batch layer has completed the analysis with a data set. Consequently, real-time data analysis cannot be performed with batch and serving layers. This is done by the speed layer in combination with the serving layer. Based on the most recently arrived data of the speed layer and the analysis results of the serving layer, the analysis results can be queried together and analyzed in real time.

The technological extension within the framework of exploratory business intelligence has led in recent years to the fact that various representatives from science and consulting practice use a different term, i.e. "business analytics", whereby the term analytics is supplemented by the word business. The terms **business intelligence**, **advanced analytics** and **business analytics** have since then been frequently used together by software manufacturers, so that a distinction is hard to recognize.[263]

---

[262] Cf. März and Warren (2015, p. 18 ff.).

[263] Cf. e.g. Inform (2017).

The definition of *Gartner* puts the development of analysis and forecasting models using algorithms in the foreground:

> Business analytics is comprised of solutions used to build analysis models and simulations to create scenarios, understand realities and predict future states. Business analytics includes data mining, predictive analytics, applied analytics and statistics, and is delivered as an application suitable for a business user.[264]

Chamoni and *Gluchowski* define business analytics as follows:

> Business Analytics (BA) can be understood as a collection of different methods and technologies, which serve to gain insights from available data for business decisions to control the business processes.[265]

For Langmann, business analytics is basically an umbrella term for the automated evaluation of many different data (big data) using statistical analysis methods and algorithms.[266]

The introduction of the term business analytics is apparently intended by some authors to displace or replace the older term business intelligence, because it is distinguished in particular by the systematic application of mathematical and statistical methods and the insights, relationships and forecasts gained thereby.[267] However, if one looks closely at the definitions of business intelligence and business analytics, one will not find any significant difference. In my opinion, when introducing or using a term, the question should be asked whether the original term does not already meet the criteria of definition. Data mining was already regarded early on as a part of data warehouse and business intelligence and already includes the distinction of predictive and prescriptive forecasts using model building (see Sect. 5.5.5.3).[268] The technological possibilities to perform these forecast calculations, however, were not very powerful in this early phase without the big data technology. Only the technological push especially of the in-memory technology brought significant progress in this area. The term business intelligence, however, was not defined as technology-restrictive, but technology-open. From today's perspective, big data technology belongs to business intelligence just as much as data warehouse technology. If one uses the core idea "use of the insights or knowledge from the available data", then in my opinion the traditional BI with the data warehouse technology is responsible for far more decisions of the business processes than insights and forecasts using the big data technology. Questions, such as e.g. with which customer do I make profit or which customer groups are currently interesting, can be taken immediately from the data warehouse for secure market processing. The prognostic questions, e.g. which

[264] Cf. Gartner (2016).

[265] Cf. Chamoni and Gluchowski (2017, p. 9).

[266] Cf. Langmann (2019a, pp. 5–8).

[267] Cf. Chamoni and Gluchowski (2017, p. 8 ff.).

[268] Cf. e.g. Bissantz et al. (2000, pp. 377–407).

potential customers are interested in our products in the future, can additionally be very valuable for market processing. Nevertheless, the management will not drop the existing customers and only follow the new ideas for market processing. It will get an overall picture for market processing based on the new and past insights, and take this into account in its sales planning and initiate appropriate measures.

Due to the statistical uncertainty of the analytical findings and forecasts of big data analytics, a company will not rely exclusively on these data. The core information that is used for corporate management is derived from the structured data. Therefore, in this presentation, the original term **business intelligence** is preferred to the new term **business analytics**.[269] It will show in the future which naming prevails in science and practice. Going by the literal meaning, the second word in the term business analytics (analytics) falls short as it only covers the theory or the art of analyzing with the help of data, starting from the static processing of the available diverse data up to the technically supported data analysis.[270] The use of knowledge from data goes beyond the statistical methods and technical systems used. The management incorporates the acquired knowledge into the actions and decisions of the company. There is a connection between people (especially managers), companies (their goals and tasks) and the technology available for this purpose (here information technology).[271] The second word in the term business intelligence (intelligence) meets this linking aspect in my opinion better. The management is not a simple consumer and executor of the statistical analytical results, but it has the intelligence to integrate them into the strategic and operational framework of action. Furthermore, some key tasks cannot or can only be poorly done with big data technology, but better with traditional BI technology, especially:

- structured multidimensional analyses
- commenting and documenting the findings
- integrated corporate planning (budget, medium-term, strategic planning and forecasting)
- company-wide action derivation

The assignments of the technologies as well as analysis and forecasting methods to the terms are visualized in the following figure (Fig. 5.60).

The exploratory business intelligence thus extends the traditional BI by model-based statistical analyses, forecasts and action derivations for selected business problems to be processed with the help of the big data technology.

[269] Some authors even use both terms and the acronym BIA for Business Intelligence & Analytics. This mixture shows in my opinion how vague the terms are used. See Ereth and Kemper (2016, pp. 458–464) and Chen et al. (2012, pp. 1165–1188).

[270] See Lanquillon and Mallow (2015, p. 55).

[271] Following the argumentation of Felden (2017, pp. 1–8).

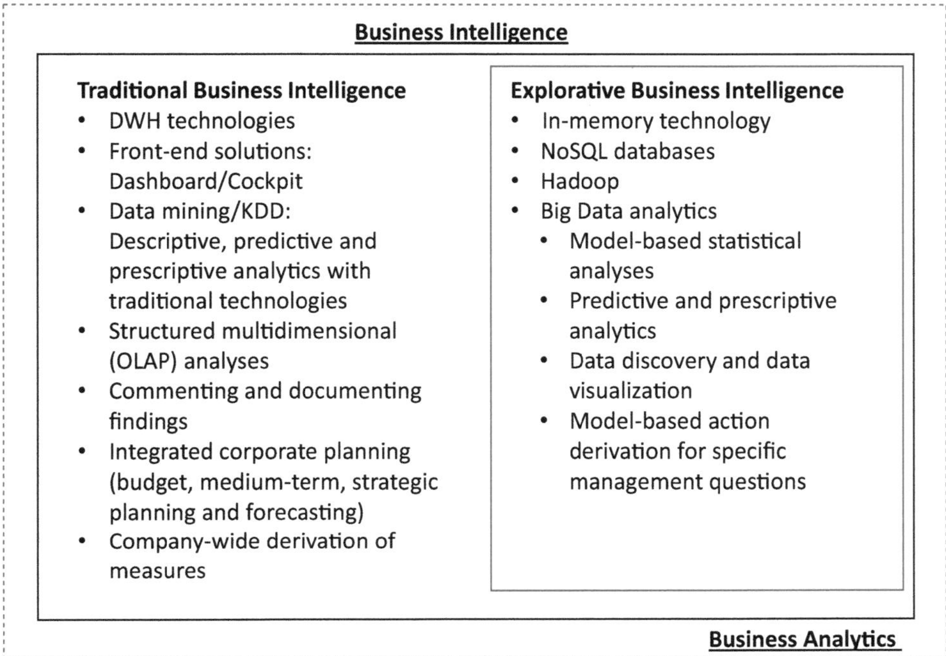

Fig. 5.60 Technologies as well as analysis and forecasting methods in use of business intelligence or business analytics

In connection with static analysis and forecasting functions of data mining or predictive analytics (see Sect. 5.5.5.3), descriptions and presentations often try to depict the analysis spectrum of BI and especially of exploratory BI. In doing so, traditional reporting is pushed into the corner of coping with the past (descriptive analysis). Figure 5.61 shows such a classification. However, these classifications are incomplete and simplify the trade-off of complexity, cost and benefit (value) for the companies. The forecasting function is wrongly considered only in the predictive and prescriptive analysis, whereas the traditional planning is simply ignored. But a complete comparison is correct. Furthermore, it should be emphasized that the analysis and forecasting types are not independent of each other. In particular, predictive and prescriptive analyses often require historical data from the traditional IT systems or from the DWH to make qualified forecasts. The planning and reporting can therefore at most be supplemented by better forecasts. It is rather an extension and not a replacement investment. Also, the value classification is unfortunate as it does not differentiate the value creation and makes value increases dependent only on the future forecasting ability.

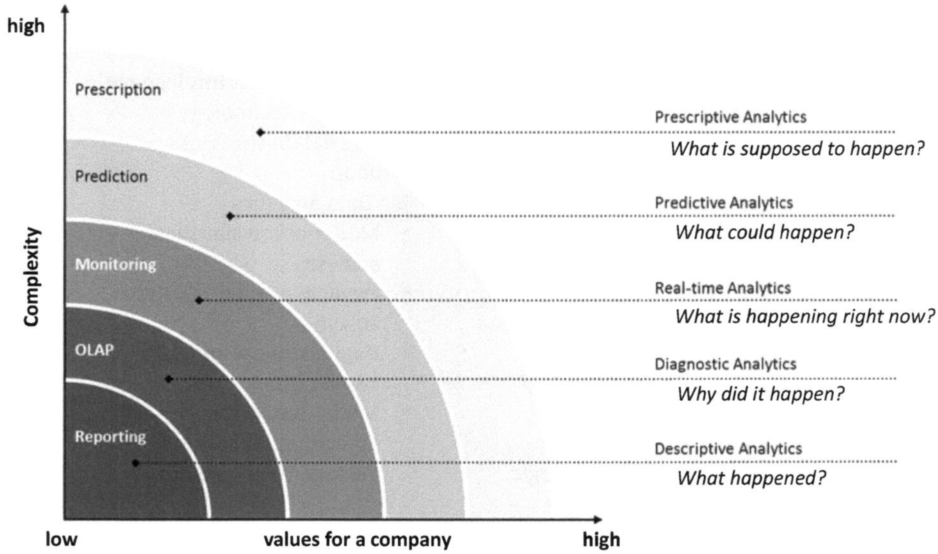

Fig. 5.61 BI analysis spectrum according to Lanquillon and Mallow based on Eckerson (*Source*: Lanquillon and Mallow (2015, p. 56) based on Eckerson 2007)

These deficits are to be improved in the following description. In the Fig. 5.62, the most important analysis and forecasting types are assigned to the analysis and forecasting phases. The phases are arranged with a time reference from the past through the present to the future. In the phases, the traditional planning/forecasting is inserted. In addition, the predictive analysis and predictive forecasting are separated to position them correctly in the time assignment.

The analysis and forecasting types first list the **traditional planning and relational reporting**. It becomes clear that companies today have tools based on traditional planning and analysis that allow them to manage business in a future-oriented way. Goals and planning targets are formed and monitored. In addition, planning and forecasting are possible which are usually based on the experience and assessment of the executives. They form the foundation for the holistic management, design and coordination in the company. The basis for this are the recording and billing systems of the companies (especially the accounting, ERP, CRM and inventory management systems) as well as the tools used for planning and analysis, often also Excel (see Sect. 5.4).

With the **multidimensional OLAP-based planning and analysis**, the planning and analysis capability improves significantly compared to the traditional planning and analysis systems. Due to the multidimensional planning, the coordination of the planning areas can be optimized. The multi-layered analysis in the data space allows to detect anomalies and trends faster. The main efforts here are the setup and operation of the listed levels of a data warehouse (see Sects. 5.4.5 and 5.5).

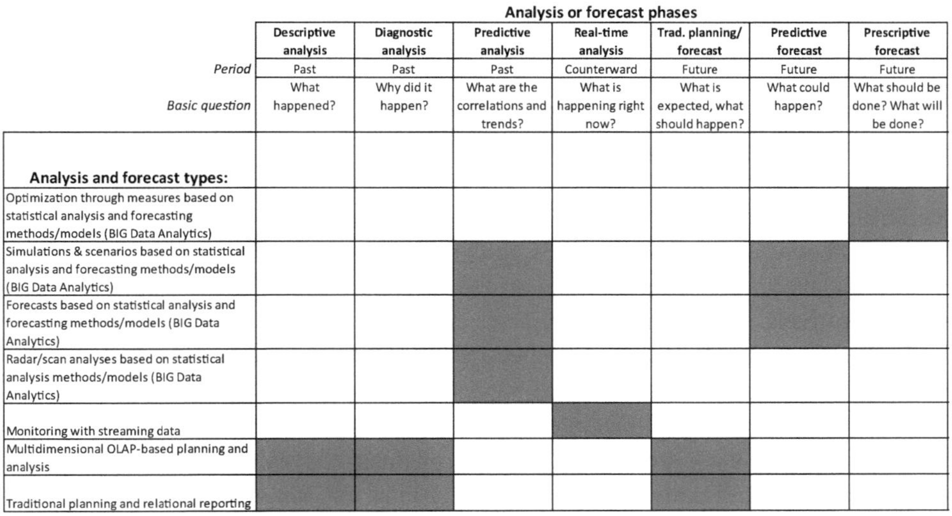

Fig. 5.62 Phase assignment of analysis and forecasting types

The **monitoring of streaming data** is interesting for business models in which it is valuable to evaluate current data in a timely manner for control purposes. Examples are price changes for gas station operators or for financial service providers. In addition to the direct data analysis of the source data via virtual layers of the DWH, exploratory BI techniques (e.g. distributed data storage, NoSQL and in-memory technology) are particularly suitable here, which also include external data sources.

Big data based radar or scan analyses are useful for the continuous observation of the business environment (competitors, customers, suppliers, markets, technologies …) and often have a strategic character. Using statistical analysis models, mainly correlations and trends regarding market behavior are detected. For this purpose, foreign data sources (e.g. web and social media data) with mostly unstructured data are evaluated. Essentially, statistical analysis methods (e.g. data, text, web mining or predictive analytics) are used, which have already been introduced in Sect. 5.5.5.3.

Big data based forecasts or predictions also rely on the statistical analysis and forecasting methods and models that are used in the context of data, text and web mining or predictive analytics. Using established value driver models, statistically evaluated own and foreign data as well as further assumptions, more "accurate" predictions can be made than would be possible with the help of traditional planning functions. Unlike "operations research", foreign and unstructured data can now also be processed in large quantities. This is made possible by the use of exploratory BI technology (e.g. Hadoop, NoSQL and in-memory). Areas of application for the digital forecasts can be individual planning areas such as sales and revenue planning.[272] But also cross-cutting planning of individual value drivers up to the operational and financial result of a company with the help of so-called value driver models are possible.

Big data based simulations and scenarios extend the forecasting functionality of the digital forecasts by simulation and scenario functions. In the simulation, alternative value-driving factors and different value expressions can be simulated with regard to the result effect. In addition, different scenarios can be calculated stochastically, something which was extremely complex with the support of previous IT. Using the Monte Carlo simulation, for example, a statement about the results to be achieved when varying value-driving factors under risk aspects can be made by specifying probabilities of occurrence. The accuracy of worst- and best-case scenarios increases as a result.

The functionality of big data based predictions, simulations or scenarios can be further optimized by setting up measures and proposals for decisions that lead to a desired predicted goal. In the extreme case, the best alternative is automatically selected and executed. This significantly reduces the latency time for using data as ideally no time losses occur during data collection or loading, analysis, decision making and implementation.[273] Typical examples are well-structured decision models such as order scheduling for inventory optimization based on cash register information or the initiation of maintenance when identifying anomalies in components within the framework of IoT (Internet of Things) or Industry 4.0. Complex business decisions, such as strategy changes, site closures and new product development, are currently not conceivable and cannot be mapped in terms of models. Here, only additional information can be provided in the sense of an early warning system, which must be translated into decisions and instructions for action by experts or executives.

In summary, one can see the good analysis and forecasting enhancement possibilities that result from explorative BI. Ultimately, it remains a sober investment decision in IT and know-how. Based on the current state of IT-supported planning and reporting of the company, the extensions or changes by new analysis and forecasting types have to be weighed up in terms of effort and benefit. In the following Fig. 5.63 complexity and benefit assessments for the analysis and forecasting types are listed. Here, more plus signs were assigned if the complexity or the benefit were estimated higher. The assessments of effort and benefit have to be made individually in each case. Changed technical possibilities can lead to a shift in the evaluation over time.

The effort mainly involves investments in technology (e.g. databases, storage techniques and software for the frontend) and in the know-how of the employees in general. Specifically for digital analysis and forecasting types, efforts for the tools and the know-how of the employees regarding the application of statistical methods and models are incurred. The latter is scarce in many companies.

---

[272] See Möller et al. (2016, pp. 509–518).

[273] Cf. Kemper et al. (2010, p. 9).

	Analysis or forecast types						
	Trad. planning and reporting	OLAP-supported planning/analysis	Monitoring with streaming data	Radar/Scan analysis	Forecasts	Simulations & Scenarios	Optimization/ Actions
Complexity							
Use of technology and know-how in general	Relational databases, software	OLAP databases and software	Traditional (DWH) and exploratory BI technology (Big Data analytics) with distributed data management, NoSQL, in-memory technology, and software				
Tools	+	++	+++	+++	+++	+++	+++
Know-how	+	++	+++	+++	+++	+++	+++
Use of statistical methods and models	not mandatory, but possible			a. o. data, text, web mining/ predictive analytics			
Tools	+	++	+	+++	+++	+++	+++
Know-how	+	++	+	+++	+++	+++	+++
Benefit foundation							
Documentation	++	+++	+	+			
Coordination	++	+++	+	+	+	+	+
Design	++	+++		+	+	+	+
Forecast	+	++			+++	+++	+++
Target setting	++	+++			+	+	+
Target achievement	++	+++	+	+			
Overall control	++	+++			+	+	
Selective control	+	++	+	++	++	++	+++
Innovation	+	++	+	+++	+++	+++	+++
Response advantage	+	++	++	+++	+++	+++	+++

Fig. 5.63 Effort and benefit assessment of analysis and forecasting types

Regarding the benefit creation for controlling and management tasks, it is noticeable that central planning, reporting and controlling functions such as documentation, goal setting, design and coordination from the perspective of the entire company are better supported by relational or multidimensional solutions as these are better suited to build up extensive planning and reporting areas for an integrated corporate management. On the other hand, the explorative analysis and forecasting types tend to be more suitable for selective management questions, which can be better mapped with the help of certain stochastic methods and compact value driver models based on large amounts of data. Advantages of these stochastic analysis and forecasting types are especially the forecasting accuracy, the response speed and the innovation power.

Due to the many different data sources and targets of traditional and exploratory BI, there are still developments that insert a virtual layer (**data virtualization**) between the second and third level of data analysis and use (see Fig. 5.58), a layer which provides a virtual access to the actual physical data sources and targets for the analysis and planning tools. This layer is optional, but has the advantage that the development of the analyses and plans is based on a virtual uniform, combinable and integrated data basis. The data access and calculation of the data is still done in the data sources and targets of the first and second level. Due to the technology used here, the virtual accesses have to be designed in a performant way.

Some BI software vendors have increasingly used the in-memory technique (e.g. with Qlik View and Jedox) in recent years and build directly on the data of the source systems without creating a physical intermediate data storage. They need, instead of the physical intermediate data storage, a virtual view on the primary data sources. This data virtualization is an extended approach of virtual data management, which deviates from the traditional data provisioning (via ETL or ELT) via a physical intermediate data storage. At query time, the data is provided directly from the distributed source systems.[274]

The data virtualization is done step by step like the ETL or ELT process. The connections to the respective required data sources are established via connectors. In the virtualization layer, the data is transformed, integrated and provided for the analysis application.

The advantage is the low or no longer required storage space and the access to the source data in real time. An improvement of the data quality and integrity is only given if the source data already meet these criteria, which is often not the case in practice. The data in the source systems are also changed and have no change history. Since the business requirements have to be mapped also in the virtual levels, the effort, flexibility and dynamics of the rules are similar to those of non-purely virtual models, except for the intermediate storage.[275]

A further disadvantage is that slow source systems can severely limit the speed of the virtual data integration. Sometimes, existing source systems are not suitable for a large dynamic data query at all. Low response speeds of the source systems also affect the performance in the source and analysis systems.[276] The data virtualization is possible for simple queries from the aforementioned advantages. For complex business queries on different primary sources, there are difficulties that cannot be easily created with the help of the virtual queries in the sense of a user-friendly self-service BI. Here, traditional data warehouse systems have clear advantages due to the ETL process in order to prepare a structured data set for analysis purposes with a physical intermediate storage.

It is also interesting to observe the development of the industry leader, SAP AG. The strategic orientation of the products in the ERP and BI area aims at the use of the SAP-owned database SAP HANA and the associated advantages of the in-memory technique. SAP HANA is a database technology developed by SAP, which is based on an in-memory technology and represents a coordinated combination of hardware and software. It is independent of the data source used, i.e. it also supports data sources from third-party providers.[277] When the new ERP software is used with the new SAP HANA database, it is called S/4HANA. For the creation of procedures, SAP uses its own scripting language, SAP HANA SQLScript. In addition, there is the possibility to use the programming language R for statistical analyses.[278] For start-up companies that have not used SAP before, SAP HANA offers the speed advantage of in-memory technology and a reduced range of functions compared to the traditional SAP NetWeaver technology. For long-time users of SAP ERP and SAP BW systems, a switch will not be so easy because they

---

[274] Cf. Luber and Litzel (2019).

[275] See also Luber and Litzel (2019), Krishnan (2013, pp. 191–195) and Iffert and Bange (2018).

[276] See also Luber and Litzel (2019) and Iffert and Bange (2018).

[277] Cf. Berg and Silvia (2013, pp. 33–35).

[278] Koglin (2016, p. 61 ff.).

do not want to lose their elaborately built structured data storage. Here, SAP offers to use their data warehouse SAP BW with SAP HANA (SAP BW/4HANA or BW on Hana). In this case, the use of InfoCubes is avoided and direct access to the data of the Data Store Objects in the DWH is possible.[279] It can be assumed that many companies will still need a structured data storage for their business information, which stores the data of the source systems harmonized by transformation processes in intermediate layers.[280] With the SAP Hana technology, there will be direct data accesses that access the data source or low layers of data storage virtually via various views. The three most important views that SAP HANA provides are "attribute", "analytic" and "calculation" views. Attribute views resemble two-dimensional table queries. With the help of analytic views, multi-dimensional queries can be set up similar to the traditional DWH. The main sources of the calculation views are attribute and analytic views and the underlying database tables. With the help of the calculation views, further complex calculations can be performed, which contain e.g. case and if statements. To create the views, SQL scripts or graphical tools can be used. The user-friendliness in terms of self-service BI is significantly lower than with traditional analysis front-end tools.[281] The technological basis of some classic DWH objects such as InfoCubes are replaced with SAP Hana by new modified objects such as the so-called "advanced data store objects". However, there will still be an area for structured data storage in data management with data-carrying or virtual layers similar to the classic data warehouse, which has a great importance for business information. In this case, a complete waiver of one or more intermediate layers of data management is not to be expected, as it will always be necessary for business information to structure and harmonize the data with regard to a business intelligence-supported controlling.[282] The access to the stored data in SAP Hana is either done with a collection of apps via the SAP Fiori UX Launchpad, which accesses the data storage of SAP HANA directly, or via the various SAP BI front-ends (WebIntelligence, SAP BusinessObjects Dashboard, SAP BusinessObjects Design Studio, Crystal Reports, BusinessObjects Explorer, SAP Lumira, etc.).[283] It will be exciting to follow this trend at SAP AG but also at other software vendors in the next few years. Since it is not to be expected that companies will simply write off the investments made in classic DWH solutions and switch to the new in-memory technology, it is rather to be assumed that the systems, similar to Fig. 5.58 (Traditional and explorative BI), will use a parallel technological basis of traditional DWH and big data technology which are connected to each other.

In summary, it becomes clear that with big data technology, especially an extension of the traditional BI to an explorative BI is useful as far as it can be built up economically

[279] Cf. Merz et al. (2015, pp. 153 ff. and 277).

[280] Cf. Haupt (2011).

[281] Cf. Merz et al. (2015, p. 259 ff.).

[282] Cf. @tfxz-Blog (2014).

[283] Cf. SAP SE (2013), Merkt et al. (2015) and Kessler et al. (2014, pp. 31–37).

appropriately. Here, the technology is used efficiently and effectively, something which is most sensible for the respective analysis and forecast of the data. Advantages of the powerful SQL and MSQL queries of traditional BI are complemented by performant big data analytics of the explorative BI:

- increased expandable (scalable) data base of diverse source systems
- increased speed of data processing
- increased availability of data through parallel processing
- increased model-based analysis and forecasting capability.

While in the beginnings of data warehouse technology, the hope was to have found a **Single Point of Truth (SPoT)** for the data storage, processing and use of information for business decisions, today it is agreed that there is no single technical platform for this, but rather it depends on the collaboration of sensibly used technologies. For this, the term "**logical SPoT**" is also used. It expresses that there will not be a single valid DWH architecture, but rather a overall architecture of data storage and use.

If one positions further currently discussed topics such as cloud computing, sandboxing and self-service BI (SSBI) in the explorative business intelligence approach, the picture shown in Fig. 5.64 is the result.

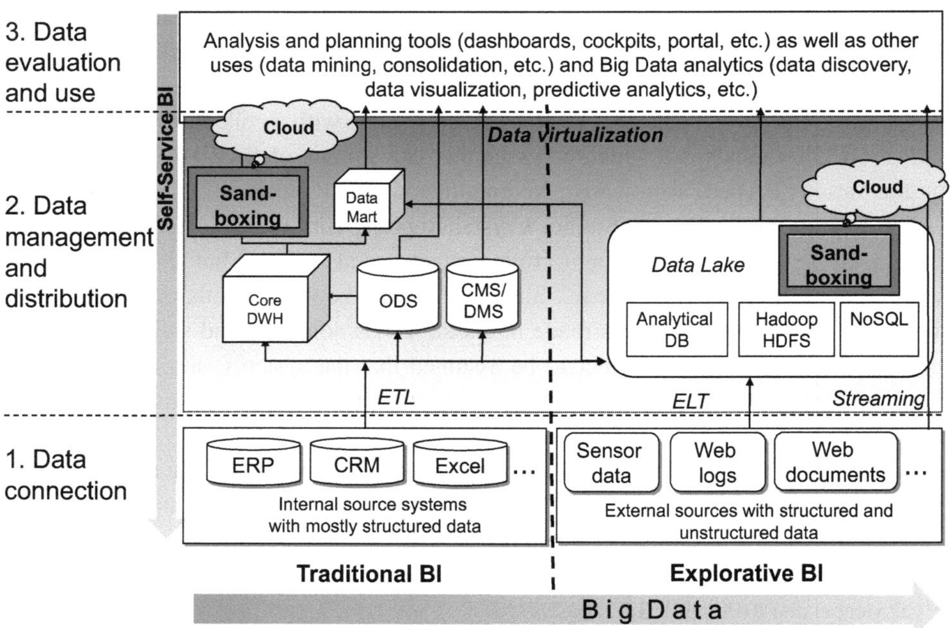

Fig. 5.64 Explorative BI, cloud computing, sandboxing and self-service BI

As already mentioned in Sect. 4.1.5.4, the term **Self-Service BI** implies that business users can create and use analyses and planning without relying on the know-how of IT departments and experts. For this purpose, the business users need explorative **data discovery** or **visual discovery tools**, which greatly simplify proprietary data connection, data transformation and modeling and with which analyses can be created user-friendly (see Sect. 5.5.5.3). The term **data discovery** stands for a business user-driven iterative process that tries to obtain insights (data patterns, relationships, anomalies) of important business relationships in the context of analyses and forecasts for business management not only with historical data, but also with external data from various sources.[284] For this purpose, data visualization is integrated in the analysis and forecast of the data discovery process. The BI software applications in the field of data discovery and data visualization use the in-memory technology and apply statistical models and algorithms in the sense of big data and predictive analytics. For the graphical presentation, these BI systems use not only the classic chart types (e.g. pie, column and line charts), but also modern chart types that visualize classifications and statistical relationships (e.g. with sunburst, treemap charts or histograms), which were already introduced in Sect. 3.8.4 and in the context of predictive analytics in Sect. 5.5.5.3. In the BI trend monitor of BARC (2017), data discovery and data visualization are highlighted as the most important BI trend.[285] In the BARC GUIDE 2020, data discovery still ranks second among the BI trends. It can be assumed that levels of the tools for self-service BI (SSBI) starting from the third to the first level are used both in the traditional and explorative BI.

If the domain users or experts are able to derive insights from the data analysis (including exploratory data) that are structured and can be used sustainably, it makes sense to integrate these data into the data warehouse. This achieves an improvement of the development process and a reduction of the development times of a BI system, which is also discussed under the term **agile BI**.[286] With the help of **sandboxing**, data areas are made available to the domain users, but also to external stakeholders that are aligned to the respective group depending on the access authorization. Thus, the users receive only those data areas that are intended for their use, taking into account data protection and data security. The potential use of sandboxing is possible in both traditional and exploratory BI, just like cloud computing. Big data services are mainly obtained from the cloud in practice projects due to the exploratory data and the desired high availability of the data.[287] Which variant of cloud computing is more suitable for the use of BI-supported reporting and which variants can be excluded is explained in the following section.

[284] Extended from Iffert (2017).

[285] See BARC (2017) and BARC GUIDE (2020, p. 11).

[286] Cf. Zarinac (2016, p. 140 f.).

[287] Cf. Buschbacher et al. (2014, p. 90).

5.9 Cloud Computing

As a technical platform for data storage, besides the classic data storage on the server in the company, cloud computing is an option. Cloud computing refers to IT services and services that can be outsourced. Traditionally, companies use their software **on premises**, that is, they install and operate their hardware and software solutions in the premises of the company. A corresponding usage and license model is concluded with the software manufacturers, in which, for example, a certain defined number of users can use the software for a one-time license fee. Hardware and software costs represent an investment. In addition, annual maintenance costs usually accrue as a percentage of the license costs. With cloud computing, hardware and software systems are used in one or more external data centers. For this, usage fees are agreed upon similar to a rent. The investment risk is significantly lower.

A great advantage of cloud computing arises from the distribution of large amounts of data using replications of the data on different servers/clusters. The replications achieve a very high availability of the data, which is very important for online commerce and social networks, for example.[288] The worldwide location of the data helps to reduce access times for local regions. In addition to the high availability and the improvement of access times, further advantages of cloud computing can be identified. Many customers and companies can share the infrastructure and services. Unused capacities can be better utilized. Due to the scale effect, the IT services can be offered more cost-effectively through a central management. This also applies to the services regarding data protection and data backup, which can be created more extensively and securely. In addition to conventional security technologies for defense against and combat of digital threats and attacks, new technologies such as network forensics, user data analysis and the isolation of software applications (Sandboxing) can be used here. These measures would not be economically feasible for one company alone. Moreover, the companies lack the technical know-how for this.

If applications are only needed occasionally, the use via the cloud can be cheaper because they are billed according to demand. Furthermore, the rental contracts can be terminated more quickly.

As an important problem area in cloud computing, data protection and data security have emerged after the NSA affair. Due to different national regulations regarding data protection and data security, e.g. in Germany at least the General Data Protection Regulation (GDPR) in conjunction with the new Federal Data Protection Act (BDSG) as well as the Telemedia Act and the Copyright Act, these have to be taken into account when considering the location of servers and granting access.[289] Although technical means such as firewalls, virus scanners, encryption of data before transmission with various

[288] Cf. Giegerich (2014, p. 321 f.).

[289] Cf. Bitkom (2013, p. 24 ff.).

encryption techniques (e.g. HTTPS and SSL) or network solutions (e.g. Virtual Private Network—VPN) can help to increase the security and protection of data, a residual risk remains.

Furthermore, the processes for IT security also have to be considered. Certificates such as ISO 27001 or SAS70 specify the requirements for development, commissioning, use, monitoring, maintenance and further development of an IT security management system including the control of risks in the company. In addition to technical potentials, these also help to create processes and organizational prerequisites to achieve a higher level of data security. The user himself poses a high risk potential for data security, for example, if data is stolen or misappropriated by employees or if third parties gain access to computers or application programs due to lack of care. Password security concepts, employee training, back-up creation, access cards, dongle or biometric recognition are some measures that can prevent misuse here.

The following service models are distinguished in cloud computing:[290]

- **IaaS (Infrastructure as a Service):**
  The customer, for example a company, is provided with IT infrastructure resources, e.g. computers/servers, storage space, network and other basic resources and associated services (maintenance etc.).
- **PaaS (Platform as a Service):**
  The customer, for example a company, is provided with operating systems and systems for development (including the necessary IaaS) and associated services, on which the customer can use his own development environment for creating and running applications.
- **SaaS (Software as a Service):**
  The customer, for example a company, is provided with a full-fledged application (including the necessary IaaS) and associated services.

For all three service models, so-called service level agreements (SLA) are concluded, which specify the services (e.g. availability of the systems, location of data storage, support according to service level, transmission paths etc.) to be delivered for the rental price.

Furthermore, in cloud computing, the transparency regarding the locality of the IT resources provided must be distinguished. Here, the following forms are distinguished:[291]

[290] Cf. Mell and Grance (2011, pp. 2–3) and Duisberg (2011, p. 49).

[291] Cf. Birk and Wegener (2010, p. 642).

- **Public cloud:**
  In a public cloud, the customer has no transparency over the physical location and distribution (fragmentation) of the IT resources provided (for example, domestic/foreign).
- **Private cloud:**
  In a private cloud, the customer receives transparency over the physical location and distribution of the IT resources provided (for example, location of a server in a foreign data center or in the own company).
- **Hybrid cloud:**
  The hybrid cloud is a mixed form of public and private cloud. The customer receives, for example, transparency over the physical location of the IT resources provided, but not over the distribution. Another example would be that only certain software applications and parts of the data can be used in a private cloud and other parts of the data in a public cloud.

In relation to application software in the area of planning and reporting, **IaaS** means that the company brings its (own or purchased) software and only rents and uses hardware.

The **PaaS** is not relevant for companies that use planning and reporting software as the software solutions themselves usually provide suitable generators for creating reports and planning forms. PaaS would possibly be interesting for software companies that develop their own reporting and planning applications.

SaaS is potentially relevant for the companies again if planning and reporting software systems can be rented and used via an SLA. The SLA would have to regulate the data connection of the source systems in this case if they are not also included in the SLA.

A pure **public cloud model** for reporting and planning of a company is not feasible as the data relevant for control are highly sensitive and the distribution of data without knowledge of the location is not justifiable due to data protection and data security as long as different national regulations apply. This applies especially to personal, to control-relevant (e.g. patents, offers, recipes etc.) and to strategic data, which are contained in the reporting or the planning.

With a **private cloud model**, the company can contractually secure transparency over the location and distribution of the data in the SLA. In addition, the provider must comply with data protection and data security guarantees according to the respective national legal situation and include any additional ones in the SLA. Contract breaches and penalties must also be regulated. It must be checked whether the risk of loss, embezzlement or espionage of data as well as the associated fraudulent actions are adequately covered by compensation payments. As with a loan (trust), a repayment (compensation) is only possible if the creditworthiness of the borrower is given. In addition, instruments for controlling data loss or misuse of data are necessary and must be established. Here, the cloud provider has the upper hand. Moreover, it remains uncertain how high the rate of undetected cases of abuse is.

The **hybrid cloud model** can be interesting for companies for planning and reporting when non-protective or security-relevant information can also be made available in a public cloud area, for example to enable better, i.e. more cost-effective and faster access etc. for dealers and branches of the company worldwide. To protect control-relevant information, it remains internal or in the private cloud.

In summary, cloud computing remains an investment decision between in- or outsourcing for the company. All advantages and disadvantages have to be weighed up for a model to be selected. For planning, IaaS and SaaS models in combination with private cloud or hybrid cloud are possible as described above. For larger cloud projects, it is important that the costs and services are transparent for the company through the SLA. However, the practice looks different. As in other industries, e.g. insurance, the contracts (services, costs, terms, rebate and termination rules, penalties etc.) are complex and opaque. For this reason, the examination and analysis support by a cloud consultant or the support by a cloud broker can be useful to get a better market overview and to receive valuable information for the investment decision and risk assessment. In addition, when making investment decisions regarding outsourcing via cloud computing, the dependencies in the event of a later change or insourcing must also be reconsidered. Application tests and references help in an introduction phase to get to know the unknown, to identify risks and deficiencies and to check the know-how and services of the provider.

5.10 Mobile Computing

The mobile phone has been established in our society for quite some time and has become an essential part of our lives. It gives us the security of being reachable at any time and the freedom to receive or pass on information regardless of our location. Mobile computing deals in this sense with the *"[...] totality of devices, systems and applications that provide a mobile user with the meaningful information and services related to his location and situation"*.[292]

Especially since the development of the BlackBerry by the company RIM (Research in Motion), mobile computing has also become a fixed component within the business world. Another innovative milestone in mobile computing was the introduction and intensive marketing of the iPhone by the company Apple in 2007, as a result of which the public interest in so-called smartphones increased enormously. Modern tablet PCs such as the iPad also developed by Apple, represented a further innovation step.

In order to assess the limits and potentials with regard to mobile reporting and mobile planning, it is necessary to consider the various mobile devices that are suitable for this purpose with regard to the different functionalities and characteristics. Furthermore,

[292] Cf. Bollmann and Zeppenfeld (2010, p. 4).

wireless networks are another decisive factor for the effective use of mobile reporting and must therefore be differentiated.

5.10.1 Mobile Devices

Mobile devices provide the technological basis for mobile reporting and mobile planning. Their functionality and performance are therefore crucial for effective use. In recent years, the performance of these devices and the range of functions offered have increased enormously.[293] In addition, the market offers a great variety of different mobile devices. These devices can be mainly assigned to the categories of notebooks, tablet PCs and mobile phones. Furthermore, technical developments such as smart glasses, smart watches or other wearable objects or devices (wearables and gadgets) such as clothing or digital assistants (such as Siri from Apple or Alexa from Amazon) can be observed. Each object or device category offers specific advantages and disadvantages in its use as a mobile platform. The following sections will look at the currently most frequently used mobile devices (notebooks, mobile phones and tablet PCs).

5.10.1.1 Notebooks

Notebooks were the first to provide a nearly full-fledged replacement for stationary PCs and also enable location-independent work. Their performance and functionality are approximately equivalent to that of a PC and, due to the mostly uniform operating systems, compatibility of applications is largely ensured. Modern communication technologies such as WLAN (Wireless Local Area Network) and UMTS (Universal Mobile Telecommunications System) or LTE (Long Term Evolution 4G or 5G) also enable a connection to wireless networks. With a relatively large and clear screen and comfortable input options (mouse and keyboard), notebooks are quite suitable for mobile use and are already used in this sense. However, they have only limited mobility compared to other mobile devices due to their size, weight and usually low battery life. Furthermore, the waiting time for connecting, booting up and finding the right information is usually much longer than with mobile phones and tablets. Smaller notebooks are also called netbooks. Particularly thin notebooks, so-called ultrabooks, try to combine the advantages of tablets and notebooks.[294] Some manufacturers offer computers that allow the screen to be detached or flipped from the notebook, so that the computer can be operated like a tablet PC. The following special notebook types are distinguished:[295]

---

[293] Cf. Bollmann and Zeppenfeld (2010, pp. 87–111).

[294] Ultrabook is a registered trademark of Intel.

[295] NN (2020, p. 9).

- Detachable: The notebook has a detachable keyboard and can thus be converted into a tablet.
- Ultrabooks: Notebooks with this designation must meet requirements set by the chip manufacturer Intel for weight, performance and a long battery life.
- Convertibles: The display of the notebook can be rotated 360 degrees, so that the notebook turns into a tablet.

In addition, there are notebook types that are specifically tailored to their user clientele, e.g. gaming or business notebooks.

5.10.1.2 Mobile Phones

The use of mobile phones (cell phones) has gone beyond making phone calls for a long time. Mobile phones now also provide the user with a variety of other functions. These functions can include, for example, various multimedia applications, calendar and organizer, contact lists, calculator, camera, internet browser and GPS (Global Positioning System). Alternatively to the touchscreen, the applications of the phones can also be controlled by voice control via a built-in microphone. In addition, modern mobile phones have several different data transmission options (including WLAN, Bluetooth, UMTS, LTE), which enable fast and unlimited communication and data transfer. The latest generation of mobile phones is what the one referred to as "smartphone". Compared to classic mobile phones, smartphones are characterized in particular by a very high performance and by operating systems that allow an extension of various applications (so-called Apps, a short form of application). For navigation or operation, a touch-sensitive screen (touchscreen) has predominantly established itself. In addition, location-based services can be offered, for example, with the help of the GPS receiver. With the introduction of the Apple iPhone in 2007, smartphones initially only represented a niche product. Meanwhile, smartphones are very popular and can also be found in lower price segments offered by different manufacturers. In companies, BlackBerry devices from *RIM* were often used in the first generation of smartphones. These devices differed from the common smartphones today by a QWERTY keyboard over which the input was made. In addition to Apple's IOS operating system (iPhone), the operating systems Android by Google, OS by BlackBerry and Windows Phone by Microsoft have also established themselves for smartphones and tablets of different manufacturers (Samsung, HTC, Microsoft, Huawei etc.). With regard to mobile reporting, the smartphone represents a possible, albeit smaller platform in addition to the tablets. It offers the necessary connection options to establish a fast internet connection and the functionality to use both extensive browser-based and native applications, the latter of which were created directly for the operating system environment.

Furthermore, smartphones are characterized by their user-friendliness and have a much higher mobility than notebooks due to their size and battery life. Waiting times for switching on and booting are very low. However, different operating systems make compatibility between the systems difficult. Furthermore, the display area has only very limited dimensions to comfortably display or analyze detailed reports or analyses. However, the small screen surface is sufficient for compact evaluations and quick key figure analyses. For planning, the smartphones are too small; here, for example, the tablet PCs or notebooks are preferable in the mobile application. Hybrid forms between tablet and smartphone, so-called Smartlets or Phablets, can also only partially remedy this shortcoming. If visual holograms are projected into the room with the help of smartphones, but also other mobile devices, and can be controlled interactively, new potentials for planning and reporting are created. The size disadvantage is then eliminated.

5.10.1.3 Tablets

Tablets often have the same technical features that we have already listed for mobile phones, such as calendar, organizer, calculator, camera, phone, internet browser, GPS, fast booting and the possibility to use many applications as apps. Compared to the mobile phone, the main advantage for the analysis and planning functions is the size of the displays. In contrast to notebooks, a tablet has a touch-sensitive screen that allows input, for example, via an electronic pen or with the fingers of the hand. Therefore, the integration of a keyboard as in notebooks is usually dispensed with. In addition, tablets usually have the same functionality as notebooks or stationary PCs. Tablets were only used by a small group of users in the past. Only since the introduction of the iPad by Apple, this category has moved into the focus of mobile devices in terms of business application. This new generation of tablet computers, such as the Samsung Galaxy Tab and the apple iPad, are characterized by an intuitive and easy operation via a capacitive touchscreen and are therefore very well suited for presentation purposes. The gap in functionality compared to classic tablets and notebooks is getting smaller and smaller. Due to the easy analysis and navigation possibilities, the short waiting times when switching on and the very good mobility characteristics (e.g. long battery life), these devices are very well suited for general mobile use and in particular for mobile analysis and planning functions.

5.10.2 Wireless Networks

Wireless networks enable flexible access to information and are therefore a prerequisite for "mobile computing". However, these networks differ greatly in their transmission speed and location independence. While WLAN (or WiFi) has a very high data transmission rate, these networks are still strongly location-bound. Mobile networks, in contrast to locally bound WLAN networks, are widely available, but their transmission rate is much lower. Mobile networks have developed enormously in recent years and still have great development potential.[296] The first digital mobile network GSM (Global System

[296] Cf. Bollmann and Zeppenfeld (2010, pp. 87–111).

for Mobile Communication) is only very limitedly suitable for data transmission. With a connection-oriented data transmission with GSM, a permanent connection is only possible via an uninterrupted access with corresponding resource consumption. Moreover, the speed is very low despite acceleration techniques. Only the extension GPRS (General Packet Radio Service) is better suited for data transmission not only because of the packet-oriented transmission, but also thanks to a higher speed. UMTS (Universal Mobile Telecommunications System) is the successor of GMS and GPRS and offers a much higher transmission speed of several 100 Kbit/s or, in connection with HSPA (High Speed Packet Access), even transmission rates in the Mbit range. The fourth generation of mobile standards with the name LTE (Long Term Evolution) achieves significantly higher download rates than the predecessor UMTS with transmission rates of several 100 megabits up to 1 gigabit per second. With the new standard 5G, the transmission rate of the 4G-LTE standard will be 10 to 100 times higher (10,000 MBits/s = 10 gigabits/s).[297]

Only with the development of modern mobile networks, mobile reporting was made possible and therefore serves in connection with the mobile devices as a basic requirement for the mobile use of information. Technologies such as UMTS and LTE with 4G and 5G standard offer sufficient speed for sending and receiving extensive reports. The almost nationwide availability and the low costs for the use enable an almost unlimited access to the internet. Information is thus available anywhere and at any time. However, this requires a mobile information supply on the part of the companies.

5.10.3 App – Application

While traditionally, users could only access their software solutions via graphical user interfaces, such as the SAP GUI, software manufacturers developed so-called web interfaces in the course of the intensive use of the internet, where users can access the programs via generally accessible web browsers. With the spread of mobile data and program use via mobile devices, however, calling up the programs via web pages became uncomfortable, so that the software manufacturers used the app as a starting point for the software application. A characteristic of many apps is that they have a condensed intuitive operation of the apps via tiles or icons by the user, according to the limited possibilities of the device. The user can customize his or her navigation view and, for example, develop a personalized homepage. Furthermore, many complex software solutions that run on desktop computers, such as the SAP ERP system, have the problem that simple navigation through the program is only possible for experienced users. Here, too, the app technology can help by making the menu navigation completely or partially via apps.

[297] Cf. Lossau (2018, pp. 1–5).

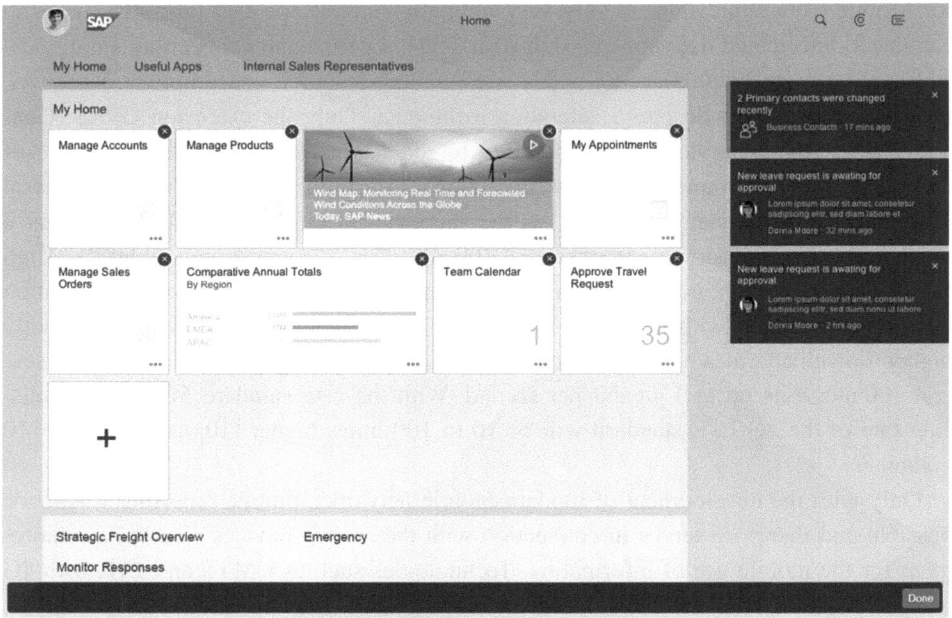

Fig. 5.65 SAP Fiori launchpad homepage in edit mode (*Source*: SAP Fiori 2017a)

The word "app" stands for mobile application and is usually executed via a tile or icon. Technically, three forms of apps are distinguished, the web apps, the native apps and the hybrid apps, which are not pursued further here.[298] The company Apple was certainly a pioneering forerunner with its iTunes store. Other manufacturers, especially Google with the Android operating system as well as Windows and BlackBerry, followed. At SAP, for example, SAP Fiori based on SAP HANA was introduced to make business figures available for mobile devices.[299] SAP Fiori uses, like other providers, increasingly a flat design with the use of the tile look for the apps and a user-individual structure of the program entry (see Fig. 5.65).

Often, complete program controls are divided into small program parts that can be accessed via apps. One of the most important advantages of the flat design and the apps for end devices is that they use less resources (CPU and GPU performance) and have lower loading times. Other important advantages of the flat design are the contemporary appearance and the adaptation of the flat design to the different end devices (see Fig. 5.66).

[298] Cf. Keist et al. (2016, pp. 109–113). The difference is similar to the one in Sects. 5.10.5.1 and 5.10.5.2 for mobile business applications.

[299] Cf. Mathew (2015, p. 1 ff.) and Engelbrecht and Wegelin (2015, p. 25).

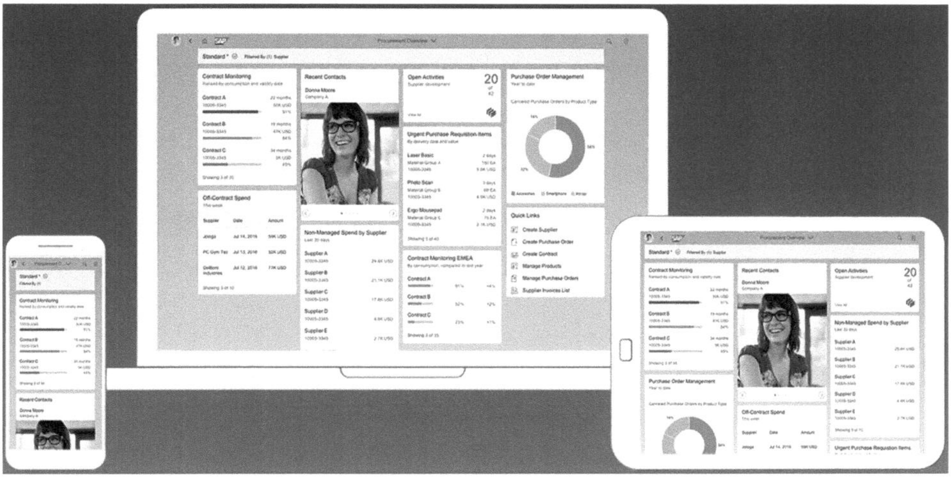

Fig. 5.66 SAP Fiori responsive design for different end devices (*Source*: SAP Fiori 2017a)

While a normal menu control via the so-called "shell-bar" in the head of the SAP Fiori launchpad homepage is possible for basic functions such as searching or logging in and out, different tile types on the screen page shape the app navigation.

There are very many different tile types, of which only some important ones are presented here:[300]

- The **standard tiles** show only content that is static, such as texts, images and icons.
- The **dynamic key figure tiles** show variable numerical values that display the current values of a key figure depending on the entry point.
- The **key performance indicator tiles with warning functions** show warnings for the selected key figures, e.g. in traffic light signal colors.
- The **graphic tiles** show selected graphic types with dynamic time-varying values for the selected key figures, such as bar or pie charts.

A example of a management dashboard with different tile types is shown in Fig. 5.67, which can be built individually with the help of Diamant/4.

Selected tiles can be grouped together, so that the user can build up a tile hierarchy through which he can call up entire program modules.

When a tile is activated by the user, applications are usually executed. The following types of apps can be distinguished:

- The transactional app launches an application program, in which tasks and functions, such as a logistic movement, a booking in accounting, a calculation or planning entry, can be performed.

[300] Cf. Krüger (2015, p. 204 ff.).

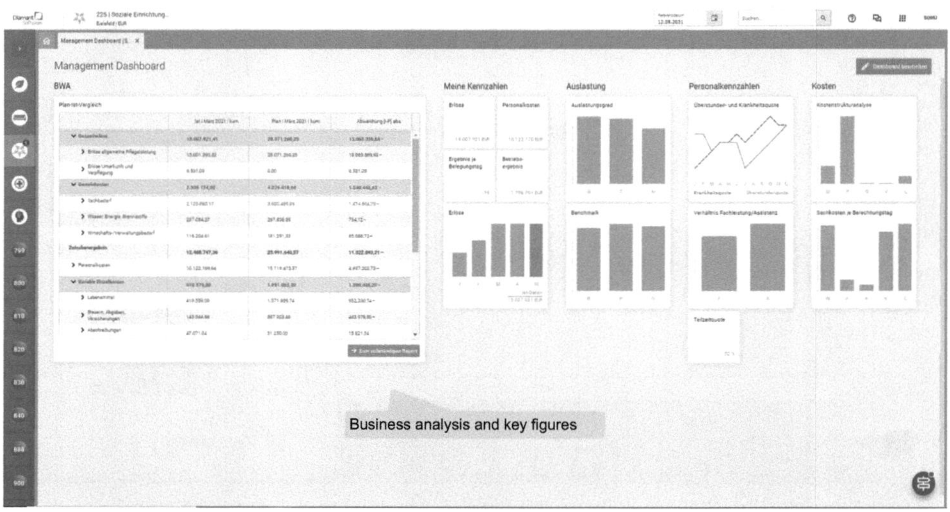

Fig. 5.67 Management dashboard with Diamant/4 (*Source*: Diamant Software 2021)

- A fact sheet app displays the results of a search, such as the number of objects.
- The analytical apps show aggregated key figures of a defined query in a report.

Planning and reporting tasks can currently only be performed to a limited extent using the app technology, unlike traditional OLAP-based frontend tools. For an analytical reporting and an integrated planning, it is obvious that the user can currently only create simple reports and planning forms in real time directly using the app technology, such as with Fiori based on SAP Hana. For complex analytical reports and plans, OLAP-based frontend tools, such as Cubeware, Jedox, SAP BusinessObjects etc. are better to use with their multidimensional data selection. The main reason for this limitation is that the toolkits, i.e. the development tools for the app technology, are not as powerful and difficult to use for users without programming skills. For example, the "SAP User Interface for HTML 5" (SAPUI5) requires very good knowledge of Javascript and HTML, while in the SAP Design Studio the creation of reports is essentially graphical by drag-and-drop of preconfigured elements. In addition, they have better visualization techniques. Another disadvantage is that the analytical apps of SAP Fiori are limited to the SAP Hana data storage. Most BI-based frontend tools are much more flexible regarding the data sources to be connected. Since SAP Fiori was primarily created for program navigation, it is not to be expected that SAP tries to eliminate this deficit with this tool. It is rather to be expected that classic OLAP frontend systems can be called via an app, so that powerful multidimensional ad-hoc analyses (e.g. with slice and dice) as well as complex planning tasks can be done here.[301] Also, the expansion of a wide range of apps does not suggest that this will capture the multifaceted analysis needs of reporting and planning.

[301] See the following source for this: SAP Fiori (2017b).

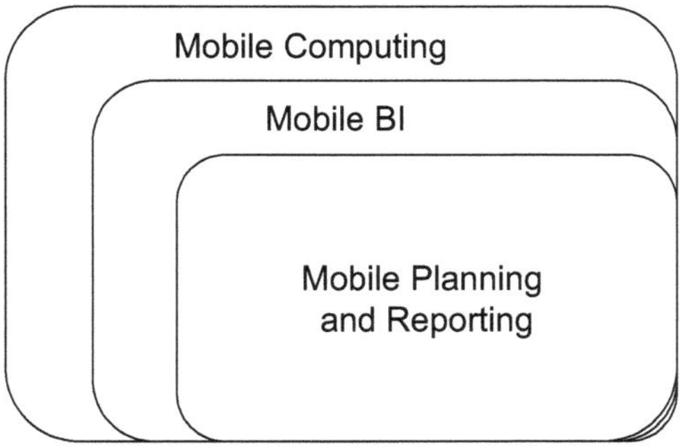

Fig. 5.68 Mobile Computing, Mobile BI and Mobile Planning and Reporting

5.10.4 Classification of Reporting and Planning in Mobile Computing

In relation to the three levels of the data warehouse, mobile computing can be classified as a technical platform for data analysis and use, i.e. specifically for analysis and planning purposes, as well as other business applications.

Mobile reporting and mobile planning can be subordinated to the research field of Mobile BI. Mobile BI deals with the access to decision-relevant data using mobile devices and with the design of the mobile BI system. The previous focus of Mobile BI lies in the mobilization of the operational reporting system.[302] Planning is hardly considered.

Mobile reporting is a core element of Mobile BI and is sometimes used synonymously. In its broadest sense, both Mobile Reporting and Mobile BI can be assigned to the framework of Mobile Computing. Mobile Computing is a very broad field that deals with the mobile use and ubiquity of computers in today's society. This research field provides the technical basis for the realization of Mobile Planning and Reporting and also covers the entire social development of mobilization. This relationship is illustrated in Fig. 5.68. The main advantages of mobility lie in the location-independent and target-group-specific information supply and the possibility of communication. This means a very fast access to important information, even if no stationary computer is available. Especially for managers and field staff who are often on the move, the mobile access offers many advantages, in particular when all important information for the upcoming tasks, such as an important contract negotiation or a decisive sales talk, are available.

5.10.5 Mobile Business Intelligence

Mobile computing has triggered a significant change in society. With modern devices, mobile networks and widespread WLAN networks, the technical basis was created to

[302] Cf. Bensberg (2008, p. 72).

access relevant information without being bound to a stationary workplace. Therefore, mobile computing can also fundamentally change the application of BI. However, a prerequisite for mobile information supply is a BI system that is extended by mobile components. The research field of mobile BI creates this framework. The concept of mobile BI offers enormous potentials for various business areas which can contribute to efficiency increases of decision processes.

In the implementation and design of a mobile BI system, many challenges arise for the companies. These challenges lie, according to *Bensberg*, especially in the adaptation of already existing BI solutions.[303] These have to meet the requirements that result from the different characteristics of the mobile environment. As described in the previous section, the mobile computing sector is characterized by a variety of mobile devices that differ greatly in their functionality. An integration of these different devices into an existing BI system is not feasible without an extension of the system. In addition, the access to the planning data for the decision maker has to be enabled via various wireless networks. The stationary BI system is therefore extended by mobile components that are supposed to enable the access of the devices via different access modes. The structure of a mobile BI system builds on the first two levels of the data warehouse architecture or the exploratory BI system (incl. big data technology) and is shown in Fig. 5.69.

The communication properties of mobile systems differ significantly from the properties of stationary systems. While access to the BI system in a "classical" distributed

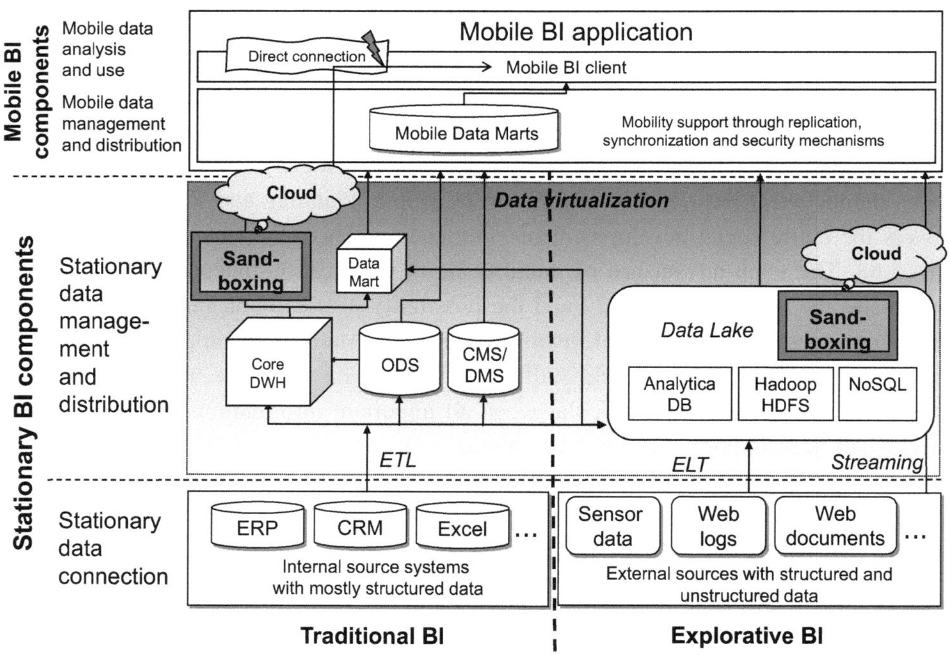

Fig. 5.69 Structure of mobile BI systems

[303] Cf. Bensberg (2008, pp. 75–79).

system, such as the local company network, can usually be established without interference, frequent changes in connection quality occur in mobile distributed systems.[304] Thus, connection failures (also for energy saving at the end device) and fluctuating data transfer rates must be taken into account in the implementation of a mobile BI system. With the extension to mobile components, the system is accordingly adapted. In order to ensure the maintenance of access to the data stocks of the DWH system or the explorative BI system (incl. big data technology) via mobile devices even in the event of a short-term connection failure, the establishment of mobile data marts is required. **Mobile data marts** represent a reduced data layer set up on the end devices. Thus, in the event of a disconnection from the system, data can still be accessed, for example, in the form of caching.[305] To realize a consistent data replication, a replication and a constant synchronization of the data after a reconnection is required. A component called middleware or also mobile server takes over this process of regular data synchronization between the stationary and the mobile database. With these disconnection and reconnection functions, mobile access to databases is enabled. In addition, the user can also work with replicated data sets without a data connection (offline) (see Fig. 5.70). Without this intermediate

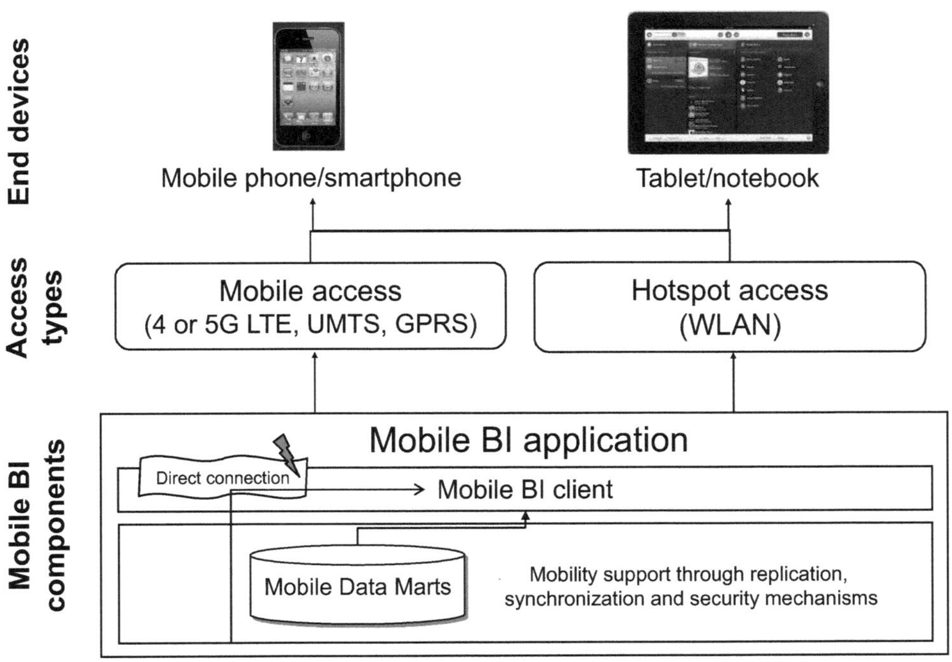

Fig. 5.70 Mobile access types and devices (Own illustration. The brands, images and symbols of the iPhone and iPad are the exclusive property and trademarks of Apple Inc.)

[304] Cf. Fuchß (2009, pp. 137–151).

[305] Cf. Schill and Springer (2007, pp. 265–271).

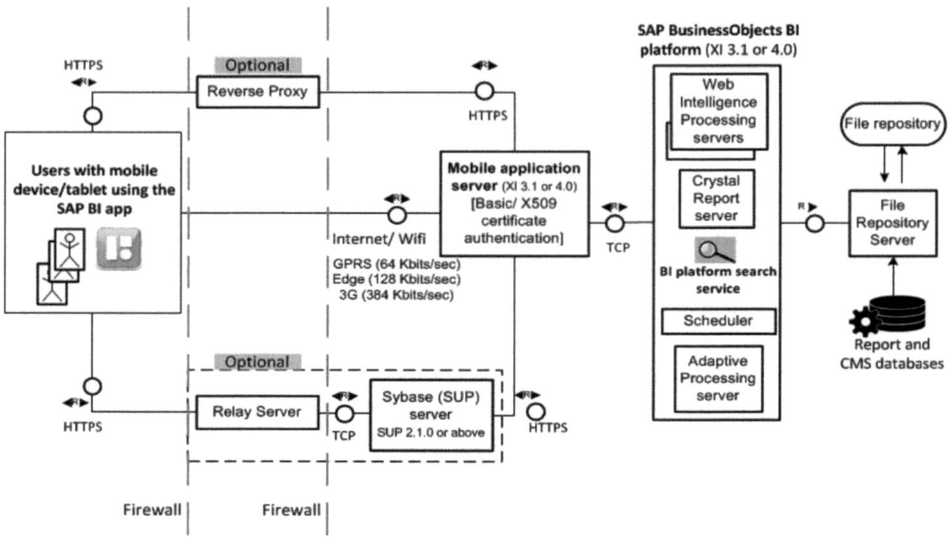

Fig. 5.71 SAP BusinessObjects mobile architecture

layer, a mobile analysis is only possible if a network connection is available. An example of an architecture for mobile BI systems is shown in Fig. 5.71 using SAP BusinessObjects as an example.[306]

The mobile access is via LTE, UMTS, GSM and GPRS or via a Hotspot access or WLAN. In addition, the mobile devices can also be connected to the company networks and thus to the stationary data warehouse via fixed network access via Ethernet or DSL.

The connection of mobile communication technology with the internet offers further possibilities. If the data storage for reporting and planning is done via the internet in a so-called cloud, the mobile devices can access mobile management information worldwide via this access. If the stability and security of cloud computing are satisfactorily solved in the next few years, this will result in very good development prospects for mobile reporting and mobile planning, especially with regard to the availability and speed of planning and decision-relevant information.

Furthermore, security mechanisms such as encryption and authentication procedures must be set up in a mobile BI system.[307] Especially with regard to the transmission to mobile phones, the sensitive data must not only be protected from eavesdropping and spying, but also be inaccessible in case of loss or theft of the device. Ideally, in the event of loss of the device, the entire data set should be deleted via remote kill or wipe functions.

306 Cf. SAP AG (2016, p. 5).

307 Cf. Bensberg (2008, p. 76).

The heterogeneity of mobile devices is another problem that arises when designing a mobile BI system. As explained before, these devices differ greatly from each other in terms of their characteristics, such as performance, operating systems and functionality. A mobile server is a possible solution for this problem. In a three-tier architecture in the sense of the Managed Query Environment (see Sect. 5.5.5.6), it can adapt the data before transmitting them to both the properties of the devices and the transmission speed and stability of the communication channels.[308] In particular, web applications can achieve compatibility with several different devices in this way. With the help of filtering and compression methods, the data volumes can also be reduced before transmission and made available to the user in different quality levels, depending on the respective application situation. The potentials of the various devices and transmission technologies can thus be fully exploited.

Mobile Device Management

For mobile computing, it makes sense to manage all mobile devices used for business purposes via a uniform software. These software solutions are summarized under the term Mobile Device Management (MDM). Similar to the conventional Local Device Management (LDM), i.e. the administration of stationary servers and PCs, special requirements are in the foreground for MDM. Due to the heterogeneous mobile devices, it is important to ensure that all devices are compatible with the typical platforms and applications. A shared private and business use on the devices is often not desired. If a shared use is allowed, the MDM and the device must ensure that private and business use are separated as best as possible. Usage agreements with the employees regulate the details.[309] The MDM also serves as a hub for updates, patches, recoveries and backup functions for the software of the devices and the platform. Here, the incident and security management plays a major role. The administrator of an MDM should be able to add, block or remove devices as easily as possible. In case of lost devices, the data on the devices must be deleted or protected as best as possible.[310] While at the beginning of the mobile hype under the motto "Bring Your Own Device" (BYOD) employees were allowed to use their private devices for business purposes and to open them for company data, for security reasons today the principle "Choose Your Own Device" (CYOD) is preferred, in which employees can choose from a list of predefined devices that have been optimized for the company's platform in advance and are usually configured with appropriate applications based on roles.[311]

Another component in the BI system, which enables mobile information supply, is the mobile BI client. This frontend application is located on the device and provides the

308 Cf. Schill and Springer (2007, pp. 274–280).

309 Cf. Donie and Raeburn (2015).

310 Cf. Kersten and Klett (2012, p. 103 ff.).

311 Cf. Hansel (2015).

user with presentation and application functions. The client takes over the visualization and can also support further application functions. The design of the clients leads to a trade-off between performance demand and server dependency. On the one hand, the dependency on a stationary server can be avoided by having the client process the data directly on the device. This method can only be realized on powerful devices due to the high resource demand. Therefore, a large part of today's frontend applications are based on web technologies that shift the preparation to the server. The two most important approaches, rich client and thin client, are explained in the following two sections.[312] Other options are hybrid applications, a mixture of rich and thin client applications, and container applications, in which the container provides access control to a middleware platform on which the actual application for mobile use is located. However, these are not considered further.

5.10.5.1 Rich Client/Native Application

A rich client application is an independent, installed program on the end device which provides access to the BI system. A native program is developed specifically for operation on a certain operating system such as iOS, Android and Windows Phone.[313] Under certain conditions, however, a rich client can also be designed platform-independent, e.g. with the programming language Java. The rich client is characterized by a large range of functions, related to the visualization techniques, the application and navigation possibilities. This allows, for example, extensive offline functionalities to be supported. Therefore, this client is especially suitable for comprehensive and dynamic mobile reporting applications that go beyond the simple viewing of static reports. Information that is not needed directly can be stored and retrieved when needed. In case of changes, the data is synchronized by a push service. Furthermore, device-specific functions such as camera, GPS location, Bluetooth etc. can be used. By using the appropriate development environment, the operating and design elements of the devices are used in the best possible way.[314] The application of a rich client demands the performance of the end devices more intensively than the other technologies, which may require an exchange of the devices. Furthermore, rich clients are usually not compatible with all operating systems of the different end devices. The investment is therefore worthwhile mainly for companies with standardized end devices. In case of updates and operating system changes, adjustments of the applications are necessary.

5.10.5.2 Thin Client/Web Application

A simple alternative to the rich client is the use of web-based applications. Web applications are conventional web pages that adapt to the screen size with the help of e.g.

[312] Cf. Bensberg (2008, p. 77).

[313] Cf. Louis and Müller (2013, p. 23).

[314] Cf. Beckert et al. (2012, p. 139 f.) and Homann et al. (2013, p. 52 f.).

HTML5, CSS3 (Cascading Style Sheet—Level 3) and JavaScript and thereby ensure an optimal presentation for mobile devices. The web browser pre-installed on the devices is usually used as the access interface. In this context, one also speaks of responsive web-design, which means that the program pages are designed so flexibly that they can be accessed via different mobile devices. Different display sizes and navigation techniques (e.g. mouse click, tapping, swiping etc.) are taken into account. This enables the use of the applications on different devices without having to make additional adjustments. In addition, expenses for installation and maintenance on the individual devices are eliminated. In the first years of rich- and thin-client development, it was disadvantageous that the thin-client development had a lower user-friendliness and the device-specific components could not be used. In the meantime, thin-client developments can access device-specific components. The applications are largely equivalent in their functionality to native applications. However, thin-client developments may still have to consider adjustments for updates by the browser manufacturers.[315] The use of thin clients ranges from accessing simple web pages that enable access to data sets, to specially prepared and interactive cockpit and dashboard solutions. The preparation via a special server is associated with additional effort. However, since navigation on web pages via mobile devices can be very cumbersome, a higher preparation effort can certainly be justified. Modern web-based applications can hardly be distinguished in their design and functionality from dedicated applications and are therefore also called rich internet applications.[316] Nevertheless, the performance of the thin clients still depends strongly on the transmission speed of the mobile network. A nearly permanent and fast network access is therefore required to ensure efficient use.

5.10.6 Mobile Reporting and Mobile Planning

Studies by the *Dresner Advisory Services LLC* on mobile business intelligence from 2010 and 2011 show a clearly growing trend towards mobile BI and especially mobile reporting in companies.[317] The first hype is over, however. In the current BI trend monitor of BARC, mobile BI ranks 11th.[318] In my estimation, however, the topic will gain in importance again due to the increasing mobility. Generally, the studies neglect the topic area of

[315] Cf. Homann et al. (2013, p. 53 f.).

[316] Cf. Kemper et al. (2010, p. 251).

[317] Cf. Dresner Advisory Services LLC: Mobile Business Intelligence Market Study, 2010 and 2011. URL: http://www.microstrategy.com/mobile/mobile-bi-landscape-dresner.pdf (accessed on 25.07.2011) and URL: http://www.informationbuilders.com/pdf/press/dresner_mobile_bi_2011.pdf [accessed on 25.07.2011].

[318] Cf. BARC (2017). Trend Monitor 2017 http://barc.de/trend-monitor. Accessed on 29.07.2017.

mobile planning. Just like in stationary BI, only reporting solutions were considered at first. Typical characteristics and functions of mobile reporting solutions are:

- view charts/reports (tabular and graphical reports)
- KPI monitoring (key figure analyses)
- alerts (alarm/warning functions with colour coding or conditional formatting)
- interactive analyses
- drill-down navigation
- drag-and-drop navigation
- data selection, data filtering
- guided analytics
- geographic analyses
- dashboards

Following the development in stationary BI, mobile planning will also expand its role in mobile BI as the independence from a stationary connection can be interesting for many planners.

Since the terms mobile reporting and mobile planning are not yet clearly defined, the following definitions should be used here.

The term **mobile reporting** refers to the time and location **in**dependent information supply and use of all control and decision-relevant information of the decision-makers of a company, whereby these are ideally optimized for the respective mobile device and coordinated with the stationary reporting.

The term **mobile planning** refers to the time and location **in**dependent information supply and use of all planning-relevant information in the form of planning forms, reports and functions for the decision makers of a company, whereby these ideally are coordinated with the stationary planning system and optimized for the respective mobile device.

Important tasks of mobile reporting and mobile planning are therefore the location-independent information supply, as well as the integration and coordination with the stationary reporting and planning system and the optimization of the output for the respective device. Here, the spectrum ranges from a static mobile reporting system, e.g. the provision of reports by e-mail or PDF files, to interactive analysis and planning functions in which the user accesses a data base that he can use arbitrarily (filter, display, compare etc.). Examples of a smartphone and tablet application are shown in Figs. 5.72 and 5.73.

Mobile reporting and mobile planning complement the classical reporting and planning with new possibilities, which are given by the time and stationindependent informa-

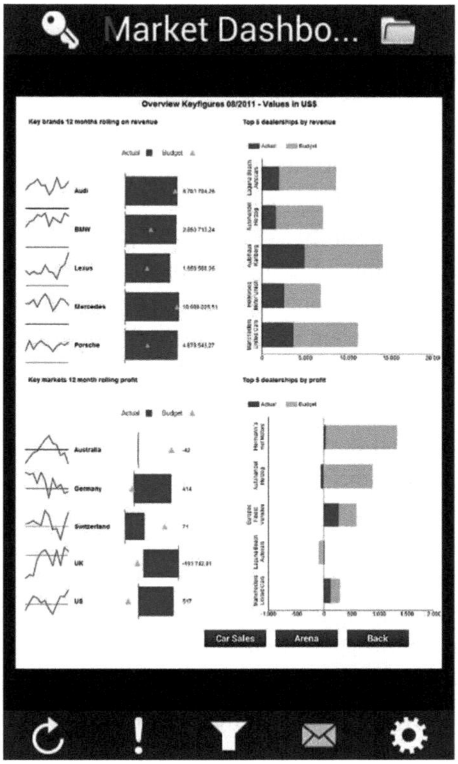

Fig. 5.72 Mobile reporting with a smartphone (C8 Cubeware) (Cf. Cubeware, http://de.cubeware.com/produkte/c8-mobile/galerie-mobile-bi-android.html. Accessed on 23.03.2015)

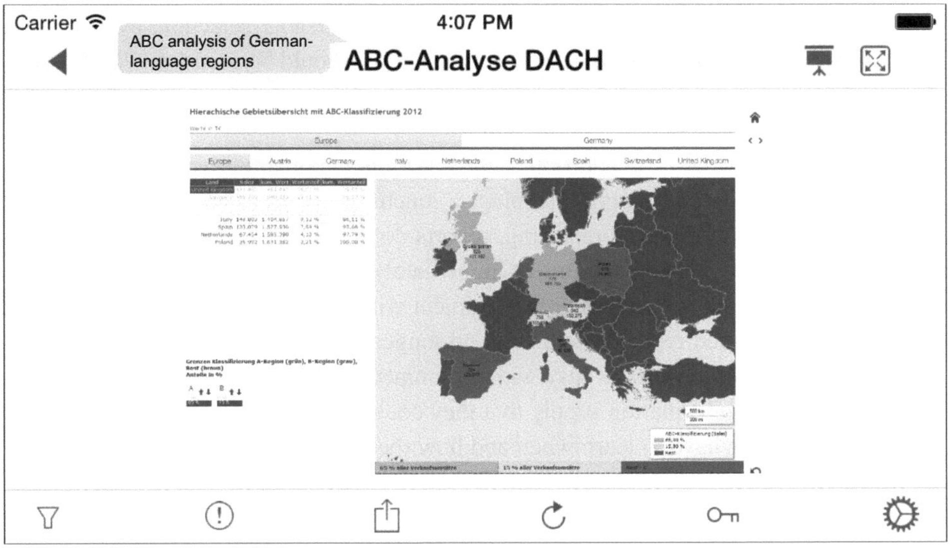

Fig. 5.73 Mobile reporting with a tablet (C8 Cubeware) (Cf. Cubeware, http://de.cubeware.com/produkte/c8-mobile/galerie-mobile-bi-ios.html. Accessed on 23.03.2015)

tion supply. Accordingly, there are also new requirements that have to be met by mobile solutions:

Usability

According to the classical reporting and planning, the reports and planning forms must also be clear in their design in mobile BI, so that the user can quickly extract the information relevant to him and carry out planning. However, these requirements are much more demanding in their implementation in mobile reporting and mobile planning, as the limited display options on the relatively small screens of the mobile devices make it difficult to present the information clearly. Therefore, the reports should not only be formatted according to the application context, but also adapted to the medium used (e.g. automatic scaling). The acceptance of the reports by the user requires a purpose- and recipient-oriented design of the information. Only in this way it can be ensured that the recipient accepts the relevant information and can also assess it.[319] Important control functions in the context of gesture recognition of a simplified input on a touchscreen display are e.g. tapping an object, double-tapping an object, swiping an object to change the page or scroll, dragging and dropping an object, enlarging objects by spreading two fingers or the complementary element, pinching the fingers to reduce an object and many other techniques. In contrast to mouse control, the right mouse button with a context menu is omitted. For this purpose, a two-finger control (press and double press) can be used if necessary. Due to different finger sizes and fine motor skills, the objects to be controlled must not be designed too small. Furthermore, it must be taken into account that depending on the device (smartphone or tablet) the positioning of the objects to be operated is within the reach of the fingers, i.e. at the bottom edge of the phone and at the two side edges of the tablet. Furthermore, simple input functions, such as calendar selection, cylinder time selection, on and off switches or plus and minus buttons, should be used (see Fig. 5.74)

Resilience

Another problem that arises in addition to the clarity is the quantitative load on the user. With mobile reporting and mobile planning, a much faster information supply is made possible. In contrast to classical reporting, reports do not have to be printed out, but can be made available to the recipient almost immediately at any time. However, this feature also entails the risk of overwhelming the recipient with too much information. From the multitude of transmitted reports, the user is no longer able to process the relevant information. Business people were so busy with the immediate availability of e-mails and the resulting permanent information supply that they could not concentrate on other tasks.[320] In addition, the users have to learn when and how they can best process the information,

---

[319] Cf. Jung (2011, pp. 207–209).

[320] Cf. Bollmann and Zeppenfeld (2010, p. 42).

Fig. 5.74 Simple exemplary input functions

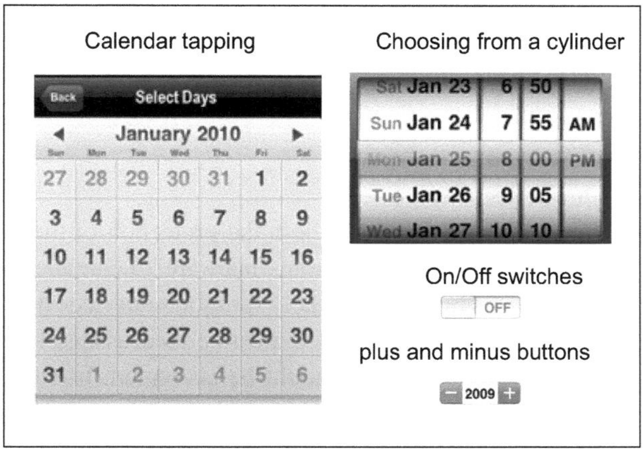

i.e. they also have to learn when not to use the mobile information supply and when it may be useful to turn off the devices.

Security

Especially the technical requirements differ strongly in mobile reporting and mobile planning from the requirements of the stationary systems for planning and reporting. As mentioned before, mobile distributed systems have to adapt to the characteristics of the wireless networks to enable a flawless transmission. Another significant requirement is the security aspect in information supply. Related to the use of mobile devices and the transmission over wireless and public networks, the security of the data can only be achieved with the help of additional measures. The importance of security in mobile computing is easy to understand as mostly sensitive and business-critical information is transmitted.[321] The loss or theft of these confidential or competitive data can lead to devastating consequences for a company. Therefore, different risks have to be considered, which are higher for mobile devices than for stationary applications. Based on these risks, different security mechanisms can be developed. These security mechanisms have to prevent unauthorized access in case of theft and espionage. For the IT, the challenge is to adapt the security mechanisms specifically to the mobile environment without restricting the user in the usage. Thus, there is a trade-off between risk minimization and user-friendliness in terms of security.

Lopez illustrates this relationship with an example. A user cannot be expected to enter a 20-digit password on a smartphone. Instead, a 6-digit alphanumeric password should be preferred, for example, to ensure the same level of security.[322]

[321] Cf. Bensberg (2008, p. 76).

[322] Cf. Lopez (2009, p. 2).

Moreover, the security requirements differ depending on which device or operating system is used. They have to be designed individually according to the device type. Due to the various risks, the security requirements can be differentiated according to three different levels:

- Device security
  Device security refers to both the protection of the data on the mobile device and the protection of the device from "malware", i.e. malicious software. In particular, theft or other loss is much more likely for mobile devices than for stationary computers. Therefore, the data on the device must be inaccessible to strangers. To ensure this security, the data on the device should be encrypted. In addition, access to the device should be protected by a password or a PIN (Personal Identification Number) or a biometric identification feature (e.g. fingerprint). In some cases, the encryption of the data is already integrated in the operating system of the mobile phone, such as in the BlackBerry. Moreover, the device must be protected from viruses, worms, trojans, etc., which could destroy the data or allow unauthorized access to the device, by firewalls and virus scanners. Furthermore, the mobile devices can be protected with a smart card, which uses its own processor for data encryption. It can be used for authentication, encryption, access control and the creation of a digital signature, among other things.
- Transmission security
  Wireless networks are used for data transmission in mobile reporting. These are much more open than wired networks and therefore freely accessible to all people. An unauthorized access in the transmission of data is thus harder to prevent than in closed, local networks. The transmission security includes in this sense the protection from foreign and unwanted access in the data transmission. For this purpose, access restrictions in the form of encryption methods must be used. In the WLAN network, the connection can be protected, for example, with the help of a WPA2 key (Wi-Fi Protected Access 2). Mobile networks are already encrypted by the network operators. However, this encryption is relatively unreliable and should be supplemented by own measures.[323]
- Network security
  Of course, the security requirements of mobile reporting also include the protection of the local company network. The opening of the internal network to the outside entails additional security risks. The mobile access to the internal data stocks must be prevented for unauthorized persons without making the access more difficult for authorized users. Accordingly, the firewalls must be adapted to the mobile use of the network and authentication methods must be set up. The use of a Virtual Private Network (VPN) is another security method with which a secure remote access to the corporate network can be enabled.

[323] Cf. Bollmann and Zeppenfeld (2010, pp. 127–130).

Among the security features of a Mobile Device Management (MDM) are, among others:[324]

- locking and deleting devices remotely
- encryption of files and data transmission
- localization of lost devices
- backup and restore functions
- jailbreak avoidance (unauthorized removal of usage restrictions on IT devices)
- password setup
- automatic locking in case of rule violation
- log-on controls and time locks
- deactivation of individual functions

Integration and implementation of BI system components

For the use of mobile reporting applications, additional components must be integrated into the existing BI system beforehand (see Sect. 5.10.4). The market offers so-called BI suites for this purpose.[325] BI suites offer all layers of the BI system in the form of a complete solution. With the use of single source authoring, reports can be generated directly from a uniform data basis such as the data warehouse for both mobile and stationary use, ideally without requiring a further adaptation phase for report design. However, practice shows that device-specific reports and planning forms have to be developed for each device due to the different frontend solutions, operating systems and display sizes. Another important criterion for the selection of the application is the speed of the software in processing and transmitting the data. Even when using standard software products, easy implementation should be ensured. Instead of using different applications, a software with a uniform interface for mobile reporting should be used. Ideally, these applications offer the same functions as comparable stationary applications and thus enable the user to have a shorter familiarization phase.

Compatibility

In addition to the implementation requirements, mobile reporting and planning applications also require a high compatibility. Mobile programs should be compatible with a variety of mobile devices, operating systems and transmission methods and ideally adapt to the individual characteristics of the platforms. Another aspect is the use of old systems and their data stocks. Accordingly, the mobile applications should be compatible with the previous system. Furthermore, the compatibility of the mobile BI system in multi-user operation is required. Many users and a high complexity of the processes should therefore not limit the performance of the mobile and stationary reporting and planning applications.

---

[324] Cf. Kersten and Klett (2012, p. 103 ff.).

[325] Cf. Bensberg (2008, p. 77).

Functionality

The same functions are expected from mobile reporting and mobile planning as have already been explained for the stationary systems. The main focus here is on user-friendly functionality that adapts to the user role. The following functions are distinguished, which are either particularly suitable or particularly **un**suitable for mobile applications:

- Functions that are suitable for mobile applications:
  - sorting and aggregation functions
  - navigation options
  - condensed standard reporting
  - exception reporting (e.g. with colour coding)
  - analysis reporting with slice and dice, drill-down, roll-up
  - condensed information (key figures, graphical information, condensed management cockpits/dashboards)
  - short comments
  - planning functions with few value inputs (e.g. for triggering simulation and scenario calculations)
- Functions that are rather unsuitable for mobile applications:
  - individual ad-hoc reporting
  - list reporting and larger tables
  - drill-through
  - extensive text and comment functions
  - planning functions with many value entries

5.10.7 Mobile Success Potentials

With the help of mobile reporting and planning applications, a time and location independent information supply and planning is enabled and thus a virtually limitless availability of information and planning possibilities is achieved. Latest technological advances, both in mobile data transmission and in the mobile devices themselves, enable an ever faster and more reliable access to decision-relevant and planning data. The dissemination of mobile applications grows with increasing performance and interactivity of the applications and the decreasing costs for the implementation. Applications for mobile planning are hardly to be found. However, it can be assumed that after the establishment of mobile reporting, mobile planning applications will also spread.

The performance of mobile reporting can be assessed based on the usage potentials of the users. As part of the study conducted by *Dresner Advisory Services*, the participants were asked about the benefits that arise in connection with mobile BI. The most important benefits of the respondents were the limitless and fast access to data sets.[326] The user is offered more spatial and temporal flexibility in the decision-making process with mobile reporting.[327]

Not only executives, but also the operational level can benefit from this greater flexibility as a result of the falling costs for mobile BI. This could be an important entry argument especially for medium-sized businesses. On the one hand, for example, field service staff in the operational business can access current data sets directly at the customer's site and advise the customer more efficiently with the help of this data. Although this application was already possible with conventional notebooks, modern smartphones and tablet PCs offer much better mobility. In this respect, a higher employee productivity and an improved customer satisfaction could be realized for a company. On the other hand, planning decisions by executives can be made much faster than before based on the immediately available data. Companies can react immediately to changes in the market and consequently better respond to changing customer needs and competitive situations. In doing so, the executives are increasingly actively involved in the reporting process and can request the information they need themselves, without being tied to a stationary workplace.

With the increasing spread of mobile reporting, the conversion of the previous reporting system to mobile BI components will therefore be inevitable for companies.

References

@tfxz-Blog. 2014. *Data warehousing on HANA: Managed or freestyle? BW or native?* https://tfxz.wordpress.com. Accessed 18 July 2014.

Alexander, S., and T. Grosser. 2017. *Analytische Datenbanken – Merkmale, Funktionen und Einsatzszenarien.* http://www.searchenterprisesoftware.de/tipp/Analytische-Datenbanken-Merkmale-Funktionen-und-Einsatzszenarien. Accessed 19 July 2017.

Alexander, S., A. Haisermann, T. Schabicki, and S. Frank. 2018. Robotic Process Automation (RPA) im Rechnungswesen und Controlling – Welche Chancen ergeben sich? *Controlling* 30(3):11–19.

Allweyer, T. 2016. *Robotic Process Automation – Neue Perspektiven durch Robotic Process Automation.* Kaiserslautern: Fachbereich Informatik und Mikrosystemtechnik – Hochschule Kaiserslautern.

Alpar, P., and J. Niedereichholz, eds. 2000. *Data Mining im praktischen Einsatz.* Braunschweig: Springer Vieweg.

Apel, D., W. Behme, R. Eberlein, and C. Merighi. 2009. *Datenqualität erfolgreich steuern.* München: dpunkt.

Azevedo, P., G. Brosius, S. Dehnert, et al. 2005. *Business Intelligence und Reporting mit Microsoft SQL-Server.* Heidelberg: Microsoft Press/dpunkt.verlag GmbH.

[326] Cf. Dresner Advisory Services LLC (2011, p. 23).

[327] Cf. Bensberg (2008, pp. 72–79).

Baars, H. 2016. Predictive Analytics in der IT-basierten Entscheidungsunterstützungl – Methodische, architektonische und organisatorische Konsequenzen. *Controlling* 28(3):174–180.

Ballhorn, C. 2017. *Data Warehouse vs. Data Lakel – Das sind die Unterschiede*. https://www.silicon.de/blog/data-warehouse-vs-data-lake-das-sind-die-unterschiede. Accessed 4 Oct 2019.

Bange, C., B. Marr, and A. Bange. 2009. *Performance Management: Eine weltweite Umfrage*. BARC-Studie August.

Bange, C., et al. 2013. Architektur. In *Data-Warehouse-Systeme: Architektur, Entwicklung, Anwendung*, eds. A. Bauer and H. Günzel, 4th edn, 5–180. Heidelberg: dpunkt.

BARC. 2014. *Big Data Analytics 2014l – Auf dem Weg zur datengetriebenen Wirtschaft*. http://www.exasol.com/fileadmin/content-de/pdf/Studien/BARC_Big_Data_Analytics_2014.pdf. Accessed 17 Dec 2014.

BARC. 2017. *BI-trend monitor 2017*. http://barc.de/trend-monitor?s=s_9_550. Accessed 12 July 2017.

BARC. 2020. *BARC GUIDE Data, BI & Analytics 2020*. Würzburg: BARC GmbH.

Barton, T., C. Müller, and C. Seel, eds. 2018. *Digitalisierung in Unternehmen. Von den theoretischen Ansätzen zur praktischen Umsetzung*. Wiesbaden: Springer Vieweg.

Bauer, H. 2020. *KI-Prozessoren: Warum die neuen Chips die Zukunft sind – PC Magazin*. https://www.pc-magazin.de/ratgeber/ki-prozessor-intel-google-huawei-apple-zukunft-forschung-3199054.html. Accessed 4 Oct 2020.

Bauer, A., and H. Günzel. 2013. *Data Warehouse Systeme; Architektur, Entwicklung und Anwendung*, 4th edn Heidelberg: dpunkt.

Baumöl, U., and P.-D. Berlitz. 2014. Big Data als Entscheidungsunterstützungl – Herausforderungen und Potenziale. In *Der Controlling-Berater: Controlling und Big Data*, eds. R. Gleich, A. Klein, M. Kirchmann, and J. Leyk, Vol. 35, 159–176. Freiburg: Haufe.

Becker, J., O. Richter, and A. Winkelmann. 2011. Analyse von Plattformen und Marktübersichten für die Auswahl von ERP und Warenwirtschaftssystemenl – Arbeitsbericht 121: Westfälische Wilhelms-Universität Münster. http://www.wi.uni-muenster.de/institut/arbeitsberichte/ab121.pdf. Accessed 17 Aug 2011.

Beckert, A., S. Beckert, and B. Escherisch. 2012. *Mobile Lösungen mit SAP*. Bonn: Galileo Press.

Behme, W. 1996. Business Intelligence als Baustein des Geschäftserfolgs. In *Das Data Warehouse-Konzept – Architektur-Datenmodelle-Anwendungen*, eds. H. Mucksch and W. Behme, 27–46. Wiesbaden: Springer Gabler.

Behme, W., and H. Mucksch. 1997. Die Notwendigkeit einer entscheidungsorientierten Informationsversorgung. In *Das Data Warehouse-Konzept: Architektur – Datenmodelle – Anwendungen*, eds. W. Behme and H. Mucksch, 2nd edn Wiesbaden: Springer Gabler.

Behme, W., J. Holthuis, and H. Mucksch. 2000. Umsetzung multidimensionaler Strukturen. In *Das Data Warehouse-Konzept – Architektur-Datenmodelle-Anwendungen*, eds. H. Mucksch and W. Behme, 4th edn, 215–242. Wiesbaden: Springer Gabler.

Bensberg, F. 2008. Mobile Business Intelligence – Besonderheiten, Potenziale und prozessorientierte Gestaltung. In *Erfolgsfaktoren des Mobilen Marketing*, eds. H. H. Bauer, T. Dirks, and M. D. Bryant. Berlin: Springer.

Berg, B., and P. Silvia. 2013. *Einführung in SAP HANA*, 2nd edn Bonn: Galileo Press.

Berry, M. W., A. Mohamed, and B. W. Yao. 2018. *Supervised and unsupervised learning for data science*. Switzerland: Cham.

Big Data Blog. 2015. *Hadoop bekommt Konkurrenz: Apache Spark ist 100 mal schneller*. https://bigdatablog.de/2015/05/07/hadoop-bekommt-konkurrenz-apache-spark-ist-100-mal-schneller/. Accessed 4 July 2016.

Birk, W., and C. Wegener. 2010. Über den Wolken: Cloud Computing im Überblick. *DuD Datenschutz und Datensicherheit* 34(9):641–645.

Bissantz, N., J. Hagedorn, and P. Mertens. 2000. Data mining. In *Das Data Warehouse-Konzept*, eds. H. Muksch and W. Behme, 4th edn, 377–407. Wiesbaden: Springer Gabler.

Bissantz, N., J. Hagedorn, et al. 2001. Data mining. In *Lexikon der Wirtschaftsinformatik*, ed. P. Mertens, 4th edn, 130–131. Wiesbaden: Springer.

BitKom. Understanding artificial intelligence as automation of decision making, https://www.bitkom.org/sites/default/files/file/import/Bitkom-Leitfaden-KI-verstehen-als-Automation-des-Entscheidens-2-Mai-2017.pdf (2017).

Bitkom. 2013. *Management von Big Data-Projekten, BITKOM-Leitfaden 2013*. http://www.bitkom.org/de/themen/79758_76511.aspx. Accessed 12 Mar 2015.

Bitkom. 2014. *Big Data-Technologien – Wissen für Entscheider*, 21–24, 45–47. http://www.bitkom.org/de/publikationen/38337_78776.aspx. Accessed 8 Jan 2015.

Bliznak, Karol. 2020. SAP Analytics Cloud und SAP Digital Boardroom: Machine Learning und intelligente Assistenten erleichtern Echtzeitsteuerung. In *Reporting und Business Analytics*, eds. Andreas Klein and Jens Gräf, 153–173. Stuttgart: Haufe Group.

Bollmann, T., and K. Zeppenfeld. 2010. *Mobile Computing – Hardware, Software, Kommunikation, Sicherheit, Programmierung*. Witten: W3L GmbH.

Brenckmann, I., and M. Pöhling. 2012. *In-Memory Computing als Treiber neuartiger Geschäftsanwendungen*. http://www.heise.de/developer/artikel/In-Memory-Computing-als-Treiber-neuartiger-Geschaeftsanwendungen-1620949.html. Accessed 15 July 2014.

Brücher, C. 2013. *Rethink big data*. Heidelberg: mitp.

Buck-Emden, R. 1995. *Die Client/Server-Technologie des SAP R/3: Basis für betriebswirtschaftliche Standardanwendungen*, 2nd edn Bonn: Addison Wesley.

Burow, L., J. Leyk, and C. Briem. 2014. Das Experten-Interview zum Thema „Controlling und Big Data". In *Der Controlling-Berater: Controlling und Big Data*, eds. R. Gleich, A. Klein, M. Kirchmann, and J. Leyk, Vol. 35, 13–20. Freiburg: Haufe.

Buschbacher, F., R. Konrad, B. Mußmann, and M. Weber. 2014. Big Data-Projekte: Vorgehen, Erfolgsfaktoren und Risiken. In *Der Controlling-Berater: Controlling und Big Data*, eds. R. Gleich, A. Klein, M. Kirchmann, and J. Leyk, Vol. 35, 83–106. Freiburg: Haufe.

Buxmann, P., and H. Schmidt. 2019. *Künstliche Intelligenz – Mit Algorithmen zum wirtschaftlichen Erfolg*. Berlin: SpringerGabler.

Caesar, D., and M. Friebel. 2011. *Microsoft SQL Server 2008 R2: Schnelleinstieg für Administratoren und Entwickler*. Bonn: Galileo Computing.

Chamoni, P., and P. Gluchowski. 2004. Integrationstrends bei Business-Intelligence-Systemen. *Wirtschaftsinformatik* 46(2):119–128.

Chamoni, P., and P. Gluchowski. 2017. Business analytics – State of the art. *Zeitschrift für Controlling Management Review* 61(4):8–17, 9.

Chamoni, P., et al. 1997. Online analytical processing (OLAP). In *Lexikon der Wirtschaftsinformatik*, ed. P. Mertens, 3rd edn Berlin: Springer.

Chamoni, P., F. Beekmann, and Bley. 2010. Ausgewählte Verfahren des Data Mining. In *Analytische Informationssysteme*, eds. P. Chamoni and P. Gluchowski, 4th edn Berlin: Springer.

Chen, H., R. H. Chiang, and V. C. Storey. 2012. Business intelligence und analytics – From big data to big impact. *MIS Quarterly* 36(4):1165–1188.

Cleve, J., and U. Lämmel. 2014. *Data mining*, 57–192. München: de Gruyter Oldenbourg.

Codd, E. F., S. B. Codd, and C. T. Salley. 1993. *Providing OLAP to user-analysts: An IT mandate*. Ann Arbor: Codd & Associates.

Dahnken, O., P. Keller, J. Narr, and C. Bange. 2003. *Software im Vergleich – Integrierte Unternehmensplanung*. München: Oxygon.

Dahnken, O., P. Keller, J. Narr, and C. Bange. 2004. *Planung und Budgetierung, 21 Software-Plattform zum Aufbau unternehmensweiter Planungsapplikationen*. München: Oxygon.

Data Academy, and R. Davenport. 2008. *White-paper: ETl vs ELT*. http://www.dataacademy.com/files/ETL-vs-ELT-White-Paper.pdf. Accessed 9 May 2015.

Davenport, T., and J. Kirby. 2016. Just how smart are machines? *MIT Sloan Management Review* 57(3):21–25.

Deloitte. 2019. *Deloitte center for process robotics. Get ready, the robots are here*. https://www2.deloitte.com/de/de/pages/innovation/contents/Center_for_Process_Robotics.html. Accessed 3 Dec 2019.

Determann, L., and M. Rey. 1999. Chancen und Grenzen des Data Mining im Controlling. *Controlling* 11(2):43–147.

Diamant Software. 2021. *Management-Dashboard und Isi-Sprachassistent mit Diamant/4*. Bielefeld: Diamant Software GmbH.

Dinnessen, F., and S. Halfmann 2018. *Wie Software-Roboter die öffentliche Verwaltung entlasten können*. https://www.detecon.com/de/wissen/wie-software-roboter-die-oeffentliche-verwaltung-entlasten-koennen. Accessed 9 Oct 2018.

Donie, P., and B. Raeburn. 2015. Mobiles Arbeiten – Verbindungsmacher. *Business Intelligence Magazin* 1:48–49.

Dorschel, J. 2015. *Praxisbuch Big Data*. Karlsruhe: Springer Gabler.

Dresner Advisory Services LLC. 2011. *Mobile business intelligence market study, 2010 und 2011*. http://www.microstrategy.com/mobile/mobile-bi-landscape-dresner.pdf. Accessed 25 July 2011.

Duisberg, A. 2011. Gelöste und ungelöste Rechtsfragen im IT-Outsourcing und Cloud Computing. In *Trust in IT – Wann vertrauen Sie ihr Geschäft der Internet-Cloud an?*, eds. A. Picot, U. Hertz, and T. Götz, 9–70. Berlin: Springer.

Düsing, R., and C. Heidsieck. 2009. Analysephase. In *Data-Warehouse-Systeme: Architektur, Entwicklung, Anwendung*, eds. A. Bauer and H. Günzel, 3rd edn, 104–127. Heidelberg: dpunkt.

Eckerson, W. 2007. *Predictive analytics: Extending the value of your data warehousing investment*. TDWI Best Practices Report, TDWI Research.

Edlich, S., et al. 2010. *NOSQL: Einstieg in die Welt nichtrationaler Web 2.0 Datenbanken*. München: Hanser.

Egger, N., J. M. Fiechter, C. Rohlf, J. Rose, and S. Weber. 2005. *SAP BW Planung und Simulation*. Bonn: Galileo Press.

Egger, N., Kästner Hastenrath, S. Kramer, et al. 2009. *Reporting und Analyse mit SAP Business Objects*. Bonn: Galileo Press.

Elmasri, R., and S. B. Navathe. 2007. *Fundamentals of database systems*, 5th edn. Bosten: Addison Wesley.

Engelbrecht, M., and M. Wegelin. 2015. *SAP Fiori: Implementierung und Entwicklung*. Bonn: SAP Press.

Engels, C. 2015. *Basiswissen business intelligence*, 2nd edn. Dortmund: W3I.

Ereth, J., and H.-G. Kemper. 2016. Business analytics und business intelligence. *Zeitschrift für Controlling* 28:458–464.

Ertel, W. 2016. *Grundkurs Künstliche Intelligenz – Eine praxisorientierte Einführung*, 4th edn. Wiesbaden: Springer Vieweg.

Farkisch, K. 2011. *Data-Warehouse-Systeme kompakt*. Heidelberg: Springer.

Fasel, D., and A. Meier, eds. 2016. *Big Data Grundlagen, Systeme und Nutzungspotenziale*. Wiesbaden: Springer Vieweg.

Feindt, B. J. 2014. *Die E-Bilanz in kleinen und mittleren Unternehmen (KMU)*. Wiesbaden: Springer Gabler.

Feindt, M., and D. Grüßling. 2014. Strategische Entscheidungen mit automatisierten Prognosen operativ umsetzen. In *Der Controlling-Berater: Controlling und Big Data*, eds. R. Gleich, A. Klein, M. Kirchmann, and J. Leyk, Vol. 35, 177–188. Freiburg: Haufe.

Felden, C. 2010. Predictive analytics. In *Analytische Informationssysteme*, eds. P. Chamoni and P. Gluchowski, 4th edn, 307–328. Berlin: Springer.

Felden, C. 2017. *Business Analytics, In Enzyklopädie der Wirtschaftsinformatik*, 1–8. http://www.enzyklopaedie-der-wirtschaftsinformatik.de/lexikon/daten-wissen/Business-Intelligence/Analytische-Informationssysteme%2D%2DMethoden-der-/Business-Analytics. Accessed 26 May 2017.

Finlay, S. 2014. *Predicitve analytics, data mining and big data: Myths, misconceptions and methods*. New York: Palgrave Macmillan.

Folkers, A. 2019. *Grundlagen des Deep Learning*. Wiesbaden: Springer Spektrum.

Forrester. 2016. *About Us. Forrester Research ist ein US-amerikanisches Marktforschungsunternehmen für den Bereich Informationstechnologie*. https://www.forrester.com/marketing/about/about-us.html. Accessed 18 July 2016.

Friedl, G. 2019. Künstliche Intelligenz im Controlling. *Zeitschrift für Controlling* 31(5):35–38.

Freiknecht, J. 2014. *Big Data in der Praxis – Lösungen mit Hadoop, HBase und Hive – Daten speichern, aufbereiten, visualisieren*. München: Hanser.

Frietsch, H. 2016. BI und Big Data in einer ausbalancierten Architektur. In *Business Intelligence erfolgreich umsetzen*, ed. M. Lang, 163–177. Düsseldorf: Symposion.

Frochte, J. 2019. *Maschinelles Lernen: Grundlagen und Algorithmen in Python*, 2nd edn. Hamburg: HANSER.

Fuchß, T. 2009. *Mobile Computing – Grundlagen und Konzepte für mobile Anwendungen*. Munich: Hanser.

Gabriel, R. 2010. Expertensystem. In *Enzyklopädie der Wirtschaftsinformatik – Online-Lexikon*, eds. K. Kurbel, J. Becker, N. Gronau, E. Sinz, and L. Suhl, 4th edn. http://www.oldenbourg.de:8080/wi-enzyklopaedie/lexikon/. Accessed 15 Jan 2011.

Gabriel, R., P. Chamoni, and P. Gluchowski. 2000. Data Warehouse und OLAP – Analyseorientierte Informationssysteme für das Management. *Zeitschrift für betriebswirtschaftliche Forschung* 52(2):74–93.

Gabriel, R., P. Gluchowski, and A. Pastwa. 2009. *Data warehouse & data mining*. Herdecke/Witten: W3I.

Gartner. 2015. *Gartner says solving ‚big data‘ challenge involves more than just managing volumes of data*. http://www.gartner.com/newsroom/id/1731916. Accessed 20 Sept 2015.

Gartner. 2016. *Business analytics*. http://www.gartner.com/it-glossary/business-analytics. Accessed 22 Oct 2016.

Gehra, B. 2005. *Früherkennung mit Business-Intelligence-Technologien. Anwendung und Wirtschaftlichkeit der Nutzung operativer Datenbestände*. Diss. Wiesbaden.

Geißner, W., and M. Wolfrum. 2015. Szenario-Analyse und Simulation: ein Fallbeispiel mit Excel und Crystal Ball. *Der Controlling-Berater* 17:241–264.

Gentsch, P. 2018. *Künstliche Intelligenz für Sales, Marketing und Service*. Wiesbaden: Springer.

Gentsch, P. 2019. *Künstliche Intelligenz für Sales, Marketing und Service*, 2nd edn. Wiesbaden: Springer Gabler.

Gesellschaft für Informatik, Dominik Klein, Phuoc Tran-Gia, und Matthias Hartmann. 2013. *Big data*. http://www.gi.de/nc/service/informatiklexikon/detailansicht/article/big-data.html. Accessed 9 Feb 2015.

Giegerich, H.-J. 2014. IT-Sicherheitsmanagement in der Cloud. *Controlling* 6:320–324.

Gleich, R. 2001. *Das System des Performance Measurement. Theoretisches Grundkonzept, Entwicklungs- und Anwendungsstand*. Munich: Vahlen.

Gleich, R., and M. Tschandl. 2018. *Digitalisierung & Controlling*. Freiburg: HAUFE.

Gluchowski, P. 2001. Business intelligence. *HDM – Praxis der Wirtschaftsinformatik* 38(222):5–15.

Gluchowski, P. 2010. Techniken und Werkzeuge zur Unterstützung des betrieblichen Berichtswesens. In *Analytische Informationssysteme*, eds. P. Chamoni and P. Gluchowski, 4th edn, 259–280. Berlin: Springer.

Gluchowski, P. 2016. Business Analytics – Grundlagen, Methoden und Einsatzpotenziale. *HMD Praxis der Wirtschaftsinformatik* 53(3):273–286.

Gluchowski, P., and P. Chamoni. 2016. *Analytische Informationssysteme: Business-Intelligence-Technologien und -Anwendungen*. Wiesbaden: Springer.

Gluchowski, P., R. Gabriel, and C. Dittmar. 2008. *Management support systeme und business intelligence*, 2nd edn. Berlin: Springer.

Goecken, M. 2006. *Entwicklung von Data-Warehouse-Systemen*. Wiesbaden: Vieweg+Teubner.

Hahne, M. 2005. *SAP business information warehouse*. Berlin: Springer.

Hahne, M. 2016. Architekturkonzepte und Modellierungsverfahren für BI-Systeme. In *Analytische Informationssysteme: Business Intelligence-Technologien und -Anwendungen*, eds. P. Gluchowski and P. Chamoni, Vol. 5, 147–184. Berlin: Springer.

Hammann, P., and B. Erichson. 2006. *Marktforschung*, 4th edn. Stuttgart: UTB.

Hammer, R. M. 1998. *Strategische Planung und Frühaufklärung*, 3rd edn. Munich: de Gruyter Oldenbourg.

Hanning, U., ed. 2008. *Vom data warehouse zum corporate performance management*. Ludwigshafen: Institut für managementinformationssysteme e. V.

Hansel, S. 2015. Langsamer Abschied vom Laisser-faire. *Wirtschaftswoche* 12:58–62.

Haupt, J. 2011. *SAP netweaver BW 7.30: LSA data flow templates series*. http://scn.sap.com/people/juergen.haupt/blog. Accessed 15 Mar 2015.

Hesseler, M. 2009. Costumizing von ERP-Systemen. *Zeitschrift für Controlling & Management* 53(Sonderheft 3):48–55.

Hesseler, M., and M. Görtz. 2007. *Basiswissen ERP-Systeme*. Witten: Herdecke.

Heuer, A., et al. 2001. *Datenbanksysteme in Büro, Technik und Wissenschaft*. 9. GI-Fachtagung Oldenburg, Berlin/Heidelberg, 7.–9. März 2001.

Homann, M., H. Wittges, and H. Krcmar. 2013. *Entwicklung mobiler Anwendung für SAP*. Bonn: Galileo Press.

Hortonworks. 2013. *Apache Hadoop Anwendungsmuster*. http://bigdatacongress.com/files/2013/09/Hortonworks-ApacheHadoopPatternsOfUse-v1-0_German.pdf. Accessed 5 Jan 2017.

Horváth, P. 2008. *Controlling*, 11th edn. Munich: Vahlen.

Hotho, A., A. Nürnberger, and G. Paaß. 2005. A brief survey of text mining. *Zeitschrift für Computerlinguistik und Sprachtechnologie* 20(1):19–62.

Humm, B., and F. Wietek. 2005. Architektur von Data Warehouses und Business Intelligence Systemen. *Informatik Spektrum* 23(2):3–14.

Hundertmark, S. 2020. *Chatbot Implementation – Ein paar hilfreiche Experten-Tipps für den Start*. https://morethandigital.info/chatbot-implementation-ein-paar-hilfreiche-experten-tipps-fuer-den-start/. Accessed 23 June 2020.

IBM Consulting Services. 2003. *SAP Berechtigungswesen*. Bonn: SAP Press.

IBM Institute for Business Value und Säid Business School. 2012. *Analytics: Big Data in der Praxis – Wie innovative Unternehmen ihre Datenbestände effektiv nutzen*. http://www935.ibm.com/services/de/gbs/thoughtleadership/download-bigdata.html. Accessed 12 Dec 2014.

IDC. 2011. *IVIEW, extracting value from chaos*. http://www.emc.com/collateral/analyst-reports/idc-extracting-value-from-chaos-ar.pdf. Accessed 9 Feb 2015.

Iffert, L. 2017. *Data Discovery – Definition und Marktüberblick*. 24.03.2017. Hannover. http://barc.de/events/downloadfile/id/XrVBGmwzckM54OF54Ff4. Accessed 3 July 2017.

Iffert, L, and A. Bange. 2018. *Datenvirtualisierung – Eine lohnenswerte Technologie für datengetriebene Unternehmen*. https://www.bi-scout.com/datenvirtualisierung. Accessed 5 Oct 2019.
Inform. 2017. *Qlik business intelligence*. https://www.inform-software.de/produkte/qlik-business-intelligence. Accessed 26 May 2017.
Institut für Business Intelligence. 2013. *Studie Competing on Analytics 2013: Herausforderungen – Potentiale und Wertbeiträge von Business Intelligence und Big Data*. Ludwigshafen: Steinbeis TransferinstitutInstitut für Business Intelligence (IBI).
de Intelligence. 2017. *Analytische Datenbanken*. https://www.intelligence.de/news/analytische-datenbanken-uberblick-zur-datenbanktechnologie.html. Accessed 19 July 2017.
Jedox. 2016. *Jedox social analytics showcase*. http://jedox-social-analytics.com/. Accessed 7 July 2016.
Jetter, W. 2004. *Performance management*, 2nd edn. Stuttgart: Schäffer-Poeschel.
Jordan, C., and D. Schnider. 2011. *Data warehousing mit oracle: Business intelligence in der praxis*. Munich: Hanser.
Jung, H. 2011. *Controlling*, 9th edn. Munich: de gruyter Oldenbourg.
Kaufmann, M. 2014. Die Geister die wir riefen. In *Big data*, eds. D. Fasel and A. Meier, 383–400. Wiesbaden: Springer Vieweg.
Keist, N.-K., S. Benisch, and C. Müller. 2016. Möglichkeiten und Grenzen der plattformübergreifenden App-Entwicklung. In *Mobile Anwendungen in Unternehmen*, eds. T. Barton, C. Müller, and C. Seel, 109–119. Wiesbaden: Springer Vieweg.
Kemper, H. G., H. Baars, and W. Mehanna. 2010. *Business Intelligence – Grundlagen und praktische Anwendungen*, 3rd edn. Wiesbaden: Springer Vieweg.
Kersten, H., and G. Klett. 2012. *Mobile device management*. Heidelberg: mitp.
Kessler, T., T. Hügens, F. Delgehausen, M. Abdel Hadi, and V. G. Saiz Castillo. 2014. *Reporting mit SAP BW® und SAP BusinessObjects™*, 2nd edn. Bonn: Galileo Press.
King, S. 2014. *Big Data – Potential und Barrieren der Nutzung im Unternehmenskontext*. Wiesbaden: Springer VS.
Klein, D., P. Tran-Gia, and M. Hartmann. 2013. *Big Data*, Würzburg. http://www.gi.de/nc/service/informatiklexikon/%20detailansicht/article/big-data.html. Accessed 6 Mar 2015.
Knapp, P., and F. Kusterer. 1996. Weltweit einsatzfähige Führungsinformationssysteme, Umsetzungen und Anforderungen. In *conference volume of the 11th Deutschen Controlling Congress*, eds. T. Reichmann, 219–244. Düsseldorf: Gesellschaft für Controlling e. V.
Knöll, H. D., C. Schulz-Sacharow, and M. Zimpe. 2006. *Unternehmensplanung mit SAP BI*. Wiesbaden: Vieweg+Teubner.
Koglin, U. 2016. *SAP S/4 Hana. Voraussetzung – Nutzen – Erfolgsfaktoren*. Bonn: Rheinwerk.
Kononenko, I., and M. Kukar. 2007. *Machine learning and data mining: Introduction to principles and algorithms*. Boca Raton: Horwood.
Kreutzer, R. T., and M. Sirrenberg. 2019. *Künstliche Intelligenz verstehen*. Wiesbaden: Springer.
Krishnan, K. 2013. *Data warehousing in the age of big data*. Waltham: Morgan Kaufmann – Elsevier Inc.
Krudewig, W. 2012. *E-Bilanz-gerecht kontieren und buchen*. Freiburg: Haufe.
Krüger, J. 2015. *SAP simple finance an introduction*. Bonn: Rheinwerk.
Krystek, U., and R. Moldenhauer. 2007. *Handbuch Krisen- und Restrukturierungsmanagement. Generelle Konzepte, Spezialprobleme, Praxisberichte*. Stuttgart: Kohlhammer.
Kumar, P., and A. Tiwari. 2017. *Ubiquitous machine learning and its applications*. Hershey: IGI Global.
Küsters, U. 2001. Data Mining Methoden: Einordnung und Überblick. In *Handbuch Data Mining im Marketing – Knowledge Discovery in Marketing Databases*, eds. H. Hippner, U. Küsters, M. Meyer, and K. D. Wilde, 95–130. Braunschweig: Vieweg Gabler.

Langmann, C. 2019a. *Digitalisierung im Controlling*. Wiesbaden: Springer-Gabler.
Langmann, C. 2019b. *Robotic Process Automation in Controlling und Rechnungswesen*. Beitrag vom 26.03.2019. https://www.management-circle.de/blog/robotic-process-automation-in-controlling-und-rechnungswesen/. Accessed 8 Nov 2019.
Langmann, C., and D. Turi. 2020. *Robotic Process Automation (RPA), Voraussetzungen, Funktionsweise und Implementierung am Beispiel des Controllings und Rechnungswesens (2020)*. Wiesbaden: Springer Gabler.
Lanquillon, C., and H. Mallow. 2015. Advanced analytics mit big data. In *Praxishandbuch Big Data*, ed. J. Dorchel, 55–89. Wiesbaden: Springer Gabler.
Laudon, K., and J. Laudon. 1988. *Management information systems*, 2nd edn. New York: Pearson.
Leiserson, C. E., R. L. Rivest, and C. Stein. 2010. *Algorithmen – Eine Einführung*. Munich: Oldenbourg.
Leßweng, H.-P. 2004. Einsatz von Business Intelligence Tools (BIT) im betrieblichen Berichtswesen. *Controlling* 1:41–49.
Lixenfeld, C. 2015. Tools für den Leistungscheck von BI-Systemen können nützlich sein. *Computerwoche* 11:24.
Lopez, M. D. 2009. *Successful mobile deployments require robust security*. Whitepaper. https://www.bitpipe.com/detail/RES/1242855514_161.html. Accessed 20 June 2017.
Lossau, N. 2018. *Jetzt wird das Internet 100 Mal schneller*. https://www.welt.de/wissenschaft/article176189726/Mobiles-Internet-mit-5G-10-000-Megabit-pro-Sekunde.html. Accessed 8 May 2018.
Louis, D., and P. Müller. 2013. *Jetzt lerne ich Android 4-Programmierung*. Munich: Markt + Technik.
Luber, S., and N. Litzel. 2017. *Was ist Grid Computing?* https://www.bigdata-insider.de/was-ist-grid-computing-a-629099/. Accessed 4 Oct 2019.
Luber, S., and N. Litzel. 2019. *Was ist Datenvirtualisierung*. https://www.bigdata-insider.de/was-ist-datenvirtualisierung-a-841015/. Accessed 4 Oct 2019.
Lubos, G. 2020. Künstliche Intelligenz im Controlling. *Controller Magazin* 1(2):45–49.
Manhart, K. 2011a. *Grundlagenserie Business Intelligence (Teil 1) Berichtssysteme: Grundtypen und Techniken*. http://www.tecchannel.de/server/sql/1751728/berichtssysteme_teil_1_grundtypen_und_techniken/index6.html. Accessed 28 June 2011.
Manhart, K. 2011b. *Business intelligence*. http://www.tecchannel.de/server/sql/1739205/business_intelligence_teil_2_datensammlung_und_data_warehouses/index8.html. Accessed 19 Feb 2011.
Martens, H. 2019. *In sieben Schritten zum erfolgreichem RPA-Projekt*. Beitrag vom 05.07.2018. https://www.computerwoche.de/a/in-sieben-schritten-zum-erfolgreichen-rpa-projekt,3545352. Accessed 2 Nov 2019.
Martin, W., and E. von Maur. 1997. Data warehouse. In *Lexikon der Wirtschaftsinformatik*, ed. P. Mertens, 3rd edn, 105–106. Berlin: Springer.
März, N., and J. Warren. 2015. *Big data – Principles and best practices of scalable realtime data systems*. Shelter Island: Manning.
Mathew, B. 2015. *Beginning SAP Fiori*. Bangalore: Apress.
McTear, M., Z. Callejas, and D. Griol. 2016. *The conversational interface – Talking to smart devices*. Cham.: Springer International.
Mehler, A., and C. Wolf. 2005. Einleitung: Perspektiven und Positionen des Text Mining. *Zeitschrift für Computerlinguistik und Sprachtechnologie* 20(1):1–18.
Meier, M., W. Sinzig, and P. Mertens. 2003. *SAP Strategic Enterprise Management™/Business Analytics – Integration von strategischer und operativer Unternehmensführung*, 2nd edn. Berlin: Springer.

Mell, P., and T. Grance. 2011. *The NIST definition of cloud computing*. Gaithersburg: National Institute of Standards and Technology.

Merkt, S., H. A. Müller, and J. Tscherkaschina. 2015. *SAP BusinessObjects Design Studio – Das Praxishandbuch*. Bonn: SAP Press.

Mertens, P. 2002. Business Intelligence – Ein Überblick, Arbeitspapier Nr. 2/2002, Universität Erlangen-Nürnberg. http://www.wi1-mertens.wiso.uni-erlangen.de/veroeffentlichungen/download/Business_Intelligence-ein_Ueberblick_Arbeitspapier_der_Universitaet_Erlangen-Nuernberg.zip. Accessed 23 July 2011.

Mertens, P., and D. Barbian. 2019. Erreicht Künstliche Intelligenz auch das Controlling? *Controlling & Management Review* 63:8–17.

Mertens, P., and J. Griese. 1988. Informations- und Kontrollsysteme. In *Industrielle Datenverarbeitung*, eds. P. Mertens and J. Griese, 2nd edn. Wiesbaden: Betriebswirtschaftlicher Verlag Dr. Th. Gabler.

Mertens, P., and J. Griese. 2002. *Integrierte Informationsverarbeitung 2. Planungs- und Kontrollsysteme in der Industrie*, 9th edn. Wiesbaden: Springer.

Merz, M., T. Hügens, and S. Blum. 2015. *SAP BW auf SAP – HANA: Implementierung und Migration*. Bonn: Galileo Press.

Mohr, M. 2006. *HCC-Einführungsschulung zum SAP Business Information Warehouse – Grundlagen, Reporting und Analyse, Modellierung und Staging*. Munich: SAP University Competence Center at the Technical University of Munich at the Chair of Information Systems - Prof. Dr. H. Krcmar.

Möller, K., F. Federmann, S. Pieper, and M. Knezevic. 2016. Predictive Analytics zur kurzfristigen Umsatzprognose. *Zeitschrift für Controlling* 28(8–9):509–518.

Mori, H., A. Yasuda, and M. Araki. 2017. An evaluation method for system response in chat-oriented dialogue system. In *Dialogues with social robots: Enablements, analyses, and evaluation*, eds. K. Jokinen and G. Wilcock. Singapore: Springer.

Mucksch, H., and W. Behme. 2000. Das Data Warehouse-Konzept als Basis einer unternehmensweiten Informationslogistik. In *Das Data Warehouse-Konzept*, eds. H. Mucksch and W. Behme, 4th edn, 3–80. Wiesbaden: Gabler.

Müller, S. 2014. Die neue Realität – Erweiterung des Data Warehouse um Hadoop, NoSql & Co. In *Big data*, eds. D. Fasel and A. Meier, 447–457. Wiesbaden: Springer Vieweg.

Müller, S., and C. Keller. 2015. *Pentaho und Jedox*. München: Hanser.

Navrade, F. 2008. *Strategische Planung mit Data-Warehouse-Systemen*. Wiesbaden: Springer Gabler.

NN. 2020. Notebook-Typen im Überblick. *IMTEST* 3(2020):9.

NN/w3resource. 2015. *NoSQL*. http://www.w3resource.com/mongodb/nosql.php. Accessed 28 Jan 2015.

Obermaier, R. 2019. *Handbuch Industrie 4.0 und Digitale Transformation*. Wiesbaden: Springer Gabler.

Oehler, K. 2000. *OLAP, Grundlagen, Modellierung und betriebswirtschaftliche Lösung*. Munich: Hanser.

Oehler, K. 2006. *Corporate performance management*. Munich: Hanser.

Oehler, K. 2020. Ist maschinelles Lernen in der Planung einsetzbar? *Controlling & Management Review* H3:22–30.

Oppelt, R. U. G. 1995. *Computerunterstützung für das Management*. Munich: Oldenbourg.

Pariser. 2011. *Eli Pariser: The filter bubble: What the Internet is hiding from you*. New York: Penguin Press.

Pastwa, A. 2010. *Serviceorientierung im betrieblichen Berichtswesen. Entwicklung eines Architektur- und Vorgehensmodells zur konzeptionellen Gestaltung von Berichtsprozessen auf Basis einer SOA und XBRL.* Frankfurt a. M.: Lang.

Pends, N., and R. Creeth. 1995. *The OLAP report. Market share analysis.* http://www.olapreport.com/market.htm. Accessed 28 Jan 1999.

Peper, I. 2018. Effizienzsteigerungen im Controlling durch Robot Process Automation – Eine Analyse am Beispiel des Monatsabschlussprozesses der Kathrein Automotive Group. *Controlling* 30(3):20–27.

Petersohn, H. 2005. *Data Mining – Verfahren, Prozesse, Anwendungsarchitektur.* Munich: Oldenburg.

Plattner, H. 2011. *In-memory data management: An inflection point for enterprise applications.* Heidelberg: Springer.

Portal, F. 2020. Wie Automatisierung und maschinelles Lernen die Finanzfunktion neu gestalten. *Controller Magazin* 3(4):69–72.

Pütter, C. 2011. *Keine Hilfe für BI-Projekte.* http://www.cio.de/2239428. Accessed 12 Mar 2011.

QlikTech. 2016. *Predictive analytics with qlikview and qlik sense.* http://go.qlik.com/PredictiveanalyticsQ2_Registration_LP.html. Accessed 25 May 2016.

QlikTech. 2017. *Qlikview governance dashboard.* http://global.qlik.com/at//~/media/Files/resource-library/global-us/direct/da-tasheets/DS-Practical-Data-Governance-As-You-Deploy-EN.ashx. Accessed 7 June 2017.

QlikView. 2011. *Qlikview-demo sales analysis.* http://demo.qlikview.com/QvAJAXZfc/opendoc.htm?document=Executive%20Dashboard.qvw&host=Demo10&anonymous=true. Accessed 20 July 2011.

R. 2016. *Statistische Datenanalyse mit R.* http://www.r-statistik.de/. Accessed 7 July 2016.

Reichmann, T. 2006. *Controlling mit Kennzahlen und Managementberichten*, 7th edn. Munich: Vahlen.

Reichmann, T. 2011. *Controlling mit Kennzahlen, Die systemgestützte Controlling-Konzeption mit Analyse- und Reportinginstrumenten*, 8th edn. Munich: Vahlen.

Reinsel, D., J. Gantz, and J. Ryding. 2018. *The digitization of the world – From edge to core.* https://www.seagate.com/files/www-content/our-story/trends/files/idc-seagate-dataage-whitepaper.pdf. Accessed 20 Aug 2020.

Richter, M. 2003. Fallbasiertes Schließen. In *Handbuch der Künstlichen Intelligenz*, eds. G. Görz, C. R. Rollinger, and J. Schneeberger, 4th edn, 407–430. Munich: Oldenbourg.

Rogers, H. 1967. *Theory of recursive functions and effective computability.* New York: McGraw-Hill.

Rouse, M. 2014. *MapReduce.* https://www.computerweekly.com/de/definition/MapReduce. Accessed 15 Aug 2019.

Ruf, R., and W. Schwab. 2016. Visual analytics. *Controlling* 28(8/9):495–501.

Runkler, T. A. 2010. *Data Mining – Methoden und Algorithmen intelligenter Datenanalyse., 96.* Wiesbaden: Springer Vieweg.

Ruprecht, J. 2003. Zugriffskontrolle im Data Warehouse. In *Data warehouse management*, eds. E. von Maur and R. Winter, 113–147. Berlin: Springer.

Sack, R. 2013. *Big Data – Strategischer Vorteil im internationalen Wettbewerb.* http://www.wi.hswismar.de/~laemmel/Lehre/WA/Artikel1306/Sack_BigData.pdf. Accessed 16 Dec 2014.

SAP AG. 2003. *SAP SEM, Scorecard-Analyse und Ursache-Wirkungs-Analyse.* Screenshot aus dem System SAP SEM. Walldorf: SAP AG.

SAP AG. 2008. *SAP BEx Analyser. Screenshot aus dem System SAP BEx Analyser.* Walldorf: SAP AG.

SAP AG. 2011. *SAP ERP CC System – Standardbericht. Selektionsmaske und Planungsformular aus einem IDES-System*. Walldorf: SAP AG.

SAP AG. 2016. *SAP businessobjects mobile for iOS – Document version: 6.3.10 – 2016-02-09*. Walldorf: SAP AG.

SAP Fiori. 2017a. *Design guideline*. https://experience.sap.com/fiori-design-web/home-page/. Accessed 28 May 2017.

SAP Fiori. 2017b. *Integration von businessObjects-tools*. https://blogs.sap.com/2016/04/13/integrating-sap-fiori-and-sap-businessobjects-bi/. Accessed 22 May 2017.

SAP SE. 2013. *SAP Fiori UX*. http://help.sap.com/fiori_bs2013/helpdata/de/d7/8c41524b66b470e10000000a423f68/content.htm?current_toc=/de/84/154353a7ace547e10000000a441470/plain.htm%26show_children=true. Accessed 16 Dec 2014.

SAP-Portal. 2009. *SAP-Portal*. http://help.sap.com/saphelp_sem60/helpdata/de/5d/7c4b52691011d4b2f00050dadfb23f/frameset.htm. Accessed 26 Jan 2009.

SAP-Portal. 2010. *SAP-Portal*. http://help.sap.com/saphelp_sem60/helpdata/de/39/100c38e15711d4b2d90050da4c74dc/frameset.htm. Accessed 20 July 2010.

Sauer, K., and J. Sturm. 2019. Digitale Assistenten in der Unternehmenssteuerung? *Controlling & Management Review* 63(4):34–44.

Schäfer, M. 2014. SAP beschleunigt die Sortimentsoptimierung. *IS-Report* 8:22–23.

Scheer, A. W. 1990. *EDV-orientierte Betriebswirtschaftslehre*, 4th edn. Berlin: Springer.

Scheuer, D. 2020. *Akzeptanz von Künstlicher Intelligenz*. Wiesbaden: Springer Vieweg.

Schill, A., and T. Springer. 2007. *Verteilte Systeme – Grundlagen und Basistechnologien*. Berlin: Springer.

Schinzer, H. D. 1996. *Entscheidungsorientierte Informationssysteme. Grundlagen, Anforderungen, Konzept, Umsetzung*. Wiesbaden: Springer.

Schinzer, H. D., C. Bange, and H. Mertens. 1999. *Data Warehouse und Data Mining, Marktführende Produkte im Vergleich*, 2nd edn. Munich: Vahlen.

Schinzer, H., C. Bange, and H. Mertens. 2000. Wachstum, Trends und gute Produkte – Neue BARC-Studie zum OLAP- und Business Intelligence-Markt. *is report* 1(4):10–17.

Schmitz, U. 2015. Nutzung von In-Memory-Technologie in der BI. In *Handbuch Business Intelligence – Potenziale, Strategien, Best Practices*, ed. M. Lang, 233–247. Düsseldorf: Symposion.

Schmitz, Robert, BI Scout: BI Trends 2019: Integrating Artificial Intelligence in Analytics Scenarios, https://www.bi-scout.com/bi-trends-2019-integration-vonkuenstlicher-intelligenz-in-analytics-szenarien (2018).

Schneider Steffen, Reporting Pulse Check.—What topics move the experts from Controlling & Finance?, https://www.haufe.de/controlling/controllerpraxis/reportingpulse-check-das-bewegt-finanzexperten_112_471720.html (2021).

Schön, D. 2004. Moderne Planungskonzepte und Reportingtools. In *conference volume of the 19th Deutscher Controlling Congress*, ed. T. Reichmann, 287–337. Dortmund: Gesellschaft für Controlling e. V.

Schön, D. 2011. *Ergebnisse zur empirischen Untersuchung: Business Intelligence für Reporting und Planung im Mittelstand*. The complete results of the study are available for download via the following link. http://www.fh-dortmund.de/de/studi/fb/9/personen/lehr/schdie/103020100000206873.php. Accessed 1 June 2011.

Schön, D., and R. Müller. 2010. Mittelstandscontrolling für Inhaber und Manager. In *conference volume of the 25th Deutscher Controlling Congress*, ed. T. Reichmann, 123–165. Dortmund: Gesellschaft für Controlling e. V.

Schön, D., and M. Pook. 2015. Bedarfsgerechte Steuerung durch IT-gestütztes Berichtswesen. *innovative Verwaltung* 37(10):10–13.

Schön, D., M. Ständer, and M. Liebe. 2013. Unternehmens- und Projektsteuerung mit Hilfe eines BI-gestützten Controlling-Cockpits am Beispiel der Securiton GmbH. *Zeitschrift für Controlling* 25(4/5):252–260.

Schonschek, O., and J. Haas 2020. *Was Unternehmen über Chatbots wissen müssen*. https://www.computerwoche.de/a/was-unternehmen-ueber-chatbots-wissen-muessen,3329735. Accessed 14 Sept 2020.

Schrödl, H. 2009. *Business Intelligence mit Microsoft SQL Server 2008*, 2nd edn. München: Hanser.

Schroeck, M. R., R. Shockley, J. Smart, D. Romero-Morales, and P. Tufano. 2015. *Analytics: Big Data in der Praxis*, ed. IBM Institute for Business Value in cooperation with the Saïd Business School at the Universität Oxford, 1–20. Ehningen: IBM.

Schubert, S. 2013. Why people and process matter, in addition to great technology. *Predictive Analytics*. http://blogs.sas.com/content/subconsciousmusings/2013/01/11/why-people-and-process-matter-in-addition-to-great-technology-in-predictive-analytics/. Accessed 9 May 2015.

Search Business Analytics. 2016. *embedded BI*. http://searchbusinessanalytics.techtarget.com/definition/em-bedded-BI-embedded-business-intelligence. Accessed 15 June 2016.

Seiter, M. 2017. *Business Analytics: Effektive Nutzung fortschrittlicher Algorithmen in der Unternehmenssteuerung*. Munich: Vahlen.

Sellmair, R., M. Demary, A. Hierlemann, and B. Mutschler 2019. *Robotic process automation*. Beitrag vom 11.04.2019. https://ereignishorizont-digitalisierung.de/intelligente-maschinen/robotic-process-automation/. Accessed 2 Nov 2019.

Seufert, A. 2014. Das Controlling als Business Partner: Business Intelligence & Big Data als zentrales Aufgabenfeld. In *Der Controlling-Berater: Controlling und Big Data*, eds. R. Gleich, A. Klein, M. Kirchmann, and J. Leyk, Vol. 35, 23–45. Freiburg: Haufe.

Seufert, A., and K. Oehler. 2009. Grundlagen Business Intelligence. In *Business intelligence & controlling competence*, Vol 1. Stuttgart: Steinbeis-Edition.

Shearer, C. 2020. The CRISP-DM model: The new blueprint for data mining. *Journal of Data Warehousing* 5:13–22.

Shortliffe, E. H. 1976. *Computer-based medical consultations: MYCIN*. New York: Elsevier.

Siegel, E. 2013. *Predictive analytics, the power to predict who will click, buy, lie, or die*. Hoboken: Wiley.

Sinz, J. E., and A. Ulbrich vom Ende. 2010. Architektur von Data-Warehouse-Systemen. In *Analytische Informationssysteme*, eds. P. Chamoni and P. Gluchowski, 4th edn, 175–196. Heidelberg: Springer.

Smeets, M., R. Erhard, and T. Kaußler. 2019. *Robotic Process Automation (RPA) in der Finanzwirtschaft. Technologie – Implementierung – Erfolgsfaktoren für Entscheider und Anwender*, 8th edn. Wiesbaden: Springer Gabler.

Sperner, L. 2020. *Warum Expertensysteme nicht als Teilgebiet von Künstlicher Intelligenz*. Innovationsblog de DZ Bank Gruppe. https://innovationsblog.dzbank.de/2017/10/18/warum-expertensysteme-nicht-als-teilgebiet-von-kuenstlicher-intelligenz-wahrgenommen-werden/. Accessed 15 July 2020.

Spitzner, J., and M. L. Schneider. 2015. *Simulation in der Unternehmenssteuerung: Studienergebnisse*. Munich: Spitzner Consulting GmbH.

Stephan, S. 2020. *Die Arten von Chatbots*. https://cloud-blog.arvato.com/die-arten-von-chatbots/. Accessed 16 Sept 2020.

Taschner, A. 2013. *Management Reporting. Erfolgsfaktor internes Berichtswesen*. Wiesbaden: Springer Gabler.

TECChannel – IT im Mittelstand. 2014. *Datenmengen explodieren durch Sensordaten*. http://www.tecchannel.de/storage/news/2056615/datenmengen_explodieren_durch_sensordaten/. Accessed 16 Dec 2014.

Tripathi, A. M. 2018. *Learning robotic process automation – Create software robots and automate business processes with the leading RPA tool UiPath*. Birmingham: Packt.

Turban, E., J. E. Aronson, and T. P. Liang. 2004. *Decission support and intelligent systems*, 7th edn. Upper Saddle River: Prentice-Hall, Inc.

University of Helsinki and Reaktor. 2020. *The Elements of All, Onlinekurs*. https://course.elementsofai.com/de/. Accessed 18 June 2020. https://innovationsblog.dzbank.de/2017/10/18/warum-expertensysteme-nicht-als-teilgebiet-von-kuenstlicher-intelligenz-wahrgenommen-werden/. Accessed 8 Sept 2020.

Vaduva, A., and T. Vetterli. 2001. Metadata management for data warehousing: An overview. *International Journal of Cooperative Information Systems* 10(3):273–298.

Walker-Morgan, D. 2010. *NoSQL im Überblick*. http://www.heise.de/open/artikel/NoSQL-im-Ueberblick-1012483.html. Accessed 14 Dec 2014.

Warner, D. 2007. *SQL für Praxis und Studium*. Poing: Neubert.

Wartala, R. 2012. *Hadoop: zuverlässige, verteilte und skalierbare Big-Data-Anwendungen*. Munich: Open Source Press.

Weber, J. 2013. Verhaltensorientiertes Controlling. Plädoyer für eine (nicht) ganz neue Sicht auf das Controlling. *Zeitschrift für Controlling* 25(4/5):217–222.

Weber, H. W., and U. Strüngmann. 1997. Data Warehouse und Controlling – Eine vielversprechende Partnerschaft. *Controlling* 1:30–36.

Weigend, A. 2017. Data for the people: Das implizite explizit machen! *mobile Business* 4(5–6):16–19.

Welker, P. 2015. Big Data versus Data Warehouse Koexistenzperspektive. *Business Intelligence Magazin1* 1:42–44.

Werner Roland, PwC Germany: Reporting 5.0, https://www.pwc.de/de/pressemitteilungen/2018/pwclauncht-reporting-5-0.html (2018).

Werner, Roland. PwC: Reporting 5.0, https://www.pwc.de/de/im-fokus/finance-transformation/future-ofsteering/reporting-5-0-ki-revolutioniert-das-reporting.html (2021).

Wild, J. 1981. *Grundlagen der Unternehmensplanung*, 3rd edn. Opladen: Westdeutscher.

Winter, J. 2018. Künstliche Intelligenz und datenbasierte Geschäftsmodellinnovationen – Warum Unternehmen jetzt handeln sollten. In *Service business development*, eds. M. Bruhn and K. Hadwich, Vol. 1, 60–79. Wiesbaden: Springer Gabler.

Winterstein, A., and E. Leitner. 1998. An Informationen nicht ersticken. *Client Server Computing* 3:34–39.

Wullenkord, A. 2018. Künstliche Intelligenz im Rechnungswesen und Controlling. In *Erfolgreiches Controlling – Theorie, Praxis und Perspektiven; Prof. Dr. Thomas Reichmann zum 80*, eds. A. Wiesehahn and M. Kissler, 113–129. Baden-Baden: Nomos.

Wuttke, L. 2020. *Machine Learning: Definition, Algorithmen, Methoden und Beispiele*. https://datasolut.com/was-ist-machine-learning/. Accessed 13 July 2020.

Zarinac, T. 2016. Technologien, Architekturen und Prozesse. In *Agile business intelligence*, eds. S. Trahasch and M. Zimmer, 131–142. Heidelberg: dpunkt.

Summary and Outlook

6

Contents

Based on the four analysis fields of the research framework, the most important results and future trends for planning and reporting in BI-supported controlling will now be summarized. The most important future topics are listed in Table 6.1 in bullet points:

Technical content

- **Integrated planning and reporting in the BI-supported controlling cockpit**
  Planning and reporting will increasingly grow together and no longer be viewed in isolation from each other. As part of the integrated planning, **strategic, medium-term, operational and anticipating planning tasks** are interlinked. As a tool for strategic integration, the **Balanced Scorecard** or the extended Balanced Chance and Risk Card will prevail, in which especially the planning and control of strategic projects in relation to the medium-term and operational impact on results play an important role. The operational plans are continuously updated by **forecasts or rolling plans**, so that it will be possible to base dispositive decisions for resource management on current plan values. The planning areas in an integrated planning calculate modeled dependencies of the individual planning areas, e.g. starting from the sales planning to the balance sheet and financial planning. For example, payment terms (discount granting and payment behavior of customers) and payment dates (installments) as well as tax payments (sales tax liability) are taken into account for the financial planning. Furthermore, more and more important success drivers for the company are recognized, with which more accurate forecasts, simulations and

D. Schön, *Planning and Reporting in BI-supported Controlling*,
https://doi.org/10.1007/978-3-658-41044-5_6

Table 6.1 Future development trends in planning and reporting

Technical content: – Integration of planning and reporting in the BI-supported controlling cockpit – Technical concepts – Expansion of planning and analysis functions – Standard solutions and mapping of individual requirements	**IT support**: – Powerful frontends – Data discovery/data visualization – Big data, data mining or predictive analytics, hybrid traditional and exploratory BI – AI-supported analytics and chatbots – Mobile systems – Cloud computing and IT security – Self-service BI – Embedded BI and BI portal integration – BI monitoring or BI on BI – Open-source products
Organization: – Strengthening the organization for planning and reporting – Efficient authorization and access systems – Target group-oriented preparation – Distribution and collaboration	**Processes**: – General process support – Data quality check and control – Bilateral integration of BI and ERP planning and reporting processes – Alternative and circular processes – Substitution of manual report creation processes by RPA

sensitivity analyses can be determined, which are the basis for successful recommendations and measures. Fig. 6.1 shows a "**value driver-based planning**" of personnel costs, where the total personnel costs can be simulated and forecasted by changing cost-driving factors, such as new hires and personnel cost increases. Such and similar value driver models will become even more powerful in the future as AI-supported models can improve both the success drivers as input factors and the impact models in terms of output. In the personnel example, the relationship between the click rate in e-recruitment and an improved hiring process in terms of qualitative selection as well as faster contract closures for new employees could be recognized and applied.

Due to the model extensions and the higher usage intensity of the data, better scenario analyses are also possible. These include, for example, scenario planning for worst-, middle- and best-case scenarios or gap analyses to close a strategic gap as a guide for management decisions.

The reporting system serves, on the one hand, as a planning basis (e.g. via orientation values) and, on the other hand, as an **analysis and control medium for target achievement** and, if necessary, for planning and course correction. Analysis and planning functions are not performed separately from each other but more and more together and continuously.

All planning and reporting areas are ideally summarized in a BI-supported controlling cockpit. It forms the center for all important control information for management

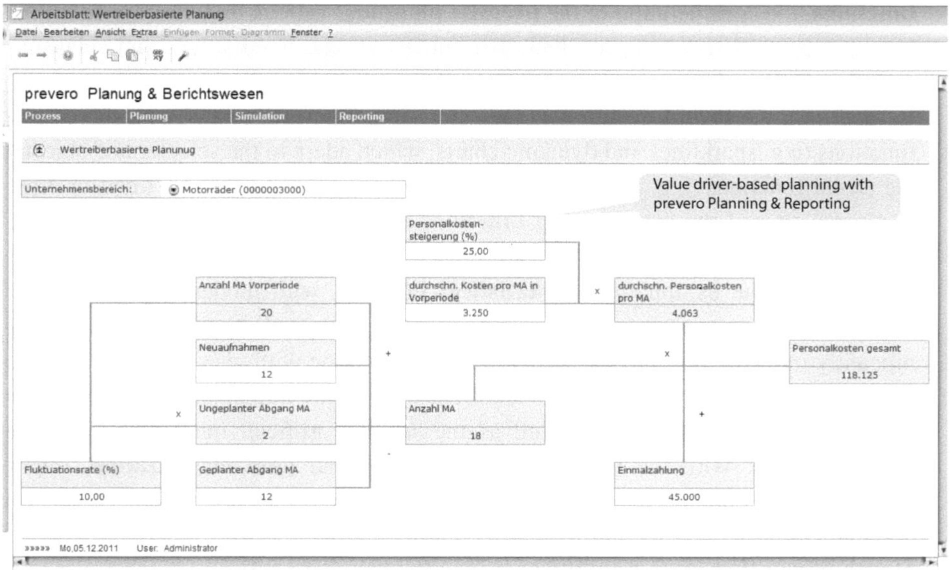

Fig. 6.1 Value driver-based planning with prevero planning & reporting (Taken from the presentation slides of Hein 2011)

decisions at all levels of management. It will also increasingly become an early warning system to avoid risks and exploit opportunities for the company.

- **Technical concepts**

  Due to the increasing complexity of control and the management orientation in the corporate management, planning and reporting systems become an important control instrument. The executives in the company will be able to react faster to market changes, recognize trends and align their company to the new requirements at an early stage. Competitive advantages are achieved by using BI-supported controlling systems. Historically grown planning and reporting solutions should be revised from scratch to eliminate content and other deficiencies. For this purpose, technical concepts will be necessary which consider the future control concept of the company holistically. The content of the technical concept determines the technical implementation. The range of technical concepts extends from the operational to the strategic controlling modules, starting from the start cockpits - with their key performance indicators up to the basic variables and indicators of the detailed reports (see Sect. 3.9). The technical concepts form the basis for the subsequent technical implementation concepts as well as for checking the ongoing data quality up to the individual document level of the upstream systems.
- **Expansion of planning and analysis functions**

  To support the executives, a further expansion of the planning and analysis functions is required. In the planning tasks, the main features are **simulation and scenario**

functions, the consideration of the **premise planning**, various **distribution functions** (e.g. season shift regulator, residual distributions around fixed values and splashing) and **plausibility aids** (e.g. reconciliation reports, validation rules).

In reporting, the **layout design** becomes more important, especially with the graphic functions (e.g. sparklines and dynamic charts, which adapt to the selection of objects), the signal functions (e.g. alerts, colour-coding, conditional formatting, trend or direction arrows)[1] and the **commenting functions** (e.g. comment reports, aggregation of comments, audio and video comments).

Furthermore, the exploitation of **statistical methods** with modern forecasting and trend calculations as well as the use of **data mining or predictive analytics** will significantly improve the analysis quality of the information with regard to their interpretation of data patterns, relationships and trends. Here, the integration and exploitation of the **big data technology** as well as the specific artificial intelligence plays an increasing role. With the help of **data discovery** and **data visualization**, important economic relationships and trends are identified, which might otherwise have remained hidden.

- **Standard solutions and mapping of individual requirements**

  To save costs, development and implementation time, standard solutions are used in areas where many companies have similar interests, especially in the medium-sized sector. Examples of establishing standard solutions are:
  - integration of profit, balance sheet, income statement and financial planning as well as consolidation
  - Balanced Scorecard and risk management
  - sales and revenue planning as well as derivation of the downstream resource plans
  - single project and multi-project planning
  - performance measurement systems such as the ROI tree or performance measurement systems for value-oriented corporate management
  - industry solutions

  So far, only solution providers that cover individual areas of responsibility can be found on the software market. Especially in the area of profit, financial and balance sheet planning as well as partly in consolidation, there are suitable solutions. Standard solutions are less common in the upstream operational planning areas, e.g. for sales and resource planning or purchasing planning.

  However, standard solutions can never be so pronounced that they cover all individual control requirements of the companies. Therefore, the systems must be open and flexible to map the individual requirements of the respective company. It is to be expected that the standard solutions will expand further and that sub-areas will be more integrated than before. Here, templates for the individual reporting and planning areas must be provided, which can be individually adapted.

[1] See Unrein (2016, p. 217 f.).

Organization

- **Strengthening the organization for planning and reporting**
  To enforce efficient planning and reporting systems, the institutionalization and organizational anchoring will have to be strengthened.
  The **controlling** assumes the central technical coordination. In the **IT**, in addition to the previous knowledge of the data source systems (e.g. ERP systems) and their connections, now also advanced knowledge is necessary in the fields of data warehousing, business intelligence, big data, and possibly RPA, artificial intelligence and chatbots. In addition, there are the administration tasks in mobile computing with the help of a mobile device management and, in the case of distributed data storage, cloud computing. Controlling and IT staff establish a **business intelligence organization**, which works together with the management and the departments on the continuous development of the planning and reporting system. Instead of setting up a separate BI organization with a BI competence center, it is advisable, in my opinion, for medium-sized and also for larger companies, to use a BI team with representatives from the areas of controlling, IT and the departments under the leadership of controlling (see Sect. 4.1.5.3).
- **Efficient authorization and access systems**
  Decision support and the prevention of data misuse are ensured by efficient authorization and access systems. Role-based assignments of functions and object values must become easier. Security gaps in the control system are to be consistently closed. Mobile and local device management systems help, for example, to block persons and devices with regard to data misuse.
- **Target group-oriented preparation**
  The results of planning and reporting are relevant today for many internal and external target groups, so the support of a target group-related assignment in terms of level of detail and scope of information is necessary. For this purpose, suitable tools such as navigation systems and storyboards are to be provided (see Sect. 4.2.3.3). With the help of chatbot technology, e.g. by voice recognition, the pull reporting from the user's perspective will improve as preferences of information requests are recognized and thus the information delivery can be designed in a more targeted and compact manner.
- **Distribution and collaboration**
  The target-oriented distribution of planning tasks and reports via automated assignment and dispatch functions (push reporting) and the retrieval of provided information (pull reporting) will gain in importance, via the central controlling cockpit or via further specialized planning and reporting servers (e.g. via management query environments), via portal solutions or via document and content management systems. As already mentioned in the context of target-oriented preparation, chatbot technology, e.g. through speech recognition, will intensify pull reporting.
  In the organization of planning and reporting, further functions for better collaboration are to be expected. These include scheduling, reserving analysis meetings, chats,

etc. regarding location, participants and the technical equipment and documents to be provided. This applies internally within the company but also across companies if, e.g., national or international business partners are to be involved. Presentation and communication possibilities within companies, in social networks and across companies with other collaboration systems will continue to establish themselves.

Processes

- **General process support**
  The entire planning and reporting process is currently still little supported by suitable instruments. Here, **workflow, status and tracking systems** could gain in importance. A system-supported workflow graphically displays the individual subtasks of the entire planning process. An integrated usable planning calendar schedules and coordinates the subtasks. Activity and status reports help to control and monitor the planning and reporting process.
- **Data quality check and control**
  Every planning and reporting system is only as good as the data quality of the system. The data quality is to be checked and improved across the board. Problems with data quality arise throughout the data processing process. They often start already at data entry and range from transformation to data analysis. The data quality criteria (completeness, correctness, relevance, comprehensibility, consistency, uniqueness, timeliness, etc.) are generally to be met for IT, especially for business intelligence and big data. Causes for poor data quality, such as incorrect postings, data redundancies, duplicates or no time for data maintenance (lack of resources), are to be remedied. In order to increase data quality sustainably, a data quality management is to be implemented that continuously checks and corrects data quality. Here, modern tools of data quality assurance such as data catalogs, data profiling, data cleansing as well as key performance indicators and reports for measuring data quality in BI monitoring are to be used. Furthermore, rules for data sovereignty and standards for data maintenance (e.g. in master data management) are to be established.
- **Bilateral integration of BI and ERP planning and reporting processes**
  Due to the different data management concepts in BI and ERP systems, a collaboration is advisable. This was already shown in the ETL process in the direction of the BI systems (see Sect. 5.5.2), but also applies vice versa. This is especially the case for computationally intensive planning and reporting tasks, such as the resolution of the quantity and value structures in the MRP runs or the internal cost allocation. These computationally intensive programs can be left in the ERP system, instead of being replicated in the BI system. In this case, an integration of BI and ERP processes is advisable in order to exploit the advantages of both system worlds. In addition, the ERP systems may need the results of the values achieved in the BI planning in order to have a complete data basis (e.g. for the transfer price determination in the cost center accounting and valuation in the calculation).

- **Alternative and circular processes**
  To simplify the planning, graded or alternative planning processes are useful, in which important information is planned in detail analytically and less important information is simplified, e.g. extrapolated from the past. For this purpose, the systems must offer graded planning processes, which can be used optionally.
  Certain planning processes take place circularly. For example, guidelines from the management level are broken down to the detail areas. The decentralized units then plan their services bottom-up. The results have to be checked and verified. There are correction needs and new planning rounds until a final **release and approval** of the budget or the forecast is reached. These quality assurance and approval processes need better support, which provide, among other things, plausibility aids and analysis functions regarding the deviations and approval process aids.
- **Substitution of manual reporting processes by RPA**
  With the help of RPA technology, inefficient manual reporting and maintenance processes can be replaced. This intermediate technology is only interesting for companies that have little automation in planning and reporting so far and mostly use very heterogeneous software products. In this case, a rationalization by RPA is worthwhile for frequently recurring, standardizable and rule-based processes which were previously performed by users doing manual, labor-intensive and monotonous work on several different software interfaces. If a company already uses a BI-supported controlling cockpit, the analysis and planning processes, such as report creation and distribution, are already automated, so that RPA is no longer necessary.

IT support

- **Powerful frontends**
  The trend for frontends is moving away from vendor-specific (proprietary) to web-based frontends. Cockpits and dashboards can be easily accessed via web browsers without the need to install additional graphical user interfaces. With some software vendors, the proprietary frontends are still more powerful than the web frontends and have increased functionality. In addition, there are still some software vendors that use an integrated Excel interface as an Excel-add-in solution for planning and analysis.
- **Data discovery/data visualization**
  With the help of **data discovery**, external data is evaluated in addition to the structured data stock of the DWH within the framework of exploratory BI in an analysis process. The aim is to identify business-relevant relationships and make better forecasts. The BI systems use statistical models and algorithms with the support of big data technology, artificial intelligence, data mining and predictive analytics (see following points). With the help of **data visualization**, the users have graphic types at their disposal that can better visualize these data relationships and weights. In addition to the classic chart types (e.g., pie, column and line charts), these are mainly

modern chart types that visualize classifications and statistical relationships (e.g., with sunburst and treemap charts or histograms). Technologically, these software solutions use the in-memory technique.[2] The tools for data visualization will continue to grow in their performance range. New forms and techniques of visualization facilitate the analysis of the data.[3]

- **Big Data, data mining or predictive analytics, hybrid traditional and explorative BI**
  The diverse and huge amounts of data from the internet and social media platforms, the sensor data from our mobile devices and arbitrary objects such as machines, vehicles, components, buildings, etc. provide data around the clock that can be of high interest for the companies. This treasure of information has to be exploited with big data technology and interpreted with data mining or predictive analytics regarding data patterns, relationships and trends to provide better forecasts for the future (see Sect. 5.7).
  With the help of in-memory computing, calculations and analyses of huge unstructured data sets, but also of structured data from traditional DWH systems, will be achieved with a significantly higher speed.
  Traditional and explorative BI will use the new possibilities of big data technology in hybrid overall IT system solutions (see also Sect. 5.8).[4]
  While some manufacturers expand their BI systems to comprehensive **BI suites** that include both traditional and explorative BI, other providers specialize in specific BI areas and focus their offer on subareas of BI, e.g., in the frontend, in the DWH or in the context of big data analytics. In this case, the companies can assemble the **Best of Breed** from the individual components.[5]
- **AI-supported analytics and chatbots**
  With the help of narrow artificial intelligence, improved models for identifying success drivers (advanced diagnostic analytics) as well as better models for forecasting can be generated and used (predictive analytics). This way, recommendations for actions to achieve targeted goals can be derived better (prescriptive analytics). A significant boost will be given to the BI-supported controlling by the use of chatbot technology. AI-supported chatbots will recognize the preferences of the management regarding desired information through an ever-improved text and speech recognition and provide all the required reports, indicators and background information (see Sect. 5.7).

[2] Cf. BARC (2016,2017).

[3] Cf. BARC (2016,2017).

[4] See Gluchowski (2014, p. 240 f.).

[5] See Lünendonk (2015, p. 18).

- **Mobile Systems**
  The spread of smartphones and smartpads in society and in businesses means that planning and reporting systems need to be more oriented toward these mobile devices than before (see Sect. 5.10). The rich-client/thin-client technology extends the traditional thin clients with powerful control elements, configurable event-driven methods and additional communication models.[6] However, the planning and reporting content must be designed in such a way that it is optimally adapted to the size of the screen and the resolution. There must be a content and layout coordination between mobile and stationary reporting. Mobile systems, especially tablets, will tend to prevail over phones in reporting and planning as analyses and planning require larger input surfaces. Phones and small mobile portable devices such as glasses, watches, clothing (wearables) or digital assistants (such as Apple's Siri or Amazon's Alexa) will rather be used for short information messages and quick analyses. Furthermore, it will be interesting to see which of the currently available operating systems and standards from Apple, Android, Windows or the general standards such as HTML5 will prevail technologically. A modern mobile device management (MDM) manages the many different devices of the mobile users. Access to software and data as well as blocking of devices are just as much part of the tasks as installing updates or backup functions.
- **Cloud computing and IT security**
  The trend away from on-premises software inside the company toward cloud computing continues. Cloud computing involves the outsourcing of software, IT services, platforms and infrastructure (see Sect. 5.9).
  Providers of Software as a Service (SaaS) market software solutions as service providers over the web; these can be used instead of the own software.
  For planning and reporting, these outsourcing or distribution processes seem to be only trend-setting insofar as secure and stable access can be guaranteed and data misuse can be excluded. If this is guaranteed, the cloud offers an unprecedented worldwide mobile availability of planning and control-relevant management information. For this reason, appropriate systems and tools (e.g. encryption techniques) will have to be developed through IT or cybersecurity, which provide the best possible protection for the sensitive corporate data.
- **Self-service BI**
  For planning and reporting, a further development of the creation tools for reports and planning forms is expected, which will enable even the IT layman to develop forms and reports without much programming knowledge. This development is discussed under the keyword self-service BI.[7] Here, the further developments of the wizard- or assistant-guided generators and development suites for creating reports, planning

[6] See Neumann (2011, p. 2).

[7] Cf. Lünendonk (2015) and BARC (2017).

forms and statistical forecasts are particularly noteworthy. But also in the area of data modeling, suitable tools should be available for non-IT specialists. The company *pmOne* offers, for example, the possibility to create the structures of the data model intuitively with a mind map using the module OneMind.[8] The BI systems can be administered by the business user and the controlling department themselves. The support of central IT departments or IT experts is only necessary for particularly difficult problems.

- **Embedded BI and BI portal integration**
  Established ERP system providers try to extend their solutions by embedded BI functions. Here, for example, individual BI objects such as reports or complete dashboards can be integrated.[9] Furthermore, BI or data warehouse solutions are integrated into portals. This creates a central entry point for the user. Contents from document or content management systems, such as videos, audios and images, are integrated via mash-ups. For planning and reporting, unstructured documents in various output formats (video, audio and image files etc.) can thus be linked with classic data analyses in the form of tables and graphs.
- **BI monitoring or BI on BI**
  With the introduction of BI monitoring or BI on BI, companies gain the ability to check their BI system themselves. In addition to controlling and managing the performance of the system itself, the aim is to analyze the user behavior and to collect and evaluate the ratings of the users. In addition, the data quality in BI monitoring has to be analyzed. The goal is to gain insights into the performance, data quality and usage rate of the BI system. This way, measures can be taken to continuously further develop the BI system so that it can be used more effectively in the company. With the help of AI-supported chatbots, the user behavior can be analyzed even better as the requests of the users can be evaluated for information gaps or redundant information.
- **Open-source products**[10]
  An increase in open-source products can be observed among the providers of BI software, in which the author of a software defines terms of use in the licenses that usually leave the software to the user for "free" use without major restrictions. An important prerequisite for the free usability is the license fee freedom and an openly accessible source code. Whether the open source products will prevail over the conventional systems in the distant future is currently not foreseeable.

---

[8] Cf. pmOne (2011).

[9] Cf. Search Business Analytics (2016).

[10] Cf. for further technical developments in the field of business intelligence, among others, Kemper et al. (2010, p. 249 ff.).

References

BARC. 2016. BI trend monitor 2016. http://barc.de/ele-ments/download/id/V3T4hapj7kB9F-hj0hGGd. Accessed: 23 July 2016.

BARC. 2017. BI-trend monitor 2017. http://barc.de/trend-monitor?s=s_9_550. Accessed: 12 June 2017.

Gluchowski, P. 2014. Aktuelle Trends in der Business Intelligence. *Controlling* 26(4/5): 235–243.

Hein, A. 2011. *prevero Unternehmen und Lösungen – Unternehmenssteuerung mit prevero*. Vienna: prevero Austria Lecture script.

Kemper, H. G., H. Baars, and W. Mehanna. 2010. *Business Intelligence – Grundlagen und praktische Anwendungen*, 3rd edn. 249. Wiesbaden: Springer Vieweg.

Lünendonk. 2015. Marktstichprobe 2015 – Der Markt für Business Intelligence und Business Analytics in Deutschland. http://luenendonk-shop.de/out/pictures/0/lue_bi-studie_f170815_fl.pdf. Accessed: 9 Mar. 2016.

Neumann, T. 2011. Richt Thin Clients für Web-Anwendungen. http://www.sigs.de/publications/os/2005/01/neumann_OS_01_05.pdf. Accessed: 18 July 2011.

pmOne Produktinformationen. 2011. http://www.pmone.de/de/produkte-loesungen/bi-technologie/onemind/onemind-im-schnelldurchlauf-innovative-datenbankmodellierung/. Accessed: 19 July 2011.

Search Business Analytics. 2016. embedded BI. http://searchbusinessanalytics.techtarget.com/definition/em-bedded-BI-embedded-business-intelligence. Accessed: 15 June 2016.

Unrein, D. 2016. *Excel im Controlling*. Munich: Vahlen.

MIX
Papier aus verantwortungsvollen Quellen
Paper from responsible sources
FSC® C105338

If you have any concerns about our products,
you can contact us on
ProductSafety@springernature.com

In case Publisher is established outside the EU,
the EU authorized representative is:
Springer Nature Customer Service Center GmbH
Europaplatz 3, 69115 Heidelberg, Germany

Printed by Libri Plureos GmbH
in Hamburg, Germany